THE HANDBOOK OF TASK ANALYSIS FOR HUMAN-COMPUTER INTERACTION

THE HANDBOOK OF TASK ANALYSIS FOR HUMAN-COMPUTER INTERACTION

Edited by

Dan Diaper
Bournemouth University

Neville A. Stanton
Brunel University

LEA LAWRENCE ERLBAUM ASSOCIATES, PUBLISHERS
2004 Mahwah, New Jersey London

Editor:	Bill Webber
Editorial Assistant:	Kristin Duch
Cover Design:	Kathryn Houghtaling Lacey
Textbook Production Manager:	Paul Smolenski
Full-Service Compositor:	TechBooks
Text and Cover Printer:	Hamilton Printing Company

This book was typeset in 10/12 pt. Times Roman, Bold, and Italic.
The heads were typeset in Americana, Americana Bold, and Americana Bold Italic.

Lawrence Erlbaum Associates, Inc., Publishers
10 Industrial Avenue
Mahwah, New Jersey 07430
www.erlbaum.com

Library of Congress Cataloging-in-Publication Data

The handbook of task analysis for human-computer interaction/edited by Dan Diaper & Neville A. Stanton.
p. cm.
ISBN 0-8058-4432-5 (casebound : alk. paper) – ISBN 0-8058-4433-3 (pbk. : alk. paper)
1. Human-computer interaction—Handbooks, manuals, etc. 2. Task analysis—Handbooks, manuals, etc.
I. Diaper, D. (Dan) II. Stanton, Neville, 1960–

QA76.9.H85H365 2003
005.1—dc21 2003012434

Contents

Preface

"Dan, do you remember that conference on task analysis that John Annett organized at Warwick University in the early 1990s? The one where we couldn't believe people were talking about the same thing."

"Yeah, Neville, ..."

The most widely cited reference on task analysis has been *Task-Analysis for Human-Computer Interaction*, edited by Dan Diaper, who must take the main blame for this new *Handbook of Task Analysis for Human-Computer Interaction*, as his motive was in part to stem the trickle of requests from around the world for chapters from the out-of-print 1989 book. We, the editors, wanted to produce the definitive reference on task analysis for human-computer interaction (HCI). In this we have failed, and the new handbook provides merely a comprehensive sample of the current research on and use of task analysis. We expect, however, for it to still be the best reference source available. Indeed, our hope is that this first edition is so successful that the completely revised second edition that we'll prepare in 3 or 4 years will be nigh definitive.

Our other reasons for editing this handbook, apart from self-aggrandizement (money certainly wasn't a motive), are as follows: The whole field of task analysis, it seems to us, remains fragmented and poorly understood by many, and a belief that task analysis is at the very heart of virtually all HCI work, although readers should regard Diaper's pathetic excuses concerning chapter 1—about coverage, precision, clarity, and combining two chapters into one—as inadequate for his being self-indulgently prolix, he does explain our beliefs about the importance of task analysis.

The editorial process for this handbook's production was based on an egalitarian ethic. First we advertised for abstracts of possible contributions. We then selected most of them and put them into a tentative structure. Once we'd received all the chapters, we sent the authors comments and suggestions, usually quite extensive. We always intended to defer to the expertise of the authors, but we are proud to report that there was virtually no controversy between authors and editors over these comments and suggestions. The chapter authors have been a pleasure to work with. Thank you. Dan then edited all the chapters and cross-referenced them so it would be a proper book, not just a collection of separate chapters, like a conference proceedings.

From the start, we intended to produce an electronic version of the handbook to go with the paper one. Our publisher, Lawrence Erlbaum Associates (LEA) and TechBooks who handled its production, were wonderful during production of the book, especially in dealing with chapters from so many different people in different formats. LEA has more than lived up to its promises after winning the feeding frenzy that the editors' proposal invoked amongst the dozen major scientific publishers they had initially contacted.

The handbook's five sections are prefaced by an explanation of their rationale and a brief summary of the chapters' contents and relationships. Each chapter starts with an abstract. We have tried to be thorough in cross-referencing the chapters and indexing the contents. It is therefore not necessary to repeat information about the chapters here.

We anticipate that people will pick up this book for different reasons. Some will be newcomers to task analysis. For these we recommend diving straight into chapter 2 and getting some experience doing task analysis. Start small and build up from there. When you have some experience, try tackling chapter 1 or some of the other chapters in part I, as they should start to make sense and will give you an overview of the field. For the skimmer, we recommend reading the section introductions first and then going to chapter 1 if you still haven't found what you are looking for. For the hardcore task analyst, we recommend going straight to the section that interests you. That means part II for industrialists, part III for psychologists, and part IV for computer scientists. Despite this advice, we hope all readers will take a detour from their home disciplines and explore the contents of the book further. Try wandering into some foreign territory; you never know, you might enjoy it. For those that just want a doorstop, use another book.

—Dan Diaper
—Neville Stanton

List of Contributors

Dan Diaper
School of Design, Engineering and Computing
Bournemouth University
Dorset, United Kingdom BH12 5BB
ddiaper@bournemouth.ac.uk

Neville A. Stanton
Department of Design
Brunel University
Egham, Surrey, United Kingdom TW20 OJZ
Neville.Stanton@brunel.ac.uk

Mourad Abed
LAMIH-UMR CNRS 8530
University of Valenciennes
F-59313 Valenciennes Cedex 9, France
Mourad.abed@univ-valenciennes.fr

John Annett
Dept. of Psychology
University of Warwick
Coventry, United Kingdom CV4 7AL
J.Annett@warwick.ac.uk

Jonathan Arnowitz
PeopleSoft
San Francisco, California 94109
jonathan_arnowitz@peoplesoft.com

Chris Baber
Electronic, Electrical and Computing Engineering
The University of Birmingham
Birmingham, United Kingdom B15 2TT
c.baber@bham.ac.uk

Sandrine Balbo
Dept. of Information Systems
The University of Melbourne
Victoria, Australia 3010
sandrine@unimelb.edu.au

Philip J. Barnard
MRC Cognition and Brain Sciences Unit
15 Chaucer Road
Cambridge, United Kingdom CB2 2EF
philip.barnard@mrc-cbu.cam.ac.uk

John M. Carroll
Computer Science Department
The Pennsylvania State University
University Park, Pennsylvania, USA

Brian Casey
Hyperion Solutions Corp.
Stamford, Connecticut, USA 06902
brian_casey@hyperion.com

Steven Clarke
Microsoft Corp.
Redmond, Washington, USA 98052-6399
stevencl@microsoft.com

William Cockayne
William Cockayne
Change Research
San Francisco, California, USA 94117
bill@cockayne.com

Jose Coronado
Hyperion Solutions Corp.
Stamford, Connecticut 06902
jose_coronado@hyperion.com

Rudy P. Darken
Dept. of Computer Science
Naval Postgraduate School
The MOVES Institute
Monterey, California. USA 93943-5118
darken@nps.navy.mil

Helmut Degen
Competence Center, User Interface Design
Siemens AG, Corporate Tech CT IC 7
Munich, Germany 81730
hdegen@acm.org

Alan John Dix
Computing Department
Lancaster University
Lancaster, United Kingdom LA1 4YR
alan@hcibook.com

Kentaro Go
Dept. of Computer Science and Media Engineering
University of Yamanashi
Kofu, Japan 400-8511
go@media.yamanashi.ac.jp

Saul Greenberg
Dept. of Computer Science
University of Calgary
Calgary, Alberta, Canada T2N 1N4
saul@cpsc.ucalgary.ca

Clare-Marie Karat
IBM T.J. Watson Research Center
Hawthorne, New York, USA 10532
ckarat@us.ibm.com

John Karat
IBM T.J. Watson Research Center
Hawthorne, New York, USA 10532
jkarat@us.ibm.com

David Kieras
Electrical Engineering & Computer Science Department
University of Michigan
Ann Arbor, Michigan, USA 48109-2110
kieras@eecs.umich.edu

Christophe Kolski
LAMIH-UMR CNRS 8530
University of Valenciennes
F-59313 Valenciennes Cedex 9, France
christophe.kolski@univ-valenciennes.fr

Quentin Limbourg
Institute d'Administration et de Gestion
Université catholique de Louvain
Louvain-la-Nueve, Belgium
limbourg@isys.ucl.ac.be

Shijian Lu
CSIRO Mathematical and Information Sciences
North Ryde, Australia NSW 1670
Shijian.Le@csiro.au

Jon May
Dept. of Psychology
University of Sheffield
Sheffield, United Kingdom S10 2TP
jon.may@shef.ac.uk

Tom McEwan
Social Computing Research Group
Napier University
Edinburgh, United Kingdom EH10 5DT
T.McEwan@napier.ac.uk

Thomas Ormerod
Dept. of Psychology
Lancaster University
Lancaster, United Kingdom LA1 4YF
t.ormerod@lancaster.ac.uk

Nadine Ozkan
Lanterna Magica
Montreal, Quebec, Canada H2V 2P3
nadine@lanternamagica.com

Cécile Paris
CSIRO Math and Information Sciences
North Ryde, Australia NSW 1670
cecile.paris@csiro.au

Fabio Paternò
ISTI-C.N.R.
Pisa, Italy 56100
fabio.paterno@isti.cnr.it

Sonja Pedell
Competence Center, User Interface and Design
SIEMENS AG, Corporate Tech CT IC 7
Munich, Germany 81730
pedell@asm.org

Chris Phillips
Institute of Information Sciences and Technology
Massey University, Turitea Campus
Palmerston North, New Zealand
c.phillips@massey.ac.nz

Devina Ramduny-Ellis
Computing Department
Lancaster University
Lancaster, United Kingdom LA1 4YR
Devina@comp.lancs.ac.uk

Karen Renaud
Dept. of Computing Science
University of Glasgow
Glasgow, United Kingdom G12 8RZ
karen@dcs.gla.ac.uk

Chris Scogings
Institute of Information and Mathematical Sciences
Massey University, Albany Campus
Albany, Auckland, New Zealand
c.scogings@massey.ac.nz

Andrew Shepherd
Synergy Consultants Ltd
c/o 3 Outwoods Drive
Loughborough
Leicestershire, United Kindom LE11 3LR
andrew.shepherd@synergy-ergs.com

Rick Spencer
Microsoft Corp.
Redmond, Washington, USA 98052
ricksp@microsoft.com

Frank Spillers
Experience Dynamics
Portland, Oregon, USA 97232
frank@experiencedynamics.com

Dimitri Tabary
LAMIH-UMR CNRS 8530
University of Valenciennes
F-59313 Valenciennes Cedex 9, France
dimitri.tabary@univ-valenciennes.fr

Jean-Claude Tarby
Laboratoire TRIGONE
Universite des Sciences et Technologies de Lille
Villeneuve d'Ascq Cedex, France
Jean-Claude.Tarby@univ-lille1.fr

Phil Turner
Social Computing Research Group
Napier University
Edinburgh, United Kingdom EH10 5DT
p.turner@napier.ac.uk

Martijn Van Welie
Dept. of Software Engineering
Vrije Universiteit
Amsterdam, The Netherlands 1081 HV
martijn@cs.vu.nl

Keith Vander Linden
Dept. of Computer Science
Calvin College
Grand Rapids, Michigan, USA 49546
kvlinden@calvin.edu

Jean Vanderdonckt
Institute d'Administration et de Gestion
Université catholique de Louvain
Louvain-la-Nueve, Belgium B-1348
vanderdoncktj@acm.org

Tobias J. van Dyk
Dept. of Computer Science and Information Systems
University of South Africa
Pretoria, South Africa 0003
vdyktj@unisa.ac.za

Gerrit C. van der Veer
Information Management and Software Engineering
Vrije Universiteit
Amsterdam, The Netherlands
gerrit@cs.vu.nl

John Vergo
IBM T.J. Watson Research
Hawthorne, New York, USA 10532
jvergo@ibm.com

Julie Wilkinson
School of Computing and Engineering
University of Huddersfield
Queensgate, Huddersfield, United Kingdom HD1 3DH
j.wilkinson@hud.ac.uk

Peter Windsor
Usability Limited
Cambridge, United Kingdom CB1 7UG
peter@usability.co.uk

B.L. William Wong
Dept. of Information Science
University of Otago
Dunedin, New Zealand
william.wong@stonebow.otago.ac.nz

THE HANDBOOK OF TASK ANALYSIS FOR HUMAN-COMPUTER INTERACTION

I

Foundations

Believe it or not, task analysis has foundations going back to the turn of the last century. One of the reasons for its longevity is that task analysis methods have adapted. Evolutionary changes in its associated disciplines have been accompanied by developments in task analysis. This has given rise to over one hundred methods, along with permutations of those methods, that call themselves task analysis. Obviously we cannot represent all of those methods here, and to do so would do more harm than good. Rather, we have taken to present contemporary thinking about task analysis together with a representative set of methods.

A multi-authored handbook such as this one needs an introductory section that provides readers with an overview of the topic's foundations. The seven chapters in part I form a book within a book. That is, they introduce, or at least exemplify, most of the main concepts, methods, and techniques discussed in more detail in the remaining chapters. Part I has a rational, linear structure, which is outlined below. The first and final chapters, chapters 1 and 7, are in one sense poles apart, as the first is primarily concerned with the theoretical issues underlying task analysis and the last describes a particular task analysis method. These two chapters, however, close a cycle and dovetail in their concern for a systemic approach (and much else), as evidenced by the considerable number of cross-references in each.

Does one start with a simple introductory tutorial or present the theory first? Egocentric or not, the editors have chosen to start with the theory (chap. 1) and then provide the tutorial (chap. 2). The rationale for this decision is that readers who have some experience of task analysis need the theory first to put their own expertise and that expressed in all the other chapters into context. On the other hand, students and other readers new to task analysis are advised to start with the chapter 2 tutorial before tackling chapter 1 or any of the other chapters.

Chapter 1

Understanding Task Analysis for Human-Computer Interaction

Dan Diaper

Chapter 1 is visionary. It is also long, although it can be treated as two chapters, one theoretical (sections 1.2 to 1.4) and the other practical (section 1.5). The chapter has a primary mission, to establish a case that task analysis is at the heart of virtually all HCI activities, even if its centrality isn't always recognized. The chapter is no mere review but extends the very definition of the task concept, relating it to the achievement of work that satisfies goals, although goals are so characterized that they need not be human. Indeed, even the concept of a hierarchy of goals, behaviors, or whatever is not safe from the chapter's systematic and systemic reformulation of what is task analysis. The chapter introduces the discipline of human-computer interaction (HCI) and its twin, software engineering, and locates task analysis within both an extremely broad definition of HCI and a far narrower one associated with user-computer interfaces. The chapter concludes with an integrative perspective on task analysis that hints at research directions for the future.

Chapter 2
Working Through Task-Centered System Design
Saul Greenberg

If you know little or nothing about task analysis, then start by reading chapter 2. The chapter summarizes a well-established, undergraduate-orientated exercise that can be done by students in around 4 hours. Admittedly the task-centered system design (TCSD) method on which the example is based is a lightweight one, but that is surely appropriate for someone just starting. Although the form of task analysis demonstrated is a little rough and ready, it will introduce the reader to some of the basic premises. Saul argues that the approach is good enough to get started, and we agree that you should start small, and build up to bigger and better things.

Chapter 3
Hierarchical Task Analysis
John Annett

Including a chapter on Hierarchical Task Analysis (HTA) by one of its progenitors in the "Foundations" section was a "no-brainer" decision. Invaluably, the chapter starts by charting the history of task analysis over most of the last century. For many people who work with task analysis, a hierarchical task analysis is right at the center of their task analysis activities. Indeed, HTA is the most widely cited task analysis method in this handbook. What the chapter makes abundantly clear, however, is that the HTA method itself involves the top-down (i.e., deductive) decomposition of *goals* and that this is not the same as decomposing activities, behaviors, or whatever. (This is not appreciated by some in the task analysis practitioner community.) Furthermore, HTA is irrevocably committed to solving problems and is thus the very essence, given its simplicity, of an engineering method. Despite the apparent simplicity, it takes considerable practice to become a competent analyst.

Chapter 4
GOMS Models and Task Analysis
David Kieras

A close runner-up to HTA in the Most Cited Methods Stakes is the family of Goals, Operators, Methods, and Selection Rules (GOMS) methods, and chapter 4 provides a sound introduction to the main principles of a GOMS analysis. GOMS methods are intended to be used after carrying out a basic task analysis such as an HTA, and, based on a cognitive psychological model of users, GOMS can claim considerable success, over several decades, at predicting time, satisfactory performance, and, more recently, tasks errors.

Chapter 5
Scenario-Based Task Analysis
Kentaro Go and John M. Carroll

When "future systems" are mentioned, the term "scenario" is frequently found in close proximity. Chapter 5 provides a brief introduction to Scenario-Based Design by summarizing more detailed treatments published elsewhere (e.g., Carroll, 2000). In contrast to Carroll's usual sober future design examples, Chapter 5 provides a glimpse of a more distant future and how families may interact electronically. Although much task analysis work in HCI does involve the analysis of existing systems and prototypes, this chapter provides an important corrective by showing how a task analytic approach can make major contributions to highly creative design activities.

Chapter 6
Comparing Task Models for User Interface Design
Quentin Limbourg and Jean Vanderdonckt

The preface suggests that there is too little agreement about task analysis, even among the cognoscenti. One desperate need, therefore, is for the analysis and classification of different task analysis concepts, methods, and tools. In consequence, this handbook contains not one but several task taxonomies, notably in this chapter and in chapters 22–25 and 28. The editors placed chapter 6 in the "Foundations" section because it provides a comparison of nine typical but different task analysis methods. The chapter ingeniously borrows from software engineering a single representational approach, that of entity relation attribute modeling, to describe and allow comparison between all of the nine methods.

Chapter 7
DUTCH: Designing for Users and Tasks from Concepts to Handles
Gerrit C. van der Veer and Martijn van Welie

As a book within a book, part I deserves at least one major example of a task analysis method. The editors deliberately chose the designing for users and tasks from concepts to handles (DUTCH) method because it (a) can be used across the entire span of the software life cycle; (b) has been successfully used in a reasonably large range of different applications in both academe and, much more importantly, the software industry; and (c) is supported by software tools. In the example presented, by using a Computer Supported Co-operative Work (CSCW) example, it deals with a genuinely complex, difficult, and, to be honest, not fully understood application domain, and one of the strengths of the method is that it provides a means of describing both the current task model and future task representations.

REFERENCES

Carroll, J. M. (2000). *Making use: Scenario-based design of human-computer interactions.* Cambridge, MA: MIT Press.

1

Understanding Task Analysis for Human-Computer Interaction

Dan Diaper
Bournemouth University

Task analysis is at the core of most work in human-computer interaction because it is concerned with the performance of work, and this is what, crucially distinguishes it from other approaches. A systemic approach is adopted to describe the properties of the models used in task analysis. The roles of task analysis in HCI and software engineering are introduced before practical issues, concerning how to perform the typical, early stages that are common to most task analysis methods, are addressed. The main example used throughout the chapter involves air traffic control and the theoretical and practical issues are finally illustrated within a method called systemic task analysis.

The goal of this chapter is to provide and illustrate the fundamental concepts that underlie task analysis so that its readers will subsequently find that task analysis is easy to understand.

1.1 TASK ANALYSIS IS EASY

The claim that "task analysis is potentially the most powerful method available to those working in HCI and has applications at all stages of system development, from early requirements specification through to final system evaluation," is Diaper's (1989a) conclusion. Although not disputing the potential of task analysis Anderson et al. (1990) suggested that "confusion reigns" and that "designers and human factors specialists fumble with the concept" of task analysis. If you agreed with Anderson et al. a dozen years ago, then, until you've read this chapter, you will probably still agree with them. The purpose of this chapter is to demonstrate that task analysis is easy to understand. This, of course, is a lie.

If you really want to understand the task analytic perspective on HCI, then one must have a reasonable competence in at least philosophy, psychology, sociology, ergonomics, and of course computing. On the other hand, although these disciplines may be part of the essential professorial armory, there is only a limited global need for theoretically driven academics, notwithstanding how necessary it is to have a few of them, so that most people interested in task analysis, and virtually all practitioners who want to use task analysis, do not need the arcane baggage of the theorizing academic. It is desirable, however, to recognize where such arcana fit into task analysis, and this chapter marks at least some places where some truly complex issues are relevant.

This chapter is unashamedly systemic, although this is not a novel approach. Shepherd (2001), for example, starts his book on hierarchical task analysis (HTA) by presenting a systems approach, and both the HTA and Goals, Operators, Methods, and Selection Rules (GOMS) methods are located within a systemic approach in the chapters of this handbook that describe them (chaps. 3 and 4, respectively). A particular style of systems model based on Checkland's (1981) soft systems methodology (SSM) conceptual models (Diaper, 2000a; Patching 1990), is used throughout the chapter in conjunction with Dowell and Long's (1989; Long, 1997) general HCI problem. Although a number of different examples are used in this chapter, the main one involves air traffic control (ATC).

1.1.1 Chapter Summary

If this chapter were a sentence, there would be a semicolon between sections 1.4 and 1.5. The first half of the chapter introduces the architecture of a systemic model (section 1.2); describes how this must be extended to model the performance of work, which is the core concept for task analysis, and explains why task analysis is at the heart of nearly all HCI activities (sec. 1.3); and then places task analysis in the context of HCI and software engineering (section 1.4). The second half of the chapter (section 1.5) addresses the use of task analysis in computing-related projects (section 1.5.1) and then introduces the common first stages of most task analyses, although some methods do use these stages in more or less an abbreviated form (section 1.5.2). A new task analysis method, systemic task analysis (STA), is introduced at the end of the chapter (section 1.5.3) to illustrate both the theoretical and practical issues discussed throughout.

Task analyses produce one or more models of the world, and such models describe the world and how work is performed in it (section 1.2). Descriptive models of the world consist of two types of entity: things and the relationships between things. Things can be physical or intangible, and most things are parts of other things (section 1.2.1). Intangible things include mental and software processes, emergent systemic properties, and models that are information rather than physically based. Task analysis models are among the intangible things.

Relationships between things in models are of two sorts: conceptual and communicative (section 1.2.2). Conceptual relationships (section 1.2.2.1) typically concern classifying things. Many taxonomic systems, including those used in task analysis, use a hierarchical structure, although, it is argued, much of the natural world is not truly hierarchic and would be better modeled as a heterarchy, using a "level of abstraction" concept while allowing things to have multiple memberships in other things. Communicative relationships in descriptive models of the world (section 1.2.2.2) involve how different things affect each other. Adding these relationships to taxonomic ones reinforces the suggested advantages of heterarchies over hierarchies and allows the introduction of the basic systemic model, based on SSM, used throughout the rest of the chapter.

What makes task analysis in HCI distinctive is its primary concern with the performance of systems (section 1.3.1), and performance in task analysis is fundamentally about doing work to achieve goals (section 1.3.2). (Dowell and Long's 1989; Long 1997) general HCI problem distinguishes between the work system and its application domain. A work system in HCI typically is made of people, including users, and computer systems and usually other things as well. Work is performed by the work system so as to change its application domain, and it is successful if the changes meet the goals of the work performed. Complex systems will contain numerous overlapping work systems and application domains (section 1.3.3), and it is necessary for task analysts to appropriately identify the work system and application domain of interest.

This chapter defines HCI as an interdisciplinary engineering discipline that is a subdiscipline of ergonomics (section 1.4). Furthermore, it suggests that the historical division between HCI and software engineering is unfortunate, as both study the same sort of systems for similar engineering purposes. It then introduces two views of HCI, one broad and the other narrow (section 1.4.1), and relates these views to different definitions of work systems and application domains. Task analysis needs to have models of people's psychology within the systems that are studied, and the chapter therefore discusses various styles and types of cognitive model (section 1.4.1.1). Task analysis must also consider the psychology of those who are involved in computing-related engineering projects, as they are either the final end users of task analysis methods or users of the results from task analyses (section 1.4.1.2).

Software engineering's solutions to the software crisis have been primarily anthropocentric because the crisis is one of human productivity (section 1.4.2). Its solutions have been to develop methods for working on software projects and to support many of these methods with computer-assisted software engineering (CASE) tools. The roles of task analyses need to be located within the software life cycle. Within projects, task analyses vary in their fidelity (i.e., how well they relate to the assumed real world) and are best thought of as always studying simulations or selected examples of their world of interest (section 1.4.2.1). The chapter emphasizes the importance of the concept of agent functionality and discusses the allocation of function between different things in a work system (section 1.4.2.2). It also points out the need for task analytic representations to be device independent in the context of abstracting task representations and discusses the possible role of goals in such abstraction (section 1.4.2.3).

The second half of the chapter starts by arguing that task analysis can be of use in most stages of any computing project (section 1.5). Stressing that any practical method must involve considerable iteration (section 1.5.1.1), the chapter suggests five questions that should precede undertaking a task analysis:

1. Which project stages will use task analysis? (section 1.5.1 and 1.5.1.1)
2. What knowledge is the task analysis intended to provide? (section 1.5.1.2)
3. What is the desirable task analysis output format? (section 1.5.1.3)
4. What data can be collected? (section 1.5.1.4)
5. What task analysis method? (section 1.5.1.5)

Following discussion of these five questions, the chapter describes the firsts two stages of most task analysis methods: (a) data collection and representation (section 1.5.2.1) and (b) activity list construction and the classification of things (section 1.5.2.2). It is after these stages that different task analysis methods diverge (section 1.5.2.3).

To illustrate these early common stages, it is necessary to choose a method and notation. Systemic task analysis (STA) is used because it has been developed from the ideas exposed in the first half of the chapter (section 1.5.3). In describing its first stage (section 1.5.3.1), the chapter uses the ATC example previously introduced but adds more project-specific details. The chapter describes how video data might be collected in an ATC tower (section 1.5.3.2), constructs a sample fragment of an activity list from these data, and proposes that sometimes such construction is sufficient task analysis in some projects or project stages (section 1.5.3.3). Although STA uses a formal, heterarchical modeling method, the chapter offers the start of a hierarchical analysis of the fragment of the activity list (section 1.5.3.4) to illustrate how task analysis methods might proceed following the fairly common early stages.

1.2 DESCRIBING THE WORLD

Does the world exist? The philosophical tradition of solipsism denies that this question can be answered, and even Descartes' proof of his own existence, encapsulated in his famous phrase "Cogito ergo sum" (I think, therefore I am), has for many years been recognized as a conclusion based on a faulty argument (Watling, 1964). Science and engineering, however, traditionally require belief in a real world external of the mind. The position of the author, which he calls *pragmatic solipsism*, is an intermediate one. It treats the world's existence as an assumption, and one corollary is that all that can ever be known about the world are merely models. Such models of the world can be compared, and some will be preferred over others for some purposes. In science, for example, Ockham's razor might be applied to choose the simplest of two equally good competing theories, although this begs questions of what is meant by "simplest" and "good." This chapter is based on the pragmatic solipsist assumption and argues that one useful class of models of the world for engineering purposes consists of the task analytic ones. It then attempts to elucidate the common nature of this class of model while recognizing that no model of the world can be proved to be true. It thus admits alternative models, which might be called *perspectives* (see also, e.g., chap. 7), and allows that some will be better suited for some purposes than others.

At the core of any task analytic model there are two things:

1. A description of the world.
2. An account of how work is performed in the described world.

It is the latter of these that differentiates task analysis from approaches that are primarily descriptive. These two are sometimes combined, but it is almost certainly easier to maintain a conceptual separation between them, as this chapter does.

A model is a description of the assumed real world. A painting or a poem can be a model of the world, as can the sorts of diagrammatic, logical, or language-based models used in HCI and software engineering. Design requires two models of the world: a current one and a future one. Design is a goal-directed activity involving deliberate changes intended to improve the current world, so the need to model the future in design is unquestionable. In practice, models of possible future worlds need to be based on models of the current world because the world is a very complicated place and accurately predicting the future must accommodate its complexity (see also chap. 30). In addition, constructing a model of the current world is easier because the current world can be observed and because the model can often be scientifically tested to determine if what the model predicts to be the case is actually so. For example, an analyst might construct a model of the current world that predicts that on web pages animated images surrounding text will interfere with the task of searching the text for information, as Zhang (1999) proposed. This prediction could be tested using typical realistic web pages, as Diaper and Waelend (2000) did. They found that the prediction was false. As Alavi (1993) stated, "Most research studies have been conducted in laboratory settings that limit the generalizability of their findings to more complex organizational settings."

In contrast to laboratory-based science, task analysis creates its models of the current world by studying it in situ. Famously, it involves the observation of people performing their tasks in an environment as realistic as possible. The performance may be documented using video cameras or by taking notes. Task analysis can also make use of other data collection methods (section 1.5.1.4; Diaper, 1989a) such as interviews (e.g., chap. 16), immersive methods such as those used in ethnography (see chap. 6 and 14), or various forms of invented scenarios (see chap. 5).

Unfortunately, since descriptions of the world in HCI include people in complex social and organizational environments, truly accurate prediction of any postdesign future is virtually impossible, not least because people often (a) adapt their behavior to designed changes, (b) alter other aspects of the world to accommodate what has been designed, and (c) change and use what has been designed in ways unanticipated by the designers. As Sharples (1993) commented, "The way designers intend technology to be used very often differs from the actual users' behavior" (p. 67).

Prototyping approaches (section 1.4.2) offer only a partial solution to the problem of predicting postdesign worlds (Carroll, 1991), and the underlying rationale for providing the best possible model of the current world is firstly one of cost, as the fewer cycles of prototyping required to achieve a satisfactory design, the cheaper will be the overall design exercise. Perhaps even more importantly, design is less likely to fail the more accurately the future world can be predicted, and a model of the future will almost certainly be more accurate if the model of the current world is accurate.

Descriptive models of the world tend to be constructed from two general classes of entity:

1. Things.
2. Relationships.

The next two subsections introduce these basic components. The discussion ignores time and even sequence and therefore the performance of work, which is introduced in the next main section (1.3).

1.2.1 Things

A model represents the world as being made of something, variously called things, objects, entities, items, components, pieces, parts, attributes, properties, and so forth. In HCI the sorts of things that are used in its models include people, computer systems, and usually other things as well. Things can be parts of other things, so that a user-computer interface might be considered to consist of things called computer output and input devices, such as a screen and loudspeakers and a keyboard and mouse.

Although it is reasonable from a systemic perspective to treat people as just another sort of thing, they have a special status—and not just because they are far more complicated than anything else, to the point that they do not even understand themselves (Diaper, 1989b, 2002a). People individually possess moral attributes, and cannot be treated like other things. For example, unlike computers, they cannot be legally turned off (killed) or intentionally damaged. In addition, people as individuals must ultimately be considered as moral atoms, even though moral issues often involve two or more people. For example, we might consider a particular political system to be immoral, but our reason for viewing it as immoral is likely to be its harmful effects on some people.

Most models of the world contain some things that have an assumed tangible, physical reality (i.e., they can be perceived by the five human senses), and usually such things provide a starting point for model building. Many things in HCI models, however, are intangible, most obviously things that are mental or are software. Similarly, the concept of intangible properties is central to systemic models because, as Patching (1990) stated, "It is not possible to see, hear or touch a social, or political, or industrial relations system" (pp. 9–10). Of course, this leads to some fundamental philosophical problems that while of interest to the theorizing academic, are not germane to the practice of task analysis. From a pragmatic point of view, what is important is not the absolute reality of things but their psychological reality (i.e., whether people believe in them). Analysts' models of the world are one example that fall into this category of intangible thing. Similarly, software can also believe things about the world, although the tendency is to blame the programmers. For example, prior to the 2K revisions, many programs believed that "no person is older than 100 years" and were automatically going to assign the age of 1 year to people born in 1899 upon the arrival of the new millennium.

Many of the things used in HCI and task analysis are described as information rather than physically. Although, for convenience, it is common for physical things to be described as containing information, care needs taking with the concept of information, as this is not really a property of a physical thing but of some other thing, usually an agent, that perceives the physical object. This is obvious when one considers the different information supposedly contained in a document perceived by a person who understands the language in which the document is written and by a person who does not. Apart from physical and informational descriptions, things can be described in other ways, depending on the perspective chosen by the analyst (e.g., money, power, or responsibility). These alternatives, however, tend to be used in only a small percentage of task analyses. That still leaves a wide range of different perspectives to choose from. This chapter assumes that psychological, sociological, and other such perspectives are basically information-processing ones, even when their focus is on, for example, affective issues (see chap. 29), because affect changes a person's view of the world, as Nussbaum (2001) argues. Section 1.5.3.3 presents an example in which the psychological property of trust between coworkers is important. Chapter 14 describes other sorts of relevant psychological properties, such as fatigue effects, and chapter 28 reviews psychological theories and models in detail. Although Checkland's SSM is celebrated for its multiple-perspectives modeling of the world, the integration of different perspectives remains a problem that, except in some specialized cases such as physical reductionism (Diaper, 1984), is epistemologically intractable and only solvable by the application of the analyst's craft skills (Long, 1986; Long & Dowell, 1989).

Just what is a thing is a complex metaphysical question (e.g., Lowe, 2002). Ignoring metaphysics, however, in practice task analysis traditionally divides its things into two main types, objects and actions (Diaper, 1989c). Often objects are further categorized as agents or things used by agents. Actions, of course, are what agents do, alone or with their objects and other agents. An activity list line, which can be used to describe a task step (as recommended in section 1.5.2.2), typically has a single agent, a main action and perhaps some subordinate ones, and the recipients of the action (i.e., objects and other agents). People generally seem to understand what is meant by an agent, and some methods in the task analysis literature describe actions and objects as verbs and nouns, respectively, although this is only an approximate metaphor. Still ignoring the truly complex metaphysical issues, the key property of agents is their ability to initiate behavior whereas other objects are passive (see also chap. 26). For example, a person putting words down on paper is an agent performing the action of writing, and the passive objects are the pen, ink, and paper. If the paper is given to someone else to read, then its transmission is another task, initiated by one agent and with another as the recipient, even though some properties of the paper (e.g., its location) have also changed. Probably the real difference between agents and other objects is that the latter have processes that are adequately describable by the physical sciences (physics, chemistry, some biology, etc.) whereas agents have processes that, because they involve emergent properties (section 1.2.1) or cognitive or computer information processing (e.g., chap. 15), require other sciences to be brought in. One conceptual difference between using a word processor rather than pen and paper is that with the word processor the writing is to another agent—the computer system. Thus, dialogue-based models are common in computing applications whereas they are not used for objects describable by the physical sciences alone. Other sorts of classifications of things are possible (e.g., some things can be classified as triggers; see chaps. 14 and 19).

1.2.2 Relationships

All things in a model have at least one relationship to another thing and often more than one relationship with other things. Isolates that have no interaction with anything can have no effect on the rest of the system and so can be legitimately ignored.

Although the following distinction may not be absolutely true theoretically, it is useful to classify relationships between things in models as of two types, conceptual and communicative.

1.2.2.1 Conceptual Relationships

Conceptual relationships are typically taxonomic (see chap. 20 for a discussion of taxonomies and classification schemes). Probably the simplest type of conceptual relationship concerns what things are made of, and these are typically modeled hierarchically (Diaper, 1984). Thus the user interface mentioned in section 1.2.1 is_made_of a computer output and input system that is_made_of the screen and loudspeakers (output) and the keyboard and mouse (input). Similarly, the computer system is_made_of the user interface and other unspecified computer hardware and software things and the complete system includes a computer and a user and the environment in which the user and computer system operate. Figure 1.1 illustrates this hierarchy using a tree diagram and a Venn diagram, which is a graphical representation of a formal, logical model based on set theory, the foundation of all modern mathematics (Ross & Wright, 1988). The two representations are logically identical (Diaper, 2001a); that is, they contain exactly the same information but in a different graphical format.

Other types of relationship may be used for taxonomic purposes, for example, kinship or the "is_similar_to" relationship. The latter can be a complex relationship if things are judged to vary on more than one attribute (an attribute is a property of a thing as well as a thing itself). Bowker and Star (2000) present a compelling case that taxonomic systems are common and important in most areas of human endeavour. Furthermore, not only do classification systems get used alone, they also form the basis for more complex types of models of the world, including task analytic models.

Conceptual relationships do not have to be specified hierarchically, although this is common in many HCI and software engineering methods and justified on the basis that people find hierarchies naturally easy to understand (see de Marco, 1979, on the design of data flow diagrams (DFDs)); Paterno, in chapter 24, claims that people's understanding of hierarchies is "intuitive." Diaper (2000a, 2001a, 2001b) has suggested the neologism *heterarchy* to describe models in which a thing is related to more than one other thing (see Fig. 29.1 for a simple example). In a tree-style model, such as Fig. 1a, heterarchical conceptual relationships can be represented either by repeating nodes in the tree or by allowing child nodes to have more than one parent node. This is not, however, an elegant solution, as is discussed below.

Although most things engineered by people do have a hierarchical design, it is far less clear that the natural world and our social worlds are arranged hierarchically (Diaper, 2000a). Biologists, for example, have found many an organism that is "*incertae sedis* meaning that its position in the classificatory system is uncertain" (Clark & Panchen, 1971). Diaper (2001a) demonstrated that even a simple office system's organization of people's roles needs to be modeled heterarchically when roles are shared. Thus, although hierarchically arranged DFDs, for example, may be appropriate if used to describe some types of software, it is of concern that methods such as HTA and many of the task analysis methods classified in chapters 6, 22, 24, and 25 may be forcing an inappropriate hierarchical organization on to their model of the world. In contrast, from its inception in the early 1980s, task analysis for knowledge descriptions (TAKD; Diaper, 1989c; Diaper & Johnson, 1989), used a heterarchical representation of a tree with nodal repetition, although it was mislabeled a hierarchy for a long time by its developers (Diaper, 2001b).

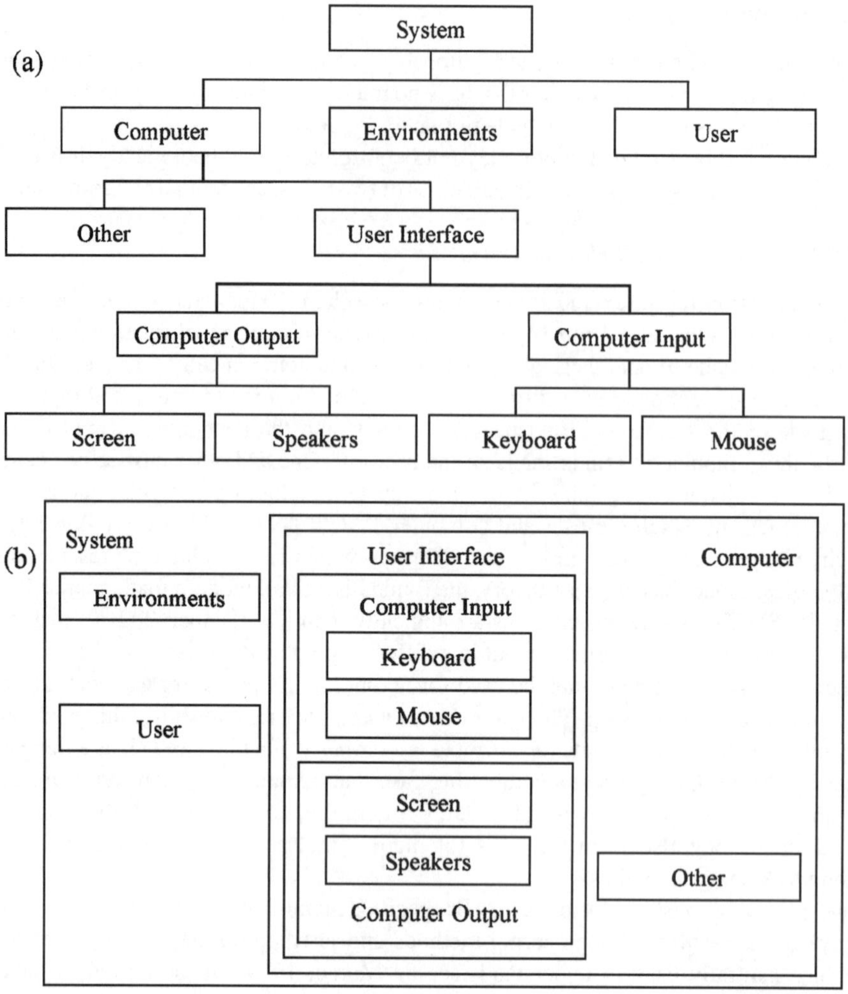

FIG. 1.1. Both 1(a) and 1(b) show the logically identical conceptual relationships between some components of a system that involves a computer and a user. Figure 1(a) represents this as a tree diagram and 1(b) as a Venn diagram. Note that the size and shape of the sets represented in a Venn diagram are not meaningful.

Given the familiarity that most people have with hierarchical models, then the currect practical advice is to start with a hierarchical structure but, when confronted with something *incertae sedis*, to consider moving to a heterarchical model. Using hierarchical models does have the advantage of forcing analysts to determine whether some thing is of one sort or another and thus encourages careful thought about things. There are times, however, when classification decisions have to be contrived, and heterarchicies provide a logical alternative that need not compromise a model's validity for the sake of maintaining a hierarchical structure.

1.2.2.2 *Communicative Relationships*

In a communicative relationship, two or more separate things communicate with or affect one another is some way. Communicative interactions require another type of thing, as all interactions between things are mediated by something. For instance, light from a computer's screen is the physical thing that links the screen to the user's eyes, and mechanical energy is

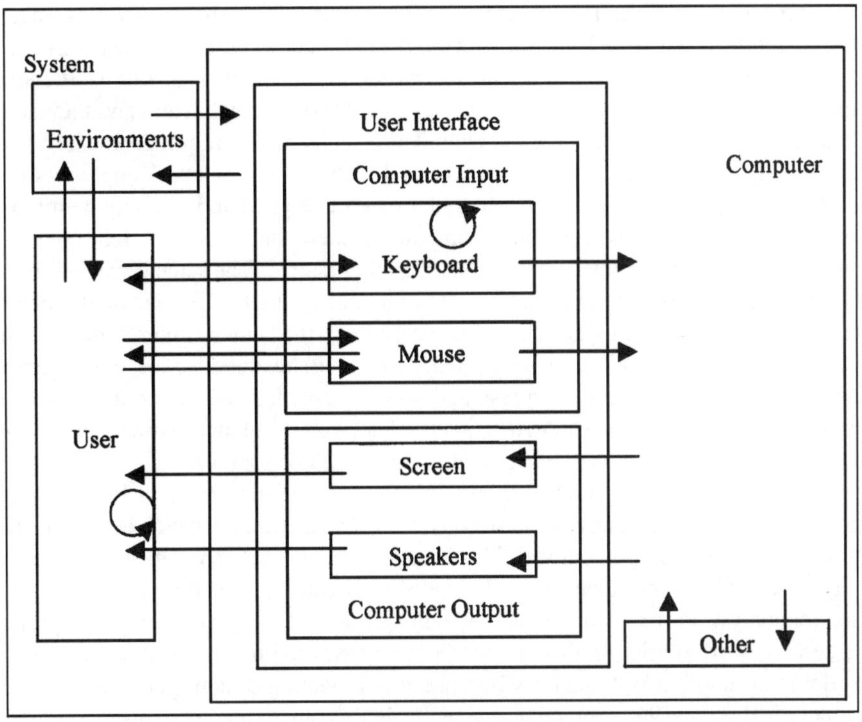

FIG. 1.2. A Venn diagram model of a system consisting of a user and a computer. The diagram shows the communicative relationships between the things in the model.

transferred between fingers and keys on a keyboard. In HCI, it is usual but not universal to represent communicative things (mediators of communicative relationships) as information that is being transmitted.

There are clear advantages to using a Venn diagram (Fig. 1.1.b) for modeling purposes rather than a tree diagram (Fig. 1.1.a). Among the pluses, communicative things can be added easily to the taxonomic structure using a quite different graphical notation to that representing conceptual relationships (see Fig. 1.2). In Fig. 1.2, the communicative things are represented by unlabeled arrows that overlay the hierarchical representation of Fig. 1.1.b. Thus, Fig. 1.2 is no longer a hierarchy but a heterarchy. That is, the communicative relationships can be within or between hierarchic levels, although only those that jump levels are shown in Fig. 1.2.

Not to label and describe arrows is too common a sin in HCI and software engineering modeling activities. The problem is that arrows often represent different types of thing. Diaper (2000a) argued that if defining arrows is a necessary requirement for simpler, single-perspective models such as DFDs, it is even more necessary for multimedia and multiple-perspective models. Illustrating such semantic differences, in Fig. 1.2 the input devices provide immediate kinaesthetic feedback to the user as they are touched or moved. The mouse is modeled as a two-function input device (movement is separated from button clicking), but whether there are two types or only a single type of feedback may depend on many things, including the nature of the task and the expertise of the user. The computer does not provide feedback to the input devices (i.e., the devices do not "know" whether they are connected to the computer or not). The keyboard's Caps Lock and Num Lock keys are represented by a circular arrow to indicate that they affect what the keyboard sends to the computer but do not themselves send information to the computer. Figure 1.2 does not show a relationship from the user to the computer output devices. The rationale for this is that, although the user may engage in attention-directing, orientating behaviors toward the output devices, nothing is communicated to the devices (i.e., a screen or loudspeaker does not know whether anyone is attending to its output). In many cases it is important not to forget these user behaviors and they can also be indicated, as in Fig. 1.2, by a circular arrow. The circular arrow might represent behaviors, such as head and eye movements, that are potentially observable by a task analyst as well as human attentional processes (see chap. 15) that are psychological and can only be inferred.

Users do not interact only with their computer systems but inhabit a rich, complicated environment made of other things, often including other people. The same is true of computer systems. Fig. 1.2 has not decomposed the environment of either the user or the computer system. It should be noted, however, that some aspects of the "environments" may be shared by both user and computer and that some, usually most, will be unique to just one of these. If a decomposition using a hierarchical representation is attempted, it will fail or at best be clumsy because of the shared things in the environments. Heterarchical models, which allow things to be components of more than one higher level thing, avoid this problem.

Figure 1.2 is potentially a systems model. It follows the conventions of SSM conceptual models but is drawn to stricter criteria, based on DFDs, than is common in SSM (Diaper, 2000a). It is, however, a static, atemporal model. That is, it describes the world independently of time or even sequence and therefore independently of events, because events must occur in time. It is common for analysts to run such a static model in their minds similar to the way programmers run the code they are writing in their minds (in both cases to predict run-time performance). This is doing an implicit task analysis because it is modeling system performance. Diaper, McKearney, & Hurne, (1998) argued that in realistically complicated systems an implicit task analysis is always likely to be done poorly, and they presented their Pentanalysis Technique as a method for turning implicit task analyses into explicit ones.

1.3 PERFORMANCE AND WORK

Task analysis focuses on the performance of systems. Annett and Stanton (1998; see also Diaper, 2002b, 2002c) suggest this when they state that task analysis is the collective noun used in the field of ergonomics, which includes HCI, for all the "methods of collecting, classifying and interpreting data on human performance" (p. 1529). Perhaps because both Annett and Stanton have a psychology background, they emphasize "human performance." From a systems perspective, it would be preferable to alter the definition to read thus:

> **Task analysis is the collective noun used in the field of ergonomics, which includes HCI, for all the methods of collecting, classifying, and interpreting data on the performance of systems that include at least one person as a system component.**

The two key concepts in this definition are performance and data. Performance is how a system behaves over time. A datum is the description of some part of a system at a point in time, and data may be collected over time and from different parts of the system.

1.3.1 Performance

The performance of a system is how it behaves over time. Systemic time can be characterized as a series of instants each representing the system in a particular state. Systems that have human and computer components have many more instants than can possibly be observed, so what is available to analysts is a series of snapshots of the state of the system with missing states between these, like the frames of a cinema film. In practice, what is available to analysts as data is the representation of the state of only one or a very small number of system components. To extend the cinema film analogy, the analyst cannot see all of any frame but only one or a few areas of it. Thus, all but the most trivial systems will appear to operate in parallel. That is, at each instant more than one component of the system can be expected to have changed in some way, and the analyst cannot know the states of all the components that do change from instant to instant. The problem is the same one that bedevils testing any large software system but worse, that a system state is possible that prevents the expected operation of a part of the system being observed and so making it difficult to identify the cause or causes of the unexpected system state.

Figure 1.2 can be used to illustrate this. The user can be observed pressing keys on the keyboard and we could test the ASCII string that is sent from the keyboard to the computer and so on until each character is echoed on the screen. Usually, as far as the user is concerned, this echoing loop from keyboard to screen is instantaneous, but sometimes it is perceptibly slow, sometime frustratingly so, for example, when the other computer components place a high demand on computer-processing resources.

One obvious question for user interface designers in the above scenario is, How slow can echoing be, in particular situations, without causing the user some sort of problem? This type of question is at the very heart of task analysis; as Annett, in (chap. 3), proclaims of HTA, the focus should be on solving problems. The issue is how to define "satisfactory performance" for a system and its components. The concept of work is critical to defining "satisfactory" in this context.

1.3.2 Work

Dowell and Long (1989; Long, 1997) have produced perhaps the most sophisticated framework for the discipline of HCI. Their general HCI design problem can be characterized at a high level as having two system components—a work system and an application domain—in an environment. Figure 1.3 illustrates the basic model, which can be described so:

> **Work is achieved by the work system making changes to the application domain. The application domain is that part of the assumed real world that is relevant to the functioning of the work system. A work system in HCI consists of one or more human and computer components and usually many other sorts of thing as well. Tasks are the means by which the work system changes the application domain. Goals are desired future states of the application domain that the work system should achieve by the tasks it carries out. The work system's performance is deemed satisfactory as long as it continues to achieve its goals in the application domain. Task analysis is the study of how work is achieved by tasks.**

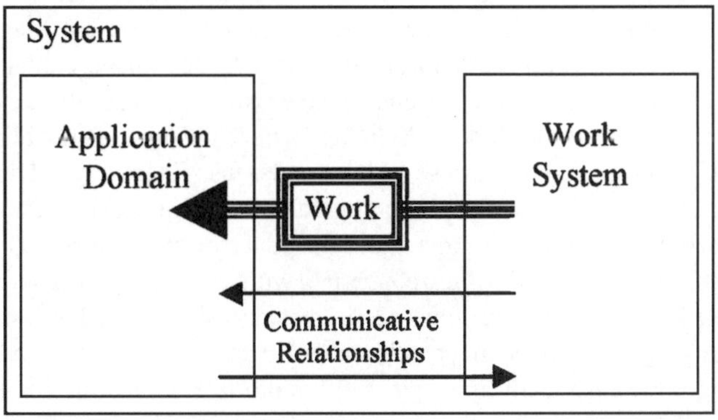

FIG. 1.3. The general model of work, based on Dowell and Long (1989; Long, 1997).

Dowell and Long's conception is that tasks are not performed within the work system but by the system as it acts on the application domain. Furthermore, the communicative relationships between the work system and application domain may be what changes the application domain, but they are not the same as the work performed, because work is defined in terms of the achievement of goals. Consequently, describing the communicative relationships is not sufficient for describing a task. Annett (chap. 3) comes to a similar conclusion when he states that "simply to list actions without understanding what they are for can be misleading." What is needed for task analysis is an understanding of how the communicative relationships arise in the work system and how they cause changes within the application domain—and, of course, vice versa (i.e., how feedback from the application domain arises and then affects the work system).

Throughout this handbook, the concept of goals is closely tied to that of tasks, and goals are nearly always described as being psychological, in that it is people who have them. In contrast, in the systemic version of Dowell and Long it is the work system that has goals, which specify the intended changes in the application domain. The advantage of this view of goals is that it recognizes that nonhuman things can have goals. In SSM, nonhuman goals reflect one aspect of what are called "emergent properties" of a system. Organizational systems, for example, can have goals that may be different from those of any individual person within the system. This is not a controversial claim in systemic approaches such as SSM or in any approach that uses an organic metaphor for organizational life. Certainly very few, if any, employees of a large business have a personal goal such as "increase market share," although this might well be a good partial explanation of the business's behavior with respect to its external environment (i.e., its application domain). Furthermore, this view of goals allows parts of an organization and individuals within it to have different goals and even ones that conflict. Indeed, a great deal of collaborative work is focused, not on achieving common goals, but on resolving conflict (Easterbrook, 1993; Shapiro & Traunmuller, 1993). Similarly, this view of goals makes it reasonable to suggest that things like computer systems can have goals separate from those of their users, a point explicitly made in chapter 26. Thus a business computer might have as a goal "to maintain client database accuracy" and attempt to achieve this by forcing users to operate in a way that they resist, perhaps because it requires more effort from them than they wish to expend. Although it has been suggested that goals in a computer system are simply those of its designers (Bench-Capon & McEnery, 1989a, 1989b), the alternative is that large computer systems possess properties, perhaps emergent ones, unanticipated by their many designers (Barlow, Rada, & Diaper, 1989; see also sec. 1.4.1.2).

It is probable that one of the main confusions that has arisen in task analysis over the years has been the assignment of different types of goals to individual people rather than to work systems. One goal of a data entry clerk, for example, might be "to achieve all the data entry work assigned as quickly as possible with a detectable error rate acceptable to management so as to obtain a bonus payment." In this case, the computer system (and perhaps the management checking function) is the clerk's application domain, and the clerk, the documents containing the data and other things, and the management, office, and social environments constitute the work system. Confusion arises, however, if the clerk is also taken to have the organizational goal "to increase market share." This is clearly not the case (if it were, bonus systems would not be necessary). Note that this argument is based on the psychological assumption that possession of a goal provides motivation, or willingness to act to achieve the goal (for further discussion, see chap. 15).

Teleology is that branch of philosophy concerned with causes, reasons, purposes, and goals. Goals are included because they are future desired states that require the same causal model. In the case of a goal, its achievement, which will occur in the future, can be caused by events happening now. Task analysis is generally mono-teleological in that behavior, particularly that defined through low-level, detailed descriptions, of behavior, is commonly represented as being caused by a single goal. Most of the chapters in this handbook promote a mono-teleological philosophy. A hierarchy of goals, as used in HTA, consists of multiple related goals, but a person can also perform an action on the basis of unrelated goals. Furthermore, unrelated goals that nonetheless motivate the same behavior cannot be simply prioritized in a list, because different goals have more or less motivational potency depending on their specific context.

For example, a chemical plant operator's unrelated goals for closing a valve might be (1) to stop the vat temperature rising, (2) to earn a salary, and (3) to avoid criticism from the plant manager. The first might concern the safety of large numbers of people, the second is sociopsychological and might concern the operator's family responsibilities, and the third is personal and might concern the operator's self-esteem. These three goals correspond to different analysis perspectives, the sociological, the sociopsychological, and the personal psychological; and there are other possible perspectives as well. Furthermore, people might have different goals within a single perspective.

Of course, the principle that any given behavior is likely to be caused by multiple goals can be extended to any complex work system, as discussed above. At the least, task analysts should be cautious whenever the explanation of the behavior of a work system is ascribed to a single goal, because there will probably be many, within and across different perspectives, and a set of goals will be prioritized differently in different contexts. Nor is a first-past-the-post threshold model likely to be satisfactory, as a combination of goals, rather than a single one, will most often reach some notional threshold and so trigger behavior. In the chemical plant operator's case, the usual main reasons to close the valve might be to avoid management criticism and ensure receiving a salary, but after an argument with the manager about salary, the operator's main goal might become protection of the health of people at risk from a chemical accident. If the latter now more strongly drives the operator than the other two goals, we might want to be quite sure that the operator is aware of the safety issues, which he or she might not be if the main motivating goals had previously been to avoid management criticism and ensure a salary.

Figure 1.4 provides a concrete but highly simplified example in traditional air traffic control (ATC). Chaps. 13 and 25 describe more modern computerized ATC systems. The application domain contains things like aircraft, weather, and so forth. The goal of ATC is to facilitate the safe and expeditious movement of aircraft, and this is both an organizational goal and one held by many of the workers. It is doubtful, however, whether it is one held by the canteen staff, for example, even though they could affect the accomplishment of the organizational goal if they fail to achieve their own goals, for examples concerning food hygiene. The simplified ATC work system in Fig. 1.4 consists of various people, the flight strip system, radar, radio and normal telephones, and so forth. The ATC work system carries out tasks intended to meet the ATC work system goals of the safe and expeditious movement of aircraft in the application domain. The concept of a goal is further explored in chap. 30, which takes a more radical position on the concept than that presented here.

One important representation of each aircraft in an ATC work system is the flight strip, which records the status of each aircraft and is updated by the ATC officer (ATCO) or the flight chief, who between them have responsibility for the aircraft represented by the flight strips in front of them. Sample flight strips are shown in chapter 19. Updating a flight strip is not a task of the ATC work system. It's not an ATC work system task because its performance does not directly change the application domain; that is, updating a flight strip does not change the real world of aircraft. A particular flight strip can only be correct or incorrect, satisfactory or not, with respect to other representations of the information within the ATC work system. The flight strip has no direct connection with its aircraft. It is conceptually related to the aircraft but not communicatively, except via other things. (Note that the conceptual relationship is not shown on Fig. 1.4, but it could be if this was the analyst's focus). If the ATCO updates a flight strip

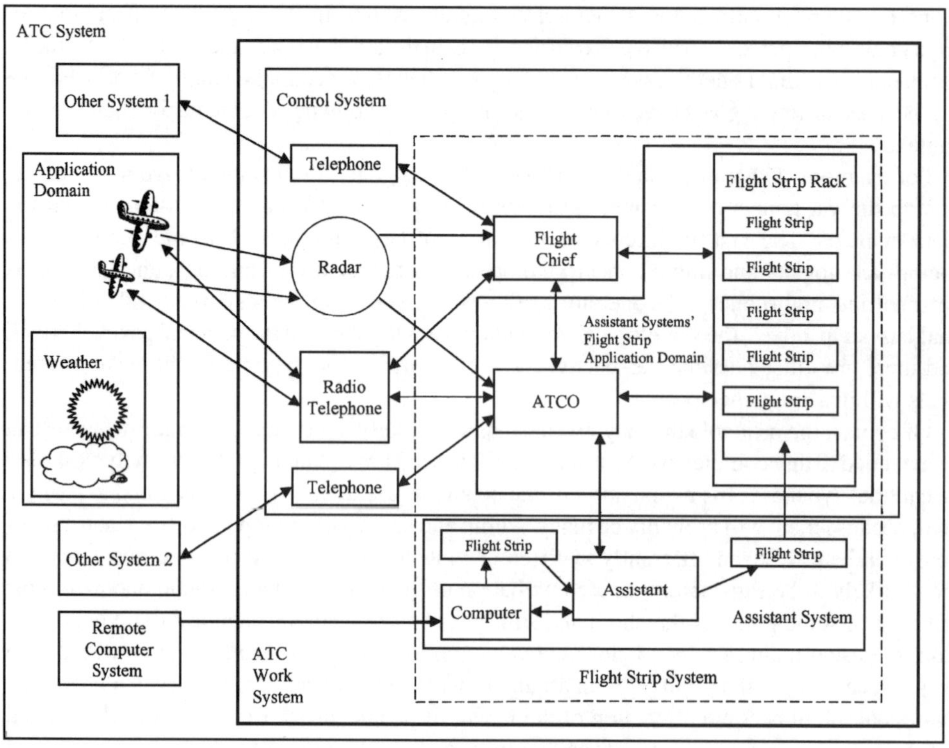

FIG. 1.4. A simplified model of a traditional air traffic control (ATC) system that uses paper-based flight strips.

with information that is incorrect with respect to the aircraft that the strip purports to represent but is correct with respect to the source of the information, say, incorrectly displayed aircraft information on a radar screen, then what is incorrect in the system is not the flight strip but the radar. In safety critical applications such as ATC, there is usually considerable deliberate information redundancy so that, on the principle of double-entry bookkeeping, information mismatches have a high probability of being detected. Provided a problem does not escape from within the ATC work system and that there are no time or other types of penalties incurred, then it has no effect on the satisfactory performance of work achieved in the application domain of aircraft movements.

The above example, in which updating a flight strip is shown not to be an ATC work system task, would appear to present a conundrum, as obviously, if a flight strip is not updated or is changed incorrectly, the result could be a real world "incident," what used to be called a "near miss" between aircraft in the application domain. The solution to this apparent problem with the Dowell and Long (1989) definition of work involves establishing what to compare the state of a flight strip with and using a multiple systems model that always has a work system and application domain for each of its embedded models.

1.3.3 Multiple Models

The final piece in this part of the jigsaw puzzle is the concept of multiple work systems and application domains. In theory, any set of components that have some communicative relationship, directly or by other things, can be designated by the analyst as constituting either a work system or an application domain. In practice, analysts will tend to assume that most of the possible work systems and application domains are not sensible. The minimal criterion to apply, however, is whether a definition of work is possible.

There are a couple of little theoretical twists that may help in understanding the ideas but do not usually have great practical relevance. First, with complex application domains, it is possible to consider reversing the work system and application domain. Although this would be insane in normal ATC, as it would mean that the purpose of the aircraft movements was to satisfactorily perform work by changing the ATC work system, it is a conceivable method of testing a new ATC system using real aircraft. Equally insane, at least superficially, is to suggest that a simple tool like a hammer performs work by changing its user, although the idea that people adapt to their tools does imply that using a hammer has effects on the hammerer, and this could be the goal of a tool's designer. Educational systems, of course, do explicitly have user-changing goals. The cynical view that real people are driven by bureaucracies or by computer systems rather than the reverse can be easily understood by switching application domain and work system. This concept is most useful when there are chains of work systems and application domains, but care needs to be taken, as often the application domain of work system_n is not quite the same as that for work system_{n+1}.

The second twist is that, although many analysis methods use a hierarchical structure, it is not necessary to do so, and in many cases it is more valid to use heterarchical models (section 1.2.2). A heterarchy will greatly increase the number of work systems and application domains that can be considered for analysis. In practice, it is therefore necessary to be careful in defining both the application domain and the work system of interest.

Returning to the ATC example, the ATC work system is modeled as containing three of many possible work systems, the control system, the assistant system, and the flight strip system. The flight strip system, delineated by dashed lines, includes the assistant system and part of the control system, and provides an example of a heterarchical system.

The assistant system has a computer, a human assistant, and flight strip components. The computer receives its input from some other, remote computer and prints a flight strip, which the assistant then removes and places in an appropriate empty slot in the flight strip rack. The new flight strip is "cocked" (it sticks out a bit from the other ones), and there is often some communication, verbal or nonverbal, between the ATCO and the assistant. The assistant system can be considered as a work system, and its application domain is the flight strip application domain, which is that part of the control system (including the ATCO and flight strip rack) that the assistant system communicates with so as to perform work in this application domain. The work goals of the assistant system are to correctly place flight strips and to ensure the ATCO is aware of new strips in the flight strip rack. The critical point is that the work of the assistant system is not directly related to real aircraft or, in this case, to the flight chief, who is not responsible for checking the arrival of new flight strips.

It is usually an error, and probably a common one, for analysts to jump from identifying a failure to achieve the task goals with respect to one application domain and work system to concluding that a similar failure has also occurred with respect to a different, often more general, application domain. In other words, analysts unjustifiably assume that if a failure occurs in one part of a system, it will cause a problem in some other part of the system, usually a problem in a real world application domain, such as that part of the ATC system that includes aircraft flying around the sky (the application domain in Fig. 1.4). Only in a zero redundancy system is this assumption guaranteed to be correct. Zero redundancy virtually never occurs in real systems involving people because such systems are inherently noisy (i.e., human performance is variable and prone to error).

To avoid problematic assumptions, analysts need to study the relevant work system with respect to the goals they wish to refer to. Incorrect information on a flight strip, for example, may have no effect on the movement of aircraft if such errors are trapped and rectified before they affect the ATC work system goal (i.e., safe and expeditious movement of aircraft). Such errors may well cause concern to those involved in ATC system design. If only the communicative relationships between ATCOs and flight strips have been analysed, then rational efforts at solving work problems, (i.e., based on analysis), can only be directed at these relationships (e.g., attempts to redesign the ATCO–flight strip interface to reduce the probability of input errors). If the more general ATC system has been analyzed, then other design options rationally become available, such as improving the ATC work system's ability to detect and correct errors or reducing the number of possible transmission and transcription errors.

1.4 HUMAN-COMPUTER INTERACTION AND SOFTWARE ENGINEERING

The general assumption throughout this handbook is that the purpose of performing a task analysis is to improve computer systems and make them better tools for people's benefit. HCI's genesis as a distinct discipline arose in the early 1980s, and from this time HCI has been recognized as being inherently interdisciplinary and as requiring consideration of both psychological and computing issues as well as issues in many other areas (Diaper, 1989d, 2002a, 2002b).

HCI is a specialized subdiscipline of ergonomics. It is so for two related reasons. First, HCI restricts itself to the study of systems that have a computer component. Second, HCI's emphasis has been on human psychology far more than in traditional ergonomics, which started with an emphasis on matching human and device physical capabilities (see also chap. 28). This shift of emphasis is related to the view that computers are tools to augment human mental abilities whereas nearly all earlier tools augmented human physical abilities. In its early days, HCI was primarily concerned with human cognition, to the extent that Norman's (1986) classic chapter is titled "Cognitive Engineering," although wider issues related to organizational psychology were pioneered, for example, by Mumford (1983), and issues of social psychology have been promoted since the development of the field of computer supported cooperative work (CSCW) in the early 1990s. Furthermore, as chapter 29 argues, consideration of cognitive psychological issues is not sufficient, even within the psychology of individuals, and emotion (affect), life style, fashion, and so on, as important aspects governing human perception and behavior, also need to be taken into account (see also section 1.2.1).

HCI is an engineering discipline rather than a science because its goals are inherently practical and involve satisfying design criteria (Diaper, 1989d, 2002a). HCI design criteria, however, although they often involve aspects of computer systems, may also involve altering human cognition (e.g., by training people) and affect (e.g., by increasing their pleasure and hence motivation) as well as revising organizational processes and other aspects of application domains or work systems. As an example of the latter, improved ATC work systems have changed the ATC application domain so that more aircraft can be safely accommodated within the same airspace.

Although the roots of computer science are in mathematics (Diaper, 2002a), and there is scope for formal methods within HCI (Harrison & Thimbleby, 1990) and task analysis (e.g., chap. 11 and most of the chapters in part IV), HCI is most closely related to the computing field of software engineering. Diaper et al. (1998; Diaper, 2002a) go so far as to suggest that no distinction should ever have been made between software engineering and HCI because both are engineering disciplines concerned with the same types of systems and their difference is merely one of emphasis, with software engineering focusing more on software and HCI more on people.

1.4.1 Human-Computer Interaction

The discipline of HCI continues to be defined in two different ways, broadly and narrowly (Diaper, 1989d, 2002a). The broad view of HCI is that it is concerned with everything to do with people and computers. The narrow view is that it is concerned with usability, learnability, intuitiveness, and so forth, and is focused on the user-computer interface. These two views of HCI fit neatly with quite different definitions of work systems and application domains (section 1.3). In the narrow view, the application domain is the computer system, and the work system is the end user and other things in the user's environment. The users' tasks, in the narrow view, are to satisfactorily change the state of the computer system. The consequences of the computer's state change for other parts of the whole system are outside this narrow view of HCI. Hammer (1984) refers to interface design as a second-order issue and claims it is only relevant when two computer systems possess equivalent functionality (section 1.4.2.2; but see also chap. 30). Hammer's definition of functionality is firmly grounded in the real-world consequences of the automated office systems he discusses. User interface design issues are a major part of HCI, but a broader definition of the application domain and work system is needed if real-world consequences are to be taken into account.

There is a moral imperative associated with the broad view of HCI (Pullinger, 1989). It is certainly necessary for some discipline to consider the myriad effects that computers have on people, individually and collectively, and the supporters of the broad view claim that this discipline is and should continue to be HCI. The broad view treats user interface design as an important but subsidiary part of the whole HCI enterprise. Task analysis is at the core of HCI because of its emphasis on changes made to application domains, which are often in the real world (i.e., outside of the work systems). Such changes inherently concern system performance, which is what distinguishes task analysis from other more static, atemporal systems models (section 1.2.2.2).

1.4.1.1 The Psychology of Users

Notwithstanding the need in HCI to consider affective, social, organizational, and other such issues, most of the psychology in HCI and in current approaches to task analysis focuses on human cognition, and it is human cognition that is the main ingredient of user models in HCI. The point to recognize is that the cognitive psychology of people is much more complicated than, for example, the information-processing abilities of computer systems and that this creates a fundamental problem for task analysis. If an analyst cannot understand the operation of a basic system component (such as the human element), then it is nigh impossible to predict how the various things in a system will interact and produce the behavior of the system. A cognitive user model of some type is essential, however, as the Skinnerian-based behaviorist approach (chap. 3), which treats the mind as a black box and merely attempts to relate its inputs and outputs, has since the early 1960s generally been recognized as inadequate for either the explanation or prediction of human behavior. Fortunately (and provided that HCI is treated as engineering rather than science), it is possible to make use of psychological models of users that are good enough for engineering purposes, even if they are inadequate scientifically.

A vast range of psychological user models (see chap. 28 for a review) are employed across the range of task analysis methods available (for reviews of task analysis methods, see chap. 6, 22, 24, and 25). Admittedly, at one extreme the user models are implicit and resemble behaviorist models in their concentration on human behavior in a task context rather than on how mental operations cause such behaviors. The cost of ignoring mental operations is that, as soon as analysts move from describing behavior within an existing system to predicting behavior within new systems, they must rely on their craft expertise in human psychology. In one sense, all people are expert psychologists because they inhabit a rich, complex social world in which they must be able to understand and predict both their own behavior and that of other people. On the other hand, people do not do this very well, for otherwise a science of psychology would be unnecessary or at least could be simpler and more understandable.

Scientific cognitive psychology is not an easy subject to comprehend, even though it has concentrated on describing the mental architecture common to everyone (usually modeling the mind as an information-processing device) rather than on the contents of people's minds. As an example, it focuses on how memory works (architecture) rather than what is remembered (content). Chapter 15 provides an example of a sophisticated information-processing model of human cognition that can be used in task analysis. Although this interacting cognitive subsystems (ICS) model for cognitive task analysis (CTA) well illustrates the complexity and hence analytical power of scientifically based cognitive models and has demonstrated reasonable predictive adequacy over a quarter of a century of development, it is by no means universally accepted by professional cognitive psychologists, and there are alternative models, such as ACT-R (Anderson & Lebiere, 1998), that make similar claims to analytical and predictive capability. Furthermore, radically different cognitive architectures, such as those based on parallel distributed information processing (Hinton & Anderson, 1981), also have their supporters among professional cognitive psychologists.

A useful contrast can be made between the ICS approach and that described in chapter 16, where it is assumed, probably incorrectly (Diaper, 1982, 1989b), that people have access to their own mental processes and can describe these when interviewed appropriately. As chapter 19 points out, there is often a considerable difference between what people say and what they do (see also Diaper, 1989a, and chap. 28). Although unsatisfactory from a scientific perspective, the assumption made in chapter 16 may well be adequate for some HCI engineering purposes, as people's beliefs about their own minds are likely to affect their behavior even if such beliefs can be experimentally demonstrated to be inadequate or even incorrect. Chapter 16 reports a considerable degree of accuracy in the recall of relevant events in some of the interviews collected on critical incidents.

Between the extremes represented by implicit psychological models and models like ICS are the more popular cognitive models associated with such task analysis methods as HTA (chap. 3) and GOMS (chap. 4). Note that HTA is based on the description of tasks as a hierarchy of goals and that, as Annett says in chapter 3, it "differs radically from earlier methods of task analysis by beginning, not with a list of activities, but by identifying the goals of a task." It is unfortunate that some users of HTA continue to confound psychological goal states with descriptions of behavior. Consider Annett's first example: "the goal hierarchy for an acid distillation plant operator's task" (Fig. 3.1) This hierarchy has a high-level goal. ("Inspect instruments & locate fault") and lower level ones (e.g., "1.2.4 Start standby pump"). Given that a videotape could be made of the operator starting the standby pump it is easy to understand why some people forget that in HTA the object is to describe psychological goals and not behavior. Remember Annett's warning, already quoted in section 1.3.2: "Simply to list actions without understanding what they are for can be misleading" What is really needed in HTA and in task analysis generally is a language that differentiates psychological things from observable behavior so as to avoid this potential source of confusion. Note that one alternative would be to assent to philosophical behaviorism (Atkinson, 1988), which should not be confused with the Skinnerian sort. Philosophical behaviorism denies that correctly specifying psychological things is possible, and it thus can allow observable behavior to stand for unknown psychological states (Diaper & Addison, 1991). As Diaper (2001b) pointed out, philosophical behaviorism has not become popular. Nonetheless, it is discussed further, from a more radical perspective, in chapter 30.

In chapter 4, Kieras states that GOMs is intended as a method to be used after a basic task analysis has been carried out and that its purpose is "to provide a fomalized representation that can be used to predict task performance well enough that a GOMS model can be used as a substitute for much (but not all) of the empirical user testing needed to arrive at a system design that is both functional and usable." GOMS employs hierarchical decomposition so as to identify, among other things, "primitive operators," which may be internal (cognitive) or external (observable behavior). It then assigns an estimated execution time to each primitive operator, usually 50 ms for internal ones and longer durations, based on task observation data, for external ones. GOMS uses a "production system model for human cognition." This model, although undoubtedly a misrepresentation of human psychology, predicts task performance time by simply summing the time associated with each primitive operator. That is, it treats task processes as independent of each other, although GOMS-CPM (Critical Path Method; see chap. 22) is claimed to incorporate some parallel cognitive processing. GOMS has enjoyed considerable success over the years at predicting both task performance time and errors. The GOMS psychological model of users is thus good enough, usually, for engineering purposes.

In summary, there is no escaping the need for task analysis to include a psychological model of the human parts of a system. The model used, however, will be adequate if it merely approximates the real, albeit still unknown, psychology of people. After all, HCI is an engineering discipline, and its practitioners must do their best to satisfy the practical criteria associated with particular projects. A number of chapters in this handbook (e.g., chaps. 3, 7, and 17) suggest that task analysis requires expert practitioners. One reason to believe that it does is that selecting and then using one or more psychological models requires experience and craft skills (Diaper, 1989e) and should not be confused with merely following the steps of a method, no matter how well specified or complicated the method is.

1.4.1.2 The Psychology of Computing Industry People

Software engineers are people too! So are systems analysts, designers, programmers, and even managers. A specialized branch of HCI was developed in the 1980s and became known as the psychology of programming (PoP), which is something of a misnomer these days, as the field covers all sorts of people involved in the computing industry. PoP has the potential to become a vast field of study, but it remains fragmented and underresearched (Diaper 2002a). An understanding of computing industry people (CIPs) is of importance to HCI and for task analysis for two reasons. First, as the producers of computer systems, CIPs have some influence over the computer systems that are delivered. Second, CIPs are users of methods and CASE tools and of the results of these.

Norman's (1986) classic HCI model identifies models of two types: users' models of the system image and designers' models of the users' models of the system image. In reverse order, the system image is everything that the end users can perceive about a computer system by interacting with it, including its manuals, any training provided, and so on. These users construct their users' models on the basis of the system image (see also chaps. 18, 19, and 24). The computer system's designers have their own models of the users' models of the system image, and Norman suggests that one major source of problems with delivered computer systems is that there has been a mismatch between the designers' models and the users' models.

Furthermore, some of CIPs' influence on the development of a computer system is implicit; that is, people make decisions and perform actions without understanding their basis or perhaps even recognizing that they've made a decision. Various different styles of working, whether in systems analysis, design, programming, or management, provide numerous examples of consequences of development that are style dependent and not understood by the workers. Carroll (2000) makes a case that design in particular is difficult, is not routine, and is creative, and just the last of these implies that no two designers or design teams will ever produce quite the same thing. Indeed, the dispute between Bench-Capon and McEnery (1989a, 1989b) and

Barlow et al. (1989) (see sec. 1.3.1.2) hinges on the latter's view that delivered computer systems have many properties that are implicitly caused by those who do the designing and building of them, including how these people are managed. The importance of this for HCI and task analysis is that it partially explains why the performance of computer systems is more complicated than ever described in their design and development documentation. This chapter has already suggested that people, as system components, are very complex (section 1.2.1), and perhaps the complexity of any reasonably sized computer system has also been generally underestimated.

The second reason for introducing CIPs into this chapter is that they are users of methods and CASE tools, including task analytic ones, and users of the results of the methods, perhaps because some methods, including task analytic ones, need to be carried out by those with the appropriate craft expertise (sections 1.4.1.1 and 1.6 and chaps. 3, 4, 7, and 17). Diaper (1989e, 2002a) suggested that there is a problem delivering HCI methods to the computing industry. One of the primary reasons TAKD failed (Diaper, 2001b) is that it was too complicated to do; certainly a CASE tool was essential (see also sections 1.4.2 and 1.5.1.5). Diaper has even suggested that, when designing a method, the designers should determine the method's design requirements by modeling its CIP users as much as by looking at the domain of the problem that the method addresses. TAKD, for example, was originally intended to identify the knowledge people required to perform IT tasks and so form the input to IT syllabus design. Unfortunately, it gave too little consideration to TAKD's other potential users. Undoubtedly this caused TAKD to be too complicated to be successfully delivered to the computing industry, notwithstanding its flexibility and analytical power. CASE tools may be a good way to facilitate the industrial uptake of methods, they may sometimes be necessary, perhaps for methods such as TOOD (chap. 25), but they are not sufficient. The delivery of task analysis to the computing industry is further discussed in chap. 30.

1.4.2 Task Analysis in the Software Life Cycle

The discipline of software engineering developed from the recognition in the late 1960s that there was a "software crisis," namely, that the demand for quality software had far outstripped the capabilities of those able to produce it. The software crisis remains with us today. Most software engineering solutions have been anthropocentric (Diaper, 2002a) and have attempted to increase software engineering productivity and improve the volume and quality of the software produced by providing methods of working, organizational as well as software-oriented ones, and building CASE tools to support some of these methods. Chapters 23 and 24 review a number of task analysis CASE tools, and most of the chapters in part IV of this handbook, along with others, such as chapter 7, describe CASE tools that support particular task analysis methods.

The most widely cited early attempt to produce a structured software development method is now usually referred to as the "classic waterfall model" (e.g., Pressman, 1994; Sommerville, 1989). This model, describes the major stages of software development using labels such as requirements, design, coding, testing, and delivery. The waterfall aspect is the strict sequencing of these main stages. Although it is always possible to return to an earlier stage to make changes, the development must again cascade through all the stages below that one. It is generally accepted, after much practical software engineering experience, that the earlier in the waterfall any changes are made, the greater is the additional cost to a project. Often the costs of change are suggested to be an order of magnitude greater for each stage further back in the sequence. The waterfall model is not now used in most software engineering projects, primarily because it fails to describe how software engineers actually work, but it continues to provide a convenient vocabulary to describe the software life cycle.

The waterfall model was followed by many structured software development methods, such as Jackson Systems Design (JSD; e.g., Cameron, 1983) and subsequently, the overly complicated Structured Systems Analysis and Design Method (SSADM; e.g., Ashworth & Goodland, 1990; Downs, Clare, & Coe, 1988; Eva, 1994; Hares, 1990). A method of this type defines one or more notations and provides relatively simple descriptions of how to use these. One general problem is that methods are difficult to describe. Methods involve processes but most natural languages are declarative, that is, they are much better at describing things than how things are transformed (Diaper, 1989b, 2001b; Diaper & Kadoda, 1999).

As an alternative to structured methods, various prototyping approaches have been developed. In a prototyping approach, an incomplete version of the software is developed and tested in a series of iterations. The hope, of course, is that a satisfactory system will eventually result (see section 1.2 and chap. 7).

A question that is always relevant is, 'Where will task analysis be used in the software life cycle?' The old HCI slogan, "Get the human factors into the earliest stages of a project," still applies (e.g., Lim & Long, 1994). In particular using task analysis early and repeatedly throughout a project can improve quality and reduce costs. It has also been argued (e.g., Diaper, 1989c) that the cost of using task analysis is greatly reduced when it is used throughout a project, as the initial effort of producing the systemic description of the world can be amortised over the life of the project. For example, if task analysis has been used at the requirements stage, it should be cheap to use the same model, with appropriate minor changes to reflect the development of the new computer system, for evaluation purposes, and quality should be improved because similar problems are being addressed at both these stages (see section 1.5.1.1).

1.4.2.1 Task Simulation and Fidelity

Howsoever one collects the data for a task analysis of a current system (sections 1.2, 1.5.1.4, and 1.5.2.1), all one ever has is a simulation of the real tasks of interest (Diaper, 2002b; chap. 30). There are three reasons for this. First, as Carroll (2000) has pointed out, generally there are potentially an infinite number of tasks carried out by different people that can be studied, but only a small number can be selected for analysis. Second, as mentioned in section 1.3.1, only a small part of a system can ever be observed so that the task data are always incomplete. Third, there is nearly always a Heisenberg effect: the act of collecting the data alters what is being studied. The only circumstances in which a Heisenberg effect would not occur is where the people or other things being observed are not affected (and, in the case of people, not even aware they are being observed). Thus, observation and other ways of collecting data about tasks are not easy and require considerable craft skill (section 1.5.1.4; Diaper, 1989a). In the case of a proposed future system (section 1.2), of course, the tasks can only be simulations, as broadly define in Life, et al. (1990).

It is therefore necessary to realize that task analysis always deals with simulations. *Fidelity* (a closely related term is *validity*) refers to the degree to which a simulation models the assumed real world (Section 1.2). (From a solipsist perspective, it is the degree of agreement between two models of the assumed real world.) Diaper (2002b) argued that fidelity varies widely across task analyses. Fidelity can be reasonably high when an existing system or prototype and its real end users in their natural environment are studied. Fidelity can be very low when use scenarios (short prose descriptions of the performance of a putative future system) are used (Carroll, 2000; chap. 5), which is not to say that these are not helpful, particularly in the earlier stages of the software life cycle. Between these extremes are approaches that are intermediate in fidelity and that are usually recognized as and called simulations. Many of these are paper based or use only a user interface without the back-end application software (e.g., Diaper's, 1990a, Adventure Game Interface Simulation Technique [AGIST]). The Wizard of Oz simulation technique provides higher fidelity simulations and has been widely used, for example, for studying natural language–processing computer systems that cannot yet be built (e.g., Diaper, 1986; Diaper & Shelton, 1987, 1989). In a Wizard of Oz simulation, the natural language–processing capabilities of a future computer system are, unknown to the users tested, simulated by a person in another location, and with sufficient care in design, such a simulation can be very convincing to users. Undoubtedly there is a strong correlation between fidelity and the point at which task analysis is used in the software life cycle, with fidelity usually lower the earlier task analysis is used (see, e.g., Benyon & Macaulay, 2002).

1.4.2.2 Functionality

Within the computing disciplines, the term *function* and its derivatives, such as functional and functionality, are polysemous (they have more than one meaning). Functionality, for instance, refers to features available on the user interface and to capabilities of the back-end software that support the achievement of goals. These two types of functionality share an asymmetric relationship. The first is really a user interface design (UID) issue, supported by the narrow view of HCI (section 1.4.1), and it is important because a computer system's back-end functionality is useless if people cannot access and use it easily. On the other hand, the back-end functionality has to be there first. Questions about what a computer system can do as part of a work system are crucial to HCI and are more important than interface issues (section 1.4.1). Although Sutcliffe (1988) is right that good UID can save a functionally sound computer system, he is wrong to suggest that better user interfaces can save poor software. In HCI, functionality is a property of any thing in a work system, but it is usually ascribed to a work system's agents (section 1.2.1; see also chap. 30).

Functionality is typically thought of as static, as a property of a work system's agents and not of its performance. Although this may be a helpful way to think about functionality, it is obviously wrong, as there will often be functionality performance issues. For example, there are many things that cannot be computed simply because the program would take too long to run; encryption algorithms rely on this. At the other end of the spectrum, the functionality of a real-time control system is severely constrained by the sequence and timing of events as well as complexities in the application domain beyond the system's direct control. Given that task analysis is primarily concerned with the performance of work, one of its potential contributions is to get analysts to realize that functionality applies to performance as well.

One specialized use of task analysis is for functional allocation (see chaps. 6, 22, and 24), which involves dividing the labor among the agents and determining which other things are used and how. Automating everything possible and leaving the remainder to people was soon recognized by ergonomists as a recipe for disaster.

As an example of how functionality can be allocated differently, consider where the intelligence to perform tasks is located in work systems that include expert systems and work systems that include hypertext ones (e.g., Beer & Diaper, 1991). Both types of computer system can be formally described by graph theory as having an architecture of nodes connected by links that can be traversed. In an expert system, the intelligence involved in link-traversal tasks resides in the expert system's inference engine, and the users are reduced to being suppliers of data to the expert system. In a hypertext system, however, the intelligence required to traverse links is the users', and the computer part of the system is passive (section 1.2.1). Web search engines provide one means of allocating some of the link-traversal task functionality to the computer system.

Functionality is obviously such an important concept in software engineering, HCI, and task analysis that it is odd that it has not be more fully discussed within the research literature. Although the term functionality is not absent from this handbook's index, for example, it is not one of the most widely used terms herein. Chapter 30 returns to the topic of functionality and takes a more radical stance.

1.4.2.3 Implementation Independence, Abstraction, and Goals

David Benyon and the author of this chapter have long been engaged in a friendly public discussion about whether and in what way task analytic descriptions are independent of any particular design or implementation (Benyon, 1992a, 1992b; Benyon & Macaulay, 2002; Diaper, 2002c; Diaper & Addison, 1992). Similarly, several chapters of this handbook disagree as to whether task analysis should represent realistic, concrete examples of tasks (e.g., chaps. 2 and 3) or generalized, abstract, higher level task descriptions (e.g., chaps. 4 and 7). Most task analysis methods are capable of doing both, of course, but there appears to be little agreement as to how close a task analysis should be to a specific implementation in a particular environment, and so on. Obviously, when task analysis is used prior to design, there must at least at some low level of abstraction and some independence between the task analytic representation and the design to be subsequently produced. When there is an existing system or prototype on which to base a task analysis, however, the question arises as to how closely the task analysis should be tied to use of the system (i.e., what degree of the fidelity the task simulation should possess; section 1.4.2.1). This issue has almost certainly caused confusion in the task analysis literature. Part of the confusion can be resolved by considering where in the software life cycle task analysis is to be used (sections 1.4.2 and 1.5.1) and what its purpose is (section 1.5.1.2). When used to evaluate a current system, a task analysis might be very concrete, describing specific things and behaviors in a detailed way. In contrast, in a creative design situation (e.g., the family-wear scenarios in chap. 5), the task analyses must be at a higher level of abstraction because the design has not yet been fleshed out. Furthermore, some writers, such as Benyon and Macaulay (2002), have argued that further abstraction is desirable. Diaper (2002b), however, suggested that scenarios such as those of Carroll are not proof against premature commitment to some detailed design. In contrast, for example, although Task Analysis for Error Identification (TAFEI; chap. 18) is able to evaluate the design of devices prior to their implementation, the method is highly device dependent.

The concept of goals, at least people's goals, occurs in every chapter of this handbook, although only this chapter (section 1.3.2) and chapter 26 state that things other than people can have goals. Although there is a strong case for considering goals to be essential for explaining the observable behavior of system components (sections 1.3.2, 1.4.1.1, and 1.5.1.5), there appears to be a second, generally implicit argument that describing goals is a means of abstracting away from an implementation-dependent description. The argument here is that describing what people want to do (their goals) is easily separable from what they have to do (the tasks they must perform) to achieve these goals. Section 1.4.1.1 suggested that in HTA (and probably many other task analysis methods) one source of confusion is that descriptions of psychological states such as goals are not sufficiently different from descriptions of observable behavior. A related concern is that if task analysts believe that by describing goals they are safe from making premature design commitments, they might make such commitments unwittingly. Furthermore, if people naturally think in relatively concrete scenarios, as proposed by Diaper (2002c), then unwitting premature design commitments are quite likely. Overall, the lower the level of abstraction (i.e., the more detailed the task analysis), the greater the chance of such problems occurring. The issue of the abstraction role of goals is dealt with in chapter 30.

1.5 DOING A TASK ANALYSIS

There are many types of projects that involve the design, development, delivery, and maintenance of computer systems. All these projects have a human element in that the projects themselves are staffed by people (section 1.4.1.2), but all also have a broader HCI element, as delivered computer systems always have an impact on some people's lives. Furthermore, given that task analysis, because of its concern with system performance issues, is at the core of HCI (sections 1.1 and 1.4.1), virtually all computer-related projects will involve some form of task analysis. This conclusion is not invalidated just because the majority of computing projects do not explicitly recognize the premises of the argument: Any computing project involves people inside and outside the project; such people are a main focus of interest for HCI; HCI is concerned with performance and hence task analysis. It would undoubtedly facilitate clarity of thought, however, if the premises and conclusion were accepted in every computing project so that the various roles of task analysis could be identified in each project.

One reason, of many, for identifying the roles of task analysis in computing projects is that task analysis cannot be fully automated. Analysts must make many decisions that are subjective (neither right or wrong) but may be more or less useful. Task analysis CASE tools (sections 1.4.1.2 and 1.4.2 and chaps. 23 and 24) can make analysts' decision making easier in many ways, but some "judgment calls" are unavoidable. Perhaps a long way in the future, approaches such as Subgoal Templates (chap. 17) might lead to some artificial intelligence–based subjective decision making, but this is far beyond current AI capabilities and may never be deemed entirely desirable. Thus, given that computing projects will involve task analysis and that task analysis will be iffy and introduce indeterminacy (aka noise) in project management, then identifying the use of task analysis in projects must be a valuable aid to project management and project cost estimation. All this boils down to the proposition that computing projects will, nearly always, involve task analysis, so they might as well recognize task analysis, and if they are going to recognize it, then they might as well call it task analysis as well.

This handbook, though by no means exhaustive, contains a considerable number of different task analysis methods suitable for different purposes at different stages in projects. What most distinguishes these methods are their later analysis stages (see section 1.5.2).

1.5.1 What For and How?

Whether well specified or not, all software engineering projects involve some form of development method. Obviously, the first question is, 'Where could task analysis be used in the project?' but answering this may be far from simple. Many activities in software engineering methods are not task analytic, so perhaps the first real question for each step in a method (section 1.4.2) is to ask, What can task analysis directly contribute? Sometimes the answer will be "nothing," as is usually the case when encoding a design and for some forms of code testing. This answer is reasonable, because task analysis only has an indirect contribution to such project stages via other stages, such as requirements and design specification and functional and user testing.

Having identified where task analysis can contribute to a project, analysts commonly leap to the question, What task analysis method? This is an error, and it can make the results of the task analysis unusable. The author has been as guilty of this mistake and in his case it resulted in many years of research trying, for example, to bridge the gulf between requirements and design (e.g., Diaper, 1990a, 2001b), because the requirements representations from the task analyses were incompatible with design ones. The solution, once this problem is recognized, is to work backwards at the project level and iterate around at the individual stage levels.

1.5.1.1 Working Backwards and Around

Computing projects are going to produce something because they are engineering projects, although the something produced may not be a computer system or software but instead a training program, manual, or management system. Indeed, at the start of some projects, even the type of solution may not be known, such as when using SSM to address soft problems rather than hard ones. Once the stages of a project have been identified so that the "Where could task analysis be used in the project?" question can be addressed, it is best to start with the relevant final stage of the project and work back through the stages asking this question. This strategy is the opposite of current practice, but it is sensible because in most projects the relevant, nearly final stage, will involve some form of testing of the performance of the project's product. Performance, of course, is central to task analysis, so this stage will almost certainly have a yes answer to the question. Furthermore, in an ideal development method, everything specified in the earlier stages of a computing project ought to be evaluated before the product is released, which is a consequence of the principle that everything specified at one project stage should be covered in subsequent project stages. The same principles apply to prototyping cycles as well.

Having identified the stages to which task analysts can contribute, analysts can address a whole host of other questions at each stage. Ultimately, the project stage that will require the most extensive, detailed, and expensive task analysis should determine how task analysis will be used in other stages so that the costs can be amortized over these stages (section 1.4.2). Although a common assumption is that the stage needing the most extensive task analysis is a requirements one, it might just as plausibly be a near final performance evaluation one, and it could be even some other stage, depending on the type of project.

Starting with the stage that will require the most extensive task analysis, the critical questions are as follows:

1. What do you want to know from the task analysis?
2. What is the most appropriate task analysis output format?
3. What data can be collected?
4. What task analysis method?

Although it is tempting (and common practice) to ask these question as they arise in the carrying out of the work, it is better to ask them initially in the order given above. The questions are not independent, so it is essential to iterate around them to refine the answers to each. Just as one should never conduct a scientific experiment without knowing how the data will be analyzed, one should never conduct a task analysis without addressing these questions. In the worst case cost scenario, a great quantity of task analysis data will be collected and analysed but never used. This is costly for two related reasons. First, collecting data is itself expensive. Second, large volumes of data not only take a long time to analyze but generally generate confusion among the analysts. It is easy to suffer two combinatorial explosions, of data collection and of analysis, sometimes to the extent that what is produced is less useful than what would have been produced if a more modest but better planned approach had been adopted.

Once the above questions have been addressed for the project stage that will make the most extensive use of task analysis, the same questions can be addressed for the other stages. Generally, the answer to question 4 should be the same for each project stage, and likewise for question 3, for the same data collection method should be used at each stage even if the emphasis on the data and the degree of detail vary. Questions 1 and 2 are stage specific. The rationale for using a consistent task analysis approach is that it will (1) amortize the costs over the project, (2) maintain or improve the relationship between project stage, and (3) transfer expertise across the stages.

Planning across the whole project is the key, but such planning needs to be approached iteratively, and the analysts need to be prepared to return to each project stage and modify their planning. No doubt sometimes a single task analysis approach cannot be adopted across an entire project, but use of a single approach should be a goal that is abandoned reluctantly. In many real computing projects, different personnel work on different project stages, and insufficient communication and management across the stages is a common problem. In a typical case, the design and implementation team hand over their prototype to the human factors team for usability testing, but the latter use an explicit task analysis method that radically differs from the, perhaps implicit, method used for requirements and design specification, making the results of the usability testing partially if not totally inapplicable to the design. One of the strengths of Coronado and Casey's use of task analysis (chap. 8) is that a similar task analysis is used iteratively throughout the stages of a project.

The following four sections address in more detail the four key questions discussed above.

1.5.1.2 What Do You Want to Know From the Task Analysis?

For each stage in any well planned computing project, this must be the critical question. Given the vast range of projects and the variety of development methods, offering general advice is nigh impossible. Worse, this is a very difficult question to answer, and the suspicion must exist that in real projects it is often dodged on the basis that the task analysis is exploratory and will highlight interesting issues. Unfortunately, in general task analyses won't have this result unless the analyst looks for "interesting issues" in the analysis. Admittedly, some issues may arise serendipitously, as the act of performing any sort of analysis often enforces a discipline on the analyst that leads to insights that would not otherwise occur, as also argued in chapter 4. On the other hand, basing an engineering approach on hoped-for serendipity is not a trustworthy strategy. The trick is to be resigned to repeated refinement of the answers to this question as the others are considered. In other words, start with this question but don't expect to get it right the first time.

1.5.1.3 What Is the Most Appropriate Task Analysis Output Format?

One of the strengths of structured approaches to development in software engineering is that different stages use one or more different representations. For example, among other representations, DFDs, Entity Life Histories (ELHs), Entity Relation Diagrams (ERDs), and Process Outlines (POs) are used in SSADM (section 1.4.2), and use cases, activity diagrams and interaction diagrams, are used in UML (see chaps. 7, 11, 12, 19, 22, 23, 24, and 27). It would therefore seem desirable for task analysis output to be in the form of one or more of such software engineering representations or be easily related to them. This desideratum is not met by many task analysis methods, although, in this handbook, methods developed for industrial application (part II) and those discussed in part IV have been designed or adapted to relate to software engineering representations. One potential disadvantage of task analysis methods that easily relate to software engineering representations is that they often require a specific software engineering method to be adopted or demand changes, often considerable, in the chosen software engineering method. In either case, the needed revisions may be unacceptable or costly for companies that have invested heavily in a particular software engineering approach not explicitly supported by task analysis, although such companies might benefit in the long term by switching to methods that do have an explicit task analytic component.

The core of the problem is the historical separation of software engineering and HCI (section 1.4). Many task analysis methods were developed by researchers with a psychological background, and these methods and their outputs often do not integrate well with those of software engineering. Furthermore, adapting task analysis methods post hoc so that their output is in a software engineering representation seems to be difficult. Diaper (1997) reported that an attempt was made to modify TAKD so that its output would be in the form of ELHs, but this attempt was subsequently recognized as a failure (Diaper, 2001b). On the other hand, the Pentanalysis technique has been successfully used to directly relate task analysis to DFDs (Diaper et al., 1998).

Although asking how task analysis output contributes to a computing project is essential, there are many cases where, even though the output does not easily fit with the software engineering approach, undertaking one or more task analyses is still valuable. One reason is that software engineers and systems analysts, for example, often do task analysis implicitly and poorly (section 1.2.2.2; Diaper et al., 1998), and doing a proper task analysis explicitly is nearly always bound to lead to an improvement in quality (at a cost, of course). Another reason is that the act of doing a task analysis can have benefits not directly related to the collected data. First, preparing for a task analysis and building models of the world can improve the similar models that are essential in software engineering. Second, collecting task analysis data forces analysts to look at their world of interest with care and often in more detail than is common in software engineering projects. Third, the analysis of the task analytic data usually causes further reappraisal of the data or the collection of new data. The author of this chapter has frequently commented (e.g., Diaper, 2001b) that in more than 20 years of doing task analyses he has never got the task analysis data representation of activity lists (section 1.5.2.2) correct the first time, and TAKD's method explicitly recognized this in its analysis cycles. In summary, no one wants to expend effort, time, and money conducting task analyses that cannot significantly contribute to a project, but the contribution of a task analysis may be considerable even if it is indirect and the analysis output does not directly map to the software engineering representations that the project uses.

1.5.1.4 What Data Can Be Collected?

One of the myths about task analysis is that it involves the detailed observation of tasks. Although in some cases this may be so, task analysis should be able to use and integrate many types of data (Diaper, 1989c, 2001b; Johnson, Diaper & Long, 1984; chap. 3). Integrating different types of data is facilitated by treating data from a performance-based perspective (section 1.3). Of course, disentangling the identification of what data can be collected from the methods of collecting it is not easy. The effort is worthwhile, however, so that the analysts are at least aware of what sorts of data might have been missed.

Apart from the observation of performance, task analysis data can be gathered from interviews and questionnaires, from all sorts of documentation, from training programs, and from consideration of existing systems. Part of the skill of the task analyst is in recognizing how data conflict and being able to see, for example, that the official account of how something works (e.g., according to training systems or senior management) is at odds with actual practice. Resolving such conflicts requires social skills and often raises ethical issues. Task analysis, because its primary focus is performance, tends to be socialist. After all, it is the workers, those who actually perform the tasks of interest, who are the task experts and who will usually be those most affected by changing the current system. Indeed, many IT system developments have failed because the workers were not properly carried along. Bell and Hardiman (1989) reported on one such case:

> We even have one example of a highly successful knowledge-based system which is saving its owners many millions of dollars a year. The designers consulted some of the potential users of the system when the system was being built, but as the designers only worked 9 am to 5 pm they only consulted the first- and second-shift users. When the system went live, the night-shift workers refused to have anything to do with it: the system was only being used by two-thirds of the work-force. (p. 53)

Where there is conflict between the views of managers and workers, negotiating the difference is essential, as a new or revised system will likely fail if it is based on an incorrect model of the world (note that the official model will often be less correct). Managers do have a right and duty to manage, of course, but some react defensively to any suggestions that their view of the world is not the only possible one. The author's experience working as a consultant to industry is that the workers who perform the tasks of interest are often enthusiastic, sometimes overly so, about providing data for task analyses because they perceive it as a way to advantageously influence their future work. It is also worth noting, in passing, that people are usually quite enthusiastic about talking about their job anyway, because they spend hours a day at it and it tends to be a taboo subject outside of the work environment and often with everyone except their immediate coworkers.

Although there are always general issues concerning the fidelity of data (section 1.4.2.1), one particularly important aspect of data fidelity is the data's level of detail. Obviously, collecting highly detailed data that are not used is hideously expensive, but having to return to data sources to collect more data can also be very expensive (Diaper, 1989a). There are two sets of related problems having to do with the fidelity of data. One arises for analysts who use a deductive (top-down) task analysis decomposition method such as HTA, and the other confronts those who use inductive (bottom-up) methods such as were characteristic of TAKD.

First, because HTA is a deductive analysis method, even though it has its P × C stopping rule (chap. 3), the analysts do not know the required level of data detail until they have done the HTA analysis. Unless the analysts are highly inefficient and collect quantities of data that remain unanalysed, they will have to switch back and forth between data collection and analysis. The problems of such iteration are relatively obvious when the data are derived from interviews or questionnaires. In such cases, if a topic is not raised, then the data required will be absent, and the analysts will have to return to their data sources. There are expenses, however, even with data sources that are relatively complete, such as video recordings of tasks or extensive documentation (e.g., training manuals), as these sources have to be returned to and reanalysed. To take the case of video recordings, where missed data are available on the recordings (and sometimes the data are simply not captured), the recordings have to be watched again. Making video recordings of tasks is easy, but it is time consuming and tedious to extract data from them, say, about 10 hours of analysis for every hour of recording. Note that a video recording is not itself an observation but merely a medium that allows analysts to observe events outside of their real-time occurrence (Diaper, 1989a).

A related problem that plagues inductive methods is that the level of data detail must be selected prior to the data's analysis. The only safe and efficient approach here is to try to ensure that the initial data collected are as detailed as is likely to be necessary. Although collecting detailed data is expensive, the advantage of working from details upward is that only the pertinent data need be analyzed. Inductive methods are therefore particularly appropriate for analyzing data from media such as video recording.

Deductive and inductive analysis methods also face task sampling problems (chap. 30). One substantial problem for inductive methods is that, until the analysis has been done, empirically based principles to guide sampling are lacking. There are two types of sampling problem given a range of tasks have been sampled, (1) tasks not sampled from within the range, and (2) possibly relevant tasks outside the sampled range. Although neither problem is amenable to a guaranteed solution, solving the latter is particularly difficult and must rely on the cunning and imagination of the analyst; the former sometimes can be dealt with using some form of statistical analysis. Chap. 8 describes commercial situations where analysts' access to people and their tasks are very limited, and chapter 2 prioritizes the desirability of sources of task analysis data, recognizing that it is not always possible to adequately sample tasks with the ideal people and in the ideal environments.

With deductive methods, it is theoretically possible to establish what data to sample as the analysis progresses to lower levels of detail. What Green (1990) calls "viscosity" however, comes into play: as the analysis progresses, it becomes increasingly hard and expensive for the analysts to change earlier parts of the analysis. Viscosity is less of a problem with induction, and is particularly well illustrated by machine induction, as it is widely recognized that most things that can be validly induced from a data set are not what is wanted. The equivalent problem with deduction is not widely recognized, however, perhaps because deduction engines such as program compilers are easy to produce. Of course, they are easy to produce because we don't care exactly what is deduced but only that it is a valid version. That is, apart from the issue of functionality (e.g., that some compilers, are optimized for execution efficiency, others for debugging), we don't care that different compilers produce different machine codes. In contrast, in task analysis we do care what is deduced at each level of an analysis, and thus similar sampling and analysis problems occur with both inductive and deductive approaches.

Although in a minority, there are application areas where a complete set of correctly performed tasks can be analyzed, but usually not a complete set of incorrectly performed ones. There are two such application areas: (a) where the primary tool to be used is of such limited functionality that it can only be used in one way to perform a task (e.g., Baber and Stanton's digital watch, discussed in chap. 18); and (b) in some safety critical systems, defining safety very broadly (Diaper, 1989d) where correct task performance is enforced. Note that enforcement is only likely to be close to successful when functionality is relatively limited, that is, not in complex, flexible safety critical environments such as ATC (section 1.3.2 and chap. 13).

Carroll (2000; see also Diaper, 2002b, 20002c) is correct however, that there are potentially an infinite number of ways that different people may perform a range of tasks more or less successfully and efficiently, and thus it is necessary to sample tasks for analysis. There are two basic approaches to task sampling: using opportunity samples and doing selective sampling. Both may be used in a project, but they do have different consequences for analysis, as noted below.

In an opportunity sample, which is most frequently used with some form of relatively high fidelity data recording, such as video recording, the analyst, having carefully arranged everything beforehand, of course, simply arrives and records whatever tasks are going on in the workplace. This is a particularly good approach when the tasks or subtasks of interest are of a relatively short duration and repetitive. ATC might be a good example, although even when we used opportunity sampling a decade ago in the Manchester ATC tower, we did try to ensure that, in the 1 hour of live video recording that we took, the workload did vary, as we were aware that ATCOs and flight chiefs work slightly differently depending on their current workload. For example, when they are busy, they are more likely to give instructions to an aircraft to fly to a radio beacon rather on a specific geographical heading. The great advantage of opportunity sampling is that it allows an analysis of the frequency of tasks and task components. The disadvantage, of course, is that some critical but rare tasks or subtasks may not be observed, and information on these must always be sought from the task experts.

Although selective sampling of tasks has the disadvantage of making frequency data unavailable, it does result in a principled coverage of the range of tasks sampled. The sampling can be any combination of frequent tasks, complex tasks, important tasks, and error-prone tasks. Obviously the task analyst needs to understand the possible task range before collecting the data, and the most common source of such information consists of those who currently perform the tasks or might do so in the future. Usually one gets this sort of information by interviewing people, but care needs to be taken with such interviews because people do not usually classify the tasks they do in the way the analyst is likely to. Critical incidence reports (chap. 16) constitute one rich source of complex, important, and error-prone tasks, although there are potential problems with how well people recall critical events (section 1.4.1.1 and chap. 28). Another useful source of task samples consists of training schemes, which generally cover frequent and important tasks well but are often less helpful for identifying complex tasks (because subtasks interact) and error-prone task and error recovery tasks (which are sometimes not recognized at all in training schemes).

The final warning on task data collection is that, whatever method is used (e.g., generating scenarios, ethnography, interviews, and task observation), adequate collection of data depends on the craft expertise of the task analyst. One weakness in the contents of this handbook is that, unlike its predecessor (Diaper, 1989a), it does not contain a chapter that explicitly addresses task data collection methods, although many chapters provide advice and examples relevant to the analysis methods they describe. There are lots of tricks of the trade for any of these data collection methods, but the key to success is always thorough preparation in advance of data collection.

1.5.1.5 What Task Analysis Method?

As noted in section 1.5.1.1, the question of what task analysis method to use should be the last one addressed and not the first, but the questions covered in the subsections above should be iterated, as they are not independent. A task analysis method in this context is a method that converts task-orientated data into some output representation for use in a project. A major goal of chapter 22 is to help analysts select task analysis methods appropriate for their purposes.

A not unreasonable suspicion is that, in practice, people either choose a task analysis method with which they are familiar or they use something that looks like HTA. As chapter 3 makes clear, there is more to HTA than merely the hierarchical decomposition of task behaviors, and there it is argued goals rather than behaviors are what are decomposed in HTA (see also sections 1.3.2 and 1.4.1.1). Indeed, this handbook's editors rejected more than one chapter because it simply glossed over the relevant analysis method, often referring to it merely as a "basic task analysis," although this concept does still appear in a few chapters that address other aspects of task analysis in detail (e.g., chaps. 4 and 7). Section 1.5.2 does attempt to describe what may be thought of as a "basic" task analysis in that it tries to describe the initial steps involved in virtually any task analysis before a specific analysis method is applied to the data.

Although analysts understandably would prefer to use a task analysis method with which they are familiar, giving in to this preference is generally a mistake and may be based on a misunderstanding of what is complicated about task analysis. The actual methods used to analyze data are quite simple, and it is their baggage, much of it essential, unfortunately, that adds the complications. For example, one of the disadvantages of many of the task analysis CASE tools described in part IV is that (a) they demand inputs of certain sorts that may have to meet data completeness criteria that in a specific project situation are not relevant and (b) they are designed to produce very specific types of output. Chapter 4 also comments on the additional effort that a CASE tool may require (see also section 1.5.1.3).

None of the various chapters that purport to classify task analysis methods and CASE tools (e.g., chaps. 6 and 22–25) entirely succeed at providing a mechanism for selecting a particular method across all the situations in which task analysis might be applied in projects (which is not to say that they lack valuable information about the methods they cover). The general advice has to be to know the input data and the desired output and let these drive the selection of the analysis method. On balance, it is probably the required type of output that is the most important selection criterion. The reason for this is the lack of integration of many task analytic and software engineering methods (section 1.5.1.3).

As with many software engineering methods, in task analysis the representations used for analysis are usually the method's output representations, although some methods do have a translation stage so that they output a representation used in a software engineering method, such as a UML representation (section 1.5.1.3 and chaps. 7, 11, 12, 19, 22–24, and 27). Task analysis methods have stages and a small number of analysis operations within each stage. One aspect of HTA that makes it a simple task analysis method is that it has only a single main analysis representation. Most task analysis methods have a relatively small number of stages and representational forms, such as Scenario-Based Design (Carroll, 2000; chap. 5) and ConcorTask-Trees (chap. 24). Methods such as GOMS (chap. 4) and TAFEI (chap. 18) are partial task analysis methods in that the they deal with the later stages of what a task analyst does. TOOD (chap. 25) is an example of a complicated task analysis method that is probably only realistic to use, unless you are its inventors and intimates, when supported by a CASE tool.

Methods such as HTA are recognized as being difficult in that analysts need training and possibly years of experience to acquire expertise (chaps. 3 and 28). Indeed, many task analysis methods are rather messy in their description and in practice. In some cases, analysts must work around a method's different stages, although methods such as TOOD (chap. 25), control this rather better than most, partly as a consequence of its being supported by CASE tools. Diaper (2001b) discussed how building a CASE tool to support TAKD (the LUTAKD toolkit) made clear to the method's inventor the need to simplify the model of how analysts visit TAKD's stages. Feedback from task analysts trying to use TAKD indicated that stage navigation was a difficulty and, although it is only a small program, one of the LUTAKD toolkit's most important facilities was a navigation system that forced analysts to make very simple decisions concerning what analysis stage to do next.

Task analysis methods do up to three things with their input data: extract it, organize it, and describe it. Although task analysis data may come in many formats, the most common is a prose description (a "task transcript" or "scenario") of one or more tasks. Other formats, such as check sheets (Diaper, 1989a), interview notes, ethnography reports, and so forth, can be treated like prose task descriptions. The extraction, organization, and description processes vary widely across different methods, but however these are done, they tend to have the general features described below.

Extraction typically involves two operations: Representing each task as a sequence of short sentences, known as an activity list, a task protocol, or an interaction script (Carroll, 2000; Diaper, 2002b), and classifying things. As noted in section 1.2.1 and, for example, by Diaper (1989c), the things tend to be organized into objects and actions, and often the objects are separated into agents and things used by agents. Classifying things in other ways has developed over the last dozen years or so, as triggers, for example (chaps. 14 and 19).

Organization also involves two typical operations: integrating different task descriptions if more than one description of a task is used and categorizing the information extracted. The most common organization is a hierarchical one, although section 1.2.2.1 argues for heterarchical models, and some methods, such as simplified set theory for systems modelling (SST4SM) (Diaper, 2000a, 2001a), can use a flat, nonleveled model to produce temporary alternative classification schemes. The most common consequence of the categorization process is the division of task descriptions into subtasks.

Naturally, almost all task analysis methods claim to be able to combine descriptions of a task performed by different people in different ways. Quite a few methods are able to combine different tasks into a single task representation. On the other hand, there are some outstanding problems with combining even the same task performed in different ways. This is most obvious in task analyses that use high-fidelity data capture methods, usually video recording, to document a specific realistic task. Given more than one recording of the same task, the different ways it is carried out ideally need combining in a single task representation. Readers of this handbook and the other task analysis literature should be alert to the various ways that most task analysis methods fudge this issue. Diaper (2002b) described Carroll's (2000) method of noting alternatives at the bottom of the activity list as quite crude, but at least the issue is recognized by Carroll; it is not in some of this handbook's chapters. The most common fudge is to have a single ideal task description that limits the description of use to an optimized set of subtasks and operations. Although many methods use this approach, TAFEI (chap. 18) provides a particularly clear example because it explicitly exploits the approach to look for deviations from the ideal task model (errors). Nor is it clear that abstracted task models (section 1.4.2.3) help solve this problem. In many cases, to represent different tasks, analysts must employ craft skill, and chapter 30 argues that this is one problem area that ought to be recognized and dealt with.

The task analysis description operations are what primarily distinguish different task analysis methods. The variety of these operations is vast, and a great deal of this handbook is devoted to describing the operations used by the task analysis methods addressed. The description operations aid in further extraction (Carroll, 2000; chap. 5) and organization or add properties to objects and actions (e.g., those having to do with sequence or preconditions). The next section describes the typical first stages of task analysis before the analysts use whatever description processes their chosen method specifies.

1.5.2 The First Stages of Task Analysis

Like so much in life, preparation is the key to doing a good task analysis. The analysts should start by addressing the following five questions:

1. Which project stages will use task analysis?
2. What do you want to know from the task analysis?
3. What is the most appropriate task analysis output format?
4. What data can be collected?
5. What task analysis method?

To obtain reasonable answers to these questions, the analysts will have to have done most, if not all, of the intellectual preparation for the task analysis. Questions 1, 2, and 3 concern the role of the task analysis in the project. Answering questions 4 and 5 will require some basic background understanding of the tasks to be analyzed. What remains to be sorted out is the logistics, who will do what tasks when. The logistics may be trivial, as in the case of a small, well-integrated design team generating scenarios, but they can be very complicated, as when video recordings are to be made in the workplace. With experience, making video recordings of a wide variety of tasks in different environments is quite easy, but there is a lot of craft trickery in dealing with ambient light and sound, camera angles, and so on, beyond knowing how to operate the camera. Indeed, there is a lot of craft trickery in all data collection methods, which is one more argument for the suggestion that task analysis should be done by those with expertise in it (see also section 1.6).

The following subsections attempt to describe the first steps in nearly all task analyses. Different task analysis methods use different terminology, of course, and some of what is described below is skipped or is implicit in some methods. The description is independent of the data collection method, but the default method used is making video recordings of a range of tasks by a number of different people in their real work environment. This is a fairly high-fidelity data collection method (section 1.4.2.1), but the same issues arise with other methods such as interviews and even with low-fidelity task simulations such as scenarios (Diaper, 2002b).

1.5.2.1 Data Collection and Representation

By the time analysts are ready to start collecting data for a task analysis, they will have a systemic model of the work system and application domain (sections 1.2 and 1.3). It may only be in their minds, but they must have identified the things that will be recorded. This is obvious for video data but is even true, at the other end of the fidelity continuum, for scenario generation, for it is necessary to know, in general terms, what the scenarios are to be about. Note that the act of writing a scenario is equivalent to making a video recording of a task. The systemic task analysis (STA) approach introduced in section 1.5.3 argues that if the analysts are required to have a systemic model, they should start by making it explicit.

Task selection involves identifying the tasks that will be used to supply data (section 1.5.1.4). One needs to identify the tasks, their number, who will do them, and whether similar tasks are to be done more than once and by different people, for example. In general, the higher the fidelity of the data collection method, the more expensive, in all sorts of ways, the data collection exercise. It is obviously usually much easier to generate some more scenarios than to have to revisit a workplace to make more video recordings. One important decision to make concerns the level of detail of the data that will be collected, as too high a level will require further data collection and too low a level will be more expensive and may cause the analysts confusion (section 1.5.1.1).

The result of a task data collection exercise will be a set of records that describe how a set of tasks were performed. The records might be video- or audiotapes, written scenarios, ethnography reports, interview notes, or summaries of document sources such as training manuals. Interestingly, high- and low-fidelity data collection methods, such as video recording and scenario generation, respectively, result in records that more closely follow task structure than intermediate methods such as interviewing, even when the interviews are task performance orientated (e.g., Diaper, 1990b; chap. 16; but see sections 1.4.1.1 and 1.5.1.4).

A scenario as a story about use (Carroll, 2000; chap. 5) is a task transcript, that is, a prose description of task performance. One of the minor myths about task analysis that seems to have survived for decades is that if a transcript is lacking, creating one is the next step in the task analysis. One justification, more common in industry than academe, for producing a task transcript, say from a video recording, is that this is an administrative task that can be done by labor cheaper than a task analyst. An argument against this arrangement is that doing transcription is a good way for analysts to understand their data and their analysis. For video transcription, the rough estimate, on which there seems to be a consensus, is that transcription takes about 10 times longer than the recording. A better alternative when a transcript is not available is to take the records, whether interview notes, video or audio recordings, or whatever and immediately convert them into the activity list format that is central to virtually all task analysis methods. If one does have a task transcript, then producing an activity list is the next step.

1.5.2.2 Activity Lists and Things

An activity list, sometimes called a "task protocol" or an "interaction script" (Carroll, 2000, chap. 10), is a prose text description of a task that has the format of a list. Each item on the list represents a task step. The heuristic that this author has promulgated for years (Diaper, 1989c, 2001b, 2002b) is that each line of an activity list should have a single main agent that performs a main action that affects other things (i.e., objects and agents). Depending on the level of detail, there may be short sequences of subsidiary actions specified, such as those that never vary (e.g., pressing the [Return] key after a command line input), and the main action will often entail quite lengthy strings of task components (e.g., when entering text using a word processor). People do carry out tasks or subtasks in parallel (e.g., simultaneously browsing their PC's diary while arranging a meeting with someone on the phone), and the advice here is to treat the two related tasks separately in the activity list and represent each of the related task components on adjacent activity list lines. Switching between two or more unrelated tasks can be treated in the same way, or separate activity lists can be constructed, if the interleaved tasks don't affect each other. The end result of the activity list construction process should be an ordered list of the steps performed to carry out each task recorded.

It is important to note that, to follow the above heuristic, it is necessary to have identified the things that will be represented in the activity list and to at least have classified them as actions, agents, and other objects (see sections 1.2.1 and 1.5.2.1). One might chose to extend this classification while constructing the activity list (e.g., identifying triggers, as proposed in chap. 19) or this might be done at a later analysis stage.

Generally the analyst will want to give each line on the activity list a unique identifier, and the obvious thing to do, it seems to most people, is to number the activity list lines. If you do this starting 1, 2, 3, 4, . . . , then, in practice, you are almost bound to run into difficulties. As I've mentioned, after 20 years of doing task analysis I've still never got an activity list right first time and always have to return to it once I've done some further analysis. If I numbered the lines, I would need to renumber them each time I made a change. Even if software is being used that propagates changes to activity list line numbers throughout all the analyses, such changes still give analysts difficulty. After all, the point of identifying each activity list line is so that it can be easily referred to and found. When I number activity list lines these days, I prefer to use intervals of 100 or 1,000 so that I can add lines subsequently without renumbering.

With observational data collection methods, it is possible to collect time data. Most analog and digital video cameras allow the recording to be time stamped, usually to the nearest second. Recording task performance on paper and using a stopwatch is quite a difficult task for the analyst, and the timing is probably only reliable to within 3–5 seconds. Whenever possible I always try to use video rather than pen and paper. If collecting the time data is free, as with video recording, then it is still an issue whether to put it on the activity list. The temptation is to do it because the data are available, but my experience is that time stamping an activity list is a time-consuming business and should be done only if it is essential to what will be analyzed subsequently. It usually has to be done when efficiency issues are the primary concern of the task analysis. One advantage of using GOMS after producing an activity list is that GOMS provides estimates of the time it takes to perform task components (chap. 4), so time doesn't have to be recorded on the activity list. Nontheless, it is a good idea to check the GOMS estimates for a sample of tasks or subtasks against the actual durations recorded.

Many HCI tasks involve a dialogue between a small number of agents. Where there are two, three, or four agents (four is probably the upper limit), a good format for an activity list is to represent each agent in a separate column. This is easy with two agents, typically a user acting as the agent of the work system and a computer acting as the application domain's agent (sections 1.3.2 and 1.3.3). One advantage of this format is that the 3Ps (perceive, process, and perform) can be applied to each agent's column, similar to the cognitive model advocated by TOOD (chap. 25; see also chap. 28). Another advantage is that the format allows representation of behavior by one agent and different processing and responses by other agents. For example, in ATC, the ATCO may type an input to a flight strip control system while being observed by the flight chief sitting next to the ATCO; the computer system and the flight chief will obviously treat the ATCO's behavior differently.

An activity list describes a single instance of one or more tasks by listing what occurs in sequence, if not in time, although it may be more or less concrete (i.e., more or less detailed; section 1.4.2.3). The problems that remain concern how to deal with parallel tasks, how to deal with interrupted and interleaved tasks, and how to combine activity lists to produce generic task descriptions. These problems are properly part of the task analysis processes discussed in the next subsection, and some of them are illustrated in section 1.5.3.4 (see also chap. 30).

1.5.2.3 Analysis in Task Analysis

To start with a true but apocryphal story, something was obviously learned since the first British HCI conference in 1985, because the next year, at HCI'86, there was a stock answer to most questions from the audience: "Oh, we did a task analysis." In most cases, this meant that they had stood over the user's shoulder and watched for a few minutes. Although something may be better than nothing, they did have go and find a user, after all, this really is not good enough, and it gives task analysis a bad name. One option, perhaps understandably not often mentioned by those who produce articles and books on task analysis methods, including many of the coauthors of this handbook, is to settle for producing activity list descriptions and to decide that no further analysis needs to be done. Producing activity list descriptions of tasks really does force an understanding of the individual tasks (see also chap. 4). The question is, how much more do the subsequent analyses add? Sometimes a lot, of course, but there are cases, for example, in requirements analysis, where what is learned from activity list construction is the most useful contribution of the task analysis. In addition, sometimes when evaluating a prototype, for example, it is easy to spot task performance problems directly from individual activity lists.

See section 1.5.3.4 for a discussion of the general sorts of operations involved in different task analysis methods. As noted in section 1.5.2.2 and chapter 30, there are issues to be resolved concerning how to integrate task descriptions and deal with parallel, interleaved, and alternative versions of tasks.

1.5.3 Systemic Task Analysis

The discussion in section 1.5.2 is intended to apply to task analysis in general, even though the advice about activity lists is quite specific because some such representation is common in most task analysis methods, although formats vary. In contrast, this section illustrates with a microscopically small example how the first stages of a task analysis can be carried out. This requires a particular notation and method, and what will be introduced is a new task analysis method, currently dubbed systemic task analysis (STA). This method takes the ideas about systems and work performance introduced in sections 1.2 and 1.3 as its basis. A full description of the STA method is not provided here, but just the representations that STA uses at each of its initial main stages.

The main stages in STA, which all iterate in something like a classic waterfall-type model (section 1.4.2), are as follows:

1. Static systemic modeling.
2. Data collection.
3. Activity list construction.
4. Dynamic systemic modeling.
5. Simplified set theory for systems modeling (SST4SM).

1.5.3.1 Static Systemic Modeling and the Example Scenario

Section 1.5.2.1 argues that at the start of a task analysis exercise the analyst must have some form of systems model, however implicit, so STA makes a virtue of this by having its first step make such a model explicit. Using ATC as an example (see section 1.3.2), a systems model such as shown in Fig. 1.4 is the sort of thing that can be quickly obtained by looking around a control tower and interviewing a few people after perhaps having done some background reading. Such a system model identifies the things in the system (section 1.2.1), models both their conceptual and communicative relationships (sections 1.2.2.1 and 1.2.2.2, respectively), and identifies various goals that start to define the work to be achieved (section 1.3.2). It is a static model, however, in that it does not represent either time or sequence. For this reason and because there is no explicit representation of performance, it is not a task analytic model, although the analysts will probably run the model in their minds (section 1.2.2.2).

For the example, let us further assume as was the case with our Manchester Airport study (Section 1.3.2), that the primary interest is in a requirements capture exercise which is starting by modelling how the current paper flight strip system performs. The work, in the early 1990s, was carried out with about a ten year planning horizon for replacing the paper based system with an electronic flight strip system.

1.5.3.2 Data Collection

Having gained access to the Manchester ATC tower, not an easy thing, as there are major safety considerations, we set up a video camera to record the ATCO sitting in front of the flight strip system. Despite the cacophony of a real ATC tower, what the ATCO and the flight chief, who sits next to the ATCO, say is audible on the video recording (getting good sound required some careful, pre-researched placement of the camera microphone). We also had an extension feed to the video camera to record both sides of the radio telephone communication between the ATCO and the aircraft (also researched in advance). The video camera is left to run for an hour, and as expected the workload increases during the session. The ATCO is interviewed immediately after the session using a post-task walk-through. In this walk-through, the ATCO stops and starts the recording of the session and explains what is going on, either spontaneously or prompted by us. We record the walk-through by pointing the camera at the TV screen showing the original task video while the ATCO makes comments. The time stamp on the original video is easily readable on the walk-through recording, which is also time stamped. The walk-through of the 1-hour Manchester Airport task video takes about 3 hours, a typical amount of time.

1.5.3.3 Activity List Construction and Analysis

Whatever the format of the task data collected, an activity list is a fundamental (perhaps the fundamental) representation of a task in task analysis. Table 1.1 shows a fragment of an activity list that might have been created from the Manchester Airport videos. Producing it would involve first watching the task video and identifying the actions and other things with their times, then going through the walk-through video. Table 1.1 represents what an activity list might look like after several cycles of development during which the analyst has refined the descriptions of the task steps and added comments. The irregular activity list line numbering reflects this cyclical development (section 1.5.2.2). The activity list could be further refined, as discussed below.

The activity list Table 1.1 follows the systems model shown in Fig. 1.4 by separating the work system, whose primary agent is the ATCO, from the application domain, which is the flight strip rack. In a paper-based flight strip system, the flight strip rack is an object and not an agent, as the rack has no processes (sections 1.2.1 and 1.4.2.1 and chap. 26). The focus of the task analysis in Table 1.1 is on what the ATCO does, and the application domain descriptions are represented at a higher level and less completely than the descriptions of the ATCO's behavior.

At the first activity list line, 4800, the ATCO is in radio communication when the assistant places a new flight strip in the rack; the ATCO finishes this task at line 5700. At line 5800, the ATCO starts to scan the flight strip rack. The subject of the comment attached to this line (i.e., the trigger for this task) is the sort of thing an expert task analyst would have asked about during the walk-through. In a further refined activity list, it might be represented by its own activity list line specifying the ATCO's cognitive operations. Line 6100, for example, does indicate the cognitive rule the ATCO claims to use. Having decided to deal with a particular aircraft at line 6100, the ATCO then drops into a subtask concerning when the aircraft's flight strip arrived on the flight strip rack. The ATCO queries the flight chief about the flight strip before returning to the main current task of dealing with the selected aircraft at line 6700.

TABLE 1.1
Fragment of an ATC Activity List

ID	*Time*	*Work System*	*Application Domain*	*Comments*
...				
4800	14:24	Assistant places new flight strip for aircraft AC 234, cocked, on rack.	New flight strip AC234, cocked, on rack.	*... ATCO on radio to aircraft AC 207. Note input from the assistant system to the work system.*
...				
5500	15:08	ATCO ends radio dialogue to aircraft AC 207.	Radio turned off.	
5600	15:10	ATCO updates flight strip AC 207 with pen.		
5700	15:12		Flight strip AC207 updated.	
5800	15:15	ATCO scans rack of live flight strips.		*Cognitive trigger: "Do at end of each task and do often."*
5900	15:22	ATCO scans rack for new, cocked flight strips.		*Cognitive trigger: "Do after live flight strips and if no immediate tasks."*
6000	15:24	ATCO reads flight strip AC 234.		
6100		ATCO decides flight strip AC 234 needs dealing with as its arrival time is within 5 minutes.		*Aircraft AC 234 is on a diverted flight.*
6200	15:28	ATCO takes flight strip AC 234 off rack and holds it in hands.	Flight strip AC 234 removed from rack.	*Physical ATCO behavior.*
6210	15:28	ATCO doesn't know exactly when flight strip AC 234 arrived on rack.		*ATCO says this "only a mild concern" as knew he'd looked at the whole rack "3 or 4 minutes ago" (Actual = 5m:15s ago).*
6250	... 15:28	Flight Chief is looking at radar.	Radar display is normal.	*Flight Chief's current activity.*
6270	15:28	ATCO is looking at whether the Flight Chief is "free."		*This is a complex work- and social-based decision but done frequently.*
6300	15:30	ATCO shows flight strip AC 234 to Flight Chief.		*The flight strip is moved to be within easy reading distance for the Flight Chief while still being held by the ATCO.*
6301	15:30	ATCO says to Flight Chief, "When did this arrive?"		
6450	15:33	Flight Chief reads flight strip AC 234.		
6550	15:36	Flight Chief attempts to recall when last saw Assistant near rack.		*This is not an official Flight Chief responsibility.*
6650	15:40	Flight Chief says to ATCO, "About 5 minutes ago."		*This triggers the ATCO to return to dealing with the AC 234 flight strip.*
6700	15:43	ATCO focuses attention on rack.		
...				

The activity list in Table 1.1 could be more detailed or less detailed. For example, some activity list lines could be combined, such as 5500 and 5560; 5800–6000 and 6700; and 6200–6301 and 6450–6650. The level of detail chosen for Table 1.1 was primarily driven by the author's perception of its understandability by this chapter's readers, but it is the sort of level of description he has often found useful when analyzing important tasks. ATC is highly safety critical, so some depth of analysis is usually warranted.

Section 1.5.2.3 suggests that producing an activity list can sometimes be sufficient for some projects' purposes. To produce one such as in Table 1.1 does require the analyst to understand, quite thoroughly, each task analyzed. Given that the main project that the Manchester Airport study was to contribute to involved designing computer replacement systems for the paper-based flight strip rack, there are two major design issues that arise just from the activity list fragment in Table 1.1: (a) new flight strip arrival time information, and (b) ATCO and flight chief cooperative work and communication and their estimates of each other's current workload.

At line 6100, the ATCO notices that an aircraft is due to become "live" in a few minutes and realizes that he doesn't know when the flight strip appeared in the rack. With a paper system, an obvious solution to this problem would for the assistant to write on each flight strip the time it is placed on the rack. The ATCO expresses only "mild concern" at this problem, although we might consider looking in more detail at the difference between the ATCO's estimates of when the rack was last scanned and the actual time between scans (e.g., is the difference between "3 or 4 minutes" and 5:15 minutes significant?). With an electronic flight strip system, there are a range of design issues. Apart from merely recording the time a flight strip arrives on the rack, each flight strip could have one or more clocks or countdowns associated with it. For example, each flight strip could display when it was last modified, how long until the next action is due, and when it was last attended to. The latter could be achieved by selecting individual strips, whole blocks of them, or all strips after scanning them (see chap. 13). Any such design is likely to have consequences for cognitive actions like the triggers identified in lines 5800 and 5900. One measure of subsequent design success might be the reduced probability of the ATCO and flight chief needing to discuss such problems, particularly when they are both busy.

Sitting next to each other in the current ATC work system, the ATCO and the flight chief talk a lot, and when both are not busy, not all of the talk is narrowly task focused. Some is even social persiflage, which may be quite important, as the two have to work in a highly coordinated fashion, trusting each other. Note that these are important psychological things about these people (Section 1.2.1). Even assuming that a decision has been made to keep the ATCO and the flight chief next to each other in front of the electronic system, lines 6250 to 6650 still highlight a number of design issues. They are the sort of issues that task analysis tends to be better at identifying than other HCI and software engineering methods. First, the ATCO has to be able to estimate what the flight chief is doing, so he needs some way to get an overview quickly, and presumably the same is true for the flight chief. Second, the ATCO wants the flight chief to read a particular strip but does not want to give the flight chief control of it (information garnered from the post -task walk-through). These two issues create computer-generated display requirements concerning the general observability and specific readability of flight strips by someone not directly in front of the display. Interestingly, even after extensive task analyses, the actual new ATC system, described in chapter 13, had initial problems with display readability. A really creative task analyst might also foresee some design issues concerning less variety of style in both the display (e.g., the loss of variation in hand written things an flight strips when they are computer generated) and the ATCO and flight chief's behaviors, which will make estimating what the other person is doing harder in the new electronic system.

The above is a lot of design- and evaluation-relevant stuff to get from a few lines of a single task's activity list. Analyze more lines and more tasks and a great deal more design requirements can be taken directly from the task activity lists. Why do more analysis?

1.5.3.4 Abstract and Dynamic Systemic Modeling

Why do more task analysis after constructing an activity list? The most obvious answer has to do with the volume of information. In many, probably most, task analyses, the analyst winds up with many tens and often hundreds of pages of activity lists. What task analysis methods solve, more or less successfully, is the problem of reducing this volume by combining tasks, representing them at higher levels of description so as to allow different task instances to be combined, and representing alternative and optional tasks (e.g., an error-recovery subtask). Most task analysis methods do other things as well, and these things partially distinguish them from each other.

HTA is one of the hardest methods to do well. It is a popular method in industry but is not always done well (Ainsworth & Marshall, 1998) because it is poorly specified and relies on considerable craft expertise (chaps. 3 and 28). Although they look more complicated because they are better specified, particularly the task analysis methods supported by CASE tools, many methods are actually easier to carry out than HTA because the analyst is guided through the process (section 1.5.1.5). Notwithstanding my personal preference for heterarchical models (section 1.2.2.1), Fig. 1.5 shows where one might start producing a hierarchical model from the activity list fragment in Table 1.1. An inductive, bottom-up approach (section 1.5.1.4) has been used, and at this stage the specific activity list lines have been left in as hierarchy terminators below some initial task classifications. Whether the resulting hierarchy is actually an HTA diagram as described by Annett (chap. 3) depends on how these task classifications are defined; in an HTA "Deal with new flight strips," for example, would be an ATCO goal and not a description of behavior. As more activity list lines become incorporated in the hierarchical analysis, the model will change. For example, not all the subtasks that follow the "Identify

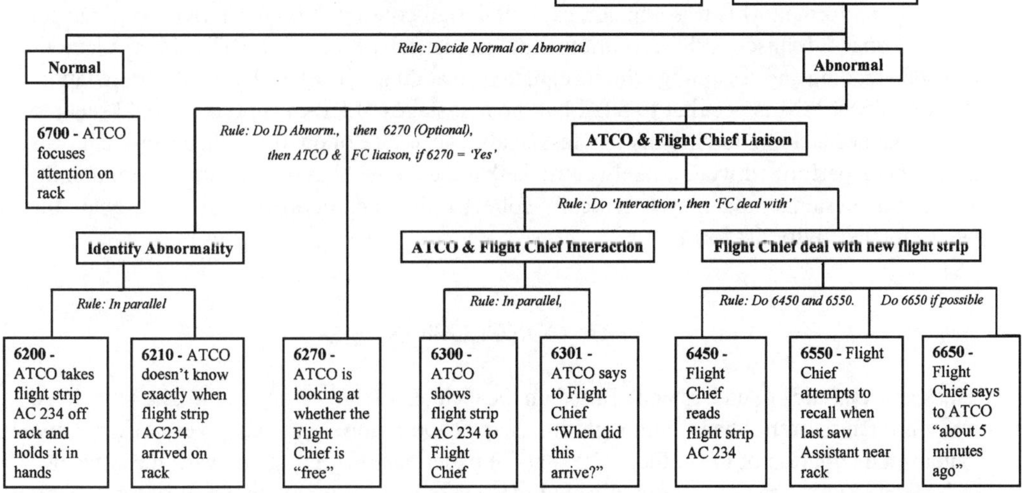

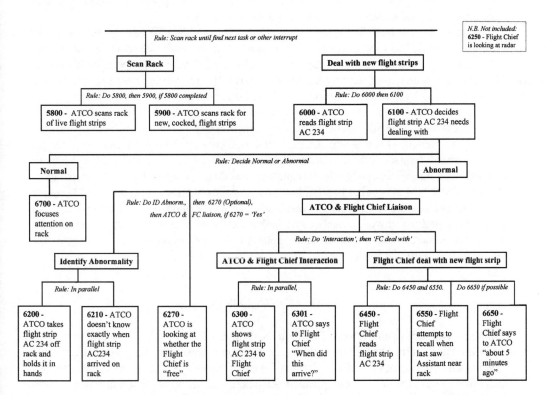

FIG. 1.5. The start of an example hierarchical model based on the activity list in table 1.1.

abnormality" subtask will involve "ATCO & flight chief liaison," so incorporating more activity list lines will require some structural alteration to the diagram. One of the things that does drive this sort of analysis is that combining activity list lines generally forces abstraction (section 1.4.2.3). Obviously, if the comment for line 8000 is "ATCO reads flight strip AC321," which is very close to the comment for line 6000, then these two lines can simply be combined by deleting or making general the aircraft ID. In contrast, line 6100 might have a quite complex extension, with more activity list line analysis. For example, there may be alternative strategies to what "needs dealing with" means, using a variant like "needs dealing with soon" which might cause the ATCO to continue scanning the rack but to return to a strip as soon as possible after completing the scan.

One major difficulty that HTA shares with other task analysis methods and systemic models of all kinds is that, although a model can be invalid (i.e., not match what actually happens; sections 1.4.2.1, 1.5, and 1.5.1.4), in most situations there can be a number of different valid models. Making decisions on how to build a model is one of the core craft skills of virtually all analysts, whether in task analysis, IT systems analysis, software engineering, and so on.

STA attempts to deal with the above problem by providing a great deal of flexibility in the structure of its systems model. Starting with a systems model such as in Fig. 1.4, activity list lines can be added to the diagram by directly locating the lines (numbers) on things, including the communcative relationships. This is how the Pentanalysis Technique (Diaper et al., 1998) works, but there the systems model is a hierarchical DFD–based one rather than a heterarchical one. The systems model will frequently have to be refined, often by dividing things into smaller things. The end result of this analysis process in STA is a systems model with all the routes the different tasks take through it. The model thus combines the tasks and represents all the alternatives discovered. The remaining problem for STA, and a fundamental one for all HCI task analysis methods, is how to use the results of the analysis to support the project. Note that a good analyst would already have decided this (section 1.5.1.2).

STA uses a formal modeling system based on set theory called Simplified Set Theory for Systems Modeling (SST4SM). SST4SM is simplified in the sense that it is designed to be understandable, and even usable, by the "mathematically challenged" (Diaper, 2000a, 2001a). SST4SM's equations cause activity list lines to be rewritten in terms of the sets involved in the systems model and can have properties of each set or subset represented as set elements. With sensible set names, SST4SM's equations can be rewritten as English-like sentences, so the method provides one mechanism of automatic abstraction by producing prose activity list lines in a common format. This was a major, explicit design consideration for TAKD (e.g., Diaper 2001b; Diaper & Johnson, 1989). As briefly illustrated in Diaper (2000a), SST4SM has methods for understanding and reasoning with its equations that do not involve algebraic manipulation. The use of SST4SM is peculiar to STA, but the first stages of STA's approach would seem to cover what needs doing in virtually any task analysis exercise of quality. When a task analysis has not been performed well, a likely cause is that the analyst has not understood (and built) an adequate systems model, systematically collected the data, or adequately represented the data in some activity list form.

1.6 CONCLUSION

Is task analysis easy to understand? Based on the chapter's word count, you might be inclined to say no. On the other hand, although this chapter took more than half a year to write and contains a distillation of the author's 20 years of experience with task analysis, the concepts, each taken in turn, are not that complicated. Also note that the whole first half of the chapter is devoted to exposing, in a rational order, the many issues that underlie task analysis and which most writings on the subject address only partially if at all. The second half then tries to illustrate the practical aspects of the first half's theorizing. Section 1.1 describes the chapter's contents and so a summary will not be repeated here.

In a number of places in this chapter and in other chapters in this handbook, it is suggested that task analysis should be carried out by experts. There are two general arguments supporting this suggestion: (a) Task analysis is so complicated that only experts can understand it, and (b) like many things in life, doing task analysis well requires all sorts of craft skills that people really only learn through practical experience. Although agreeing with (b), this chapter attempts to refute (a). Although design is clearly a craft, chapter 11 makes explicit the value of using task representations that nonexperts can understand, and chapter 10 describes a commercial software development environment where task analysis is to be carried out by people who are not HCI experts.

Most of this handbook's chapters are cross-referenced in this chapter. Yet, because the field of task analysis is divers and fragmentary, as stated in the preface, the editors have refrained from standardizing the models, concepts, and terminology that occur in the other chapters. Readers may find that attempting to relate the contents of the other chapters to the ideas in this one is an informative exercise and will help elucidate what each of the other chapters actually covers. Whether editorial standardization might be enforceable in future editions of this handbook remains to be seen. As a prerequisite, people must agree on solutions to the problems of task analysis discussed in the final chapter of this handbook.

Finally, for those readers who are still struggling to understand the underlying concepts of task analysis, the next chapter provides a practical task analysis tutorial aimed to get students going at doing task analysis.

REFERENCES

Ainsworth, L., & Marshall, E. (1998). Issues of quality and practicality in task analysis: Preliminary results from two surveys. *Ergonomics, 41*, 1607–1617.

Alavi, M. (1993). An Assessment of electronic meeting systems in a corporate setting. In G. De Michelis, C. Simone, & K. Schmidt (Eds.), *Proceedings of the Third European Conference on Computer Supported Cooperative Work.*

Anderson, J. R., & Lebiere, C. (1998). *The atomic components of thought.* Hillsdale, NJ: Lawrence Erlbaum Associates.

Anderson, R., Carroll, J., Grudin, J., McGrew, J., & Scapin, D. (1990). Task analysis: The oft missed step in the development of computer-human interfaces: Its desirable nature, value and role. In D. Diaper, D. Gilmore, G. Cockton, & B. Shackel (Eds.), *Human-computer Interaction: Interact, '90* (pp. 1051–1054). Amsterdam: North-Holland.

Annett, J., & Stanton, N. A. (Eds.) (1998). Editorial [Special issue on task analysis]. *Ergonomics, 41*, 1529–1536.

Ashworth, C., & Goodland, M. (1990). *SSADM: A practical approach.* New York: McGraw-Hill.

Atkinson, M. (1988). Cognitive science and philosophy of mind. In M. McTear (Ed.), *Understanding cognitive science* (pp. 46–68). Chichester, England: Ellis Horwood.

Barlow, J., Rada, R., & Diaper, D. (1989). Interacting WITH computers. *Interacting With Computers, 1*, 39–42.

Beer, M., & Diaper, D. (1991). *Reading and writing documents using Headed Record Expertext.* In M. Sharples (Ed.), *Proceedings of the Fourth Annual Conference on Computers and the Writing Process* (pp. 198–207). University of Sussex, Brigaton, UK. Reprinted in AISB Newsletter, (1991), *77*, 22–28.

Bell, J., & Hardiman, R. J. (1989). The third role: The naturalistic knowledge engineer. In D. Diaper (Ed.), *Knowledge elicitation: Principles, techniques and applications.* (pp. 47–86). Chichester, England: Ellis Horwood.

Bench-Capon, T., & McEnry, A. (1989a). People interact through computers not with them. *Interacting With Computers, 1*, 31–38.

Bench-Capon, T., & McEnry, A. (1989b). Modelling devices and modelling speakers. *Interacting With Computers, 1*, 220–224.

Benyon, D. (1992a). The role of task analysis in systems design. *Interacting With Computers, 4*, 102–123.

Benyon, D. (1992b). Task analysis and system design: The discipline of data. *Interacting With Computers, 4*, 246–259.

Benyon, D., & Macaulay, C. (2002). Scenarios and the HCI-SE design problem. *Interacting With Computers, 14*, 397–405.

Bowker, G. C., & Star, S. L. (2000). *Sorting things out: Classification and its consequences*. Cambridge, MA: MIT Press.

Cameron, J. R. (1983). *JSP & JSD: The Jackson approach to software development*. IEEE Computer Society Press, Los Angeles.

Carroll, J. M. (1991). History and hysteresis in theories and frameworks for HCI. In D. Diaper & N. Hammond, (Eds.), *People and computers VI* (pp. 47–56). Cambridge: Cambridge University Press.

Carroll, J. M. (2000). *Making use: Scenario-based design for human-computer interactions*. Cambridge, MA: MIT Press.

Checkland, P. (1981). *Systems thinking, systems practice*. New York: Wiley.

Clark, R. B., & Panchen, A. L. (1971). *Synopsis of animal classification*. London: Chapman & Hall.

De Marco, T. (1979). *Structured analysis and system specification*. Englewood Clifts NJ: Prentice-Hall.

Diaper, D. (1982). *Central backward masking and the two task paradigm*. Unpublished Ph.D. dissertation, University of Cambridge.

Diaper, D. (1984). An approach to IKBS development based on a review of "Conceptual Structures: Information Processing in Mind and Machine" by J. F. Sowa. *behavior and Information Technology, 3*, 249–255.

Diaper, D. (1986). *Identifying the knowledge requirements of an expert system's natural language processing interface*. In M. D. Harrison & A. F. Monk (Eds.), *People and computers: Designing for usability* (pp. 263–280). Cambridge: Cambridge University Press.

Diaper, D. (1989a). Task observation for Human-Computer Interaction. In D. Diaper (Ed.), *Task analysis for human-computer interaction* (pp. 210–237). Chichester, England: Ellis Horwood.

Diaper, D. (1989b). Designing expert systems: From Dan to Beersheba. In D. Diaper (Ed.), *Knowledge elicitation: Principles, techniques and applications* (pp. 15–46). Chichester, England: Ellis Horwood.

Diaper, D. (1989c). Task Analysis for Knowledge Descriptions (TAKD): The method and an example. In D. Diaper (Ed.), *Task analysis for human-computer interaction* (pp. 108–159). Chichester, England: Ellis Horwood.

Diaper, D. (1989d). The discipline of human-computer interaction. *Interacting With Computers, 1*, 3–5.

Diaper, D. (1989e). Giving HCI away. In A. Sutcliffe & L. Macaulay (Eds.), *People and computers V* (pp. 109–120). Cambridge: Cambridge University Press.

Diaper, D. (1990a). Simulation: A stepping stone between requirements and design. In A. Life, C. Narborough-Hall, & W. Hamilton (Eds.), *Simulation and the user interface* (pp. 59–75). London: Taylor & Francis.

Diaper, D. (1990b). Analysing focused interview data with Task Analysis for Knowledge Descriptions (TAKD). In D. Diaper, D. Gilmore, G. Cockton, & B. Shackel (Eds.), *Human-computer interaction: Interact, '90* (pp. 277–282). North Holland: Elsevier.

Diaper, D. (1997). Integrating human-computer interaction and software engineering requirements analysis: A demonstration of task analysis supporting entity modelling. *SIGCHI Bulletin, 29*(1).

Diaper, D. (2000a). Hardening soft systems methodology. In S. McDonald, Y. Waern, & G. Cockton (Eds.), *People and computers XIV* (pp. 183–204). New York: Springer.

Diaper, D. (2001a). The model matters: Constructing and reasoning with heterarchical structural models. In G. Kadoda (Ed.), *Proceedings of the Psychology of Programming Interest Group 13th Annual Workshop* (pp. 191–206). Bournemouth: Bournemouth University.

Diaper, D. (2001b). Task Analysis for Knowledge Descriptions (TAKD): A requiem for a method. *behavior and Information Technology, 20*, 199–212.

Diaper, D. (2002a). Human-computer interaction. In R. B. Meyers (Ed.), *The encyclopedia of physical science and technology* (3rd ed.; vol. 7; pp. 393–400). New York: Academic Press.

Diaper, D. (2002b). Scenarios and task analysis. *Interacting With Computers, 14*, 379–395.

Diaper, D. (2002c). Task scenarios and thought. *Interacting With Computers, 14*, 629–638.

Diaper, D., & Addison, M. (1991). User modelling: The Task Oriented Modelling (TOM) approach to the designer's model. In D. Diaper & N. Hammond (Eds.), *People and computers VI* (pp. 387–402). Cambridge: Cambridge University Press.

Diaper, D., & Addison, M. (1992). Task analysis and systems analysis for software engineering. *Interacting With Computers, 4*, 124–139.

Diaper, D., & Johnson, P. (1989). Task Analysis for Knowledge Descriptions: Theory and application in training. In J. Long & A. Whitefield (Eds.), *Cognitive ergonomics and human-computer interaction* (pp. 191–224). Cambridge: Cambridge University Press.

Diaper, D., & Kadoda, G. (1999). The process perspective. In L. Brooks, & C. Kimble (Eds.), *UK Academy for Information Systems 1999 Conference Proceedings* (pp. 31–40). New York: McGraw-Hill.

Diaper, D., McKearney, S., & Hurne, J. (1998). Integrating task and data flow analyses using the Pentanalysis Technique. *Ergonomics, 41*, 1553–1583.

Diaper, D., & Shelton, T. (1987). Natural language requirements for expert system naive users. In *Recent developments and applications of natural language understanding* (pp. 113–124). London: Unicom Seminars Ltd.

Diaper, D., & Shelton, T. (1989). Dialogues With the Tin Man: Computing a natural language grammar for expert system naive users. In J. Peckham (Ed.), *Recent developments and applications of natural language processing* (pp. 98–116). London: Kogan Page.

Diaper, D., & Waelend, P. (2000). World Wide Web working whilst ignoring graphics: Good news for web page designers. *Interacting With Computers, 13*, 163–181.

Dowell, J., & Long, J. (1989). Towards a conception for an engineering discipline of human factors. *Ergonomics, 32*, 1513–1535.

Downs, E., Clare, P., & Coe, I. (1988). *Structured systems analysis and design method: Application and context.* Englewood Cliffs, NJ: Prentice Hall.

Easterbrook, S. (Ed.). (1993). *CSCW: Cooperation or conflict?* New York: Springer-Verlag.

Eva, M. (1994). *SSADM version 4: A user's guide* (2nd ed.). New York: McGraw-Hill.

Green, T. R. G. (1990). The cognitive dimension of viscosity: A sticky problem for HCI. In D. Diaper, D. Gilmore, G. Cockton, & B. Shackel (Eds.), *Human-computer interaction: interact '90.* (pp. 79–86). Amsterdam: North-Holland.

Hammer, M. (1984, February). The OA mirage. *Datamation*, pp. 36-46.

Hares, J. (1990). *SSADM for the advanced practitioner.* New York: Wiley.

Harrison, M., & Thimbleby, H. (1990). *Formal methods in human-computer interaction.* Cambridge: Cambridge University Press.

Hinton, G. E., & Anderson, J. A. (Eds.). (1981). *Parallel models of associative memory.* Hillsdale, NJ: Lawrence Erlbaum Associates.

Johnson, P., Diaper, D., & Long, J. (1984). *Tasks, skills and knowledge: Task analysis for knowledge based descriptions.* In B. Shackel (Ed.), *Interact '84:* First IFIP Conference on Human-Computer Interaction (pp. 23–27). Amsterdam: North Holland.

Life, A., Narborough-Hall, C., & Hamilton, W. (1990). *Simulation and the user interface.* London: Taylor & Francis.

Lim, K. Y., & Long, J. (1994). *The MUSE method for usability engineering.* Cambridge: Cambridge University Press.

Long, J. (1986). People and computers: Designing for usability. In M. Harrison, & A. Monk (Eds.), *People and computers: Designing for usability* (pp. 3–23). Cambridge: Cambridge University Press.

Long, J. (1997). *Research and the design of human-computer interactions or "What Happened to Validation?"* In H. Thimbleby, B. O'Conaill, & P. Thomas (Eds.), *People and computers XII* (pp. 223–243). New York: Springer.

Long, J., & Dowell, J. (1989). Conceptions of the discipline of HCI: Craft, applied science, and engineering. In A. Sutcliffe & L. Macaulay, (Eds.), *People and computers V* (pp. 9–34). Cambridge: Cambridge University Press.

Lowe, E. J. (2002). *A survey of metaphysics.* Oxford: Oxford University Press.

Mumford, E. (1983). *Designing participatively.* Manchester, England: Manchester Business School Press.

Norman, D. (1986). Cognitive engineering. In D. Norman, & S. Draper (Eds.), *User centered system design: New perspectives on human-computer interaction* (pp. 31–61). Hillsdale NJ: Lawrence Erlbaum Associates.

Nussbaum, M. C. (2001). *Upheavals of thought: The intelligence of emotions.* Cambridge: Cambridge University Press.

Patching, D. (1990). *Practical soft systems analysis.* London: Pitman.

Pressman, R. S. (1994). *Software engineering: A practitioner's approach* (European ed.). London: McGraw-Hill.

Pullinger, D. J. (1989). Moral judgements in designing better systems. *Interacting With Computers, 1*, 93–104.

Ross, K. A., & Wright, C. R. B. (1988). *Discrete mathematics* (2nd ed.). Englewood Cliffs NJ: Prentice-Hall.

Shapiro, D., & Traunmuller, R. (1993). CSCW in public administration: A review. In H. E. G. Bonin (Ed.), *Systems engineering in public administration* (pp. 1–17). New York: Elsevier.

Sharples, M. (1993). A study of breakdowns and repairs in a computer-mediated communication system. *Interacting With Computers, 5*, 61–78.

Shepherd, A. (2001). *Hierarchical task analysis.* London: Taylor & Francis.

Sommerville, I. (1989). *Software engineering.* (3rd ed.). Reading, MA: Addison-Wesley.

Sutcliffe, A. (1988). *Human-computer interface design.* London: Macmillan.

Watling, J. L. (1964). Descartes. In D. J. O'Connor, (Ed.), *A critical history of western philosophy* (pp. 170–186). London: Macmillan.

Zhang, P. (1999). Will you use animation on your web pages? In F. Sudweeks, & C. T. Romm (Eds.), *Doing business on the Internet: Opportunities and pitfalls* (pp. 35–51). New York: Springer.

2

Working Through Task-Centered System Design

Saul Greenberg
University of Calgary, Alberta

This chapter presents a how-to tutorial to a version of Lewis and Rieman's Task Centered System Design methodology. Using an example of an interface being developed for a catalog store, we show in detail how a practitioner can identify key tasks, use those tasks to do a rudimentary requirements analysis and how one can evaluate prototype designs through a task-centered walkthrough.

2.1 INTRODUCTION

In 1993, Clayton Lewis and John Rieman introduced task centered system design (TCSD), a highly practical discount-usability engineering methodology. In its essence, TCSD is a process where designers do the following:

- Articulate concrete descriptions of real-world people doing their real-world tasks.
- Use these descriptions to determine which users and what tasks the system should support.
- Prototype an interface that satisfy these requirements.
- Evaluate the interface by performing a task-centered walk-through.

Because TCSD is simple to learn and apply, I have been teaching it for almost a decade within an introductory human-computer interaction (HCI) course for computer scientists. I could only devote about four classes to this method, so I reworked the Lewis and Rieman material into a short form that provides students with a terse explanation of the process as well as a worked example that illustrates how to apply it to a problem. Students then use TCSD in their first assignment to analyze a real-world problem of their choosing and to develop and evaluate an interface that solves this problem. Most students find TCSD to be an eye-opener in terms of its simplicity and effectiveness—they are surprised at how well it informs their interface designs and how well it lets them evaluate the nuances of their designs.

This chapter summarizes my reworked approach to task-centered system design. It both paraphrases and adds to the Lewis and Rieman material. The first part of this chapter details the main steps of the TCSD process. The second part applies this process to an actual example.

2.2 TASK-CENTERED SYSTEM DESIGN

The version of TCSD described in this chapter is divided into four phases. Each phase is described below.

2.2.1 Phase I: Identification

In the first phase of TCSD, you identify specific users of the system and articulate example realistic tasks that they would do. Your goal is to produce a manageable list of representative users and tasks that give realistic coverage of who would use the system to do what kinds of tasks. To achieve this goal, you need to first discover what tasks users do, then write these up as task descriptions, and finally validate the descriptions to make sure they represent reality. These steps are detailed below.

2.2.1.1 Discovering the Tasks That Users Do

TCSD strives for realism. This means you should discover how real people do their real tasks. Yet depending on your situation, you may or may not be able to access these real people. Consequently, you should select the approach below that best fits your situation.

The Ideal: Observing and/or Interviewing the Real End Users. Get in touch with current or potential users. These users may now be employing paper methods, competing systems, or antiquated systems for doing their tasks. Observe them as they do their task activities, and interview them about what they are doing (Beyer & Holtzblatt, 1998). For example, if you are interested in customers who purchase items in a store, you should observe and talk to store customers as they move about the store. These interviews and observations are critical. They transform "the user" from an abstract notion into real people with real needs and concerns. It will allow you put a face on the faceless and help you understand where they are coming from.

Second Best: Interviewing the End-User Representative. If you absolutely cannot get in direct contact with end users, you can carefully select and interview end-user representatives as stand-ins. These *must* be people who have direct contact with end users and have intimate knowledge and experience of their needs and what they do. It is crucial that the representatives have a deep and real (rather than idealized) understanding of what end users actually do. People who work "in the trenches" with the end users are the best bet. For example, you can talk to a store's front-line sales staff about their customers if you cannot observe or talk to the customers directly. A better option is to interview these front-line staff as they deal with the customers; this way you can observe firsthand what customers do.

When All Else Fails: Make Your Beliefs about the End Users and the Task Space Explicit. If you cannot get in touch with real end users or their representatives, use your team members to articulate their assumptions about end users and tasks. Because this runs the serious risk of producing end-user and task descriptions that bear no relation to reality, you should do this only as a last resort. Still, you will at least produce a diverse list of expected end users and their tasks (because you have a diverse team), and it will put everyone's beliefs and assumptions on the table. You can always show these to your clients later, and hopefully you will be able to compare them to real-world situations to see if these tasks indeed reflect what real end users do.

No matter which approach you choose, you have to determine your stopping conditions. That is, you need to decide when you should stop gathering data and generating user and task descriptons (see also chap. 3 and 17). In practice, you will notice ever increasing repetition as you do your observations and interviews. You simply stop when new types of people and tasks are rare and when it is no longer cost-effective to continue.

2.2.1.2 Developing Good Task Descriptions

You must write up the results of your observations and interviews as good task descriptions. Each task description should adhere to five very important rules.

It describes what the user wants to do but does not say how the user would do it. The description should not include any interface mechanics about how the task is actually carried out. That is, you do not want to detail task steps that are peculiar to the system being used. What you are really doing here is identifying the users' goal, as Hierarchical Task Analysis (HTA) does (chap. 3), as well as the concrete steps they would take to achieve this goal no matter what system was being used (Cooper, 1999). This is important, because you will want to use these tasks to generate several alternative designs that let the user accomplish the task in quite different ways. Similarly, you will use these tasks to compare different interface design alternatives in a fair way.

It is very specific. A description is concrete. It says exactly what the user wants to do, including the actual items the user would eventually want to input (somehow) into the system and what information or results the user will want out of it. This is important because it provides concrete rather than imaginary data for the types of information the system must handle.

It describes a complete job. The description should flow through all aspects of the task, starting at the very beginning and concluding at the very end. This is important because the complete description forces you to consider how interface features will work together. You can also use the complete description to contrast how information input and output are carried through a particular interface design. That is, you can ask, Where does information come from? Where does it go? What has to happen next?

It says who the users are and reflects their real interests. The description should name real people and should include what they know or don't know about performing the task and using computers. This is important because the success of a design is strongly influenced by what the users know. Because you need to observe real people to do this, you will tend to find tasks that illustrate required functionality in a person's real-world context. If the task names real people, you can go back to them and ask them about any information you are missing. You will eventually use this information to see if people realistically have the desire, knowledge, and/or capabilities to accomplish their task using your system design.

As a set, the task descriptions identify a broad coverage of users and task types. Collectively, the descriptions should identify the typical "expected" users, the occasional but still important users, and the unusual users. Similarly, they should identify typical routine tasks, infrequent but important tasks, and unexpected or odd tasks. This is important because you will need a way to decide the coverage of your system design—which tasks and user groups must be included in the design and which could be left out. Because your initial set of descriptions may have similar entries, you should reduce it by choosing particular user and task stories (see chap. 5) that best represent the expected classes of users and tasks. In practice, you should end up with a manageable number of descriptions that still give good coverage.

2.2.1.3 Validating the Tasks

Your final step in the identification phase is to get a reality check of your task descriptions. You can do this by circulating the descriptions back to the end users or the end-user representatives. These people should see if the descriptions fairly summarize the activities. Specifically, they should check to see whether the set of descriptions adequately covers the potential end users of your product, whether the descriptions really represent what people do, and whether their details are realistic. You should ask for details omitted from the original task descriptions as well as corrections, clarifications, and suggestions. Rewrite the descriptions as corrected task descriptions.

This step is critical if you used end-user representatives or if your team just made up what they thought were good descriptions. Although it may not be possible for you to interview and observe many real clients, you can probably get a few to comment on a compiled list of prototypical tasks.

2.2.2 Phase II: User-Centered Requirements Analysis

You will rarely design a system that handles all possible users and tasks equally well. This could be because you do not have the budget to develop an all-encompassing system, because the diversity of possible users and tasks is too high to be handled by a common system, or because you cannot cost-justify certain features. As a rule of thumb, most systems are considered successful if they have about 90% coverage, that is, if 90% of the people can do 90% of the tasks reasonably well. This also means that these systems exclude 10% of the people and tasks. The next phase in TCSD is for you to decide what people and tasks will be included or excluded from your design. This list will become the basic user-centered requirements analysis of your system design.

2.2.2.1 Deciding which User Types to Include

You need to identify which user types will be supported by your design. Because each description identifies a representative user, you can separate users into user types. Although you want to be careful not to overstereotype people, you will likely find that some of the groups are clearly separate, with quite distinct needs and goals. In a store setting, for example, the store customers would make up one group and the sales clerks another. As well, you may find that some users differ considerably from each other even though the tasks they wish to accomplish are similar. You may find people with different levels of computer experience or with different levels of knowledge and experience of the actual task. You should go through the list and make some hard decisions about whom to include in your system design.

- *Absolutely must include.* The system design must support these user types. They are the basic audience, and leaving them out would seriously undermine the purpose of the entire system.
- *Should include if possible.* These user types are of lesser importance or are perhaps somewhat atypical. The system design should strive to accommodate them if possible. However, it is acceptable if they must do somewhat more work to use the system, if they are excluded altogether (perhaps because other workarounds exist), or if their inclusion is deferred to the next system release.

- *Exclude.* These user types are rare, unimportant, or quite different from the core users; their inclusion cannot be justified from a cost perspective; or they have workarounds that allow them to do without the system. Although these users may be able to use the existing system, the system design should not go out of its way to accommodate them.

2.2.2.2 Deciding Which Tasks to Include

Similarly, you need to identify what tasks will be handled efficiently and effectively by your design. Because each description is task centered, you can order the task descriptions by the following criteria.

- *Absolutely must include.* These are key tasks and identify the essential things that people would do with the system. They are typically frequent and important tasks.
- *Should include if possible.* These tasks should be included if budget and time permit. Although still important, they are perhaps somewhat rarer than the "must include" tasks. If they are not included in version 1 of the system, they should be included in version 2.
- *Could include.* These are lesser tasks that could be supported by the system but only if they could be included "almost for free," That is, if the necessary features can be easily added without impacting the rest of the interface or if the task can be accommodated simply by the way the system supports the other tasks, then it can be included.
- *Exclude.* These tasks are unimportant and/or so rare that no effort should be made to include them in the system.

2.2.3 Phase III: Design Through Scenarios

With the descriptions and requirements in hand, you can now start thinking about the interface design. Each description creates the character and plot of a story. You generate design possibilities by exploring how specific designs support the telling of this story. Each design should take into account how its features work together to help users accomplish their real work. It should also reflect the expected users' knowledge and motivation as well as the real-world context of its potential use (Carroll, 2000; chap. 5).

As design ideas unfold, you can judge and quickly modify your interface by seeing how well it supports the story told by your core set of user-task descriptions. That is, you can perform an "in the small" task-centered walk-through (discussed in section 2.2.4) to see how well your interface and its features support particular user types and tasks.

2.2.4 Phase IV: Evaluate Via Task-Centered Walk-Throughs

A *usage scenario* combines an interface design with one of your user-task descriptions. In this phase, you choose a scenario and perform a *task-centered walk-through* of it (Nielson & Mack 1994). With a walk-through, you tell a concrete story about what a particular user would do and see step by step when performing his or her particular task using the interface (Carroll, 2000; chap. 5).

Walk-throughs are an excellent low-cost way to evaluate your interface, for you will quickly discover trouble spots as you move through the task. At its cheapest, you can do it by yourself, with no need for end-user involvement. However, walk-throughs tend to produce richer results when they are performed with others on your team, particularly if other members have perspectives that differ from yours e.g., designers, implementers, and end users (Bias, 1994).

Lewis and Reiman's (1993) algorithm for performing a task-centered walk-through is surprisingly simple and easy to do:

Select one of the task scenarios.
For each of the user's steps/actions in the task:
  Can you build a believable story that motivates the user's actions?
  Can you rely on the user's expected knowledge and training about system?
  If you cannot, you have located a problem in the interface.
  Note the problem and any comments or solutions that come to mind.
Once a problem is identified, assume it has been repaired.
Go to the next step in the task.

To make this walk-through algorithm work effectively, you must put yourself in the mind and context of the end user. You are essentially role-playing. You must stay true to the spirit of what the person is trying to accomplish, what the person knows, and what is reasonable for the person to do. The story you tell must be complete and must ring true. The story should start at the very beginning of the task, perhaps even before the person touches the computer. For each expected step in the task as prescribed by the interface design, you have to ask if the person will know what to do next, whether the person will know how to use the interface controls to do it, and whether the person can comprehend the feedback provided by the system. You should continue through the task to the bitter end even when you discover that your design is so terrible that it should be abandoned. This is because discovering other problems now may help you avoid them in successive designs and may even give you insights into new designs.

2.3 A WORKING EXAMPLE: CHEAP SHOP

This sections steps through a working example to illustrate how this process can be applied.

2.3.1 The Situation

Cheap Shop is a catalog-based department store known for its low-cost merchandise. A customer shops by browsing one of the paper catalogs scattered around the store. As the customer finds each desired item, he or she enters its item code from the catalog onto an order form. The customer then gives this form to a sales clerk at the front counter. After a modest time (3–8 minutes), the warehouse clerk delivers the items from the back room to the sale clerk at the front counter. The sales clerk passes them to the customer. The customer checks the items and pays the sales clerk for the items he or she wants. A sample catalog listing is shown in Fig. 2.1 and a filled-in form is shown in Fig. 2.2.

Cheap Shop has contracted you to evaluate an in-store computer system that it has prototyped. This prototype system is to be used by customers to indicate and buy the items they want. Once requests are made, the system sends them to the warehouse so that the items can be brought to the front counter for final processing by a clerk. If the prototype has major problems, you can either suggest how it can be repaired or propose a completely new design.

2.3.2 The Cheap Shop Prototype

This prototype is intended to be available on all Cheap Shop Department Store computers. Shoppers in the store decide on the items they want by browsing the catalog and can then purchase the items by entering the relevant information into the screens shown in Figs. 2.3 and 2.4.

FIG. 2.1. The catalog entry for the stroller.

FIG. 2.2. The filled-in order form for the stroller.

To order the first item:

- Shoppers follow the sequence on screen 1 to enter their personal information and their first order.
- Text is entered via keyboard, and the tab or mouse is used to go between fields.

To order additional items:

- Shoppers fill in screen 2 after clicking *Next Catalog Item* (can be repeated).

To complete an order:

- Shoppers click *Trigger Invoice.*
- The system automatically tells shipping and billing about the order.
- The system returns to a blank screen 1.

To cancel the order:

- Shoppers do not enter input for 30 seconds (as if they walked away).
- The system will then clear all screens and return to the main screen.

Input checking:

- All input fields checked when either button is pressed.
- Erroneous fields will blink for 3 seconds and will then be cleared.
- The shopper can then reenter correct values in those fields.

2.3.3 Producing User and Task Descriptions for Cheap Shop

We collected descriptions by monitoring customer activity at the Cheap Shop store. We validated each description by interviewing the customers and by asking them if it reflected what they did. We also sat behind the counter with the store clerks, where we observed what they did and how customers and clerks talked to each other. Later, we gave the complete set of descriptions to the store clerks (the client representatives) and asked them if the descriptions

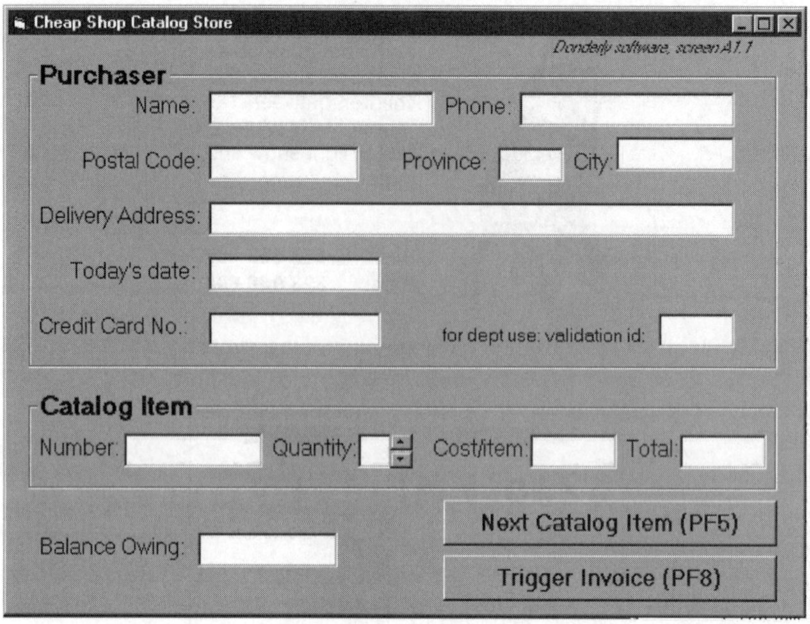

FIG. 2.3. Prototype design, screen 1.

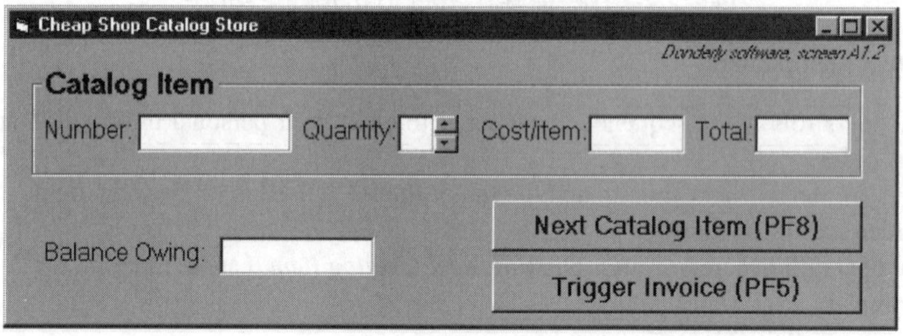

FIG. 2.4. Prototype design, screen 2.

typified what they saw in terms of customer requests. Three of our descriptions are included below. Notice that they follow the rules for good task descriptions identified earlier.

2.3.3.1 Task 1

Fred Johnson, who is caring for his demanding toddler son, wants a good-quality umbrella stroller (red is preferred but blue is acceptable). He browses the catalog and chooses the JPG stroller (cost $98, item code 323 066 697). He pays for it in cash and uses it immediately. Fred is a first-time customer to this store, has little computer experience, and says he types very slowly with one finger.

Fred has many properties of our typical expected user. Many customers are first-time shoppers, and a good number have no computer experience and are poor typists. Similarly, the task type is routine and important. Many people often purchase only one item, and a good number of those pay by cash. As with Fred, people often have a general sense of what they want to buy but decide on the actual product only after seeing what is available.

2.3.3.2 Task 2

Mary Vornushia, an elderly arthritic woman, is price-comparing the costs of a child's bedroom set consisting of a wooden desk, a chair, a single bed, a mattress, a bedspread, and a pillow all made by Furnons Company. She takes the description and total cost away with her to check against other stores. Three hours later, she returns and decides to buy everything but the chair. She pays by credit card and asks for the items to be delivered to her daughter's home, which is the basement suite at the back of the house at 31247 Lucinda Drive.

Like Mary, a reasonable number of store customers are elderly and have infirmities that inhibit their physical abilities. A modest number of them also enjoy comparison shopping, perhaps because they have more time on their hands or because they are living on a low income. Although this would be considered a "major" purchase in terms of the total cost, the number of items purchased is not unusual. Delivery of large items is the norm, and many customers pay by credit card for larger orders.

2.3.3.3 Task 3

John Forham, the sole sales clerk in the store, is given a list of items by a customer who does not want to use the computer. The list includes 4 pine chairs, 1 pine table, 6 blue place mats, 6 "lor" forks, 6 "lor" tablespoons, 6 "lor" teaspoons, 6 "lor" knives, 1 "tot" tricycle, 1 red ball, and 1 "silva" croquet set. After seeing the total, the customer tells John he will take all but the silverware and decides to add 1 blue ball to the list. The customer starts paying John by credit card but then changes his mind and decides to pay cash. The customer tells John he wants the items delivered to his home the day after tomorrow. While this is occurring, six other customers are waiting for John. John has been on staff for 1 week and is only partway through his training program.

The third task introduces the clerk as a system user. Although every store will have a few clerks, they are vastly outnumbered by the number of customers using the system. Because the store has high staff turnover, new employees such as John are common. Thus John reflects a "rare" but important group of users. The task that John is asked to do by the customer, though complex, is fairly typical, as people making large numbers of purchases often ask a clerk to help them. Similarly, clerks mention that customers often change their mind partway through a transaction, changing what they want to buy or how they want to pay for it. Customers, however, rarely give specific delivery dates (most want delivery as soon as possible). Lineups for clerks do happen during busy times.

2.3.4 The Task Walk-Through

Because we already had an interface in hand, we began by performing a task-centered walk-through of it. The example in Table 2.2 below shows our walk-through analysis performed with a scenario that combines our first task description (Table 2.1) with this interface.

TABLE 2.1
The Walk-through Scenario

Interface: Cheap Shop prototype 1.

Description 1. Fred Johnson, who is caring for his demanding toddler son, wants a good-quality umbrella stroller (red is preferred but blue is acceptable). He browses the catalog and chooses the JPG stroller (cost $98, item code 323 066 697). He pays for it in cash and uses it immediately. Fred is a first-time customer to this store, has little computer experience, and says he types very slowly with one finger.

TABLE 2.2
From Entering the Store to Finding the Computer

Task Step	*Knowledge? Believable? Motivated?*	*Comments/Solutions*
a. Enters store.	Okay.	
b. Looks for catalog.	Okay if paper catalog is used, but what if the catalog is online?	Finding paper catalogs is not a problem in the current store. However, we were not told if the paper catalog would still be used or if the catalog would be made available on line. ***Note:*** Ask Cheap Shop about this. If they are developing an electronic catalog, we will have to consider how our interface will work with it. For now, we assume only a paper catalog is used.
c. Finds red JPG stroller in catalog.	Okay.	The current paper catalog has proven itself repeatedly as an effective way for customers to browse Cheap Shop merchandise and to locate products.
d. Looks for computer.	Modest problem.	As a first-time customer, Fred does not know that he needs to order through the computer. Unfortunately, we do not know how the store plans to tell customers that they should use the computer. Is there a computer next to every catalog (so its association can be inferred) or are there a limited number of computers on separate counters? Are there signs telling Fred what to do? ***Note:*** Ask Cheap Shop about the store layout and possible signage. ***Possible solution***: Instead of screen 1, a startup screen can clearly indicate what the computer is for (e.g., "Order your items here" in large letters).

For reporting purposes, each walk-through report is preceded by a description of the scenario that identifies which interface is being analyzed. This report is contained in Table 2.1, whereas Tables 2.2-2.1 tells the story, for they record the step-by-step results of the task walk-through algorithm. The first column describes each task step in sequence. The second column asks if that person has the knowledge or training to do this step and if it is believable that the person would be motivated to do what is asked. The final column records problem details, comments, and (optional) solutions to any problems.

The sequence in Table 2.2 illustrates the value of starting the walk-through at the very beginning of the task. Notably, we see that we are missing information about whether paper or electronic catalogs will be used, how the computerized system is situated in the store environment, and whether signage or other instructional material will tell customers what to do. While doing task-centered design, you should examine what information is missing and list what assumptions you are making. You should validate these assumptions, as incorrect ones can profoundly affect how the interface will perform in the actual setting. Even when assumptions reveal issues "outside" the actual interface being designed, they can be critical to its success.

The sequence in Table 2.3 illustrates fundamental problems that arise as one walks through task step details, such as how Fred selects and moves between fields and how Fred enters input. It also illustrates how the inspector can do a reality check on walk-through steps in the large, such as whether the expected sequence of activities matches Fred's goals and whether Fred is

TABLE 2.3
Entering Personal Information

e. Enters name.	No motivation to do this!	Fred's task is to buy the stroller, but the scenario shows that the system is asking him for his name. Fred may be reluctant to enter his name if (say) he believes that he will be added to a mailing list without his permission. ***Note.*** Ask Cheap Shop why they are asking for the customer's name and other contact information.
f. Selects the name field.	Knowledge lacking: Fred does not know how to select a field.	To enter his name, Fred is expected to click and type into the first text field on this form. Yet Fred has little computer experience, and thus he may not know what to do. He may also be reluctant to experiment with the system. ***Possible solutions***: a. Have the first field preselected, with the cursor in it. b. Have a poster next to the computer describing these basic acts.
g. Types his name.	Knowledge lacking: Fred types poorly, does not know name format.	Because Fred types poorly, text entry will be slow and tedious. This further dampens Fred's motivation, as he is entering information that is unimportant to the task. Fred is uncertain about formats: Does he type his name as "Fred Johnson" or "Johnson, Fred"?
h. Moves to phone field.	Knowledge lacking.	Fred may not know how to tab or mouse over to the next field because of his unfamiliarity with computers.
i. Fills in phone, postal code, province, and city.	Poor motivation; poor format knowledge.	If Fred can complete steps e–h, he will be able to continue with the following fields. However, motivation will decrease even further as Fred painfully types unnecessary information into the system. Fred continues to have formatting concerns about how he should enter information. Should the phone number include the area code, spaces, and/or hyphens? Should he spell out the province or use the abbreviation? Should he leave a space in the postal code?
j. Enters delivery address.	Violates the task.	Fred will use the stroller immediately, but the system asks for his delivery address. Fred may incorrectly assume that he is filling in the wrong form and give up. We also noticed that the order of the contact information does not follow the typical flow. That is, one would expect "Name, Address, City, Province, Postal Code, Phone" rather than the odd order shown in the form.
k. Enters today's date.	No motivation.	This is an odd field. Why should Fred enter the date when the system already knows it? Can he skip it? If he does fill it in, he would be quite lucky to enter a recognizable date format.
l. Enters credit card information.	Violates the task.	Fred is paying by cash, and thus he is unwilling to enter his credit card number. He is also concerned that others may see his credit card information as he types it onto the screen. Finally, this seems an odd place to ask for payment information. Most stores ask for it at the end of transaction, not at the beginning.
m. Ignores validation ID.	Okay.	Although Fred will likely do the right thing, this field should not be here. It has nothing to do with Fred's task. ***Possible solution***: Remove it.
Steps e–m.	Not needed for task.	***Possible solution***: This entire part of the interface is not needed or is at best optional (e.g., if it is for getting onto the mailing list). Delete it entirely or move it into a very secondary area that can be filled in after the transaction is completed.

TABLE 2.4
Buying the Stroller

n. Enter item number for the JPG red stroller.	Is motivated, but has problems; error-prone.	The item number for Fred's stroller is written in the paper catalog as 323 066 397. Because catalogs are common, he may be able to figure out what he has to do. However, the format is a bit mysterious—should he include spaces or not? If the paper catalog is in an awkward place, Fred will have to rely on his memory to enter the number or he will constantly be running back and forth between the catalog and the computer. Because Fred is a poor typist, he may have difficulty typing the number correctly.
o. Enter quantity.	Knowledge low, motivation high.	Fred wants one stroller only. However, this "spinner widget" is somewhat mysterious to Fred. Because he does not know computers, he will likely not know that he can type the amount directly or just click the arrows to select the quantity. ***Partial solution***: Have the spinner show a 1 by default.
p. Enter cost/item.	Motivation low.	Why should Fred enter the cost? Surely the system knows this. If this field is actually used to display the cost, then it has the wrong visual affordance, as it looks like a text box. Perhaps Fred would be willing to enter a deeply discounted cost, but this will probably be treated as a system error.
q. Enter total.	Motivation low.	See above point.
r. Enter balance owing.	Motivation low, knowledge low.	See above point. Fred will also be uncertain about how this field differs from the "Total" field.
s. Click "Trigger Invoice" or press PF5.	Knowledge low.	Being inexperienced with computers, Fred may not recognize or know how to use a clickable button. The "PF5" label is also mysterious, as Fred does not recognize it as a keyboard shortcut. Fred will find the meaning of "Trigger Invoice" cryptic, as it is in the language of the database system rather than in his language. This may leave him at a loss of what to do. ***Possible solutions***: Remove the "PF5" label. Change ''Trigger Invoice" to something more meaningful.

willing to enter the expected information. In this case, we see several serious problems that must be repaired.

The sequence in Table 2.4 illustrates how poorly the interface handles the most critical part of the task, where Fred specifies the item he wants to buy and tries to complete the transaction. We also see that the interface is full of jargon and unneeded or poorly designed interface components.

As with the opening task sequence, the sequence in Table 2.5 illustrates how the closing of the task must recognize factors that go beyond the interface. In this case, we see a serious problem in how the electronic part of the task flows through to the physical completion of the task.

Table 2.5 walks through the correct task sequence. What happens when errors or other events occur that disrupt this sequence? Table 2.6 demonstrates that walk-throughs are also effective for discovering how the system design deals with problems and with events arising from the end user's real-world context.

Walk-throughs of the interface using the other task descriptions (sections 2.3.3.2 and 2.3.3.3) yield other problems. In the second description, we identify a user (Mary) who cannot use the mouse or type because she is arthritic. What should be done with her? We also see serious

TABLE 2.5
Picking up the Stroller and Paying for It

t. Waits for item at sales counter.	Knowledge low, motivation high.	Fred has to go to the sales clerk and wait for the item to appear. Yet he may not know this, especially because the computer returns to the initial empty screen. Has the transaction completed successfully? Is there signage that says what has to happen? ***Possible solution***: Provide a final screen that tells Fred what he has bought and what he has to do next.
u. Gets item from sales clerk.	Knowledge low, motivation high.	If other items are appearing aside from Fred's, he may not know which items are his unless the boxes are clearly labeled or the box sizes and shapes give it away. Similarly, the sales clerk has no easy way to identify whose items have appeared unless the name given in the name field is somehow attached to the items. ***Possible solution***: After "Trigger Invoice" is clicked, the system could print out a sheet listing the chosen items, which Fred can then give to the sales clerk.
v. Pays for item in cash.	Okay.	Although this is straightforward, there is a question about how the clerk will tally up this bill. This is the clerk's problem, but we don't want Fred to wait excessively.
w. Uses it immediately.	Okay.	

problems when multiple items are ordered: The system gives no feedback on what was entered, and there is no way to correct errors without reentering everything from scratch. There is also no easy way for Mary to price-compare (since no printout is provided), nor is there any way for her to recall information that she previously entered. Details are also a problem: There is no easy way for her to tell the system about the unusual delivery location (a basement suite at the back of a house). As for the third description, we see that the system is completely unsuitable for use by the store clerk. Item entry is too slow, and the clerk will not be able to keep up with the changes requested by the customer. The slowness also affects other customers waiting in line, and we see that the system cannot accommodate delayed delivery.

The bottom line is that this design is a disaster and should be completely revamped.

2.3.5 Determining User-Centered Requirements

Before redesigning the interface, we revisited the descriptions and made decisions on which users and tasks should be supported by the design. We split the main user types and tasks into the categories and dimensions shown in Table 2.7.

As regards user types, we decided that the system design must include first-time and repeat customers who are computer-knowledgeable and who are willing to use the computer. These customers make up a large customer base, and they have skills that should allow them to use the system with minimal effort. The requirements should also include customers who are computer naive or are nontypists, for these too make up a large part of the customer base. However, the system can exclude those who may have disabilities that interfere with their ability to enter their data in the system; those people are fairly rare, and they can simply ask a sales clerk to help them. We also recommended that a different system be designed for sales clerks, because their needs are unlike those of typical customers.

TABLE 2.6
Interruptions, Errors, and Exceptions

1. Event: interruptions and timeout.		
a. Deals with toddler.	Knowledge high, motivation high.	Fred's toddler starts demanding his attention part way through this task (say, after he has entered the item number). Fred comforts his child.
b. Deals with timeout.	Knowledge low, motivation low.	Unfortunately, this took more than 30 seconds, which means that the system has cleared the screen. Fred has to reenter all this information, which he will likely not do. Note that a similar problem will occur if Fred lingers too long on any step in this task.
2. Error: incorrect item number		
a. Recognizes error message.	Knowledge low.	If Fred enters an incorrect item number, the system will blink that field for 3 seconds and then clear it. It is extremely unlikely that Fred will know what this means.
b. Enter corrected item number.	Knowledge low, motivation medium.	Even if Fred realizes that he has made a mistake entering the item number, he will be uncertain about what he did wrong (since the number is no longer there) or how to correct it.
3. Event: red stroller unavailable		
a. No red stroller is in stock.	Knowledge low.	If there is no red stroller in stock, how does Fred find this out? Will the sales clerk tell him (in which case the clerk needs to identify Fred)?
b. Reenters all information for blue stroller.	Motivation low.	We cannot imagine that Fred would be willing to go through this whole process again, especially because his demanding toddler is likely losing patience.
		Possible solution: When the customer selects an item, the interface should clearly indicate if it is in stock.

As regards tasks, the system design must let people choose and easily modify a list of merchandise comprising at least one to seven items and be able to quickly determine individual and total costs. They should be able to pay by cash, credit card, or debit card and take the merchandise away with them. These are all important and very frequent tasks that customers do. The system should also let people buy more than seven items, although making this possible can be deferred, as the large majority of people buy seven items or less at Cheap Shop. The system could include the ability to let people comparison shop (e.g., by printing out the screen) or retrieve previous orders. The system can exclude delivery or payments by invoice, as these are fairly rare and are best handled by the sales clerk.

2.3.6 Interface Design

From these requirements and from our knowledge of the problems detected in our walk-through, we developed the prototype described below (this nonfunctional prototype was created in about 45 minutes). It differs considerably from the previous version. Because people may not know how to type or use a mouse, we use only a bar code reader and a touch screen for input. Paper catalog descriptions of items will be modified to include bar codes. Next to the computer is located a "getting started" instructional poster that pictorially shows people how to use the bar code reader and tells them the sales clerk would gladly help them if they have any problems. Because we wanted this system to let people easily select and purchase just a few items (perhaps changing their mind on the way), the interface portrays itself as a dynamic shopping list. Because people may be unfamiliar with the system, context-sensitive on-screen instructions prompt the user every step of the way.

TABLE 2.7
Categories of User Types and Tasks

USERS

Customers
- First-time vs. repeat customers
- Computer knowledgeable vs. computer naive
- Typists vs. nontypists
- Willing to use the computer vs. unwilling
- People with disabilities who may have trouble with fine motor control

Sales clerks
- Experienced and trained
- New staff member; has passed introductory training session

TASKS

Choosing merchandise
- One item
- Multiple items
- Modifying the selected list of items

Pay by
- Cash
- Credit or debit card
- Invoice

Reviewing cost
- Individual item cost
- Total costs
- Comparison shopping

Merchandise pickup
- Immediate
- Delivery

To illustrate how this works, let's revisit the first task in Table 2.1. In the new initial screen (Fig. 2.5), the "wizard" tells Fred how to add an item to the shopping list. Fred finds his stroller in the catalog and scans the bar code next to it. The second screen (Figure 2.6) shows the result: A picture of the stroller, its description, and its cost appear, and it is automatically added to the shopping list. Because the stroller comes in several colors, the wizard tells Fred he can change the current color (chosen by default) by touching the color he wants (this would automatically modify the shopping list). Fred wanted red, but sees the red stroller is out of stock, so he leaves blue as the color. At this point, the system lets the customer scan in other items or delete or modify items already in the list by touching the appropriate button located at each item's right. If the modify button is touched, the item reappears in the "What you selected" box. Fred just wants the stroller, so he touches the "Place your order" button at the bottom. A copy of the

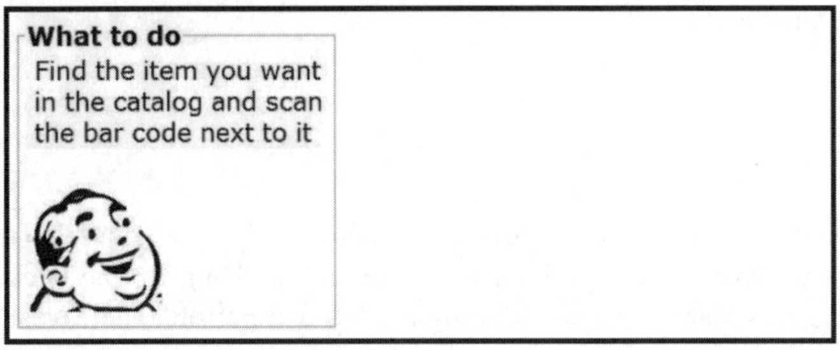

FIG. 2.5. The initial screen of the new prototype.

FIG. 2.6. The filled-in screen of the new prototype.

shopping list prints out (including its own unique bar code), and the wizard tells Fred to give it to the sales clerk at the front counter. The clerk takes the order from Fred and uses it to collect the item when it appears from the warehouse. Fred is ready to pay by cash, so the clerk displays the order on his screen by scanning the bar code on the printout and processes it through his own system.

This prototype should handle other tasks as well. The comparison shopper in Task Description 2 (section 2.3.3.2) can choose to print the list without placing the order. If she comes back later to order the items, she simply scans the bar code on the printout, which puts her shopping list back onto the screen (instructions on the form tell shoppers that they can do this). This particular shopper may not be able to do this because of her arthritis, but scanning would be possible with mild cases of the disease.

Does this interface work? Do a detailed task-centered walk-through to discover where it succeeds and where it fails. There are certainly flaws in it! Although we created this solution to address some of the problems identified in the task analysis, we don't really know how good these fixes are. We do know that major redesigns such as this one will likely introduce new problems, both large and small.

2.4 CONCLUSION

Task-centered system design is a very effective discount usability engineering method suitable for many interface development problems. It offers a big "bang for the buck," allowing the achievement of a reasonable design with only modest extra effort. The caveat is that it is not perfect. Compared to other more precise techniques, you will likely miss a task or user group or may overlook a task nuance during the walk-through. I would not use TCSD to design an air traffic control system (see chap. 13), but I would use it if I had to produce a noncritical system under a limited budget. I would also supplement it with other evaluation techniques, such as heuristic evaluation (Nielsen & Mack, 1994) and user testing (Dumais & Redish, 1993).

This chapter should be sufficient to get you started on your own task-centered system design project. If the method excites you, you should also read Lewis and Reiman's (1993) book on the topic. They go into further detail and describe how the TCSD method relates to other aspects of interface design and evaluation.

There are other good sources on related topics. Beyer and Holtzblatt (1998) describe the contextual interview, which is an excellent way to discover the information needed for detailed user and task descriptions. Carroll (2000) goes into great detail on scenario-based design. Cooper (1999) presents a variant on TCSD called goal-centered system design. Nielsen and Mack (1994) collect readings about other discount interface inspection methods. Finally, my own teaching materials on this topic are available at www.cpsc.ucalgary.ca/~saul/hci_topics/topics/tasks.html. Materials include PowerPoint presentations, student assignments, and teaching tips.

ACKNOWLEDGMENTS

The National Sciences and Engineering Research Council of Canada (NSERC) provided funding. This material has been refined over years of teaching University of Calgary undergraduate students. I owe them my gratitude.

REFERENCES

Beyer, H. & Holtzblatt, K. (1998) *Contextual design: Defining customer-centered design.* San Francisco: Morgan Kaufmann.

Bias, R. (1994). The pluralistic usability walkthrough: Coordinated empathies. In J. Nielsen & R. Mack (Eds.), *Usability inspection methods* (pp. 63–76). New York: Wiley.

Carroll, J. (2000). Making use: Scenario-based design of human-computer interactions. Cambridge, MA: MIT Press.

Cooper, A. (1999). *The inmates are running the asylum: Why high-tech products drive us crazy and how to restore the sanity.* SAMS.

Dumas, J., & Redish, J. (1993). *A practical guide to usability testing.* Norwood, NJ: Ablex.

Lewis, C., & Reiman, J. (1993). *Task centered user interface design: A practical introduction.* Boulder, CO: University of Colorado. Shareware book available from ftp.cs.colorado.edu/pub/cs/distribs/clewis/ HCI-Design-Book/

Nielsen, J., & Mack, R. (1994). *Usability inspection methods.* New York: Wiley.

3

Hierarchical Task Analysis

John Annett
University of Warwick

Hierarchical task analysis is one of the most familiar of the analytical methods employed by human factors specialists in the UK. This chapter reviews the historical origins of HTA and explains the underlying principles and rationale. It is best regarded not so much as a strict procedure but a generic approach to the investigation of problems of human performance within complex, goal-directed, control systems, including those involving computers. The main section comprises a 7-step procedural guide with illustrative examples. The chapter concludes with a discussion of the variety of uses and a brief review of evaluation studies.

3.1 HISTORICAL BACKGROUND

3.1.1 The Search for a Unit of Analysis

Hierarchical task analysis (HTA) was developed at the University of Hull in the late 1960s in response to the need for a rational basis for understanding the skills required in complex non-repetitive operator tasks, especially process control tasks such as are found in steel production, chemical and petroleum refining, and power generation. The principal analytic tools available up to that time were derived either from classical work study (Gilbreth, 1911; Taylor, 1911) or skill taxonomics based on psychometric constructs. Work study provided a set of simple units such as "select", "grasp", and "assemble" descriptive of repetitive manual operations, but proved inadequate for "mental" work such as monitoring, controlling, and decision making. Some of Gilbreth's units, such as "select," implied a cognitive process, but no serious attempt was made to elaborate or analyze the underlying mental activity.

The "programmed instruction" movement inspired by B. F. Skinner's (1954) behaviorist theory of learning was also based on the notion that complex behavior, including intellectual tasks such a doing mathematics, comprised a set of simple stimulus-response (S-R) units. Intelligent behavior could, according to Skinner, be reduced to a chain of individually reinforced S-R connections. To "program" a course of instruction, the teacher had to identify these components and reinforce each individually in a carefully planned sequence in which complex behavior was "shaped" in a series of small steps progressively closer approximations to the desired behavior. In the absence of any systematic theory of cognition, the shaping of behavior could only be empirically determined, and so analysis was seen, like the learning process itself, as a matter of trial-and-error.

The alternative arising out of psychometric studies of motor skills (Seashore, 1951) sought to describe work in terms of the skills and abilities required for successful task performance. Tasks were executed by the deployment of skills such as "coordination," "dexterity," "verbal knowledge," "spatial relations," and so on, as defined by patterns of performance on psychometric tests. The analysis of tasks in these terms might be accomplished by correlating task performance with test results but in practice was rarely attempted. There was the added complication that these patterns were observed to change with practice (Fleishman & Hempel, 1954).

The classification of tasks in terms of supposed underlying cognitive processes continued to provide a focus for analysis prompted by the notion of taxonomies of learning objectives such as that of Bloom, Engelhart, Furst, Hill, & Krathwohl (1956) or Gagné (1965). Gagné's taxonomy comprised a hierarchy of cognitive skills, with simple S-R connections and behavioral chains at the lowest level rising through the ability to make discriminations, to form concepts, acquire general principles, and solve problems as progressively more complex skills. Although such schemes attributed greater significance to cognitive processes than did Skinner's behaviorism, they suffer from the problem of all taxonomies, namely, that they are intrinsically "fuzzy" and susceptible to changes in definition and fashions in the use of language.

HTA constituted a radically new approach based on functional rather than behavioral or psychometric constructs and using a fundamental unit called an *operation.* A functional task analysis begins by defining *goals* before considering actions by which the task may be accomplished. Complex tasks are defined in terms of a hierarchy of goals and subgoals nested within higher order goals, each goal, and the means of achieving it being represented as an operation. The key features of an operation are the conditions under which the goal is activated (see chap. 19) and the conditions which satisfy the goal together with the various actions which may be deployed to attain the goal. These actions may themselves be defined in terms of subgoals. For example, thirst may be the condition which activates the goal of having a cup of tea, and subgoals are likely to include obtaining boiling water, a teapot with tea, and so on. By analyzing tasks in this way, we not only avoid the awkward problem of behavioral taxonomies but focus attention on what really matters, that is, whether the task is actually accomplished and, if not, why not. In short, task analysis is seen not so much as a describing of actions or even cognitive processes as such but as a systematic method of investigating problems of human performance (see chap. 1).

3.1.2 Modeling Human Performance

3.1.2.1 Systems Theory

The origin of this functional approach lay in systems theory and information-processing models of human performance (see, e.g., Morgan, Cook, Chapanis, & Lund, 1963). Systems theory sees task performance as the interaction between human and machine, the latter becoming increasingly complex as computers and automation developed. Each have their respective strengths and weaknesses, and overall system performance can be optimized by correct allocation of function between human and machine consistent with the performance capabilities of each (see chaps. 6 and 22). System performance can also be analyzed in terms of the sum of individual sources of variation, or error, in system output. Chapanis (1951) pointed out that the relative importance of the error contributed by uncorrelated sources of variance increases quadratically with relative size. In other words, the most effective means of improving overall system performance is to identify and deal first with factors generating the largest error variance. These could be either the machine or the human—or indeed any identifiable feature of their interaction, such as the visibility of displays, the dynamics of the controls, the length of the sequence of relevant actions, the complexity of the decisions required, and many other physical and psychological factors. HTA takes this top-down approach by first examining high-level goals and decomposing the task into subgoals and subsubgoals, looking for those subgoals which are more difficult to attain (i.e., generate more error) and that therefore restrict or even prevent overall goal attainment.

3.1.2.2 Feedback Theory

The functional analysis of a task into a hierarchy of operations within which are nested suboperations is compatible with the kind of information-processing model of human performance advanced by early "cognitive" theorists such as Broadbent (1958) and Miller, Galanter, and Pribram (1960). Both, in contrast to the behaviorist view, envisaged human behavior as goal-directed and essentially controlled by feedback. Although Broadbent's theory concentrated on the capacity limitations of the principal feedback loop, Miller et al. expounded the hierarchical structure of feedback loops and especially the concept of a *plan*. They proposed a basic behavioral unit comprising a *test* of the goal state and an *operate* function connected in a recursive loop such that if the goal state tests negative, then *operate* is triggered, but if the goal state tests positive, then exit from the loop is permitted. This basic feedback loop was called a Test-Operate-Test-Exit (T-O-T-E) unit. Their classic example is a T-O-T-E for hammering a nail: The test is "Is the nail flush?" If negative, then "hammer" (operate), then test again. If positive, then "exit." Complex behavior, say, constructing a bookshelf from a self-assembly kit, may be described as a nested hierarchy of T-O-T-E units such as "Is the kit unpacked?" (if no, unpack it), "Are all components present?" and so on. Tasks such as controlling an acid purification plant (Duncan, 1972) can be similarly decomposed into subgoals such as ensuring the plant is started up, is run within safety limits, and is shut down under certain conditions, and each of these is further "unpacked" to whatever level of detail the situation might demand, such as testing for the presence of faults and adopting suitable remedial procedures. The detailed description of the goal hierarchy is termed a *plan*, and this designation is adopted in HTA. Some plans, such as cooking recipes, are simple, often comprising a set of subgoals, such as obtaining ingredients, preparing them, and heating them, arranged in a simple sequence. Others, such as the plans required to detect, diagnose, and remedy a fault in a nuclear power plant, are much more complex, but they are plans nonetheless. Examples of the variety of plans will be given later in the chapter.

3.2 RATIONALE AND PRINCIPLES

3.2.1 Analysis Versus Description

The rationale for HTA springs from the ideas outlined in the previous section. The principles on which the method is based were all present or implicit in the original statement of HTA (Annett, Duncan, Stammers, & Gray, 1971), but they are here elaborated for clarification and to make it easier to understand when and how to use HTA. Analysis is not just a matter of listing the actions or the physical or cognitive processes involved in carrying out a task, although it is likely to refer to either or both. Analysis, as opposed to description, is a procedure aimed at identifying performance problems (i.e., sources of error) and proposing solutions. This distinction between *task description* and *task analysis* was made by R. B. Miller (1962; see also chap. 24), and emphasizes the purpose of analysis as providing solutions to initially specified problems. The problem might be to design a suitable interface or perhaps decide what kind of training to provide, and in each case the course of the analysis might be different, depending on what kinds of information are most relevant to the question being asked.

3.2.2 Tasks and Goals

A task is "any piece of work that has to be done" (*Shorter Oxford Dictionary*, 1973). Every task may therefore be defined in terms of its *goal* or *goals*. HTA differs radically from earlier methods of task analysis by beginning, not with a list of activities, but by identifying the goals of the task. It aims to provide a *functional* analysis rather than a behavioral description. In routine repetitive tasks, actions vary little while the environment and purpose remain constant. In complex tasks, the same goals may be pursued by different routes and different means, depending on circumstances peculiar to each occasion. Simply to list actions without understanding what they are for can be misleading. Complex systems, including those involving humans and computers, are designed with goals in mind, and understanding how a system attains or fails to attain its designated goal is the primary purpose of the analysis.

A *goal* is best stated as *a specific state of affairs*, formally a *goal state*. The goal state can be an event or some physically observable value of one or more variables that act as criteria of goal attainment and ultimately of system performance. At any one time, a goal may be *active* or *latent*. Active goals are those being currently pursued, whereas latent goals are those that may be pursued under conditions that might arise. The importance of this distinction will become apparent when we consider the concept of a *plan*.

3.2.3 Decomposition and Redescription

Goals are often complex; that is, they are defined by more than one event or by values of more than one variable. Where these can be individually identified, the analysis should specify these component goal states by the process of *decomposition*. HTA envisages two kinds of decomposition. The first comprises identifying those goal states specified by multiple criteria, for example, to arrive at a destination (an event) having expended minimum effort (variable 1) and with no injury (variable 2). The second kind of decomposition comprises the identification of subgoals in any routes that may be taken in attaining the overall goal state. Goals may be successively unpacked to reveal a nested hierarchy of goals and subgoals. This process of decomposition, also referred to as *redescription*, has the benefit, according to the general principle proposed by Chapanis (1951), of comprising an economical way of locating sources of general system error (actual or potential) in failure to attain specific subgoals.

3.2.4 Operations, Input-Action-Feedback

An *operation* is the fundamental unit of analysis. An operation is defined by its goal(s) (or goals). It is further specified by (a) the circumstances in which the goal is activated (the *I*nput), (b) the activities (*A*ction) that contribute to goal attainment, and (c) the conditions indicating goal attainment (*F*eedback); hence operations are sometimes referred to as *I-A-F units*. As shown in the introduction, operations are equivalent to T-O-T-E units (Miller et al., 1960) in that they are feedback (or servo) loops. Just as goals may be decomposed into constituent subgoals, so operations may be decomposed into constituent suboperations arranged in a *nested hierarchy*. Suboperations are included within higher order (or superordinate) operations, the attainment of each subgoal making a unique contribution to the attainment of superordinate goals. The suboperations comprising a superordinate operation should be mutually exclusive and collectively comprise an exhaustive statement of the subgoals and superordinate goals.

An *action* can be understood as an injunction (or instruction) to do something under specified circumstances, as illustrated by the T-O-T-E for hammering a nail into a piece of wood. Input and feedback both represent *states* or *tests* in the Miller et al. formulation. These states register either error, therefore requiring action, or the cancellation of error, signaling the cessation of that action ("operate" in Miller et al's [1960] terminology). An action can be understood formally as a *transformation rule* (Annett, 1969, pp. 165–169), that is, a specification of how a servo responds to an error signal and its cancellation. For example, in a manual tracking task, the transformation rule can be specified by an equation, known as a *transfer function*, which quantifies the control output required to correct for an error signal of given direction and magnitude (McRuer & Krendel, 1959). This is, however, a special case, and normally, for example in self-assembly kits, computer software handbooks, and cookbooks, instructions are specified using commonly understood verbs. Some verbs (e.g., *chamfer*, *defragment*, and *marinate*) form part of a technical vocabulary and may need to be redescribed in simpler terms (how *does* one chamfer wood, defragment a computer disc, or marinate meat?), and each redescription comprises a set of suboperations.

3.2.5 Plans

As indicated above, suboperations collectively redescribe their superordinate operation, but typically we need to know not only the constituent suboperations but the order, if any, in which they should be carried out (e.g., "To chamfer, first secure the piece to be chamfered, then obtain a suitable file . . ."). The specification of the rule, or rules, governing the order in which suboperations should be carried out is called a *plan*. Plans can be of various types, the commonest being simply a *fixed sequence* or *routine procedure*, such as "do this, then this, then this," and so on. Another common type of plan specifies a *selective rule* or *decision*: "If *x* is the case, do this. If *y* is the case, do that. "These two types of plan are significant because they imply knowledge on the part of the operator. The required knowledge may be simple procedural knowledge or extensive declarative knowledge of the environment, the limits and capabilities of the machine, safety rules, and much else besides. In this respect, HTA anticipated the requirement for what is now known as *cognitive task analysis* (Schraagen, Chipman, & Shalin, 2000; chap. 15).

A third distinct type of plan requires two or more operations to be pursued in parallel. In other words, the superordinate goal cannot be attained unless two or more subordinate goals are attained *at the same time*. Such a plan is known as a *time-sharing* or *dual task* plan, and this type of plan also has significant cognitive implications for the division of attention (e.g., chaps. 14 and 15) and in the case of team operations the distribution of information between team members acting together. When a goal becomes active, its subordinate goals become active according to the nature of the plan. For example, in a fixed sequence, the goal of each suboperation becomes active as the previous subgoal is attained. Where the plan involves a selective rule, only those goals become active that are specified by the application of the rule; the rest remain latent. In a time-sharing plan, two or more goals are simultaneously active.

3.2.6 Stop Rules

The decomposition of goal hierarchies and the redescription of operations and suboperations might continue indefinitely without the use of a *stop rule* specifying the level of detail beyond which no further redescription is of use. The ultimate stop rule is just that: Stop when you have all the information you need to meet the purposes of the analysis. However, since the general purpose of HTA is to identify sources of actual or potential performance failure a general stop rule is common: Stop when the product of the probability of failure (p) and the cost of failure (*c*) is judged acceptable. This is known as the *p* × *c criterion* (Annett & Duncan, 1967), and its prime benefit is that it keeps the analytical work down to the minimal amount and focuses the attention of the analyst on those aspects of the task that are critical to overall system success. In practice, lack of empirical data may mean that *p* and *c* can only be estimated, but it is the *product* of the two that is crucial to the decision to stop or continue decomposition. The obvious reason for stopping is that the source of error has been identified and the analyst can propose a plausible remedy in terms of either system design, operating procedures, or operator training (i.e., by redesigning the cognitive task).

3.3 PROCEDURAL GUIDE TO HTA

HTA is a flexible tool that can be adapted to a variety of situations and needs. Data may be derived from any number of different sources (chap. 1), the analysis can be continued to any desired level of detail, and there is no rigid prescription of how the results may be used. HTA can nevertheless be carried out in a number of different ways that may involve greater or lesser attention to individual steps in the fundamental procedure outlined below. In general, the benefits of HTA, and its reliability and validity, are proportional to the effort that goes into following this procedure. The analyst is nevertheless entitled to trade off effort for value by shortening or adapting the procedure to suit specific needs. Some ways of doing this are mentioned. The steps are summarized in Table 3.1.

3.3.1 Step 1: Decide the Purpose(s) of the Analysis

The purpose of the analysis has important implications for the way in which it is carried out, including the preferred data collection procedures, the depth of the analysis, and the kinds of solutions (results) that can be offered. Typical purposes are designing a new system, troubleshooting and modifying an existing system, and developing operator training all of which involve the design or redesign of the operator's tasks.

3.3.1.1 System Design

Special considerations may include the design of the equipment interface and operating procedures and manning with the aim of optimizing workload and minimizing errors. Few designs are totally novel, and, as suggested by Lim and Long (1994), an analysis of a comparable extant system may prove useful in identifying difficulties to be avoided in the new design. In the case of an extant design, relevant data may be collected from records of performance, errors, and accidents; from the views of expert users, supervisors, and managers; and by direct observation. Depending on the observed, reported, or even anticipated symptoms of failure, the analyst's attention may well focus on particular aspects of the task, such as displays, communications, complex decision rules or heuristics to be employed, and the number and roles of operators required for successful system performance. Where relevant data are not available, limited hierarchical decomposition, as recommended by Ormerod (2000), may be helpful in drawing attention to possible sources of error. Some error prediction methods, such as Task Analysis for Error Identification (TAFEI) by Baber and Stanton (1994; chap. 18), use HTA as a first stage, but the accuracy with which errors can be predicted will depend on the quality of the information available for the analysis. Particular attention should be paid to operations that play a critical role in system performance, and reference to published ergonomic design standards may be valuable. In general, the analysis and report should address the sponsor's questions.

TABLE 3.1
Principal Steps in Conducting HTA

Step Number	*Notes and Examples*
1. Decide the purpose(s) of the analysis	1. Design of system/interface/operating procedures/manning. 2. Determine training content/method.
2. Get agreement between stakeholders on the definition of task goals and criterion measures	1. Stakeholders may include designers, managers, supervisors, instructors, operators. 2. Concentrate on system values and outputs. 3. Agree performance indicators and criteria.
3. Identify sources of task information and select means of data acquisition	1. What sources as are available? e.g. direct observation, walk-through, protocols, expert interviews, operating procedures and manuals, performance records, accident data, simulations.
4. Acquire data and draft decomposition table/diagram	1. Account for each operation in terms of input, action, feedback and goal attainment criteria and identify plans. 2. Sub-operations should be (a) mutually exclusive (b) exhaustive 3. Ask not only what *should* happen but what *might* happen. Estimate probability and cost of failures.
5. Re-check validity of decomposition with stakeholders	1. Stakeholders invited to confirm analysis, especially identified goals and performance criteria. 2. Revert to step 4 until misinterpretations and omissions have been rectified.
6. Identify significant operations in light of purpose of analysis	1. Identify operations failing p × c criterion 2. Identify operations having special characteristics, e.g. high work-load, requiring teamwork, specialist knowledge etc.
7. Generate and, if possible, test hypotheses concerning factors affecting learning and performance	1. Consider sources of failure attributable to skills, rules and knowledge. 2. Refer to current theory/best practice to provide plausible solutions. 3. Confirm validity of proposed solutions whenever possible.

3.3.1.2 Training

The intended product of the analysis is also important in determining the appropriate stop rule. To produce a fully documented training program for novices, the level of detail may need to be capable of generating very specific how-to instructions in plain language. If the purpose is to identify the type of training required, the analysis should identify operations and plans of particular types that are thought to respond to particular training methods. For example, where inputs are perceptually or conceptually complex, special recognition exercises may be required, or where procedures are especially critical, operating rules and heuristics and system knowledge will need to be learned (Duncan, 1972). In summary, the analysis should anticipate the kinds of results which would provide answers to the original questions such as design recommendations, training syllabi and so on.

3.3.2 Step 2: Definition of Task Goals

Task performance, by definition, is goal-directed behavior, and it is therefore crucial to the analysis to establish what the performance goals are and how one would know whether these goals have been attained. A common mistake is to take this question to be about observed operator behavior, such as using a particular method, rather than performance outcomes, such as frequency of errors and out-of-tolerance products. Bear in mind that the effort of analysis is ultimately justified by evidence of the outcomes of performance, an issue that is taken up again in the section on validity.

Different stakeholders (designers, trainers, supervisors, and operators) can sometimes have subtly different goals. It is better to identify problems of this kind early in the analysis by thorough discussion with all the relevant stakeholders. If goals appear to be incompatible, the analyst can sometimes act as a catalyst in resolving these issues but should not impose a solution without thorough discussion (see also chap. 1). As the decomposition proceeds, more detailed goals and more specific criterion measures are identified, and it can emerge that different operators with ostensibly the same overall purpose have slightly different plans (ways of doing things) that may imply different subgoals. Objective performance measures provide the opportunity to compare methods in terms of superordinate criteria.

For any goal, the key questions, which may be asked in many different forms, are these: (a) What objective evidence will show that this goal has been attained? and (b) What are the consequences of failure to attain this goal? Answers to the first question can form the basis of objective performance measures that may subsequently be used in evaluating any design modifications or training procedures proposed on the basis of the analysis. Answers to the second question may be used to evaluate the $p \times c$ criterion and hence the degree of detail of the analysis. Answers to both questions form the essential basis for agreement about the system goals. If goals cannot be stated in these objective terms, then the sponsors and stakeholders are unclear about the purposes of the system, and the validity of the entire analysis is called in question.

3.3.3 Step 3: Data Acquisition

Classical work study methods were typically based on direct observation. However, HTA is concerned with the functional aspects of the task, and actual behavior may not always be the best guide. If the purpose is to look for ways to improve operator performance on an existing system, records of actual operator performance, including both the methods used by operators and measures of success or failure, will be important. Errors may be rare, but critical incident data can provide useful insights into the origins of performance failure (chap. 16). If the purpose is to make recommendations on a new design, then data relating to comparable (e.g., precursor) tasks may be helpful, but the designer's intentions are critical. In the absence of actual performance data, the analyst should challenge the designer with what-if questions (see chaps. 5 and 21) to estimate the consequences of performance failure. Sometimes data concerning performance on preexisting systems or comparable tasks may prove useful.

Preferred sources of data will clearly vary considerably between analyses. Interviewing the experts is often the best way to begin, particularly if the interviews focus on system goals, failures, and shortcomings. Direct observation may provide confirmatory information but, especially in the case of nonroutine tasks, may yield relatively little information concerning uncommon events that may be critical. Formal performance records such as logs and flight recorders may be available, especially in safety-critical systems, but these are often designed primarily with engineering objectives in mind, and the human contribution to system performance can sometimes be difficult to determine from a mass of recorded data. Doing focused interviews with recognized experts aimed at identifying performance problems is often the only practicable method of obtaining estimates of the frequency and criticality of key behaviors. In some cases, where the informants are unclear about what would happen in certain circumstances and what would be the most effective operator strategy, it may be helpful to run experimental trials or simulations. In one (personally observed) case where even skilled operators were not clear about the cues used in reaching an important decision, an experiment was run in which the effects of blocking certain sources of information were observed. In

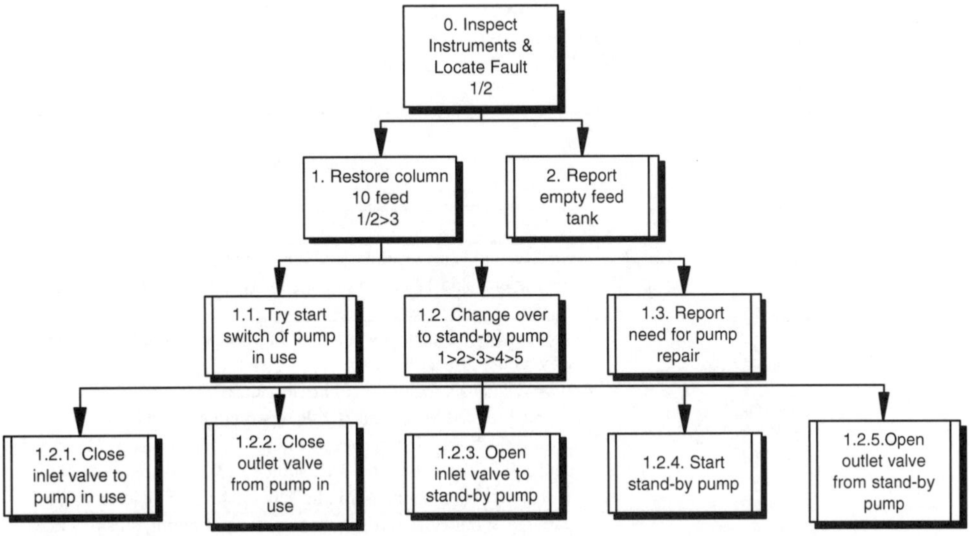

FIG. 3.1. Section of the goal hierarchy for an acid distillation plant operator's task.

this instance, it was determined that the operators were unconsciously using the sound of the machinery rather than the sight of the product to reach a key decision.

3.3.4 Step 4: Acquire Data and Draft a Decomposition Table or Diagram

In general, the more independent sources of data consulted, the greater the guarantee of validity of the analysis. It is all too easy to think of operations simply as actions. The critical goal of HTA is to be able to relate *what* operators do (or are recommended to do) and *why* they do it and *the consequences if it is not done correctly*. Only when this is thoroughly understood is it possible to create a meaningful diagram or table (see Fig. 3.1 and Table 3.2). A useful logical check on the validity of a proposed decomposition table is that all suboperations must be (a) mutually exclusive and (b) exhaustive (i.e., must completely define the superordinate operation).

The use of standard notation generally helps in the construction of tables and diagrams as well as the interpretation of results, and it can also improve communication between analysts and stakeholders. The recommended notation system provides for each identified operation a unique number that may be used in both diagrams and tables. The notation should also specify plans and indicate stops. Stops represent the most detailed level of the analysis, typically the level that is most relevant to the results of the search and the recommendations for further action.

Both tabular and diagrammatic formats have their advantages, and typically both are used. The diagrammatic format (see Fig. 3.1) often helps to make clear the functional structure of the task whereas the tabular format (see Table 3.2) is more economical of space and facilitates the recording of supplementary notes, queries, and recommendations. Individual operations are normally numbered with 0 standing for the top-level goal, in effect the title of the task, and suboperations being numbered in the order of description (which is not necessarily the same as the order of execution). Thus operations 1, 2, and 3 would be the three principal suboperations of task 0; 1.1, 1.2, 1.3 would be the three suboperations into which operation 1 is decomposed; and 2.1 and 2.2 would be the two suboperations into which operation 2 is decomposed. Each additional digit indicates a new level of decomposition. In the diagrammatic

TABLE 3.2
Extracts from the Tabular Format Identifying those Operations Likely to Cause Problems and Offering Recommendations

Operation	*Problems and Recommendations*
0. Inspect instruments & locate fault. 1/2	*Input*: Need to learn location and function of up to 50 control panel instruments and 128 hand-operated valves throughout plant and recognise fault pattern. *Feedback*: Only available when fault indicators restored to normal following correct diagnosis. *Action*: Diagnosis required within 3 minutes. *Plan*: restore feed to column 10 OR report empty feed tank. *Recommendations*: (1) improve labelling of control valves. (2) Provide decision tree to optimise search strategy. (3) Simulation of possible fault patterns for training. Methods could be (a) rote learning of patterns & remedies or (b) 'technical story' training plus practice.
1. Restore column feed. 1/2 + 3	*Plan*: try start switch of pump in use OR change over to stand-by pump AND report need for pump repair.
1.2. Change over to stand-by pump. 1 > 2 > 3 > 4 > 5	*Action*: Essential to follow correct sequence due to danger of acid spill. *Plan*: follow strict sequence (sub-operations 1–5) *Recommendation*: Correct sequence to be thoroughly memorised.

format, a vertical line descends from the superordinate operation to a horizontal line covering the set of suboperations into which it is expanded.

Plans are implicit in the decomposition structure, but they can be made more explicit by adding the appropriate algorithm to the diagram . A succinct way of presenting four basic types of plan in the diagram was employed by Annett, Cunningham, and Mathias-Jones (2000); it is shown in Fig. 3.1. The symbol ">" is used to indicate a sequence, "/" to represent an either/or decision, "+" to represent dual or parallel operations, and ":" to represent multiple operations in which timing and order are not critical. The lowest or most detailed level of decomposition, being often the most important level of description, is typically indicated in the diagram by some feature of the box, such as the double sidebars in Fig. 3.1, and may be similarly indicated in the table by a typographic variant, such as boldface.

3.3.5 Step 5: Recheck Validity of Decomposition with Stakeholders

Stakeholders, especially those unfamiliar with HTA, sometimes change their minds as the analysis proceeds. By thinking about the task, they may realize that events do not always occur in a standard way or that something that has never been known to happen just could. For this reason, an iterative process is recommended wherever possible, with the full analysis being developed over a series of interviews and by cross-checking between sources such as training manuals and experienced operators (see chap. 2). The process of cross-checking not only provides the best guarantee of the reliability of the completed analysis but encourages the stakeholders to develop a sense of ownership of the analysis and consequently to share responsibility for recommendations arising from the results.

3.3.6 Step 6: Identify Significant Operations

It is common in decomposition methods used by large organizations, such as the military, to adopt a fixed number of levels, such as *job*, *task*, and *subtask*, but this means that some parts of the task may be described in unnecessary detail whilst others require more. As a general rule, the recommended criterion (the stop rule) is to stop the analysis of any given operation when the probability of performance failure multiplied by the cost to the system of performance failure ($p \times c$) is acceptably low. The rationale for this general rule is that the analysis is essentially a search process aimed at identifying significant sources of actual or potential system failure, so when no more sources can be identified, then clearly the analysis should cease. However, modifications to this general rule may be made in certain circumstances. For example, when the analysis is part of the design process, it may be desirable to stop at a level that is device independent, specifying the component operations functionally but leaving open exactly how these operations will be implemented in terms of equipment or software (see Lim & Long, 1994; Ormerod, Richardson, & Shepherd, 1998).

Different stopping rules may serve different purposes. For example, Crawley (1982) used a form of HTA, referred to as overview task analysis (OTA), in a study of air traffic control. The analysis was stopped at tasks that could be readily identified and judged by controllers as having certain characteristics, such as being particularly demanding or especially satisfying, that provided a basis for deciding whether automation should be considered. Annett et al. (2000) analyzed naval antisubmarine warfare command team tasks in sufficient detail to identify certain types of teamwork such as operations depending critically on intra-team communication or discussion, active collaboration, or the synchronization of actions.

3.3.7 Step 7: Generate and, If Possible, Test Hypotheses Concerning Task Performance

HTA is principally carried out on the assumption that it is a tool to be used by a specialist who is looking for particular classes of problem and has a range of optional recommendations available. The analysis provides a means of generating hypotheses concerning the likely sources of actual or potential failure to meet overall task goals and to propose appropriate solutions. It must be clearly understood that HTA as a method does not itself include a set of diagnostic categories nor a set of acceptable solutions to the problems identified. These will depend on the predilections and technical capabilities of the analyst. HTA simply provides an efficient procedure for identifying sources of actual or potential performance failure. However, Reason's (1990) classification of human error, based on Rasmussen's (1983) taxonomy of *skill-based*, *rule-based*, and *knowledge-based* performance, may be helpful in this context. The analyst is prompted to develop hypotheses to account for failures and propose practical solutions, but these are to be regarded as hypotheses that should be put to the test, since testing is the only way to guarantee that the analysis is valid.

3.4 ILLUSTRATIVE EXAMPLES

3.4.1 A Process Control Task

This analysis was carried out by Keith Duncan as part of the University of Hull Task Analysis project and is described more fully in Duncan (1972). The tabular format has been changed from the original to something closer to the notation recommended here. The problem was to devise a training program to teach process operators to locate and cure faults in an acetic acid distillation plant. Figure 3.1 shows a section of the goal hierarchy, and Table 3.2 presents some of the notes derived from the analysis. The presence of a fault is normally indicated by one or more clear alarms, and the physical actions, such as throwing a switch or closing a valve, are mostly simple. A core problem was learning the identity and function of around 50 instruments in the control room plus a further 128 control valves located throughout the plant and learning a large number of possible symptom patterns and their remedies. This was made especially difficult by variable time delays between corrective action and the restoration of normal function and by the possibility of compounding failure due to faulty diagnosis. Three minutes was regarded as the maximum allowable time for diagnosis. Note also a changeover to a standby pump was potentially dangerous if the necessary steps were not carried out in the correct order. As regards training solutions, there was a clear need for some form of simulation because it could take many months of on-the-job experience to be exposed to a representative range of faults and variable feedback delay in the confirmation of diagnosis, to say nothing of the danger of misdiagnosis. Three possible training solutions were suggested and subsequently tested experimentally, and in addition suggestions were offered concerning a labeling system to make it easier to identify the control valves.

3.4.2 A Naval Team Task

Table 3.3 shows in tabular form part of the analysis of a task carried out by the antisubmarine warfare team in the operations room of a warship. It is described more fully in Annett et al. (2000). The team comprises the principal warfare officer (PWO), the active sonar director (ASD), and the action picture supervisor (AcPS), all located in the ship's operations room. They are also in touch with the officer of the watch (OOW) and the electronic warfare director (EWD) elsewhere on the ship and in other platforms, such as the helicopter (Helo) and maritime patrol aircraft (MPA). The overall aim of the team is to defend a highly valued unit (e.g., a troopship) by identifying and responding appropriately to submarine threats picked up by the ship's sensors. The purpose of this analysis was to be able to identify and provide objective measures of key team skills. The team members were already proficient in their individual skills, and the table shows only some of the critical team operations. The notes indicate the goal of each suboperation and criterion measure and include a plain language statement of the nature of the teamwork involved. To identify threats (1.1), the team must, at the same time, scan for possible threats (1.1.1) and classify threats (1.1.2), of which more than one could be present at any given time. One possible classification is an Immediate Threat (1.1.2.1.).

Having identified these and other critical team behaviors, their occurrence was observed during simulator-based exercises. An objective scoring system was then based on five types of team behavior, including sending and receiving information, discussing ambiguous data and courses of action, collaborating in the execution of a plan, and synchronizing team actions. At the time of writing (spring 2003), the Royal Navy is in the process of evaluating the method as a means of objectively assessing collective performance.

3.5 EVALUATION

3.5.1 Uses of HTA

HTA has been used in a wide variety of contexts for the full range of problems that confront human factors practitioners. Shepherd (2001) cites a range of examples including simple procedural tasks, such as changing a printer cartridge, using a word processor and the supermarket checkout task, through fine motor skills of minimal access (keyhole) surgery to air traffic control and management tasks. In a survey of 30 task analysis studies in the defence industry Ainsworth and Marshall (1998) found 2 cases of its use in system procurement, 7 for

TABLE 3.3
Extracts from an HTA(T) of an Anti-submarine Warfare team. From Annett et. al. (2000)

Team Operation	*Notes on Teamwork*
1.1.1. Scan for threats	*Goal*: To ensure no potential threat goes undetected. *Measure*: Reporting response time of individuals to whom 'stimuli' have been presented. Some contacts ignored because they are thought to be irrelevant (e.g., an aircraft track) *Teamwork*: All members alert to sources of information. Sources include sonar contacts or visual sightings and from own ship or other platform. The ASD and the AcPS monitor sonar contacts and are best able to take action on a submarine contact or a torpedo. EWD monitors ESM, AcPS monitors for 'riser/sinker' and with Helo for 'feather' and team members should monitor voice circuits for Helo or MPA reports. PWO ultimately responsible for ensuring vigilance and team members awareness of possible threats. *Plan*: Scanning is continuous throughout the exercise.
1.1.2. Classify threats [1/2 > 3]	*An immediate threat, an approaching torpedo has to be identified as such. A contact from whatever source is usually ambiguous and team must put together available data to arrive at the most likely identification. Classifying at certain levels (possub1, possub2 etc.) has consequences for subsequent actions . Grading system may be used incorrectly.* *Classification categories are possub lo 1, lo 2, hi 3 etc. hi4, probsub (needs more information—e.g., acoustic information) then to certsub when a qualified observer actually sees it.* *Goal*: Identify immediate threat. Correctly classify all contacts. *Measure*: Frequency of correct classification. Time to make correct classification. *Teamwork*: Normally classification is a matter of discussion between PWO, ASWD and CO. Incorrect or missing information may result in late or incorrect classification. Plan: If threat is immediate (torpedo) go to [1.1.2.1], else go to chart check. [1.1.2.2], then [1.1.2.3]
1.1.2.1. Identify Immediate Threat [>1.2]	*Goal*: Recognition of an immediate threat—e.g., a torpedo track—and give correct range & bearing. Measure: Immediate threat correctly recognised and correct bearing/range given. *Teamwork*: PWO must be adequately informed by other team members. In case of Torpedo Counter Measure (TCM) OOW is responsible. *Plan*: Requires immediate response (e.g., TCM). Go to [1.2].

manpower analysis, 9 for interface design, 5 for operability assessment and 2 instances of its use in specifying training.

Duncan (1972) described the use of HTA in the design of training for process control operators in the petrochemical industry (see section 3.4.1). Other uses include assessing workload and manning requirements (Fewins, Mitchell, & Williams, 1992; Penington, Joy, & Kirwan, 1992). The cognitive task problems identified by the analysis were dealt with by a combination of interface design, changes in staffing levels, and automation. Crawley (1982) used an abbreviated version of HTA to identify critical air traffic control tasks, and Annett et al. (2000) used a variant of HTA to identify critical team functions. Shepherd (2001) outlined the application of HTA to production teams, supervision of an automatic railway, and collaboration between members of medical teams. HTA has also been used for the purpose of hazard assessment and error prediction (Baber & Stanton, 1994; Penington, 1992; Reed, 1992).

HTA was recommended as a precursor to systematic design by Lim and Long (1994) in their Method for Usability Engineering (MUSE). Ormerod (2000) also advocated a modified version of HTA, called the Sub-Goal Template (SGT) method (chap. 17), as part of the design process. The decomposition proceeds down to the level at which the specification of an operation is independent of any specific equipment or interface design. This approach frees the designer to consider a range of design possibilities.

The principle of hierarchical goal decomposition first developed in HTA has become widely used in a number of well-known HCI methods, including Goals, Operators, Methods, and Selection rules (GOMS; Card, Moran, & Newell, 1983; chap. 4) and knowledge analysis of tasks (KAT; Johnson & Johnson, 1991). In addition, task analysis for knowledge description (TAKD; Diaper, 2001; Diaper & Johnson, 1989) explicitly incorporate the essentials of HTA into an extended methodology (see also chaps. 12, 15, 19, and 27).

3.5.2 Usability, Validity, and Reliability

It is reasonable to expect that the same standards of usability should apply to the methods used by human factors specialists as apply to the objects of their studies. Ainsworth and Marshall's (1998) survey found that training in the use of HTA is somewhat variable, and it appears that some practitioners neglect some of the important steps outlined in this chapter. Ainsworth and Marshall also noted that "it appeared that the most insightful analyses were undertaken by analysts who had the most human factors experience" (p. 1617).

Stanton and Young (1998) surveyed the use of 27 methods, including HTA, used by professional ergonomists. The consensus view was that HTA was useful but time consuming and required more training and practice than most others. These findings were confirmed by giving engineering students training and practice in 11 of the most commonly used methods, including HTA. Patrick, Gregov, and Halliday (2000) reported a study in which a small sample of students received training in the main features of HTA and were then required to draw up an analysis of painting a door or making a cup of tea. Their analyses were then scored on 13 criteria dealing with the principal features of the method. Overall performance was found to be poor, particularly in respect of the students' ability to construct an adequate hierarchy. A second study using the same population and tasks but with enhanced training generated analyses of higher quality although still not without problems. These results confirm the conclusions reached by Ainsworth and Marshall (1998) and Stanton and Young (1998)—that HTA is far from simple and takes both expertise and practice to administer effectively.

Evidence from these studies suggests that careful attention to the basic steps summarized in Table 3.1 is recommended as the best guarantee of validity and reliability. In particular keeping the purpose of the study in sight throughout is crucial to the validity of the analysis. The ultimate test of validity lies in Step 7, the empirical test of the hypotheses on which the recommendations are based. Sadly, such results are rarely, if ever, reported. Reliability rests principally on the skills of the analyst in extracting and cross-checking data from various sources (Step 3) and on consultation with stakeholders (in Step 5). A good analyst will always pursue and try to resolve apparent disagreement between informants or inconsistencies in the data. In this way reliability can be maximised. There are many practical reasons why direct evidence of the validity and reliability of HTA, in common with other analytical methods, is scarce, but perhaps the best evidence that HTA has been found a valuable tool lies in its continued use in a wide variety of contexts over the past 30 years.

REFERENCES

Ainsworth, L., & Marshall, E. (1998). Issues of quality and practicability in task analysis: Preliminary results from two surveys. *Ergonomics, 41*, 1607–1617.

Annett, J. (1969). *Feedback and human behavior*. Harmondsworth, UK: Penguin.

Annett, J., Cunningham, D., & Mathias-Jones, P. (2000). A method for measuring team skills. *Ergonomics, 43*, 1076–1094.

Annett, J., & Duncan, K. D. (1967). Task analysis and training design. *Occupational Psychology, 41*, 211–221.

Annett, J., Duncan, K. D., Stammers, R. B., & Gray, M. J. (1971). *Task analysis*. London: Her Majesty's Stationery Office.

Baber, C., & Stanton, N. A. (1994). Task analysis for error identification: A methodology for designing error-tolerant consumer products. *Ergonomics, 37*, 1923–1941.

Bloom, B. S., Engelhart, M. D., Furst, E. J., Hill, W. H., & Krathwohl, D. R. (Ed.). (1956). *Taxonomy of educational objectives*. New York: Longmans, Green.

Broadbent, D. E. (1958). *Perception and communication*. London: Pergamon.

Card, S., Moran, T. P., & Newell, A (1983). *The psychology of human-computer interaction*. Hillsdale, NJ: Lawrence Erlbaum, Associates.

Chapanis, A. (1951). Theory and methods for analyzing errors in man-machine systems. *Annals of the New York Academy of Sciences, 51*(6), 1179–1203.

Crawley, R. (1982). Predicting air traffic controller reaction to computer assistance: A follow-up study (AP Report No. 105). University of Aston, Applied Psychology Department, Aston, England.

Diaper, D. (1989). Task analysis for knowledge descriptions (TAKD): The method and an example. In D. Diaper (Ed.), *Task analysis for human-computer interaction* (pp. 108–159). Chichester, England: Ellis Horwood.

Diaper, D. (2001). Task analysis for knowledge descriptions (TAKD): A requiem for a method. *behavior and Information Technology, 20*, 199–212.

Diaper, D., & Johnson, P. (1989). Task analysis for knowledge descriptions: Theory and application in training. In J. Long & A. Whitefield (Eds.) *Cognitive ergonomics and human-computer interaction* (pp. 191–224). Cambridge: Cambridge University Press.

Duncan, K. D. (1972). Strategies for the analysis of the task. In J. Hartley (Ed.), *Strategies for programmed instruction* (pp. 19–81). London: Butterworths.

Fewins, A., Mitchell, K., & Williams, J. C. (1992). Balancing automation and human action through task analsis. In B. Kirwan & L. K. Ainsworth (Eds.), *A guide to task analysis*. (pp. 241–251). London: Taylor & Francis.

Fleishman, E. A., & Hempel, W. E. (1954). Changes in factor structure of a complex task as a function of practice. *Psychometrika, 19*, 239–252.

Gagné, R. M. (1965). *The conditions of learning*. New York: Holt, Rinehart & Winston.

Gilbreth, F. B. (1911). *Motion study*. Princeton, NJ: Van Nostrand.

Johnson, P., & Johnson, H. (1991). Knowledge analysis of tasks: Task analysis and specification for human-computer systems. In A. Downton (Ed.), *Engineering the human-computer interface* (pp. 119–144). London: McGraw-Hill.

Kirwan, B. (1992). A task analysis programme for a large nuclear chemical plant. In B. Kirwan & L. K. Ainsworth (Eds.), *A guide to task analysis* (pp. 363–388). London: Taylor & Francis.

Kirwan, B., & Ainsworth, L. K. (Eds.). (1992). *A guide to task analysis*. London: Taylor & Francis.

Lim, K. Y., & Long, J. (1994). *The MUSE method for usability engineering*. Cambridge: Cambridge University Press.

McRuer, D. T., & Krendel, E. S. (1959). The human operator as a servo system element. *Journal of the Franklin Institute, 267*, 381–403, 511–536.

Miller, G. A., Galanter, E., & Pribram, K. (1960). *Plans and the structure of behavior*. New York: Holt.

Miller, R. B. (1962). Task description and analysis. In R. M. Gagné (Ed.), *Psychological principles of system development* (pp. 187–228). New York: Holt.

Morgan, C. T., Cook, J. S., Chapanis, A., & Lund, M. W. (1963). *Human engineering guide to equipment design*. New York: McGraw-Hill.

Ormerod, T. C. (2000). Using task analysis as a primary design method: The SGT approach. In J.-M. Schraagen, S. F. Chipman, & V. L. Shalin (Eds.), *Cognitive task analysis* (pp. 181–200). Mahwah, NJ: Lawrence Erlbaum Associates.

Ormerod, T. C., Richardson, J., & Shepherd, A. (1998). Enhancing the usability of a task analysis method: A notation and environment for requirements specification. *Ergonomics, 41*, 1642–1663.

Patrick, J., Gregov, A., & Halliday, P. (2000). Analysing and training task analysis. *Instructional Science, 28*(4), 51–79.

Penington, J. (1992). A preliminary communications systems assessment. In B. Kirwan & L. K. Ainsworth (Eds.), *A guide to task analysis* (pp. 253–265). London: Taylor & Francis.

Penington, J., Joy, M., & Kirwan, B. (1992). A staffing assessment for a local control room. In B. Kirwan & L. K. Ainsworth (Eds.), *A guide to task analysis* (pp. 289–299). London: Taylor & Francis.

Rasmussen, J. (1983). Skills, rules, knowledge: Signals, signs and symbols and other distinctions in human performance models. *IEEE Transactions: Systems, Man and Cybernetics, 13*, 257–267.

Reason, J. (1990). *Human error*. Cambridge: Cambridge University Press.

Reed, J. (1992). A plant local panel review. In B. Kirwan & L. K. Ainsworth (Eds.), *A guide to task analysis* (pp. 267–288). London: Taylor & Francis.

Schraagen, J.-M., Chipman, S., & Shalin, V. (2000). *Cognitive task analysis*. Mahwah, NJ: Lawrence Erlbaum Associates.

Seashore, R. H. (1951). Work and motor performance. In S. S. Stevens (Ed.), *Handbook of experimental psychology*. (1341–1362). New York: Wiley.

Shepherd, A. (2001). *Hierarchical task analysis*. London: Taylor & Francis.

The Shorter Oxford Dictionary (1973). Third Edition, Oxford: Clarendon Press.

Skinner, B. F. (1954). The art of teaching and the science of learning. *Harvard Educational Review 24*, 86–96.

Stanton, N., & Young, M. (1998). Is utility in the eye of the beholder? A study of ergonomics methods. *Applied Ergonomics, 29*(1), 41–54.

Taylor, F. W. (1911). *The principles of scientific management*. New York: Harper.

4

GOMS Models for Task Analysis

David Kieras
University of Michigan

Analyzing a task into Goals, Operators, Methods, and Selection rules (GOMS) is an established method for characterizing a user's procedural knowledge. When combined with additional theoretical mechanisms, the resulting GOMS model provides a way to quantitatively predict human learning and performance for an interface design, in addition to serving as a useful qualitative description of how the user will use a computer system to perform a task. This chapter focusses on GOMS models as a task-analytic notation and how to construct them.

4.1 INTRODUCTION

The purpose of task analysis is to understand the user's activity in the context of the whole human-machine system, for either an existing or a future system (Chapter 1). Although understanding human activity scientifically is the goal of psychology and the social sciences, the constraints on system design activity preclude the lengthy and precise analysis and experimentation involved in scientific work. Thus a task analysis for system design must be rather more informal and primarily heuristic in flavor compared to scientific research. The task analyst must do his or her best to understand the user's task situation well enough to influence the system design given the limited time and resources available.

Despite the fundamentally informal character of task analysis, many formal and quasi-formal systems for task analysis have been proposed. However, these systems do not in themselves analyze the task or produce an understanding of the task. Rather, they are ways to help the analyst observe and think carefully about the user's actual task activity, and they provide a format for recording and communicating the results of the task analysis. Thus a task analysis methodology both specifies what kinds of task information are likely to be useful to analyze and provides a heuristic test for whether the task has actually been understood. That is, a good test for understanding something is whether one can represent it or document it, and constructing such a representation can be a good approach to trying to understand it (see chap. 1).

A formal representation of a task helps by ensuring that the analyst's understanding is more reliably communicated. Finally, some of the more formal representations can be used as the basis for computer simulations or mathematical analyses to obtain quantitative predictions of task performance, but it must be understood that such results are no more correct than the original and informally obtained understanding underlying the representation.

GOMS is such a formalized representation that can be used to predict task performance well enough that a GOMS model can be used as a substitute for much (but not all) of the empirical user testing needed to arrive at a system design that is both functional and usable. This predictive function is normally presented as the rationale for GOMS modeling (see Card, Moran, & Newell, 1983; John & Kieras, 1996a, b; Kieras, in press). However, GOMS models also qualify as a form of task-analytic representation, with properties similar to Hierarchical Task Analysis (HTA, see Annett, Duncan, Stammers, & Gray, 1971; Kirwan & Ainsworth, 1992; chap. 3) but with the special advantage of being able to generate useful predictions of learning and performance.

This chapter presents GOMS modeling as a task analysis method, emphasizing the process of analysis and construction of a GOMS model. Information about using the GOMS model for design evaluation and prediction of learning and performance is not covered here. GOMS methodology is quite detailed, especially when GOMS is used in a computer simulation of human performance, but due to lack of space, the presentation has been considerably simplified, and almost all specifics related to performance prediction have been eliminated. The interested reader should examine the cited sources and contact the author for treatments that are both more complete and more up to date. Also due to the lack of space, this chapter contains only one complete example of a GOMS model, a simple text editor (see appendix the end of the chapter). The reader might find it useful to gain a preliminary understanding of what a GOMS model is like by briefly examining this example before reading further.

4.1.1 The GOMS Model

A GOMS model is a description of the *procedural knowledge* (see chap. 15) that a user must have in order to carry out tasks on a device or system; it is a representation of the "how to do it" knowledge that is required by a system in order to get the intended tasks accomplished. The acronym GOMS stands for *G*oals, *O*perators, *M*ethods, and *S*election rules. Briefly, a GOMS model consists of descriptions of the Methods needed to accomplish specified Goals. The Methods are a series of steps consisting of Operators that the user performs. A Method may call for subgoals to be accomplished, so the Methods have a hierarchical structure. If there is more than one Method to accomplish a Goal, then Selection Rules choose the appropriate Method depending on the context. Describing the Goals, Operators, Methods, and Selection Rules for a set of tasks in a formal way constitutes doing a GOMS analysis, or constructing a GOMS model.

In the Card et al. (1983) formulation, the new user of a computer system will use various problem-solving and learning strategies to figure out how to accomplish tasks using the computer system, and then with additional practice, these results of problem-solving will become methods - procedures that the user can routinely invoke to accomplish tasks in a smooth, skilled manner. The properties of the methods will thus govern both the ease of learning and ease of use of the computer system. In the research program stemming from the original proposal, approaches to representing GOMS models based on cognitive psychology theory have been developed and validated empirically, along with the corresponding techniques and computer-based tools for representing, analyzing, and predicting human performance in human-computer interaction situations.

John and Kieras (1996a, 1996b) describe the current family of GOMS models and the associated techniques for predicting usability, and list many successful applications of GOMS to practical design problems. The simplest form of GOMS model is the Keystroke-Level Model, first described by Card, Moran, and Newell (1980), in which task execution time is predicted by the total of the times for the elementary keystroke-level actions required to perform the task. The most complex is CPM-GOMS, developed by Gray, John, and Atwood (1993), in which the sequential dependencies between the user's perceptual, cognitive, and motor processes are mapped out in a schedule chart, whose critical path predicts the execution time.

In between these two methods is the method presented in Kieras (1988, 1997a), NGOMSL, in which learning time and execution time are predicted based on a programlike representation of the methods that the user must learn and execute to perform tasks with the system. NGOMSL is an acronym for *N*atural *GOMS L*anguage, which is a structured natural language used to represent the user's methods and selection rules. NGOMSL models thus have an explicit representation of the user's methods, which are assumed to be strictly sequential and hierarchical in form. NGOMSL is based on the cognitive modeling of human-computer interaction by Kieras and Polson (Kieras & Polson, 1985; Bovair, Kieras, & Polson, 1990). As summarized by John and Kieras (1996a, 1996b), NGOMSL is useful for many desktop computing situations in which the user's procedures are usefully approximated as being hierarchical and sequential. The execution time for a task is predicted by simulating the execution of the methods required to perform the task. Each NGOMSL statement is assumed to require a small fixed time to execute, and any operators in the statement, such as a keystroke, will then take additional time depending on the operator. The time to learn how to operate the interface can be predicted from the length of the methods, and the amount of transfer of training from the number of methods or method steps previously learned. In addition, NGOMSL models have been shown to be useful for defining the content of on-line help and documentation (Elkerton & Palmiter, 1991; Gong & Elkerton, 1990).

This chapter uses a newer computational form of NGOMSL, called GOMSL (*L*anguage) which is processed and executed by a GOMS model *simulation* tool, *GLEAN3* (*GOMS* *L*anguage *E*valuation and *A*nalysis). GLEAN3 was inspired by the original GLEAN tool developed by Scott Wood (1993; see also Byrne, Wood, Sukaviriya, Foley, & Kieras, 1994) and reimplemented and elaborated in Kieras, Wood, Abotel, and Hornof (1995) and then again as summarized in Kieras (1999). Unlike the earlier versions, GLEAN3 is based on a comprehensive cognitive architecture, namely, a simplified version of the EPIC architecture for simulating human cognition and performance (Kieras & Meyer, 1997). GOMSL and GLEAN3 have been also been used to identify likely sources of errors in user interfaces and model human error recovery (Wood, 1999, 2000) and also to model the performance of teams of humans who interact with speech (Santoro & Kieras, 2001).

4.1.2 Strengths and Limitations of GOMS Models

It is important to be clear on what GOMS models can and cannot do. See John and Kieras (1996a, 1996b) for more discussion.

GOMS starts after a basic task analysis. In order to apply the GOMS technique, the task analyst must first determine what goals the user will be trying to accomplish. The analyst can then express in a GOMS model how the user can accomplish these goals with the system being designed. Thus, GOMS modeling does not replace the most critical process in designing a usable system, that of understanding the user's situation, working context, and overall goals. Approaches to this stage of interface design have been presented in sources such as Gould (1988), Diaper (1989), Kirwan and Ainsworth (1992), and Beevis et al. (1992). Once this basic level of task analysis has been conducted, constructing the GOMS model can then provide an elaborated account of how the user does the task.

GOMS represents only the procedural aspects of a task. GOMS models can account for the *procedural* aspects of usability; these concern the exact steps in the procedures that the user must follow, and so GOMS allows the analyst to determine the amount, consistency, and efficiency of the procedures that users must follow. Since the usability of many systems depends heavily on the simplicity and efficiency of the procedures, the narrowly focused GOMS model has considerable value in guiding interface design. The reason why GOMS models can predict these aspects of usability is that the methods for accomplishing user goals tend to be tightly constrained by the design of the interface, making it possible to construct a GOMS model given just the interface design, prior to any prototyping or user testing.

Clearly, there are other important aspects of usability that are not related to the procedures entailed by the interface design. These concern both lowest-level perceptual issues like the legibility of typefaces on CRTs, and also very high-level issues such as the user's conceptual knowledge of the system, e.g., whether the user has an appropriate "mental model" or the extent to which the system fits appropriately into an organization (see John & Kieras, 1996a). The lowest-level issues are dealt with well by standard human factors methodology, whereas understanding the higher-level concerns is currently a matter of practitioner wisdom and the higher-level task analysis techniques. Considerably more research is needed on the higher-level aspects of usability, and tools for dealing with the corresponding design issues are far off. For these reasons, great attention must still be given to the overall task analysis, and some user testing will still be required to ensure a high-quality user interface.

GOMS models are practical and effective. There has been a widespread belief that constructing and using GOMS models is too time-consuming to be practical (e.g., Lewis & Rieman, 1994). However, the many cases surveyed by John and Kieras (1996a) make clear that members of the GOMS family have been applied in many practical situations and were often very time- and cost-effective. A possible source of confusion is that the development of the GOMS modeling techniques has involved validating the analysis against empirical data. However, once the technique has been validated and the relevant parameters estimated, no empirical data collection or validation should be needed to apply a GOMS analysis during practical system design, enabling usability evaluations to be obtained much faster than user testing techniques. However, the calculations required to derive the predictions are tedious and mechanical; GLEAN was developed to remove this obstacle, but of course, additional effort is required to express the GOMS model precisely enough for a computer-based tool to use it (see also chap. 1).

4.2 GENERAL ISSUES IN GOMS ANALYSIS

4.2.1 Overview of GOMS Analysis

Carrying out a GOMS analysis involves defining and then describing in a formal notation the user's Goals, Operators, Methods, and Selection Rules. Most of the work seems to be in defining the Goals and Methods. That is, the Operators are mostly determined by the hardware and lowest-level software of the system, such as whether it has a mouse, for example. Thus the Operators are fairly easy to define. The Selection Rules can be subtle, but usually they are involved only when there are clear multiple methods for the same goal. In a good design, it is clear when each method should be used, so defining the Selection Rules is (or should be) relatively easy as well.

Identifying and defining the user's goals is often difficult, because the analyst must examine the task that the user is trying to accomplish in some detail, often going beyond just the specific system to the context in which the system is being used. This is especially important in designing a new system, because a good design is one that fits not just the task considered in isolation, but also how the system will be used in the user's job context. As mentioned above, GOMS modeling starts with the results of a task analysis that identifies the user's top-level goals. Once a goal is defined, the corresponding method can be simple to describe because it is simply the answer to the question "how do you do it on this system?" The system design itself largely determines what the methods are.

One critical process involved in doing a GOMS analysis is deciding what and what *not* to describe. The mental processes of the user are incredibly complex (see chap. 1 and 15); trying to describe all of them would be hopeless. However, the details of many of these complex processes have nothing to do with the design of the interface and so do not need to be worked out for the analysis to be useful. For example, the process of reading is extraordinarily complex; but usually, design choices for a user interface can be made without any detailed consideration of how the reading process works. We can treat the user's reading mechanisms as a "black box" during the interface design. We may want to know *how much* reading has to be done, but rarely do we need to know *how* it is done. So, we will need to describe when something is read and why it is read, but we will not need to describe the actual processes involved. A way to handle this in a GOMS analysis is to "bypass" the reading process by representing it with a "dummy" or "placeholder" operator. This is discussed more below. But making the choices of what to bypass is an important, and sometimes difficult, part of the analysis.

4.2.2 Judgment Calls

In constructing a GOMS model, the analyst is relying on a task analysis that involves judgments about how users view the task in terms of their natural goals, how they decompose the task into subtasks, and what the natural steps are in the user's methods. These are standard problems in task analysis (see Annett et al., 1971, Kieras, 1997b, Kirwan & Ainsworth, 1992). It is possible to collect extensive behavioral data on how users view and decompose tasks, but often it is not practical to do so because of time and cost constraints on the interface design process. Instead, the analyst must often make *judgment calls* on these issues. These are decisions based on the analyst's judgment, rather than on systematically collected behavioral data. In making judgment calls, the analyst is actually speculating on a psychological theory or model for how people do the task, and so will have to make hypothetical claims and assumptions about how users think about the task. Because the analyst does not normally have the time or opportunities to collect the data required to test alternative models, these decisions may be wrong, but making them is better than not doing the analysis at all. By documenting these judgment calls, the analyst can explore more than one way of decomposing the task, and consider whether there are serious implications to how these decisions are made. If so, collecting behavioral data might then be required. But notice that once the basic decisions are made for a task, the methods are determined by the design of the system, and no longer by judgments on the part of the analyst.

For example, in the example in the appendix for moving text in MacWrite, the main judgment call is that due to the command structure, the user views moving text as first cutting, then pasting, rather than as a single unitary move operation. Given this judgment, the actual methods are determined by the possible sequences of actions that MacWrite permits to do cutting and pasting.

In contrast, on the IBM DisplayWriter, the design did not include separate cut and paste operations. So here, the decomposition of moving into "cut then paste" would be a weak judgment call. The most reasonable guess is that a DisplayWriter user thinks of the text movement task not in terms of cut and paste subgoals, but in terms of the subgoals of first selecting the text, then issuing the Move command, and then designating the target location. So what is superficially the same text editing task may have different decompositions into subgoals, depending on how the system design encourages the user to think about it.

It could be argued that it is inappropriate for the analyst to be making *assumptions* about how humans view a system. However, notice that any designer of a system has *de facto* made many such assumptions. The usability problems in many software products are a result of the designer making assumptions, often unconsciously, with little or no thoughtful consideration of the implications for users. So, if the analyst's assumptions are based on a careful consideration from the user's point of view, they cannot do any more harm than that typically resulting from the designer's assumptions, and should lead to better results.

4.2.3 How do Users do the Task?

If the system already exists and has users, the analyst can learn a lot about how users view the task by talking to the users to get ideas about how they decompose the task into subtasks and what methods and selection rules they use (e.g., chap. 16). However, a basic lesson from the painful history of cognitive psychology is that people have only a very limited awareness of their own goals, strategies, and mental processes in general. Thus the analyst cannot simply collect this information from interviews or having people "think out loud." What users *actually* do can differ a lot from what they *think* they do. The analyst will have to combine information from talking to users with considerations of how the task constrains the user's behavior, and most importantly, observations of actual user behavior (see chap. 2). So, rather than asking people to describe verbally what they do, a better approach is having users demonstrate on the system what they do, or better yet, observing what they normally do in an unobtrusive way.

In addition, what users actually do with a system may not in fact be what they *should* be doing with it. The user, even a very experienced one, is not necessarily a source of "truth" about the system or the tasks (cf. Annett et al., 1971). As a result of poor design, bad documentation, or inadequate training, users may not in fact be taking advantage of features of the system that allow them to be more productive (see Bhavnani & John, 1996). The analyst should try to understand why this is happening, because a good design will only be good if it is used in the intended way. But for purposes of a GOMS analysis, the analyst will have to decide whether to assume a suboptimal use of the system or a fully informed one.

This situation deserves further discussion. In many task-analysis or modeling situations, especially with complex systems, the human user can perform the task in a variety of different ways, following different *task strategies* - the external structure of the task does not strongly constrain what the user must do, leaving them free to devise and follow different strategies that arrive at the same end result. In such a case, it will be difficult to determine the goal structure and the methods. One approach, of course, is to rely on empirical data and observation about what users actually do in the task. This is certainly the desired approach, but it can be extremely difficult to identify the strategy people use in a task, even a very simple one (Kieras & Meyer, 2000). Furthermore, empirical results cannot be used if the system in question is under design and so has no users to observe, or if an existing system has not been, or cannot be, studied in the required detail because of the severe practical difficulties involved in collecting usage data for complex systems.

In such cases, the analyst is tempted to speculate on how users might do the task, and as noted above, such speculation by the task analyst and interface designer is likely to be better than than haphazard decisions made by whoever writes the interface code. However, if the task is indeed a complex one, trying to guess or speculate how the user does the task can result in an endless guessing game. A better approach is to consider whether the system designers have a concept of how the system is supposed to be used, and if so, construct a GOMS model for how the user *should* do the task. This is much less speculative, and is thus relatively well-defined. It represents a sort of best-case analysis in which the system designer's intentions are assumed to be fully communicated to the user, so that the user takes full advantage of the system features. If the resulting GOMS analysis and the performance predictions for it reveal serious problems, the actual case will certainly be much worse.

An additional elaboration of this approach is to use bracketing logic (Kieras & Meyer, 2000) to gain information about the possible actual performance using the system. Construct two models of the task, one representing using the system as cleverly as possible, producing the the fastest-possible performance, and another that represents the nominal or unenterprising use of the system, resulting in a slowest-reasonable model. When performance predictions are computed, these two models will bracket what actual users can be expected to do. By comparing how the two models respond to, for example, changes in workload and analyzing their performance bottlenecks, the analyst can derive useful conclusions about a system design, or especially about the relative merits of two designs, without having to make detailed unsupported assumptions about the user's actual task strategies (e.g., Kieras, Meyer, & Ballas, 2001).

4.2.4 Bypassing Complex Processes

Many cognitive processes are too difficult to analyze in a practical context. Examples of such processes are reading, problem solving, figuring out the best wording for a sentence, finding a bug in a computer program, and so forth. One approach is to bypass the analysis of a complex process by simply representing it with a "dummy" or "placeholder" operator, such as the `Think_of` operator in GOMSL (see below). In this way the analyst documents the presence of the process, and can consider what influence it might have on the user's performance with a design. A more flexible approach is the "yellow pad" heuristic: suppose the user has already done the complex processing and has written the results down on a yellow note pad and simply refers to them along with the rest of the information about the task instance.

For example, in MacWrite, the user may use tabs to control the layout of a table. How does the user know, or figure out, where to put them? The analyst might assume that the difficulties of doing this have nothing to do with the design of MacWrite (which may or not be true). The analyst can bypass the process of how the user figures out tab locations by assuming that user has figured them out already, and includes the tab settings as part of the task instance description supplied to the methods. (cf. the discussion in Bennett, Lorch, Kieras, & Polson, 1987). The analyst uses the GOMSL `Get_task_item` operator to represent when this information is accessed.

As a second example, consider a word-processor user who is making changes in a document from a marked-up hardcopy. How does the user know that a particular scribble on the paper means "delete this word"? The analyst can bypass this problem by putting in the task description the information that the goal is to `Delete` and that the target text is at such-and-such a location (see example task descriptions below), and then using the `Get_task_item` operator to access the task information. The methods will invoke this operator at the places where the user is assumed to have to look at the document to find out what to do. This way, the contents of the task description show the results of the complex reading process that was bypassed, and the places in the methods where the operator appears mark where the user is engaging in the complex reading process.

The analyst should only bypass processes for which a full analysis would be irrelevant to the design. But sometimes the complexity of the bypassed process is related to the design. For example, a text editor user must be able to read the paper marked-up form of a document, regardless of the design of the text editor, meaning that the reading process can be bypassed because it does not need to be analyzed in order to choose between two different text editor designs. On the other hand, the POET editor (see Card, Moran, & Newell, 1983) requires heavy use of find-strings, which the user has to devise as needed. This process can still be bypassed, and the actual find strings specified in the task description. But suppose we are comparing POET to an editor that does not require such heavy use of find strings. Any conclusions about the difficulty of POET compared to the other editor will depend critically how hard it is to think up good find-strings. In this case, bypassing a process might produce seriously misleading results.

4.2.5 Generative Models, Rather than Models of Specific Task Instances

Often, user interface designers will work with *task scenarios* (chap. 5), which are essentially descriptions in ordinary language of task instances and what the user would do in each one. The list of specific actions that the user would perform for a specific task can be called a *trace*, analogous to the specific sequence of results one obtains when "tracing" a computer program. Assembling a set of scenarios and traces is often useful as an informal way of characterizing a proposed user interface and its impact on the user.

If one has collected a set of task scenarios and traces, the natural temptation is to construct a description of the user's methods for executing these specific task instances. This temptation must be resisted; the goal of GOMS analysis is a description of the *general* methods for accomplishing a set of tasks, not just the method for executing a specific instance of a task.

If the analyst falls into the trap of writing methods for specific task instances, the resulting methods will probably be "flat," containing little in the way of method and submethod hierarchies, and also may contain only the specific keystroke-level operations appearing in the trace. For example, if the task scenario is that the user deletes the file FOOBAR, such a method will generate the keystroke sequence of "DELETE FOOBAR <CR>." But the fatal problem is that a tiny change in the task instance, such as a different file name, means that the method will not work. This corresponds to a user who has memorized by rote how to do an exact task, but who cannot execute variations of the task.

On the other hand, a set of *general* methods will have the property that the information in a specific task instance acts like "parameters" for a general program, and the general methods will thus generate the specific actions required to carry out that task instance. Any task instance of the general type will be successfully executed by the general method. For example, a general method for deleting the file specified by `<filename>` will generate the keystroke sequence of "DELETE" followed by the string designated `<filename>` by followed by <CR>. This corresponds to a user who knows how to use the system in the general way normally intended. Such GOMS models are *generative*—rather than being limited to specific snippets of behavior, they can generate all possible traces from a single set of methods. This is a critical advantage of GOMS models and other cognitive-architecture models (for more discussion, see John & Kieras, 1996b; Kieras, in press; Kieras, Wood, & Meyer, 1997).

So, if the analyst has a collection of task scenarios or traces, he or she should study them to discover the range of things that the user has to do. They should then be set aside and a generative GOMS model written that contains a set of general methods that can correctly perform any specific task within the classes defined by the methods (e.g., delete any file whose name is specified in the task description). The methods can be checked to ensure that they will generate the correct trace for each task scenario, but they should also work for *any* scenario of the same type.

4.2.6 When Can a GOMS Analysis be Done?

4.2.6.1 After Implementation—Existing Systems

Constructing a GOMS model for a system that already exists is the easiest case for the analyst because much of the information needed for the GOMS analysis can be obtained from the system itself, its documentation, its designers, and the present users. The user's goals can be determined by considering the actual and intended use of the system; the methods are determined by what actual steps have to be carried out. The analyst's main problem will be to determine whether what users actually do is what the designers intended them to do, which often isn't the case (chap. 1), and then go on to decide what the users' actual goals and methods are. For example, the documentation for a sophisticated document preparation system gave no clue to the fact that most users dealt with the complex control language by keeping "template" files on hand which they just modified as needed for specific documents. Likewise, this mode of use was apparently not intended by the designers. So the first task for the analyst is to determine how an existing system is actually used in terms of the goals that actual users are trying to accomplish. Talking to, and observing, users can help the analyst with these basic decisions (but remember the pitfalls discussed above).

Because in this case the system exists, it is possible to collect data on the user's learning and performance with the system, so using a GOMS model to predict this data would only be of interest if the analyst wanted to verify that the model was accurate (see chap. 1), perhaps in conjunction with evaluating the effect of proposed changes to the system. However, notice that collecting systematic learning and performance data for a complex piece of software can be an extremely expensive undertaking. Chapter 22 suggests that it is "daunting"; if one is confident of the model, it could be used as a substitute for empirical data in activities such as comparing two competing existing products.

4.2.6.2 After Design Specification—Evaluation During Development

There is no need for the system to be already implemented or in use for a GOMS analysis to be carried out. It is only necessary that the analyst can specify the components of the GOMS model. If the design has been specified in adequate detail, then the analyst can identify the intended user's goals and describe the corresponding methods just as in the case of an existing system.

Of course, the analyst cannot get the user's perspective since there are as yet no users to talk to. However, the analyst can talk to the designers to determine the designer's intentions and assumptions about the user's goals and methods, and then construct the corresponding GOMS model as a way to make these assumptions explicit and to explore their implications. Predictions can then be made of learning and performance characteristics, and then used to help correct and revise the design. The analyst thus plays the role of the future user's advocate, by systematically assessing how the design will affect future users. Since the analysis can be done before the system is implemented, it should be possible to identify and put into place an improved design without wasting coding effort.

However, the analyst can often be in a difficult position. Even fairly detailed design specifications often omit many specific details that directly affect the methods that users will have to learn. For example, the design specifications for a system may define the general pattern of interaction by specifying pop-up menus, but not the specific menu choices available, or which choices users will have to make to accomplish actual tasks. Often these detailed design decisions are left up to whoever happens to write the relevant code. The analyst may not be able to provide many predictions until the design is more fully fleshed out, and may have to urge the designers to do more complete specification than they normally would.

4.2.6.3 During Design—GOMS Analysis Guiding the Design

Rather than analyze an existing or specified design, the interface could be designed concurrently with describing the GOMS model. That is, by starting with listing the user's top-level goals, then defining the top-level methods for these goals, and then going on to the subgoals and sub-methods, one is in a position to make decisions about the design of the user interface directly in the context of what the impact is on the user. For example, bad design choices may be immediately revealed as spawning inconsistent, complex methods, leading the designer quickly into considering better alternatives. See Kieras (1997b) for more discussion of this approach. Clearly, the designer and analyst must closely cooperate, or be the same person.

Perhaps counter to intuition, there is little difference in the approach to GOMS analysis between doing it *during* the design process and doing it after. Doing the analysis during the design means that the analyst and designer are making design decisions about what the goals and methods *should be*, and then immediately describing them in the GOMS model. Doing the analysis *after* the system is designed means that the analyst is trying to determine the design decisions that were made *sometime in the past*, and then describing them in a GOMS model. For example, instead of determining and describing how the user does a cut-and-paste with an existing text editor, the designer-analyst *decides* and describes how the user *will* do it. It seems clear that the reliability of the analysis would be better if it is done during the design process, but the overall logic is the same in both cases.

4.3 GOMSL: A NOTATION FOR GOMS MODELS

This section presents the GOMSL (*GOMS L*anguage) notation system which is a computer-executable version of the earlier NGOMSL (Kieras, 1988, 1997a). GOMSL is an attempt to define a language that will allow GOMS models to be executed with a computer-based tool, but at the same time be easy to read. An analyst can use GOMSL in an informal fashion; if performance predictions are needed, he or she can compute the results by hand, or alternatively, tighten up the GOMSL and then use a computational tool such as GLEAN3 to run the model or generate performance predictions.

GOMSL is not supposed to be an ordinary programming language for computers, but rather to have properties that are directly related to the underlying *production rule models* described by Kieras, Bovair, and Polson (Bovair et. al., 1990; Kieras & Bovair, 1986; Kieras & Polson, 1985; Polson, 1987). So GOMSL is supposed to represent something like "the programming language of the mind," as absurd as this sounds. The idea is that GOMSL programs have properties that are related in straightforward ways to both data on *human performance* and theoretical ideas in cognitive psychology. If GOMSL is clumsy and limited as a computer language, it is because humans have a different architecture than computers. Thus, for example, GOMSL does not allow complicated conditional statements, because there is good reason to believe that humans cannot process complex conditionals in a single cognitive step. If it is hard for people to do, then it should be reflected in a long and complicated GOMSL program. In this document, GOMSL expressions are shown in `this typeface`.

4.3.1 Task Data

4.3.1.1 Object-Property-Value Representation

The basic data representation in GOMSL consists of *objects* with *properties* and *values*. Each object has a symbolic name and a list of properties, each of which has an associated value. The object name, property, and value are symbols. This representation is used in several places: For example, long-term memory is represented as a collection of objects, or items, each of which has a symbolic name and a set of property-value pairs. For example, the fact that "plumber" is a skilled trade and has high income might be represented as follows:

```
LTM_Item: Plumber.
  Kind is skilled_trade.
  Income is high.
```

In this example, "Plumber" is an object in LTM that has a "Kind" property whose value is "skilled_trade" and an "Income" property whose value is "high."

Another example is declarative knowledge of an interface as a collection of facts about the Cut command in a typical text editor:

```
LTM_Item: Cut_Command.
  Containing_Menu is Edit.
  Menu_Item_Label is Cut
  Accelerator_Key is Command-X.
```

The "cut" command is described as an object whose properties and values specify which menu it is contained in, the actual label used for it in the menu, and the corresponding accelerator (shortcut) key. Likewise, a task instance is described as a collection of objects each of which has properties and values. There are operators for accessing or retrieving visual objects, task or long-term memory items, and then accessing their individual properties and values.

4.3.1.2 Working Memory

Working memory in GOMSL consists of two kinds of information: one is a *tag store* (Kieras, Meyer, Mueller, & Seymour, 1999), which represents an elementary form of working memory. The other kind is the *object store* which holds information about an object that has been brought into "focus", that is, placed in working memory and whose property values are thus immediately available in the form of a *property-of-object* construction.

The working memory tag store consists of a collection of symbolic values stored under a symbolic name or *tag*. Tags are expressed as identifiers enclosed in angle brackets. In many cases, the use of tags corresponds to traditional concepts of verbal working memory; syntactically, they roughly resemble variables in a traditional programming language. At execution time, if a tag appears in an operator argument, it is replaced with the value currently stored under the tag. An elementary example:

```
Step 1. Store "foo.txt" under <filename>.
Step 2. Type_in <filename>.
```

In Step 1, the `Store` operator is used to place the string "foo.txt" into working memory under the tag `<filename>`. In Step 2, before the `Type_in` operator is executed, the value stored under the tag is retrieved from working memory, and this becomes the parameter for the operator. So Step 2 results in the simulated human typing the string `"foo.txt"`.

The object stores correspond to working memory for visual input, task information, and long-term memory retrievals. All three of these have the common feature that gaining access to an object or item will be time consuming, but once it has been located or retrieved, further details of the object or the item can then be immediately used by specifying the desired property of the object with a *property-of-object* construction. So the operation of bringing an object or item into focus is time-consuming, but then all of its properties are available in working memory. But if the "focus" is changed to a different object or item, the information is no longer available, and a time-consuming operation is required to bring it back into focus. This mechanism represents in a simple way the performance constraints involved in many forms of working memory and visual attention. This analysis is a simplification of the very complex ways in which working memory information is accessed and stored during a task (see Kieras, Meyer, Mueller, & Seymour, 1999, for more discussion).

4.3.2 Goals

A goal is something that the user tries to accomplish. The analyst attempts to identify and represent the goals that typical users will have. A set of goals usually will have a hierarchical arrangement in which accomplishing a goal may require first accomplishing one or more subgoals.

A goal description is a pair of identifiers, which by convention are chosen to be an action-object pair in the form: `verb noun`, such as `delete word`, or `move-by-find-function cursor`. Either the noun or the verb can be complicated if necessary to distinguish between methods (see below on selection rules). Any parameters involved that modify or specify a goal, such as where a to-be-deleted word is located, are represented in the task description, and made available when the method is invoked (see below).

4.3.3 Operators

Operators are actions that the user executes. There is an important difference between goals and operators. Both take an action-object form, such as the goal of `revise document` and the operator of `Keystroke ENTER`. But in a GOMS model, a goal is something to be accomplished, whereas an operator is just executed. This distinction is intuitively based, and is also relative; it depends on the level of analysis chosen by the analyst (John & Kieras, 1996b). The procedure presented below for constructing a GOMS model is based on the idea of first describing methods using very high-level operators, and then replacing these operators with methods that accomplish the corresponding goal by executing a series of lower-level operators. This process is repeated until the operators are all *primitive operators* that will not be further analyzed.

As explained in more detail later, based on the underlying *production system* model for human cognition, each step in a GOMS model takes a certain time to execute, estimated at 50 ms. Most of the built-in mental operators are all executed during this fixed step execution time, and so are termed *intrastep* operators. However, substantially longer execution times are required for external operators, such as pointing to the target with a mouse, or certain built-in mental operators such as searching long-term memory. Thus, these are *interstep* operators because their execution time occurs between the steps, and so governs when the next step is started.

4.3.3.1 External Operators

External operators are the observable actions through which the user exchanges information with the system or other humans. These include perceptual operators, which read text from a screen, scan the screen to locate the cursor, or input a piece of speech, and motor operators, such as pressing a key, or speaking a phrase. The analyst usually chooses the external operators depending on the system or tasks, such as whether there is there a mouse on the machine.

Listed below are the primitive motor and perceptual operators whose definitions and execution times are based on the physical and mental operators used in the Keystroke-Level Model (Card et al., 1983; John & Kieras, 1996a, 1996b). These are all interstep operators. Based on the simplifying logic in the Keystroke-Level Model, operators for complex mental activities are assumed to take a constant amount of time, approximated with the value of 1,200 ms, based on results in Olson and Olson (1990). Each operator keyword in the list below is shown in `this typeface`; parameters for operators are shown in *`this typeface`*. Unless otherwise stated, an operator parameter can be either a symbol, a tag, or a *property of object* construction. An identifier enclosed in angle brackets, as in `<tag>`, is a tag name parameter. A description of the operator and its execution time for each operator is given.

`Keystroke` *`key_name`*

Strike a key on a keyboard. If the keyname is a string, only the first character is used. Execution time is 280 ms.

`Type_in` *`string_of_characters`*

Do a series of keystrokes, one for each character in the supplied string. Execution time is 280 ms per character.

`Click` *`mouse_button`*

The designated mouse button is pressed and released. Execution time is 200 ms.

`Double_click` *`mouse_button`*

Two clicks are executed. Execution time is 400 ms.

`Hold_down` *`mouse button`*

Press and continue to press the mouse button. Execution time is 100 ms.

`Release` *`mouse_button`*

Release the mouse button. Execution time is 100 ms.

`Point_to` *`target_object`*

The mouse cursor is moved to the target object on the screen. Execution time is determined by the Welford form of Fitts' Law with a minimum of 100 ms if object location and sizes are specified; it is 1,100 ms if not.

`Home_to` *`destination`*

Move the right hand to the destination. The initial location of the right and left hands is the keyboard.

Possible destinations are `keyboard` and `mouse`. Execution time is 400 ms. All of the manual operators automatically execute a `Home_to` operator if necessary to simulate the hand being moved between the mouse and keyboard.

`Look_for_object_whose` *`property is value, ...`* `and_store_under` *`<tag>`*

This mental operator searches visual working memory, (essentially the visual space) for an object whose specified properties have the specified values, and stores its symbolic name in working memory under the label `<tag>` where it is now in focus. If no object matching the specification is present, the result is the symbol `absent`, which may be tested for subsequently. Time required is 1,200 ms, after which additional information about the object is available immediately. Putting a different object in focus will result in the previous object's properties being no longer available. If the visual object disappears (e.g., it is taken off of the display screen), then it will be removed from visual working memory, and the object information is no longer available. Static visual objects and their properties can be defined as part of the GOMS model.

```
Get_task_item_whose  property is  value, ... and_store_under
 <tag>
```

This mental operator is used to represent that the user has a source of information available containing the specifics of the tasks to be executed, but the analyst does not wish to specify this source, but is assuming that it requires mental activity to produce the required task information. For example, the user could be "thinking up" the tasks as he or she works, recalling it from memory, or reading the task information from a marked-up manuscript or a set of notes (see the "yellow pad" heuristic below). The task information would be specified in a set of `Task-items`, presented below, that together define a collection of objects and properties, the task description. This operator searches the task description for a task item object whose specified properties have the specified values, and stores its symbolic name in working memory under the label `<tag>`. It is now in focus, and additional properties are now available. Time required is 1,200 ms. Putting a different task item in focus will result in the previous item's properties being no longer available. If the specified object is not found in the task description, then the resulting symbol is `absent`, and this value can be subsequently tested for.

4.3.3.2 Mental Operators

Mental operators are the internal actions performed by the user; they are non-observed and hypothetical, inferred by the theorist or analyst. In the notation system presented here, some mental operators are "built in"; these are primitive operators that correspond to the basic mechanisms of the cognitive processor, the *cognitive architecture*. These are based on the production rule models described by Bovair et al. (1990). These operators include actions like making a basic decision, storing an item in Working Memory (WM), retrieving information from Long-Term Memory (LTM), determining details of the next task to be performed, or setting up a goal to be accomplished.

Other mental operators are defined by the analyst to represent complex mental activities (see below), normally as a placeholder for a complex activity that cannot be analyzed further. A common such *analyst-defined* mental operator is verifying that a typed-in command is correct before hitting the `ENTER` key; another example would be a stand-in for an activity that will not be analyzed, such as `LOG-INTO-SYSTEM`.

Below is a brief description of the GOMSL primitive mental operators. These are all intrastep operators.

Flow of Control. Only one flow of control operator can appear in a step. A submethod is invoked by asserting that its goal should be accomplished:

```
Accomplish_goal:  goal
AG:  goal  (abbreviation)
Accomplish_goal:  goal using  pseudoargument tag list
```

This is analogous to an ordinary CALL statement; control passes to the method for the goal, and returns here when the goal has been accomplished. The pseudoargument tag list is described in a later section below. The operator:

```
Return_with_goal_accomplished
RGA  (abbreviation)
```

is analogous to an ordinary RETURN statement. A method normally must contain at least one of these.

There is a branching operator:

```
Goto  step_label
```

As in structured programming, a `Goto` is used sparingly; normally it used only with Decide operators to implement loops or complicated branching structures.

A decision is represented by a step containing a `Decide` operator; a step may contain only one Decide operator and no other operators. The `Decide` operator contains one or more IF-THEN conditionals and at most one ELSE form. The conditionals are separated by semicolons:

```
Decide:  conditional
Decide:  conditional;  conditional; ...  else-form
```

The IF part of a conditional can have one or more predicates. The THEN part or an else-form has one or more operators:

```
If  predicates Then  operators
Else  operators
```

The predicates consist of one or more predicates, separated by commas or comma-and. If all of the predicates are true, then the operators are executed, and no further conditionals are evaluated. An else-form may appear only at the end; its operators are executed only if all of the preceding IF conditions have all failed to be satisfied. Normally, `Decide` operators are written so that the if-else combinations are mutually exclusive. Note that nesting of conditionals is not allowed. Here are three examples of `Decide` operators:

```
Step 3. Decide: If <id_letter> is "A" Then Goto 4.

Step 8. Decide:
  If appearance of <current_thing> is "super duper", and
          size of <current_person> is large,
   Then Type_in string of <current_task>;
  If appearance of <current_thing thing> is absent, Then RGA;
  Else Goto 2.
Step 5. Decide:
  If <button_label> is ACCEPT Then Keystroke K1;
  If <button_label> is REJECT, Then Keystroke K2;
  Else Keystroke K3.
```

If there are multiple IF-THEN conditionals, as in the second and third example above, the conditions must be mutually exclusive, so that only one condition can match, and the order in which the If-Thens are listed is supposed to be irrelevant. However, the conditionals are evaluated in order, and evaluation stops at the first conditional whose condition is true.

The `Decide` operator is used for making a simple decision that governs the flow of control within a method. It is not supposed to be used to implement GOMS selection rules, which have their own construct (see below). The IF clause typically contains predicates that test some state of the environment or contents of WM. Notice that the complexity of a `Decide` operator is strictly limited; only one simple `Else` clause is allowed, and multiple conditionals must be mutually exclusive and independent of order. More complex conditional situations must be handled by separate decision-making methods that have multiple steps, decisions, and branching.

The following predicates are currently defined:

```
is is_equal_to  (synonyms)
is_not is_not_equal_to  (synonyms)
is_greater_than
is_greater_than_or_equal_to
is_less_than
is_less_than_or_equal_to
```

These predicates are valid for either numeric or symbol values. If the compared values are both numeric, then a comparison between the numeric values is performed. Otherwise both values are treated as symbolic, and the character string representations of the two are compared in lexicographic order-if necessary, a number is converted to a standard string representation for this purpose.

Memory Storage and Retrieval. The memory operators reflect the distinction between *long-term memory* (LTM) and *working memory* (WM) (often termed *short-term memory*) as they are typically used in computer operation tasks. WM contents are normally stored and retrieved using their *tags*, the symbolic name for the value being stored in working memory. These tags are used in a way that is somewhat analogous to variables in a conventional programming language.

```
Store  value under  <tag>
```

The value is stored in working memory under the label *<tag>*.

```
Delete  <tag>
```

The value stored under the label *<tag>* is deleted from working memory.

```
Recall_LTM_item_whose  property is  value, ... and_store_under
 <tag>
```

Searches long-term memory for an item whose specified properties have the specified values, and stores its symbolic name in working memory under the label *<tag>* where it is now in focus. Time required is 1,200 ms, after which additional information is available immediately. Putting a different task item in focus will result in the previous item's properties being no longer available.

Consistent with the theoretical concept that working memory is a fast scratchpad sort of system, and how it is used in the production system models, the execution time for the

`Store` and `Delete` operators is bundled into the time to execute the step; thus they are intrastep operators. The `Store` operator is used to load a value into working memory. The `Delete` operator is more frequently used to eliminate working memory items that are no longer needed. Although this deliberate forgetting might seem counter-intuitive, it is a real phenomenon; see Bjork (1972). For simplicity, working memory information does not "decay" and so there is no built-in limit to how much information can be stored in working memory. This reflects the fact that despite the considerable research on working memory, there is not a theoretical consensus on what working memory limits apply during the execution of procedural tasks (see Kieras et al., 1999). So rather than set an arbitrary limit on when working memory overload would occur, the analyst can identify memory overload problems examining how many items are required in WM during task execution; GLEAN can provide this information.

There is only a recall operator for LTM, because in the tasks typically modeled with GOMS, long-term learning and forgetting are not involved. The `Recall_LTM_item` operator is an interstep operator that takes a standard mental operator execution time, but once the information has been placed in focus in WM, additional information about the item can be used immediately with the `x_of_y` argument form.

Analyst-Defined Mental Operators. As discussed in some detail below, the analyst will often encounter psychological processes that are too complex to be practical to represent as methods in the GOMS model and that often have little to do with the specifics of the system design. The analyst can bypass these processes by defining operators that act as placeholders for the mental activities that will not be further analyzed. Depending on the specific situation, such operators may correspond to keystroke-level model *mental* operators, and so can be approximated with as standard interstep mental operators that require 1.2 sec. The `Verify` and `Think_of` operators are intended for this situation; the analyst simply documents the assumed mental process in the description argument of the operator.

```
Verify  description
Think_of  description
```

These operators simply introduce a time delay to represent when the user must pause and think of something about the task; the actual results of this thinking are specified elsewhere, such as the task description. The description string serves as documentation, nothing more. Each operator requires 1,200 ms.

4.3.4 Methods

A method is a sequence of steps that accomplishes a goal. A step in a method typically consists of an external operator, such a pressing a key, or a set of mental operators involved with setting up and accomplishing a subgoal. Much of the work in analyzing a user interface consists of specifying the actual steps that users carry out in order to accomplish goals, so describing the methods is the focus of the analysis.

The form for a method is as follows:

```
Method_for_goal:  goal
Step 1.  operators.
Step 2.  operators.
...
```

Abbreviations and pseudoparameters are allowed:

```
MFG:  goal  (abbreviation)
Method_for_goal:  goal using  pseudoparameter tag list
```

Steps. More than one operator can appear in a step, and at least one step in a method must contain the operator `Return_with_goal_accomplished`. A `step` starts with the keyword `Step`, contains an optional label, followed by a period, and one or more operators separated by semicolons, with a final period:

```
Step.  operator.
Step  label.  operator.
Step.  operator;  operator;  operator.
Step  label.  operator;  operator;  operator.
```

The label is either a number or an identifier. The labels are ignored by GLEAN except for the `Goto` operator, which searches the method from the beginning for the first matching label; this designates the next step to be executed. Thus the labels do not have to be unique or in order. However, a run-time error occurs if a `Goto` operator does not find a matching label. Using numeric labels throughout highlights the step-by-step procedure concept of GOMS methods, but plan on renumbering the steps and altering `Gotos` to maintain a neat appearance.

4.3.4.1 Method Hierarchy

Methods often call submethods to accomplish goals that are subgoals. This method hierarchy takes the following form:

```
Method_forgoal:  goal
  Step 1.  operators.
  Step 2.  <operators>
  ...
  Step i. Accomplish_goal:  subgoal.
  ...
  Step m. Return_with_goal_accomplished.
Method_for_goal:  subgoal
  Step 1.  operators.
  Step 2.  <operators>
  ...
  Step j. Accomplish_goal:  sub-subgoal.
  ...
  Step n. Return_with_goal_accomplished.
  ...
```

4.3.4.2 Method Pseudoparameters

The simple tagged-value model of working memory results in WM tags being used something like variables in traditional programming languages, but because there is only one WM system containing only one set of tagged values, these “variables” are effectively global in scope. This makes it syntactically difficult to write “library” methods that represent reusable “subroutines” with "parameters." To alleviate this problem, a method can be called with *pseudoarguments* in the `Accomplish_goal` operator and the corresponding *pseudoparameters* can be defined for a method or selection rule set. These are automatically deleted when the method completed. For example:

```
  Step 8. Accomplish_goal: Enter Data using "Name",
           and name of <currentperson>.
Method_for_goal: Enter Data using <field_name>, and <data>
  Step 1. Look_for_object_whose label is <field_name>
          and_store_under <field>.
  Step 2. Point_to <field>.
  Step 3. Click <button>.
  Step 4. Type_in <data>.
  Step 5. Delete <field>; Return_with_goal_accomplished.
```

The "pseudo" prefix makes clear that these "variables" do not follow the normal scoping rules used in actual programming languages-the human does not have a run-time function-call stack for argument passing.

4.3.5 Selection Rules

The purpose of a *selection rule* is to route control to the appropriate method to accomplish a goal. Clearly, if there is more than one method for a goal, then a selection rule is logically required.

There are many possible ways to represent selection rules. In the approach presented here, a selection rule responds to the combination of a *general* goal and a specific context by setting up a *specific* goal of executing one of the methods that will accomplish the general goal. Selection rules are if-then rules that are grouped into sets that are governed by a general goal. If the general goal is present, the conditions of the rules in the set are tested in parallel to choose the specific goal to be accomplished. The relationship with the underlying production rule models is very direct (see Bovair et al., 1990). The form for a selection rule set is:

```
Selection_rules_for_goal:  general goal
If  predicates Then Accomplish_goal:  specific goal.
If  predicates Then Accomplish_goal:  specific goal.
...
Return_with_goal_accomplished.
```

A common and natural confusion is when a selection rule set should be used and when a `Decide` operator should be used. A selection rule set is used exclusively to route control to the suitable method for a goal, and so can only have `Accomplish_goal` operators in the `Then` clause, while a `Decide` operator controls flow of control within a method, and can have any type of operator in the then clause. Thus, if there is more than one method to accomplish a goal, use that goal as a general goal, and define separate methods to accomplish the more specific goals; use a selection rule set to dispatch control to the specific method. To control which operators in what sequence are executed within a method, use a `Decide`.

4.3.6 Auxiliary Information

In order to execute successfully, the methods in a GOMS model often require additional information; this information is auxiliary to the step-by-step procedural knowledge represented directly in the GOMS methods and selection rules, but is logically required for actual tasks to be executed. For example, if the user is supposed to type in an exact string from memory, this string must be specified somehow.

The syntax for specifying auxiliary information is based on describing objectlike entities with properties and values; these descriptions can appear along with methods and selection rule sets. They must not be placed inside methods and selection rule sets but can appear in any order with them and each other.

4.3.6.1 Visual Object Description

Visual objects are described outside of any methods as follows:

```
Visual_object:  object_name
  property_name is  value.
  ...
```

For example, a red button labeled "Start" would be described as:

```
Visual_object: start_button
  Type is Button.
  Label is Start.
  Color is Red.
```

A step like the following will result in `start_button` being stored in WM under the tag `<button>`:

```
Step. Look_for_visual_object_whose Type is Button, and
  Label is Start
         and_store_under <button>.
```

Subsequently, the following step will point to the button if its color is red:

```
Step. Decide: If Color of <button> is Red,
         Then Point_to <button>.
```

Note that the names for visual objects are chosen by the analyst. GLEAN3 reserves two property names, `Location` and `Size`, for use by the Visual and Manual processors. All other property names and values can be chosen by the analyst.

4.3.6.2 Long-Term Memory Contents

The contents of Long-Term Memory can be specified as a set of concepts (objects) with properties and values. Note the since the value of a property can be the name of another object, complicated information structures are possible. The syntax:

```
LTM_item:  LTM_concept
  property_name is  property_value.
  ...
...
```

For example, information about the "Cut" command in a simple text editor could be specified as:

```
LTM_item: Cut_Command
  Name is CUT.
  Containing_Menu is Edit.
  Menu_Item_Label is Cut.
  Accelerator_Key is Command-X.
```

4.3.6.3 Task Instances

A *task description* describes a generic task in terms of the goal to be accomplished, the situation information required to specify the goal, and the auxiliary information required to accomplish the goal that might be involved in bypassing descriptions of complex processes (see below). Thus, the task description is essentially the "parameter list" for the methods that perform the task. A *task instance* is a description of a specific task. It consists of specific values for all of the parameters in a task description. A set of task instances can be specified as task item objects whose property values can refer to other objects to form a linked-list sort of structure. The syntax is similar to the above:

```
Task_item:  task_item_name
  property_name is  property_value.
  ...
...
```

4.4 A PROCEDURE FOR CONSTRUCTING A GOMS MODEL

A GOMS analysis of a task follows the familiar top-down decomposition approach. The model is developed top-down from the most general user goal to more specific subgoals, with primitive operators finally at the bottom. The methods for the goals at each level are dealt with before going down to a lower level. The recipe presented here thus follows a top-down, breadth-first expansion of methods.

In overview, start by describing a method for accomplishing a top-level goal in terms of high-level operators. Then choose one of the high-level operators, replace it with an `Accomplish_goal` operator for the corresponding goal, and then write a method for accomplishing that goal in terms of lower-level operators. Repeat with the other level operators. Then descend a level of analysis, and repeat the process for the lower-level operators. Continue until the methods have arrived at enough detail to suit the design needs, or until the methods are expressed in terms of primitive operators. So, as the analysis proceeds, high-level operators are replaced by goals to be accomplished by methods that involve lower-level operators.

It is important to perform the analysis breadth-first, rather than depth-first. By considering all of the methods that are at the same level of the hierarchy before getting more specific, similar methods are more likely to be noticed, which is critical to capturing the procedural consistency of the user interface.

4.4.1 Step 1: Choose the Top-Level User's Goals

The top-level user's goals are the first goals that will be expanded in the top-down decomposition. It is worthwhile to make the topmost goal, and the first level of subgoals, very high-level to capture any important relationships within the set of tasks that the system is supposed to address. An example for a text editor is `revise document`, whereas a lower-level one would be `delete text`. Starting with a set of goals at too low a level entails a risk of missing the methods involved in going from one type of task to another.

As an example of very high-level goals, consider the goal of `produce document` in the sense of publishing-getting a document actually distributed to other people. This will involve first creating it, then revising it, and then getting the final printed version of it. In an environment that includes a mixture of ordinary and desktop publishing facilities, there may be some important subtasks that have to be done in going from one to the other of the major tasks, such as taking a document out of an ordinary text editor and loading it into a page-layout editor, or combining the results of a text and a graphics editor. If only one of these applications is under design, say the page-layout editor, and the analysis start only with goals that correspond to page-layout functions, the analysis may miss what the user has to do to integrate the use of the page-layout editor in the rest of the environment.

As a lower-level example, many Macintosh applications combine deleting and inserting text in an especially convenient way. The goal of `change word` has a method of its own; that is, double click on the word and then type the new word. If the analysis starts with `revise document` it is possible to see that one kind of revision is changing a piece of text to another, and so this especially handy method might well be noticed in the analysis. But if the analysis starts with goals like `insert text` and `delete text` the decision has already been made about how revisions will be done, and so it is more likely to miss a case where a natural goal for the user has been well-mapped onto the software directly, instead of going through the usual functions.

4.4.2 Step 2. Write the Top-Level Method Assuming a Unit-Task Control Structure

Unless there is reason to believe otherwise, assume that the overall task has a unit-task type of control structure. This means that the user will accomplish the topmost goal (the overall task) by doing a series of smaller tasks one after the other. The smaller tasks correspond to the set of top-level goals chosen in Step 1. For a system such as a text editor, this means that the topmost goal of `edit document` will be accomplished by a unit-task method similar to that described by Card et al. (1983). One way to describe this type of method in GOMSL is as follows:

```
Method_for_goal: Edit Document
  Step. Store First under <current_task_name>.
  Step Check_for_done.
  Decide: If <current_task_name> is None, Then
          Delete <current_task>; Delete <current_task_name>;
          Return_with_goal_accomplished.
  Step. Get_task_item_whose Name is <current_task_name>
          and_store_under <current_task>.
  Step. Accomplish_goal: Perform Unit_task.
  Step. Store Next of <current_task> under
        <current_task_name>;
          Goto Check_for_done.
```

The goal of performing the unit task typically is accomplished via a selection rule set, which dispatches control to the appropriate method for the unit task type, such as:

```
Selection_rules_for_goal: Perform Unit_task
  If Type of <current_task> is move,
   Then Accomplish_goal: Move Text.
  If Type of <current_task> is delete,
   Then Accomplish_goal: Erase Text.
  If Type of <current_task> is copy,
   Then Accomplish_goal: Copy Text.
  //... etc. ...
  Return_with_goal_accomplished.
```

This type of control structure is common enough that the above method and selection rule set can be used as a template for getting the GOMS model started. The remaining methods in the analysis will then consist of the specific methods for these sub-goals, similar to those described in the extended example below.

In this example the task type maps directly to a goal whose name is a near-synonym of the type, but this is not always the case. A good exercise is to consider the typical Video Cassette Recorder (VCR), which has at least three modes for recording a broadcast program; the selection rule for choosing the recording method consists not of tests for a simple task types like "one button recording", but rather the conditions under which each mode can or should be applied. For example, if the user is present at the beginning of the program, but will not be present at the end, and the length of the program is known, then the one-button recording method should be used.

4.4.3 Step 3. Recursively Expand the Method Hierarchy

This step consists of writing a method for each goal in terms of high-level operators, and then replacing the high-level operators with another goal/method set, until the analysis has worked down to the finest grain size desired. First, draft a method to accomplish each of the current goals. Simply list the series of steps the user has to do (an "activity list", Chapter 1). Each step should be a single natural unit of activity; heuristically, this is just an answer to the question "how would a user describe how to do this?" Make the steps as general and high-level as possible for the current level of analysis. A heuristic is to consider how a user would describe it in response to the instruction "don't tell me the details yet." Define new high-level operators, and bypass complex psychological processes as needed. Make a note of the analyst-defined operators and task description information as it is developed. Make simplifying assumptions as needed, such as deferring the consideration of possible shortcuts that experienced users might use. Make a note of these assumptions in comments in the method.

If there is more than one method for accomplishing the goal, draft each method and then draft the selection rule set for the goal. A recommendation: defer consideration of minor alternative methods until later; especially for alternative "shortcut" methods.

After drafting all of the methods at the current level, examine them one at time. If all of the operators in a method are primitives, then this is the final level of analysis of the method, and nothing further needs to be done with this method. If some of the operators are high-level, non-primitive operators, examine each one and decide whether to provide a method for performing it. The basis for the decision is whether additional detail is needed for design purposes. For example, early in the design of a specialized text-entry device, it might not be decided whether the system will have a mouse or cursor keys. Thus it will not be possible to describe cursor movement and object selection below the level of high-level operators. In general, it is a good idea to expand as many high-level operators as possible into primitives at the level of keystrokes, because many important design problems, such as a lack of consistent methods, will show up mainly at this level of detail. Also, the time estimates are clearest and most meaningful at this level. For each operator to be expanded, rewrite that step in the method (and in all other methods using the operator) to replace the operator with an `Accomplish_goal` operator for the corresponding goal.

For example, suppose the current method for copying selected text is:

```
Method_for_goal: Copy Selection
Step 1. Select Text.
Step 2. Issue Command using Copy.
Step 3. Return_with_goal_accomplished.
```

To descend a level of analysis for the Step 1 operator `Select Text,` rewrite the method as:

```
Method_for_goal: Copy Selection
Step 1. Accomplish_goal: Select Text.
Step 2. Issue Command using Copy.
Step 3. Return_with_goal_accomplished.
```

Then provide a method for the goal of selecting the text. This process should be repeated until all of the methods consist only of operators that are either primitive operators, or higher-level operators that will not be expanded.

4.4.4 Step 4. Document and Check the Analysis

After the methods and auxiliary information have been written out to produce the complete GOMSL model, list the any analyst-defined operators used, along with a brief description of each one, and the assumptions and judgment calls made during the analysis. Then, choose some representative task instances, and check on the accuracy of the model either by hand or with the GLEAN tool, to verify that the methods generate the correct sequence of overt actions, and correct and recheck if necessary.

Examine the judgment calls and assumptions made during the analysis to determine whether the conclusions about design quality and the performance estimates would change radically if the judgments or assumptions were made differently. This sensitivity analysis will be very important if two designs are being compared that involved different judgments or assumptions; less important if these were the same in the two designs. It may be desirable to develop alternate GOMS models to capture the effects of different judgment calls to systematically evaluate whether they have important impacts on the design.

Example in the Appendix

The Appendix contains a complete example GOMS model and a summary of how it was constructed with the above procedure. A more complete description of the construction procedure can be found in Kieras (1988, 1997a, 1999).

4.5 CONCLUSION AND NEW DIRECTIONS

GOMS was originally intended as an analytic approach to evaluating a user interface: a way to obtain some usability information early enough in the design process to avoid the expense of prototype development and user testing (see John & Kieras, 1996a, 1996b; Kieras, in press). However, as GOMS was developed, it incorporated some of the developing ideas about computational modeling of human cognition and performance, and has thus become a framework for constructing computer simulations of the subset of human activity that is especially relevant to much of user interface design. Thus, as presented here, GOMS can be defined as a task-analytic notation for procedural knowledge that (1) has a syntax and semantics similar to a traditional programming language; (2) assumes a simplified cognitive architecture based on the scientific literature that can be implemented as a computer simulation; (3) can be executed in a simulation to yield a simulated human that can interact with an actual or simulated device; (4) the static and run-time properties of the resulting simulation model predict aspects of usability such as time to learn or task execution time.

The advantage of GOMS modeling of human activity over the several alternatives (see Kieras, in press) is its relative simplicity and conceptual familiarity to software designers and developers. The key part of such simulations is, of course, the representation of how the simulated human understands the task structure and the requirements; GOMS models provide a convenient and relatively intelligible way to describe the procedural aspects of tasks. The fact that these task-analytic models have both a direct scientific tie to human psychology and also can be used in running computer simulations means that the GOMS task-analytic notation has well-grounded claims to rigor and utility beyond more informal approaches. Of course, formality has a price: constructing a formal representation is always more work than an informal one. However, GOMS models can be "sketched" in an informal manner that preserves the intuitive concepts of how the task is done, and these informal models can then be tightened up if necessary. Thus there would appear to be no disadvantages to using GOMS for representing procedural tasks: the level of formality of the notation can be adjusted to the requirements of the analysis situation.

Although GOMS as it currently stands is a useful and practical approach, it is by no means "finished." There are many unresolved issues and new directions to explore within the general approach that GOMS represents, and the specifics of GOMS as presented in this chapter. The remainder of this concluding discussion will deal with three topics undergoing development in GOMS; these are human error; interruptions, and modeling of teams.

4.5.1 Modeling Error Behavior

The GOMS models originally presented in Card et al. (1983) and subsequently dealt only with *error-free* behavior, although Card et al. (1983, p. 184ff) did sketch out what would be required to apply GOMS to errors. Historically, it has been a daunting problem to model human error and how to deal with it in design more precisely and specifically than the usual high-level general advice. However, as presented at length in Wood (2000), if attention is restricted to errors in procedural tasks, and GOMS is used to represent procedural knowledge, substantial progress can be made. That is, a remarkable thing about the extant theoretical work on human error is that it does not have at its core a well-worked out and useful theory of normal, or error-free, behavior in procedural tasks; it is hard to see how one could account for errors unless one can also account for correct performance! GOMS, in the architectural sense presented here, is such a theory of correct performance, and thus provides good starting point for usefully representing human error behavior.

Wood (2000) follows up on this insight and the original Card et al. (1983) proposal in three general ways: First, once the human detects an error, recovering from it becomes simply another goal; a well-designed system will have simple, efficient, and consistent methods for recovering from errors (see chap. 19). Thus the error-recovery support provided by a user interface can be designed and evaluated with GOMS just like the "normal" methods supported by the interface. Second, the way in which an error recovery method should be invoked turns out to be a difficult notational problem, and has a direct analogy to how error handling is done in computer programming. The modern solution for computer programming is *exceptions*, which provide an alternate flow of control to be used just for error handling, leaving the main body of the code uncluttered to represent only the normal activity of the program. Wood suggests that a similar approach would be the desirable extension to GOMS models: when an error is detected, an exception-like mechanism invokes the appropriate error-recovery method and then allows the original method to resume. However, exception handling is subtle even in computer programming, and humans may or may not work in the same way; more theoretical and empirical work is needed. Third, the ways in which humans detect that they have committed an error is currently rather mysterious, and we lack a good theoretical proposal that could be used easily in design situations. To finesse this problem, Wood (1999, 2000) proposed using a set of heuristics for examining a GOMS model and identifying what type of error was likely to occur at each method step, when the error would become visible, and what method the user would have to use to recover from it. This information in turn suggests how one might modify the interface to reduce the likelihood of the error or make it easier to detect and recover from it. For example, suppose the GOMS model includes a mental operation to compute a value from two numbers on the screen that is subsequently used in a decide operator that invokes one of two sub-methods to complete a task. A possible error is to miscompute the value, but the error might not be manifested until the decide operator had taken the user through a series of other steps to the wrong display, for which no further steps could be executed. Such an interface, with its error-prone requirements and delayed-detection properties, might compound its poor design by forcing the user to start from the beginning to correct the error. The interface could be redesigned to make the computation unnecessary if possible, make an error obvious sooner, or provide an efficient recovery procedure. Wood (2000) demonstrated the value of this heuristic analysis for error-tolerant design using a realistic e-commerce application.

4.5.2 Interruptions

The normal flow of control in a GOMS model as presented here is the hierarchical-sequential flow of control used in traditional programming languages: a method executes steps in sequence; if a step invokes a submethod, the steps in that submethod are executed in sequence until the submethod is complete, whereupon the next step in the calling method is executed in sequence. This simple control regime is why GOMS models are relatively easy to construct compared to other cognitive modeling approaches (see Kieras, in press). However, in realistic situations, humans often have to respond to interrupts of various kinds; an everyday example is responding to a telephone call, then resuming work after handling the call. Trying to account for interruptability within the confines of hierarchical-sequential flow of control is technically possible, but it is also clumsy and counter-intuitive: statements that check for interrupting events must be distributed liberally into the normal flow of processing. Computer technology rejected such an approach many decades ago with the introduction of specialized hardware in which an interrupting signal automatically forces the computer to suspend execution of whatever it is doing and start executing interrupt-handling code instead; once the interruption is dealt this, the interrupted process can be resumed.

GOMS models for many computer applications do not seem to require such interrupt processing because the analyzed task is limited to the human interacting with the computer; other devices, such as the telephone, are not included; and the activity with the computer is all user-initiated; the computer responds only to the user's activity, and in such a way that the user always waits for the computer to finish its response before continuing. The typical text-editor task fits this description, along with many ordinary computer-usage situations. However, in other tasks, the machine can present events that are *asynchronous* with the user's activity, and the task requirements can be such that the user must respond to these asynchronous events promptly, or at least not ignore them. An example of such a task situation appears in the military task modeled by Santoro, Kieras, and Campbell (2000). Here the user is supposed to monitor a radar display showing the movements of aircraft in the vicinity of a warship and perform various tasks in response to what the aircraft do. For example, if an aircraft exhibits suspicious behavior, the user is supposed to establish radio contact with the aircraft and ask for identification and clarification. Such activity can take several minutes to complete; but in the meantime, other events must be noted, even if no overt activity is performed in response. For example, another aircraft could suddenly appear on the display, and the user must note that it should be given priority for inspection and decision-making once the current activity is done. In analogy with computer programming, such checking could be done with many statements throughout the GOMS methods, but both practicality and intuition requires some kind of interrupt mechanism analogous to those used in computers. Some of the production-rule cognitive architectures (see Byrne, in press) provide a natural approach: GOMS can be extended to include a set of If-Then statements whose conditions are evaluated whenever the relevant psychological state of the user changes; these rules specify what goal to accomplish if a specific interrupting condition is present. Thus for the radar operator's task, one interrupt rule was that if a new "blip" with a red color-code appears on the display, add it to a list of high-priority blips. As part of its process for choosing the next task, the top-level method in the model checks this list and activates a goal based on what it finds there.

The interrupt-rule concept provides a natural mechanism for giving a GOMS model the ability to respond to asynchronous events while preserving the simplicity of the hierachical-sequential control structure for the bulk of the task. It also provides a potential way to represent error detection; an interrupt rule could be checking for evidence of an error, and then invoke the appropriate error-handling method. Working out the details of such an approach is a matter for future research. But in the meantime, it appears that GOMS models can successfully combine a simple program-like representation with an intuitive form of interruptability.

4.5.3 Modeling Teams of Users

Many design situations involve teams of humans that cooperate to perform a complex task. Doing more than simply acknowledging the possible incredible complexity of human interactions involved in a team is well beyond the scope of this chapter. However, there is a subset of team activity that can be encompassed with GOMS modeling: the case where the team is following a procedure consisting of specified interactions between the team members, each of whom is likewise following a set of specified procedures. Such situations are common in many military team situations, such as the *combat information center* teams analyzed by Santoro and Kieras (2001) and not dissimilar to Annett's naval example (chap. 3). In the Santoro and Kieras example, each human in the team sits at a workstation that incorporates a radar display, and has certain assigned tasks, such as making the radio contacts described above. The team members are supposed to communicate with each other, using speech over an intercom, to coordinate their activity, such as ensuring that high-priority blips get examined. In this case, a model for the team can be constructed simply as a team of models: Each team member's taks is represented by a GOMS model; part of the member's task is to speak certain messages to other team members, and respond to certain messages from other team members. The structure of the team procedures determines which messages are produced by what member, and how another member is supposed to respond to them. The interrupt capability described above is especially useful because it simplifies handling of asynchronous speech input. Once the individual GOMS models have been developed, the activity of a team can be simulated simply by running the whole set of interacting individual models simultaneously. The simulation can then show whether the team procedures result in good performance by the team as a whole. Thus the rigor and strengths of GOMS modeling and task analysis can be extended from the domain of individual user interfaces to the domain of team structure and team procedures.

APPENDIX: SAMPLE GOMS MODEL

This appendix presents a complete example GOMS model for a simplified subset of the MacWrite text editor that describes how to move, delete, or copy text selections. The model contains methods that start at the topmost level and finish at the keystroke level, along with the necessary auxiliary information for the methods to be executed (by hand or by GLEAN), and a set of four *benchmark tasks* to be performed. The GOMSL starts with the auxiliary information for a set of tasks, visual items, and items assumed to be in LTM. The methods themselves start with the Top-Level Unit Task method. One should be able to get at least of rough idea of the methods simply by reading them, even without detailed knowledge of the syntax of GOMSL; this is one of the goals of the NGOMSL and GOMSL notations.

The process of constructing the model will be summarized. A more complete step-by-step construction of a similar example can be found in Kieras (1988, 1997a, 1999). Because of the top-down expansion of methods, the methods were constructed in roughly the same order as they appear in the example. The construction started with the topmost user's goal of editing the document. Taking the above recommendation, the first piece of the model is simply a version of the unit task method and the selection rule set that dispatches control to the appropriate method. This assumes an ad hoc task representation that was refined as the construction continued.

The unit task method and its selection rule specify a set of second-level user goals. This example focuses on the methods for the goal of moving text. A first judgment call was that users view moving text as first cutting, then pasting. The method for this goal was written accordingly, initially with two high-level operators—`Cut Selection`, then `Paste Selection`—followed by a `Verify` that the cut and paste has been done correctly.

Descending a level, the high-level operators were then rewritten into `Accomplish_goal` operators and methods for cutting and pasting selected text were provided. Another judgment call is that users are aware of the general-purpose idea of text selection, so the method for cutting a selection starts not with keystroke-level actions for selecting text that would be specific to the cutting goal, but rather with another high-level operator for selecting text which then was rewritten into an `Accomplish_goal` operator. Since there are a variety of ways to select text, a selection rule specifies three different specific contexts in which text selection is needed. It thus maps the general goal of selecting text to three different specific goals, each of which has its own method. The paste method similarly has a subgoal of selecting the insertion point, but there is only one way to do it, and so only a single method was provided. Notice how this set of judgments effectively states that the user has some general-purpose "subroutines" for cutting, pasting, selecting text, and selecting an insertion point. Expressing this conclusion as goals and methods identifies some key properties of the interface (e.g., selection can be done in the same way for all relevant tasks in the text-editing application) and assumes that the user makes use of them, or *should* make use of them.

Descending another level, note that the cutting and pasting methods involve picking a command from a menu (for simplicity, assume that users do not make use of the command-key shortcuts). An important property of well-designed menu systems is that the procedure for picking a command is uniform across the menu system. Thus, the operators of `Invoke_cut_command` and `Invoke_paste_command` were replaced with a single `Issue Command` method that is given the "name" or "concept" of the desired command as a pseudoargument and makes the proper menu accesses. Thus the `Issue Command` method first retrieves from LTM which menu to open, finds it on the screen, opens it, and then finds and selects the actual menu item.

To make this example more complete, the methods for deleting and duplicating text were added. Often, writing the methods for additional goals is quite easy once the first set of methods have been written—the lower level submethods are simply reused in different combinations or with different commands; this is one symptom of a good design. After drafting the methods the analyst collected the task information that the methods require and reconciled and revised the task representation as necessary. In addition, the auxiliary information was collected and specified, such as the LTM items required by the `Issue Command` method. For the GLEAN tool to execute these methods, there needs to be some visual objects for the methods to look for and point at. In this case, these objects need only be minimal or "dummy" objects. The example also includes as auxiliary information a set of four editing tasks specified in a "linked list" form that is accessed by the top-level unit task method.

```
Define_model: "MacWrite Example"
  Starting_goal is Edit Document.

Task_item: T1
  Name is First.
  Type is copy.
  Text_size is Word.
  Text_selection is "foobar".
  Text_insertion_point is "*".
  Next is T2.

Task_item: T2
  Name is T2.
  Type is copy.
  Text_size is Arbitrary.
  Text_selection_start is "Now".
  Text_selection_end is "country".
  Next is T3.

Task_item: T3
  Name is T3.
  Type is delete.
  Text_size is Word.
  Text_selection is "foobar".
  Text_insertion_point is "*".
  Next is T4.

Task_item: T4
  Name is T4.
  Type is move.
```

```
  Text_size is Arbitrary.
  Text_selection_start is "Now".
  Text_selection_end is "country".
  Next is None.

// Dummy visual objects - targets for Look_for and Point_to
Visual_object: Dummy_text_word
  Content is "foobar".
Visual_object: Dummy_text_selection_start
  Content is "Now".
Visual_object: Dummy_text_selection_end
  Content is "country".
Visual_object: Dummy_text_insertion_point
  Content is "*".

// Minimal description of the visual objects in the editor
interface
Visual_object: Edit_menu
  Label is Edit.
Visual_object: Cut_menu_item
  Label is Cut.
Visual_object: Copy_menu_item
  Label is Copy.
Visual_object: Paste_menu_item
  Label is Paste.

// Long-Term Memory contents about which items are in which
   menu
LTM_item: Cut_Command
  Name is Cut.
  Containing_Menu is Edit.
  Menu_Item_Label is Cut.
  Accelerator_Key is COMMAND-X.
LTM_item: Copy_Command
  Name is Copy.
  Containing_Menu is Edit.
  Menu_Item_Label is Copy.
  Accelerator_Key is COMMAND-C.
LTM_item: Paste_Command
  Name is Paste.
  Containing_Menu is Edit.
  Menu_Item_Label is Paste.
  Accelerator_Key is COMMAND-V.

// Top-Level Unit Task Method
Method_for_goal: Edit Document
  Step. Store First under <current_task_name>.
  Step Check_for_done.
  Decide: If <current_task_name> is None, Then
          Delete<current_task>;Delete <current_task_name>;
          Return_with_goal_accomplished.
```

```
  Step. Get_task_item_whose Name is <current_task_name>
          and_store_under <current_task>.
  Step. Accomplish_goal: Perform Unit_task.
  Step. StoreNext of <current_task> under <current_task_name>;
          Goto Check_for_done.

Selection_rules_for_goal: Perform Unit_task
  If Type of <current_task> is move,
   Then Accomplish_goal: Move Text.
  If Type of <current_task> is delete,
   Then Accomplish_goal: Erase Text.
  If Type of <current_task> is copy,
   Then Accomplish_goal: Copy Text.
  //... etc. ...
  Return_with_goal_accomplished.

Method_for_goal: Erase Text
  Step 1. Accomplish_goal: Select Text.
  Step 2. Keystroke DELETE.
  Step 3. Verify "correct text deleted".
  Step 4. Return_with_goal_accomplished.

Method_for_goal: Move Text
  Step 1. Accomplish_goal: Cut Selection.
  Step 2. Accomplish_goal: Paste Selection.
  Step 3. Verify "correct text moved".
  Step 4. Return_with_goal_accomplished.

Method_for_goal: Copy Text
  Step 1. Accomplish_goal: Copy Selection.
  Step 2. Accomplish_goal: Paste Selection.
  Step 3. Verify "correct text moved".
  Step 4. Return_with_goal_accomplished.

Method_for_goal: Cut Selection
  Step 1. Accomplish_goal: Select Text.
  Step 2. Accomplish_goal: Issue Command using Cut.
  Step 3. Return_with_goal_accomplished.

Method_for_goal: Copy Selection
  Step 1. Accomplish_goal: Select Text.
  Step 2. Accomplish_goal: Issue Command using Copy.
  Step 3. Return_with_goal_accomplished.

Method_for_goal: Paste Selection
  Step 1. Accomplish_goal: Select Insertion_point.
  Step 2. Accomplish_goal: Issue Command using Paste.
  Step 3. Return_with_goal_accomplished.
```

```
// Each task specifies the "size" of the text involved
Selection_rules_for_goal: Select Text
  If Text_size of <current_task> is Word,
   Then Accomplish_goal: Select Word.
  If Text_size of <current_task> is Arbitrary,
   Then Accomplish_goal: Select Arbitrary_text.
Return_with_goal_accomplished.

// The task specifies the to-be-selected word
Method_for_goal: Select Word
  Step 1. Look_for_object_whose Content is
  Text_selection of  <current_task>
          and_store_under <target>.
  Step 2. Point_to <target>; Delete <target>.
  Step 3. Double_click mouse_button.
  Step 4. Verify "correct text is selected".
  Step 5. Return_with_goal_accomplished.

// The task specifies the beginning and ending word of the text
Method_for_goal: Select Arbitrary_text
  Step 1. Look_for_object_whose Content is
           Text_selection_start of <current_task>
           and_store_under <target>.
  Step 2. Point_to <target>.
  Step 3. Hold_down mouse_button.
  Step 4. Look_for_object_whose Content is
           Text_selection_end of <current_task>
           and_store_under <target>.
  Step 5. Point_to <target>; Delete <target>.
  Step 6. Release mouse_button.
  Step 7. Verify "correct text is selected".
  Step 8. Return_with_goal_accomplished.

Method_for_goal: Select Insertion_point
  Step 1. Look_for_object_whose Content is
           Text_insertion_point of <current_task>
           and_store_under <target>.
  Step 2. Point_to <target>; Delete <target>.
  Step 3. Click mouse_button.
  Step 4. Verify "insertion cursor is at correct place".
  Step 5. Return_with_goal_accomplished.
// Assumes that user does not use command-key shortcuts

Method_for_goal: Issue Command using <command_name>
// Recall which menu the command is on, find it, and open it
  Step 1. Recall_LTM_item_whose
           Name is <command_name>
           and_store_under <command>.
  Step 2. Look_for_object_whose
          Label is Containing_Menu of <command>
          and_store_under <target>.
```

```
  Step 3. Point_to <target>.
  Step 4. Hold_down mouse_button.
  Step 5. Verify "correct menu appears".

// Now select the menu item for the command
  Step 6. Look_for_object_whose
         Label is Menu_Item_Label of <command>
         and_store_under <target>.
  Step 7. Point_to <target>.
  Step 8. Verify "correct menu command is highlighted".
  Step 9. Release mouse_button.
  Step 10.Delete <command>; Delete <target>;
          Return_with_goal_accomplished.
```

REFERENCES

Annett, J., Duncan, K. D., Stammers, R. B., & Gray, M. J. (1971). *Task analysis*. London: Her Majesty's Stationery Office.

Beevis, D., Bost, R., Doering, B., Nordo, E., Oberman, F., Papin, J-P., Schuffel, H., & Streets, D. 1992. *Analysis techniques for man-machine system design* (Report No. AC/243(P8)TR/7). Brussels: Defense Research Group, NATO HQ.

Bennett, J. L., Lorch, D. J., Kieras, D. E., & Polson, P. G. (1987). Developing a user interface technology for use in industry. In H. J. Bullinger & B. Shackel, (Eds.), *Proceedings of the Second IFIP Conference on Human-Computer Interaction (Interact '87)* (pp. 21–26). Amsterdam: Elsevier Science Publishers.

Bhavnani, S. K., & John, B. E. (1996). Exploring the unrealized potential of computer-aided drafting. In *Proceedings of the CHI '96 Conference on Human Factors in Computing Systems* (pp. 332–339). New York: ACM.

Bjork, R. A. (1972). Theoretical implications of directed forgetting. In A. W. Melton & E. Martin (Eds.),*Coding processes in human memory* (pp. 217–236). Washington, DC: Winston.

Bovair, S., Kieras, D. E., & Polson, P. G. (1990). The acquisition and performance of text editing skill: A cognitive complexity analysis. *Human-Computer Interaction, 5*, 1–48.

Byrne, M. D. (in press). Cognitive architecture. In J. Jacko & A. Sears (Eds.), *Human-computer interaction handbook*. Mahwah, N. J.: Lawrence Erlbaum Associates.

Byrne, M. D., Wood, S. D, Sukaviriya, P., Foley, J. D, & Kieras, D. E. (1994). Automating interface evaluation. In *Proceedings of CHI'94*, (pp. 232–237). New York: ACM.

Card, S. K., Moran, T. P., & Newell, A. (1980). The keystroke-level model for user performance time with interactive systems. *Communications of the ACM, 23*, 396–410.

Card, S., Moran, T., & Newell, A. (1983). *The psychology of human-computer interaction*. Hillsdale, NJ: Lawrence Erlbaum Associates.

Diaper, D. (Ed.). (1989). *Task analysis for human-computer interaction*. Chicester, England: Ellis Horwood.

Elkerton, J., & Palmiter, S. (1991). Designing help using the GOMS model: An information retrieval evaluation. *Human Factors, 33*, 185–204.

Gong, R. & Elkerton, J. (1990). Designing minimal documentation using a GOMS model: A usability evaluation of an engineering approach. In *Proceedings of CHI'90: Human Factors in Computer Systems* (pp. 99–106). New York: ACM.

Gould, J. D. (1988). How to design usable systems. In M. Helander (Ed.), *Handbook of human-computer interaction* (757–789). Amsterdam: North-Holland.

Gray, W. D., John, B. E., & Atwood, M. E. (1993). Project Ernestine: A validation of GOMS for prediction and explanation of real-world task performance. *Human-Computer Interaction, 8*, 237–209.

John, B. E., & Kieras, D. E. (1996a). Using GOMS for user interface design and evaluation: Which technique? *ACM Transactions on Computer-Human Interaction, 3*, 287–319.

John, B. E., & Kieras, D. E. (1996b). The GOMS family of user interface analysis techniques: Comparison and contrast. *ACM Transactions on Computer-Human Interaction, 3*, 320–351.

Kieras, D. E. (1988). Towards a practical GOMS model methodology for user interface design. In M. Helander (Ed.), *Handbook of human-computer interaction* (pp. 135–158). Amsterdam: North-Holland.

Kieras, D. E. (1997a). A Guide to GOMS model usability evaluation using NGOMSL. In M. Helander, T. Landauer, & P. Prabhu (Eds.), *Handbook of human-computer interaction* (2nd ed.; 733–766). Amsterdam: North-Holland.

Kieras, D. E. (1997b). Task analysis and the design of functionality. In A. Tucker (Ed.) *The computer science and engineering handbook* (pp. 1401–1423). Boca Raton, FL: CRC.

Kieras, D. E. (1999). *A guide to GOMS model usability evaluation using GOMSL and GLEAN3.* Available FTP: ftp://www.eecs.umich.edu/people/kieras

Kieras, D. E. (in press). Model-based evaluation. In J. Jacko & A. Sears (Eds), *Human-computer interaction handbook* (pp. 1139–1151). Mahwah, NJ: Lawrence Erlbaum Associates.

Kieras, D. E., & Bovair, S. (1986). The acquisition of procedures from text: A production-system analysis of transfer of training. *Journal of Memory and Language, 25*, 507–524.

Kieras, D., & Meyer, D. E. (1997). An overview of the EPIC architecture for cognition and performance with application to human-computer interaction. *Human-Computer Interaction, 12*, 391–438.

Kieras, D. E., & Meyer, D. E. (2000). The role of cognitive task analysis in the application of predictive models of human performance. In J. M. C. Schraagen, S. E. Chipman, & V. L. Shalin (Eds.), *Cognitive task analysis* (pp. 237–260). Mahwah, NJ: Lawrence Erlbaum Associates.

Kieras, D., Meyer, D., & Ballas, J. (2001). Towards demystification of direct manipulation: Cognitive modeling charts the gulf of execution. In *Proceedings of the CHI 2001 Conference on Human Factors in Computing Systems* (pp. 128–135). New York, ACM.

Kieras, D. E., Meyer, D. E., Mueller, S., & Seymour, T. (1999). Insights into working memory from the perspective of the EPIC architecture for modeling skilled perceptual-motor and cognitive human performance. In A. Miyake & P. Shah (Eds.), *Models of working memory: Mechanisms of active maintenance and executive control* (pp. 183–223). New York: Cambridge University Press.

Kieras, D. E. & Polson, P. G. (1985). An approach to the formal analysis of user complexity. *International Journal of Man-Machine Studies, 22*, 365–394.

Kieras, D. E., Wood, S. D., Abotel, K., & Hornof, A. (1995). GLEAN: A computer-based tool for rapid GOMS model usability evaluation of user interface designs. In *Proceedings of the ACM symposium on user Interface Software and Technology (UIST '95)* (pp. 91–100). New York: ACM.

Kieras, D. E., Wood, S. D., & Meyer, D. E. (1997). Predictive engineering models based on the EPIC architecture for a multimodal high-performance human-computer interaction task. *ACM Transactions on Computer-Human Interaction. 4*, 230–275.

Kirwan, B., & Ainsworth, L. K. (1992). *A guide to task analysis.* London: Taylor & Francis.

Lewis, C., & Rieman, J. (1994) *Task-centered user interface design: A practical introduction.* Shareware book available FTP: ftp.cs.colorado.edu/pub/cs/distribs/clewis/HCI-Design-Book

Olson, J. R., & Olson, G. M. (1990). The growth of cognitive modeling in human-computer interaction since GOMS. *Human-Computer Interaction, 5*, 221–265.

Polson, P. G. (1987). A quantitative model of human-computer interaction. In J. M. Carroll (Ed.), *Interfacing thought: Cognitive aspects of human-computer interaction* (pp. 184–235). Cambridge, MA: MIT Press.

Santoro, T. P., Kieras, D. E., & Campbell, G. E. (2000). GOMS modeling application to watchstation design using the GLEAN tool. In *Proceedings of the Interservice/Industry Training, Simulation, and Education Conference* (pp. 964–973). Arlington, VA: National Training Systems Association.

Santoro, T., & Kieras, D. (2001 October). GOMS models for team performance. In J. Pharmer and J. Freeman (Organizers), *Complementary methods of modeling team performance*. Panel presented at the 45th Annual Meeting of the Human Factors and Ergonomics Society, Minneapolis–St. Paul.

Wood, S. (1993). *Issues in the implementation of a GOMS-model design tool.* Unpublished manuscript, University of Michigan, Ann Arbor.

Wood, S. D. (1999, August). The application of GOMS to error-tolerant design. Paper presented at the 17th International System Safety Conference, Orlando, FL.

Wood, S. D. (2000). Extending GOMS to human error and applying it to error-tolerant design. Unpublished doctoral dissertation, University of Michigan, Ann Arbor.

5

Scenario-Based Task Analysis

Kentaro Go
University of Yamanashi

John M. Carroll
The Pennsylvania State University

This chapter examines the use of scenarios in analysing tasks in human-computer interaction. The scenario-based requirements process employs scenarios, specific stories and examples of past use of existing systems, as well as future use of potential systems, as fundamental tools for capturing and articulating tasks and activities. This chapter defines basic terms for the scenario-based requirements process, which include scenarios, claims, scenario-based requirements analysis, task exploration and analysis using scenarios, claims analysis, and scenario exploration. In addition, it proposes a framework for concept and scenario evolution in which concepts in the abstraction domain and scenarios in the detail domain iteratively evolve. This framework provides another facet of scenario-based requirements analysis. The chapter also illustrates an example of the scenario-based requirements process. The concept of Familyware, communication tools that provide status information about remotely located family members using peripheral displays and devices, is developed through an example process. This example illustrates where scenarios are obtained from and how they evolve. The chapter concludes with a discussion about the characteristics of the scenario-based approach and task analysis.

5.1 INTRODUCTION

Scenarios have been used as a powerful design tool. They are employed throughout the design process and in various disciplines, including human-computer interaction, requirements engineering, object-oriented design, and strategic planning (Go & Carroll 2002; Jarke, Bui, & Carroll, 1998). Carroll (1995) collected the early research on the scenario-based approach to design in the area of human-computer interaction. More recently, several practitioners have documented efforts to use scenarios in systems design (Beyer & Holtzblatt, 1998; Cooper, 1999; Johnson, 2000; McGraw & Harbison, 1997; see also chaps. 1, 2, 4, 7, 9, 13, 18, 21–24, 26, and 27).

Scenarios facilitate design activities to provide a lightweight way of creating and reusing usage situations (Carroll, 2000). They are integrated and flexible use-oriented design representations that are easily developed, shared, and manipulated. They are applicable to many system development activities. They are fundamental design artifacts in human-computer interaction and have several roles throughout the system life cycle. For example, scenarios facilitate user-designer communication as a vehicle of knowledge. They treat unforeseen activities by users as if they were tangible design artifacts.

This chapter consists of three parts. The first part describes basic terms used in scenario-based requirements analysis. It specifies scenarios and claims and then briefly explains Rosson and Carroll's (2001) approach. It also describes a task exploration and analysis approach. The second part, which is transitional, illustrates the framework of concept and scenario evolution. It shows an alternative view of the scenario-based requirements analysis to the one described in the first part; to be precise, it models the evolution of concepts in the abstraction domain and of scenarios in the detail domain. The final part demonstrates how scenarios can contribute to the activities of task analysis. Basically this chapter describes how we develop ideas for personal communication systems called Familyware (Go, Carroll, & Imamiya, 2000). Familyware provides status information about remotely located family members using peripheral displays and devices. At the early design and analysis stage, we develop a root concept of Familyware, then gather usage scenarios from interviews and brainstorming sessions. These activities result in an envisioned scenario, which is analyzed later. At the end of this chapter, we discuss the contribution scenarios can make to task analysis.

5.2 SCENARIO-BASED REQUIREMENTS ANALYSIS: IDEAS AND CONCEPTS

This section specifies scenarios and claims. It includes a short summary of Scenario-Based Requirements Analysis (SBRA) by Rosson and Carroll (2001). SBRA employs claims analysis as a key technique. Also, the section explains scenario exploration, which uses a structured brainstorming session to produce new ideas.

5.3 SCENARIOS

A scenario is a description that contains actors, background information about them, and assumptions about their environment, their goals or objectives, and sequences of actions and events. It may include obstacles, contingencies, and outcomes. In some applications, scenarios may omit one of the elements or express it simply or implicitly.

A scenario is a shared story among various stakeholders in system design. For example, customers or project managers describe their visions in episodes. Users talk about problems they face as they happen. Designers record the rationale of a design in the form of an example and develop mock-ups to illustrate what users should do with the design. Technical writers explain the task of users in a manual and write it up as a story. These are examples of scenarios shared among stakeholders and distributed throughout the design cycle.

Scenarios are expressed in various media and forms. For example, they can be textual narratives, storyboards, video mock-ups, or scripted prototypes. In addition, they may be in informal, semiformal, or formal notation. A typical example of an informal scenario is a story, a kind of scenario frequently used for envisioning user tasks in human-computer interaction.

The following is an example of a textural narrative scenario; it is an excerpt from Carroll (2000). It envisions ideas about an interface and interaction in a video information system.

Looking for the fast-forward button

> Walter has been browsing some clips pertaining to the project manager's views of the lexical network as they developed through the course of the project. One clip in particular seems to drag a bit, and he wonders how to fast forward through the rest of it—perhaps he can just stop the playout? (Box 4.2, p. 81)

5.3.1 Claims

A designed artifact may contain various *trade-offs*. Trade-offs describe the pros and cons of an artifact. In contrast, *claims* are more specific in context than trade-offs. A claim is a description of a trade-off relating to specific usability concerns with a given artefact; in other words, it creates an instance of a trade-off. It articulates the upsides and downsides of artifact usability.

Claims enumerate implicit causal relations in a scenario. They describe trade-offs instantiated in a scenario context (and in that sense they explain the scenario). The use of claims and scenarios in combination, therefore, helps designers discuss the consequences of design moves at various levels of analysis. Consider the "looking for the fast-forward button" scenario, for instance: Video is a very rich, intrinsically appealing medium, but the data are difficult to search and must be viewed linearly in real time. Articulating this important yet implicit feature of video together with the scenario increases the scenario's effectiveness in system design and analysis.

It is useful to provide a format for trade-offs and claims in order to produce, document, analyze, and reuse them. Carroll and Rosson (1992) suggest the following practical form:

{Some design feature}
+ causes {desirable consequences}
– causes {undesirable consequences}

For example, the claims about video information related to the "looking for the fast-forward button" scenario can be described as follows:

Video information claim: video information
+ is a very rich medium,
+ is an intrinsically appealing medium,
– is difficult to search,
– must be viewed linearly in real time.

Putting claims in this form is useful for analysts because it naturally indicates some things that are missing in the analysis. For example, if there are several downsides listed in a claim about an artifact without any upsides, then analysts should consider its upsides.

The activity of investigating the claims of a scenario is called claims analysis. This is an analytic evaluation method involving the identification of scenario features that have significant positive or negative usability consequences. Articulating claims about a technology in scenarios (i.e., systematically enumerating its potential trade-offs) produces a fair analysis of its pros and cons of by moderating the tendency of usability experts to produce and address only the downsides of an artifact.

5.3.2 Scenario-Based Design

Scenario-Based Design (SBD) uses scenarios (and thereby claims) as a central representation throughout the entire system life cycle. SBD has three key characteristics: (1) It is a life cycle methodology, (2) it is strongly oriented toward inquiry (in Pott's sense, it is intended for "a series of questions and answers designed to pinpoint where information needs come from and when," Potts, Takahashi, & Anton, 1994, p. 21), and (3) it can be employed in a wide variety of ways.

SBD covers everything from requirements and visions to summative evaluation. Most task analysis methods are employed at a particular point in system development, and neither leverage nor are leveraged by any other representation or technique employed anywhere else in the system development life cycle. This lack of integration with life cycle development activities could be one reason that these other task analysis methods have had little impact, as Diaper (2002; also chap. 1) admits in his review of Carroll (2000), and are adopted only in organizations that employ structured waterfall development methods (e.g., military contractors and insurance companies).

SBD is strongly oriented to inquiry. Most task analysis methods take some sort of specification as a given and then further articulate it and perhaps refactor it. However, this is mild stuff if one is interested in discovering insights into human activity, radically new ways of doing familiar things, or entirely new types of things to do. SBD includes and addresses these other, more creative concerns. It may worry a person with a very structured task analytic perspective that SBD helps to generate all sorts of novel activity concepts, because such a person may see that as being beyond the purview of task analysis. But it clearly is useful to imagine new tasks, and in creative and rapidly emerging areas of information technology, imagining new tasks may be much more critical to success than finely analysing out-of-date task concepts.

Traditional task analysis methods such as Hierarchical Task Analysis (HTA; chap. 3; Shepherd, 1989) seek to achieve highly precise procedural specifications so that a task analyst knows exactly what steps to perform, how to perform them, and what order to perform them in. This is highly desirable, especially if the task analyst is only marginally competent or not very creative. This is where structured methods really prove their value. However, the downside of well-specified method scripts is that they lack flexibility. Thus, if circumstances are novel or if new opportunities present themselves, the highly structured methods chug ahead just as they always do. SBD is highly flexible and accommodates different practitioner styles, different contexts of application, and idiosyncratic constraints and opportunities in a particular project. An analyst using SBD can merely envision a scenario, can employ the concepts from Making Use (Carroll, 2000), can methodically follow the more structured approach of SBRA, or can even employ a hierarchical task analysis of a systematic sample of scenarios. All of these approaches belong to the family of SBD methods, and each can be employed when it is most appropriate.

5.3.3 Scenario-Based Requirements Analysis

Scenario-Based Requirements Analysis (SBRA) has been developed as the starting-point activity in the scenario-based design of human-computer interaction (Rosson & Carroll, 2001). Its framework is illustrated in Fig. 5.1.

Initiating SBRA, analysts prepare a *root concept* prior to going into the field. The root concept is a document describing the vision, rationale, assumptions, and stakeholders of the target system. It is derived from various sources. For example, the vision may come from open-ended discussions among various people related to the target project. Identifying those people—the stakeholders—is also part of the root concept. The rationale may come from discussions about the current technology and problems in the target domain. Finally, listing assumptions about the project and their impacts on it can provide helpful ideas for the analysts.

After preparing the root concept and questions about it, the analysts conduct field studies. They use several tools and techniques of task observation and recording (Diaper, 1989).

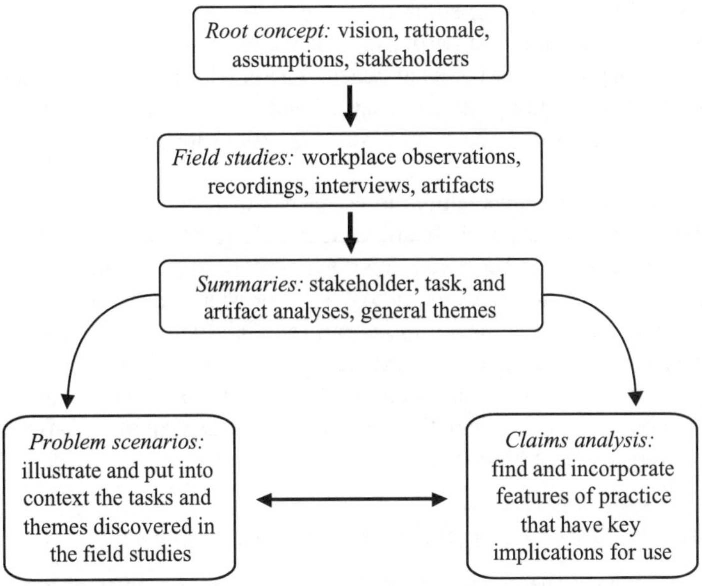

FIG. 5.1. Overview of scenario-based requirements analysis (Rosson & Carroll, 2001).

They conduct qualitative research by observing the workplace, recording the work setting and activities, interviewing stakeholders, and analyzing artifacts.

Then they summarize the collected data to identify and illustrate the stakeholders, activities, and tools or artifacts of the project. Also, they recognize the general themes or workplace themes of the project. During this process, the analysts use task analysis techniques such as HTA to decompose complex tasks into subtasks. In addition, in this summarizing step the analysts may employ a technique similar to the *affinity diagram* method of contextual design (Beyer & Holtzblatt, 1998). In this technique, the analysts jot down ideas on Post-it notes and then form them into a group and give it a title. More recently, practitioners have incorporated techniques of this kind into design projects and reported their experiences.

Having identified the basic elements from the field data, the analysts enter into iterative cycles of scenario description and claims analysis. First, they gather all the information about current practice to create *problem scenarios*. Problem scenarios contain the identified elements such as the project's stakeholders, their activities, and the tools or artifacts they use. They represent and illustrate the current practice of the project. Based on the problem scenarios, the analysts conduct a claims analysis. The iterative cycles of scenarios and claims can be supported with the scenario exploration technique.

5.3.4 Scenario Exploration

Requirements analysis can be seen as an *inquiry process*—an iterative cycle of questions and answers that aims to discover and explain hidden requirements (Potts et al., 1994). On this view, scenario exploration falls into the category of requirements analysis. It is a structured brainstorming session in which a group of analysts conduct a systematic inquiry into a usage situation. It may include actual users and customers. Also, various stakeholders in system analysis and design can contribute to the activity of scenario exploration. Usability specialists use the past experiences and knowledge of usability in addition to design heuristics and theories in human-computer interaction. Users can participate in this activity to provide their domain knowledge.

A scenario exploration session starts from a basic scenario illustrating part of the root concept of the target domain. The participants of the session put questions to the scenario. Typical questions may include who, what, when, where, why, and how questions (5W + 1H questions) and what-if questions (see also chaps. 3 and 21). For each question, the participants provide solutions as scenarios. The scenarios and questions iteratively evolve as the analysis progresses.

Scenario exploration employs simple tools: pieces of paper such as Post-it notes and a wide-open area such as a wall, whiteboard, or table. There are two types of paper (possibly color-coded): scenario slips and question slips. On a scenario slip, the analysts write a scenario; on a question slip, they write a question (a 5W + 1H or what-if question) about the scenario. Analysts put them side by side or top to bottom (possibly with a link or line) in the wide-open area. The answers to the question form scenarios, so the analysts write each answer on a scenario slip and put it next to the question slip. Each scenario in turn evokes new questions. As the analysts continue working like this, they obtain a large map of scenarios and questions in a relatively short time in a group session.

5.3.5 Framework of Concept and Scenario Evolution

In the theory of design, one considers the tension between the abstract concept of design principles and the details of specific interaction techniques. The scenarios discussed in this chapter generally refer to the detail domain: They illustrate specific actors, contexts, and uses. As an example, the next section discusses the evolution of the Familyware concept. In an envisioning Familyware scenario, a girl is excited about her father's message and expresses her feelings by shaking her teddy bear; the feelings inside her are expressed in a physical behavior that serves as the trigger to signal her father. This episode may not need to be developed because as it is, it illustrates one of the views of the Familyware concept. To develop the Familyware concept, therefore, discussion in the abstraction domain also is necessary. Fig. 5.2 illustrates the framework of concept and scenario evolution.

Concepts at the abstraction level evolve along with scenarios at the detail level. Designers develop a root concept before they go into the field. They log real-world episodes, which represent specific tasks and activities at the detail level. The episodes then contribute to strengthen

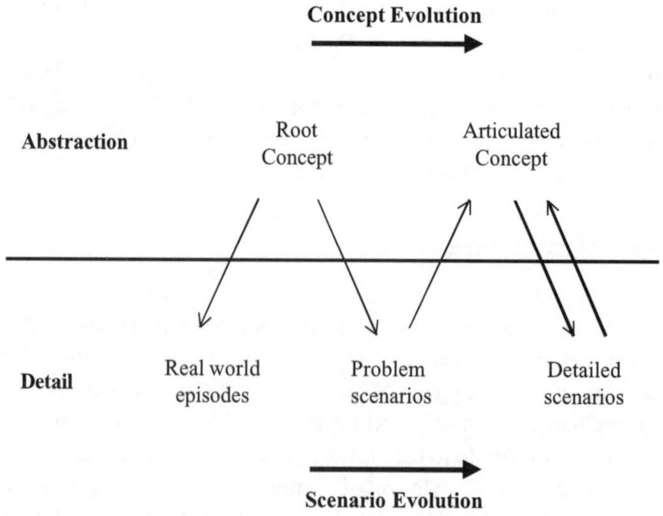

FIG. 5.2. Framework of concept and scenario evolution.

the root concept. This step also provides the themes of the concept. The episodes show helpful hints to envision problem scenarios, which are articulated at the detail level. Comparing the real-world episodes and the problem scenarios allows users to articulate the root concept and add further issues that have to be considered. The articulated concept is used to envision more detailed scenarios. Then an analysis of the produced scenarios improves the concept that is used to produce them. This iterative cycle continues to articulate the concept of a target system. The framework of concept and scenario evolution supports and provides another facet of SBRA described in the previous section.

5.4 DEVELOPING THE FAMILYWARE CONCEPT AND ITS USE

This section demonstrates a scenario-based requirements process based on the framework of concept and scenario evolution. The example involves Familyware—a collection of communication tools for distant family members and people who have a close relationship (Go et al., 2000).

5.4.1 Root Concept

The high-level goal of the Familyware project is summarized in Table 5.1. The fundamental idea of the project comes from supporting communication in a community. Unlike previous work, like the Blacksburg Electronic Village project (Carroll & Rosson, 1996), the Familyware project focuses on relatively small communities such as a family and friends. Familyware seeks to allow those community members to exchange messages. The basic rationale behind it is that digital surroundings and peripheral displays with a communication function foster a feeling of connection among family members. Table 5.1 shows parent and child as an example of a stakeholder group. (Go et al., 2000, discussed friends as another example of a stakeholder group.)

5.4.2 First-Round Interview

The purpose of the initial interview was to obtain episodes of family communication in order to verify the assumptions of the root concept. We also gathered examples of opportunities of family communication to strengthen the root concept of Familyware. We conducted an

TABLE 5.1
The Root Concept of the Familyware Project

Component	*Contributions to the Root Concept*
High-level vision	Family members exchange messages.
Basic rationale	Digital surroundings and peripheral displays with a communication function keep a feeling of connection.
Stakeholder group	
Family	
Parent	More effortless awareness.
Child	Easy to communicate.
Starting assumptions	Use everyday artifact as display. Internet is used for long-distance communication.

informal interview with a college professor who has a daughter and a son. She explained her experience with her children in the following scenarios:

> *Telescope Scenario*: Alison is a college professor and she lives with her two children. Her children did not want to go to kindergarten at first, so she stayed with them in their classroom on the first day of school. Next day, she sat in her car, watching the classroom activities through the window. Then she told her children that she would watch them from her office (on campus) through a telescope—though that is physically impossible. The children were comfortable going to kindergarten because they felt that their mother would still be with them. Now the younger child sometimes says, "I did XYZ today. Did you see me?"

This episode articulates the opportunity of family communication. Alison's younger child seeks to have a feeling of connection to her. He wants to share his experiences with her. Another episode she mentioned reveals a more complicated communication situation:

> *Phone Call Scenario*: Alison is writing a research paper for a conference. The deadline is not far away, so she has often been staying late at her office. Kaz, Alison's son, has been worrying about her: "Why doesn't Mom come home?" He felt he had to do something. In the kitchen, he found his grandmother's phone number on the wall. He called her and asked for his mom's office phone number. Then Kaz called Alison's office. She was surprised but also happy to hear from him. For the rest of the evening, Kaz called her office every 10 minutes. Alison could understand his behavior; nevertheless, she eventually became annoyed.

This episode raises the issue of availability for communication. Alison experienced a mixture of positive and negative feelings. Her primary work was to write the conference paper. Except for an emergency, she did not want to be disturbed by a direct communication device like a telephone. Using everyday artefacts to display information may take care of this issue. Table 5.2 summarizes the themes and issues obtained from these scenarios.

5.4.3 Listing Basic Tasks and Activities

Another method we used to obtain scenarios was to conduct a participatory design meeting with potential users. From our own experience, we know that it is impossible to enumerate all the tasks and activities of a college professor. It is pointless to videotape all his or her tasks and activities because it is too expensive to analyze them, not to mention the privacy issues that would arise. Instead, we held an informal brainstorming session. In less than 20 minutes, we listed more than 50 tasks of a college professor. Some of them are given in Table 5.3.

TABLE 5.2
Summary of Themes from Interview with College Professor

Theme	*Issues Contributing to the Theme*
Shared feeling of connection	Family members want to keep in touch with each other to keep a feeling of connection and share experiences.
Wider view of social activities using technology	Family members have their own primary work. Do not want to be disturbed during primary work.
Communication among the small, local, private community	Family members already have close relationships.

TABLE 5.3
Some Tasks and Activities of College Professors

Prepare class material
Participate in faculty meeting
Grade midterm and final
Go to lunch
Teach class
Hold research meeting
Make a cup of tea
Compose research paper
Develop final exam
Write grant proposal
Perform field work
Send FAX
Make a phone call
Read a research paper
E-mail
Review paper
Order materials
Discuss with students
Manage project
Conduct experiment
Hold research seminar
Plan business trip
Browse Web pages
Meet company people

The list is clearly not comprehensive. The purpose of creating it is to derive day-in-the-life scenarios as a touchstone of requirements elicitation. The participants were asked to recall and imagine a typical day of work, then each of them created a day-in-the-life scenario by picking tasks out of the task list. Because this scenario is a specific instance, tasks may require descriptions of assumptions and backgrounds, which are attached to the scenario. For example, when a college professor grades a midterm examination, this implies that the examination had been conducted. When she composes a research paper, we may assume that the paper deadline is imminent.

The two leftmost columns of Table 5.4 show a college professor's day-in-the-life scenario. Tasks are presented in chronological order, and the first column indicates the approximate time the corresponding task starts. The tasks in parenthesis ("arrive at office" and "leave office") are assumptions added by the analysts. Developing a day-in-the-life scenario like Table 5.4 reveals these assumed tasks by putting all the considered tasks in a wider context. In fact, working on a day-in-the-life scenario helps the analysts to focus on linkages between individual tasks and assumed, forgotten, or overlooked tasks that occur before or after them.

After creating the day-in-the-life scenario, we visited the professor's office to discover the context information for each task. During our visit, we added information about sites, artifacts, and participants to the original scenario. In addition, for each of the tasks, we asked ourselves what if our children want to contact to us as they did in Alison's episode. Based on this we judged the professor's availability during the tasks. Table 5.4 shows a day-in-the-life scenario with the context information. For each task, the Site column represents where the task is carried out; the Artifacts column shows the artifacts used; the Participants column lists other people involved in the task; and the Availability column specifies the priority of the task—no. 1

TABLE 5.4
A Day-in-the-Life Scenario of a College Professor with Context Information

Time	*Task*	*Site*	*Artifacts*	*Participants*	*Availability*
9:00	(Arrive at office)				
9:00	Make a cup of tea	Office	Teapot, cup, spoon		3
9:10	E-mail	Office	PC		2
9:50	Read a research paper	Office	Paper, notebook, marker, pen, PC		2
10:30	Teach class	Class room	OHP, transparency, marker, notebook	Students	1
12:00	Go to lunch	Café	Glass, knife, fork, spoon, plate, paper napkin	Friends	3
13:00	Grade midterm	Office	Answer sheet, pen, notebook, PC		2
14:30	Compose research paper	Office	PC, notebook, pen, dictionary, book, whiteboard, marker		1
16:30	Participate in faculty meeting	Meeting room	Notebook, pen, whiteboard, maker	Faculty members	1
18:30	Discuss with students	Lab	Whiteboard, marker	Students	1
19:30	(Leave office)				

does not accept interruption, no. 2 can accept interruption, and no. 3 accepts interruption. For example, at 10:30 the professor, who is the main actor in this scenario, starts teaching a class in a classroom. She makes use of an overhead projector and transparencies. She sometimes takes notes on a transparency with a marker while referring to her notebook. In the class, she teaches students: They are the main participants in teaching. When she teaches, she concentrates on the class so she does not want to be disturbed by anyone from outside except in an emergency situation. In contrast, at 12:00 she joins her friends for lunch. At the café, she uses a glass, knife, fork, spoon, plates, and paper napkin. During lunch, she does not mind if other people interrupt her: She accepts interruptions. In short, the college professor's tasks employ a PC as a primary artifact, and her availability depends on the task, not on the site. The identified elements are used to envision a problem scenario for Familyware.

The theme of mobility emerges from the analysis. Although most tasks in Table 5.4 are carried out in the professor's office, she does conduct tasks in various other places, such as a classroom, meeting room, and research laboratory. In fact, the assumed tasks, "arrive at office" and "leave office," involve movement. Furthermore, she might go on a business trip. In this situation, the travel time and distance might be longer. How to enable communicate during the moving process is a challenging theme of the root concept of Familyware. Table 5.5 shows an additional theme not included in Table 5.4. The identified themes are considered in order to create scenarios along with the root concept.

5.4.4 Scenario Development: Familyware Scenario

Based on the activity so far, we developed the following problem scenario. The tasks and themes identified in the detail domain are put in context so that they become prominently illustrated for evaluation. In this scenario, two family members, Wendy and Sean, are the main actors.

TABLE 5.5
An Additional Theme From the Day-in-the-Life Scenario of a College Professor

Theme	*Issues Contributing to the Theme*
Mobility	Family members do not remain in the same place.

They are separated by a long distance and try to share their feelings via their everyday artifacts. The scenario assumes that the Internet is used as a background communication channel.

> *Wendy Scenario*: Wendy, a 5-year-old girl, picks up her favorite teddy bear and takes it to a corner of her room. In the corner there is a TV, and when she approaches it, the TV turns on automatically. An electronic greetings card appears on the TV screen; it is from her dad. The card plays back his voice: "I am going to bring a puppy home with me!"
>
> Sean, Wendy's dad, is working at a software company in College Park, Maryland. He has been gone all week from his home in Blacksburg, Virginia. Wendy loves him and is looking forward to the weekend.
>
> Wendy is excited by the news: Dad is going to bring home a puppy! She holds her teddy bear tightly, then shakes it. These are Sean and Wendy's special actions—if she thinks of him, she squeezes and shakes her teddy bear. Though she is not aware of it, the teddy bear contains an electronic device that has a wireless connection to the Internet. The device senses being shaken and sends a signal to Sean's computer.
>
> In his office, Sean is composing a project report on his computer. On the screen there is a small window displaying a photo of Wendy. He notices that a big smile has appeared on it and understands that she is thinking of him. He also smiles and thinks about the coming weekend.

This problem scenario is used not only as a shared vision by the analysts and designers but also as key material for analysis at the detail level. A claims analysis is conducted to examine the practical features that have key implications for use in the problem scenario. The features include an electronic greetings card on TV, the Internet as a background communication channel, a teddy bear as an interactive device, and a photoframe as a window on the PC screen:

Electronic greetings card on TV
+ can be implemented by a conventional e-mail system
– but requires both the sender and receiver to have networked computers.

The Internet as a background communication channel
+ provides global access from anywhere in the world
– but might be expensive to use or might require authentication actions from the user.

Teddy bear as an interactive device
+ is an everyday object for children,
+ allows a young child to express his or her feelings by physical manipulation,
– but may not provide feedback that a message has been sent.

Photoframe as a window on the PC screen
+ is an everyday object for office workers,
+ could display photos of a parent's children,

+ could display different photos depending on the message received from the teddy bear via the Internet,

– but may not provide feedback that a message was received.

The claims add further details about the problem scenario, and the process of examining claims in the problem scenario poses questions for Familyware. For example, the claim related to the electronic greetings card on TV says that it can be implemented by a conventional e-mail system but requires both the sender and the receiver to have networked computers. This raises a general question about how to install the Familyware system at the sites. This question and others derived from the claims analysis process will be used in the next design step.

5.4.5 From Detail to Abstraction

The next step of our design activity obtains key abstraction-level design principles from the detail level. We held a design meeting on the Familyware concept to analyze the problem scenario. Table 5.6 shows the tasks and artifacts identified from the problem scenario. Most items, such as picking up a teddy bear, are directly obtained from the problem scenario, but the ones in italics are assumed tasks identified from the analysis. For example, the problem scenario does not mention how the father sends the electronic card; in addition, it does not define how to install the Familyware artefacts.

The analysis continues to derive an abstraction for each task. Table 5.7 lists pairs of abstraction and detail for the tasks. The child may install a Familyware artifact herself or may have it installed at her site for her. She and her father make a promise: They define a special action. She triggers the Familyware artifact: She picks up her teddy bear and takes it to a corner of the room. She notices a message from her father: She looks at an electronic greetings card on TV. After that, she sends a message to her father: She hugs the teddy bear and shakes it. In contrast, the parent may set up a Familyware artifact on his office computer. He and his daughter make a promise, as described above. He sends a message using the artifact: He sends an electronic card to his daughter. He has his primary work: He works on a project report using a computer. He notices a message as a change in the photo displayed on his computer's desktop.

Two issues emerged during the analysis: the symmetry or asymmetry of Familyware artifacts and consciousness of sending a message. For example, there is asymmetry in the teddy bear and the photoframe on PC system. The child uses the teddy bear as an input device and the

TABLE 5.6
Tasks and Artifacts From the Scenario Analysis

Child		*Parent*	
Task	*Artifact*	*Task*	*Artifact*
Install Familyware artifact		*Install Familyware artifact*	
Define a special action	Teddy bear	Define a special action	Teddy bear
Pick up teddy bear and take it to a corner of the room	Teddy bear	*Send electronic card*	
Look at electric card	TV monitor	Compose project report	PC
Hug teddy bear and shake it	Teddy bear	Glance at photo	PC

TABLE 5.7
Detail Versus Abstraction Analysis of the Problem Scenario

Child		*Parent*	
Detail	*Abstraction*	*Detail*	*Abstraction*
Install Familyware artifact	Install Familyware	*Install Familyware artifact*	Install Familyware
Define a special action	Make promises	Define a special action	Make promises
Pick up teddy bear and take it to a corner of the room	Trigger Familyware	*Send electronic card*	Send messages
Look at electric card	Notice messages	Compose project report	Primary work
Hug teddy bear and shake it	Send messages	Glance at photo	Notice messages

TV monitor as a display, whereas the parent uses the PC as an input and display device. A symmetrical interface is also possible for Familyware artifacts. A necklace-style peripheral display, for example, has symmetric features and uses. The necklace Familyware is intended to support communication between close friends (e.g., boyfriend and girlfriend). Each wears a necklace that can send to and receive from the partner a simple signal, such as the temperature of the mounted stone (Go et al., 2000). It can be implemented by using wireless technologies such as radio transmission or a digital cellular system to transmit the signal.

The style of communication may depend on the degree of consciousness involved in sending the message. If the child is conscious of sending a message to her father when she holds and shakes her teddy bear, then the communication will succeed after he has noticed the message. If she is not conscious of sending messages to her father, then the communication is in fact one-way communication: Only her father knows the meaning of the communication.

These issues contribute to articulating the concept of Familyware. They allow us to consider what-if questions when we create more scenarios: What if Familyware artefacts have symmetry and what if they do not involve conscious message sending? These questions lead to other groups of Familyware artefacts. For example, the necklace interface for a boyfriend and girlfriend is a design solution to the symmetry issue, and a rattle-photoframe interface handles the unconscious actions of a baby (for further discussion, see Go et al., 2000). These identified issues are used to envision further scenarios for Familyware use, together with the root concept and the issues already recognized at the initial analysis.

5.4.6 Envisioning Further Familyware Uses

The design process continues to generate new scenarios from the concept developed so far. The example here illustrates how to conduct scenario exploration. Initially, it assumes that the main actors are a company worker and his daughter, as described in the problem scenario, but other actors are added if necessary. Although the company worker's main task is to compose a project report using a PC, he notices that his Familyware artefact has changed status.

At the beginning of the scenario exploration, the analysts take notes about the basic tasks on scenario slips. In this example, there are five basis tasks: setting up Familyware, making promises, triggering Familyware, sending messages, and receiving messages. Apart from the Familyware-related tasks, the family members have their primary work.

The analysts focus first on the message-sending task. They gather episodes, examples, and anecdotes. A straightforward tactic is to describe the task based on the problem scenario. Thus, they write down how the teddy bear is being used on a scenario slip: Wendy holds her teddy bear and shakes it, then a message is sent to her father. This simple scenario involves assumptions, such as that a teddy bear is likely to be an everyday artefact for a young girl. Challenging the assumption allows designers to ask the question, What if the actor is a boy? A boy might have a teddy bear too, but there are alternatives.

The analysts examine everyday objects for boys while considering the root concept of Familyware. They identify a portable game machine as a boy's everyday object, then they compose a corresponding scenario. A boy has a portable game machine (e.g., Nintendo GameBoy). When he uses it, he sends a message to his father in much the same way as in the teddy bear case. The issue of being conscious of sending a message gives rise to another scenario at this point. The analysts move their focus to the situation in which a boy is not aware of sending a message. They set up a scenario in which a boy pedals his bicycle. The boy loves his bicycle and usually plays with it outside his house. As he pedals it, information about its wheel rotation speed is sent to his parent as a message. Another question related to the gender issue in the Wendy scenario is what gender-free artifact is possible for kids. The analysts choose shoes as possible artifacts. A small kid has a favorite pair of shoes and she wears them almost everyday. At each step, the shoes send a signal to her parent by means of wireless mobile technology.

In contrast, discussion about the message-noticing task tries to articulate an alternative use situation for a Familyware artifact in the Wendy scenario. The analysts set up a commuting situation. A father commutes by car everyday and sometimes drives his car on a business trip. He notices that his car navigation system is displaying an e-mail message from his daughter. This radical change emerges if the analysts challenge an assumption of the Wendy scenario, namely, that Wendy's father stays in his office. If he is not in his office, he may be doing different tasks. Setting up the new situation generates this straightforward question: What items can provide messages without disturbing the driver? Possible channels include sound—music from the car radio or the sound of the car's engine. Another channel is the illumination of dashboard instruments. Part of the scenarios and questions is summarized below. The indentation represents the depth of scenarios and questions.

Basic scenario: Sending a message.

Scenario: Wendy holds her teddy bear and shakes it; this causes a message to be sent to her father.

Question: What if the actor is a boy?

Scenario: A boy has a portable game machine, such as a Nintendo GameBoy; when he uses it, he sends a message to his father in much the same way as when he plays a game.

Scenario: A boy loves his bicycle. He usually plays with it outside his house. As he pedals his bicycle, information about its wheel rotation speed is sent to one of his parents as a message.

Question: What everyday artifacts for kids are gender-free?

Scenario: A small kid has a favorite pair of shoes and wears them almost everyday. At each step, the shoes send a signal to one of her parents by means of wireless mobile technology.

Basic scenario: Noticing a message.

Scenario: A girl's father commutes by car everyday. He sometimes drives his car on a business trip. He notices that his car navigation system is displaying an e-mail message from his daughter.

Question: What items do not disturb a driver?

Scenario: He is listening to music while driving. It is jazz music, but it smoothly changes to rock music. He notices this change and understands that his son has sent him a message.

Question: What if he does not listen to music while driving?

Scenario: While driving, he notices that the sound of the car's engine has become a bit higher. From this sign, he is aware of his son's message.

Scenario: The dashboard instruments are illuminated while he is driving. He notices that the illumination color has changed from light blue to light green and recognizes that his son has sent him a message.

Although the scenario exploration seeks to create as many scenarios as possible so that the analysts can pinpoint hidden requirements behind the concept, an obvious question is where to stop. A stopping heuristic may be the point where scenarios produce a general usability rationale contributing to the concept of the target system. Another stopping heuristic is related to time—because analysts and designers conduct a scenario exploration session as collaborative work, their time is expensive. The session should have a time limit. Two hours seems realistic.

After this session, the analysts sketch potential artifacts in order to develop and concretize their further possibilities. After considering the above scenario-question threads, they choose the shoes scenario for sending messages and the music-in-a-car scenario for noticing messages. They sketch simple interfaces and conduct a claims analysis.

A pair of shoes, which sends signals
+ is an everyday artefact for anybody,
+ can be used virtually anytime,
– but might be heavy for kids,
– and might cause a privacy issue.

The music in a car that conveys a kid's message
+ can be used while driving,
– but might cause a safety issue.

These issues provide feedback to the concept at the abstraction level in order to articulate it. More iterative cycles of articulating the concept and detailing the scenarios might be needed to develop them further.

Figure 5.3 summarizes the example in this section. At the abstraction level, the root concept evolved in the sense that more issues and themes were added to articulate it. At the detail level, in contrast, several scenarios were derived to implement the concept with issues and themes and generate further issues and themes. The process started with the development of the root concept. Then, in order to verify the root concept, the analysts collected real-world episodes such as the telescope scenario and the phone call scenario. The day-in-the-life scenario of a college professor was also envisioned to identify assumed tasks. These activities produced the themes of the root concept. The obtained theme and the root concept were used to create the Wendy scenario. This problem scenario was analyzed to obtain further issues and themes. These were added to the root concept to articulate it. Again, the detailed scenarios that implemented the concept with the themes were envisioned to identify further issues and themes, placing them in a new context. The intertwining process of creating scenarios, verifying concepts, and

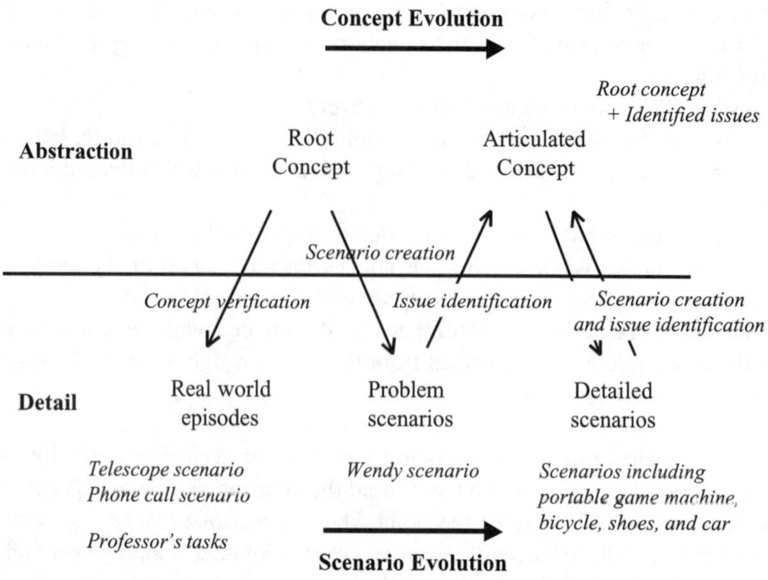

FIG. 5.3. A summary of the Familyware example.

identifying issues between the abstraction and the detail promoted the concept and scenario evolution.

5.5 CONCLUSION

Scenario-based design and traditional task analysis have common properties: They both focus on users, concrete descriptions, and specific instances. Scenario-based design in human-computer interaction puts a strong emphasis on users. It articulates who the users are, what they do with a technology and how they do it, and what the circumstances are in which they work. A scenario is a concrete description of work and activities, so it describes a specific instance and usage situation. Similarly, the user-centered view is taken by task analysis, too. Unlike system analysis, task analysis puts the stress on users and their method of use. It obtains a concrete description of tasks. For example, Hierarchical Task Analysis (HTA) creates a task hierarchy and decomposed subtasks, including plans for subtask execution. The final step of the decomposition may enumerate specific keystrokes such as pressing the "Delete" key or clicking the "Cancel" button. In this sense, task analysis also considers specific instances of tasks.

In contrast, scenario-based design and task analysis have different perspectives: Scenario-based design is more oriented toward inquiry, and task analysis is more often oriented toward to obtaining a single, correct description, although Diaper (chap. 1) argues against both the "single" and the "correct" concepts. Scenarios articulate a view from specific instances. When they are employed to describe current uses of a technology, these may be regular uses or rare ones. When they are employed to describe potential uses of a technology, these might or might not be realized in the future. Thus, in a sense, the limited number of scenarios do not describe the complete uses of a technology. Rather, scenarios focus on exemplifying a concept in the abstraction domain as a rich, concrete description in the detail domain so that designers and analysts, and even users, can share a vision, knowledge, and experience of the project. The shared scenarios provide a chance to generate question and answers regarding the situation described in the scenarios. The scenarios also allow those people to contribute to the concept—making a contribution of their vision, knowledge, and experience—because new scenarios added to the previous ones provide feedback to the concept. As the concept behind a scenario evolves, the scenario instantiating it evolves; thus, the scenario contributes to the evolution of the concept.

Task analysis is appropriate for producing a precise, correct description by analyzing the current use of an existing technology. To design a training manual for a system, correct procedures of tasks performed on the system should be described. A step-by-step description a task produced by a task analysis method offers the task information needed for the training manual. In addition, it can be used to estimate the time required to complete the task, a typical application of methods such as GOMS (chap. 4). These cases require a precise, correct description of the task. To examine knowledge transfer from a previous system to a new system, the procedures and knowledge for completing a task on the previous system should be articulated and analyzed. The knowledge description can be used to evaluate how easily the users of the previous system learn the new system. This also requires a correct description. The description produced by a task analysis method reflects the existing use of a system but may suggest future uses of it.

The scenario-based approach described in this chapter can enhance the benefits of of task analysis. Following the performance of a task analysis using HTA, for example, scenarios add various views to the results, allowing designers and analysts to explore new possibilities of use. Although this chapter concentrates on scenario elicitation through interviews, scenario development through brainstorming, and relatively heuristic analyses of scenarios, the robust and detailed data obtained from the task analysis of current use situations may be a rich source of information for the scenario development process. The scenarios created through the process can be analyzed using task analysis techniques even though, as mentioned, the envisioned scenarios are not exhaustive. Because they are not exhaustive, the scenario development activities help designers and analysts to identify implicit tasks that may be assumed, forgotten, or ignored and to take into account the relationships among tasks. Overall, scenarios help to expand and create design ideas based on the results of task analysis.

REFERENCES

Beyer, H., & Holtzblatt, K. (1998). *Contextual design.* San Francisco: Morgan Kaufmann.

Carroll, J. M. (1995). *Scenario-based design: Envisioning work and technology in system development.* New York: Wiley.

Carroll, J. M. (2000). *Making use: Scenario-based design of human-computer interactions.* Cambridge, MA: MIT Press.

Carroll, J. M., & Rosson, M. B. (1992). Getting around the task-artefact cycle: How to make claims and design by scenario. *ACM Transactions on Information Systems, 10,* 181–212.

Carroll, J. M., & Rosson, M. B. (1996). Developing the Blacksburg Electronic Village. Communications of the ACM, *39(12)*, 69–74.

Cooper, A. (1999). *The inmates are running the asylum: Why high-tech products drive us crazy and how to restore the sanity. Indianapolis,* IN: SAMS.

Diaper, D. (1989). *Task observation for human-computer interaction.* In D. Diaper, (Ed.) *Task analysis for human-computer interaction* (pp. 210–237). Chichester, England: Ellis Horwood.

Diaper, D. (2002). Scenarios and task analysis. *Interacting With computers, 14,* 379–395.

Go, K., & Carroll, J. M. (2003). Blind men and an elephant: Views of scenario-based system design. *ACM Interactions, forthcoming.*

Go, K., Carroll, J. M., & Imamiya, A. (2000). Familyware: Communicating With someone you love. In A. Sloane, & F. van Rijn (Eds.), *Home informatics and telematics: Information, technology and society* (pp. 125–140). Boston: Kluwer Academic Publishers.

Jarke, M., Bui, T. X., & Carroll, J. M. (1998). *Scenario management: An Interdisciplinary approach. Requirements Engineering, 3*(3-4), 155–173.

Johnson, J. (2000). *GUI bloopers: Don'ts and do's for software developers and web designers.* San Francisco: Morgan Kaufmann.

McGraw, K., & Harbison, K. (1997). *User-Centered Requirements: The scenario-based engineering process.* Mahwah, NJ: Lawrence Erlbaum Associates.

Potts, C., Takahashi, K., & Anton, A. I. (1994). *Inquiry-based requirements analysis. IEEE Software, 2*(11), 21–32.

Rosson, M. B., & Carroll, J. M. (2001). *Usability engineering: Scenario-based development of human-computer interaction.* San Francisco: Morgan Kaufmann.

Shepherd, A. (1989). Analysis and training in information technology tasks. In D. Diaper, (Ed.), *Task analysis for human-computer interaction* (pp. 15–55). Chichester, England: Ellis Horwood.

6

Comparing Task Models for User Interface Design

Quentin Limbourg
Université catholique de Louvain

Jean Vanderdonckt
Université catholique de Louvain

Many task models, task analysis methods, and supporting tools have been introduced in the literature and are widely used in practice. With this comes need to understand their scopes and their differences. This chapter provides a thorough review of selected, significant task models along with their method and supporting tools. For this purpose, a meta-model of each task model is expressed as an Entity-Relationship-Attribute schema (ERA) and discussed. This leads to a comparative analysis of task models according to aims and goals, discipline, concepts and relationships, expressiveness of static and dynamic structures. Following is discussion of the model with respect to developing life cycle steps, tool support, advantages, and shortcomings. This comparative analysis provides a reference framework against which task models can be understood with respect to each other. The appreciation of the similarities and the differences allows practioners to identify a task model that fits a situation's given requirements. It shows how a similar concept or relationship translates in different task usage models.

6.1 INTRODUCTION

User-Centered Design (UCD) has yielded many forms of design practices in which various characteristics of the context of use are considered. Among these, task analysis is widely recognized as one fundamental way not only to ensure some user-centered design (Hackos & Redish, 1998) but to improve the understanding of how a user may interact with a user interface to accomplish a given interactive task. A task model is often defined as a description of an interactive task to be performed by the user of an application through the application's user interface. Individual elements in a task model represent specific actions that the user may undertake. Information on subtask ordering as well as conditions on task execution is also included in the model.

Task analysis methods have been introduced from disciplines with different backgrounds, different concerns, and different focuses on task. The disciplines include:

Cognitive psychology or *ergonomics* (Stanton & Young, 1998). Task models are used to improve the understanding of how users may interact with a given user interface for carrying out a particular interactive task. Task analysis is useful for identifying the cognitive processes (e.g., data manipulation, thinking, problem solving) and structures (e.g., the intellectual skills and knowledge of a task) exploited by a user when carrying out a task and for showing how a user may dynamically change them as the task proceeds (Johnson, Diaper, & Long, 1984). It can also be used to predict cognitive load and rectify usability flaws.

Task planning and allocation. Task models are used to assess task workload, to plan and allocate tasks to users in a particular organization, and to provide indicators to redesign work allocation to fit time, space, and other available resources (Kirwan & Ainsworth, 1992).

Software engineering. Task models can capture relevant task information in an operational form that is machine understandable (see chap. 1). This is especially useful where a system needs to maintain an internal task representation for dynamic purposes, such as to enable adaptation to variations in the context of use (Lewis & Rieman, 1994; Smith & O'Neill, 1996).

Ethnography. Task models can focus on how humans interact with a particular user interface in a given context of use, possibly interacting with other users at the same time (see chap. 14).

Existing task models show a great diversity in terms of formalism and depth of analysis. They also are used to achieve a range of objectives (Bomsdorf & Szwillus, 1998, 1999):

- *To inform* designers about potential usability problems, as in HTA (Annett & Duncan, 1967; chap. 3).
- *To evaluate* human performance, as in GOMS (Card, Moran, & Newell, 1983; chap. 4).
- *To support design* by providing a detailed task model describing task hierarchy, objects used, and knowledge structures, as in TKS (Johnson, 1992; chap. 15) or CTT (Paternò, 1999; chap. 24).
- *To generate* a prototype of a user interface, as in the Adept approach (Wilson & Johnson, 1996; chap. 24).

These different objectives give rise to many interesting concepts. For example, task analysis methods used in cognitive analysis go beyond the goal level to analyze the cognitive workload, execution time, or knowledge required to carry out a set of tasks. In this respect, they are similar to user modeling. On the other hand, methods intended to support cooperative work have developed formalisms to represent how tasks are assigned to different roles, broadening the scope of task analysis with organizational concepts. This situation leads to a series of shortcomings (see also chap. 30):

Lack of understanding of the basic contents of each individual task model, including the rationale behind the analysis method, the concepts, their relationships, their vocabularies, and the intellectual operations involved.

Heterogeneity of the contents of different models due to the variety of methods employed by different disciplines. The heterogeneity encompasses different focuses, different vocabularies, and different definitions of what a task model may be.

Difficulty in matching the contents of two or more analyses done using different task analysis models. Sometimes no matching between contents can be established at all. As each method has its own vocabulary, it is difficult to relate the vocabularies of different methods to see what they cover and what they do not.

Lack of interoperability of systems. Since task-modeling tools do not share a common format, they are restricted to those task models expressed according to their accepted formats.

Reduced communication among task analysts. Owing to the lack of interoperability, a task analyst may experience trouble in communicating the results of a task analysis to another analyst or any other stakeholder. In addition, any transition between two persons may generate inconsistencies, errors, misunderstandings, or inappropriate modeling.

Difficulty in identifying frontiers of task analysis. Some methods are more focused on some specific aspects of the task to be represented whereas others tend to go beyond the limits of the task by encompassing parameters external to the task yet relevant to the context of use.

To address the above shortcomings, this chapter pursues two major goals. The first, *a theoretical goal*, is to provide a deep conceptual and methodological understanding of each individual task model and its related approach. The second, *an ontological goal*, is to establish semantic mappings between the different individual task models so as to create a transverse understanding of their underlying concepts independently of their particularities. This goal involves many activities such as vocabulary translation, expressiveness analysis, degree of details, identification of concepts, emergence of transversal concepts, and task structuring identification.

6.2 A REVIEW OF TASK MODELS

In general, any task analysis method may involve three related poles:

1. *Models*. A model captures some facets of the problem and translates them into specifications.
2. *A stepwise approach*. An approach in which a sequence of steps is used to work on models.
3. *Software tools*. A software tool supports the approach by manipulating the appropriate models.

Any task analysis method may contain representative parts that fall within each of these three poles. This comparative study is focused on the first pole, that is, on models. It is assumed that the structuring of a method's steps for modeling tasks should remain independent of the task model's contents. Therefore, the methodological part of each task model was taken to fall outside the scope of our analysis. A tool clearly facilitates the task modeling activity in hiding the model notation from the analyst and helping him or her capture it, edit it for any modification, and exploit it for future use (e.g., task simulation, user interface derivation). Most models presented below are software supported.

Lots of task models have been proposed in the literature, and some are described in this handbook: CTA (chap. 15 and 16), CTT (chap. 24; Paternò, 1999), Diane+ (chaps. 22 and 26; Barthet & Tarby, 1996; Lu, Paris, & Vander Linden, 1998), GOMS (chap. 4; John & Kieras, 1996), GTA (chaps. 7 and 24; van Welie & van der Veer, 1998; van Welie, van der Veer, & Eliëns, 1998), HTA (chap. 3; Annett & Duncan, 1967; Shepherd, 1989, 1995), MAD* (chaps. 22 and 24; Scapin & Pierret-Golbreich, 1989; Gamboa & Scapin, 1997; Scapin & Bastien, 2001), MUSE (Lim & Long, 1994), TAG (Payne & Green, 1986), TAKD (Diaper, 1989a),

TKS (Johnson, 1992; see also Chapter 15), TOOD (chap. 25; Mahfoudhi, Abed, & Tabary, 2001).

A subset of task models were selected to reflect the disciplinary variety. Each significant discipline is represented by one member. The set is also intended to reflect the geographical range of the task models. In addition, each selected task model is integrated in a development methodology as a core or side element, and each has been submitted to experimental studies to assess its validity.

After selecting the models, we analyzed the foundation papers on each. We then decomposed each model into constituent concepts using an entity-relationship-attribute (ERA) method of analysis so as to obtain a task metamodel (see the appendix at the end of this chapter). Task models invoke concepts at different levels of importance. A concept that is similar across methods can be modeled as an entity, a relationship, or even as an attribute. We decided that these concepts should be represented in a consistent manner. For example, a temporal operator was always represented as an entity to recognize an equal importance of this concept throughout the different task metamodels. Each pertinent concept was then precisely defined and commented on. The initial terminology of the originating papers was kept for naming identified concepts.

6.2.1 Hierarchical Task Analysis (HTA)

Hierarchical Task Analysis (HTA; Annett & Duncan, 1967; chap. 3) was a pioneering method of task analysis. It was primarily aimed at training users to perform particular tasks. On the basis of interviews, user observation, and analysis of existing documents (e.g., manuals, documentation), HTA describes tasks in terms of three main concepts (Fig. 6.1): tasks, task hierarchy, and plans. Tasks are recursively decomposed into subtasks to a point where subtasks are allocated either to the user or the user interface, thus becoming observable. The task hierarchy statically represents this task decomposition. The decomposition stopping criterion is a rule of thumb referred to the $p \times c$ rule. This criterion takes into account the probability of a nonsatisfactory performance and the cost of a nonsatisfactory performance (i.e., the consequences it might produce).

Since the task hierarchy does not contain any task ordering, any task should be accomplished according to a plan describable in terms of rules, skills, and knowledge. A plan specifies an ordering in which subtasks of a given task could be carried on, thus acting as a constraint on task performance.

A plan is provided for each hierarchic level. Although the plan is an informal description of temporal relationships between tasks, it is one of the most attractive features of HTA, as

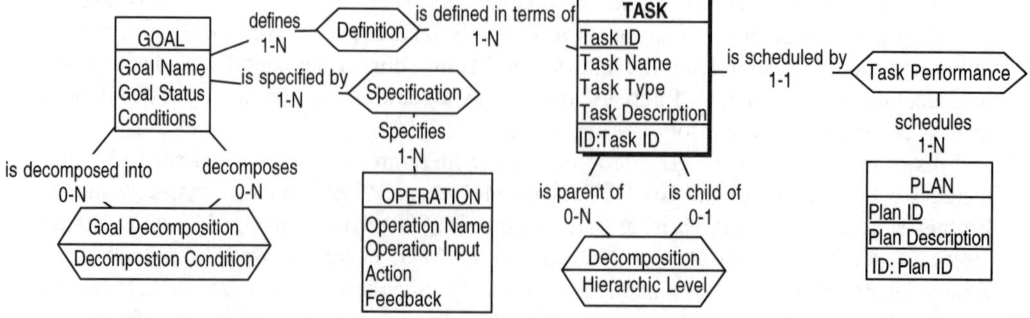

FIG. 6.1. The HTA task meta-model.

it is both simple and expressive. Plans are very close to textual description or to the activity list (chap. 1) of traditional task analysis. One advantage of plans is that they do not create any artificial tasks, as some formal notations force analysts' to do to avoid ambiguous specification. On the other hand, because plans are informal, it is not possible to apply automatic checking of properties such as consistency and reachability.

Any task can be expressed in terms of goals that are reached when the corresponding task is accomplished. Each goal has a status (i.e., latent or active) and conditions to be satisfied. The advantage here in HTA is that goals are independent of the concrete means of reaching them. Therefore, for each goal at any level of decomposition, For each goal, several different operations for reaching the goal can be imagined and specified. Each operation is consequently related to a goal (or goals) and is further specified by the circumstances in which the goal is activated (the input), the activities (action) that contribute to goal attainment, and the conditions indicating the goal has been attained (feedback).

HTA provides a graphical representation of labeled tasks and a plan for each hierarchic level explaining the possible sequences of tasks and the conditions under which each sequence is executed. HTA also supports task analysis for teamwork, as described in Annett, Cunningham, and Mathias-Jones (2000; see also chap. 3).

6.2.2 Goals, Operators, Methods, and Selection rules (GOMS)

Card et al. (1983) developed GOMS as an engineering model for human performance to enable quantitative predictions (chap. 4). By incorporating tables of parameter values that rely on a cognitive architecture, GOMS can be used as an engineering approach to task design (Beard, Smith, & Denelsbeck, 1996). The original GOMS model, referred as CMN-GOMS (Card et al., 1983), is the root of a family of models that were elaborated later (John & Kieras, 1996), such as GOMSL (GOMS language) and CPM-GOMS (Critical Path Method GOMS). Although the first uses a "mental programming language" and is based on a parallel cognitive architecture, the second uses a PERT chart to identify the critical path for computing execution time (Baumeister, John, & Byrne, 2000).

In GOMS, the concept of a *method* is essential, as methods are used to describe how tasks are actually carried out (Fig. 6.2). A method is a sequence of operators that describes task performance. Tasks are triggered by goals and can be further decomposed into subtasks corresponding to intermediary goals. When several methods compete for the same goal, a selection rule is used to choose the proper one.

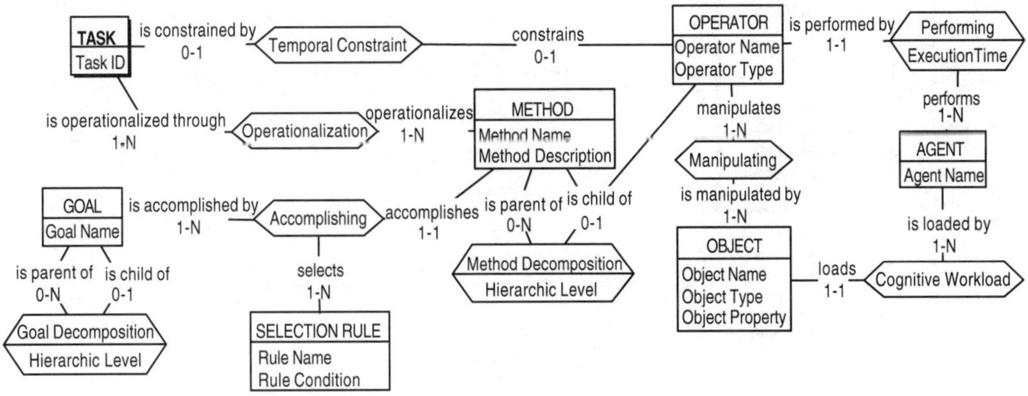

FIG. 6.2. The GOMS task meta-model.

Methods describe how goals are actually accomplished. Higher level methods describe task performance in terms of lower level methods, operators, and selection rules. The lowest level of decomposition in GOMS is the unit task, defined by Card et al. (1983) as a task the user really (consciously) wants to perform. Higher level methods use task flow operators that act as constructors controlling task execution.

GOMS makes a clear distinction between tasks and actions. First, task decomposition stops at unit tasks. Second, actions that in GOMS are termed operators are specified by the methods associated with unit tasks. Action modeling varies depending on the GOMS model and the method specification. Operators are cognitive and physical actions the user has to perform in order to accomplish the task goal. Since each operator has an associated execution time (determined experimentally), a GOMS model can help in predicting the time needed to perform a task.

Actions undertaken by the user are specified using external and mental operators. Some special mental operators are flow-control operators that are used to constrain the execution flow. Although the granularity varies according to the purpose of the analysis, it is clear that GOMS is mainly useful when decomposition is done at operational level (i.e., under the unit task level).

6.2.3 Groupware Task Analysis (GTA)

GroupWare Task Analysis (GTA) was developed by van der Veer, Lenting, and Bergevoet (1996) as a means of modeling the complexity of tasks in a cooperative environment. GTA has roots in ethnography; It is applied to design cooperative systems, and is based on activity theory. It adopts a clear distinction between tasks and actions.

An ontology describing the concepts and relations of the method has been proposed by van Welie, van der Veer, and Eliens (1998). The five central concepts are task, role, object, agent, and event (Fig. 6.3). In GTA, complex tasks are decomposed into unit tasks (Card et al., 1983) and basic tasks (Tauber, 1990). There is no indication how this relates to user interface design.

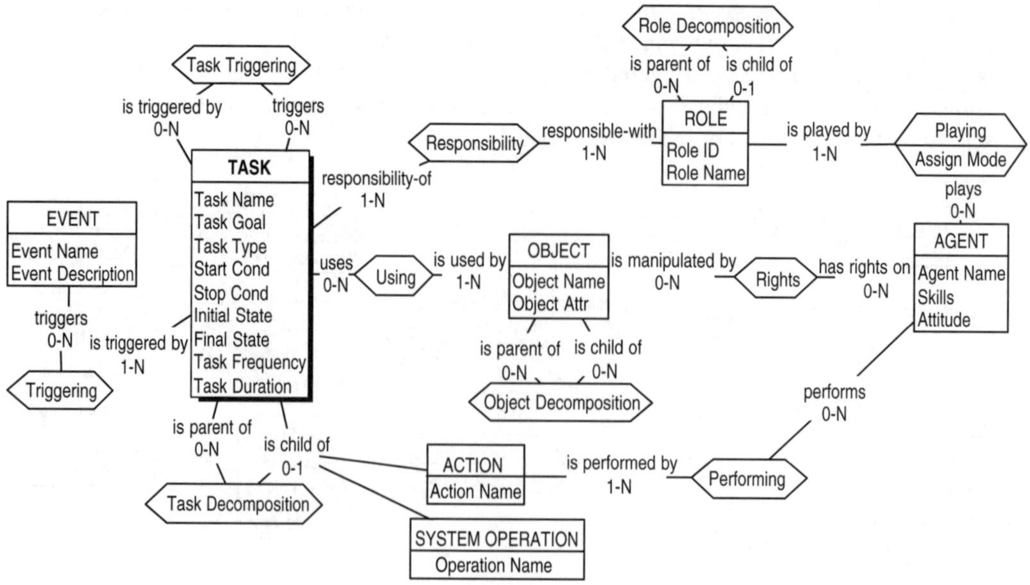

FIG. 6.3. The GTA task meta-model.

More recently, techniques have been added (van Welie, van der Veer, & Eliens, 1998), including a user action notation for the decomposition of basic tasks. An attractive feature of GTA is its capability of representing cooperative tasks (groupware). The representation is done by integrating the role concept in the task world and enabling representation of tasks sets for which a role is responsible of organizational aspects, such as how a role is assigned to different agents.

Although the ontology defined for GTA improves the conceptualization of the task world, the representation is not based on an adequate formalism. For instance, goals and actions are represented as attributes of the task and not as concepts. This is somewhat inconsistent with the fact that GTA allows for a goal to be reached in many ways. Also, since the same action can be used in many tasks, actions are better represented as a main concept. This way, object manipulation is represented as a relationship between actions and objects rather than between tasks and objects. Since actions depend on operational conditions, when different objects are used, different actions may be needed. Any task can also trigger another one. Euterpe, the software tool supporting GTA, allows the specification of constructors in a way similar to MAD*. Like in MAD*, parent-child constructors may lead to artificial tasks to satisfy the temporal constraints.

6.2.4 ConcurTaskTrees (CTT)

ConcurTaskTrees (CTT), a model developed by Paternò (1999; chap. 24), is based on five concepts: tasks, objects, actions, operators, and roles (Fig. 6.4). CTT constructors, termed *operators*, are used to link sibling tasks on the same level of decomposition. In this respect, CTT differs from previously described models, where operators act on parent-children relationships. It is also important to note that CTT has a formal definition for its temporal operators. In addition, CTT provides the means to describe cooperative tasks. To describe such a task, the task model is composed of different task trees, one for the cooperative part and one for each role that is involved in the task. Tasks are further decomposed up to the level of basic tasks, which are defined as tasks that could not be further decomposed. Actions and objects are specified for each basic task. Objects could be perceivable objects or application objects. Application objects are mapped onto perceivable objects in order to be presented to the user.

An interesting feature of CTT is that both input actions and output actions associated with an object are specified. Object specification is mainly directed toward the specification of interaction objects (interactors). The last modification made to CTT was to add the concept of platform in order to support multiplatform user interface development. A task can be associated

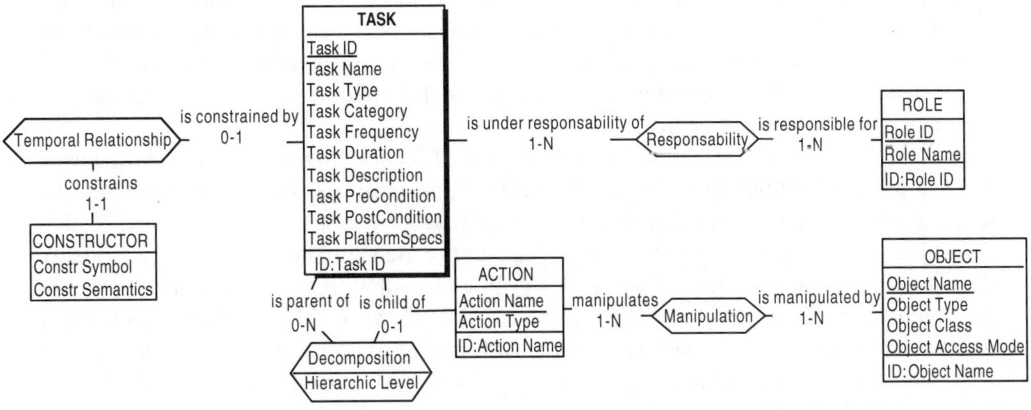

FIG. 6.4. The CTT task meta-model.

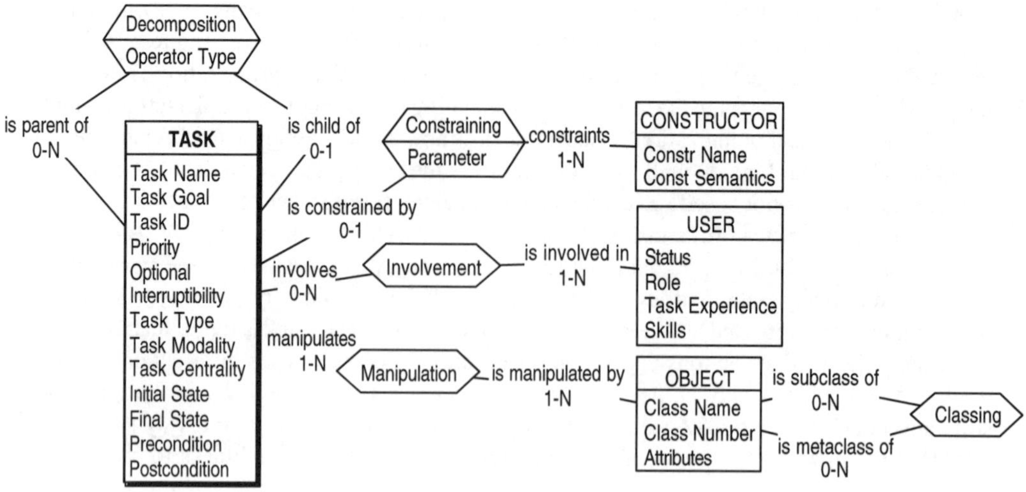

FIG. 6.5. The MAD* task meta-model.

with one or several previously defined platform descriptions. Views on the task model are obtained by filtering the model with respect to one or several platforms. CTT uses a tool for building the task model that specifies tasks, roles, and objects as well as creates a task hierarchy with temporal operators.

6.2.5 Méthode Analytique de Description de tâches (MAD*)

*M*éthode *A*nalytique de *D*escription de tâches (MAD*; Scapin & Bastien, 2001; Scapin & Pierret-Goldbreich, 1989; see also chaps. 22 and 24) provides object-oriented task specifications to support design phases. The main concepts of MAD* are task, constructor, user, and objects (Fig. 6.5). The task structure concept is implicitly represented in the decomposition relationship and the relationship with the constructor entity. As in GTA, tasks are divided into two broad categories: elementary tasks and composite tasks.

Elementary tasks are tasks that cannot be further decomposed. An elementary task contains direct reference to one or many domain objects that are manipulated by the task. A composite task is decomposed into children tasks by the use of different operators belonging to four categories: synchronization operators (i.e., sequence, parallelism, and simultaneity), ordering operators (i.e., AND, OR, and XOR), temporal operators (i.e., begin, end, and duration), and auxiliary operators (i.e., elementary or unknown). An elementary task is specified by an elementary "auxiliary" operator, and a task whose decomposition is not yet terminated is specified by an "unknown" operator.

In MAD*, a task, either composite or elementary, is characterized by several attributes: a name, a goal, an identifier (e.g., "Task 2.5.3 means third subsubtask of fifth subtask of second task"), a priority indicating preemptivity over other tasks, an optional character, a specification indicating whether the task can be interrupted, a type (i.e., sensorimotor, cognitive), a modality (i.e., manual, automatic, interactive), and a degree of centrality (i.e., important, somewhat important, secondary). The initial state specifies a state of the world prior to the execution of the task. The final state specifies a state of the world after the task execution. The goal consists of the object modifications that an agent wants to achieve. Although the goal is attached to the task as an attribute, the same goal can be reached by accomplishing different tasks. This fact motivated putting the goal concept into an entity type.

The precondition is a set of assertions constraining the initial state. Those assertions must be satisfied prior to the execution of the chosen task. Preconditions are classified as either sufficient conditions or as necessary and sufficient conditions. The postcondition is a set of assertions constraining the final state. The postcondition must be satisfied after the execution of the task. Any task can be linked to the user responsible for carrying out the task and to objects manipulated by the task. MAD* is supported by ALACIE, a task editor that allows analysts to input and manipulate their model (Gamboa & Scapin, 1997).

6.2.6 Task Knowledge Structure (TKS)

In the model known as the Task Knowledge Structure (TKS) method (Johnson & Johnson, 1989; chap. 15), the analysts manipulate a TKS, which is a conceptual representation of the knowledge a person has stored in his or her memory about a particular task (Fig. 6.6). A TKS is associated with each task that an agent (i.e., a user) performs. Tasks that an agent is in charge of are determined by the role the agent is presumed to assume. A role is defined as the particular set of tasks the agent is responsible for performing as part of his or her duty in a particular social context. An agent can take on several roles, and a role can be taken on by several agents. Even if tasks or TKSs may seem similar across different roles (e.g., typing a letter for a secretary and a manager), they will be considered as different. The "similarity" relationship is aimed to represent this situation.

The TKS for a task holds information about the task's goal, which is the state of affairs the task is intended to produce. A particular goal is accomplished by a particular task. A goal is decomposed into a goal substructure, which contains all intermediate subgoals needed to achieve it. Goal and task hierarchies are represented in Fig. 6.6 as two overlapping substructures, the first decomposing the goal into subgoals, the second decomposing tasks into subtasks. Each subgoal in the goal structure has a corresponding subtask in the task structure, and vice versa. The structure is composed either by task decomposition or by temporal or causal relationship mechanisms (*constructors*).

Constructors operate on tasks and by association on goals. The same goal can be reached by different subgoal sequencing. This leads to the concept of a plan. A plan is a particular arrangement of a set of subgoals and procedures for achieving a particular goal. As in other models, actions and objects are found at the lowest level of the task analysis. They are the

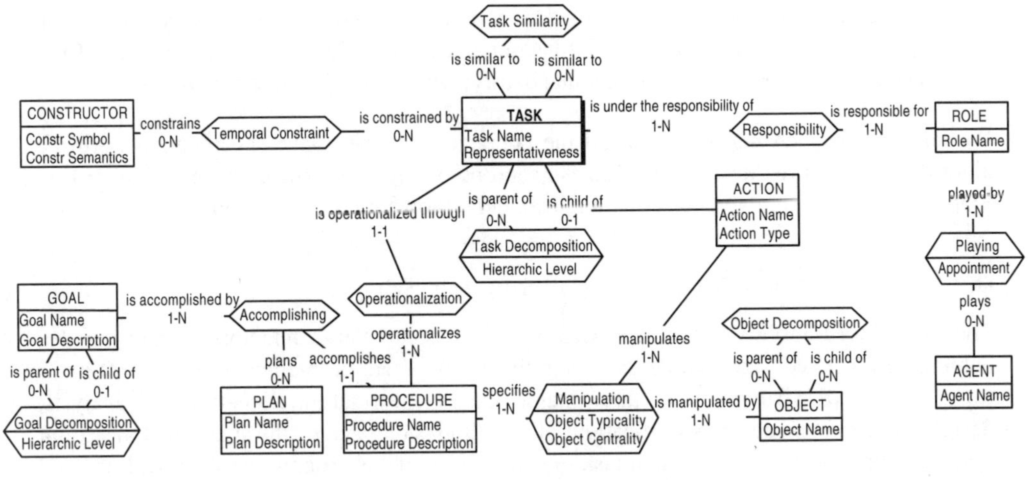

FIG. 6.6. The TKS task meta-model.

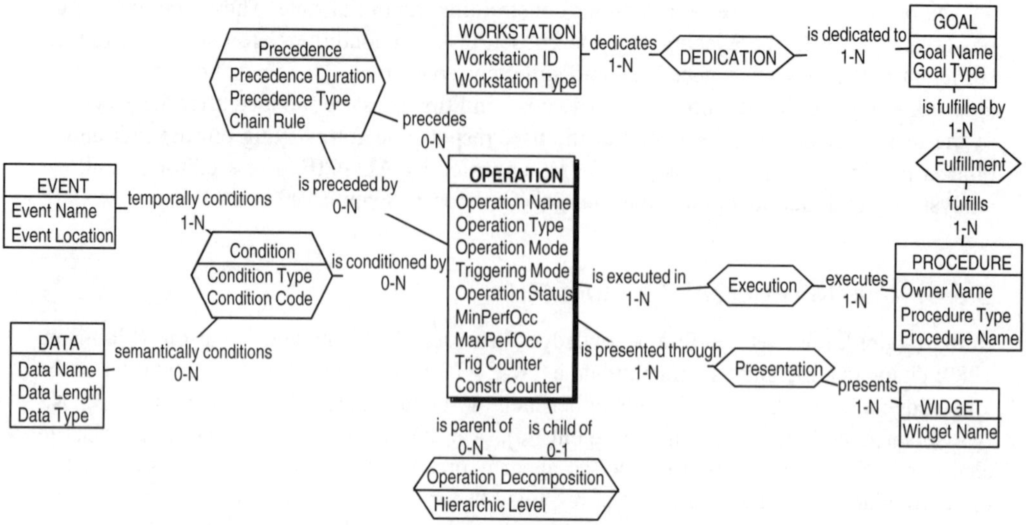

FIG. 6.7. The Diane+ task meta-model.

constituents of procedures that operationalize subtasks. One or several procedures can be linked to the same subtask. The TKS method proposes a production rule system for choosing the appropriate procedures for each context. Actions are directly related to a task tree and form its leaf level. Actions and objects have properties of interest. They can be central to the execution of a task and have typical instances. For example, the centrality and typicality of an object or action are always expressed with respect to a task or procedure that operationalizes the object or action. Note that objects are structured into a decomposition hierarchy.

6.2.7 Diane+

There are two important points to be made about the way in which Diane+ (Fig. 6.7) models a task (Barthet & Tarby, 1996; chap. 26):

1. The procedures describe only the characteristics specific to an application and do include the standard actions common to most applications, such as quit, cancel, and so on. This assumes that the supposed standard actions, previously defined, really apply to the application of interest. (If a standard action does not apply, this would be indicated.)
2. The described procedures are not mandatory; what is not forbidden is allowed.

We note that Diane+ can represent all the constraints of the above specifications. All the algorithmic structures do exist in Diane+, such as ordered sequence, unordered sequence, loop, required choice, free choice, parallelism, default operations, and so on.

6.2.8 Method for USability Engineering (MUSE)

The *M*ethod for *US*ability *E*ngineering (MUSE) is a structured human factors method aimed at helping designers consider human factors in the development of interactive software (Lim & Long, 1994). MUSE consists of three major phases: information elicitation and analysis, design synthesis, and design specification. The method is initiated by the analysis of the extant system, which results into an extant task model. This model is progressively transformed and augmented until eventually an interface model and display design is reached in the last phase.

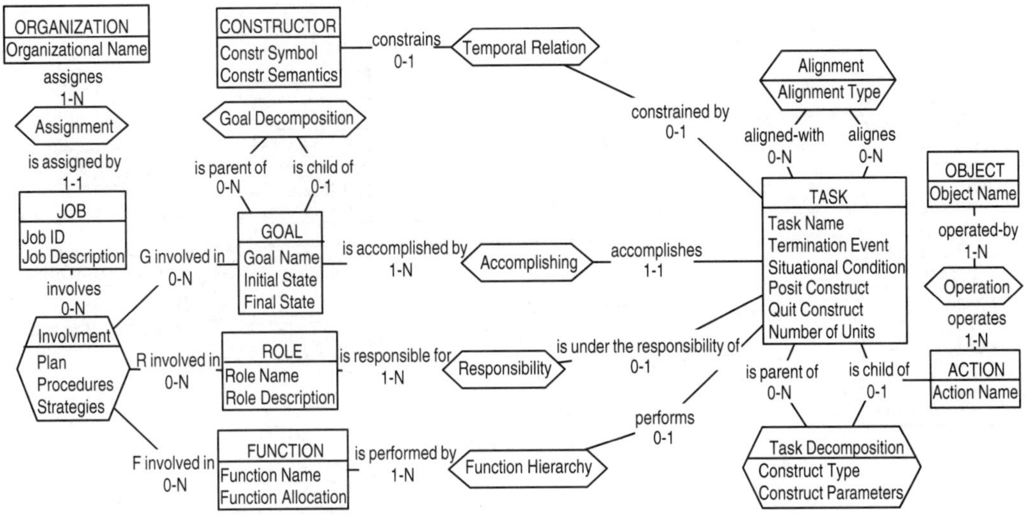

FIG. 6.8. The MUSE task meta-model.

The task model is described using task hierarchies covering several concepts. At the top of the hierarchy is the *organization*, which may be conceptualized as a supersystem to be decomposed into subsystems, called *jobs* (Fig. 6.8). Each job may involve three components:

1. A *goal hierarchy*, in which the main goal assigned to a job is recursively decomposed into subgoals, subgoals into subsubgoals, and so on. The main goal of a job describes the related subsystem as a transformation from its initial state to its final state.
2. A *role list*, in which roles assigned to a job are enumerated.
3. A *function list*, in which the specific functions required to perform the role are gathered. A function may be allocated to a human, a system, or both.

Each role and function appearing in the function lists is associated with a task, which is again hierarchically decomposed into subtasks, subsubtasks, and so on, down to the level of actions and objects. Each task is detailed with a termination event, a situational condition. The detection of unacceptable situational conditions may serve as a shortcut for terminating a path in the task tree. These constructs are used to describe uncertain events that may occur while the task is carried out. A number of units describes when a task is multiple (i.e., a instance of the same task carried out many times). Since the graphical representation of a MUSE task model is purely hierarchical, a single action composing multiple tasks or a single object operated by multiple actions may be reproduced. Other constructs related to the task decomposition and potentially requiring parameters are sequence, hierarchy, selection, iteration, concurrency, interleaving, multiplicity, posit, and quit. Figure 6.8 presents more concepts than are involved in the task modeling; the right part is the core component leading to more refined task models (Lim & Long, 1996).

6.2.9 Task Object-Oriented Description (TOOD)

Task Object-Oriented Description (TOOD) consists of an object-oriented method for modeling tasks in the domain of control processes and complex interactive systems, such as those used in air traffic control (Mahfoudhi, Abed, & Tabary, 2001; chap. 25). The method consists of

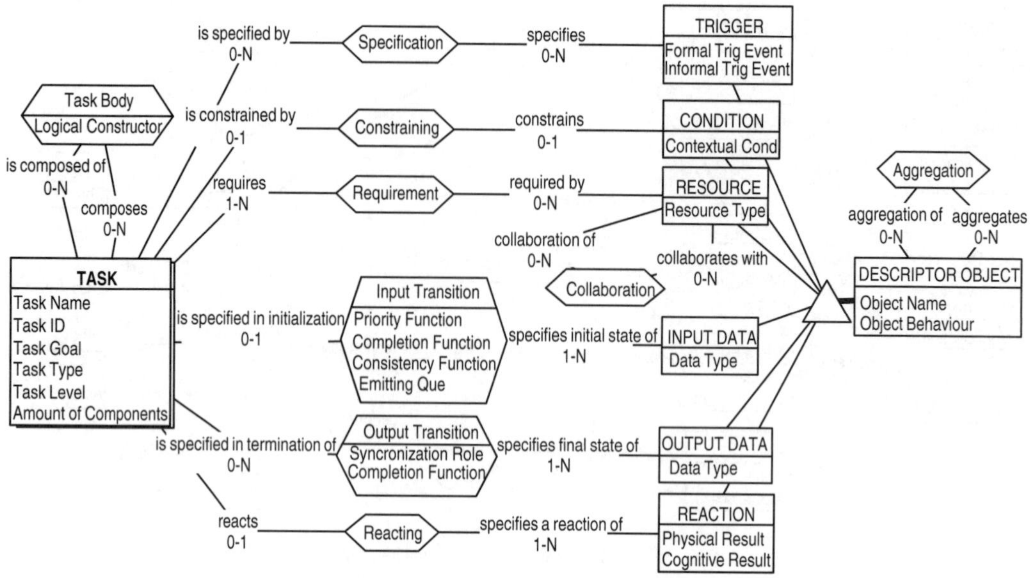

FIG. 6.9. The TOOD task meta-model.

four steps: hierarchical decomposition of tasks, identification of descriptor objects and world objects, definition of elementary and control tasks, and integration of concurrency (Fig. 6.9). Each task is treated as an instance of a task class identified by a name and an identifier and characterized by a goal, a type (i.e., human, automatic, interactive, and cooperative), the level in the hierarchy, and the total amount of task components. The task body represents a task hierarchy organized using three logical constructors (i.e., AND, OR, and XOR). Each task is then associated with a task control structure (TCS) made up of six classes of descriptor objects that are consumed when the task is carried out and they are aggregated:

1. The *triggering* class has four types of events: formal and informal events, events occuring outside and inside the system.
2. The *condition* class contains contextual conditions governing the performance of the task.
3. The *resource* class describes resources (human or system) required for the task to be performed.
4. The *input data* class specifies information items required for performance of the task. To initialize a task, an input transition expresses logical conditions on these data by sending rules and benefits from various checking functions to ensure that all conditions required to perform the task are fulfilled. For instance, the completeness function checks that all input data are available and satisfy related constraints.
5. The *output data* class specifies information items produced by the task performance. To terminate a task, an output transition expresses logical conditions on these data through synchronization rules and benefits from various checking functions.
6. The *reaction* class describes physical and cognitive results resulting from the task performance.

The combination of TOOD descriptor objects covers task hierarchy and temporal ordering. TOOD is supported by a graphical editor allowing analysts to specify instances of task classes as well as instances of their related classes.

6.3 CONCLUSION

The task models analysed in previous sections exhibit a variety of concepts and relationships. The differences between concepts are both syntactic and semantic.

6.3.1 Syntactic Differences

Syntactic differences include differences of vocabulary used for the same concept across models. The most notable syntactic differences are summed up in Table 6.1. The comparison is based on four model features: task planning (how a task is related to high-level goal to be planned), operationalization capacity (the extent to which the method provides ways to map high-level goals to low-level ways of reaching them), lowest task decomposition level (the leaf node in the task decomposition), and operational level (the task decomposition level where actions take place).

It can be observed from Table 6.1 that similar or different terms can be used to refer to the same concepts. For example, *plan, operator, constructor*, and *goal* are often used for discussing high-level task planning. Although most models do provide for task decomposition, both structurally and temporally, they do not necessarily describe how decomposition can be effectively carried out. For instance, a scenario is frequently considered to be a particular instantiation of a general task model that depends on particular circumstances within the context of use.

The two last rows describe how a task is recursively decomposed into subtasks to end up with leaf nodes. Some models do not make terminological distinction between different levels of decomposition. On the other hand, GOMS, GTA, CTT, and MUSE do separate any nonterminal level of decomposition from leaf nodes, which usually bear a special name (often, *action*).

6.3.2 Semantic Differences

Semantic differences are related to conceptual variation across models. Semantic differences can be of major or minor importance. Of major importance are the differences in entity or relationship definitions and coverage, (e.g., cases where the same concept is not defined in the same way across models). Less consequential is the variation in how an entity or a relationship is expressed. For example, constructors in GTA, MAD*, or TKS express the temporal relationship between a task and its subtasks (although the set of constructors is not identical in all models), whereas operators in CTT are used between sibling tasks. Operators used in GOMS have a dual semantics: They specify actions (cognitive and motor) performed by the user, and they are also used as syntactic constructions of the language to control task flow, similar to a programming language. Table 6.2 compares task models along the following dimensions:

Discipline of origin. The discipline of origin has an impact on the concepts that will be found in a particular model. In particular, HTA and GOMS are deeply rooted in cognitive analysis. Models rooted in cognitive psychology or related disciplines focus on cognitive concepts and avoid software artifacts, whereas software engineering models do the reverse.

Formalization. This dimension specifies whether a model is based on a formal system or not. For instance, CTT temporal operators (chap. 24) are defined in process algebra, TOOD (chap. 25) contains mathematical definitions of transitions, and Task Layer Maps (chap. 11) has a formal basis.

Collaborative aspects. In order to describe cooperative aspects, models use a role concept. A role is defined by the tasks the role is responsible for. Roles are played by agents and are

TABLE 6.1
Main Task Model Features

	HTA	*GOMS*	*GTA*	*CTT*	*MAD**	*TKS*	*Diane+*	*TOOD*	*MUSE*
Task planning	Plans	Operators	Constructors	Operators	Constructors	Plans/constructors	Goals	Input/output transitions	Goals and constructors
Operationalization		Methods/ selection rules		Scenarios	Pre- and postconditions	Procedures	Procedures		
Task tree leaves		Unit tasks	Basic tasks	Basic tasks		Actions			Actions
Operational level	Tasks	Operators	Actions/system operations	Actions	Tasks		Operations	Task	Task

TABLE 6.2
Main Semantic Variations Between Models

	Origin	*Formalisation*	*Collaborative aspects*	*Context of use variation*	*Cognitive aspects*	*System Response*	*Scope of constructors*	*Manipulated objects*
HTA	Cognitive analysis	Ø	√	Ø	Usability problems	√	Parent	Reference/task object
GOMS	Cognitive analysis	Ø	Ø	Ø	User performance	Ø	Multiple levels	Reference/task object
GTA	Ethnography, cognitive analysis, & CSCW		√ (roles/agents)	Ø	User performance	√	Parent	Object
CTT	Software engineering	√	√ (roles & cooperative parts)	√ (platforms)	Ø	√	Sibling	Detailed/task objects
TKS	Cognitive analysis & software engineering	Ø	√ (roles/agents)	√ (user)	Knowledge structures	Ø	Multiple levels	Detailed/task objects
MAD*	Psychology	√	Ø	Ø	Ø	√	Parent	Detailed/task objects
Diane+	Process control & software engineering	√	Ø	√ (devices)	Ø	√	Sibling	Object
TOOD	Process control	√	Can be expressed in transitions	Ø	Cognitive result in a reaction	√	Sibling	Resource, input data, and output data
MUSE	Software engineering & human factors	√	Ø	Ø	Ø	√	Parent	Object

Note. √ = supported, Ø = unsupported.

assigned according to organizational rules. They are represented by task trees constructed by performing several task decompositions. In the case of a cooperative role, the task hierarchy represents only the cooperative part of the activity. Since a user interface is designed for a given role, this distinction in task decomposition is useful both for user interface design and for coordinating the computer-supported cooperative work.

Context of use variation. Because the user community and the number of computing platforms and working environments are increasing, context-sensitive user interfaces are becoming more important to support task variations resulting from differences in the contexts of use. The importance given to various external elements varies from one system to another. Some approaches place an emphasis on the context of use (e.g., TKS, GTA, and, to a degree, CTT). For instance, the users' characteristics, the organization, the computing platform, and the physical environment are taken into consideration to develop a usable system. It is worth noting that several task models have been modified to enable them to characterize aspects relevant to context sensitivity. For instance, the Unified Design Method (Savidis, Akoumianakis, & Stephanidis, 2001) introduced the mechanism of polymorphic task models to integrate variations that resulted from the consideration of different design alternatives—alternatives that resulted from addressing the needs of several categories of users in different contexts of use. Similarly, CTT has been extended where particular subtasks are selected based on conditions resulting from context variations (Paternò & Santoro, 2002; Thevenin, 2001). The idea of conditional subtrees also leads to the identification and formal definition of context-dependent, context-independent, and partially context-independent decompositions, as oulined in Souchon, Limbourg, and Vanderdonckt (2002).

Cognitive aspects. This dimension concerns the incorporation and/or support of cognitive aspects in modeling activities.

System response. This dimension determines whether semantic functions from the technical system can be identified and embodied n the modeling. Some models remain open by not distinguishing any type of (sub)task (e.g., HTA), whereas others effectively pursue this goal (e.g., Diane+).

Scope of constructors. This dimension expresses the scope of the task elements on which the temporal operators work. The scope can be the *parent* or the *sibling* when any temporal operator constraint affects the ordering, respectively, between a father node in the task decomposition and its children (as in HTA) or between siblings of the same father (as in CTT). Sometimes the scope goes beyond two levels in the task hierarchy, in which case multiple levels may be required to expressed a temporal ordering (such as in Dittmar, 2000).

Manipulated objects. Although the *task model* is intended to represent how a particular task can be carried out by a particular user stereotype in a certain context of use, a *domain model* is frequently used to refer to the domain objects (or *task objects*) that are manipulated by any (sub)task at any level of decomposition. Some task models explicitly embody this information, whereas others prefer to establish a link between the task model and the domain model (*reference*). This reference may cover the entities of interest only (as in CTT), perhaps along with their relationships (as in TOOD).

6.3.3 Common Properties

The following concepts are critical for a task-based design of a user interface. First, goal and task hierarchies are essential. Second, operators must express temporal constraints between tasks. Third, a minimal requirement for dealing with cooperative aspects is role specification in terms of tasks. Fourth, objects and the actions that are performed on them make possible the detailed modeling of presentation and dialog of the user interface. More details are provided in (Limbourg, Pribeanu, & Vanderdonckt, 2001).

6.3.4 Model Usage

HTA is mainly intended as a means of training people in the use of a given interface. Although plans are attractive and precise, they are informal and are not subject to proof verification or model checking. Moreover, plans typically describe temporal constraints in a procedural way. Adding new tasks means rewriting the plans associated with the next higher level task, which may be tedious and lead to errors. Although plans are suitable for early task analysis and when information is elicited in an informal but unambiguous way, they do not provide the kind of representation able to support a (computer-aided) derivation of a user interface model.

GOMS models also assume a given user interface. There is no explicit task decomposition but only a hierarchical organization of methods used to operationalize tasks. Cognitive analysis in GOMS is done at unit task level, and it requires different levels of specification depending on the level of detail desired by the analyst. GOMS is built for the lowest level of task decomposition and thus provides no support for user interface modeling. Rather, the objective is to optimize user performance and evaluate execution time, memory workload, and learning time early in the design process. The concepts are more closely related to user modeling than to task modeling.

GTA is especially good at specifying delegation mechanisms between roles. TKS uses both plans and procedures similar to the way HTA uses plans and GOMS uses selection rules and methods. Like methods, procedures are useful for achieving a detailed specification of the elementary task when there is a need to describe actions performed on objects. Models should be declarative rather than procedural in order to support successive transforms and to be suitable for the use of computer tools (Eisenstein, Vanderdonckt, & Puerta, 2001). Task models that are primarily intended as support for evaluation and user training, like HTA and GOMS, are not suitable for supporting user interface modeling. Rather, these models require an initial design of the interface whose usability they focus on improving.

6.3.5 Expressiveness Versus Complexity

Fig. 6.10 shows that task models become increasingly complex as they become progressively expressive (e.g., they can express many different facets of task modeling and not only the task decomposition). HTA and MAD* are located at the left end of the continuum because they

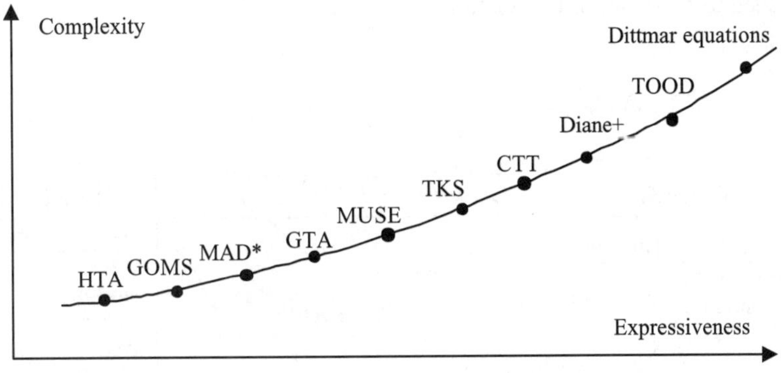

FIG. 6.10. Expressiveness versus complexity of task models.

are basically restricted to decomposing tasks into subtasks and temporal operators. Successive models refine the temporal relationships using pre- and postconditions (MUSE), abstract data types and axioms (TKS), LOTOS operators (CTT), first-order logical predicates (Diane+), mathematical functions (TOOD), and mathematical equations (Dittmar, 2000). The right side of the continuum ends with Dittmar's task model, which regulates temporal ordering by means of a set of mathematical equations, linear or nonlinear. This method is considered the most expressive (mathematical equations are very general) yet the most complex (solving a set of nonlinear, possibly conflicting, equations is a nontrivial problem).

ACKNOWLEDGMENTS

We would like to greatly thank those who helped us in this comparison of task models: Cristina Chisalita, Hilary Johnson, Peter Johnson, Kim Young Lim, John Long, Cécile Paris, Fabio Paternò, Costin Pribeanu, Dominique Scapin, Neville Stanton, Dimitri Tabary, Jean-Claude Tarby, and Gerritt van der Veer.

APPENDIX: OVERVIEW OF THE ENTITY-RELATIONSHIP-ATTRIBUTE MODEL

The figures of metamodels compared in this chapter use the notation of the Entity-Relationship-Attribute (ERA) model. This model relies on understanding the relationship between one entity and another. For example, the model in Fig. 6.11 graphically depicts the ownership of vehicles by persons. Reading the model from left to right represents the statement "An owner may own no, one, or many vehicles," whereas reading from right to left represents the statement "Any vehicle is owned by one or many owners."

ERA uses a rectangle to graphically depict an entity and a hexagon for any relationship between entities. A relationship is said to *cyclic* if the relationship has the same entity for both source and destination. Any entity or relationship can have one or many attributes. For instance, an owner may be characterized by the following attributes: the number of her identity card, her first name, her last name, and her address. If an attribute is underlined, it is an *identifier*, that is, an attribute whose values remain unique for every instance of the entity. The connectivity specifies how many instances of that entity can be associated with a single instance of the other entity via the relationship. In this example, an owner may have many vehicles, even simultaneously, and a vehicle should be owned by one or many owners. To facilitate reading, a role is added to each part of the relationship.

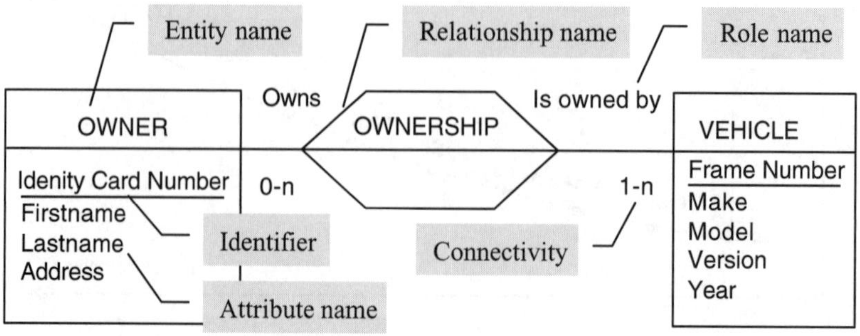

FIG. 6.11. Overview of the entity-relationship-attribute model used.

REFERENCES

Annett, J., Cunningham, D., & Mathias-Jones, P. (2000). A method for measuring team skills. *Ergonomics, 43*, 1076–1094.

Annett, J., & Duncan, K. (1967). Task analysis and training design. *Occupational Psychology, 41*, 211–227.

Barthet, M.-F., & Tarby, J.-C. (1996). *The Diane+ method*. In J. Vanderdonckt, (Ed.), *Computer-aided design of user interfaces* (pp. 95–120). Namur, Belgium: Presses Universitaires de Namur.

Baumeister, L. K., John, B. E., & Byrne, M. D. (2000). A comparison of tools for building GOMS models: Tools for design. In *Proceedings of ACM Conference on Human Factors in Computing Systems (CHI 2000)* (Vol. 1; pp. 502–509). New York: ACM Press.

Beard, D. V., Smith, D. K., & Denelsbeck, K. M. (1996). QGOMS: A direct-manipulation tool for simple GOMS models. In *Proceedings of ACM Conference on Human Factors in Computing Systems (CHI '96)* (Vol. 2; pp. 25–26). New York: ACM Press.

Bomsdorf, B., & Szwillus, G. (1998). From task to dialogue: Task-based user interface design. *SIGCHI Bulletin, 30*(4), 40–42.

Bomsdorf, B., & Szwillus, G. (1999). Tool support for task-based user interface design. A CHI '99 workshop. *SIGCHI Bulletin, 31*(4), 27–29. Available: http://www.uni-paderborn.de/cs/ag-szwillus/chi99/ws/

Card, S. K., Moran, T. P., & Newell, A. (1983). The psychology of human-computer interaction. Hillsdale, NJ: Lawrence Erlbaum Associates.

Diaper, D. (1989a). Task analysis for human-computer interaction. Chichester, England: Ellis Horwood.

Diaper, D. (1989b). Task Analysis for Knowledge Descriptions (TAKD): The method and examples. In D. Diaper (Ed.), *Task analysis for human-computer interaction* (pp. 108–159). Chichester, England: Ellis Horwood.

Dittmar, A. (2000). More precise descriptions of temporal relations within task models. In P. Palanque & F. Paternò (Eds.), *Proceedings of Seventh International Workshop on Design, Specification, and Verification of Interactive Systems (DSV-IS 2000)* (Lecture Notes in Computer Science, Vol. 1946; pp. 151–168). Berlin: Springer-Verlag.

Eisenstein, J., Vanderdonckt, J., & Puerta, A. (2001). Model-based user-interface development techniques for mobile computing. In J. Lester (Ed.), *Proceedings of ACM International Conference on Intelligent User Interfaces (IUI 2001)* (pp. 69–76). New York: ACM Press.

Gamboa, F., Scapin D. L. (1997) Editing MAD* task descriptions for specifying user interfaces, at both semantic and presentation levels. In M. D. Harrison & J. C. Torres, (Eds.), *Proceedings of Fourth International Workshop on Design, Specification, and Verification of Interactive Systems (DSV-IS '97)* (pp. 193–208). Berlin: Springer-Verlag.

Hackos, J. T., & Redish, J. C. (1998). *User and task analysis for interface design*. New York: Wiley.

John, B. E., & Kieras, D. E. (1996). The GOMS family of user interface analysis techniques: Comparison and contrast. *ACM Transactions on Computer-Human Interaction, 3*, 320–351.

Johnson, P. (1992). *Human-computer interaction: Psychology, task analysis and software engineering*. London: McGraw-Hill.

Johnson, P., Diaper, D., & Long, J. (1984). Tasks, skill and knowledge: Task analysis for knowledge based descriptions. In *Proceedings of First IFIP Conference on Human-Computer Interaction (Interact '84)* (Vol. 1; pp. 23–28). North-Holland: Elsevier Science Publishers.

Johnson, P., & Johnson, H. (1989). Knowledge analysis of task: Task analysis and specification for human-computer systems. In A. Downton, (Ed.), *Engineering the human-computer interface* (pp. 119–144). London: McGraw-Hill.

Kieras, D. E., Wood, S. D., Abotel, K., & Hornof, A. (1995). GLEAN: A computer-based tool for rapid GOMS model usability evaluation of user interface designs. In *Proceedings of the ACM Symposium on User Interface Software and Technology (UIST '95)* (pp. 91–100). New York: ACM Press.

Kirwan, B., & Ainsworth, L. K. (1992). *A guide to task analysis*. London: Taylor & Francis.

Lewis, C., & Rieman, J. (1994). *Task-centered user interface design: A practical introduction*. Available at http://www.acm.org/~perlman/uidesign.html#chap0

Lim, K. Y., & Long, J. (1994). *The MUSE method for usability engineering*. Cambridge: Cambridge University Press.

Lim, K. Y., & Long, J. (1996). Structured task analysis: An instantiation of the MUSE method for usability engineering. *interacting With Computers, 8*(1), 31–50.

Limbourg, Q., Pribeanu, C., & Vanderdonckt, J. (2001). Towards uniformised task models in a model based approach. In C. Johnson (Ed.), *Proceedings of Eighth International Workshop on Design, Specification, Verification of Interactive Systems (DSV-IS 2001)* (Vol. 2220; pp. 164–182). Berlin: Springser-Verlag.

Lu, S., Paris, C., & Vander Linden, K. (1998). Towards the automatic generation of task models from object oriented diagrams. In *Proceedings of Seventh IFIP Working Conference on Engineering for Human-Computer Interaction (EHCI '98)* (pp. xxx). Dorolrecht, Netherlands: Kluwer Academic.

Mahfoudhi, A., Abed, M., & Tabary, D. (2001). From the formal specifications of user tasks to the automatic generation of the HCI specifications. In A. Blandford, J. Vanderdonckt, & P. Gray, (Eds.), *People and computers XV* (pp. 331–347). London: Springer.

Paternò, F. (1999). *Model based design and evaluation of interactive applications.* Berlin: Springer-Verlag.

Paternò, F., & Santoro, C. (2002). One model, many interfaces. In C. Kolski & J. Vanderdonckt (Eds.), *Proceedings of Fourth International Conference on Computer-Aided Design of User Interfaces (CADUI 2002)* (pp. 143–154). Dordrecht: Kluwer Academics.

Payne, S. J., & Green, T. R. G. (1986). Task-action grammars: A model of the mental representation of task languages. *Human-Computer Interaction, 2*(2), 93–133.

Savidis, A., Akoumianakis, D., & Stephanidis, C. (2001). *The unified user interface design method.* In C. Stephanidis (Ed.), *User interfaces for all* (pp. 417–440). Mahwah (NJ): Lawrence Erlbaum Associates.

Scapin, D. L., & Bastien, J. M. C. (2001). *Analyse des tâches et aide ergonomique à la conception : l'approche MAD*.* In C. Kolski (Ed.), *Analyse et conception de l'IHM: Interaction homme-machine pour les systèmes d'information* (Vol. 1; pp. 85–116). Paris: Edition Hermès.

Scapin, D., & Pierret-Golbreich, C. (1989). Towards a method for task description: MAD. In L. Berlinguet & D. Berthelette (Eds.), *Proceedings of Work with Display Units (WWU '89)* (pp. 27–34). North-Holland: Elsevier Science.

Shepherd, A. (1989). Analysis and training in information technology Tasks. In D. Diaper (Ed.), *Task analysis for human-computer interaction* (pp. 15–55). Chichester, England: Ellis Horwood.

Shepherd, A. (1995). Task analysis as a framework for examining HCI Tasks. In A. Monk & N. Gilbert (Eds.), *Perspectives on HCI: Diverse approaches* (pp. 145–174). London: Academic Press.

Smith, M. J., & O'Neill, J. E. (1996). Beyond task analysis: Exploiting task models in application implementation. In *Proceedings of ACM Conference on Human Aspects in Computing Systems (CHI '96)* (Vol. 2; pp. 263–264). New York: ACM Press.

Souchon, N., Limbourg, Q., & Vanderdonckt, J. (2002). Task modelling in multiple contexts of use. In P. Forbrig, Q. Limbourg, B. Urban, & J. Vanderdonckt (Eds.), *Proceedings of Ninth International Workshop on Design, Specification, and Verification of Interactive Systems (DSV-IS '2002)* (Vol. 2545; pp. 59–73). Berlin: Springer-Verlag.

Stanton, N., & Young, M. (1998). Is utility in the eye of the beholder? A study of ergonomics methods. *Applied Ergonomics, 29*(1), 41–54.

Sutcliffe, A. (1989). Task analysis, systems analysis and design: Symbiosis or synthesis? *Interacting With Computers, 1*(1), 6–12.

Tauber, M. J. (1990). ETAG: Extended task action grammar, a language for the description of the user's task language. In D. Diaper, D. Gilmore, G. Cockton, and B. Shackel (Eds.), Proceedings of the 3rd IFIP TC 13 Conference On Human-Computer Interaction Interact 90, pp. 163–168, Amsterdam, 1990, Elsevier.

Thévenin, D. (2001). *Adaptation en interaction homme-machine: Le cas de la plasticité.* Unpublished doctoral dissertation, Université Joseph Fourier, Grenoble, France.

van der Veer, G. C., Van der Lenting, B. F., & Bergevoet, B. A. J. (1996). GTA: Groupware task analysis-modeling completity, *Acta Psychological*, 91: 297–322, 1996.

van Welie, M., & van der Veer, G. C. (1998). EUTERPE: Tools support for analyzing cooperative environments. In T. R. G. Green, L. Bannon, C. P. Warren, & J. Buckkleys (Eds.), *Cognition and co-operation* (pp. 25–30). Roquencourt INRIA.

van Welie, M., van der Veer, G. C., & Eliëns, A. (1998). An ontology for task world models. In P. Markopoulos & P. Johnson (Eds.), *Proceedings of the Fifth International Workshop on Design, Specification, and Verification of Interactive Systems (DSV-IS '98)* (pp. 57–70). Vienna: Springer-Verlag.

Wilson, S., & Johnson, P. (1996). Bridging the generation gap: From work tasks to user interface designs. In J. Vanderdonckt (Ed.), *Computer-aided design of user interfaces* (pp. 77–94). Namur, Belgium: Presses Universitaires de Namur.

7

DUTCH: Designing for Users and Tasks from Concepts to Handles

Gerrit C. van der Veer
Vrije Universiteit Amsterdam

Martijn van Welie
Vrije Universiteit Amsterdam

Designing Groupware systems requires methods and tools that cover all aspects of Groupware systems. We present a method that utilizes known theoretical insights and makes them usable in practice. In our method, the design of Groupware systems is driven by an extensive task analysis followed by structured design and iterative evaluation using usability criteria. Using a combination of multiple, complementary representations and techniques, a wide range of aspects of Groupware design is covered. The method is built on our experiences and is used in practice by several companies and educational institutes in Europe. We define the design process, the models needed, and the tools that support the design process.

7.1 INTRODUCTION

Designing Groupware is a complex activity. Methods for designing a Groupware system need to address many relevant aspects of the system, including the users, their tasks, and the software but also the physical and social environment of the system. Methods from the human-computer interaction (HCI) and computer-supported cooperative work (CSCW) literature individually address some of the relevant aspects, but combining their insights in practice remains difficult. Moreover, the gap between theoretical ideas on design and applying them in practice is often large, rendering the theoretical ideas virtually useless. Based on our experiences in both industrial and educational projects, we developed a practical method for designing Groupware. Despite the fact that we found that a theoretical foundation was necessary to solve certain problems in the design process, the method remained very practical. This chapter describes our method, called *DUTCH* (Designing for Users and Tasks from Concepts to Handles), presents the theory behind it, and discusses the representations that it uses.

DUTCH is a design method for complex interactive systems. Such design projects require multidisciplinary teams in which each of the various members has the responsibility for some aspect of design (for other examples, see chaps. 8, 9, and 11). The division of labor poses constraints on the use of documentation, representation, and communication within the team and in relation to clients and users. The needed representations vary in character, from formalisms to scenarios and highly informal sketches.

Based on our experience of applying DUTCH in teams in educational settings and industry, we have come to the conclusion that the combination of representations does not limit creativity but, on the contrary, stimulates creative input from designers, clients, and prospective users. Informal and open scenarios are built on the basis of formal representations of task models, and evaluation is done by systematic heuristic walk-throughs as well as by acting out the scenarios.

Designing complex interactive systems is rarely an individual activity. As soon as different user roles can be distinguished and technological innovations affect roles, procedures, and other organizational aspects, individuals from various disciplines have to be involved, and they have to work together effectively in a structured design team. In industry, complex interactive systems are typically designed by groups of 10 to 30 people, and subgroups are formed to handle particular activities. Our method is intended for the development of such systems. We suggest allocating complementary activities to small specialized groups of three to five people. The groups have to work together to complete the design. Breaking up the work is necessary when a large project is undertaken. When groups work together, the division of labor puts constraints on the way communication is handled. We found that a combination of formal and informal representations is well suited to communication between groups.

7.1.1 A Typical Example

One of our former students has been working in a company that develops designs for a large international company in information technology and electronics systems. The task domain we were involved in can be characterized as design of safety and security systems—systems intended to help protect access and promote safety in banks, systems used to monitor industrial processes, and systems used for safeguarding railroad traffic. The systems are typical Groupware systems and are embedded in complex organizations where people have different roles and responsibilities.

The company has been providing these types of designs for over 10 years. Currently the company is frequently asked to redesign existing systems as well as to design systems for new situations that resemble the situations of systems already in use elsewhere. In the past, the company used many different methods and techniques, each accommodating some particular aspects of the problem domain. However, as many problems were repeatedly encountered, the company felt it needed a better design method. It wanted a method that adequately addressed all phases of design, including the following:

- Collecting insights on current situations and describing them and analyzing them. The managers of design projects were often not even aware of the importance of this activity.
- Considering the future situation in which a new design (or redesign) would be implemented, including the new system and the new work organization and procedures. Based on previous experiences, the company felt this was necessary to do early and not after implementation of finished designs into full-blown systems.
- Relating the detailed design of a technology back to global analysis of the intended new work situation. Systematic evaluation can, when done in time, redirect the detailed design before solutions are completely implemented.
- Evaluating system usability both early and late in a design cycle. Evaluation procedures had always been used, but there has never been a clear view on what should be evaluated at which moment in the design cycle. Consequently, evaluation was driven by the availability

of commercial tools and by the standards set by the clients, who generally did not have clear views on usability even though they almost always had views on safety and reliability (as far as hardware and software were concerned).

Since the company utilized design teams extensively, it was interested in using representations that were suitable for other tasks besides describing Groupware systems. Representations were needed for purposes such as the following:

- Analyzing design knowledge in cases where designers had to collaborate with experts from different disciplines.
- Proposing and discussing global and detail solutions both internally and externally.
- Evaluating systems in different phases of the design process.
- Transfering decisions to those responsible for implementation, which usually meant handing over specifications to builders in a different company.

It was also important to have tools capable of supporting the process and the techniques. The purpose of such tools was to aid the designers in controlling the process, in iteration, and in backtracking of decisions and also in producing solutions and proposals to be elaborated by others or analyzed, tested, and evaluated. It was soon clear no "off-the-shelf" method was appropriate. Most methods only covered some of the important aspects, and many methods turned out to be difficult to use in practice.

7.2 DUTCH DESIGN

Over the past years we have taken useful bits of theories and combined them into a coherent practical method for designing complex interactive systems. DUTCH has been used successfully in both industry and education and has shown its practical value. From our experiences working with DUTCH, we learned that, for a design method to be useful, the constructors of the method must define a clear process; define the models and representations, including their semantics; and support the method and models with tools.

DUTCH is task based, which means that the user tasks drive the design process. The goal is to design both usable and useful systems. We think it is important to base the design on the work that has to be done by the users. Therefore, the users play an important role in the gathering of information about their work and in usability testing. Our process consists of four main activities: (a) analyzing a "current" task situation, (b) envisioning a future task situation for which information technology is to be designed, (c) specifying the information technology to be designed, and (d) evaluating the technology (which makes the process cyclic). Figure 7.1 gives an overview of the entire design process, including all activities and sources of information.

DUTCH requires the design team to work in groups because of the different types of expertise needed. Typically a design team comprises members from a variety of disciplines, such as computer scientists, psychologists, ethnographers, graphical designers, and industrial designers.

The design process starts with an extensive task analysis using the Groupware Task Analysis (GTA) method (van der Veer, Lenting, & Bergevoet, 1996). A task analysis of this type includes a description of the work, the work situation, and the users and other stakeholders of the system to be designed. We distinguish two task models. The first task model made is a descriptive task model, and it is used for analyzing the current task situation. The second task model is a prescriptive task model for the system to be designed. Both task models are usually

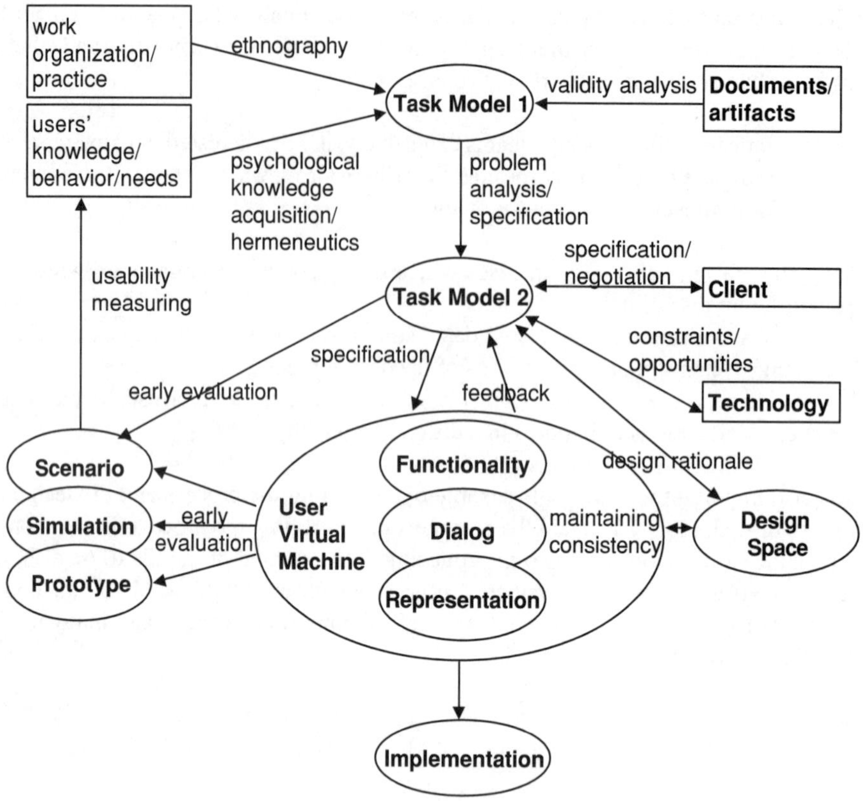

FIG. 7.1. The DUTCH design process.

constructed by task modeling specialists (e.g., psychologists and ethnographers). However, all design disciplines should participate in constructing the second task model by contributing ideas, which may later be incorporated into the model by the task modeling group.

7.3 GROUPWARE TASK ANALYSIS (GTA)

The design process starts with an extensive task analysis using the GTA method. As mentioned, the first task model made is descriptive and is used for analyzing the current task situation. The second is a prescriptive task model for the system to be designed.

7.3.1 Analyzing the Current Task Situation

In many cases, the design of a new system is triggered by an existing task situation. Either the current way of performing tasks is not considered optimal or the availability of new technology is expected to allow improvement over current methods. A systematic analysis of the current situation may help in formulating design requirements and may later help in evaluating the design. Whenever a "current" version of the task situation exists, it pays to model this. We use a combination of classical HCI techniques such as structured interviews (Sebillotte, 1988) and CSCW techniques such as ethnographic studies and interaction analysis (Jordan, 1996).

7.3.2 Envisioning the Future Task Situation

Many design methods in HCI that start with task modeling include a number of phases. Once the current situation is described (task model 1), the task structure is redesigned to include technological solutions for problems and technological answers to requirements. Johnson, Johnson, Waddington, and Shouls (1988) provide an example of a systematic approach in which a second task model is explicitly defined in the course of design decision making. Task model 2 will in general be formulated and structured in the same way as task model 1, but it is not considered a descriptive model of the users' knowledge, although in some cases it might function as a prescriptive model of the knowledge an expert user of the new technology should possess.

7.3.3 A Conceptual Framework

For describing the task world, we developed a broad conceptual framework that is based on comparisons of different approaches and on an analysis of existing and proposed systems for HCI and CSCW (see van der Veer, van Vliet, & Lenting, 1995). To design Groupware systems, it is necessary to widen the notion of a task model to include descriptions of many more aspects of the task world than just the tasks. The framework is intended to structure task models 1 and 2 and hence to guide the choice of techniques for collecting the information to be used in constructing task model 1. Obviously, task model 2 design decisions have to be made based on problems and conflicts that are represented in task model 1 and on the requirement specifications formulated in interaction with the client. For a discussion of these design activities, see van der Veer et al., (1995).

A task model for a complex situation needs to be composed using three different perspectives–those of the agents, the work, and the situation. Applying these perspectives allows designers to read and to design from different angles while using design tools to safeguard consistency and ensure completeness. The three viewpoints applied in our approach are a superset of the main focal points in the domain of HCI and CSCW. Both of these design fields consider agents ("users" vs. "cooperating users" or user groups) and work (activities or tasks, respectively the objectives of "interaction" and cooperative work). Moreover, especially CSCW focuses on situations in which technological support has to be incorporated. In HCI, this is only sometimes considered, and then mostly implicitly.

7.3.3.1 Agents

The term *agents* usually refers to people, either individuals or groups, but it may also refer to systems. Agents are considered in relation to the task world. Hence, we need to make a distinction between actors (acting individuals or systems) and the roles they play. Moreover, we need the concept of organization of agents. Agents have to be described by means of relevant characteristics (e.g., in the case of human actors, the language they speak, their typing skills, or their experience with Microsoft-Windows). Roles indicate classes of actors to whom certain subsets of tasks are allocated. By definition, roles are generic in the task world. More than one actor may perform the same role, and a single actor may have several roles at the same time. *Organization* refers to the relation between actors and roles with respect to task allocation. Delegating responsibilities, including switching them from one role to another, is part of the organization.

7.3.3.2 Work

We consider both the structural and the dynamic aspect of work. We take task as the basic concept and allow that each task can have several goals. We also make a distinction between a task and an action. Tasks can be identified at various levels of complexity. The unit level of tasks needs special attention. We make a distinction between (a) the lowest task level that people want to consider in referring to their work, the "unit task" (Card, Moran, & Newell, 1983), and (b) the atomic level of task delegation that is defined by the tool used in performing work, such as a single command in command-driven computer applications. This last type of task we call a basic task (Tauber, 1988). Unit tasks will often be role related. Complex tasks may be split up between actors or roles. Unit tasks and basic tasks may be decomposed further into user actions and system actions, but these cannot really be understood without a frame of reference created by the corresponding task (i.e., the actions derive their meaning from the task). For instance, hitting a return key has a different meaning depending on whether it concludes a command or confirms the specification of a numerical input value in a spreadsheet.

The task structure will often at least partially be hierarchical (see chap. 1). On the other hand, the effects of certain tasks may influence the procedures for other tasks (possibly with other roles involved). Therefore, we also need to understand task flow and data flow over time as well as the relation between several concurrent flows. In this context we use the concept of an event. An event is a triggering condition for a task, even if the triggering could be caused by something outside the task domain we are considering.

7.3.3.3 Situation

Analyzing a task world from the viewpoint of the situation means detecting and describing the environment (physical, conceptual, and social) and the objects in the environment. The object description includes an analysis of the object structure. Each thing that is relevant to the work in a certain situation is treated as an object by task analysis, even the environment. In this framework, objects are not defined by "object-oriented" methods. Objects may be physical things or conceptual (nonmaterial) things like messages, gestures, passwords, stories, and signatures. The task environment is the current situation for the performance of a certain task. It includes actors with roles as well as conditions for task performance. The history of past relevant events in the task situation is part of the actual environment if this features in the conditions for task execution.

7.4 THE TASK WORLD ONTOLOGY

In order to put the theory into practice, the three viewpoints (of the agents, the work, and the situation) must be expressed in a task world ontology (van Welie, van der Veer, & Eliëns, 1998b). The ontology defines the basic concepts and relationships between them that we regard relevant for the purpose of task analysis. "Basic" here indicates that we are able to describe all other relevant concepts and relations by using the basic concepts and relations. The ontology is of importance because it is the conceptual basis of all information recorded and the way it is structured and may be represented. Our ontology is derived from the three viewpoints and incorporates relevant aspects of several other task analysis methods.

7.4.1 Relationships

The basic concepts from GTA (task, object, agent, role, and event) are related in specific ways. In this section, we sketch the relationships that we are using now. For each relationship the first-order predicate definition is given and explained. Figure 7.2 shows all the concepts and relationships. The set of relationships have in practice proven to be sufficient for dealing with

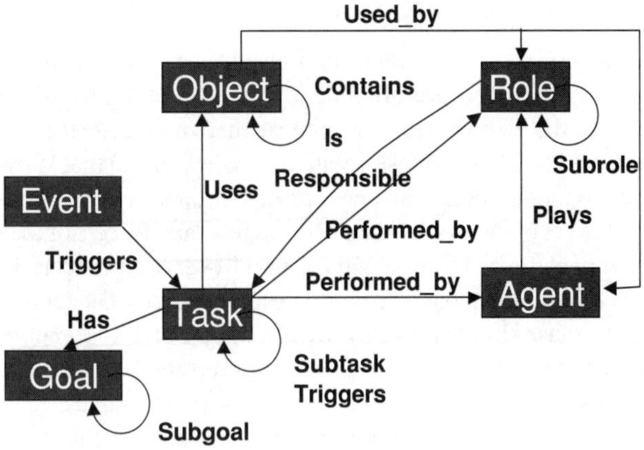

FIG. 7.2. Task World Ontology.

most design cases. Keep in mind that there are other relationships in our ontology that allow the representation of the additional concepts (e.g., tasks have attributes like complex/unit/basic as well as start-and-stop conditions). For a complete specification of our ontology, see van Welie, van der Veer, and Eliëns (1998).

Uses. The "uses" relationship specifies which object is used in executing the task and how it is used. The "uses" relationship typically changes the state of the object.
Triggers. The "triggers" relationship is the basis for specifying task flow (see chap. 19). It specifies that a task is triggered (started) by an event or another task and how it is triggered. Several trigger types are possible (e.g., OR, AND, and NEXT) to express choice, parallelism, or sequences of tasks.
Plays. Every agent should play one or more roles. The "plays" relationship also indicates how the relevant role was obtained (e.g., by delegation, by mandate, or for a socially determined reason).
Performed_by. The relationship "performed by" specifies that a task is performed by an agent. This does not mean that the agent is also the one who is responsible for the task, because this depends on the agent's role and the way it was obtained. When it is not relevant to specify the agent that performs the task, a role can also be specified as the performing entity.
Has. The "has" relationship connects tasks to goals. Each task has one or more goals that define the reason for performing the task. A goal could be either personal or a business goal.
Subtask/subgoal. The "subtask/subgoal" relationship describes the task/goal decomposition (see chap. 3).
Subrole. The "subrole" relationship brings roles into a hierarchical structure. The "subrole" relationship describes a role as encompassing other roles, including the responsibility for the tasks within the subroles. When a role has subroles, all the task responsibilities belong to the role.
Responsible. The "responsible" relationship specifies a task for which the role is responsible.
Used_by. The "used by" relationship indicates who used which object and what the agent or role can do with it. The agent's rights regarding objects can be of existential nature, indicate ownership, or indicate daily handling of objects.

7.4.2 Representations for Task Models

Our experiences have shown us that representations are very important for the effectiveness of a task analysis. Design projects are often done in teams, and a team's effectiveness often depends on the ability of the members to communicate the knowledge they have acquired. Task modeling is concerned with collecting task-related knowledge, and that knowledge needs to be documented. In task modeling, more than one representation is needed to capture all important aspects. We have built a set of "views" to describe the relevant aspects of the task world. These views resemble the work models of Contextual Design (Beyer & Holtzblatt, 1998) but our representations are intended for use in large real-world design projects. The representations have become more complex because simple models proved not to be effective enough. The ontology is used for the abstract structure of the data and allows various representations that can be regarded as views of this data. In most case, a representation shows a particular aspect of the information. Figure 7.3 presents some of the representations that we use and for which we have implemented tool support. It contains a workflow editor (top-left corner), a task tree, and a task template.

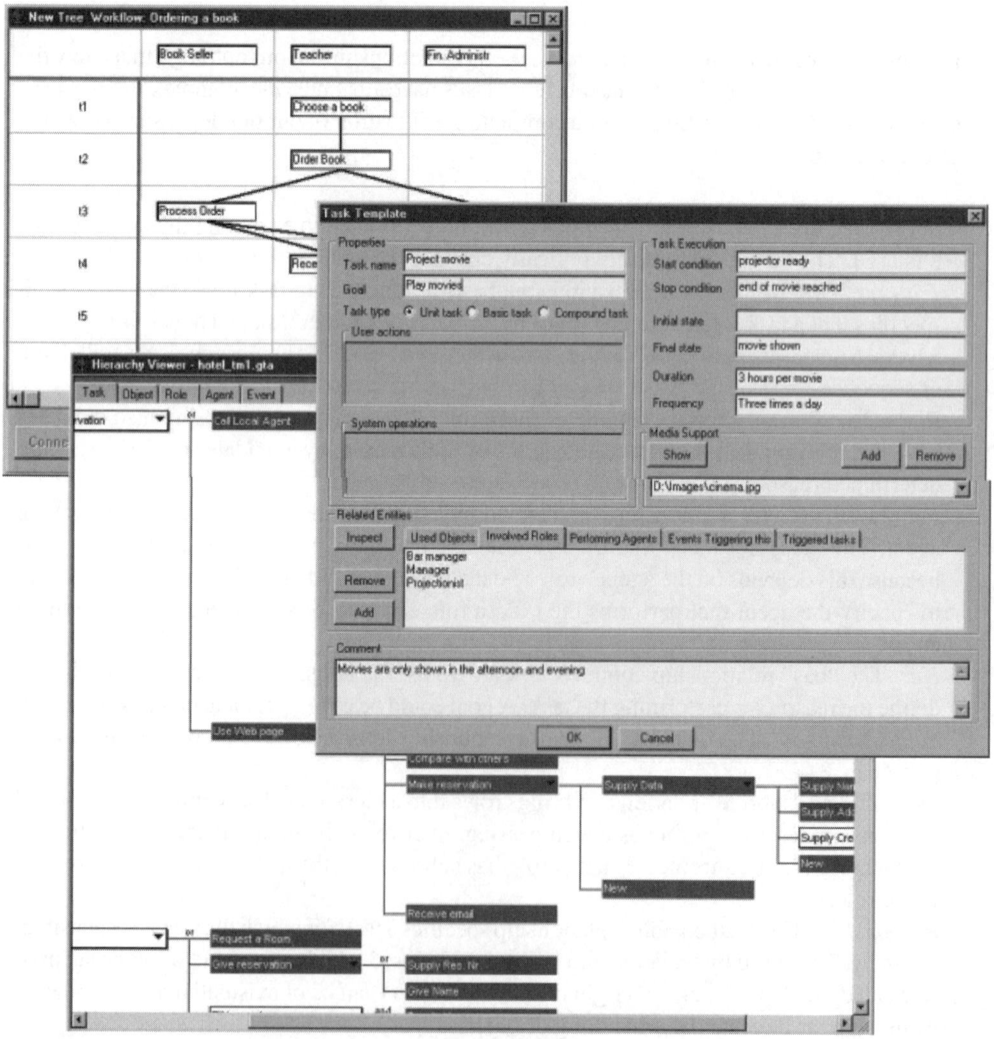

FIG. 7.3. Some representations including a task tree, task flow, and a task template.

7.4.2.1 Representing Work Structure

The purpose of modeling the work structure is to represent how people divide their work into smaller meaningful pieces in order to achieve certain goals. Knowing the structure of work allows the designers to understand how people think about their work, to see where problems arise, and to discover how tasks are related to the users' goals. The relation between tasks and goals helps the designers identify which tasks need to be supported by the system and why (i.e., identify which user goals are independent of the technology used).

Work structure is usually represented using task trees that show a hierarchic decomposition of the work. In a task tree, tasks and goals are distinguished. Often some timing information is added using constructors such as SEQ, LOOP, PAR, and OR. The constructors cannot always be used, especially when the task sequence uses a combination of sequential and optional tasks (van Welie, van der Veer, & Eliëns 1998b). Details of the task can effectively be described using templates we have developed. Details include state changes, task frequency and duration, triggering conditions and start-and-stop conditions.

7.4.2.2 Representing Work Dynamics

Work dynamics involve the sequence in which tasks are performed in relation to the roles that perform them. Workflow or activity models are needed to capture these aspects and should include the possibility of modeling parallel and optional tasks. Workflow diagrams usually describe a scenario or use case (chap. 5). A scenario is triggered by some event and usually starts with some important goal being activated. The scenario usually ends when the goal is achieved, but other goals may have been activated in the course of the tasks performed and may not be reached yet. Work dynamics can thus be modeled in an event-driven way. Case studies such as van Loo, van der Veer, & van Welie (1999) show that the event-driven dynamic aspect of cooperative work can be very important.

Another important aspect of work dynamics is collaboration and communication. Especially when multiple roles are involved in a certain task, timing and changes in control are essential to model. Roles pass objects when they communicate or collaborate, which cannot be represented well by a task tree. A workflow model can show work in relation to time and roles. The model gives the designer insight into the order in which tasks are performed and how different people are involved in them. Additionally, it can show how people work together and communicate by exchanging objects or messages. Typically, a workflow model describes a small scenario involving one or more roles, showing how the work is interleaved.

Figure 7.4 shows a variation of the UML activity diagram that we use to model work dynamics. The activity diagram focuses on how roles work together in performing tasks and how they communicate and collaborate. Additionally, goals and event are included to facilitate deeper analysis. With each task, a new goal can become active, remaining so until it is reached in a later task.

7.4.2.3 Representing Tools and Artifacts

The work environment itself usually contains many objects (a hundred or more is not unusual), some of which are used directly in tasks and others that may be "just lying around." The objects can be tools that people use, either software or hardware, but other objects may be directly manipulated in tasks. For some of these objects, it may be relevant to describe them in detail. The descriptions may include their structure, type, and attributes. For the object structure and type, we use class diagrams but without the object oriented–specific parts such as methods. Other task details, such as object attributes and the relations between the objects and their users, are described using templates.

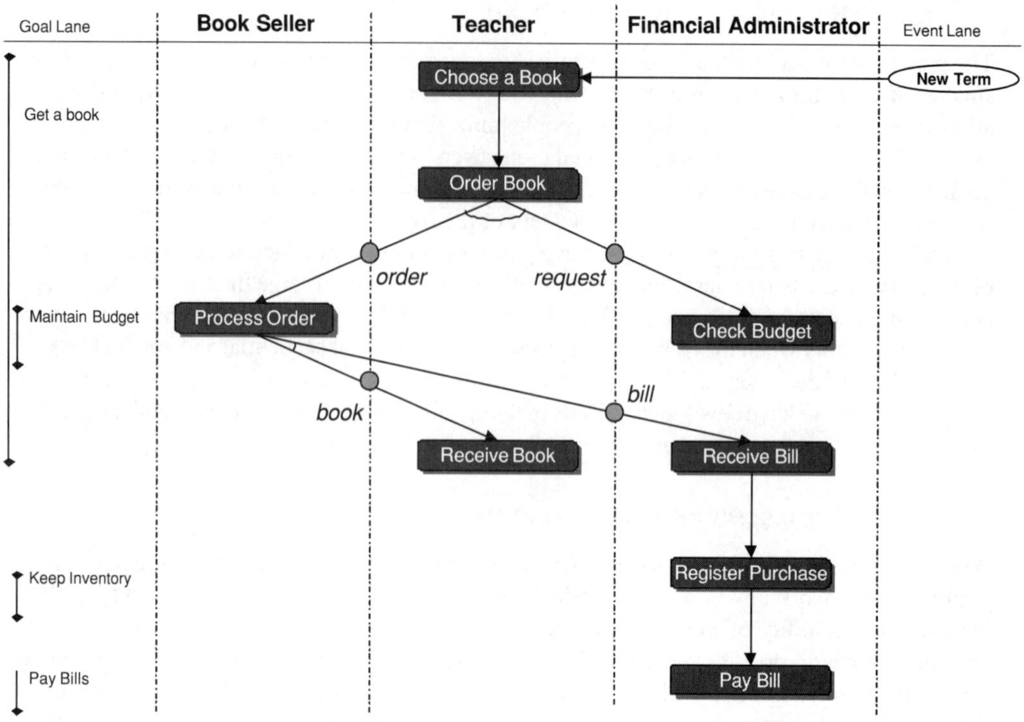

FIG. 7.4. An example of a Flow diagram.

7.4.2.4 *Representing the Work Environment*

In the past, most task analysis methods focused on modeling one user and that user's tasks. However, in current applications, group aspects are becoming more important. Classic task modeling methods lack the power to deal with group situations, modeling only the static part of the task world by identifying roles. This neglects other parts of the organization and dynamic aspects of the task world. People rarely perform their work in solitude. They work together, and share offices, and form social groups.

One aspect of the work environment is the actual physical layout. How big is the room? Where are objects positioned and what are their dimensions? Pictures, drawings, maps, and video clips can capture some of this information. Usually maps or drawings are annotated with comments on the object, such as on their reachability. Most objects that appear in such representations are also represented in task models.

Every work environment has its own culture, which defines values, policies, expectations, and the general approach to work (see chap. 19). The culture determines how people work together, how they view each other socially, and what they expect from each other. In user interface design, taking the culture into account may influence decisions on restructuring work, including roles or their responsibilities. Roles are usually used to describe the formal work structure, although "socially defined" roles are sometimes noted. In practice, roles such as "management" or "marketing" influence each other and other roles. Influence relationships of this type are part of the work culture. Describing the work culture is not straightforward, but at least some influence relationships and their relative strengths can be modeled. The work culture also encompasses policies, values, and identity.

7.4.2.5 Dealing with Multiple Representations

When multiple representations are used, it is important to integrate them. We therefore define each representation as a view on the data that is structured by our ontology. As a result, concepts can appear in several representations at the same time without confusing the semantics of the representations. For instance, a task is only specified once, but it can be part of a task tree, a task flow representation, and a role template. EUTERPE, a tool we developed (van Welie, van der Veer, & Eliëns, 1998a), can help designers who use different representation by safeguarding consistency across representations. The ontology remains hidden within the tool, and the designers are just editing representations, which become "integrated" without extra effort. This saves time, and designers can concentrate on modeling rather than editing activities.

Besides representations based on the ontology, we also capture data from ethnographic studies. These data include video fragments, sound clips, and images of objects. EUTERPE can be used to link the data to one or more of the concepts of the ontology. For instance, a short video clip can give an impression of how the work is actually done in the current situation. We discuss tools more in section 7.7.

7.5 DETAIL DESIGN AND THE USER VIRTUAL MACHINE (UVM)

After the task modeling activity, the Groupware system needs to be designed and specified. Task model 2 describes the envisioned task world in which the new system will be situated, but the details of the technology and the basic tasks that involve interaction with the Groupware system still need to be worked out. This activity consists of three subactivities that are strongly interrelated: specifying the functionality, structuring the dialog between the users and the system, and specifying the way the system is represented to the users.

7.5.1 The Process

The first activity focuses on creating a detailed description of the system as far as it is of direct relevance to the end users. We use the term *User Virtual Machine* (UVM; Tauber, 1988) to refer to the totality of user-relevant knowledge about the technology, both semantics (what the system offers the users for task delegation) and syntax (how task delegation has to be expressed by the users). The object is similar to that of essential modeling (Constantine & Lockwood, 1999), in which the tasks of the users are modeled without using the solutions of particular technology. In actual design, frequent iterations will be needed between the specifications of the tasks models and the UVM specifications. This iteration is an explicit part of the method and essential for the development of a usable system. In the transition from task model 2 to designing the UVM, the tasks and objects are used to sketch the application. The task or object structure is then used to create the main displays and navigational structure. From there, the iterative refinement process takes off.

7.5.2 Representations for Detailed Design

In the detailed design phase, the design is worked out in enough detail that a prototype can be made out of the specifications, enabling a full-blown implementation. Representations for this phase include sketches, screenshots, interactive prototypes, hardware mock-ups, and NUAN (the DUTCH version of a complete formal representation). Especially in this phase, the formal and informal representations work together well. Creative ideas are born in the informal representations and then become more detailed at each iteration. At the end of the

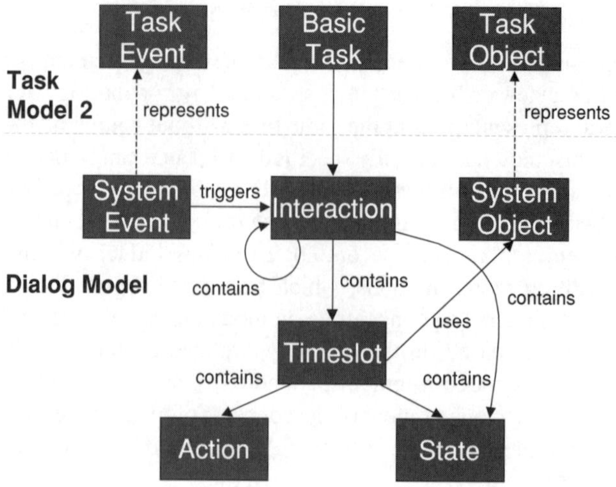

FIG. 7.5. Ontology for linking task models and NUAN.

process, a detailed specification is handed to the implementers, and this specification needs to be unambiguous. If not, the final product may not have the intended look and behavior.

For representing the dialog structure, we developed a variation of User Action Notation (UAN; Hix & Hartson, 1993; see also chaps. 22–24 and 27). UAN diagrams describe the dialog between the user and the system. Constantine's Essential Use Cases (Constantine & Lockwood, 1999) are very similar but are not able to describe the link with the internal functionality. Our variant of UAN includes both extensions that allow event-driven behavior to be specified more easily and extensions for describing mental actions and preconditions. In addition to the UAN diagrams, sketches of screen designs are used to show their representation. We found that using UAN diagrams without sketches was not desirable. The UAN diagrams describe the dialog and part of the functionality, and the sketches cover the presentational aspects of the UVM.

We use a revision of UAN, NUAN, which we explicitly relate to the task world ontology, as shown in Fig. 7.5. A basic task is decomposed into interactions, which in turn are described by a sequence of timeslots. The timeslots contain actions of the user and the system and also states describing the interface state or the application's state. Concerning actions of the user, we make a distinction between physical and mental actions. Physical actions are actions that are "steps" in the dialog structure. Mental actions are cognitive actions that the user performs. Mental actions are included to evaluate knowledge the user needs to access, possible from the system's screen or from past or current output. Again, the designers are not confronted with the ontology directly, but EUTERPE uses it to link concepts together. The designers only perceive it in the tool's functionality. In order to facilitate the transition from a task model to the specification of the UVM, our tool can link a task model with a NUAN model. The basic tasks of the task model then become the top-level interactions in the NUAN diagram, making it more visible which tasks are supported in the NUAN specification. Figure 7.6 shows an example of a NUAN interaction specification.

The three detail design aspects (functionality, dialog, and representation) are mutually dependent, and it is necessary to keep the models consistent. In addition, all the design decisions need to be documented. We use the QOC (MacLean, Young, Bellotti, & Moran, 1991) method to record the design space and the rationale for design decisions. At the moment, support for design rationale has not been integrated in our tools.

Figure 7.7 shows some hardware mock-ups used to determine the basic shape and texture of a handheld device. The prototypes were made of sponge, clay, and polystyrene foam.

Interaction: New Email			
About		**Interface Pre-state**	
This interaction describes the arrival of a new email, in case all previous mails have been read. If another message arrives during this dialogue, the dialogue would re-start immediately after the current one is finished.		1. Email running as a background process though not represented at the interface. 2. All previous emails have been read (unread_mail = false until the new arrival).	
User Actions	**Interface Actions**	**Interface State**	**Connection to Computation**
	! MESSAGE("New email has arrived")		unread_mail => true
	! ASK("Read now", [Yes,No])	Previous dialogue postponed, not aborted	
CHOOSE(task to continue, from the set of previous tasks plus new_email task)			
POINTERTO(<yesbutton>) CLICK(<yesbutton>) || POINTERTO(<nobutton>} CLICK(<nobutton>)	MOVEPOINTER(<yesbutton>) SHOW_EMAIL([latest]) MOVEPOINTER(<nobutton>)		unread_mail => false unread_mail => true
	HIDE_MESSAGE()	Previous dialogue enabled to continue	

FIG. 7.6. A NUAN description.

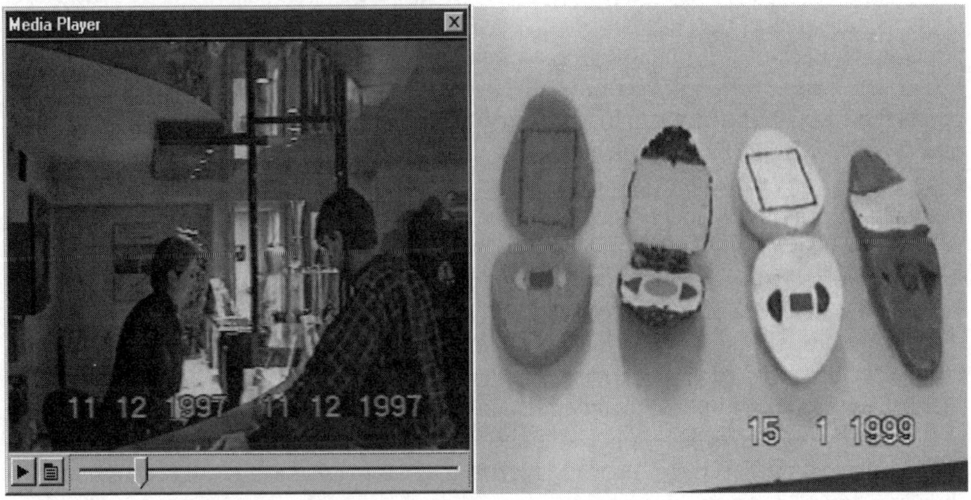

FIG. 7.7. A video fragment and some hardware mock-ups.

7.5.3 Scenarios

Scenario analysis (see chap. 5) is a technique that can be used during the development of task model 2 as well as for the evaluation of detail specifications, both early and late. Scenarios are informal descriptions of tasks that are likely to occur together in a specified situation. Some scenarios can be heuristically evaluated by a group of designers, possibly together with future users. On the other hand, in some cases scenarios show their value when actually acted out. When scenarios are acted out, the actors are instructed to try many variations within their task description.

It is useful to have some of the scenarios acted out by intended users who in fact are sceptical or afraid of using the future technology. These actors will both show how the design could fail and how to maneuver around the supposed problems. To make sure nothing is overlooked, the whole scenario is videotaped. After several scenarios have been acted out, a claims analysis is done. For each scenario, claims have been identified, and this analysis evaluates whether the claims really hold.

7.6 EVALUATION AND USABILITY TESTING

During the entire process, evaluation activities can take place. As soon as an initial task model is available, it can be evaluated using scenarios and use cases. Later on, when some initial sketches for the new system have been done, mockups and prototypes can be used for early evaluation of the design concepts (see chap. 1). Each evaluation activity can cause another iteration. For instance, a task model may turn out to be incomplete after a mockup is evaluated in a critical scenario. In that case, the designers need to go back to the task model and rethink their model.

7.6.1 Evaluating Usability

If a part of the design is worked out in detail, usability testing with a prototype and users can begin. Early evaluation can be done by inspecting design specifications or by performing walk-through sessions with designers and/or users. For early evaluation, we developed a usability framework (van Welie, van der Veer, & Eliëns 1999; see Fig. 7.8).

On the highest level, the ISO definition of usability is followed. This provides three pillars for looking at usability and is based on a well-formed theory (Bevan, 1994). The next level contains a number of usage indicators that can actually be observed in practice when users are at work. Each of these indicators contributes to the abstract aspects of the higher level. For instance, a low error rate increases effectiveness, and good performance speed indicates good efficiency—and hence can be an observable goal for design. The usage indicators are measured using a set of usability metrics.

One level lower is the level of means. Means can be used in "heuristics" for improving one or more of the usage indicators and are consequently not goals by themselves. For instance, consistency may have a positive effect on learnability, and warnings may reduce errors. On the other hand, high adaptability may have a negative effect on memorability while having a positive effect on performance time.

Each means can have a positive or negative effect on some of the indicators. The means need to be "used with care," and a designer should take care not to apply them automatically. The highest usability results from an optimal use of the means, that is, where each means is at a certain "level" somewhere between "none" and "completely/ everywhere/all the time." In order to find optimal levels for all means, the designer has to use the three knowledge domains (humans, design, and task). For example, design knowledge such as guidelines should include how changes in use of the means affect the usage indicators.

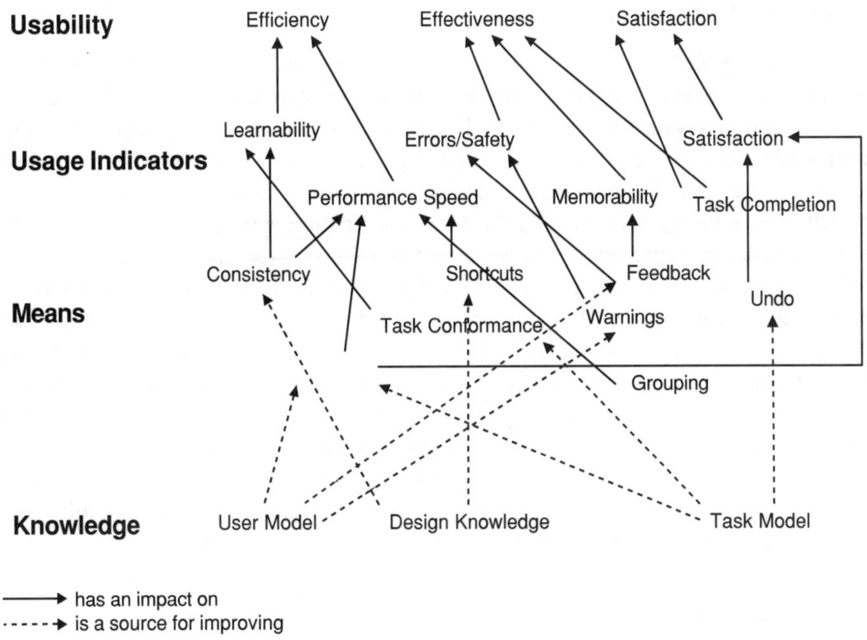

FIG. 7.8. A framework for usability.

7.6.2 Improving Usability

When evaluation shows that the usability needs to be improved, the problem is to find out which means need to be changed and how they need to be changed. As mentioned earlier, means can have a positive effect on one usage indicator and a negative effect on another. In some cases, the designer has to take a step back and look at the knowledge domains again. For instance, when the task conformance is seen as a problem, the task model can give the designers information about what is wrong with the task conformance. Similarly, the user model may give information about the memory limitations, which may require the design to have more or better feedback of user actions. Obtaining extra data may require the task model to be extended to include previously undescribed data.

7.7 SUPPORTING THE DESIGN PROCESS

Tools help to structure the process of design and aid designers in understanding their data (see the chapters in part IV for examples). Representations and diagrams are an integral part of many methods. Tools that allow these representations to be edited help reduce design costs and time. A problem of many tools is their availability. Many descriptions of methods from academia refer to tools that support the methods, but often these tools are not publicly available or not developed far enough in order to be of any practical use.

Our tool EUTERPE (van Welie et al., 1998a; see also chaps. 6 and 24) was developed to support the process outlined in the previous sections. It is being constantly revised in response to comments of users in the field and is freely available to anyone. It has been used for several years in both industry and education. Because it was designed to deal with multiple representations, it is explicitly based on an ontology that semantically links representations.

7.7.1 Deriving Representations

The ontology only defines a structure for the task model data and does not limit or dictate any representation. The tool is based on a repository that contains the data of a design project. All representations are views on the repository. The task world ontology is specified in a logic programming language (Prolog, in our case) and is the main data structure for the repository. EUTERPE offers several different representations, and all the representations are coherent because each is built up on the fly out of the same information specified using the ontology. For instance, a task tree representation does not exist in the logical model, but the structure is derived from the specified relationships of subtasks within tasks. By issuing queries to the Prolog engine, all the relationships can be inspected. Naturally, EUTERPE allows most representations to be modified as well, in which case the views need to assert the right facts in the Prolog engine. For instance, when a new subtask is added by editing the task tree view, a new fact, *subtask(X,Y)*, is asserted. This way the users of EUTERPE can work with the representations without having to deal with the logic representation underneath.

The current version of EUTERPE supports representations for task modeling and NUAN diagrams. Hierarchical structures and templates can be created for all concepts. Additionally, the models can be analyzed semiautomatically (van Welie et al., 1998a).

7.7.2 Documenting a Task Analysis

EUTERPE has two ways of producing documentation. First, it can use the printing functionality. Task trees and object hierarchies as well as lists of events and agents can be printed. If a tree does not fit on one sheet of paper, the printing automatically tiles it on multiple pages. The second way of producing documentation is by exporting the specification to HTML. EUTERPE can generate a set of HTML pages, including an applet containing a task tree that allows browsing of the task analysis results. These pages are a "read-only" view of the model, as changes are not propagated back to the Prolog engine. All the pages together can be seen as a hyperlinked task analysis document because for each concept referenced a hyperlink is added. For instance, a reference to an object used in a task becomes a link to the description of that object and vice versa. Links to images and video fragments are also generated. For navigation purposes and for getting a better overview, a simple Java applet is also included. The applet shows the trees graphically, and when a node is selected, the browser jumps to the corresponding entity. When large design teams actually used EUTERPE, the produced HTML documents were put on a Web server, and these constituted the main reference document for all members of the design team.

Another important aspect is the integration with common office applications. Designers typically write reports in which they need to include some of the design representations, such as task trees of parts of UVM specifications. A tool must therefore be able to produce output in formats that can be used in typical office applications. EUTERPE uses Windows Metafiles as the exchange format for representations. Such representations can be arbitrarily scaled or modified in office applications without loss of image quality.

7.8 DESIGNING IN TEAMS

In DUTCH, team-based design is a requirement. Teams need to work together in order to make the right decisions and to obtain a high-quality end product. For example, in a typical project we establish the following teams:

- task analysis group
- detail design group

- evaluation group
- scenario and prototyping group
- management group.

We found that it is important to have one independent group that manages all the other groups. The management groups guards the time schedules, deals with delays and group conflicts, and supervises the overall project documentation. The management's main purpose is to guard the dialogue between disciplines as well as between design phases. The management group monitors the use of representations to communicate between design subdisciplines as well as between the design team and the users, the stakeholders, and the client.

Designing in teams can work well and potentially has several advantages. First of all, working in groups on separate issues creates a competition effect. Each group depends on each other and is being judged by other groups. If one group delivers poor work the other groups immediately remind them. On the other hand if a group has developed really new ideas, it is very disappointing if the other groups just discard their creativity or fail to see its merits. Not surprisingly, it is not unusual that certain groups get really upset by the work of others. However, this even leads to qualitative improvements of the final work as it forces the management as well as the whole team to reconsider and argue about misunderstandings and rejected proposals. Another advantage is that in this way the expertise people have is used optimally. In educational settings, the management team forms the groups and designers have to apply for a position. This way everyone gets to determine what he or she does, within some limits of course. In industry, a comparable "marketing" situation and resource management is in fact common practise.

7.9 DUTCH IN USE

DUTCH has been used in several projects over the last few years in Europe, both in industry (van der Veer, Hoeve, & Lenting, 1996; van Loo et al., 1999) and in education. The method is taught to computer science and psychology students in four Dutch universities, at least one Spanish University, at least one British University, two Romanian universities, and the Open University in the Netherlands.

One of the things we have learned was that it is considered valuable to explicitly use a method that structures the design process from task analysis to usability testing. For designers, it gives them a structure for activities and contributes positively when they are negotiating with higher management (e.g., trying to get support for data-gathering activities). During the analysis phase, the GTA conceptual framework proved to be of great value. It worked as a kind of checklist to focus attention on things that matter in performing tasks. Some activities, such as ethnographic studies and interaction analyses, are new to analysts, but after the initial hurdle, the benefits are clear. It helps to create a common vocabulary in order to simplify discussions.

The representations have been evolving but now seem powerful enough for practical use. However, tool support is essential for creating and managing these representations. Because of the required iterations in the design process, the tools save designers a lot of time. When using our tool, we saw that it was important to keep editing functionality simple so that the more advanced features are not directly visible.

However, what sometimes remains difficult is to make people understand what it means to perform task analysis at the customer site and how it should be integrated into the existing system design culture. To the majority of traditional IT personal and organizations, these ideas are radically new. Their acceptance can be helped along by (partly) applying the method in current projects.

7.10 CONCLUSION

Although design methods exist that claim to cover similar design aspects, we found that an integrating method is what many designers would like. Our approach differs from others on three main points:

- DUTCH uses multiple representations integrated through use of an ontology. The representations have been used frequently and have proven to be sufficient for most design cases.
- DUTCH has a wider scope than most other methods throughout the entire development process, from initial task analysis and envisioning to detailed design and evaluation. It considers both multiple and individual users.
- DUTCH is supported by tools that are publicly available and tested in practice.

An integrated method is desirable when the method is applied in practice. For example, when we work with industry, we find that having tool support is an important issue. Being able to manage data and design representations using a tool can save a lot of time. Such benefits can help persuade higher management that it is important to do structured task-based design.

DUTCH is a design method that uses both formal and informal representations. We found the combination of these representations to be useful and desirable. The formal representations give precision when needed whereas the informal representations can effectively communicate ideas within design groups as well as to clients. DUTCH method defines a clear process, models, and representations to be used and tools that support the design process, which may explain why it is used in several countries in Europe, both in industry and education.

REFERENCES

Bevan, N. (1994). *Guidance on usability* (ISO 9241-11; Ergonomic Requirements for Office Work With VDTs). International Organization for Standarisation, Geneva, Switzerland.

Beyer, H., & Holtzblatt, K. (1998). *Contextual design*. San Francisco: Morgan Kaufmann.

Card, S. K., Moran, T. P., & Newell, A. (1983). *The psychology of human-computer interaction*. Hillsdale, NJ: Lawrence Erlbaum Associates.

Constantine, L. L., & Lockwood, L. A. D. (1999). *Software for use*. Reading, MA: Addison-Wesley.

Hix, D., & Hartson, H. R. (1993). *Developing user interfaces: Ensuring usability through product and process*. New York: Wiley.

Johnson, P., Johnson, H., Waddington, R., & Shouls, A. (1988). Task-related knowledge structures: Analysis, modeling and application. In D. M. Jones, & R. Winder, (Eds.), *People and Computers IV* (pp. 35–62). Cambridge: Cambridge University Press.

Jordan, B. (1996). Ethnographic workplace studies and CSCW. In D. Shapiro, M. J. Tauber, & R. Traunmueller (Eds.), *The design of Computer Supported Cooperative Work and Groupware systems* (pp. 17–42). Amsterdam: North-Holland.

MacLean, A., Young, R., Bellotti, V., & Moran, T. (1991). Questions, options, and criteria: elements of design space analysis. *Human-Computer Interaction, 6*, 201–250.

Sebillotte, S. (1988). Hierarchical planning as a method for task-analysis: The example of office task analysis. *behavior and Information Technology, 7*, 275–293.

Tauber, M. J. (1988). On mental models and the user interface. In: G. C. van der Veer, T. R. G. Green, J.-M. Hoc, & D. Murray, (Eds.), *Working with computers: Theory versus outcome* (pp. 89–119). London: Academic Press.

van der Veer, G. C., Hoeve, M., & Lenting, B. F. (1996). Modeling complex work systems: Method meets reality. Paper presented at the Eighth European Conference on Cognitive Ergonomics (EACE), Granada, Spain (pp. 115–120).

van der Veer, G. C., van Vliet, J. C., & Lenting, B. F. (1995). Designing complex systems: A structured activity. In *Symposium on Designing Interactive Systems (DIS '95)* (pp. 207–217). New York: ACM Press.

van der Veer, G. C., Lenting, P. F., and Bergevoet, B. A. J. (1996), GTA: Groupware task analysis-Modeling complexity, *Acta Psychologica* 91(1996), pp. 297–322.

van Loo, R., van der Veer, G. C., & van Welie, M. (1999). Groupware task analysis in practice: A scientific approach meets security problems. Paper presented at the Seventh European Conference on Cognitive Science Approaches to Process Control, Villeneuve d'Ascq, France .

van Welie, M., van der Veer, G. C., & Eliëns, A. (1998a). EUTERPE: Tool support for analyzing cooperative environments. Paper presented at the Ninth European Conference on Cognitive Ergonomics, Limerick, Ireland.

van Welie, M., van der Veer, G. C., & Eliëns, A. (1998b). An ontology for task world models. In *Proceedings of the Fifth International Workshop on Design, Specification, and Verification of Interactive Systems (DSV-IS '98)* (pp. 57–70) Vienna: Springer-Verlag.

van Welie, M., van der Veer, G. C., & Eliëns, A. (1999). Breaking down usability. M. A. Sasse and C. Johnson (eds.) Human-Computer Interaction INTERACT '99. IOS press, Amsterdam, Netherlands (pp. 613–620).

II

IT Industry Perspectives

Perhaps the biggest problem that continues to bedevil HCI researchers is that of "giving HCI away" (Diaper, 1989). The Task Analysis for Knowledge Descriptions (TAKD) method was finally publicly abandoned by its champion primarily because, although of great power, it was too complicated for realistic use in the software industry, even when supported by a CASE tool (Diaper, 2001). Diaper (2002) called this the "delivery-to-industry problem," and he currently recommends that method developers should balance method requirements that arise from the problem space, application domain, and so on, against an equivalent set of requirements concerning their method's end users and their environment, tasks, work practices, and so on. Properly, the end users of task analysis methods should be the people in the software industry who collectively are responsible for delivering computer systems.

This part is not for the faint-hearted. We can imagine many academics screaming at the liberties that have been taken in the name of the twin commercial pressures of finance and time. Nonetheless, by reading the chapters in this part, such academics may gain some appreciation for the constraints that the commercial world is under.

So serious do the editors consider the delivery-to-industry problem that they have deliberately chosen to promote the chapters on IT industry perspectives to a favored location—immediately after part I, which lays out the fundamentals of task analysis. Readers are invited to study the chapters in part II to better understand the use of task analysis in commercial IT projects and also to discover some of the important constraints on its use.

Chapter 8

A Multicultural Approach to Task Analysis: Capturing User Requirements for a Global Software Application

Jose Coronado and Brian Casey

Chapter 8 opens part II because it offers a beautiful example of international teams working on global, commercial software projects. Project constraints, particularly regarding time, are fierce, so efficiency, as well as effectiveness, is important. The chapter describes a development process that is highly iterative, has quite short development cycles, and uses task analysis within each cycle.

Chapter 9
JIET Design Process for e-Business Applications
Helmut Degen and Sonja Pedell

Hauntingly, the project constraints described in Chapter 9 resemble those of the immediately preceding chapter. The chapter shows how task analysis is integrated into the Just-in-e-Time (JIET) method, which uses many techniques and at first looks complicated. Nonetheless, the project requirements for efficiency allow the entire JIET method to be completed in only one month, with most activities taking mere days to complete. The authors offer an analysis of the time needed to conduct different activities as part of the design process.

Chapter 10
User-System Interaction Scripts
Rick Spencer and Steven Clarke

User-System Interaction Scripts provide a task representation that is basically a form of activity list (chap. 1), in other words, a prose description of a task split up into steps. Chapter 10 describes user-system interaction scripts and how they are used at Microsoft by development staff who are not HCI experts and who have but hours to undertake this form of task analysis in the software development process the company uses. The chapter shows that even under such project constraints, task analysis is still able to provide a valuable contribution to the development process. The chapter thus provides a demonstration of the falsity of the myth that all task analysis involves a time consuming, detailed analysis of tasks (chap. 1).

Chapter 11
Task Layer Maps: A Task Analysis Method for the Rest of Us
Jonathan Arnowitz

As chapter 11 recognizes, the use of task analysis in commercial IT development environments faces the problem that many of the people involved in a project may have difficulty understanding the diagrammatic representations common in task analysis methods. Task Layer Maps, which have a basis in a formal representation of a task model, provide a means to reduce the complexity of representations, increase the understanding of project team members, and enhance their contribution to task analysis exercises. Task Layer Maps are used to sequence tasks by temporal order and remove redundancy in task transitions to allow the most parsimonious task description possible.

Chapter 12
Task Analysis for e-Commerce and the Web
Tobias van Dyk and Karen Renaud

The heterogeneity of the World Wide Web's users presents a particular problem for commercial IT projects involving the design of e-commerce Web sites. Chapter 12 argues that the resource constraints on such projects require an approach that is efficient as well as effective and proposes that usability metrics overlaying a task-based description can provide a commercially realistic solution. The authors offer a proceduralized approach for calculating Web usability, so you only need follow their instructions.

Chapter 13
Using Task Data in Designing the Swanwick Air Traffic Control Centre
Peter Windsor

Along with plant process control (e.g., chap. 3 and 17), air traffic control (ATC) is one of the most widely reported application domains for the use of task analysis; there are four chapters in this handbook that use ATC examples: this chapter and chapters 1, 19, and 25. Readers unfamiliar with ATC will greatly benefit from chapter 13's description of ATC tasks. All readers should benefit from the chapter's final sections, which indicate the awesome complexity of the ATC application domain. Indeed, the suspicion should be entertained that ATC is so complicated, with so many exception cases, special circumstances, and so on, that anything even close to good design is nigh impossible. Probably no task analysis method described in this handbook can fully cope with the complexity of ATC, although the need for using task analysis in this domain is self-evident.

The approach to task analysis described in chapter 13 is an embedded one, task analysis being an integral part of virtually all the HCI activities undertaken in the design of computerized ATC systems. In this, it supports the claim in chapter 1 that task analysis is at the heart of virtually all HCI.

REFERENCES

Diaper, D. (1989). Giving HCI away. In A. Suteliffe & L. Macaulay (Eds.), *People and computers V* (pp. 109–120). Cambridge: Cambridge University Press.

Diaper, D. (2001). Task Analysis for Knowledge Descriptions (TAKD): A requiem for a method. behavior and Information Technology, *20*, 199–212.

Diaper, D. (2002). Human-computer interaction. In R. B. Meyers (Ed.), *The encyclopaedia of physical science and technology* (3rd ed.; Vol. 7; pp. 393–400). New York: Academic Press.

8

A Multicultural Approach to Task Analysis: Capturing User Requirements for a Global Software Application

José Coronado
Hyperion Solutions Corporation

Brian Casey
Hyperion Solutions Corporation

Complex business software applications present different challenges to the product design and development teams. Software development companies are often faced with difficult choices in meeting delivery deadlines. If the product is designed in isolation, there is a risk of misrepresenting business processes, making incorrect assumptions about the users, or simply overlooking specific user requirements. Software development teams should reach out to subject matter experts and end users to understand how the product supports the business process and meets user requirements.

The authors will discuss in detail their experience designing user interfaces for complex financial software applications. They describe a task analysis process they have employed to solve specific interaction problems. Specifically, they will talk about how they used it to work with international panels of users and subject matter experts. The results of these activities provided the design team with a better understanding of the target user audience, the global business processes, and key regional and legal requirements.

In addition, the authors show how this information was transformed into product design alternatives. They provide details about UI design opportunities identified and how the international panel of participants became ad-hoc members of the development team throughout the design phase. They also highlight how the findings were later validated through usability sessions conducted in North America and Europe.

8.1 INTRODUCTION

Companies in the business of providing technology solutions to broad markets are often forced to make trade-offs between product features and a successfully met deadline (see also chap. 9). Entire features may be dropped and optimal process flows might be modified in favor of an easier-to-implement system to shorten development times and keep to an aggressive schedule. These product changes usually are misaligned with the original user requirements that were gathered under more ideal circumstances. They may also not fit the users' needs or expectations or conform to standard industry practices. If a product does not meet needs or standards, the company may have to reengineer its own business processes to match the product, a quite different result than having the product complement these processes in their original form.

Software development teams are always at risk for making inaccurate assumptions about the user community for which they build applications. The pressure these teams are under to meet executive demands, revenue targets, and market expectations increases this risk the likelihood of affecting specific user groups is high. When requirements are modified or dropped from a product. For example, if a software product is not fully localized in French because resources would not allow this to happen in an acceptable time frame, it could not be sold in France (this would be mandated by the Toubon Law, Legendre, Perrut, Deniau, & Fuchs, 1994). In other cases, international user groups may reside in locales with specific legal requirements that could affect the decision whether to buy a product. For instance, in financial reporting, there are differences in the reporting requirements and laws governing business practices in the United States and European countries. If a vendor neglects to investigate and abide by these key requirements, its business success could be jeopardized.

Software development teams should reach out to subject matter experts and end users to understand how their product should support their user audience. A systematic and scientific approach to task analysis enables a product development team to understand the business processes it must support, validate requirements and iterative user interface designs, and gather valuable information from the user community.

8.1.1 Hyperion Solutions

Hyperion Solutions Corporation is a U.S.-based company and one of the leading providers of Business Performance Management (BPM) application software. More than 4,000 customer companies use Hyperion's BPM technology and application products worldwide. Most products are large scale and are used for managing business performance through multi-dimensional analysis, including financial analysis, performance measurement, and business modeling and reporting. Hyperion has a direct market presence in over 20 countries and generates about one third of its revenue from its international markets.

In 1998, Hyperion initiated the development of a suite of products for the business analysis market. These suites primarily provided financial analysis, management reporting, and planning and forecasting functionality. Hyperion was in the position of needing to expand its existing knowledge base and quickly and effectively gather information and refine the requirements for these new products. The next section explains the methodologies used to achieve those goals and the benefits derived throughout the development process.

8.1.2 User-Centered Design (UCD) at Hyperion

User-centered design methods have been in use at Hyperion since 1994. Professionals from three different areas compose the virtual UCD team. User interface engineers who are integrated into the product teams deal primarily with the user interface design and layout and the behavior of the product's front end. Visual designers are responsible for supporting all products teams and providing them with innovative, consistent, and easy-to-use graphical screen elements. In some instances, they are required to support the graphical elements of print and online documentation. Lastly, usability engineers employ systematic and scientific methods to collect data that help determine the ease-of-use of Hyperion products. The user-centered design team as a whole assists the development teams in delivering easy-to-use products that effectively support the client's business processes.

As described by Coronado and Livermore (2001), user interface engineers at Hyperion are integrated with the product teams. This allows close collaboration between the development disciplines and the user interface engineers. On the other hand, usability engineers are part of a centralized organization intended to support any product team in need of its services. This alignment allows for a virtual user-centered design team to collaborate across different areas of the company. As part of their strategic work, user-centered design engineers have developed and evolved guidelines for WEB and GUI design. These guidelines, among other things, address accessibility issues and include requirements for user interface internationalization and localization for all products. Through usability services and methodologies, user-centered design also supports data collection for activities such as task analysis and usability evaluation performed in conjunction with customers and subject matter experts.

8.2 TASK ANALYSIS

Task analysis is understood in the organization as the study of how people perform their jobs to accomplish a given goal. It details the tasks and steps that users go through to achieve a work objective and may not necessarily be tied to a particular user interface or system. However, the subject matter of the analysis would be specifically related to the business processes for an identifiable market niche.

Task analysis is probably one of the most efficient data-gathering methodologies used at Hyperion to enable its product teams to better understand users and their business processes. It produces instant feedback while the facilitator helps the participants make progress in a fun, relaxed team atmosphere. It is a tool that has helped capture in detail the procedures that finance- and accounting-rooted users follow to accomplish their goals. By involving a broad range of participants, from end- users to subject matter experts, user interface engineers, product managers, and software developers, product teams can obtain a big picture of a task flow in just a few hours. Within a couple of days, this same team can collaborate to formulate a detailed understanding of all the intricate steps of the specific business process. Unfortunately, task analysis is often overlooked in business software development and its potential benefits remain unavailable to the makers and the users of the product.

8.2.1 Why Is Task Analysis Important for a Software Development Organization?

All too often, a software product and its primary interaction with its users are based on a programmer's mental model of how the system should work. Unless the programmer has a firm understanding of his or her audience, the product is likely to cause unnecessary user frustration, confusion, and error. Other probable consequences include longer learning periods, significantly higher training costs for the end users, and higher "help desk" costs for the manufacturing company.

Task analysis can help development teams avoid these pitfalls altogether. Task analysis enables the professionals involved in the product design cycle to learn how users perform their routine tasks and enables the development team to have a common understanding and framework on which to build the system. By sharing an understanding of the users, the development team has a foundation on which to begin the task of defining product requirements.

8.2.2 When in the Development Cycle Is Task Analysis Conducted?

Kicking off the development phase with task analysis workshops sets a precedent for employing a systematic approach to satisfying user requirements throughout the development cycle (see Fig. 8.1). This will also help when it comes time for making those necessary enhancements later, since a tested methodology has already been put into place. The end result should be a product that, from the very beginning, was driven by those who will ultimately be the beneficiaries of the solution the development team provides and that should closely meet users' expectations and achieve very high customer satisfaction levels. However, the recent emphasis on rapid development dictates that some data collection activities will be deferred in favor of shorter release cycles.

When the release cycle is condensed, the user-centered design professionals must adapt to the surrounding environment and ensure that important data collection activities occur and have an effect on subsequent releases. Usually toward the end of the "current release" cycle, activities like task analysis can be performed in parallel with the product team processes for releasing the product commercially without having any impact release time-frame (see Fig. 8.2).

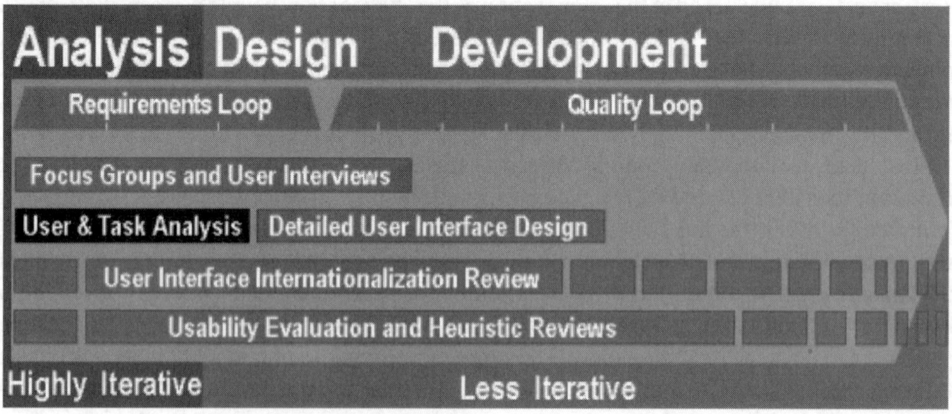

FIG. 8.1. Usability activities mapped within the ideal development process.

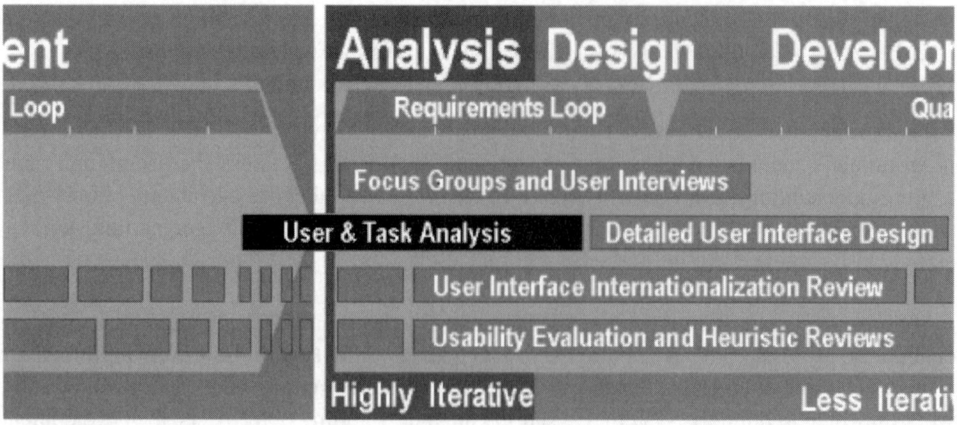

FIG. 8.2. Task analysis conducted at the end of one release cycle in order to benefit subsequent release.

TABLE 8.1
Typical Task Analysis Participants

Subject matter experts
Real end users
Potential and existing clients
Architects/systems engineers
Marketing representative
Product management (decision maker)
User-centered design (facilitator and UCD expert)
Developers
Technical writers
Trainers

8.2.3 Who Are the Participants?

A participatory task analysis methodology gets many stakeholders involved. It encourages participants to work collaboratively as peers and ensures equity among contributors. A typical group may include six to eight active participants. Some participants (see Table 8.1) may participate primarily as observers, contributing only when their specific expertise is desired.

Stakeholders such as software architects and developers should participate when the user groups need to validate the feasibility of a design idea. A product manager will benefit from feedback gathered on product requirements. The user-centered design engineer on the product team may receive the most valuable information needed to understand the business processes and the users. This knowledge is then put to work in the construction of the product.

8.2.4 How is a Task Analysis Session Conducted?

Task analysis at Hyperion utilizes the CARD methodology. The participants are asked to brainstorm and write on index cards a verb and a noun that relate to the business process under analysis. The brainstorming phase of the session encourages uninhibited contributions, as participants are free to write down any verb-noun pair. The submissions often refer to very high level activities (e.g., "Prepare Financial Report") and lower level tasks (e.g., "Enter Data Adjustment Comment").

The brainstorming initially is done by the participants independently. When they exhaust their ability to generate ideas individually, the facilitator encourages them to brainstorm in small groups. The groups then come together to share the individual and small-group word pairs. As the ideas are shared, duplicates are consolidated, but only after each submitter agrees that the meanings of the two word pairs are equivalent. During this phase, some trends may be apparent, and it is the role of the facilitator to solicit and validate any trends at this time. If the discussion becomes more focus group–like, the facilitator moderates the discussion and maintains balance, not allowing any single contributor to dominate the discussion.

Discussions can also be handled in focus group style, with the facilitator moderating the discussion, or take place in small groups in which the participants explore some of the business process ideas at hand. The facilitator moves from group to group to help the participants, to answer any questions that they may have about the methodology, and to ask them questions when there are doubts about a process or a task. It is important to solicit and guarantee balanced contributions from all.

8.2.5 What Is the Output of a Task Analysis?

The initial output provides the big picture, and some high-level processes are easily identifiable. The processes are captured on large flip chart paper on which adhesive note cards are arranged in a logical order. Visual and detailed output is discussed by all the participants. Groups of tasks, changes, and refinements are talked about, and some level of consensus is reached. If there are clear differences in certain processes, these need to be documented. In some instances, the environment (e.g., the legal system) could dictate certain requirements; in others, the particular industry might employ unique business processes (e.g., some processes are unique to insurance and do not occur in banking, etc.).

The most common types of deliverables at the end of a task analysis session are these: (a) the task flows for the tasks or areas chosen, (b) identified problem areas within the flows, and (c) the blue-sky or ideal task flows or task support.

All task flows are built jointly and then presented to the group, which could include additional stakeholders. The presentations help the participants recapitulate the discussion, validate the results, and uncover areas that may not make sense—in other words, check for completeness and clarity.

Users can also identify potential problem areas and issues that would need to be addressed. The discussion helps determine the priority of the different issues uncovered. If developers are participating in the meetings, they can talk about the feasibility of addressing certain user issues.

The participants also engage in blue-sky or ideal-world speculation. Users brainstorm on how to improve current task flows without any constraints and think on those things that would be nice to have in the flows. The input that fuels this discussion consists of the participants' knowledge of how they work and complete tasks, any relevant documentation or artifacts, and the current task flows. The goal of the blue-sky discussion is to design and document optimized (future) task flows.

The product team uses the output of task analysis to define task objects, build paper prototypes, and perform usability tests.

8.2.6 Iterative Design Process and Usability Evaluation

Task analysis on its own is a very powerful methodology, but the authors believe that its power and effectiveness can be enhanced by complementing it with other user-centered design methodologies or processes. Iterative design processes and usability evaluations are key to guaranteeing that a successful product will be delivered.

8.2.6.1 Iterative Design Process

The iterative design process allows the designers to continually evolve and refine the product during the development process. Using short and fast cycles of "analysis, design, and evaluation," the designers and some members of the development team can refine the design idea based on user requirements and feedback as well as implementation technology issues or restrictions. Practical applications of the iteration could focus on a small portion of a module or a whole section of a product (see Fig. 8.3).

8.2.6.2 Usability Evaluation

During a usability session, end users perform a series of real-world business tasks using a prototype or a functional version of the product. By means of a systematic data collection approach, the product team gathers performance (objective) and opinion-based (subjective)

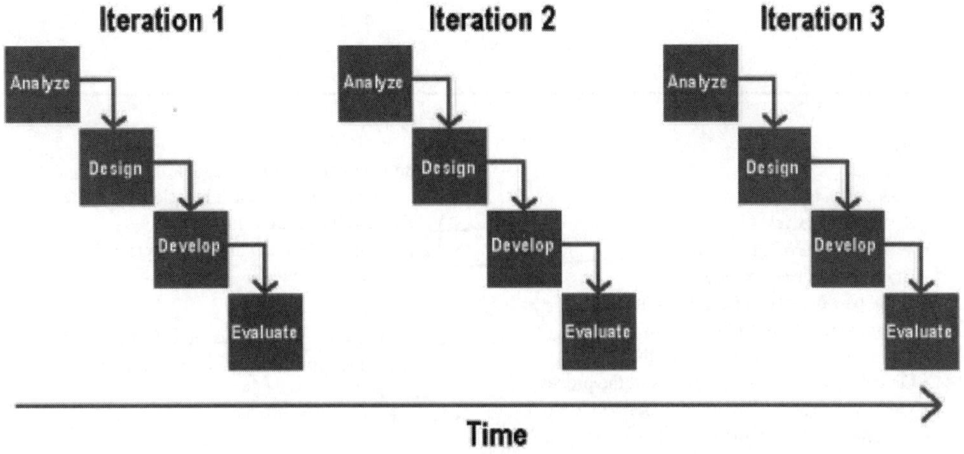

FIG. 8.3. Iterations may have a different emphasis: functionality, usability, acceptability, scalability, viability, and so on.

data that allow it to determine the ease-of-use of the user interface and uncover potential problems in the task flow or the user interface elements.

Usability sessions can also help determine how global the design is. Different cultures may have different interaction patterns or user needs for similar processes. Usability evaluation helps the product team avoid potential problems such as these:

- The product has an interface that is difficult to use.
- The product contains icons that some cultures will find offensive.
- A process flow within the product does not match specific practices in international locales.

In the following case study, the authors describe in detail the methods used to design an application and explain how the design was validated throughout the development process.

8.3 CASE STUDY: A FINANCIAL ANALYTICAL APPLICATION

From the moment that user-centered design processes were integrated into the best practices of the development life cycle, product development teams started considering the user as a member of the design team. As Bailey (1989) explained in *Human Performance Engineering*, the design team involves the user as a refiner of the idea, not as the designer of the product.

This case study shows how a group of user interface and usability engineers, in collaboration with the product managers, conducted task analysis, design, and usability sessions with clients and subject matter experts from North America and Europe. Some of the key similarities and differences between the designers' and the users' perspectives are covered. After these sessions, the product team was able to better understand the community of users and their business processes and also learned new product requirements to ensure the optimization of the financial application system.

TABLE 8.2
Task Analysis Structure

Day 1	Day 2
• Agenda. • Task analysis overview. • Session objectives. • Why do we have this process? • Brainstorm ideas individually. • Make participants share their point of view. • Organize the ideas on Post-it and paper. • Define the high-level picture of the process. • Divide or identify the big pieces of the process. • Identify the relation/interaction between each piece. • Document with comments while still fresh in participants' minds (facilitators).	• Recap where the process was left. Did we finish? • Go into more detail on each big piece identified. • Validate (walk-through) flows. • Agree on next piece to "drill" into (prioritize). • Document. Day 3 • Recap where the process was left. Did we finish? • Validate again (probably with developers). • Wrap up. • Document.

8.3.1 The Problem

In 1998, Hyperion was embarking on the creation of a new set of packaged applications. Up to this point, Hyperion was the leader in multisource consolidation, management reporting, and planning and forecasting applications. These applications are still very powerful and rich in features, but they did not address the needs of companies requiring a system architecture that could be easily deployed to a large number of users in a quick and reliable manner. The requirements of these companies demanded that the product teams reach out to the users to determine and clarify what their needs were and how they would be translated into product design.

An adjustment module for one of these new applications was the main target defined by product management. User-centered design engineers devised a three-phase plan that included task analysis, design, and usability. The task analysis would enhance the designers' understanding of the module, the business process, and the users; the design phase would cover the transformation of this understanding into user interface Design alternatives; and the usability phase would help the designers refine the idea before its implementation.

The sessions covered informal discussions, brainstorming, and walk-throughs of the different business processes within the module. Table 8.2 shows the structure of the 3-day session.

8.3.2 Understanding the Users

The user-centered design group worked with an international panel that included participants from the United States, Canada, and the United Kingdom. These participants were subject matter experts and came from industries such as civil engineering, construction, banking and finance, manufacturing, and consulting.

The participants discussed the users of this business module from three perspectives. They first looked at who the users would be (Table 8.3), then at the roles they would have and the types of tasks they would perform (Table 8.4), then at the user skills associated with different roles (Table 8.5).

By recruiting participants from different geographic regions and different industries, the design team was able to determine commonalities for the user profile of this financial application. The participants discussed and identified the professional roles within the business process for the adjustment module. At the same time, they began an initial discussion of the tasks involved in this business process and about some of the instruments used.

TABLE 8.3
Some Users for the Adjustment Module

Management accountants
Statutory consolidation accountants
Head office person
Controller in headquarters
Sr. and Jr. accountants
Auditors
Analysts
Output users (report users)

TABLE 8.4
Some User Types and Their Tasks

Senior accountant
- Liaison with sites; receives information
- Talks to auditors
- Head office person: reviews and approves
- Controller in headquarters

Analysts
- Review output/reports
- Talk to controller

Output users (i.e., of reports)

Auditors (external)
- Review output from accounting systems
- Propose adjustments
- Talk to the controller

Junior accountant
- Inputs data
- Reviews and raises issues
- Publishes data

TABLE 8.5
Users and Their Skills

Senior accountant
- High professional skills
- Ten or more years of professional and computer experience—computer on their desk
- Does not need to learn new computer skills
- Does not write "code," complex logic or formulas
- Comfortable using the system
- High level of awareness of what the system does, but accountant does not necessarily have to do the task him- or herself

Junior accountant
- Medium to high computer skills
- Less amount of experience with computers but possibly more years of education with them
- Less experienced professional than senior account

8.3.3 Understanding the Business Process

The participants understood that the main objective of the task analysis session was to define and describe the adjustment business process that clients and consultants use without tying it to a specific system or product. The discussion, which was focus group–like in format, included brainstorming, open discussion, and consensus building. In addition, the users presented the findings of each section to themselves and to other stakeholders who participated as observers. These presentations and task flow walk-throughs helped the discussants understand the logic behind the groupings and the chronological order of the tasks (based on the users' interpretation of the business process). They also helped validate the task flows with the end users themselves, the product managers, and the development team.

Since the information gathered is closely related to a financial business process and relevant only to the product design teams, the authors will not discuss it in detail. However, for illustration purposes, the results of the first part of the task analysis are described below.

Looking at the process from a high-level perspective the participants identified these major areas (Fig. 8.4):

- Analysis and system setup.
- Periodic process.
- Maintenance.
- Year-end process.

Using a mid level point of view, the participants considered the process to have the following areas (Fig. 8.5):

- Preadjustment process.
- Adjustment process (input, review, approve, publish).
- Reporting (loops back to the adjustment process)
- Final working papers.
- Maintenance.

8.3.4 Transforming New Understandings Into Product Design

The task analysis session for the adjustment module produced very valuable information. The product manager was able to uncover new product requirements and refine the existing ones. The designers were able to identify and understand key user interface issues and design opportunities and transform the findings into product design alternatives. In addition, some of the participants in the task analysis became ad hoc members of the design team and participated in design reviews and usability sessions.

Table 8.6 lists the major areas and task processes discussed with the users and shows how each was translated into a task-oriented process in the product. The structure and the task groups associated with it went through several iterations of design. Its interpretation in the user interface as well as all the elements needed to address the product requirements developed and evolved into several working prototypes of potential solutions.

The design team followed up with end users and subject matter experts, who became closely involved with the implementation of the design as development partners. Video and audio conferences and one-on-one meetings, including usability evaluation sessions in North America and Europe, helped refine the user interface design delivered to the development team.

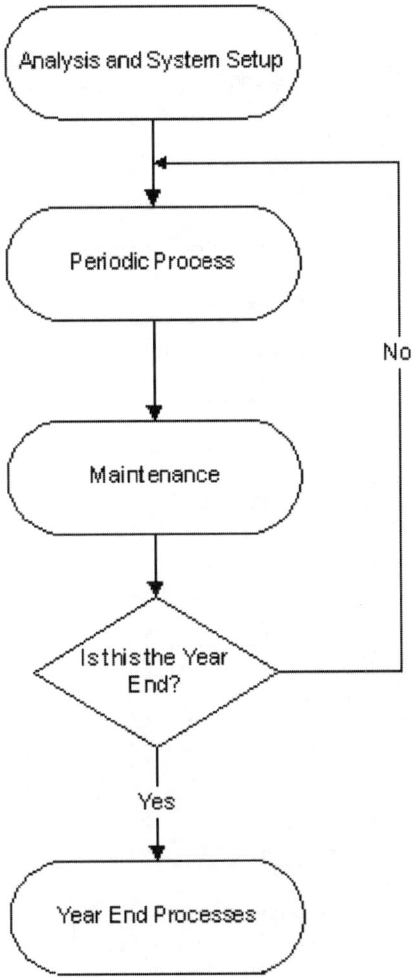

FIG. 8.4. The 50,000-foot view of the periodic process from the users' perspective.

8.3.5 Usability Evaluation Sessions

In order to obtain the most appropriate performance and preference data, the design team defined the criteria for the usability participants as follows:

- A minimum of *x* years or months of computer experience (experience with Windows and/or the Web).
- Knowledge of the business process.
- Experience with similar business process applications.
- A specific need to work with the module to be evaluated or an appropriate background for evaluating it.

Each participant individually spent about 2 hours per session helping the evaluation team by performing specific business-oriented tasks using the proposed user interface of the new modules. The session covered several tasks in the areas listed in the "Module Structure" column of Table 8.6.

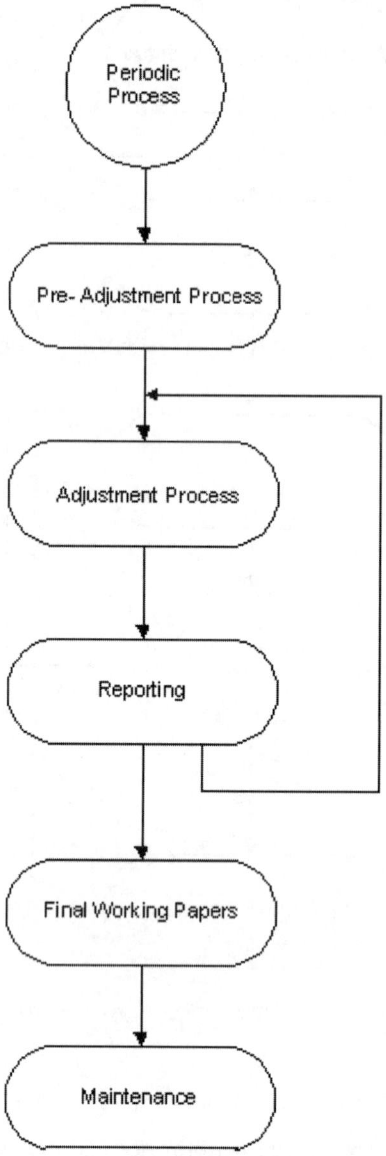

FIG. 8.5. The 20,000-foot view of the periodic process from the users' perspective.

TABLE 8.6
High-Level Task-Oriented User Interface Structure

Module Structure	*Task Process*
System setup	Set flags & security → other setup
Periodic process	Enter & validate → approve → publish/report
Automatic process	Approve → publish/report
Maintenance	Periods → system → templates

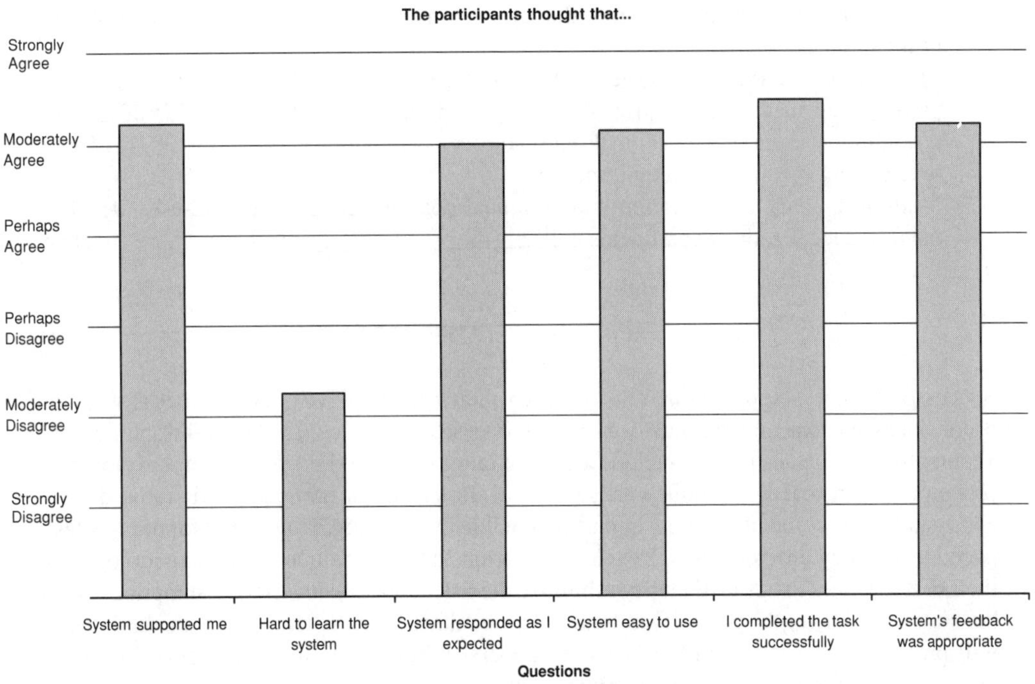

FIG. 8.6. Usability session: subjective results.

A total of 12 potential users participated in the usability phase of this module. The sessions took place in North America and Europe. Since some of the usability participants were also involved in the task analysis and design process, they had a high level of expectation and a certain degree of design ownership. However, most of the participants were seeing the usability prototype for the first time.

The results of the sessions and the feedback provided by the participants were very positive (Fig. 8.6). The participants tended to "moderately agree" that the system supported them to doing the task, was easy to use, responded as they had expected and delivered appropriate feedback.

The results of the usability phase were very positive. Since then, the development team has continued usability testing on later iterations of the product. The validation sessions, held in both North America and Europe, have ensured that the user interface of the product released to the public has evolved and been optimized.

8.3.6 Lessons Learned

As described in this chapter, the product design team is tested when working on a complex business software application. Here are the most important lessons the product team learned in the design process for the adjustment module:

- Be prepared to be flexible, as some of the methods you use need to be adapted to specific situations. Time is essential in a rapid application development environment, and the development team will expect quick turnaround from the data collection and validation methods employed.
- Integrate and work closely with the development team. Do not be afraid to share the ownership of design ideas. A participatory approach will make the design process easier

and increase the potential influence that the user interface designer can have in the design of the product.

- Make the users partners in the design process. Customer and subject matter experts enjoy participating in the design process. Acting in partnership reduces the risk of misrepresenting business processes, making incorrect assumptions about the users, or simply overlooking specific user requirements.
- Multidisciplinary collaboration and continual communication with end-user, development, and localization departments will help the product design process be more effective.

8.4 CONCLUSION

Successful business applications need to align their product development process with the users' needs and expectations and the standard processes and practices of the relevant industry. Technology companies need to include a diverse range of methodologies in their development process to ensure that their products are successful. Task analysis, participatory iterative design, and usability are some of the most important flexible methodologies available for today's fast-paced technology environments. Development teams need to include these methodologies in their best practices to ensure that they minimize the risk of making incorrect assumptions about business processes or the potential end users. The results to be gained from these methodologies will provide the design team with a better understanding of the users, the global business process, and key regional and legal requirements.

REFERENCES

Bailey, R. (1989). *Human performance engineering: Using human factors/ergonomics to achieve computer system usability.* Englewood Cliffs, NJ: Prentice Hall.

Coronado, J., & Livermore, C. (2001). Going global with the product design process: Does it make business sense? interactions Magazine, Oct. 2001, *8*(6), pp. 21–26.

Hackos, J., & Redish, J. (1998) *User and task analysis for interface design.* New York: Wiley.

Heidelberg, M. (1998). *UCD task analysis workshop.* Stamford, CN: Internal Publication. Hyperion Solutions.

Holtzblatt, K., & Beyer, H. (1997) *Contextual design: A customer-centered approach to systems designs.* San Francisco: Morgan Kaufmann.

Livermore, C., & Coronado, J. (2001) *Growing pains of designing global products.* In D. Day, & L. Dunckley (Eds.), *Designing for global markets 3. Third International Workshop on Internationalisation of Products and Systems* (pp. 47–60). Milton Keynes, England: Open University Press.

Legendre, J., Perrut, P., Deniau, X., & Fuchs, J. (1994, August 5). LOI no. 94-665 du 4 août 1994 relative à l'emploi de la langue française (La Loi Toubon [The Toubon Law]). *Journal Officiel, 180*, 11392.

9

The JIET Design Process for e-Business Applications

Helmut Degen
Siemens AG

Sonja Pedell
Siemens AG

In this chapter, we present an entire design process for e-business projects and within a special technique for the task analysis. E-Business projects in the industrial practice are characterized by heavy resource constraints concerning time and budget. For every step (e.g., requirements gathering) just a few days or even only one day is available. This requires a usable approach, that is, it is effective, efficient and satisfying for all involved people (designer, customers, clients, and others). To ensure a high result quality inspite of these constraints every step itself has to be optimized regarding the available resources and has to match very well to the next one. This applies accordingly for the task analysis that is embedded in the whole design process. The task analysis is part of the use context analysis step and is carried out in two substeps: First identifying goals, tasks, core tasks, and scenarios via structured interview (we call it "Activity Profile"), and second identifying the related process via a template method (we call it "Process Profile"). With this separation we meet the resource constraints because very often only the first substep (Activity Profile), can be carried out. The proposed approach was developed and succesfully applied in many e-business projects within the Siemens AG.

9.1 INTRODUCTION

When designing customized products, the designers should consider the wishes, goals, ideas, and expectations of the customers during the product-engineering process (Degen, 2000, Norman & Draper, 1986). In projects, two parties exist: the owners of the process and the experts executing the process. The process owners are the clients of the experts (e.g., user interface designers). The experts have to meet the clients' expectations by convincing the clients of the efficiency of the process they are following. Therefore, customer orientation is important for products and processes alike.

Usually the product-engineering process starts with market research and product definition. Next comes the attempt to design an ideal user interface, followed by construction of the technical outline, implementation, shipment of the product, and maintenance. This chapter focuses on the process for designing user interfaces for e-business applications. The expert responsible for the definition and prosecution of the design process will be referred to as a *user interface designer*.

According to Hewett et al. (1992), the field of human-computer interaction can be divided into four areas: use and context, human, computer, and development process. In the "human" and the "use and context" areas, the designers deal with customers and consumers. One of the things they might do, for example, is analyze the language and the interaction models of the users. In the "computer" area, they focus on Windows- and Web-based applications. The "development process" area covers the whole development process, from analysis to evaluation.

The mode of execution of e-business projects in industrial practice varies from project to project. For each project, the process and its complexity depend primarily on the starting conditions and the project's objective. This chapter describes how to carry out a task analysis in the context of various industrial e-business projects. The next section describes the characteristics and demands of modular and adaptable design processes and gives an overview of a particular design process, called JIET (Just in e-Time). Then come two sections describing use context analysis and task analysis, followed by the presentation of three case studies that illustrate how the JIET design process has been used in real-life projects. The chapter concludes with a brief discussion of the JIET design process.

9.2 JUST IN e-TIME

9.2.1 Adaptability of the Design Process

Experience in industrial practice shows that besides other things, e-business projects are characterized by starting conditions that are very different from those of other types of projects. These conditions directly affect the work of the user interface designer. Generally, the starting conditions can be distinguished according to the following starting parameters:

- Lead-in phase. This is the phase in the product-engineering process in which the user interface designer is needed. An essential piece of information for the user interface designer is whether user interfaces and/or an implemented technology already exist.
- Project goal. This is the goal for the user interface designer and includes the tangible results, that have to be delivered.
- Available information. This includes information on the market, the product (e.g., the e-business Web site), the provider of the product, preceding projects, and future users, including their use context.
- Available resources. These include the available time until the requested delivery date for the results and the financial resources available for the design process.
- Type of the application. Typical e-business applications are business-to-business (B2B) applications and business-to-consumer (B2C) applications. Designing a B2B application in a customer-oriented and usable way generally requires a larger expenditure than for a B2C application because the users have a large amount of expert knowledge. This knowledge has to be acquired by the user interface designer within the use context analysis (see section 9.4).

The starting parameters highlight the fact that a customer-oriented design process cannot be a rigid design process. In order to be successful and achieve good customer satisfaction, a user interface designer must use an adaptable design process. Although rigid design processes

FIG. 9.1. Just in e-Time. *Note.* Heinz Bergmeier, © Siemens.

like those of Beyer and Holtzblatt (1998) and Constantine and Lockwood (1999) put single method modules in the center, a customer-oriented design process for e-business applications should emphasize the starting parameters of the project. Suitable method modules can be selected and applied only if the lead-in phase, project goal, available information, available resources, and type of the application are known.

Therefore, it is necessary to have a process framework that contains individual method modules. These modules can be selected based on the starting parameters and can lead to a customized and efficient design process. In order to be able to offer such a design process, especially for e-business projects, the Competence Center User Interface Design of Siemens AG has developed JIET, which takes the time criticality of e-business projects into account (Fig. 9.1). The individual method components are used "just in e-time" in order to determine customer requirements efficiently and to construct user interfaces based on them.

The JIET design process consists of two phases, an analysis phase and a design phase. The analysis phase is divided into three stages: market analysis, use context analysis, and requirements gathering. The design phase is divided into four stages: user interface design, evaluation, redesign, and design specification. The focus of this chapter is on the analysis phase.

9.2.2 Usable Design Process

The academic literature on user interface design processes often makes assumptions that are rarely true in industrial practice. Above all, the principle "know your user" is often difficult to follow because of the unavailability of users.

In B2C projects, almost everyone is a potential user. Therefore, the population is practically infinitely large, and it is relatively easy to find appropriate users for the required user profile. In B2B projects, which constitute the majority of industrial projects, this is definitely not the case. The potential users are experts. They are active specialists in the customer organizations. Therefore, the recruitment of these users is very difficult, as they make up a very small population, in some cases only a few hundred people or even less.

Because one might lose customers, it is sometimes not advisable to directly approach customer organizations. Business and other strategic issues also play a major role in limiting access to potential users. Even if potential users are accessable, they are available for only a very limited time (e.g., for a single day). This has been proven to be true in many different e-business projects. Thus, potential users constitute a very limited and precious asset for the user interface designer, and the designer needs to obtain from meetings with these users as much relevant information on the topics of interest as possible.

Given the need for customer orientation of the design process as well as the bottlenecks that can occur during processing, the general requirements for a design process might be formulated as follows:

- Effectiveness. The objective can be achieved by the design process.
- Efficiency. The design process is effective, and the objective can be achieved through optimal utilization of resources.
- Satisfaction. The design process is efficient and leads to the satisfaction of all involved persons, including the user interface designer, the clients (e.g., the contact person, account manager, IT experts, business experts, and content experts), and the customers (e.g., potential users and stakeholders). Futhermore, the satisfaction covers all phases of the design process, from the first acquisition meeting to the preparation of the tender and its execution up to the presentation of the result.

These three general requirements correspond to the usability criteria associated with ISO 9241-11. Therefore, a design process that fulfills these three requirements can be called a "usable design process."

The efficiency requirement is highly correlated with the amount of time and the required financial means for the implementation of the design process (the fourth starting parameter, available resources). It also plays an important role in the achievement of customer satisfaction. The quest for efficiency affects not only the conduct of each step but its preparation and interpretation. In practice, design process activities need a considerable amount of effort and money.

9.2.3 Related Methods

Numerous descriptions of design processes can be found in the literature. Well-documented processes include those of Beyer and Holtzblatt (1998), Constantine and Lockwood (1999), Hackos and Redish (1998), and Mayhew (1999).

Beyer and Holtzblatt (1998) put the observation of the user at his or her working place at the center of their process (Contextual Inquiry). Based on our professional and personal experience, we concluded this method would not be effective for e-business projects. However, e-business product designers do need a separate view of information and tools. The artifact model (Beyer & Holzblatt, 1998, pp. 102–107) treats tools (e.g., time systems) and information (appointments) alike. In addition, some of the suggested models for presenting results, such as the flow model (pp. 92, 93), the sequence model (p. 98), and the cultural model (pp. 109, 113, 114), are difficult to understand for clients and customers and lead to low customer satisfaction. Different, more comprehensive forms of presentation, on the other hand, require more preparation and interpretation effort.

Constantine and Lockwood (1999) present a design process called Usage-Centered Design. The target audience of this process are the developers, who also become active as user interface designers within this design process. Constantine and Lockwood assume that the developers have enough domain knowledge already to be able to arrange the user interface. They present "models" with which the developers obtain a user interface, similar to a recipe. Usage-Centered Design might be suitable for applications that have already been utilized for decades in the industry (Constantine & Lockwood, 2002). However, e-business applications are relatively new for our clients. Even the business experts themselves only partially know the requirements of the potential users, as we found out during usability tests. Thus, Usage-Centered Design is not efficient and effective for the e-business domain.

Hackos and Redish (1998) present single components designed for task analysis, and these can be arranged to form a usable design process. Indeed, we frequently browsed the work of these two authors for assistance in composing the JIET design process. However, the disadvantage of the individual components is that it is unclear which component has to be used for what and which component can be omitted. Consequently, the method is not well suited to the requirements of e-business projects.

The same applies to Mayhew (1999), who presents a usability engineering life cycle. This life cycle contains an analytical part and a design part. The analytical part can be used as a framework, but the textual requirements of e-business projects (e.g., the analysis of business processes and market information as well as information flow) are not appropriately considered.

All these approaches are based on the use of a sequence with fixed method components. The JIET design process works in a different manner. It enables the user interface designer to choose appropriate method modules from the provided set (framework) based on the starting parameters of the e-business project. The JIET process does not contain significant new method components but arranges already existing components in such a way that a usable design process is made available. Table 9.1 presents an outline at the JIET process.

9.3 JIET DESIGN PROCESS

9.3.1 Analysis Phase

The objective of the analysis phase is to develop use scenarios for relevant user roles. The analysis phase consists of the following three stages: market analysis, use context analysis, and requirements gathering.

The purpose of the *market analysis* is to understand the future provider of the e-business application and the provider's business (enterprise profile of the provider, product profile). Further, the relevant suppliers and customer organizations are identified (customer profile, supplier profile). Within these organizations all job roles relevant to the project are identified (e.g., decision maker, administrator). The (different) business processes in which the provider is included are then identified (business process profile), as are the provider's competitors (competitor profile). One required result of the market analysis is the identification of key customers by their company names. Another is the creation of a list of representatives for each job role within these organizations (note that it is possible for one individual to hold more than one job role). With this pool of knowledge, these persons can be invited as representatives of the intended users for the next step: the use context analysis.

The purpose of the *use context analysis* is to understand the use context of potential users of the e-business application. First, a market profile of selected customer is created in order to understand that customer's organization and business. This profile corresponds to the market analysis of the provider in the previous step but now the focus is on the customer. The potential users are then analyzed, and the analysis identifies their goals, responsibilities, job roles, and hands-on computer experience (user profile). In case there are different job roles, the kind and number of the different roles has to be determined. An analysis of activities follows in which tasks, goals, and Web-relevant scenarios are studied (activity profile). For selected scenarios, a task analysis can be performed (process profile). The use context analysis can be further extended by investigating content objects, documents (object profile), and tools (tool profile). Eventually, all user roles can be identified (user role profile).

TABLE 9.1
JIET Design Process

Phases and Stages	*Estimated Effort*	*Required Inputs From Previous Step*	*Activity*	*Result*
Analysis phase Market analysis (provider)	Up to 2 person days (1 day in total)		Understanding the client's business, it customers and suppliers; identifying customer classes, selecting key customers and interview partners	Enterprise profile (provider) Product profile Customer profile Supplier profile Business process profile Competitor profile
Use context analysis	Up to 4 person days (2 days in total); for each object 1 additional hour; for each tool, 2 additional hours	Customer profile Business process profile Selected key customers (representatives of the customer classes) Selected persons (representatives of the job roles)	Identifying user roles and scenarios, in some cases of context scenarios for gathering requirements	Market profile (customer) User profile Activity profile Process profile Object profile Tool profile User role profile
Requirements gathering	2 person days for 5 scenarios (1 day in total)	User role profile Activity profile	Identifying use scenarios and user-specific requirements	Use scenarios (intended state) User interface requirements
Design phase User interface design	Up to 27 person days for 30 templates (15 days in total)	Activity profile Process profile User interface requirements	Designing user interfaces	User interface in the form of Interaction concept Interaction design Visual design Prototyping
Evaluation	Up to 10 person days (up to 10 days in total)	User interfaces Activity profile Selected persons (representatives of the user roles)	Conducting a usability evaluation	Improvement potentials
Redesign	Depends on effort to change; usually 3 person days (3 days in total)	User Interfaces Improvement potentials	Optimizing user interfaces	Optimized user interfaces
Design specification	5 person days (5 days in total)	Optimized user interfaces	Compiling a design specification	Design specification

Note. Estimate for a project for one country, one user role, and a B2B-Web site with 30 templates (a template is a web page for one product description that can be used within the same Web site for many products). We assume optimal working conditions, namely, that there is fine-tuned collaboration with the client, and the user interface experts work in one office area. The activity and process profiles (results of the use context analysis) together form the task analysis.

The objective of the *requirements gathering* stage is to explore requests and requirements concerning the characteristics of the e-business application. The scenarios identified in the use context analysis are used for this. The requirements can be divided into use scenarios (sequence of web pages) and specific user interface requirements (expected content and functions). Ideally, the user interface designers work with representatives of the identified user roles during this stage, which completes the analysis phase.

9.3.2 Design Phase

The objective of the design phase is to design usable user interfaces and, if desired, constract a design specification for these user interfaces. This phase consists of four stages: user interface design, evaluation, redesign, and design specification.

The objective of the *user interface design* stage is to design the user interface. Results from previous stages are used here: scenarios; results from the requirements gathering; and possibly information on content objects, documents, and tools. The products created during this phase can have increasing levels of quality:

- An *interaction concept* is a text document in which the main features of the e-business application are listed as text or by means of tables. The document contains a list of all web page templates, storyboards specific to the scenarios, a grid layout of the homepage, and a content plan for each web page template.
- An *interaction design* is a refined version of the interaction concept. The most important pages are depicted using simple means (e.g., paper prototypes), and the click-through path is outlined graphically.
- A *visual design* is a refined version of the interaction design. All pages are finally designed by a visual designer.
- *Prototypes* can be developed to create an interactive version of the user interfaces. The results of the interaction design and/or the visual design contribute to the development of prototypes.

The purpose of the *evaluation* of user interfaces is to identify potential for improvement. The evaluation can be carried out with or without the participation of users (i.e., by means of usability testing or usability inspections).

In the *redesign stage*, the user interfaces are optimized. Results from all previous stages contribute to the redesign, especially results from the evaluation. If desired, a design specification of the optimized user interfaces is generated. This will be handed over to the developers and serves as a basis for implementation.

9.4 USE CONTEXT ANALYSIS

Because the task analysis is part of the use context analysis we focus here are this phase.

9.4.1 Individual Steps

The object of the *use context analysis* is to understand the context in which the potential users will use the e-business application. This is the design process stage that will produce the relevant information about the users, the users' working environment, the tools used, and the contents. This information is necessary to obtain a high-quality, usable interface design.

The use context analysis involves constructing seven profiles: a market profile (customer), a user profile, an activity profile, a process profile, an object profile, a tool profile, and a user role profile (see Table 9.2). These profiles are described below.

TABLE 9.2
Use Content Analysis Profiles

Profile	*Inputs*[a]	*Setting*	*All Possible Results*[b]	*Instruments*	*Participants*
Market profile (customer)	Customer profile (provider) Business Process Profile	Office, customer workshop	Enterprise profile (customer) Business process profile (refined) Competitor profile	Interview guide (Word)	Customer, account manager, UI designer, and business, marketing, and possibly IT personnel
User profile	Business process profile (refined)	Customer workshop	Job profile (B2B), consumer profile (B2C) Goals Tasks Core tasks Responsibilities Needs Job role Team Business process (customer) Contact persons Computer profile Computer and Internet experiences Hardware and software	Interview guide (Word)	Customer, account manager, UI designer, and business, marketing, and possibly IT personnel
Activity profile	Core tasks Job Roles	Customer workshop	Scenario profile Current state Intended state Problems and improvements Competitor profile Provider Competitors	Interview guide (Word)	Customer, account manager, UI designer, and business, marketing, and possibly IT personnel

Continued

[a]Minimum requirements. These inputs show the minimum procedures for conducting the use context analysis.
[b]All possible results when all method modules are used. For each result, a corresponding method module exists. In this chapter, only the method modules for task flow analysis are shown.

TABLE 9.2
(Continued)

Profile	*Inputs*[a]	*Setting*	*All Possible Results*[b]	*Instruments*	*Participants*
Process profile	Job roles Scenarios (current and intended state)	Customer workshop	Process profile Context scenarios (current state) Context scenarios (intended state)	Paper forms for steps and results Interview guide (Word) Result presentation for context scenarios	Customer, account manager, UI designer, and business, marketing, and possibly IT personnel
Object profile	Scenarios (current state) Context scenarios (current state)	Client workshop	Content object description Document description Content plan and specific requirements	Interview guide (Word)	Client (not customer), UI designer, and business, marketing, and possibly IT personnel
Tool profile	Scenarios (current state) Context scenarios (current state)	Client workshop	Tool description Tool specific use scenarios and specific requirements	Interview guide (Word)	Client (not customer), account manager, UI designer
User role profile	Job roles Scenarios (current state, intended state)	Office	User roles	Interview guide (Word)	UI designer

[a]Minimum requirements. These inputs show the minimum procedures for conducting the use context analysis.
[b]All possible results when all method modules are used. For each result, a corresponding method module exists. In this chapter, only the method modules for task flow analysis are shown.

The purpose of constructing *market profiles (customer)* is to get to know each customer organization. For this purpose, an organization profile is drawn up. This profile contains the organization's size, products and services, customers, competitors, sites, and so on. It will also describe the business goals and the targets of the e-business project.

The *user profile* describes the potential users and their professional contacts. This profile focuses on their jobs, including their goals, tasks, responsibilities, and needs. The respondents classify themselves into job roles that have been predefined in cooperation with the organization. Furthermore, the users give information on their working environment, their sociodemographic characteristics, and their business contacts.

The *activity profile* describes the activities of the potential users. The designers list the tasks and possibly put them in concrete form. All the tasks that could be supported by an e-business application are selected. These are known as the core tasks. For all core tasks, concrete scenarios are formulated (chap. 5). A sample core task would be "management of telephone installation fault reports." The corresponding scenario might run as follows: "On April 3, colleague Meier-Schmitt reported a problem with his telephone, which has the number 4444. I am currently handling this problem." A scenario is an instance of a core task. Usually, existing problems are investigated, and suggestions for improvement are generated.

The purpose of the *process profile* is to show how the scenarios will be performed. Although the activity profile identifies the core tasks and includes scenarios, the process profile analyzes the task flow and the context scenarios. The task flow itself is abstract and will be made concrete in the form of a context scenario. Just as there is a concrete scenario for every core task, there is a concrete context scenario for every task flow (see Table 9.3).

To analyze the context scenarios, every single process step is examined. The concrete sequence of process steps within a scenario is called a *context scenario* (the sequence describes how the scenario will be executed). The instruments used to execute the task analysis are described in the next section. For every step of a task flow, the actor, the intention of the

TABLE 9.3
Relation Between Abstract and Concrete Levels

Abstract Terms or Expression	*Concrete Examples for Abstract Terms/Expression*
Job role	Individual person
Analyst	Jack Smith (analyst of Factory Ltd.)
Recommender	Kate White (recommender of Factory Ltd.)
Decision maker	Dagobert McDonald (decision maker of Factory Ltd.)
	Joan Shoeman (Siemens account manager)
Core task	Scenario
Ordering a telephone system	Ordering a Hicom 150 H Office with 150 devices for site "San Francisco" of Factory Ltd.
Task flow	Context scenario (example):
1. Needs analysis	1. Discussing with Dagobert the need to buy a new telephone system; result: decision to buy
2. Requesting a proposal	2. Calling Ms. Shoeman at Siemens to communicate requests
3. Getting the proposal	3. Obtaining the proposal (Word document) from Ms. Shoeman
Etc.	Etc.

step, the input, the activity itself, the needed working time, the tools used, and the output are analyzed. The activity profile and process profile are very closely connected. Constructing the activity profile is absolutely necessary. The process profile should be created only if the user interface designer does not have sufficient domain knowledge of the process.

The purpose of constructing the *object profile* is to understand all the objects used in a task flow. Objects are defined here as digital content objects and documents. A precise description of the content objects and the documents used will make the design of a user interface easier.

The *tool profile* has a similar purpose. An e-business application usually aims at standardizing a heterogeneous IT landscape, which can be analogous or digital. The goal is to avoid a separation of data and to save work steps. The tool profile aids in understanding currently used tools and in transfering successful design solutions and improving less successful design solutions.

The *user role profile* identifies the user roles. A user role is characterized by at least one job role and the relevant scenarios and context scenarios. It is possible that every job role needs its own version of a user interface, but usually one version of a user interface corresponds to more than one job role. For example, a director (job role: decision maker), a secretary (job role: administrator), and a project manager (job role: recommender) might use the same version of a project management tool, in which case there are three different job roles but only one user role (user role: project user). To avoid the expense of designing unnecessary versions, designers must identify the requested number of user roles.

9.4.2 Instruments for Use Context Analysis

For the use context analysis, we have prepared interview guides available as Word documents. Owing to space restrictions, not all instruments can be presented here. We concentrate on instruments especially helpful for doing a task analysis, from the rough gathering of tasks to the detailed description of steps in the process chain. The process chain is the core structure containing information about problems and improvements as well as the object profile and the tool profile.

The selection of the correct media for assessment is at least as important as the assessment procedure itself. In general, using a laptop computer and a projector to do interviews has turned out to be an effective and efficient method. All participants can watch and comment on the projected results directly. This creates an atmosphere of confidence and gives the interviewees a feeling of having control over the interview.

Among other things, task gathering starts during construction of the user profile. First we describe all tasks, list their objectives, and prioritize them. Then we identify the tasks for which an e-business application is desired. These tasks are the core tasks, and they will be elaborated in more detail in the next step (the assessment of core tasks).

9.4.2.1 *Task Gathering*

- Number and name of task.
- Task description.
- Objective of the task.
- Prioritization of tasks.
- e-Business support desired?

In the activity profile, we identify all core tasks i.e., all tasks for which an e-business application is desired. We also identify all possible communication channels for the core tasks. Examples of communication channels include face-to-face, contact, fax machines, and call centers. We prioritize all core tasks for the Web communication channel. Then we list priorities for all core tasks from the users' point of view.

9.4.2.2 *Assessment of Core Tasks*

- Number and name of core task.
- Current communication channel (face-to-face contact, fax machines, call center, mobile devices, Web).
- Prioritization of core tasks in regard to Web communication channel.

We register the average time spent on each core task. Both duration at present and duration in the future are of interest, as they indicate the potential for improvement. Also, there are two types of duration of importance. The time frame is the duration from the beginning of the core task to its termination. In contrast, working time is the duration of the actual work. For the core task "order products," the time frame might be 2 weeks but the real working time might be only 4 hours.

9.4.2.3 *Expenses per Core Task*

- Number and name of core task.
- Actual state.
  - Duration (time frame, working time).
  - Frequency of transaction (per time unit).
  - Number of objects (per transaction).
- Target state.
  - Duration (time frame, working time).
  - Frequency of transaction (per time unit).
  - Number of objects (per transaction).

9.4.2.4 *Assessment of Scenarios*

- Number and name of core task.
- Number and name of the scenario.
- Description of scenario.
- Result of scenario.

The assessment of scenarios constitutes an important step in the use context analysis. (As a reminder, a scenario is a concrete task). If it is not possible to conduct further surveys, the use context analysis may be terminated right here, in which case the scenarios may serve as guideline for the requirements gathering, the user interface design, and the evaluation. But if there is enough time left, the scenarios should be assessed in more detail. To aid in the assessment, numerous pieces of information get collected for each scenario (Table 9.4).

The final step in constructing the activity profile is the assessment of future scenarios. Here, the target state from the users' point of view is assessed for each future scenario.

9.4.2.5 *Expenses per Scenario*

The following represent the necessary expenses associated with a scenario or step

- Number and name of a scenario.
- Intended steps (of the scenario).

TABLE 9.4
Tool for Gathering Detailed Information About a Scenario

Identification
- Scenario name
- Scenario number

Scenario objective
- What is the objective you want to achieve with this scenario? What do you expect?

Execution
- Actor: Who is involved executing this scenario?
- Responsible executor: Who is responsible for executing this scenario?
- Single activities: What exactly is this person doing? What are the individual steps?
- Tools: Which tools is this person using?
- Requirements: Do you have specific requirements for improving the execution?

Scenario end
- Results: What is the output information/result?
- Information: What is the information relevant for the decision?
- Recipient: Who is the recipient of the output information/result?

Scenario start
- Trigger event: What or who starts the step?
- Input information: What is the required input information?
- Delivering person: Who delivers the available input information?

- Intended result (for each step).
- Intended tecipient (for each step).
- Intended tools (for each step).
- Requirements (for each step or scenario).

9.4.2.6 Context Scenario Templates

Context scenarios (task flows) are assessed as part of the process profile. For their assessment, we developed a template method, which we describe below. We differentiate between the template for a step (Fig. 9.2) and the template for the result of a step (Fig. 9.3). A step corresponds to a step in a context scenario. In cooperation with the customer, we reproduce the steps and their results from the context scenario by linking the steps and their results. As a rule, the results from one step form the input of the next step. In addition to the templates, we have Word documents for the collection of detailed information (Table 9.5).

The advantage of this procedure for revealing context scenarios is that it requires only a small amount of time. We do not, for example, use Contextual Inquiry (Beyer & Holtzblatt 1998) because it is too time consuming. In our procedure, each single step and the result of the context scenario is developed right in front of the customer. Because the steps and results are linked together, inconsistencies in the context scenario can be easily discovered and corrected just by adding new templates or by moving templates. Only after the whole context scenario has been completed can the individual templates be glued on to a piece of paper and used later as a basis for open questions. Furthermore, it is easy to optimize the current state of a context scenario while working together with the customer to construct its intended state (Fig. 9.4).

The upper half of Fig. 9.4 represents a context scenario (current state). The time flow moves from left to right. The templates are glued on to a piece of paper that is divided in two areas so that it can be read more easily. The upper area portrays only the customer activities (customer), the lower area only shows the provider activities (Siemens). It is recommended to use the real

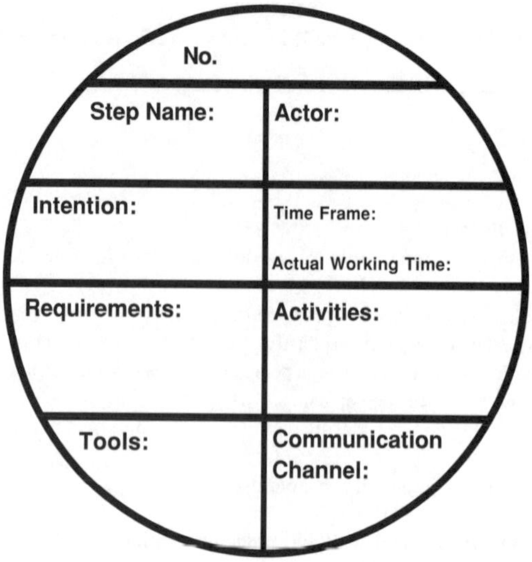

FIG. 9.2. Template for one step. "No." indicates the number of steps. Every step is unambiguously identified. "Step Name" contains the name of the step. "Actor" states the executor. "Intention" describes the intention of this step. "Requirements" describes missing features concerning this step. "Tools" mentions the tools that were used for this step. "Time Frame" records the total time, from kick-off to completion of the results. "Actual Working Time" describes the time that was actually spent carrying out this step. (e.g., a step can be spread over 2 weeks, while the step itself actually only takes 3 hours of work). "Activities" describes the possible substeps of this step. "Communication Channel" indicates which medium (e.g., face-to-face, e-mail, fax, telephone, Web etc.) was used for the execution of this step.

customer names here. Activities that will be carried out by both parties (e.g., communicating with each other) are represented by placing the circle on the center line. The lower half of Fig. 9.4 represents the context scenario (intended state). The step highlighted in gray should be supported by the future e-business application. Therefore, within the framework of the requirements gathering, a use scenario will have to be made.

The context scenarios are documented in a cumulative form. For this purpose, the relation between the target, tasks, problems and suggested improvements, scenarios (results from the activity profile stage), context scenarios (current state and intended state), and the identified user roles (results from the process profile and the user role profile) are represented on a PowerPoint sheet (Fig. 9.5).

Within the framework of the activity profiles, the targets and tasks of the job roles are investigated. These can be assigned to a basic workflow, here consisting of "Pre Sales," "Sales," and "Post Sales." Furthermore, problems and improvement potentials are identified. Within the framework of the process profiles, the context scenarios (current state and intended state) belonging to the identified scenarios (here, "buying telephone system HICOM 150 for 96 devices") are constructed. For the current state, the involved job roles (here, "Manager Telephone," "Director IT," "Siemens Account Manager," and "Siemens Service Manager") are listed. For every step in the context scenario, the relevant active job role is identified (represented by an arrow). For the intended state, the context scenario is optimized, and it is explained how this should be supported by the future e-business application. In addition, the user roles are displayed. In this example, the job roles "Manager Telephone" and "Director IT" are joined together as "Analyst Enterprise" because both job roles perform identical scenarios, and a separation for the user interface is therefore pointless. The identified context scenario shows which

No.
Result Name:
Creator:
Previous Step (No.):
Next Step (No.):
Description:
Material:

FIG. 9.3. Template for one result. "No." indicates the number of results. Every result is unambiguosly identified. "Result Name" contains the name of the result. "Creator" describes the producer of the results. "Previous Step (No.)" indicates the step in which these results have been produced. "Next Step" indicates which step uses these results as input. "Description" is the description of the results. "Material" indicates in which form (e.g., paper, fax, e-mail, Word document) the results are documented.

steps of a scenario for which users should be supported by the e-business application (here, the steps "Delivery Proposal," "Order Product," and "Tracking/Tracing"). For the identified steps, so-called use scenarios are developed in the requirements gathering stage.

9.4.3 Efficiency of the Use Context Analysis

For most of the method modules of the use context analysis, we created Word documents to aid in gathering the required information. These documents can be used without preparation time. During the interview session, they are filled in by the analysts and the interviewees. An additional interpretation effort is not necessary that is, after the data collection, when each single step is finished, the next step can be started.

For constructing the context scenarios, we created the steps and the results templates and the corresponding Word documents for collecting detailed information. This step does not require any preparation time either. The results presentation (Fig. 9.6) is used to arrange the gathered information in a customer-friendly manner. This kind of presentation displays the results in a compact and clear manner. Practice has shown that this presentation method works well and is appreciated by the clients.

TABLE 9.5
Guideline for Filling in the Templates

Identification
- Step name
- Step number

Purpose of step
- Intention: Why do you do it? What do you expect?

Execution
- Actor: Who is involved executing the step?
- Activities: What exactly is this person doing?
- Tools: Which tools is this person using?
- Actual working time: How much time does it take (only working time)?
- Time frame: How much time does it take in total (from start until end, duration)?
- Requirements: Do you have specific requirements for improving the execution?

Scenario end
- Results: What is the output information/result?
- Recipient: Who is the recipient of the output information/result?
- Next Step: What is the next step?
- Decision: How did the results help you to continue?

Scenario start
- Trigger event: What or who starts the step?
- Previous step: What is the previous step?
- Input information: What is the required input information?
- Delivering person: Who delivers the available input information?

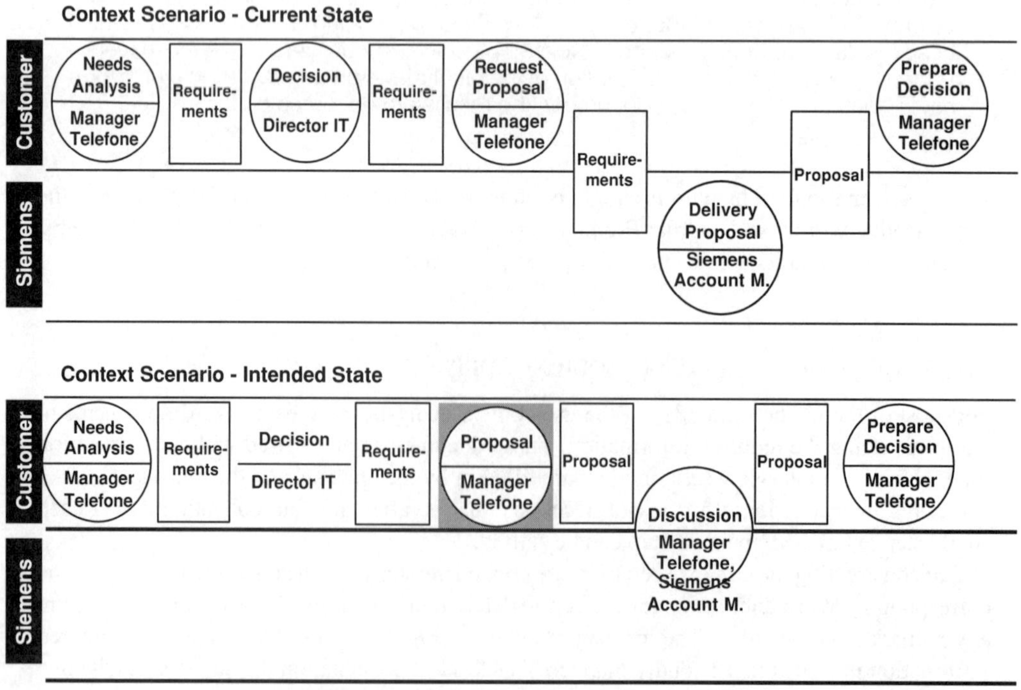

FIG. 9.4. Context scenarios with templates.

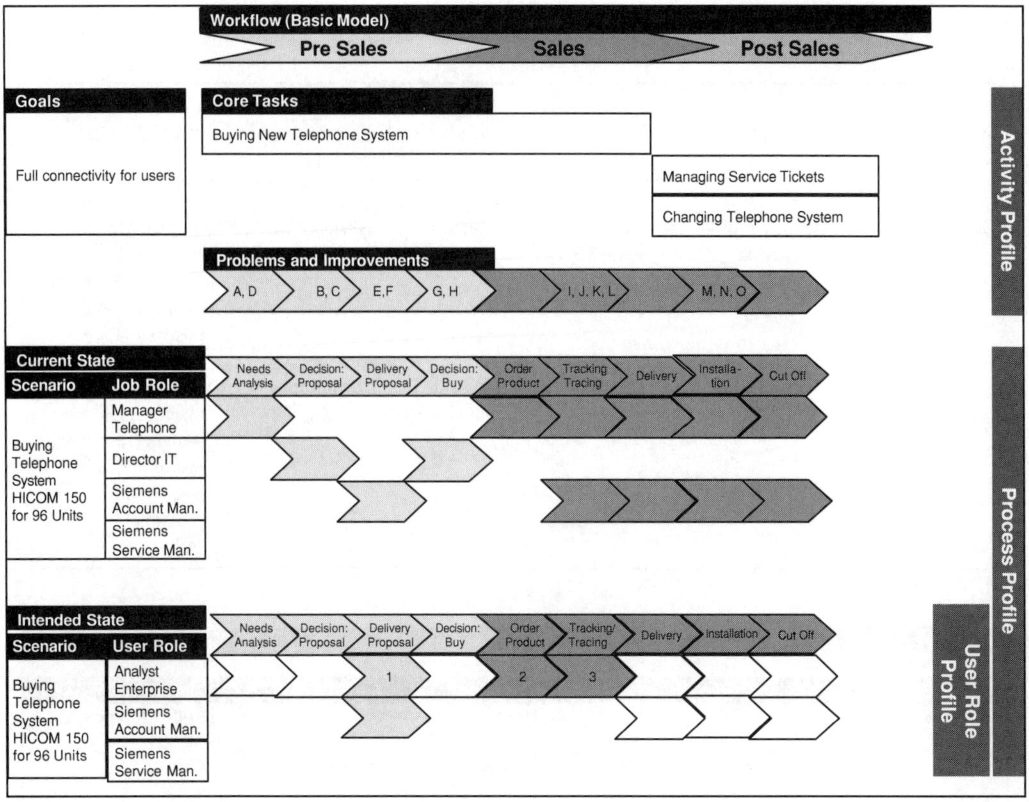

FIG. 9.5. Result presentation for the process profile.

9.5 REQUIREMENTS GATHERING

9.5.1 Individual Steps

The objective of the requirements gathering is to get user-specific requirements for the design of user interfaces. In order for the process to proceed, the following results are required: For every scenario of a user role, users sketch a use scenario by hand. On the basis of these sketches, the user-specific requirements for the design are investigated and noted in the relevant requirements form (Table 9.6).

Although the use context analysis covers the context of the potential users, the requirements gathering bridges the gap between the use context and the user interface (Wood, 1998). If possible at all, the requirements gathering should be conducted with potential users. With small effort and expenditure, it is thus possible to avoid wrong concept decisions during the design phase. The results of the requirements gathering stage are the use scenarios and user-specific requirements.

A use scenario (intended state) is understood as an episodic description of work activities that a user needs for the completion of a core task using the future dialogue system. A use scenario is possibly provided for an identified user role, a certain scenario, and eventually on the basis of an optimized context scenario (intended state). The core of the use scenario is the dialogue sequence, which is frequently presented in the form of storyboards.

Besides the dialogue sequence, there are user-specific requirements for the user interfaces. These can stress different aspects, such as functional aspects, content aspects, layout aspects, and so on. In order to be able to collect the user-specific requirements systematically, we

TABLE 9.6
Overview of the Requirements Gathering

Step	*Inputs*	*Method*	*Results*	*Instruments*	*Participants*
Use scenarios	User role profile scenarios	Customer workshop	Use scenarios	Sketch form	Customer, account manager, UI designer, business, marketing, and possibly IT personnel
User-specific requirements	Use scenarios	Customer workshop	User-specific requirements	Requirements form	Customer, account manager, UI designer, business, marketing, and possibly IT personnel

Scenario			
Scenario No. and Name:		Step No. and Name:	
Webpage			
No. (previous webpage)	Webpage No.:	Heading of the webpage:	No. (following webpage):
	What is the Intention of the webpage:	Decision relevant information:	
User acitivities:		Requirements: Content: Media (e.g. text, icon, sound): Wording: Layout: Links (with wording and no. of following pages): Functions:	
Examples			

FIG. 9.6. Requirements form.

identified categories consisting of relevant aspects of user interfaces for Web-based e-business applications. We call these categories factors. The following factors for user-specific requirements have been found to be favorable:

Content. Which content is requested to appear on a particular web page?
Media. Which media (e.g., text, graphic, photo) is preferred for presenting the content?
Wording. Which expression (e.g., "Order," "Shop," "Product Catalogue," Browse," "Buy," "Online Order") is preferred for labeling the content?
Layout. How should the different user interface elements be organized? Which color scheme, color contrast, and fonts should be used?

Function. Which functions are needed and should be assigned to which content elements?
Links. Which links should be placed on a particular web page?
Click-through. How should the web pages be connected to each other?

The factors are used for the systematic collection of requirements, and they certainly assist in the categorizing of the requirements. During the evaluation phase, these factors are used to establish error categories in order to simplify the task of eliminating identified usability errors.

9.5.2 Instruments for Requirements Gathering

The objective of the requirements-gathering stage is to identify use scenarios and user-specific requirements for the user interfaces. This stage is the bridge between the analysis and the user interface design. Because it is essential for efficiency reasons to arrive as close as possible at the desired design of the user interfaces, at this point we use a specially developed paper prototype procedure named NOGAP (A*no*ther Method for *Ga*thering Design Requirements in e-Business User Interface Design *P*rojects). At the core of the procedure are the factors already introduced for the categorization of requirements. This categorization is retrieved from a form for requirements gathering (Fig. 9.6) and a form on which users outline their ideas (Fig. 9.7).

During this entire data collection stage, the users only work with the sketch form. Each user develops and sketches ideas and wishes under guidance from the user interface designer. The result is a direct image of the user's mental model (Pedell, 1998). This procedure works very well. The "screens" constitute a concrete representation of the requirements and therefore supply the user interface designers with detailed input on how to name and arrange the required

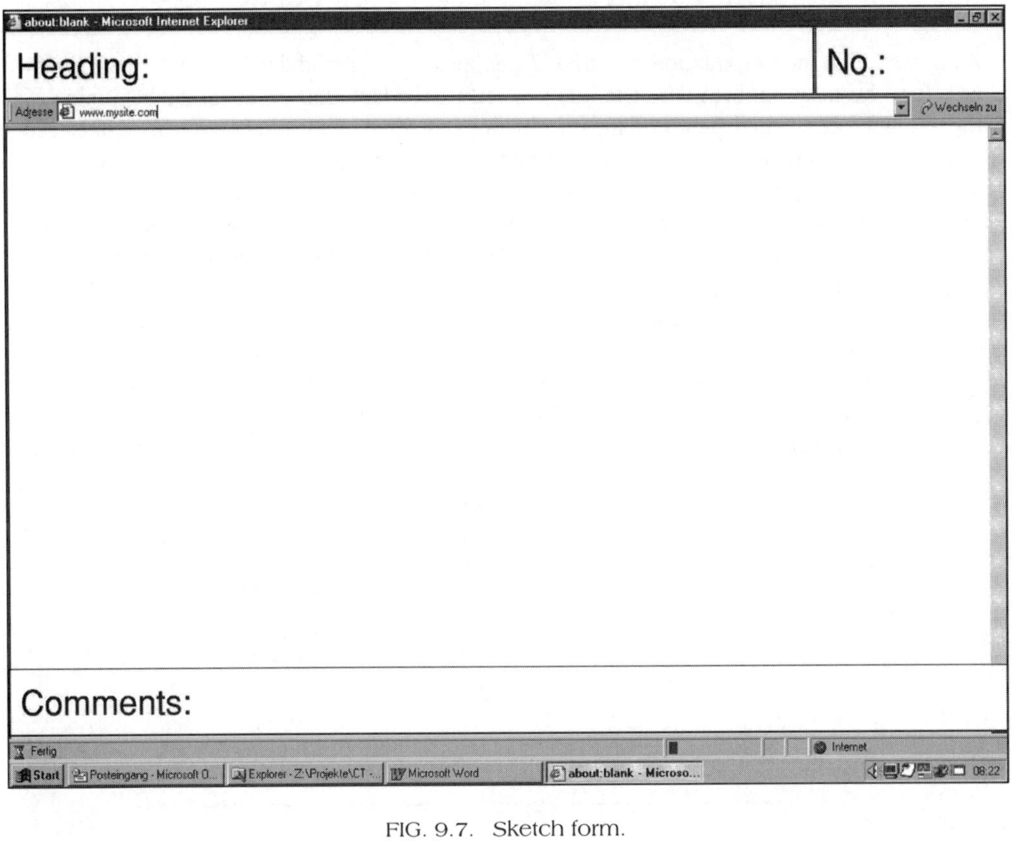

FIG. 9.7. Sketch form.

contents. Even users with little design experience start to sketch their ideas after a short warm-up period (approximately 5 minutes). The user interface designers record the ideas on the requirements form, which they also use as a guide for their questions. It is recommended to carry out this type of interview procedure with two user interface designers, one to ask the questions, the other to record the requirements. In this situation, using a transparent way of recording the results is especially important, as mentioned before. The simplest and most efficient way is to record them directly using a laptop computer. During the entire time, all participants can view the records via projection. This procedure increases the confidence of the participants and at the same time prevents mistakes. The results of the requirements gathering are used as a basis for the design of the user interfaces.

9.5.3 Efficiency of the Requirements Gathering

For requirements gathering, the sketch form and the requirements forms are used as described. The scenarios are either already available from the use context analysis or will be uncovered during the requirements gathering. Thus, no preparation is necessary for the requirements gathering. The interpretation of the requirements gathering results is reduced to the electronical scanning of the sketches and the generation of the click-through. This work can be accomplished in a few hours. The gathered requirements are already present in Word document, and no extra work is necessary at this point. Practical application of the procedure has shown that it is fast and easy to do. Thus the two method modules of the requirements gathering fulfill the required efficiency criteria.

9.6 ADAPTABILITY OF THE JIET DESIGN PROCESS

As described in the introduction, the JIET design process is a modular, adaptable design process for e-business applications. The first possibility of adaptation occurs at the beginning of the design process and depends on the information delivered by the client to the user interface designer. If the company profile, the product profile, the business processes, and the customer, supplier, and competitor profiles are available, the use context analysis can be started immediately. If the job roles, tasks, core tasks, scenarios, and context scenarios as well as the user roles of the use context are known, it is time to begin gathering information on the user-specific requirements as well as missing information on objects, documents, and tools. If all the information on the requirements is available, the design of the user interfaces can begin. The evaluation can be started once the user interfaces, the user roles, and the scenarios are available. If the usability evaluation has uncovered potential for improvement, the redesign can be initiated.

The five starting parameters lead to different start conditions, which are reduced here to two dimensions. As shown in Table 9.7, four cases are determined by whether users are involved

TABLE 9.7
Case Study Matrix

	Users not involved	*Users involved*
User interfaces already exist	Case 1	Case 3
User interfaces do not exist	Case 2	Case 4

TABLE 9.8
Adaptability of the JIET Design Process

Case 1	*Case 2*	*Case 3*	*Case 4*
Starting conditions Users are not involved User interfaces exist	**Starting conditions** Users are not involved User interfaces do not exist	**Starting conditions** Users are involved User interfaces exist	**Starting conditions** Users are involved User Interfaces do not exist
Analysis phase Market profile Constructed with client Scenarios Received from client User roles Received from client Context scenarios Process expert involved if available Use scenarios Not constructed	**Analysis phase** Market profile Constructed with client Scenarios Received from client User roles Received from client Context scenarios Process expert involved if available Use scenarios Constructed with business experts	**Analysis phase** Market profile Constructed with client Scenarios Received from client User roles Received from client Context scenarios Process expert involved if available Use scenarios Not constructed	**Analysis phase** Market profile Constructed with client Tasks, core tasks, and scenarios Constructed with users User roles Constructed with users Context scenarios Only if the UI designer has no knowledge about context Use scenarios Constructed with users
Design phase User interface design No (already done) Evaluation Usability inspection done by UI designer and business and IT experts Redesign UI designer (with business and IT experts) Design specification Probably not What should be done if little time is available UI designer together with business and IT experts inspect and redesign the interfaces	**Design phase** User interface design UI designer Evaluation Usability inspection done by UI designer and business and IT experts Redesign UI designer (with business and IT experts) Design specification If ordered	**Design phase** User interface design No (already done) Evaluation Usability test done with users (with involvement of business and IT experts) Redesign UI designer (with business and IT experts) Design specification Probably not What should be done if little time is available UI designer together with users and business and IT experts inspect and redesign the interfaces	**Design phase** User interface design? UI designer Evaluation? Usability test done with users (with involvement of business and IT experts) Redesign? UI designer (with business and IT experts) Design specification? If ordered
Quality of result Structural UI concept User interfaces with a high error probability	**Quality of result** Goal-oriented UI concept User interfaces with a high error probability	**Quality of result** Structural UI concept User interfaces with a low error probability	**Quality of result** Goal-oriented UI concept User interfaces with a low error probability

in the design process and also by whether user interfaces exist prior to the start of the design process. In addition, in the following discussion two assumptions are made: the projects are under normal (i.e., high) time pressure and any user interfaces existing were not designed by user interface designers. That is, instead of a user model, a system model formed the basis of the design (Norman, 1988). See Table 9.8 for a summary of how JIET can be adapted to each of the four situations identified in Table 9.7.

In the first case, the user interface designer has few design options, as user interfaces already exist and users are not participating in the design process. This means the user interfaces will have a high probability of error. In addition, this type of project is typically under great time pressure. Only slight changes can be made (e.g., in the wording), and usually it is impossible to incorporate other or new functions or a completely different user prompt. Since user interface designers did not participate in the design phase, the interfaces have a structural design, (e.g., they have the structure of the technical software architecture). The goals and tasks of the users were not adequately considered, and thus, it is reasonable to do a market profile in conjunction with business and IT experts. The scenarios and job roles are already visible in the user interfaces that the client has supplied. In order to obtain more information about the processes, a process expert should be consulted. If the project is running out of time, no use scenarios should be gathered, but instead the user interface designer should carry out a usability inspection working with business and IT experts. The biggest usability problems can be identified and eliminated before delivery of the final product.

In the second case, the user interface designer has many design possibilities. In order to proceed, the user interface designer, together with IT and business experts, should perform all the important steps of the analysis phase. As a result, the user interface concept will be goal oriented, but because there is no user participation planned, the user interfaces will still have a high probability of error.

In the third case, users are involved but user interfaces already exist. The user interface designer should perform all the important steps of the analysis phase, do a usability test, then embark on a redesign of the user interfaces. If the time frame does not allow a usability test, the user interface designer, together with the users and IT and business experts, should inspect and redesign the interfaces in one session. If the main usability problems are taken into consideration, the resulting user interface will have a low probability of error, but the user interface concept will be structure oriented and not correspond with the ideas of the customers.

In the fourth case, the users participate and the user interface designer is involved from the very beginning (i.e., following the product definition). From the point of view of the user interface designer, this is the ideal situation. The analysis and the evaluation phases will then be carried out with the users, and consequently the user interfaces will have a low probability of error and a goal-oriented concept.

9.7 CASE STUDIES

9.7.1 Case Study 1: Travel Worldwide (B2C)

9.7.1.1 The Project

The first case study describes an international research project. The objective of the research project is to elaborate design similarities and differences in Chinese, American, and German versions of a Web site for a (fictitious) travel agency. The Web site targets private customers. (See Table 9.9 for a summary of this and the following two case studies.)

TABLE 9.9
Case Study Summary

	Case Study 1: Travel Web Site (B2C)	*Case Study 2: Audiological Order Form (B2B)*	*Case Study 3: Marketplace for Communication Systems (B2B)*
Market analysis	X	–	X
Enterprise profile (provider)	X	–	X
Product profile	X	–	X
Customer profile	X	–	X
Supplier profile	X	–	X
Business process profile	X	–	X
Competitor profile	–	–	X
Use context analysis			
Market profile (customer)			
Enterprise/customer profile	X	X	X
Business process	–	–	X
Competitor profile	–	–	X
User profile			
Job/consumer profile	X	–	X
Computer profile	X	–	X
Activity profile			
Scenario profile	X	X	X
Problems/improvements	–	X	X
User role profile	–	–	X
Process profile			
Context scenarios profile	X	X	X
Object profile			
Content objects profile	–	–	X
Document profile	–	–	X
Tool profile	–	–	X
Requirements gathering			
Use scenarios	X	–	X
User-specific requirements	X	X	X
Design			
Interaction concept	X	–	X
Interaction design	X	–	X
Visual design	X	–	–
Prototype	–	–	X
Evaluation			
Usability inspection without users	–	–	–
Usability inspection with users	X	X	–
Usability test	–	–	X
Redesign			
Interaction concept	–	–	X
Interaction design	–	–	X
Visual design	–	–	–
Prototype	–	–	X
Design specification	–	–	–

Note. X = step was conducted; – = step was skipped.

9.7.1.2 Start Conditions

The following start conditions existed for this project:

- Lead-in phase (during the product development process): after the product definition and before starting the design phase.
- Project objective: design of user interfaces (no prototype).
- Available information: portal idea.
- Financed steps: market analysis, use context analysis, user interface design, evaluation.
- Application type: Web portal for travel offers, B2C.
- Project type: users are involved; user interfaces do not exist.

9.7.1.3 Conducting the Project

We went through the analysis and design phases of the JIET design process. In the analysis phase, the following steps were carried out:

1. Market analysis (provider): 3 hours, UI designers only.
2. Use context analysis: 30 minutes with users.
   - User profile: consumer profile (B2C), computer profile: 15 minutes with users.
   - Activity profile: scenario profile: 15 minutes with users.
3. Requirements gathering: 2 hours with users.
4. Use scenarios and user-specific requirements: 2 hours with users.

The market analysis was accomplished during the project preparation. As the user interface designers, we put ourselves in the place of a fictitious travel agency. The consumer profiles and the computer profile were constructed by working together with representatives of the user group. The scenarios were already prepared, owing to user interface the designers' experience, and they were evaluated by the participants. Likewise, the use scenarios and the user-specific requirements were identified by working with the users.

The design phase was partly conducted with the users. Two steps were carried out:

- User interface design: 3 days, user interface designers only.
- Evaluation: half an hour for each user, four users total.

At the end of the project, a visual design of the user interface was developed. Together with the users (the same persons involved in the requirements gathering), the user interface designers conducted a usability inspection.

9.7.1.4 Discussion

Using the JIET design process, it was possible to provide user interfaces for a commercial Web site in 4 days. Although the project was done for the purpose of research, the required time was nevertheless very short. Process analysis of the task flow was not necessary, since the user interface designers themselves belonged to the user group, having the knowledge how to plan and book a trip. If more time had been available, we would have uncovered problems and suggestions for improvement and would have created a competitor profile.

9.7.2 Case Study 2: Audiological Technology (B2B)

9.7.2.1 The Project

The second case study describes an international project (based in Germany and the United States) in the area of audiological technology. The objective was to gather requirements for an electronic order form for ordering hearing aids and accessories. At present, orders are made almost exclusively by using a paper form. The electronic order form targets audiologists and dispensers (United States) and acousticians (Germany).

9.7.2.2 *Starting Conditions*

The following starting conditions existed for this project:

- Lead-in phase (during the product development process): after the product definition and before starting the design phase.
- Project objective: requirements for the electronic order form.
- Available information: market profile (provider), user profile, scenario "order" prototype, and U.S. online form.
- Financed steps: use context analysis and requirements gathering.
- Application type: order form, B2B.
- Project type: users were involved and user interfaces already existed (a prototype and U.S. online form were used as information sources only).

9.7.2.3 *Conducting the Project*

Because of previous projects, a lot of information was known about the work of an acoustician. Therefore, only the analysis phase was conducted—indeed, only the steps needed to reveal the order process. The following steps were performed:

1. Use context analysis: 4.5 hours with users.
    - Activity profile: problems and improvements: 2 hours with the users.
    - Process profile: context scenario (current state and intended state) for the scenario "Order": 2.5 hours with the users.

Thus, the process profiles were developed first and used as a warm-up for a brainstorming phase to reveal problems in and ways to improve the activity profile. After reconstructing the order process, the audiologists had the individual steps clearly in mind and were able to generate the associated problems easily.

1. Requirements analysis: 1.5 hours with the users.
    - User-specific requirements: 1.5 hours with the users.

Together with the users, we gathered the requirements based on the existing prototype and online order form. The audiologists inspected the prototype and the order form by means of the order scenario and formulated their "likes" and "dislikes."

9.7.2.4 *Discussion*

In the overall process, developing the task flow initiated the requirements gathering. Owing to the audiologists' experience with the existing products (the prototype and online form), gathering the requirements by means of a usability inspection was the obvious choice. This case study shows how flexible and adaptable design processes have to be.

9.7.3 Case Study 3: Marketplace for Selling Communication Systems (B2B)

9.7.3.1 *The Project*

The third case study describes an international project whose objective was the development of a marketplace for selling communication systems (e.g., telephone systems and call centers).

The project was based in the United States and Germany. The Web site targets enterprise customers and distributors. In this case study, only the project section for enterprise customers is described.

9.7.3.2 Start Conditions

The following start conditions existed for this project:

- Lead-in phase (during the product development process): after the product definition and before starting the design phase. User interfaces already existed, but there was plenty of room to design new interfaces.
- Project objective: design of usable user interfaces (prototype).
- Available information: project sketch.
- Financed steps: market analysis (provider), use context analysis, requirements gathering, user interface design, evaluation, and redesign.
- Application type: transactional e-commerce Web site, B2B.
- Project type: users were involved; user interfaces already existed (used as information sources).

9.7.3.3 Conducting the Project

The following steps were carried out:

1. Market analysis (provider): enterprise profile (provider), product profile, customer profile, supplier profile, business process profile, competitor profile: 4 hours with business experts.
2. Use context analysis: 1.5 days with account managers, 1 day with users.
   - Market profile (customer): enterprise profile, business process, competitor profile: 3 hours with account manager, 1 hour with users.
   - User profile: job profile, computer profile: 1 hour with account managers, 1 hour with customers.
   - Activity profile: scenario profile, problems/improvements, competitor profile: 3 hours with account managers, 3 hours with customers.
   - User role profile: 4 hours, user interface designers only.
   - Process profile: 5 hours with account managers, 2 hours with customers.
3. Requirements gathering: 1 day with users.
   - Use Scenarios and user-specific requirements: 1 day (with users).

The use context analysis and requirements gathering took place on different days with different participants. In order to be best prepared for the customer workshops, for each customer we first interviewed the relevant account manager. In general, this strategy allows the user interface designers not only to become acquainted with the application domain but also to obtain knowledge about the customers. The importance of finding out about the customers should not to be underestimated, for it aids in satisfying them and also causes the customers to view the designers as truly professional. The interviews with the account managers took 1.5 days, the customer interviews took only 1 day. During the analysis phase, the user role profile is extremely important. At the project start, the client specified three different job roles (analyst, recommender, and decision maker) for five different customer classes (international enterprise customer, large national enterprise customer, medium national enterprise customer, small national enterprise customer, and distributor) in three countries (the United States, Germany, and the United Kingdom). Altogether 45 different user roles were possible (3 job roles × 5 customer classes × 3 countries = 45). One objective of the analysis phase was to identify the number of user roles. Just a quick reminder: For each user role, one interface version is needed. The analysis showed that there were only two different user roles; this means only two different interface versions were needed.

9.7.3.4 Discussion

Using the JIET design process, it was possible to carry out the interviews with the account managers and customers in a relatively short amount of time. During this project, a task flow analysis was necessary (i.e., for identifying the context scenarios). This was because the user interface designers knew little about the application domain and its processes at the beginning of the project. The representation of the context scenarios with the template technique revealed the potential for improvement, thus influencing the user interface design later on.

9.8 CONCLUSION

In order to complete user interface design projects successfully and in a customer-oriented manner, a usable design process is needed. A design process is usable if it is effective and efficient and leads to the satisfaction of all parties involved. Such a design process must be adaptive because of the different start conditions that different projects will have.

The JIET (Just in e-Time) design process is such an adaptive design process. On the one hand, it considers the boundary conditions in industrial practice; on the other hand, it is oriented toward the usability engineering process described in ISO 13407. The strength of the JIET design process lies in its modularity. Depending on the project, some method components can be omitted without seriously affecting the quality of the user interface design.

The adaptability extends to the task analysis. In design projects, knowledge of the user roles and the scenarios is a must. However, the level of detail of the scenarios depends on the previous knowledge of the user interface designers and on the type of project (B2B or B2C). If further information about the scenario is needed, an analysis of context scenarios is recommended. The method for accomplishing this described in this chapter has been applied many times in industrial practice.

Another strength of the JIET design process is its expandability. If, for example, the project objectives require the use of Contextual Inquiry (Beyer & Holtzblatt, 1998), this method module can be incorporated into the JIET framework. Thus, the JIET design process is prepared for future challenges.

If the JIET design process is changed, it is important to ensure that the new method components are compatible with the starting parameters. Furthermore, not only the collection of the data but also their interpretation should be effective and efficient.

9.9 ACKNOWLEDGMENTS

The authors would like to thank Arnold Rudorfer, Ilke Brueckner-Klein, and Madhusudhan Marur for their timely and valuable comments. We appreciate and acknowledge the creative illustration of the JIET process by Heinz Bergmeier (Fig. 9.1). Our special thanks go to Mark Hogg and Eduard Kaiser for permitting us to use the project data. Last but not the least, we thank Dr. Stefan Schoen, head CT IC 7, Siemens AG, for his constant support and for letting us to publish this method.

REFERENCES

Beyer, H., & Holtzblatt, K. (1998). *Contextual design. Defining customer-centered systems*. San Francisco: Morgan Kaufmann.

Constantine, L. L., & Lockwood, L. A. D. (1999). *Software for use: A practical guide to the models and methods of usage-centered design*. Reading, MA: Addison-Wesley; New York: ACM Press.

Constantine, L. L., & Lockwood, L. A. D. (2002). Industry brief *Interactions, 9*(2), 69–71.

Degen, H., (2000): Performance model for market-oriented design of software products. *International Journal of Human-Computer Interaction, 12*, 285–307.

Hewett, T. T., Baecker, R., Card, S., Carey, T., Gasen, J., Mantei, M., Perlman, G., Strong, G., & Verplank, W. (1992). *Curricula for Human-Computer Interaction*. ACM SIGCHI, [Online], 21.05.02. http://www.acm.org/sigchi/cdg/

ISO 9241-10: *Ergonomic requirements for office work with visual display terminals (VDTs). Part 10: Dialogue principles*.

ISO 9241-11: *Ergonomic requirements for office work with visual display terminals (VDTs). Part 11: Guidance on usability*.

ISO 13407: *Human-centered design processes for interactive systems*.

Hackos, J. T., & Redish, J. C. (1998). *User and task analysis for interface design*. New York: Wiley.

Mayhew, D. J. (1999): *The usability engineering lifecycle*. San Francisco: Morgan Kaufmann.

Norman, D. A. (1988). *The psychology of everyday things*. New York: Basic Books.

Norman, D. A., & Draper, S. W. (Eds.) (1986). *User-centered system design. New perspectives on human-computer interaction*. Hillsdale, NJ: Lawrence Erlbaum Associates.

Pedell, S. (1998). *Die Bedeutung der Gestaltgesetze Ähnlichkeit und Nähe für die Gestaltung von Benutzeroberflächen*. Berlin: Diplomarbeit. Unpublished.

Wood, L. E. (ed.). (1998). *User Interface Design. Bridging the Gap from User Requirements to Design*. CRC Press, Boca Raton, FL.

10

User-System Interaction Scripts

Rick Spencer
Microsoft Corporation

Steven Clarke
Microsoft Corporation

Though task analysis is generally considered a desirable part of user interface design, current task analysis methodologies are often considered complex and exacting and, even for human-computer Interaction specialists, the costs are often perceived to outweigh the benefits. Furthermore, the majority of software design is not done by human-computer Interaction specialists. For these reasons task analysis is rarely practiced in a formal way. This chapter presents a task analysis methodology called user-system interaction scripts that is designed to be usable by anyone, not just human-computer Interaction specialists. The chapter includes directions for using the methodology and a brief discussion of why we think it can be used by anyone.

10.1 INTRODUCTION

In the most general sense, a task analysis is a description of the human actions on a system necessary to accomplish a goal with that system (Kirwan & Ainsworth, 1992). In the fields of software engineering and human-computer interaction (HCI), task analysis is generally used to describe the actions that a user must perform with the user interface of specific software programs, such as mouse clicks, typing, and so on, in order to accomplish a goal with that piece of software.

We work in a large division of a software development company. Our division's primary focus is creating developer tools, which are computer programs that programmers use to create other computer programs. We believed that task analysis could benefit our organization by making the design process easier through providing a systematic approach to the design of human interaction with the software, ultimately resulting in more usable software and increasing customer satisfaction.

There were a total of four HCI specialists in this large division. We occasionally used task analysis while designing a human-computer interaction to help optimize the human performance in the system being designed. The general belief, which we share, is that if the human component of a human-computer interaction is explicitly designed and not merely an artifact of software engineering, the human component of the system will be more tuned to human capabilities, resulting in a more usable system (John, 1995).

We also instituted the practice of usability inspection using the Streamlined Cognitive Walkthrough technique (Spencer, 2000). We needed to create task analyses as inputs into the Streamlined Cognitive Walkthroughs as well as other usability inspection methods that we used.

10.1.1 Cost of Using Existing Methods Outweighs the Benefits

There is also a general recognition in the field of HCI that task analysis methodologies are rarely used in real-world application development. We believe that this is because existing task analysis methodologies are complex and exacting, and the cost-benefit ratio is not always perceived as justifying the performance of an analysis.

When we tried to apply task analysis methods, like various GOMS techniques, we found that the work required to complete an analysis was copious. We found that, for anything but very small interactions, we could not complete a thorough analysis in time to use it as input in design decisions. In other words, design decisions happened too fast to apply existing task analysis methods.

Finally, in order to get the full benefits, including the benefits of usability inspections on early designs, we need to do task analysis early in our development cycle. After a certain point, when designs are fleshed out and code is being written, the investment in task analysis is not likely to pay off for us.

10.1.2 Design Owners Are Not Always HCI Specialists

Furthermore, in our division, responsibility for the design of a human-computer interaction only rarely falls on someone with any training or formal knowledge of HCI. Much of our software design and development is done by teams with no HCI specialist at all. The design of the system is generally left up to project managers or computer programmers. As HCI specialists, we found ourselves spread thinly across the organization. In short, for all the task analysis that should be done for the purpose of optimizing human performance with our software systems, we simply could not possibly do it all, even if we thought it was worth the time.

We realized that if we wanted our organization to benefit from using task analysis early and often in our design process, we would need to motivate other people in our organization to learn a new way of working. We decided that program managers would have to learn task analysis.

Program managers in our division are responsible for the specification of the human-computer interaction, the definition of how the software will be implemented, and the schedule for ensuring that the software will be completed on time. Consequently, we thought there would be some obstacles to training program managers to do task analysis. Based on our experience working in our organization, we believed the following:

- Managers and engineers would not spend more then 30 minutes learning how to do task analysis.
- Managers and engineers would not do any task analysis that added noticeably to the time it took for them to complete their specifications and other job responsibilities.
- Managers and engineers would have a low tolerance for applying any methodology that seemed unnecessarily exacting or required new notations, concepts, or vocabulary.

10.1.3 Requirements for a Successful Methodology

We responded by creating a task analysis methodology we call User/System Interaction Scripts. We believe that this is a methodology that can be practiced by anyone and that we could institutionalize its practice early in the development cycle by making it part of the specification phase of our development process. Our goals for the system were these:

- User/System Interaction Scripts can be taught without introducing new language or new notational systems.
- Interaction Scripts can be learned from rules of thumb and examples.
- Interaction Scripts are informal and flexible and could be learned and practiced by someone without any formal HCI training.

The following two sections consist of a document that we created for training program managers in our organization. It is meant to stand on its own as an instructional guide to creating User/System Interaction Scripts.

10.2 USER/SYSTEM INTERACTION SCRIPTS

An interaction script (an "activity list" in chap. 1) is a written document that walks the reader through the steps that a user would make to accomplish some goal. It's an easy and clear way to communicate how the user interface for your software will be used.

An interaction script has two main parts:

1. A description of where the user is in the task.
2. A series of steps that the user would have to do to accomplish the goal and the system responses to each of those steps.

10.2.1 Sample Interaction Script

User Goal: The user is done using the computer so he or she is going to turn it off.

Step: Click the button labeled "Start" in the lower left hand corner of the screen.

System Response: A menu with about 10 items with icons pops up. The bottom menu item is "Shut Down." The other items include "Run," "Help," "Search," "Settings," "Documents," and "Programs."

Step: Click "Shut Down."

System Response: The menu disappears, and the whole screen gets grayed out. A dialog pops up that says, "What do you want your computer to do?" The dialog contains a drop down that is set to "Shut Down." There is a line of descriptive text about the Shut Down item. There are three buttons labeled "OK," "Cancel," and "Help."

Step: Click the OK button.

System Response: Everything on the screen disappears, and a dialog comes up that says, "Please wait" and "Windows is shutting down" After about 20 seconds, another dialog pops up that says, "It is now safe to turn off your computer."

Step: Click power button on the front of the computer.

System Response: All the lights go off on the front of the computer, and the monitor goes black.

Task Completed.

10.3 HOW TO CREATE AN INTERACTION SCRIPT

10.3.1 Step 1: Describe the Goal

The first step is to describe what the user is going to try to accomplish with the software. You need to communicate a few basic things in a goal description:

- where the user is in the task
- what the user wants to achieve
- additional details necessary to understand why the user would want to accomplish the goal
- additional details necessary to understand certain steps that the user will do.

10.3.1.1 The Right Level of Detail for a Goal Description

You need to include enough information in your goal description so that someone who is not intimately familiar with the software that you are describing will understand it. This will often be determined by where the task description is. In a very detailed specification document, few details may be necessary inside the goal description itself, since the rest of the specification document will provide context and details for readers.

You should feel free to vary the level of detail depending on how long the task is that you are trying to describe and how many features you are trying to describe with it. For instance, if you want to describe how to set certain settings, you will probably want to include using those settings as part of the user goal.

10.3.1.2 Sample Goal Descriptions

User Goal: The user's computer is running, but no applications are open. The user wants to write a letter, print it out, and mail it to a friend.

User Goal: The user has just created an e-mail account on a Web-based e-mail system. The user wants to write an e-mail and send it to a friend. The user knows the friend's e-mail address.

User Goal: The user has just installed an application using the apps setup program. The app relies on a component that runs on a computer elsewhere on the network. The user wants to configure the application so that it can use the component on the network.

10.3.2 Step 2: List All the Steps

Write down each step that the user has to do at the right level of detail. Your sequence of steps should only contain:

1. The **correct** steps that a user who knew exactly what to do go into the description. Don't try to use it to show how someone learning the system might use it.
2. One **singular** way of accomplishing the task. Don't try to show conditionals or alternate ways of doing things in one task description. If you want to show multiple ways of doing the same task, then just do multiple task descriptions. Of course you can reuse parts of different task descriptions that are the same.

10.3.2.1 The Right Level of Detail for User Steps

The user steps should have enough information in it so that someone who is not intimately familiar with the software can understand what the user is trying to do. A good rule of thumb is to get about as much information in each step as you would expect to see in a help document that is telling you how to use a feature. Another rule of thumb is to provide about as much detail as you would if you were giving directions to someone over the phone.

Not every step has to be at exactly the same level of detail. Multiple small actions in the use of common controls are generally rolled up into one user step. For instance, to show someone how to select the print command, you could list two steps:

Step: Click on the word "File" in the menu bar.
Step: Click on the word "Print" in the menu.

However, since a menu is such a common control, you might as well combine the steps:

Step: Select Print from the File Menu.

On the other hand, before Windows 95, buttons did not normally pop up menus. Therefore, the steps for selecting from the start menu should have been broken out:

Step: Click the button labeled "Start" in the lower left hand corner of the screen.
Step: Click "Shut Down."

10.3.2.2 Examples

The following examples show too much detail, not enough detail, and about the right level of detail for steps for the same user goal.

User Goal: The user is done writing their document, and is going to print it.

Sample Sequence With Too Much Detail.

Step: Point mouse cursor to File menu.
Step: Click mouse button.
Step: Point mouse cursor to Print command.
Step: Click mouse button.
Step: Point mouse cursor to the OK button on the Print dialog.
Step: Click mouse button.

Sample Sequence With Not Enough Detail.

Step: Print the document.

Sample Sequence With Right Level of Detail.

Step: Choose the Print command from the File menu.
Step: Click the OK button on the Print dialog.

10.3.3 Step 3: List The System Responses

You have to write down what the system does after each step the user does. You can do this after each step, or you can do it all at once at the end. The System responses should contain:

- changes that occur in the interface, such as windows or dialogs opening and closing, controls becoming enabled or disabled, etc.

- descriptions of dialog text
- substantial lags in system response
- cursor changes and other feedback.

10.3.3.1 The Right Level of Description for System Responses

You should provide enough detail so that someone not intimately familiar with the software that you are describing would be able to understand how the software is responding and how the user would know to do the next step. Generally, focus on the things that change in the interface, including new buttons or links that are available to click. You may want to summarize blocks of writing, etc. If you have a bitmap, even a rough one, it is often very helpful for communicating a lot of information about the system responses in a small space. However, it should not replace a good description of the system responses.

10.3.3.2 Examples

The following examples show too much detail, not enough detail, and about the right level of detail for system responses.

Sample System Responses With Too Much Detail.

Step: Choose the Print command from the File menu.
System Response: After 100 ms a dialog appears at coordinates 140×140. The dialog's title bar has the system setting for color and style. The dialog has a "?" button, and an "x" in the right side of the title bar. In the left-hand portion of the title bar is the word "Print" in the current system font and color. The dialog is divided vertically into three regions. The top region is bound by a single groupbox. In the upper left of the group box, the word "Printer" interrupts the groupbox. Inside the groupbox...

Sample System Responses With Not Enough Detail.

Step: Choose the Print command from the File menu.
System Response: A print dialog appears.

Sample System Responses With Right Level of Detail.

Step: Choose the Print command from the File menu.
System Response: The standard print dialog appears. It says "Print" in the title bar, and the dialog contains controls related to printing. There is an OK and a Cancel button.

10.4 USING INTERACTION SCRIPTS IN SPECIFICATIONS

10.4.1 How Do You Choose the Scenarios to Include in Specifications?

It's probably not possible to do an interaction script for every user goal that is supported by any one specification, so you need to choose what to describe. An average specification will usually have 1–3 interaction scripts, whereas a large specification may have 3–5 interaction scripts.

Naturally, you will use your own judgment in choosing which user goals you will write an interaction script for. The interaction scripts that you choose to include will communicate to readers what you think are the most important aspects of your specification. Consider the following criteria to choose what to describe with an interaction script:

- Importance to user. From the users' point of view, is the most critical functionality in the specification?
- Strategic objectives. What are your team's objectives in terms of new functionalities or technologies that you want users to adopt?
- Realism. What do users really do now, and what do you think they will really do with your product? Will they be focusing exclusively on your new features or will they be using the product pretty much the way they are now?
- What is unknown? There probably is not much point in describing a well-known functionality that has not changed for many versions and is still not changing.
- Coverage of features. An interaction script that covers multiple functional areas of your specification will demonstrate how multiple features work and how the features fit together.

10.4.2 What About Bitmaps?

Bitmaps are a very nice addition to interaction scripts. If you have them or have time to create them, you can add bitmaps after the system responses to visual demonstrate how the user interface changes as users interact with it. If you only want to create a few bitmaps, target those system responses where really big changes occur, like a new dialog appearing.

10.5 STUDYING THE EFFECTIVENESS OF INTERACTION SCRIPTS

We had two research questions that we wanted to answer:

- Would experienced program managers at Microsoft be able to use interaction scripts to describe a feature that they work on and are familiar with?
- Would experienced program managers at Microsoft be able to use interaction scripts to describe a feature that they do not work on and are not familiar with?

We designed and ran two simple studies. In the first study, we concentrated on the ability of program managers to use interaction scripts to describe a feature that they are familiar with. We describe the study and its results in the next section. We then describe the second study.

10.5.1 Are They Simple Enough?

The first study focused on determining whether people would be able to create interaction scripts using our original set of instructions. We identified a number of research questions that we wanted to answer:

- Will program managers in our company be able to create interaction scripts in their feature specifications?
- How long will program managers in our company take to create an interaction script for one of their features?

- Will program managers in our company be likely to use this format in their specifications?
- How do program managers in our company feel the interaction script format compares to writing scenarios in feature specifications?

To answer these questions, we recruited five participants to take part in the study. All of the participants were Microsoft program managers with at least 6 months' experience working at Microsoft and had written at least 10 feature specifications throughout their Microsoft career. None of the participants were HCI specialists. We gave participants a set of instructions for creating interaction scripts. Participants were asked to read the instructions and then to attempt to use the instructions to create an interaction script for any of the features they were currently working on. They were asked to spend no more than 45 minutes creating the script. After creating the script, the participants completed a short online questionnaire. The questionnaire asked participants to rate the difficulties of describing the user goal, each of the steps and system responses in the script, the effort involved in creating the script, and how worthwhile the exercise seemed. Participants were also asked to compare their experiences using the interaction script to their experiences describing scenarios in their feature specifications.

10.5.2 Results

We found that participants were able to create an interaction script for their feature using the current version of the instructions but that some participants had difficulty describing the steps and system responses at the appropriate level of detail. All participants in the initial study were able to complete at least one simple task description using the version of the instructions they were supplied with. The time taken to complete the interaction script varied from 6 minutes to 30 minutes. However, participants experienced certain difficulties while attempting the task.

One participant described difficulties he had when describing the steps and system responses: "It's hard to know how much detail to use because on one hand you're trying to let anyone unfamiliar with the process know how it works, yet you don't want to create volumes of text." In our judgment, the participant's interaction script (which follows) was not detailed enough:

> System Response: User is navigated to a details page that describes how the transaction has been handled and gives the user the option of changing whether the downloaded transaction matches an entered transaction or whether it should be a new transaction.
>
> Step: Confirm correctness of data, click the Done button.
>
> System Response: Navigate back to the register, item remains highlighted.
>
> Step: Mark the selection as read, click Review.
>
> System Response: Highlight and markings are removed from item. The next unread item is then highlighted.

Another participant said that he had encountered some problems in creating the script: "It requires too many steps. In the spec I already have, I have covered this with one paragraph of text and one bitmap." In this case, it appears that the participant thought he had to create a script covering every possible task supported by the feature. The instructions did not make it clear to the participant that each participant could choose a subset of tasks to describe.

We gave participants the opportunity to provide us with free-form feedback on the interaction scripts. In general, participants reacted favorably. The interaction script format for most people was reasonably straightforward to work with, and for at least one participant it highlighted some breakdowns in the way that she had been designing her feature (note that she refers to "simple task description," which was our original name for interaction scripts): "Simple task description seems much clearer than scenarios. I also found that this method catches potential problem areas better. It was difficult to know how to break tasks into groups and what level of detail to include. Better documentation on this would be great!"

This study encouraged us to continue to develop the interaction script format and to proceed with a more detailed study. We reacted to the findings from the initial study and attempted to improve the instructions we had created describing how to create an interaction script. We described how to use bitmaps to describe feedback to the user and explained that it is not necessary to describe a whole feature with this format, only those components of a feature that it makes sense to describe, such as a new component or a redesigned component.

In another study, an informal comparison of the scripts created by participants revealed differences between the interaction script format and the more narrative style of description chosen by most participants who did not use interaction scripts to create their scripts. The first example shown below was created by a participant who used the interaction script format to describe how the user would create and save a search. The second example is of the same task but was created by a participant who did not use the interaction script format.

User Scenario: The user wants to create a search for documents whose names begin with "Usability Report" in both their hard drive and on a network share. They want to save the search in a directory called "My Searches" inside the "My Documents" folder.

Step 1: From the Start Menu choose Find: Files and Folders.

Response: The Find dialog (described above) appears.

Step 2. Enter "Usability Report*" in the Named: edit control.

Response: N/A.

Step 3. In the "Look in" enter "c:\; [path to network share]".

Response: N/A.

Step 4. Click Find.

Response: An hourglass icon below the find button will begin spinning and a listview will appear below the tabs. Documents containing "Usability..." will begin progressively appearing in the list.

Step 5. Choose Save Search from the File Menu.

Response: The Windows File Save dialog will appear. There's a listview in the middle showing the files in the "My Documents" folder, and below that there are two dropdowns, one for the "File Name" and one for "Save as Type."

Step 6. Open the "My Searches" folder, enter a name in the "File Name" combo and click "Save," keeping the default name.

Response: The Save dialog closes.

Here is a free-form narrative text describing the same task.

To find the file the users would type the file name in the named box in name and location. They would choose browse for the look in and multi-select the local drive and on a network share. There may be some confusion on the multi-select and also the fact that the network share may not be in the list. Choose find to find the file. Once the result is back they would do a save. That saves the file to the desktop—this would be confusing. Users may expect a standard save dialog. Once the file is on the desktop they could move it to a folder called my searches in my documents. They could rename by right clicking the file and choosing rename. An easier implementation may be to just use the standard save as dialog.

We believe that the program managers who used the interaction script format were significantly more systematic in describing the user interaction with the user interface under consideration. Furthermore, the interaction scripts that they created could be used for a usability inspection method (e.g., cognitive walk-through).

Since we have deployed User System Interaction Scripts directly in an applied setting, we have not been able to do any kind of formal comparison between the same design created with interaction scripts and with the existing specification functionality. It has simply not been feasible to invest the time to create and evaluate two designs for the same user experience. However, we believe strongly that including task analysis early in the design process ultimately improves designs, and we believe that we have created a task analysis methodology that can be employed effectively by people who are not specifically trained in HCI.

10.6 CONCLUSION

We have tried to create a task analysis method that was as simple and straightforward as possible to make it much more attractive to our colleagues. The studies have shown that we have been successful at this and that program managers are significantly more systematic in task-based design efforts using the interaction script format.

Since we created the instructions for creating interaction scripts, many program managers have been using in them their specifications. When we followed up with them about why they continued to use the method, they said that improvement in readability was a major advantage over their old format, since the developers and quality assurance staff on their teams were more likely to read and understand those areas of their specifications related to user interaction. Furthermore, they think that using the interaction script format ultimately improves their designs because they see usability problems early in the specification process.

REFERENCES

Kirwan, B., & Ainsworth, L. K. (1992). *A guide to task analysis*. London: Taylor & Francis.

John, B. E. (1995). Why GOMS? *Interactions, 2*, 80–89.

Spencer, R. (2000).The Streamlined Cognitive Walkthrough Method: Working around social constraints encountered in a software development company. In *Proceedings of the (CHI 2000)*. New York, NY USA: ACM (pp. 353–359).

11

Task Layer Maps: A Task Analysis Technique for the Rest of Us

Jonathan Arnowitz
PeopleSoft

This chapter discusses a task analysis technique with a higher abstraction level that is intended for multidisciplinary usage. The technique is also designed to be quickly created from data of a user study as befits a commercial setting in less than ideal circumstances where resources are scarce and demands high. Based on a simple mathematical program, task layer maps will give the entire development team a quick and easy overview of the system being developed.

11.1 INTRODUCTION

This book is in part a survey of the broad range of best practices in task analysis for human-computer interaction design. This chapter looks at task analysis a bit differently. In particular, it considers how other parties look at a task analysis and what possible techniques could make a task analysis helpful to HCI professionals and nonprofessionals who must work together. In the past years, the HCI design team has become larger and larger, making it harder and harder to justify project expenses for task analysis just for the user interface designer, especially as team members from other disciplines back don't really know what to do with a task analysis. The Task Layer Map technique is an attempt to address this problem by reducing the complexity of task analysis design to essential information that can be customized by each discipline. The resulting drawings have been shown to be easy enough to understand to be of help to even nontechnical people during both requirements reviews and acceptance testing.

A task analysis is often considered the most essential step in designing a quality user interface. For the purposes here, I use the general definition of the task analysis given by Hackos and Redish (1998): "User and task analysis is the process of learning about ordinary users by observing them in action" (p. 7). There are many methods used for attaining the information for a task analysis as well as many different models for representing them. Whether the starting point is psychology, anthropology, or architecture, all the disciplines agree on the importance of a task analysis. How that task analysis looks and what information it contains can be as varied as the theories and methods. The problem with most task analysis diagrams is that they are often too specialized and/or obscure to be widely understood. The misunderstanding that results, along with the lack of communication, presents itself as a problem particularly (but by no means solely) when designers want to translate a task analysis into a design. If everyone cannot understand the task analysis, the basis for creating a good design is compromised. This inhibits the designers (and other non-HCI professionals involved in the project) from making full use of the task analysis, often to the detriment of the resulting product.

The first section below discusses the current problem with task analysis diagrams. The next section introduces Task Layer Maps and describes their philosophical and theoretical background. The final section presents a scenario showing how Task Layer Maps work in a software development project (with the names changed to protect the innocent).

11.2 THE CURRENT PROBLEM

In my experience (as well as in the experience of others, such as Bomsdorf & Szwillus, 1998, 1999, and van Setten, van der Veer, & Brinkkemper, 1997), one problem remains constant: In a multidisciplinary design team, there is no satisfactory technique for representing a task analysis and translating that task analysis into an interaction design. Diagram styles often have a line drawing technique that works quite well for some and yet is confusing for others. Some users demand UML; others, data flow techniques; still others, scatter diagrams; and the list goes on. Each diagram style requires its own particular perspective, its own way of doing things, and its own way of conceptualizing things.

This chapter, as must now be apparent, focuses on the physical representation of the task analysis, which we will call the *task analysis diagram*. Task analysis diagrams are often clear to technicians but mystifying to the uninitiated, or the other way around. For example, model-based development methods such as TADEUS and KAP (Bomsdorf & Szwillus, 1996; Puerta, Chengm, Ou, & Min, 1999; see also chap. 25) often include diagrams that are difficult for the lay-person to understand. CASE tools, on the other hand are created from a software engineering perspective (see the chapters in Part IV) and often are difficult to translate by designers. GOMS analysis frequently requires someone who understands the GOMS methods and thinks procedurally (Kieras, 1997; chap. 4). Even simpler hierarchical diagrams as suggested by Mayhew (1999) and Hackos and Redish (1998) can mystify users who do not understand the context of the hierarchy. Owing to the fact that HCI professionals usually work with people from a great variety of backgrounds, using specialized diagrams is not fair to at least some members of the project team. During task analysis sessions, many team members are left with folded hands because they don't understand exactly what is going on. Therefore, a new diagram technique is called for. That is not to say that I would recommend ignoring, replacing, or even modifying tried-and-true methods, many of which are discussed in this handbook. The problem I am addressing is not the methods but their representation techniques. Task Layer Maps is intended to be a technique usable with any task analysis method to communicate the resulting task analysis to the entire design team, especially where full-strength HCI details are not needed or team members may want to add other details that they find more important.

The specific problems with the current task analysis diagrams can be summed up as follows. The drawings or documentation techniques do not

1. address the needs of a multidisciplinary team
2. lend themselves to an execution technique sufficiently quick for commercial projects
3. present the design team with an easy-to-read overview of the task structure
4. give clues on how to implement a dialog design

5. provide a technique for verifying the concept against the task analysis
6. assist the creative aspects of user interface design
7. provide a flexible technique that can fit into other methods.

11.2.1 Multidisciplinary Team Needs

Because not everyone can interpret task analysis diagrams, terms often fail to get the full benefit of the team members collective talents. Worse, misunderstandings can go undetected, leading to potential mistakes in both the task analysis and the resulting product. Misunderstandings can also causes delays as team members are taught the skills needed to co-create or understand the task analysis. This problem also inhibits the ownership among all the team members.

Furthermore, task analysis diagrams often contain details that are not interesting for everyone. For example, a graphical designer does not necessarily need to know what data sources are involved in a task. As another example, the user's object can be a source of confusion and misunderstanding between object-oriented programmers and HCI designers, who frequently use the same terms to mean different things. A task analysis diagram must be able to speak to all members of the project team.

11.2.2 Quick Execution

Another major problem is that companies often view a task analysis as a nice to have but not essential for the user interface. Therefore, design teams are under pressure to produce something that is both fast to produce and easy to understand. Complex task representations can cost a lot of time to produce. Furthermore, a task analysis is often difficult to understand by people not in the design team. Consequently, clients do not understand the value of the task analysis. Even representations that are not complex require many supporting or supplementary documents. For example, Hackos and Redish (1998) list 16 different possible supporting documents. Therefore, a task analysis diagram should be quick to produce and give added value to diverse stakeholders in a project, especially those managing the project.

11.2.3 Overview

Many drawing techniques do not give the design team an overview of the tasks of an application and how they fit together. Many task design methods are spread over a cumbersome number of flip charts or padded into a thick stack of use cases or lost in a plate of spaghetti-like task diagrams. A good diagram provides an easy-to-read overview of the application.

11.2.4 Dialog Design Tool

With most task analysis representations, it is not always clear where to begin and where to proceed with dialogue design. To convert a task analysis into dialogue design (Chap. 27 addresses this explicitly), almost every designer has developed his or her own black magic that makes perfect sense to no one else. This translation problem is another barrier preventing other team members from participating in the design process. A diagram technique should help the designers explain this transformation and support their design rationale.

11.2.5 Design Verification Tool

Another problem, especially when designers work with clients in a consulting capacity, is that clients personally cannot use the task analysis for anything. Nor do they know how to test whether the resulting design is a valid result of the task analysis. If the task analysis diagram could also be helpful to the clients, they would be motivated to ensure that the task analysis was performed. Furthermore, they would get added value by having a tool to check whether the designers have done their work well. A task analysis diagram should also support naive clients in their acceptance testing.

11.2.6 Encouragement of Creativity

Very often, HCI designers can get lost in the methodology. Particularly when using complex, rigorous methods, HCI professionals forget the creative side of HCI design. A task analysis diagram should not only enable creativity but support it. (see also chap. 5).

11.2.7 Method-Independent Technique

Clients use a wide variety of development methods. There are many software development methods, project management methods, rapid application methods, and, as this book attests, HCI development methods. Thus, a task analysis diagram should be flexible enough to adapt to many different environments.

11.3 PARTIAL ORDERING

The Task Layer Map technique described in this chapter was developed during work on several projects with diverse clients. Using Task Layer Maps, everyone on the project team could more easily add their own ideas and creativity into the design process. The increase in creativity was considerable when compared with other projects using more complicated task analysis diagrams and/or flip charts. Moreover, with an easy-to-read diagram, the HCI professional becomes less the expert on task analysis and more the facilitator, a role that will lead to more input and greater creativity in a project.

The improved project results were due neither to added information, nor added terminology, nor a minimalist approach to the information. Instead our approach tries to change the presentation using techniques discussed by Tufte (1990). It also follows two rules discussed by Tufte: First, reduce visual noise, and second, use separation and layering.

With differing shape types and line styles all with specific meanings, drawings such as UML can become very "noisy" (i.e., not simple to read). Even with keys or legends, these drawings are still obscure. Task Layer Maps reduce this visual noise by using a single presentation for a task and its relations. This simple presentation style ensures that the diagram poses little or no translation problems for users. With just a single box and lines with arrows in one direction, users can concentrate on the total picture instead of trying to decipher the different symbols and lines. As users became familiar with the diagram, we discovered that coloring the background of the task boxes was a successful way of differentiating groups of tasks without recreating the visual noise problem. If we ever ran out of distinguishable colors, we found we were overspecifying.

If the information was presented in a separated and layered manner, it became even easier to read and much more understandable. An overview became easier to accomplish, as tasks could be aggregated to a higher level. How this works is the subject of the last section of this chapter. The current section focuses on the theoretical basis for the separation technique. It contains a discussion on parallel programming written largely by Duco Fijma, one of the original creators of the diagram method. The discussions on parallel programming provided in the following section originate from him and part of an earlier version of this article which had appeared at the DIS2002 conference. The information on parallel programming is essential for understanding Task Layer Maps, and I hope users already familiar with Task Layer Maps will forgive the duplication here.

The Task Layer Maps technique is based on the mathematical concept (important in parallel processing) of a *partial ordering.* This is a sound formal concept, yet it is easy

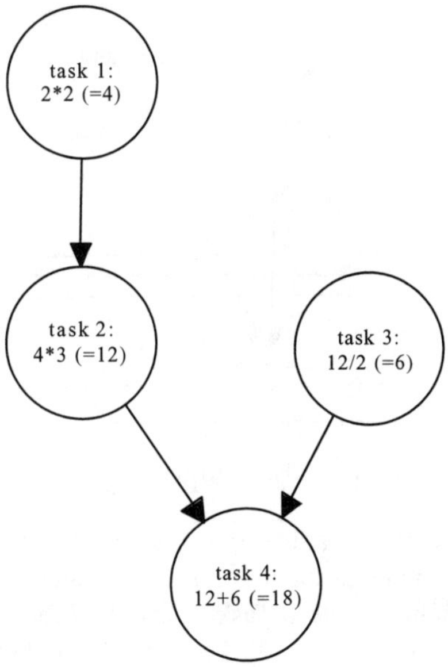

FIG. 11.1. Partial ordering of a simple expression.

to understand in an informal way, as demonstrated in the appendix at the end of the chapter.

More interesting than the formal definitions in the appendix, however, is the way we can use this concept to model a set of *tasks* together with the *dependencies* between them. Any kind of task, including user goals, elementary calculation steps in a computer algorithm, and the directions in the recipe for your favorite cheesecake, can be modeled this way.

Of course, this chapter concentrates on user goals. However, to give you a feeling for what we can do with partial orderings, it first discusses a small example borrowed from the analysis of a computer algorithm (Fijma, 1994). In this example, a task corresponds to a basic instruction a computer executes while running a program.

Consider the expression $(2 \times 2 \times 3) + (12/2)$. The tasks executed by the computer while it calculates an answer to this problem can be described by the partial ordering in Fig. 11.1. Tasks are drawn as dots or circles, and the dependencies between tasks are drawn as arrows pointing to the dependent tasks. Looking at Fig. 11.1 we observe that it describes only the essential dependencies between tasks. Task 1 must take place before Task 2, and although both Tasks 2 and 3 have to be ready before Task 4 can start, there is no inherent mutual dependency between Tasks 2 and 3.

That this relationship is made explicitly clear in the partial ordering is very attractive from a design perspective. We have, or the computer has, the freedom to execute tasks 2 and 3 in any order or to execute the tasks in parallel. Choosing beforehand for any ordering between 2 and 3 is overspecification.

As an alternative to the partial ordering presentation style, consider the alternative specification of what the computer does in some Pascal-like procedural program (which is similar to the way a GOMS model might be specified; chap. 4):

$$x = 2 \times 2; // \text{Task 1}$$

$$y = x \times 3; // \text{Task 2}$$

$$z = 12/2; // \text{Task 3}$$

$$\text{result} = y + z // \text{Task 4}$$

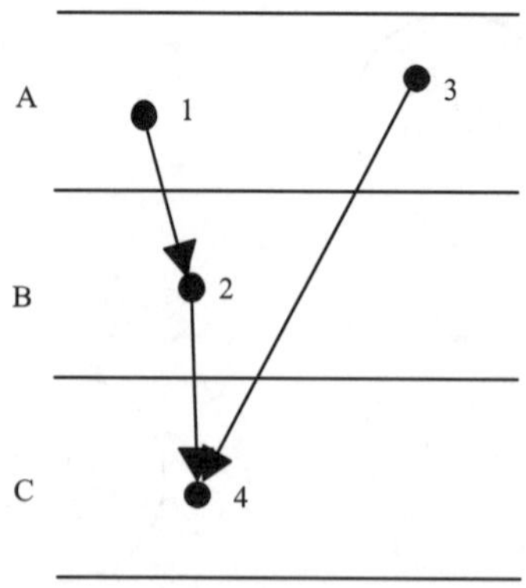

FIG. 11.2. Layered Diagram of a Partial Ordering.

This specification of what the computer has to do contains too much detail that is not relevant for the design which is also a major problem for many task analysis diagrams. The semicolons express nonessential sequencing between Tasks 2 and 3. If this specification were given too early in a design process, we would have blocked, or at least obscured, the possibility for the computer to execute Tasks 2 and 3 in parallel.

Because a partial ordering is transitive, we do not need to show direct dependencies between tasks if these are implied by other dependencies. Although Task 1 has to be performed before Task 4, we do not have to draw an arrow between 1 and 4 because it is implied by the arrows from 1 to 2 and from 2 to 4. Partial orderings, therefore, greatly reduce the number of arrows in the diagram, making the specification easier to read and the relations clearer.

Because a partial ordering is antisymmetric, a diagram cannot contain cycles. This is essential for our task-layering technique. Because of antisymmetry, we can present a partial ordering as a *layering diagram* (Fig. 11.2).

A layering diagram is essentially an allocation of the tasks to be performed in the "time slots" in which they can be performed earliest. All tasks in one layer, or time slot, can be executed in parallel when all tasks in earlier layers are done. This interpretation of layers also allows an informal understanding of why the absence of cycles is so important: Time does not flow backwards.

The *width* of a layering diagram shows the amount of parallelism in the job. In the example, at most two processors or human beings could be set to work efficiently. Even then, one of these would sit idle during execution of layers C and possibly B (assuming Task 3 could be finished before Tasks 1 and 2). Only during layer A would they would be both busy.

11.3.1 The Task Layer Map Technique

The Task Layer Map technique uses layers in three dimensions. The first two dimensions consist of a single aggregate level of tasks and/or goals. (The level of aggregation can be determined by the designer based on task complexity). The goal is to have an appropriate aggregate level to maintain an easy-to-read diagram and then zoom in on an aggregate element and view the task. For example, for a complex application to assist object-oriented programmers, you may want to use an aggregation level of user goals such as "Maintain object tree," "Maintain business objects," "Maintain tasks," and so on. Then the designer could zoom in on the "Maintain object tree" goal and create another diagram for the tasks and/or actions.

For a simpler application—the registration of telephone calls, for example—the lower level subtasks can be used as a starting point. In general, I use this criterion: If the drawing will not fit on a single page, then the aggregation level is too low for an easy overview. The fact that these drawings are quick and easy to perform (and will be quicker if a tool is developed to automate the process) means that the overview is quick and easy to perform once a task analysis has been conducted. Furthermore, the level of detail required varies between project personnel. Marketing professionals, for example, may care only about the high-level diagram and in meetings only want to discuss that. On the other hand, a meeting between developers and designers to discuss implementation of a single task may concentrate on that task's diagrammed actions.

11.3.2 How to Create a Task Layer Map

In our application of the layering technique to task analysis, task can be interpreted to mean task or subtask or goals. It is completely dependent on what will work for the application being studied, as mentioned above. Task Layer Maps are constructed in three steps:

1. Construct task set.
2. Peel task set into layers.
3. Finalize layer map.

To understand the method better, consider a simple data entry application intended to support a call center in which employees need to look up existing numbers and addresses of customers who have put in a call. They search for these in a one of many customer data files. If they find them, they will need to state which telephone number or which address they are using for their phone call. If neither a phone number nor an address is present, they need to add one. Then they have to make the phone call and record the results. Duplication and error checking are essential throughout the application. Note that the example covers only a small segment of the application and is not intended to fully represent the entire application.

11.3.2.1 Construct Task Set

This is the easy part and has not much to do with partial orderings or layering.

We started with a flip chart (you could also use your favorite charting tool), and using our favorite method of task analysis creation, we created an application flow diagram. It is essential that at least the tasks and relationships (e.g., dependencies) appear in this flow diagram.

In our example, we identified all tasks and drew them on a flip chart. Then we checked all tasks for dependencies. For each task that was dependent on another, we drew an arrowed line going from the nondependent task to the dependent task. The resulting drawing looked something like Fig. 11.3.

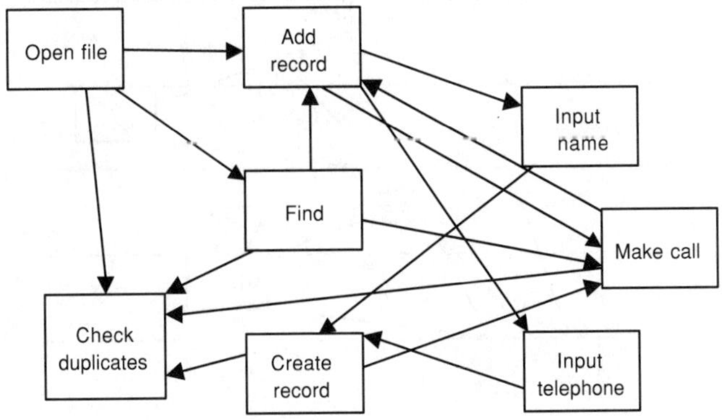

FIG. 11.3. Task set in an application flow chart.

11.3.2.2 Peel Task Set

In our technique, constructing a partial ordering from the initial task set and presenting it in layers is not performed in separate steps. In the example, we started with the construction of the layering diagram. During this process, we ran into problems (as we often do) because the initial task set was not a partial ordering (i.e., there were cycles). We intended to straighten out these cycles of the task set on the fly.

Having the task set ready and starting with a new blank drawing, we divided the drawing into as many layers as we thought we would need. Additional layers could be easily added during in the process. Then we started constructing the layers.

The first (earliest) layer was numbered 0. We placed all tasks that had no incoming arrows on this layer.

We constructed the subsequent layers so that each layer contained tasks that had incoming arrows only from previous layers. Layer 1 contained all tasks that only had an incoming arrow from tasks in layer 0. Layer 2 contained tasks with incoming arrows from 1 and possibly 0. In other words, we put tasks in the *earliest* possible layer while ensuring that no layer contained horizontal connections.

In Fig. 11.4, the task "Open file" is in Layer 0 because it does not have any incoming arrows in Fig. 11.3. The task "Find" can be added to Layer 1 because it depends only on the single task in Layer 0. Task "Check duplicates" depends on tasks in Layers 0, 1, 4, and 5. The earliest possible layer for this task is thus Layer 6 (the next layer after 5, the last of the layers containing a task that "Check duplicates" is dependent on).

If any cycles in the task's dependencies arise, the process of constructing the partial ordering stalls. In the example, this occurred after Layer 1 had been finished. At that point, there were not any tasks left that depended solely on tasks in Layer 0 or 1.

Somehow we had to "cut the cycle" to escape from this deadlock situation. For example, one of the cycles in Fig. 11.3 is the mutual dependency between "Add record" and "Make call." When we removed the arrow from "Make call" to "Add record," we allowed the process of constructing the layer diagram to be restarted by placing the "Add record" in Layer 2. Note that when an arrow is deleted, we do remove information from the flow chart (e.g., Fig. 11.3), for we want to keep this view of the tasks as well.

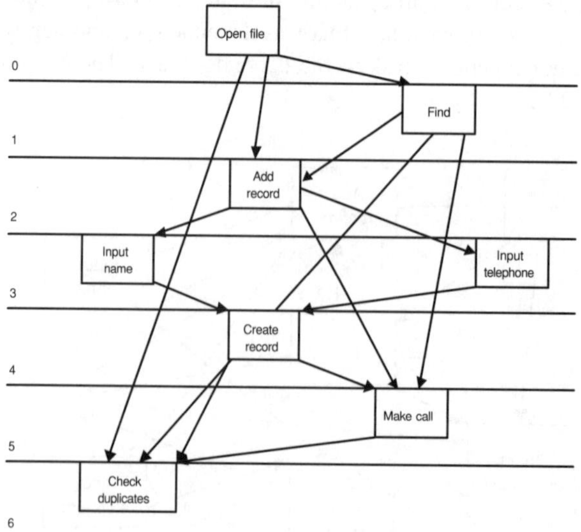

FIG. 11.4. Layering result after "peeling."

From a mathematical point of view, there is no objective rule that tells us which arrow to delete. Instead, the selection of which arrow to delete is a creative decision that the design team must make. The decision reflects which of the two arrows is the "weakest" from the user's point of view. In our example, we consulted our user information and saw that the arrow going from "Make call" to "Add record" exists because the user wants to add a second record after making a call using the first record. From this part of the user study, we determined this was a weaker dependency than that represented by the arrow going from "Add record" to "Make call" because a call cannot be made without a record. Deleting that arrow allowed us to continue our layering until we hit the next cycle. After all cycles were resolved and all tasks were assigned to a layer, the drawing we constructed looked like Fig. 11.4.

The advantage of this step is that it not only helps the design team clarify weak links in the user study but also automates items where no creativity is required. Making the tasks transitive is a purely mechanical process, whereas resolving cycles requires creativity from the design team.

11.3.2.3 Removing Redundant Dependencies

Figure 11.4 already looks a lot like a partial ordering in layer diagram format. The only thing that remains to be done is to delete redundant dependency arrows. For example, when we looked at our flip chart version of Fig. 11.4, we saw that the arrow between "Open file" and "Add record" could be deleted because it was already implied by the combination of the arrows from "Open file" to "Find" and the arrow from "Find" to "Add record." Therefore, we marked this line as deletable. We repeated this step for each arrow. In each case, the team discussed if the relation was correct and why. Normally there was not much discussion, but when relations or tasks were unclear, the misunderstanding or conclusion could be solved on the spot. This process contributed greatly to the mutual understanding of the task analysis. When we were finished, we redrew the diagram (Fig. 11.5).

The deletion of arrows in this step does not lead to information loss like the deletion of arrows in the previous step. We do not really delete them: They are demoted from being "explicit" to being "implicit" arrows in order to accommodate the transitivity of the partial ordering. Consequently, their deletion is not a creative activity and could easily be automated in a software tool that supports the construction of Task Layer Maps. (An important issue to consider is whether the gain in time is preferred over the gain in discussion with team members.)

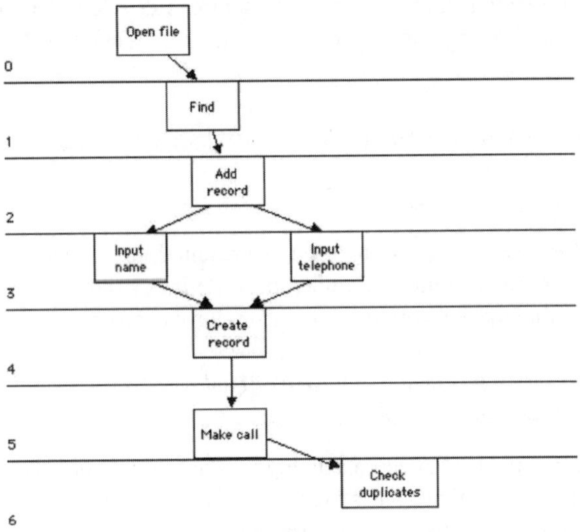

FIG. 11.5. Task Layer Map.

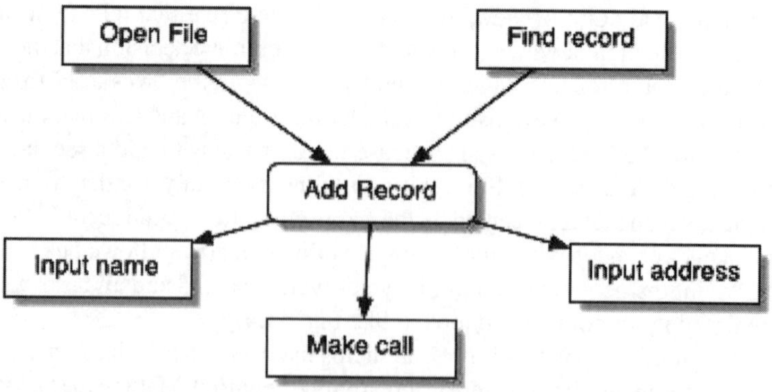

FIG. 11.6. A task star diagram for the "add record" task.

Figure 11.5 shows very clearly what are the central tasks by their clusters. Furthermore, because the tasks are now all indirectly related, application navigation is clearer, and dialog analysis can take place. Designers can use the image to explain their design decisions. Other team members can use the diagram to check for completeness in dialog design.

11.3.2.4 Task Star Map

There is one complementary task diagram that we have discovered to be useful in special situations: a Task Star Map (Fig. 11.6). The limitation of using parallel programming for our diagramming technique is that "time can never run backwards" but tasks sometimes can and do. When faced with recurring, reusable tasks throughout an application, we had in the past recommended not using the Task Layer Map technique. However, a simple Task Star Map can address the problem. Whenever a task has multiple connections, you can represent such a task on a Task Layer Map as a square with rounded corners (showing it is different from the rest of the tasks) and then link a drawing to this task that shows the relations between the task and its immediately connected tasks.

Using this system allows a developer and designer to look at a Task Layer Map and see right away that a task has been already used or will be used again somewhere else. Given the intimate knowledge of the application that design team members develop, for many this is enough. However, for very complex applications, the zoom in on the Task Star Map is very helpful.

11.4 TASK LAYER MAPS IN ACTION

11.4.1 Scenarios

To illustrate how Task Layer Maps are used, I present four scenarios from four different projects. I would like to thank Jouke Verlinden of TNO Industrial Design for contributing the fourth scenario. Note that in each scenario, the names have been change to protect the innocent.

11.4.1.1 Scenario 1: Dialog Design Tool and Design Verification Tool

The interaction designer begins to design a dialog for the "Add record" command. She uses the Task Layer Map to make sure that access to all possible following commands are represented. She also checks to make sure that unnecessary commands (commands that are only needed later) are not represented in the dialog.

The client, in turn, is delivered the beta application for the first steps in user acceptance testing. The client has key users look at the represented task analysis and compares the tasks with the application itself, checking each task to ensure that the connections represented in the Task Layer Map are indeed reflected in the software itself.

11.4.1.2 Scenario 2: Needs of a Multidisciplinary Team

The project team is busy with a complex development tool designed to develop 4GL database applications using an object-oriented paradigm. The requirements are huge and changing, the technical members are familiar with object-oriented principles, but the designers are desperately trying to understand them, and the marketing people, although think they understand them, lack a visualization of an object-oriented system and don't really understand its workings. At the first team meeting after the user research has been translated into a task analysis, the interaction designer unveils the overview of the application as Task Layer Maps.

The project team see an overview of the application for the first time. One marketing person asks, "That's all we're making?" The visual designer asks, "Is it really that simple?" The interaction designer, seeing the highest level overview, can begin exploring possible conceptual models with the team and see if they work. The team starts using construction, architecture, and film production metaphors. Because the Task Layer Map makes interaction clear, everyone joins in. During the meeting, the visual designer finally figures out how certain concepts hang together and later comes up with some sketches for a visual language addressing the different conceptual models. The drawings help with brainstorming, and during one session a metaphor is chosen. Then rough sketches for dialog and visual design are produced. These sketches set a direction that the designers can now work out. Although there is still much refinement to do, but the creative stimulation of the sessions has a lasting effect. The application of the new metaphor can always be checked against the Task Layer Map to see if it will really work or if adjustments are needed.

11.4.1.3 Scenario 3: Quick Execution and Method Independence

The creation of Task Layer Maps is based on mathematics, so their creation requires little discussion after user research has found some way of identifying serial and hierarchical connections (data flow, time, dependencies, etc.). The project team can come into a brainstorming session and review the user research and the envisioned system, and within moments after that the diagram can be completed. Since it is a mathematical system, the completion can even be shortened if a simple drawing tool was created to support the generation of the diagrams.

During the design team's brainstorming, some people may argue for certain types of terminologies, such as those for use cases, tasks, goals, user requirements, business objectives, and so on. The Task Layer Map, because it shows the flow of actions or tasks in the system, is already understood to be an abstraction. A business analyst can use the diagram as a basis for business object modeling. An object-oriented developer can use the diagram as a model for use cases. All parties agree that when they change their models, they will update the Task Layer Diagram so everyone can understand the ramifications of the changes they made to their particular part of the project work. In fact, as long as a task analysis method supports some kind of flow, that flow can be used as a basis for building Task Layer Maps. This is possible since the layers in the diagrams themselves, although concrete and visual, are abstract in the sense that they do not have to represent a point in time or a point in data flow; they can mean whatever the team wants them to mean. Task Layer Maps can become a lingua franca for the design and development team.

11.4.1.4 Scenario 4: Insurance Project

Our objective was to design a user interface for insurance brokers. The application was a workflow-based system. The purpose of the system was to allow a broker to create a price quotation for a clients and then guide the client through the process of generating a policy.

The project group consisted of information analysts, business experts, programmers, software architects, and end users. An object-oriented development method was employed, in combination with a specialized business objects framework.

When we were hired, the design challenges included these:

- The only specifications that covered the human-computer interface were use case specifications (more than 35 pages). The use cases were part of a large report. Although the use case specifications were written in natural language, it took a lot of effort to verify whether they covered all tasks and task sequences. Furthermore, the use case specifications encompassed detailed dialog designs. Including such specifications at this preliminary phase obstructed our work more than it helped.
- Our assignment was to design a user interface concept as soon as possible, so no time could be spent on analysis (especially because a lot of resources had already been spent on getting the technical specifications in place).
- The end user panel was losing interest in the project, primarily because the focus was so much on the new technology involved.

In order to deal with these design issues, we carried out the following steps:

1. We did a quick scan of all project documentation and conducted short interviews to develop an early user profile.
2. We met in a half-day workshop to construct the task set.
3. We translated the tasks into a Task Layer Map and then immediately constructed a first user interface design.
4. We used paper prototyping to improve the user interface.

Getting an Early User Profile. By studying the use case specifications and all other project documentation, we identified the scope of the application and the target users. However, the documentation proved to be too technical to use as the only source for performing the task analysis. Brief interviews with future users gave more insight into the responsibilities and the existing workflow management strategies.

Workshop on Constructing the Task Set. One of the most important findings of the previous step was that there were too many inconsistencies in the specifications. In order to share the same view among all relevant project members, we held a workshop to kick-start the task analysis. Attendees included the user panel, the information analyst (who had been responsible for the use case diagrams as well), and the team coordinator.

The workshop focused on getting as many of the tasks and relationships explicated as possible. First, all tasks were written down on sticky notes. After plenary clustering and the removal of redundancies, we started to stick the remaining notes on a whiteboard. Next, all participants were invited to draw the relationships between them. As a final exercise, we identified the most important start points and endpoints. The resulting diagram was incredibly complex, yet all workshop participants agreed on the contents of individual tasks and the relationships among them.

Peeling the Diagram. Translating the original task set into a Task Layer Map brought many observations to light. For example, "Risk assessment," initially thought of as a starting task, shifted to become a child of "Process incoming applications." The resulting Task Layer Map represented a coarse impression of the tasks. Yet it was the first time all project members could use the notion of user-centered tasks as a major requirement for their designs.

Furthermore, the resulting diagram provided a foundation for creating a preliminary interface design. For example, the diagram clearly showed the starting points and the task clusters. Some tasks had so many branches that the most obvious solution was to include all those subtasks in the same window (e.g., "Offer quotation," "Risk assessment," and "Process workflow item"). Making this information accessible at a simple glance helped in the development of interface concepts and provided a quick verification tool for the other project members.

Paper Prototyping. Based on the above diagram, a paper prototype of the application was made. During the actual prototyping sessions, task aspects were further refined, and some errors in the initial task analysis were diagnosed and fixed.

The Task Layer Map proved to be a great help in this project. Although the task analysis was not properly finished after the initial specification, the translation to dialog design was easily understood by all project members. The creation of the Task Layer Map also led to improvements in related documentation (such as the use case specifications).

11.4.2 Drawbacks to Task Layer Maps

11.4.2.1 Too Many Drawings

From project to project, we have seen a need to resort to use coloring in order to further differentiate tasks. The coloring scheme has changed based on the needs of the design group. Sometimes we have used colors to represent user groups. (One color per group), and other times we have used them to represent user objects. We have often used dotted lines to deemphasize some links or used other color lines. The design team sometimes wants several different types of images. When they do, we are forced to constantly switch between images—once to look at the user-group view and once to look at the user-object view.

We also found it necessary to take good notes to document things to make sure we did not end up in a do, undo, redo loop. We would like to develop a tool that would allow us to add this documentation to the diagram itself. We would also like to find a way of packing more information in the diagram without losing the reading and communication ease. Changing shapes and line styles has created more disadvantages than advantages.

11.4.2.2 Too Simple

Another problem with the technique is that it is a little too simple. We have a good technique for communicating the task analysis in general outline from any type of user study as long as there is some idea of relationship flow. However, getting a complete picture of an application depends other factors as well, such as user characteristics, environmental factors, user-object definitions, and so on. Fitting this information into the image has proved challenging. For now, we have resorted to different iterations of the diagram but that can be very confusing. We are also looking into an overlay technique in which a single image might have different overlays for environment, objects, grouping, and so on. The overlays could work similarly to layers in Photoshop.

11.4.2.3 Complex Applications

In complex applications, the multiple diagram problem becomes even greater. With extremely complex applications, we have to resort to using different levels of aggregation for the Task Layer Maps. This results in the team having to zoom in and out to get an overview or to go to the necessary level of specification. Each zoom-in would need another Task Layer Map. The overall Task Layer Map would then work like the map of a state or country supplemented with maps of individual cities showing city streets. The advantage here is that, as with a map, the context is clear. The disadvantage is that it increases the number of diagrams.

11.4.2.4 Applications Without Clear "Flow"

As mentioned above in our definition of *transitivity*, relations that are not transitive cannot be represented using this technique. In less formal terms, Task Layer Maps cannot be used in applications where dependencies or flows are indefinable. Without a stream of data flow, workflow, or dependencies, this technique is of little value. In applications where tasks have no dependencies or have no predictable structure, a Task Layer Map would be impossible to draw.

11.5 CONCLUSION

The Task Layer Map technique can help the entire design team in understanding the task analysis and how it relates to the dialogue design. In particular, it addresses the seven problems with task analysis modeling techniques that were listed in section 11.2.

This technique offers the members of a multidisciplinary team a simple image of the task analysis. This simple image is constructed by incorporating data the members understand—data on relationships—to form a partial ordering of the applications tasks. Its simplicity and comprehensibility mean that everyone, regardless of background, can understand the application flow.

The technique separates areas that do not need to be discussed by employing a mathematical formula to automate everything except inconsistencies or areas of concern. Typically, design teams can create and execute complex Task Layer Maps in a half day.

By presenting all tasks in a single view with a relationship flow, the technique makes it easy get the big picture.

The simpler application overview and the task clustering make it simpler to see the connection between dialog design and the task analysis. Clustering, diagram width, and other visual clues help guide the designers in their dialog creation.

The work instrument also helps developers, clients, and designers check whether their dialog design is complete (e.g., that for each window the window gets accessed from all tasks related *to* the task(s) in that window and the window gives access to all tasks that are related *from* task(s) in that window).

Task Layer Maps give team members a better overview of the application flow. The improved communication and application overview can help inspire solutions to interface problems.

The Task Layer Map technique is independent of the task analysis or user study method. It allows the use of any method as long as the relationships can be partially ordered.

Task Layer Maps facilitate the design process by giving the entire team a reference point for discussing the application. This reference point helps the designers in two major ways. First, the entire design team can understand the application, in one overview. The goals of the application and its ultimate purpose are now no longer cluttered with other functional and technical requirements, which often lead to misunderstandings within the design team. Second, Task Layer Maps expose the design rationale for the user interface design. Knowing the design rationale helps the design team understand the constraints and challenges confronting the user interface designer. Enhanced understanding of the application and knowledge of the design rationale tend to reduce costly arguments among the design team members.

This technique is still in development. The use of colors, shapes, arrow types, and so on, needs to be improved. Nevertheless, we believe this technique can be useful to product design teams now. Its rough edges keep the technique flexible. In addition, the technique should not be made too complex—quite the reverse. The greatest strength of this technique is its readability. Further development of this method should always keep this important point in mind. At this point, our main goal is to formalize the various drawing alterations (e.g., coloring the task boxes, using colored lines, etc.). We also would like to develop an overlay tool allowing the Task Layer Maps to show more information without recreating the visual noise problem that plagues other drawing techniques.

APPENDIX: FORMAL DEFINITIONS OF THE PROPERTIES OF PARTIAL ORDERING

Formally, a partial ordering is a relation with the properties of *antisymmetry, transitivity*, and *reflexivity* (Schmidt, 1988). These terms require formal definitions. (Please note the difference between equivalence ($<=>$) and implication ($=>$) in the following formal definitions.)

A relation over a certain domain, let's call it "$<<<$", is *antisymmetric* if either $a <<< b$ or $b <<< a$ but not both (except when $a = b$). Informally, these equations can be read as "*a* precedes *b*," "*b* precedes *a*," or "a is equal to b" (this latter is related to the reflexivity of the relation; see below). Formally, antisymmetry is represented as follows:

$$\text{(for all } a, b \text{ in domain)}:$$
$$(a <<< b \text{ .and. } b <<< a) <=> (a = b)$$

A relation is *transitive* if it can be concluded that $a <<< c$ from the fact that $a <<< b$ and $b <<< c$. Formally, transitivity is represented as follows:

$$\text{(for all } a,\ b,\ c \text{ in domain)}:$$
$$(a <<< b\text{.and. } b <<< c) => (a <<< b)$$

Transitivity is a standard property of dependency or flowlike relations. If c follows b and b follows a, c follows a. Not all relations are transitive. Think of "is a friend of," which cannot be treated in the same logical manner as the relations above. Consequently, nontransitive of relations (and nontransitive applications) cannot be treated with this technique.

A relation is *reflexive* if the elements of a given domain are always related to themselves:

$$\text{(for all } a \text{ in domain)}: a <<< a$$

Admittedly, this does not sound like a very interesting property, but formally it is needed as a kind of 0 in the algebra of relations. An expression like

$$\text{(for all } a, c):$$
$$(a <<< a \text{ .and. } a <<< c) => (a <<< c)$$

serves the same purpose as expressions like $0 + a = a$ and $1 \times a = a$ in calculus.

REFERENCES

Bomsdorf, B., & Szwillus, G., (1996). Early prototyping based on executable task models. In M., Tauber, Human Factors in Computing Systems: CHI 96 Conference Extended Abstracts, pp. 254–255, New York, ACM Press.

Bomsdorf, B., & Szwillus, G., (1998). From task to dialogue: Task-based user interface design In Compendium, C. Karat, & A. Lund, Human Factors in Computing Systems: CHI 98 Conference Extended Abstracts. pp. 210–211. New York, ACM Press.

Bomsdorf, B., & Szwillus, G., (1999). Tool support for task-based user interface design, In M. Williams, & M. Altom, et al., Human Factors in Computing Systems: CHI 99 Conference Extended Abstracts, pp. 169–170, New York, ACM Press.

Hackos, J., & Redish, J. (1998). *User and task analysis for interface design*, New York: Wiley.

Kieras, D. (1997). A guide to GOMS model usability evaluation using NGOMSL. In M. G. Helander, T. Landauer, & P. Prabhu, (Eds.), *Handbook of human computer interaction* (2nd ed.), pp. 733–767, Elsevier, Amsterdam, The Netherlands.

Mayhew, D. J. (1999). *The usability engineering lifecycle*. San Francisco, Morgan-Kaufmann.

Puerta, A., Chengm, E., Ou, T., & Min, J. (1999). MOBILE: User-centered interface building, In Williams, M., & Altom, M. et al., Human Factors in Computing Systems. In Tool support for task-based userinterface design. CHI 99 Conference Proceedings, pp. 426–433, New York, ACM Press.

Schmidt, D. A. (1988). *Denotational semantics*. Dubuque, IA: Wm. C. Brown.

Tufte, E. R. (1990). *Envisioning information*. Cheshire, CT: Graphic Press.

van Setten, M., van der Veer, G., & Brinkkemper, S., (1997). Comparing interaction design techniques: A method for objective comparison to find the conceptual basis for interaction design. In Designing Interactive Systems Conference Proceedings, p. 349, New York, ACM Press.

12

Task Analysis for e-Commerce and the Web

Tobias van Dyk
University of South Africa

Karen Renaud
University of Glasgow

This chapter offers additional perspectives on web-based tasks, and also introduces a way of using task-analysis to improve web-based usability. A task-based weighting scheme for evaluating these sites is proposed. This extended evaluation metric scheme and provides a more finely-tuned mechanism for assisting developers to improve usability of e-Commerce Web sites, since the user's task is included in the formulation of the guidelines. In conjunction with this, a novel usability metric prioritising scheme is used to yield information, which is noticeably different from unweighted values.

Developers of e-commerce applications are often sceptical about Website usability guidelines. User testing is also usually not carried out because it is expensive in terms of time and expertise. The main reason for developers neglecting current evaluation practices is that they are often vague, and in the case of user testing, too difficult to do effectively. The e-commerce shopping process has been analyzed from a task-based point of view, and a set of task-weighted metrics to be used by developers in evaluating their sites has been proposed. These metrics have been applied to four popular sites and the results from the evaluation have been analyzed.

12.1 INTRODUCTION

The Internet and its hosted applications remain a remarkably immature technology and application domain, given the exponential growth in Web sites and users. The combination of an immature technology, a distributed delivery and interaction mechanism, an immensely heterogeneous client base, and a usability-naive web developer community has led to both unpredictable and unique usability problems, with an impact that far exceeds anything previously experienced.

Many researchers rate ease of use as being of critical importance to the e-commerce process (Miles & Howes, 2000; Tilson, Dong, Martin, & Kieke, 1998; Vulkan, 1999). Bad usability is often blamed for causing the failure of sites (Helander & Khalid, 2000; Rohn, 1998). Jahng, Jain, and Ramamurthy (2000) argued that the uncertainty and ambiguity of badly designed sites makes users feel uncomfortable, something that is unlikely to result in a sale.

Nielsen (2000) emphasized one significant reason for the importance of usability for the commercial web. He referred to an *inversion* of the usability experience for the web when compared with other products. Web users, experience usability at a much earlier stage than conventional computer users, and bad impressions will therefore potentially impact more severely on the company providing the site. Research has indicated that online customers will seldom stay around to investigate a site that puzzles them or is difficult to learn to use. They tend to simply switch to another page on another site (Nielsen, 2000; Nielsen & Norman, 2000). Bad usability could also reduce brand loyalty and lead to the relocation of buying preference.

The prevalence of low-usability Web sites is partly due to Web developers' ease of access to the application domain, a consequence of the relatively minor barriers to entry. This leads to many usability-naive and inexperienced developers operating in the domain. In this context, various discount engineering methods offer a solution. There are a number of approaches to validating the usability of Web sites. Three of the most popular are empirical usability testing with a number of users, examination of web logs, and usability metrics. The first is expensive and probably not an option for most web developers. The second is problematical for two reasons: (a) It is essentially a postmortem analysis, and the results may come too late for remedial action to make a difference, and (b) even if it is not too late, the web logs are notoriously difficult to interpret (Nielsen & Norman, 2000). The third is a low-resource (time, money, and skills), high-yield (usability improvement) option and is thus suitable for use by inexperienced web developers.

Metrics have been used successfully by Kirakowski and Claridge (2002), who developed a commercial metric-based product called the Web Analysis and MeasureMent Inventory (WAMMI) usability questionnaire, which originated from a tool for evaluating desktop applications. WAMMI, although web-specific, is not tuned for e-commerce site evaluation and does not consider the uniqueness of the e-commerce task. Ivory, Sinha, and Hearst (2001) have worked on quantifying the measurement of information-centric Web pages in terms of presentation. They developed a tool called TANGO (Tool for Assessing NaviGation and information Organisation); (Ivory, Sinha, & Hearst, 2001), which was used to evaluate over 400 Web sites. Their tool is highly automated because it evaluates simple page elements rather than semantic content and context, both of which are addressed by the metrics introduced in this chapter.

The rest of this chapter discusses a combination of a task-based and metric-based approach to enhancing web usability. This is a viable yet novel technique for providing inexperienced developers with a tool that can be used to improve the quality of their sites.

12.1.1 Why Focus on e-Commerce Tasks?

The Internet is the largest information source ever created. Users of the Internet are said to *surf* the net, a euphemism for browsing and searching for sites of interest. Indications are that initially observed rather aimless behavior is changing and maturing. Studies by the Radio and Television News Directors Foundation and the Pew Research Center show that people want to *accomplish* something online. They are not aimless surfers anymore, looking for a novel site or anything of passing interest. Instead, the average Net user seems to be becoming more goal oriented and interested in finding *specific* information and communicating with others (Carton, 2000).

Belew (2000) examines the web browsing and searching process and finds that three steps are involved: *asking a question, constructing an answer*, and *assessing the answer*. (Note that people may revisit steps in any order while engaged in browsing and searching). Although the usability of solely information specific sites is important, the fact that the usability of e-commerce sites could impact negatively on a company's regular retail trade as well as their e-commerce trade makes it vital for designers to produce usable e-commerce sites ("Best

Practices," 2001). It can be argued that for these strongly goal-directed applications, *the bulk of a system's usability is accounted for by task focus*. Travis (2000) recommended the following:

> Allocate your design activity to reflect the needs of end-users, and put the bulk of your effort into task focus. Users will thank you because they will be able to work more effectively. You know a system has task focus when you get a warm feeling that the person who designed this system knew what they were doing. You find you are able to use the system to do exactly what you want.

It is important to note Diaper and Addison's (1991) claim that a task will be strongly dependent on the user's mental model of the computer documentation (and, for this study, online, documentation). Task decomposition (perhaps too mechanistic), knowledge-based techniques, and entity-relationship–based analysis (now replaced by UML) are three different but overlapping approaches to task analysis (Dix, Finlay, Abowd, & Beale, 1998), and of these the second is especially relevant to e-commerce. Knowledge-based descriptions of tasks are important because they identify

- the plan for carrying out the task
- the knowledge or concepts required
- the interaction between different kinds of knowledge.

In other words, they help the designer gain an understanding of the commonalities that exist between tasks in terms of knowledge requirements and anticipated plans for execution (Johnson, Diaper, & Long, 1984). In this regard, Brinck, Gergle, and Wood (2002) noted that task analysis is essential for providing an easy to use and learn system, while not exceeding human limitations. In addition, the high-level goals specified in the task analysis make explicit the functionality that is built into the system (see also chap. 1). Thus, there is little confusion about the intended purpose of the site. Because it makes explicit the procedural knowledge expected from the user, it clarifies learning requirements and can provide the basis for training materials.

It can be expected that a task analysis should yield, inter alia, the following benefits when applied to the e-commerce interaction session (Maguire, 1997):

- It provides knowledge of the tasks that the user wishes to perform (i.e., find, select, and purchase.
- It is a reference against which the value of the system functions and features can be tested.
- It is a cost-saving exercise because failure to allocate sufficient resources to the task analysis activity increases the potential for costly problems to arise in later phases of development.
- It makes it possible to design and allocate tasks appropriately and efficiently within the new system.
- It enables a more accurate specification of the functions to be included within the system and the user interface.

Most users of e-commerce systems will not have been trained in their use. The user interface will therefore have to be designed with great care so that the user can discover everything that is task supportive from the system based on the observable system state (for instance, Norman's [1986] "system image"; see chap. 18). The designer of the user interface must be sure to bestow rational behavior on the application; that is, ensure that the application behaves in a way that is reasonable and intelligible. By concentrating on the e-commerce task, the developer can move closer to a system that the user can use intuitively.

Substandard e-commerce sites currently focus on available technology rather than the user's task. Shifting the focus from *technology* to *customers* will deliver the following usability benefits, among others (Travis, 2000):

- *An improved conversion rate* (lookers: bookers ratio) and hence improved sales. On a high-volume e-commerce site, raising the conversion rate by one tenth of a percent can add thousands in revenue per month.
- *Increased hits.* A satisfied customer provides free word-of-mouth exposure, bringing more customers to the site.
- *Avoiding leakage to competitors.* One bad customer experience could cause a customer to abandon a site permanently and will possibly result in bad publicity.
- *Focusing developers on important business metrics.* These metrics include conversion rates, revenue per order, acquisition cost per new customer, percentage of repeat buyers, percentage of revenue from repeat buyers, and abandoned shopping carts.
- *Reducing support staff costs* by allowing customers to carry out maintenance tasks online.
- *Helping developers to stay focused on customers' tasks and goals* rather than on the web as a technology tool.
- *Highlighting wastage of development time* and establishing guidelines for the next move.

The next section identifies two distinct phases of the purchase task of the e-commerce shopping experience, and the following analyzes the e-commerce task with these phases in mind.

12.1.2 Two Phases of the e-Commerce Task

Some work has been done in analyzing user behavior with respect to e-commerce sites. Guttman, Moukas, and Maes (1998) identified six stages of customer purchasing behavior: need identification, product brokering, merchant brokering, negotiation, purchase and delivery, and service and evaluation. O'Keefe and McEachern (1998) proposed a model with only five processes: need recognition, information search, evaluation, purchase, and after-purchase evaluation. Singh, Jain, and Singh (1999) broke up the e-commerce process into three activities: identifying and finding a vendor, purchasing, and tracking. We examine only one of Singh's processes, namely, the one that everyone refers to as the *purchase* task. This task can be split up into two distinct phases, as shown in Fig. 12.1.

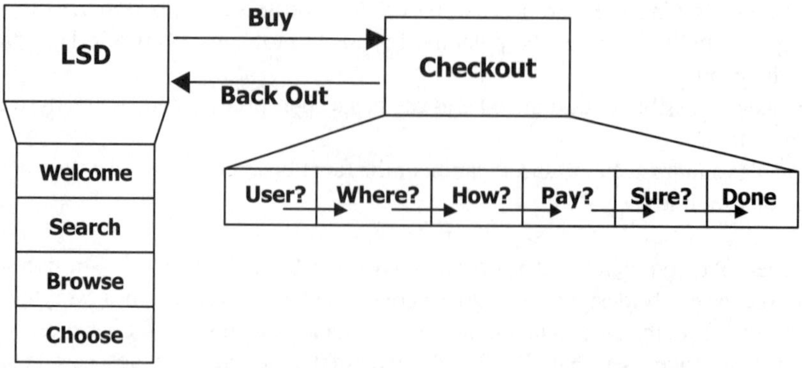

FIG. 12.1. The two phases and ten stages of the purchase task. *Note.* From *A Mechanism for Evaluating Feedback of e-Commerce Sites,* by K Renaud, T. van Dyk, and P. Kotzé, 2001.

1. *Look, See, and Decide (LSD)*. This stage will typically be used to look at available products, compare them, and make a decision whether to purchase products. This may be done one or more times until the consumer has found products that satisfy his or her needs. This phase is intensely user driven because the user is looking at and assimilating information continuously. It has the following substages, which can be traversed iteratively and in varying sequences: Welcome; Search; Browse; Choose.
2. *Checkout*. When users trigger this stage, they have made their choice of offered products and have decided to make a purchase. They now have to provide certain details, such as their address and credit card details. This stage is system driven and changes the paradigm of the interaction process from user initiative to system initiative. Feedback is of critical importance during this stage. Users who feel that they have lost control can simply leave the site without any embarrassment, unlike a user who is standing at a checkout till in a supermarket. This stage is typically composed of at least the following steps, which should be navigated in a logically serial fashion: User? → Where? → How? → Payment? → Sure? → Done.

Some Web sites will have all these stages integrated into one page (e.g., www.amazon.co.uk), but the implied functionality is the same: Each of these categories of information must be provided so that the transaction can be carried out. Brinck et al., (2002) combine UML use case analysis (see also chap. 7, 11, 19, 22, 24, and 27) and hierarchical task analysis (HTA; chap. 3) into a powerful technique. They identify two use cases for a book purchasing scenario, namely, "Buy Book" and "Complete Order," which coincide with the LSD and checkout phases identified above.

The user drives Belew's (2000) searching and browsing process during the LSD phase, asking question after question until a particular item has been located. The process is somewhat different during the checkout phase, because the *system* drives the process, asking the user questions and receiving and assessing the answers. Most web design guidelines do not take these diametrically opposing operating paradigms into account, even though the principle of dialog initiative and system versus user preemption is well established (Dix et al., 1998).

12.2 PRELIMINARY TASK ANALYSIS

A simple computer operating model may serve as an effective basis for an understanding of the goal-directed nature of an e-commerce task execution (Fig. 12.2). This model can also serve to further highlight the tasking difference between the two phases of the e-commerce shopping process. A definition for task analysis that is suitable within the context of this application domain is that offered by Dix et al. (1998); they describe task analysis as the *identification and description of the interactive system user's problem space, in terms of domain, goals, intentions, and tasks*. A more extensive but not inconsistent definition is provided in chapter 1.

The focus of this part of the study will be on identifying differences between the nature of the task during the LSD phase (which includes searching and browsing, selection, and the shopping cart) and the checkout phase of the e-commerce purchase interaction process.

The nature of the shopping task differs significantly during the LSD and checkout phases. The LSD phase is, in essence, a user-driven iterative browsing and selection task with (possibly) less well defined goals and a larger number of possible actions. The checkout phase is a system-driven, predefined, linear task with well-defined goals and subgoals and a smaller number of predefined actions.

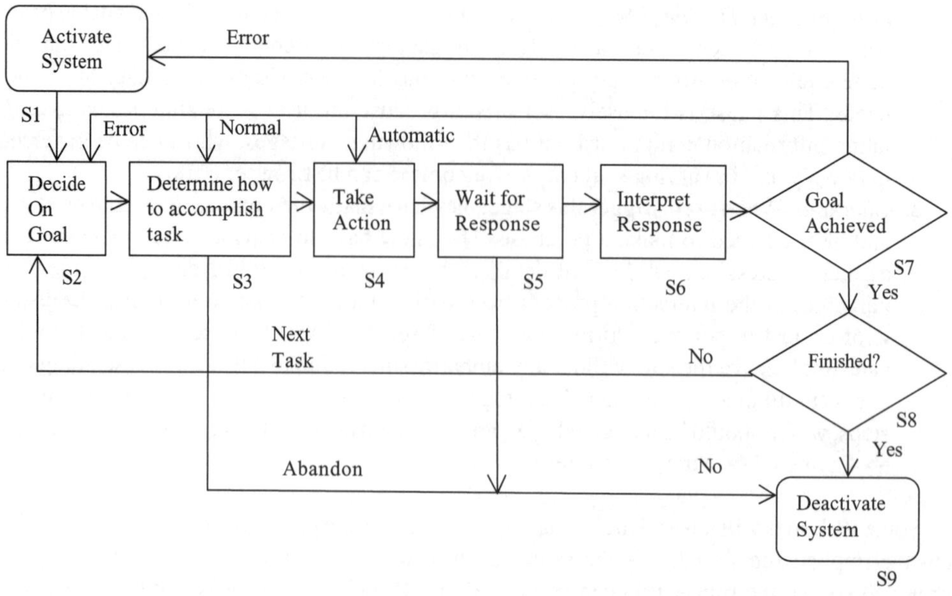

FIG. 12.2. Task analysis based on a simple iterative computer operating model. *Note.* Adapted from *Human Factors Design Handbook* (2nd ed.) by W. E. Woodson, B. Tillman, and P. Tillman, 1992, New York: McGraw-Hill.

During the LSD phase, there will be three types of goals:

1. browsing (searching) for the shopping objects
2. categorizing the shopping objects
3. specifying (selecting) shopping objects for the shopping cart.

The nature of the interaction is such that the customer should be kept interested in the results of the search-type goals, thus retaining them onsite. The focus is on discouraging user dropouts through abandonment of the goal or by linking them offsite (Dix, 1999). System errors and poor response times during this phase are perceived to be less serious by the user (but not by the Web site owner, since they may result in shopping cart abandonment).

The checkout phase has a single goal: completing the financial transaction (as defined by the contents of the customer's shopping cart and shipping preferences) as quickly and securely as possible with the minimum of disruption and roadblocks ("Best Practices," 2001). Accordingly, this phase has a set of linear and intentionally fairly rigid subgoals. The focus here is not on user entertainment but on completing the transaction rapidly and securely—before the customer changes his or her mind about the shopping cart contents and the related cost. This implies that response times and clear feedback on reasons for delays are more important here than during the LSD phase. Because errors could have a more serious (security and financial) impact, a well-designed user help function and clear explanatory subsystem are required. It is also easy to provide the user with obvious and intuitive navigation clues (progress or stage indicators) as to where the user is in the process. The best strategy is to strive for the minimum number of pages or stages rather have the user scroll moderately and then click through to a larger number of small pages ("Best Practices," 2001). This is in stark contrast to having as much as possible of the relevant information immediately visible in the LSD stage (Veen, 2001). Simplifying the process will ensure that there will be a smaller incidence of user dropout and shopping cart abandonment during this phase—provided additional expenses such as shipping costs are shown as soon as possible. The effect of an error at this stage will affect the sequence of the subgoals, and will make this phase nonlinear ("loopy").

When these aspects are applied to the model as presented in Fig. 12.2, the following should be noted:

More effort will be required for system activation (S1) in the LSD phase than for the checkout phase. For example, the customer has to have an established Internet session. The customer also needs to know about the site. This is well understood by marketing professionals, and sites are often well advertised in the media. Unfortunately, this level of attention is often not paid to other aspects of the e-commerce experience.

Goal formulation (S2) may be less clear in the LSD phase than in the checkout phase. The customer may want to reevaluate options and reformulate goals based on the range, price, and availability of the shopping objects during the outcome of a set of search results.

The intermediate stages (S2–S6) are less prescribed for the LSD phase, and there will be a natural tendency to loop back to S2 during this phase, as might occur if the response at S5 takes too long, for example.

The S3 stage is often trickier for the customer to formulate in the LSD phase. The customer may have some vague idea of an item he or she needs but have difficulty formulating a query. For example, the customer may have heard about a popular autobiography by an Irish teacher who grew up in Limerick. The customer types in many different search criteria—"Irish", "teacher", "Limerick"—before perhaps finding the book *Angela's Ashes* by Frank McCourt by browsing through the list of available autobiographies.

Interpretation of the response (S6) will be more difficult during the LSD phase than the checkout phase. The customer may be presented with a range of shopping objects from which to choose, whereas there is a linear progression during the checkout phase.

Measuring the success of the task at S5 will be more difficult for the LSD phase. The customer is dealing with a goal achievement based on an electronic description rather than confirmation of a familiar financial transaction, as in the checkout phase.

The result of an error (which may occur at S7 → S1 or S4 → S2) will be deemed to be less serious during the LSD phase than during the completion of a transaction in the checkout phase.

The transition S7 → S2 may be traversed during the LSD phase without an error having been made. It could happen as a result of a reformulated goal.

The transition S7 → S8 may be traversed even though the goal has not been achieved. For example, the site may not stock the required object.

The activity distance S2 → S7 should be as short as possible for the checkout phase, with achievement of S7 always clearly visible, perhaps by means of stage indicators.

The S3 → S9 and S5 → S9 loops should be minimized by ensuring good site usability.

It is necessary to translate the discussion of this section into some set of recommendations so that developers have guidelines to follow in order to ensure that adequate feedback is provided. Veen (2001) referred to the difficulty of evaluating Web sites. Developers using traditional usability testing methods are often faced with an iterative and time-consuming evaluation process that involves a number of users and continues until every perceivable problem is solved. The provision of a set of easily applied metrics should make it easier for e-commerce site developers to profit from accumulated research results and make the evaluation process a little less daunting.

12.3 TASK-WEIGHTED EVALUATION METRICS

The previous section discussed the differences between the two different phases of the e-commerce task. It is fitting for the two phases to have different evaluation metrics as well—

metrics corresponding to their different paradigms and needs—and for these metrics to be weighted according to their impact on usability.

We previously reported on a set of suitable evaluation metrics for measuring feedback-related usability of three e-commerce Web sites (Renaud, van Dyk, & Kotzè, 2001). The evaluation criteria used were equally weighted. In certain cases, however, it may be advantageous to prioritize some of the criteria by means of a selective, unequal weighting. Examples of how to do this may be found in Levi and Conrad (1996). They described the application of Nielsen and Mack's (1994) usability guidelines to the evaluation of a set of web pages. After the evaluation they modified the list based on feedback from their two evaluation teams (HCI and Web developers) and produced a new list by assigning severity ratings to each usability violation found on a five-point scale. In addition, they prioritized on the basis of the frequency of occurrence of the usability problem. Their scale is as follows: 0 = not a usability problem; 1 = cosmetic; 2 = minor; 3 = major; 4 = catastrophic problem. They produce a list of usability violations, which contains both frequency and severity information. Nielsen (1995) earlier produced a guideline for severity ratings based on the same scale but expanded on this by noting that the severity of a usability problem was a combination of four factors: the frequency with which the problem occurs, the impact of the problem on users if it occurs, the persistence of the problem, and the market (product popularity) impact of the problem.

Along the same lines, Bastien and Scapin (1992) referred to the *amount and importance of usability problems found.* Another technique applies a *strength-of-evidence* scale to a set of evaluation criteria (National Cancer Institute, 2001). These criteria are based on the type and number of research experiments that may support, or discount, the specific criterion. The World Wide Web Consortium (2001) prioritizes in terms of (accessibility) guidelines that *must be applied*, *should be applied*, or *may be applied.*

One suitable weighting approach, which we believe to be in part novel, is to assign a task weighting to each of these previously equal-weighted criteria scores. This task weighting has two components:

1. A *task repetition* component (R). The task repetition component is an indicator of how often this task or activity will be encountered during the interaction. A weight of 0.1 indicates the task is of low occurrence and only happens in exceptional cases, a weight of 0.2 indicates that it happens very seldom, and a weight of 1.0 indicates that it occurs regularly during the interaction. A value of 0 indicates that the activity is absent and should not play a role in the calculation of the overall score.
2. A *task complexity* component (C). The task complexity component reflects the inherent degree of difficulty in executing the task or activity. A weight of 1.0 indicates that the activity is highly complex and requires extensive background and operational knowledge or a high degree of complex interaction. A weight of 0.1 indicates that the task is simple and has low interactivity. A weight of 0 is not possible because it indicates that no interaction is required.

Each of these two components amplify or attenuate the contribution of a specific usability feedback criterion to the overall score. The overall value should be a more faithful reflection of the Web site's overall usability than that rendered by unweighted metrics.

In support of our view, it should be noted that Brinck et al. (2002) distinguish between the frequency and priority of tasks and that they (correctly) assert that the starting point for their HTA should be important tasks that occur frequently. Similarly Nielsen's (1995) "impact and persistence of the problem" could make the *problem context* a high-priority task. Lastly, Bohmann (2001) has developed and tested task metrics for a quantitative usability evaluation. These metrics make it possible to calculate the usability effect of redesign efforts. The two main metrics are task time (the time to complete a task or set of tasks) and task errors (number of errors per task).

The following section evaluates four e-commerce booksellers' Web sites using this technique.

12.4 EVALUATION OF e-COMMERCE SITES

Given a list of guidelines, developers have difficulty knowing which to follow. For example, developers are told to have the most important information visible to the user without scrolling. They are also told to provide the user with enough information to keep them interested in the site, thus increasing the chance of a sale. Which of these is more important? We propose the use of a set of *metrics* that can be used by developers to evaluate each page of a Web site. Additionally, we propose a metric-weighting technique that involves the components of *task complexity* and *task repetition*, as discussed previously.

The following describes how these metrics were applied to four e-commerce sites and comments on the efficacy of the proposed evaluation mechanism. In order to evaluate e-commerce Web sites, a raw score is given for each of the questions (metrics) as follows (Ravden & Johnson, 1989):

- Never (0). The feature is never available.
- Sometimes (1). The feature is seldom there.
- Mostly (2). The feature is usually there.
- Always (3). The feature is universally available.

The first step in the evaluation is the raw scoring of the *usability metrics*. The scores are determined per e-commerce site, per phase (LSD and checkout), per stage within a phase, and also per page, as a ratio to the maximum score. The scores for each metric in each stage are calculated by adding up the score for each page making up the stage and awarding a total for each particular metric feature. The scores for each feature are then totaled to arrive at a raw score equivalent to the percentage per site per purchasing stage. Note that the evaluator should not feel constrained by the list of metrics given here; these were adapted and selected from a much larger list (11 sections and 179 metrics) developed and extensively tested by Ravden and Johnson (1989). It is likely that differences in e-commerce sites may require the evaluator to revisit this more comprehensive list and add to (or subtract from) the list of metrics given here. It is also recommended that more than one evaluation (and evaluator) be used to arrive at the raw metric scores. Three data sets can be considered to be the absolute minimum. The individual scores from the data sets should be averaged as an input to the second step.

The second step assigns values for the two task weight components (R + C) based on the evaluators' experiences with the site during the metric scoring step. These values are designed be have little effect initially on the raw scores until the evaluator develops more confidence in applying corrections. The natural tendency will be to choose median values of close to 0.5 for both the task repetition and complexity values, in which case the weighting adjustment will effectively be 1 (R + C = .5 + .5). Thus, initially the adjustment will be no worse than the unadjusted raw metric scores. Ultimately, as experience is gained in the use of the weight factors, the weight adjustments could have a large influence on the metric score. Consider a low-complexity, low-repetition value of .5 compared with a high-repetition, high-complexity value of 1.5. The metric adjusted by the first would only contribute one third as much to the overall usability score for the e-commerce site as the second metric.

The third step is to eliminate metrics with particularly low R + C values (e. g., < 0.6) from the evaluation. Their elimination would partly alleviate the problem of a tendency toward an average of 1 for all R + C values when a large number of metrics are used.

To demonstrate our approach, consider Question 9 in Table 12.4: "Is it clear what the user has to do to complete the task?" Scoring the new Amazon.com site would be done as follows:

Step 1. Each reviewer does a dry run through the checkout phase of Amazon.com and records all actions and responses for each page.
Step 2.
a. For each page, each reviewer assigns a score for the particular usability metric. For example, the first page in the Amazon.com checkout process displays a log-in script requiring the user to provide an email address and password. The reviewers would assign a score of 3 to this page because it is obvious that the user needs to enter the details and click on the "Sign In" button. The full 3 points are given because the user does not need to scroll down to find the button and no other interpretation is possible. If the user has to search for the button or if two buttons appear with similar functions, the score may be reduced.
b. The scores for each page within the stage are added together, and these scores are averaged to arrive at an assigned score (referred to as Score). In our evoluation, we arrived at a value of 7 for that particular metric. The maximum score (*MaxScore*) is 9, as this stage consists of 3 pages each of which has a maximum score of 3.
Step 3.
a. Based on the reviewers' observations during Steps 1 and 2, suitable values are assigned to the task repetition and complexity weighting for that specific metric. In our evaluation, we used a value of .5 for the repetition because the actions to be taken in these pages are neither particularly frequent nor infrequent. The complexity of this task was assigned a value of .7, indicating that this task falls between an average and a highly complex task. This is because the order review page requires the use of extensive external knowledge and interpretation in order to carry out the task correctly.
b. The value of 16 is the total of all the R + C values for any specific phase in a site. Each R + C value represents a fraction of the total usability environment for a the site. Comparing the R + C values for a particular metric with the total value indicates how important that metric is to the total site usability (*Total R + C*).
c. The final usability coefficient (UC) for the metric is calculated by applying this formula: UC = (Score/MaxScore) × ((R + C)/Total R + C).

To complete our evaluation, we needed to calculate an overall usability score for each phase per site in order to facilitate a comparison between sites. The percentage usability score based on raw (non-task-weighted) scores is

$$\text{Raw} = (\Sigma\ \text{Score})/(\Sigma\ \text{MaxScore}) \times 100$$

The percentage usability based on task-weighted scores is calculated as

$$\text{Task} - \text{weighted} = (\Sigma \text{UC}) \times 100$$

The normalized ratios, which make it easier to compare usability scores, are expressed as ratios relative to the highest scoring site. For sites other than the top site, the score is calculated as follows:

Task-weighted score/Top site's task-weighted score

A suitable list of evaluation metrics based on Ravden and Johnson's (1989) set of 179 and derived from previously reported results [Renaud et al. (2001) are shown in Tables 12.1 and 12.2 (metrics for the LSD and checkout phases respectively)]. Typical task-weighting factors for each metric are as indicated. To use these tables, you multiply the original (raw) score

TABLE 12.1
User Task Metrics for the LSD Phase

Stage	*Metrics*	*Task-Weighting Factors (Repetition + Complexity)*
S3	Is it clear how the user must search for a product?	(.8 + .5)/8.5 = 15% (>1)
S3	Are different types of information clearly separated?	(.8 + .9)/8.5 = 20% (>1)
S4	Is it clear what needs to be done to select a product?	(.1 + .6)/8.5 = 8% (<1)
S5→S9	Does the system inform the user of reasons for delays?	(.8 + .5)/8.5 = 15% (>1)
S7→ S2	Does the search engine offer alternatives if the search fails?	(.5 + .8)/8.5 = 15% (>1)
S7→S2	Can the user undo a product selection?	(.1 + .5)/8.5 = 7% (<1)
S7→S3	Does the system allow the user to explicitly check on previous searches?	(.5 + .5)/8.5 = 12% (1)
S8→S9	Is it clear how the transition to checkout can be made?	(.1 + .5)/8.5 = 7% (<1)
Average task weighting factor for this phase		8.5/8.0 = 1.06

Note. As an example, (.8 + .5)/8.5 = 15% (>1) is to be read thus: .8 + .5 = R + C = task repetition value + task complexity value. The 8.5 is the total of the task (R + C) values for the specific set of usability metrics. The 15% is the percentage contribution to the total task weight. The (>1) signifies that R + C>1, implying an increased weighting for any usability score for this metric.

TABLE 12.2
User Task Metrics for the Checkout Phase

Stage	*Metrics*	*Task-Weighting Factors (Repetition + Complexity)*
S2, S3	Are possible actions clear?	(.8 + .7)/16 = 9% (>1)
S3	Are instructions and messages concise, clear and unambiguous?	(1 + .8)/16 = 11% (>1)
S3, S5→S9	Can the user easily back out of the process?	(.1 + .8)/16 = 6% (<1)
S4	Is the required format of user actions clearly indicated?	(.5 + .8)/16 = 8% (>1)
S5→S9	Does the system inform the user of the reasons for delays?	(1 + .5)/16 = 9% (>1)
S6	Are user actions linked to changes in the interface?	(.8 + .5)/16 = 8% (>1)
S6	Is there always an appropriate response to user actions?	(.8 + .8)/16 = 10% (>1)
S6	Does the user explicitly confirm the final purchase?	(.1 + .4)/16 = 3% (<1)
S6	Does the system indicate the current stage?	(.3 + .5)/16 = 5% (<1)
S6	Can users check on inputs provided during the process?	(.1 + .5)/16 = 4% (<1)
S7→S8 or S7→S2/S3/S4	Does the system inform the user of the success or failure of their actions?	(1 + .5)/16 = 9% (>1)
S7→S2	Do error messages indicate the what, where and why and how to recover?	(.5 + 1)/16 = 9% (>1)
S7→S8	Is it clear what the user must do to complete the task?	(.5 + .7)/16 = 8% (>1)
Average task weighting factor for this phase		16/13 = 1.23

TABLE 12.3
Results of Applying the User Task Metrics for the LSD Phase

User Task	*Amazon (2002)*	*Amazon (2001)*	*Kalahari*	*Books Online*
1. Is it clear what a user must do to search for a product?	6/9(.8 + .5)/8.5 = .102	6/9(.8 + .5)/8 = .108	6/9(.8 + .5)/8 = .108	7/9(.8 + .5)/8 = .126
2. Does the search engine offer alternatives if a search fails?	9/9(.5 + .8)/8.5 = .153	9/9(.5 + .5)/8 = .125	3/9(.5 + .5)/8 = .042	0/9(.5 + .5)/8 = .000
3. Does the system inform the user of the reasons for delays?	5/9(.8 + .5)/8.5 = .085	5/9(.8 + .5)/8 = .090	3/9(.8 + .5)/8 = .054	3/9(.8 + .5)/8 = .054
4. Are different types of information clearly separated?	8/9(.8 + .9)/8.5 = .178	7/9(.8 + .8)/8 = .155	6/9(.8 + .8)/8 = .133	9/9(.8 + .8)/8 = .200
5. Is it clear what needs to be done to select a product?	9/9(.1 + .6)/8.5 = .082	9/9(.1 + .5)/8 = .075	6/9(.1 + .5)/8 = .050	6/9(.1 + .5)/8 = .050
6. Can the user undo a product selection?	9/9(.1 + .5)/8.5 = .071	9/9(.1 + .5)/8 = .075	6/9(.1 + .5)/8 = .050	8/9(.1 + .5)/8 = .066
7. Is it clear what must be done to make the transition to checkout?	7/9(.1 + .5)/8.5 = .055	6/9(.1 + .5)/8 = .050	0/9(.1 + .5)/8 = .000	9/9(.1 + .5)/8 = .075
8. Does the system allow users to explicitly check on previous searches?	6/9(.5 + .5)/8.5 = .078	0/9(.5 + .5)/8 = .000	6/9(.5 + .5)/8 = .083	0/9(.5 + .5)/8 = .000
Percentage				
Raw	59/72 = 81.9%	51/72 = 7.8%	30/72 = 41.7%	42/72 = 58.3%
Task-weighted	.804 = 80.4%	.678 = 67.8%	.520 = 52.0%	.574 = 57.4%
Normalized Ratio				
Raw	1.0	.865	.509	.712
Task-weighted	1.0	.843	.647	.714

for each metric by the factor given in the table for that metric. Of particular interest in these two tables would be criteria with associated tasks or activities that have either high combined weight factors (i.e., R + C) or very low combined weight factors. This could imply that these task components are proportionally either more or less important to the usability evaluation.

To illustrate the technique, Tables 12.3 and 12.4 list the results of four previous evaluations of three booksellers on the Internet, namely, Amazon (performed in 2001 and 2002; www.amazon.com), Kalahari (2001; www.kalahari.net), and Books Online (2001; www.uk.bol.com).

The results from Table 12.3 show the following:

- Applying the task weighting has decreased the overall usability difference between the best site, Amazon (2002), and the worst site, Kalahari. Small changes were observed for Books Online. Elimination of low-value R + C task metrics will result in larger differences between weighted and unweighted results.
- For the lowest usability site (Kalahari), applying the task weighting results in a significant increase in its overall usability score. This would imply that Kalahari does focus on better usability for important tasks compared to other metrics that have lower R + C values.
- Applying the weighting factors has emphasized the usability differences between the new and old Amazon sites. The new Amazon has improved considerably on its usability score for the LSD phase. This is mainly because of higher scores for content layout, information presentation, the provision of a history function, and more obvious navigation to next stages in the book purchase task.

TABLE 12.4
Results of Applying the User Task Metrics for the Checkout Phase

	Amazon (2002) (3 stages)	*Amazon (2001) (6 stages)*	*Kalahari (3 stages)*	*Books Online (5 stages)*
1. Are instructions and messages concise, clear, and unambiguous?	8/9(1 + .8)/16 = 0.100	12/18(1 + .8)/16.5 = 0.073	6/9(1 + .8)/16.5 = 0.073	11/15(1 + .8)/16.5 = 0.080
2. Are possible actions clear?	7/9(.8 + .7)/16 = 0.073	12/18(.8 + .8)/16.5 = 0.065	5/9(.8 + .8)/16.5 = 0.054	12/15(.8 + .8)/16.5 = 0.078
3. Is the required format of user inputs clearly indicated?	8/9(.5 + .8)/16 = 0.072	15/18(.5 + 1)/16.5 = 0.076	7/9(.5 + 1)/16.5 = 0.071	10/15(.5 + 1)/16.5 = 0.061
4. Are user actions linked to changes in the interface?	7/9(.8 + .5)/16 = 0.063	13/18(.8 + .5)/16.5 = 0.057	6/9(.8 + .5)/16.5 = 0.053	12/15(.8 + .5)/16.5 = 0.063
5. Is there always an appropriate response to user actions?	6/9(.8 + .8)/16 = 0.067	12/18(.8 + .8)/16.5 = 0.065	6/9(.8 + .8)/16.5 = 0.065	13/15(.8 + .8)/16.5 = 0.084
6. Does the system inform the users of the success or failure of their actions?	8/9(1 + .5)/16 = 0.083	14/18(1 + .5)/16.5 = 0.071	6/9(1 + .5)/16.5 = 0.061	11/15(1 + .5)/16.5 = 0.067
7. Does the system inform users of the reasons for delays?	7/9(1 + .5)/16 = 0.073	11/18(1 + .5)/16.5 = 0.056	3/9(1 + .5)/16.5 = 0.030	5/15(1 + .5)/16.5 = 0.030
8. Do error messages indicate the what, where, and why and how to recover?	4/9(.5 + 1)/16 = 0.042	9/18(.5 + 1)/16.5 = 0.045	4/9(.5 + 1)/16.5 = 0.040	11/15(.5 + 1)/16.5 = 0.067
9. Is it clear what the user has to do to complete the task?	7/9(.5 + .7)/16 = 0.058	11/18(.5 + .8)/16.5 = 0.048	5/9(.5 + .8)/16.5 = 0.044	14/15(.5 + .8)/16.5 = 0.074
10. Does the system indicate the current stage?	8/9(.3 + .5)/16 = 0.044	17/18(.3 + .5)/16.5 = 0.046	3/9(.3 + .5)/16.5 = 0.016	15/15(.3 + .5)/16.5 = 0.048
11. Can the user easily back out of the process?	8/9(.1 + .8)/16 = 0.050	10/18(.1 + .8)/16.5 = 0.030	2/9(.1 + .8)/16.5 = 0.012	3/15(.1 + .8)/16.5 = 0.011
12. Is the final purchase explicitly confirmed by the user?	9/9(.1 + .4)/16 = 0.031	18/18(.1 + .5)/16.5 = 0.036	9/9(.1 + .5)/16.5 = 0.036	15/15(.1 + .5)/16.5 = 0.036
13. Can users check on inputs provided during the process?	7/9(.1 + .5)/16 = 0.029	9/18(.1 + 0.5)/16.5 = 0.018	6/9(.1 + .5)/16.5 = 0.024	3/15(.1 + .5)/16.5 = 0.007
Percentage	94/117 = 80.3%	163/234 = 69.7%	68/117 = 58.1%	135/195 = 69.2%
Raw Task-weighted	0.785 = 78.5%	0.686 = 68.6%	0.579 = 57.9%	0.706 = 70.6%
Normalized Ratio	1.0	0.868	0.724	0.862
Raw Task-weight	1.0	0.874	0.738	0.899

- High R + C criteria include activities associated with the presentation of information and instruction-oriented actions.
- Low R + C criteria include product selection and deselection actions and undo facilities.
- The raw scores yield useful information by themselves. They provide an evaluation mechanism that can be used by developers to flag problem feedback areas. For example, the old Amazon Web site did not have a search history facility and as a result scored 0 for metric 8. This was corrected in the new Amazon site.

The results from Table 12.4 show the following:

- The task weighting has improved the score of Books Online but decreased the scores of all three other sites. The changes are smaller than those in Table 12.3 in part because of the larger number of metrics used in this table. That is, there is to some extent an averaging around the mean of the R + C value. This could be avoided by eliminating all metrics with low R + C values (e.g., < .6) from the scoring.
- The new Amazon has improved very noticeably on its usability score for the checkout phase. The improvement is due to the folding of six previous stages onto three and also because of higher scores for layout of user options, condensed information presentation, and more intuitive navigation to the next stage in the book purchase task.
- High R + C criteria include user guidance, appropriate responses, and clarity of interaction messages and information presentation.
- Low R + C criteria include the confirmation of the purchase and abort facilities.
- Red-flagged (problem) areas based on the raw (unweighted) scores in Table 12.4 include the lack of meaningful error messages, unexplained delays, and no intuitive undo facilities.

The results and especially the approach adopted (that is, prioritizing certain criteria over another set) tailor the metrics to the nature of the task. On an intuitive level, it is clear that the repetition of a task should make it more important (that is, increase its weight, that the level of interaction required should increase its weight, that the task duration should increase its weight, and that the level of knowledge required for the task should increase its contribution to the Web site's overall usability score. A more formalized analysis for task repetition values could be obtained through a frequency count of the Knowledge Representation Grammars (KRGs) and Sequence Representation Grammars (SRGs) produced from a Task Analysis for Knowledge Description (TAKD) analysis (Diaper, 1989). The method used here is less complex, as it simply counts the occurrence of these during a typical (shopping and browsing) interaction session.

12.5 CONCLUSION

e-Commerce applications are usually strongly task oriented and goal directed Web-based interactions, and thus they lend themselves to the use of structured task analysis approaches. This chapter offers additional perspectives on Web-based tasks and also introduces a way of using task analysis to improve Web-based usability. It also proposes a task-based weighting scheme for evaluating these sites. This extended evaluation metric scheme provides a finely tuned mechanism for helping developers improve the usability of e-commerce Web sites because the user's task is included in the formulation of the guidelines. The scheme makes use of a novel usability metric prioritizing scheme to yield information that can be used during both the design and maintenance phases. The prioritizing scheme yields values that can be noticeably different from more commonly applied unadjusted values.

The approach as outlined here needs to be applied to a larger sample of e-commerce sites. It is currently designed for Web sites that fit the LSD and checkout model, but because it encompasses scoring and then weighting usability metrics, it is suitable for other types of e-commerce sites, such as those that offer Internet-based banking. Using it to evaluate such sites will require fortifying or changing the metrics by replacing and redesigning some of the items. Developing a faster questionnaire that can be delivered via the Web and to which users rather than experts can respond will also be beneficial and may facilitate the partial automation of the questionnaires. It would also be necessary to obtain reliability information on the results obtained, for example, by comparing the results with those obtained by a heuristic evaluation or by comparing then with the results from other tools, such as the Web log analytics mentioned previously. It should be stressed that the approach as outlined provides relative usability comparisons across systems. In the future, it may be possible to produce both normative and absolute usability data as a benchmark to test against. Whether this occurs will depend on the collection of a sufficiently large body of results for a range of different sites. Web sites are using increasingly sophisticated methods to watch what customers are doing and to make e-commerce pay, and a growing future trend will be the use of shopper simulation (i.e., testing usability against sophisticated computer models of the site).

REFERENCES

Bastien, J. M. C., & Scapin, D. L. (1992). A validation of ergonomic criteria for the evaluation of human-computer interfaces. *International Journal of Human-Computer Interaction, 4*(2), pp. 183–196.

Belew, R. (2000). *Finding out About*: A Cognitive Perspective on Search Engine Technology and the www. Cambride University Press. Cambridge: UK.

Bohmann, K. (2000). *User performance metrics.* Available: http://www.bohmann.dk/articles/user_performance_metrics.html.

Brinck, T., Gergle D., & Wood, S. D. (2002). *Usability for the Web: Designing Web sites that Work.* San Francisco: Morgan Kaufmann.

Carton, S. (2000). *What do people want online*? Available: http://www.clickz.com/tech/lead_edge/article.php/824471

Checklist of checkpoints for Web content accessibility guidelines. (2001). World Wide Web Consortium. Available: http://www.w3.org/TR/WAI-WEBCONTENT/full-checklist.html

Diaper, D. (1989). Task analysis for knowledge descriptions (TAKD): The method and an example. In D. Diaper (Ed.), *Task analysis for human-computer interaction* (pp. 108–159). Chichester, England: Ellis Horwood.

Diaper, D., & Addison, M. (1991). User modelling: The task oriented modeling (TOM) approach to the designer's model. In D. Diaper & N. Hammond (Eds.), *People and Computers VI* (pp. 387–402). Cambridge: Cambridge University Press.

Dix, A. (1999). Design of user interfaces for the Web. In N. W. Paton & T. Griffiths (Eds.), *Proceedings of User Interfaces to Data Intensive Systems* (*UIDIS '99*) (pp. 2–11). Edinburgh, Scotland: IEEE Computer Society Publishers.

Dix, A., Finlay, J., Abowd, G., & Beale, R. (1998). *Human-computer interaction* (2nd ed.). Hemel Hampstead, England: Prentice Hall International.

Friedman, B., & Millett, L. (1995). It's the computer's fault: Reasoning about computers as moral agents. In *Proceedings of ACM Conference on Human Factors in Computing Systems (CHI '95).* (Short Papers: Agents and Anthropomorphism; Vol. 2; pp. 226–227). New York: ACM Press.

Guttman, R. H., Moukas, A. G., & Maes, P. (1998). Agent-mediated electronic commerce: A survey. *Knowledge Engineering Review, 13*, 147–159.

Helander, M. G., & Khalid, H. M. (2000). Modeling the customer in electronic commerce. *Applied Ergonomics, 31*, 609–619.

Ivory, M. Y., Sinha, R. R., & Hearst, M. A. (2001, June). *Preliminary Findings on quantitative measures for distinguishing highly rated information-centric web pages.* Paper presented at the Sixth Conference on Human Factors and the Web, Austin, TX.

Ivory, M. Y., Sinha, R. R., & Hearst, M. A. (2001, March-April). *Empirically validated web page design metrics.* Paper presented at the Proceedings of the SIGCHI Conference on Human factors in computing systems: Changing our world, changing Ourselves. ACM special interest group computer human interaction 2001, April 2002, Seattle, WA.

Jahng, J., Jain, H., & Ramamurthy, K. (2000). Effective design of electronic commerce environments: A proposed theory of congruence and an illustration. *IEEE Transactions on Systems, Man and Cybernetics: A. Systems and Humans, 30*, 456–471.

Johnson, P., Diaper, D., & Long, J. (1984). Tasks, skills and knowledge: Task analysis for knowledge based descriptions. In B. Shakel (Ed.), *Human-Computer Interaction (Interact '84)*. (pp. 499–503). London: Elsevier Science.

Kirakowski, J., & Claridge, N. (2002). WAMMI information page. http://www.ucc.ie/hfrg/questionnaires/wammi

Levi, M. D., & Conrad F. G. (1996, July-August). *A heuristic evaluation of a World Wide Web prototype. Interactions,* pp. 50–61.

Maguire, M. (1997). *RESPECT: User requirements framework handbook.* Loughborough, England: HUSAT Research Institute.

Mandel, N., & Johnson, E. (1999). *Constructing preferences online: Can Web pages change what you want?* (Working paper). Philadelphia: University of Pennsylvania.

Miles, G. E., & Howes, A. (2000). Framework for understanding human factors in Web-based electronic commerce. *International Journal of Human Computer Studies, 52*(1), 131–163.

National Cancer Institute. (2001). *Research-based Web guidelines.* Available: http://www.usability.gov/guidelines/intro.html

Nielsen, J. (2000). *Designing Web usability: The practice of simplicity*. Indianapolis, IN: New Riders Publishing.

Nielsen, J. (1995). *Severity ratings for usability problems.* Available: http://www.useit.com/papers/heuristic/severityrating.html.

Nielsen, J., & Mack, R. (Eds.). (1994). *Usability inspection methods*. New York: Wiley.

Nielsen, J., & Norman, D. (2000, January 14). Usability on the Web isn't a luxury. *Information Week.* Available: http://www.informationweek.com/773/Web.htm

Norman, D. (1986). Cognitive engineering. In D. Norman & S. Draper (Eds.), *User-centered system design: New perspectives on human-computer interaction* (pp. 31–61). Hillsdale, NJ: Lawrence Erlbaum Associates.

O'Keefe, R. M., & McEachern, T. (1998). Web-based customer decision support systems. *Communications of the ACM, 41*, 71–78.

Ragus, D. (2001). *Best practices for designing shopping cart and checkout interfaces.* http://www.dack.com/Web/shopping/cart.html

Ravden, S. J., & Johnson, G. I. (1989). *Evaluating usability of human-computer interfaces: A practical method.* New York: Wiley.

Renaud, K., van Dyk, T., & Kotzé, P. (2001, October). *A mechanism for evaluating feedback of e-commerce sites.* Paper presented at the First IFIP Conference on e-Commerce, e-Business, e-Government, Zurich.

Rohn, J. A. (1998). *Creating usable e-commerce sites. STDVIEW*: ACM Press, New York, USA. (ACM Journal on Standardization), 6, pp. 110–115.

Singh, M., Jain, A. K., & Singh, M. P. (1999). e-Commerce over communicators: Challenges and solutions for user interfaces. In *Proceedings of the ACM Conference on Electronic Commerce* (pp. 177–186). New York: ACM Press.

Tilson, R., Dong, J., Martin, S., & Kieke, E. (1998). A comparison of two current e-commerce sites. In the *ACM 16th International Conference on Systems Documentation* Quebec, Canada ACM Press (Web Navigation Series; pp. 87–92).

Travis, D. (2000). *Back to the drawing board: Why e-commerce sites need to focus on usability.* Available: http://www.system-concepts.com/articles/ecommerce.html

Veen, J. (2001). *The art and science of Web design.* Indianapolis, IN: New Riders Publishing.

Vulkan, N. (1999, February). *Economic implications of agent technology and e-commerce. The Economic Journal, 109*, pp. F67–90.

Woodson, W. E., Tillman, B., & Tillman, P. (1992). *Human factors design handbook* (2nd ed.). New York: McGraw Hill.

13

Using Task Data in Designing the Swanwick Air Traffic Control Centre

Peter Windsor
Usability Limited

The task of Air Traffic Controllers is to monitor the four dimensional profiles of aircraft, identify and resolve potential interactions between them, to communicate those resolutions to pilots and other controllers and to record actions and agreements, all in real time. In designing a computer system to support this complex task we used a range of task analysis techniques to develop a detailed understanding of the domain and of how controllers work. In this chapter, we will give examples of the kinds of problems we addressed and discuss how the various task analysis techniques were used alongside our design methods.

13.1 EN-ROUTE AIR TRAFFIC CONTROL

London Area Control Centre, Swanwick provides air traffic services to aircraft flying in Control Areas (Airways) and Upper Air Routes in the London Flight Information Region (FIR). In lay terms, the centre is responsible for aircraft flying about 15,000 feet over the southern half of the United Kingdom (to be accurate, the base of this "en-route" volume varies between 5,000 and 20,000 feet).

Every aircraft passing through the controlled airspace files a flight plan stating the route it intends to follow, the altitude at which it wishes to fly, and the time when it will take off (or arrive at the FIR boundary). The computer systems process these data along with radar information and generate a four-dimensional profile for each aircraft. For aircraft that are in the airspace and being tracked, the profile is accurate for the near future but becomes increasingly approximate the further into the future the flight is considered. Similarly, the profile is only approximate for aircraft that are still on the ground (or have not yet come into UK radar cover).

The volume is divided up into "sectors," as illustrated in Fig. 13.1. Each sector is the responsibility of a three-person team:

- The *Tactical* controller monitors the aircraft in the sector and ensures their safe passage through it; he or she talks to the pilots and issues instructions as necessary.

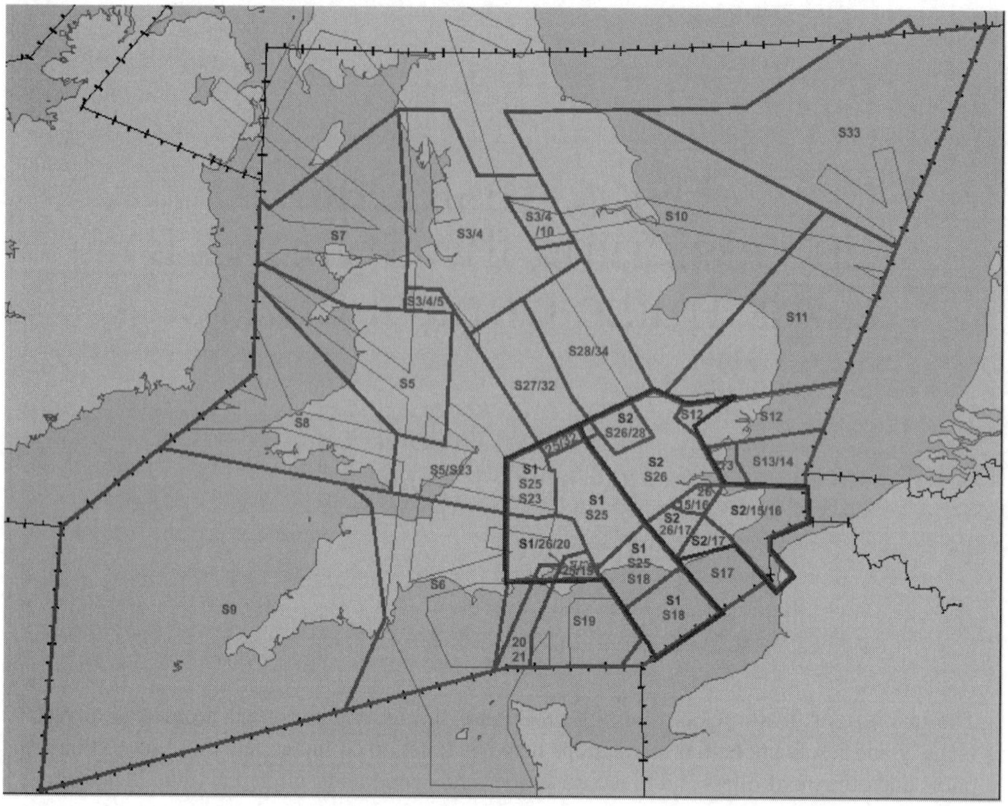

FIG. 13.1. Illustration of sectors.

- The *Planner* controller manages when and where aircraft will enter and leave the sector and "coordinates" this with the neighbouring teams.
- The sector *Assistant* assists the two controllers.

The controllers' tasks fall into four categories:

- Monitoring the situation and detecting where action is required. (The ranges of actions include responses to potential conflicts and routine actions to allow aircraft to progress.)
- Deciding what action to take.
- Communicating those decisions to the affected parties (aircraft and/or other controllers).
- Recording those decisions (and hence updating the situation being monitored).

The distinction between the Tactical and the Planner is fundamentally the time period they consider. The Tactical is concerned with the present and the next few minutes whereas the Planner is considering what will happen in 5 to 10 minutes' time.

Swanwick differs from its predecessor (West Drayton, which was similar to the Manchester ATC system described in chap. 1) in two key respects:

- At West Drayton, a single Crew Chief performed the planning tasks for several sectors; Swanwick has a separate Planner for each sector.
- The Swanwick system mediates coordination between the sectors; at West Drayton, this was done by telephone calls, with the agreements handwritten on paper flight progress strips (see chap. 19 for an example of a paper flight strip).

What Swanwick does not change is that the principal tool of the tactical controllers is a set of paper flight progress strips, which are used to provide data on the aircraft in (or expected to pass through) a sector and for recording tactical instructions.

13.2 TASK ANALYSIS AND THE DEVELOPMENT OF THE SWANWICK CENTRE

The Swanwick Centre was brought into operation in January 2002 after many years of development. The main phases of that development were as follows:

1. a competitive Project Definition in 1991 and 1992.
2. an Implementation Phase under a firm fixed price contract ending in 1997.
3. the refinement of the initial system to accommodate changed requirements (1998–2001).

My colleagues and I were involved throughout this period. We were responsible for the design of the system's user interface and were heavily involved in the design of the air traffic applications. Our approach was user centred: A group of air traffic controllers were part of the project and participated closely in the user interface and application design activities. It was also iterative, and typically three or four iterations were used to refine each part of the design from an initial concept through to a fully detailed specification.

There was no task analysis per se; rather, every piece of work included the task analysis activities necessary for its completion. This is consistent with chapter 1, which proposes that task analysis is central to virtually all HCI activities. As all the development work was done under time pressure, the analysis activities were always limited to what was necessary and sufficient to make progress with the problem at hand. Typically we would do some preliminary analysis as part of the first iteration, sufficient to allow us to come up with an initial design concept and to estimate and plan the remainder of the work. In the later iterations, we would analyse the important elements of the tasks in more detail, selecting analysis techniques that would provide the data we needed to evaluate, refine, and complete the design.

Thus, we used a variety of techniques ranging from capturing coarse task lists (especially in the Project Definition phase) through to very detailed analysis (and sometimes simulation) of critical tasks. The following sections explain the main techniques we used and illustrate them with examples.

13.2.1 The Core Tasks and the Organization of the User Interface

The first task analysis technique we applied was a broad-brush assessment to determine the key characteristics of the air traffic control tasks from which a general strategy for the user interface design could be derived. The technique is to consider:

- What are the information exchanges between the user and the envisaged system?
- What are the temporal characteristics of the tasks? That is, how are they triggered, what are their durations and frequencies, and how do they overlap?
- What happens when a task is interrupted?

In air traffic control the primary information exchange is the "scan": The controllers monitor data presented by the system and identify where intervention is necessary. This was not a novel observation, but we did confirm it for the Swanwick context and identify the specific data needed by the two controllers (see section 13.2.2). We also noted that the controllers do not remember detailed information but rather rely on it being "remembered in the world" (see chaps. 14 and 19).

These characteristics had a strong influence on the user interface design. First, they confirmed our expectation that the system needed to present information about pertinent flights in a form that supports rapid comparisons. That is, it needed tabular displays akin to the well-established paper flight progress strips.

Second, these characteristics suggested that all the data being used should be displayed all the time, as interacting with the system during the scan is unduly disruptive. Moreover, the displays needed to support pattern recognition, and so our design includes supporting information that is not directly relevant to the control tasks but aids the controller both in identifying individual flights and in classifying interactions between them.

The notion of the scan also influenced our strategy for presenting information. Our design uses three levels of alert:

1. safety-related alerts which need immediate attention and hence should interrupt the scan
2. indications of outstanding work which should not interrupt the scan but should draw attention so that they are easily identified as candidates for the next action
3. noncritical highlights (e.g., small updates to times), which should be noticeable when they are scanned past but should not perturb the scan at all.

Temporally, the air traffic control tasks fall into three groups. First, there is the scan itself and the associated decision-making tasks, which are essentially continuous. Next, there are the "executive" tasks relating to the recording and communicating of decisions. These tasks are short in duration, typically 3 to 5 seconds, and collectively of high frequency, roughly 5 to 10 per minute at busy times. The executive tasks are triggered from the scanning and decision-making tasks and are interleaved with them: The scan is not stopped during an executive task but is pushed into the background. Each individual executive task is essentially atomic, and one task will normally be completed before another is started. (An executive task, however, may involve parallel subtasks. For example, when a tactical controller gives an instruction, he or she records it on the appropriate paper strip at the same time as speaking to the aircraft.) Finally, there are "back office" tasks such as rerouting a flight. These tasks are longer in duration—they may take 1 or 2 minutes to complete—and are of comparatively low frequency, although they tend to occur in bursts (e.g., when an airport is closed due to bad weather, many flights have to be diverted).

There is also an additional set of tasks that arise from managing the configuration of the system, especially when sectors are combined together (at night when there is little traffic) and split apart (as the traffic builds up). These tasks are not inherent in air traffic control but are a consequence of how the system supports the control tasks. The configuration tasks are triggered by the daily cycle, are of medium duration (around 30 seconds), and have to be fitted in around the other tasks. They also tend to be performed incrementally and iteratively. Each total task is broken up into short pieces that can be done quickly and with intermediate states that do not interfere with the main control tasks.

The executive and back office tasks also differ in how interruptions are handled. Air traffic controllers are trained to make decisions quickly based on the situation at the time and, if a decision is deferred or reconsidered, the controller will reassess based on the updated situation. Thus, if an executive task is interrupted, it is either completed or abandoned. In the latter case, the task will be retriggered from the scan and reassessed, so there is no need for the system to retain data about the partially complete task. Back office tasks, however, take too long to be abandoned due to interruptions and so are paused part way through and picked up when the interruptions have been dealt with.

Again, these task characteristics guided our design strategy. The user interface elements used for executive tasks are accessed from the data displays used in the scan, have strong completion feedback and simple, unambiguous means of cancellation. In contrast, the parts of the user interface used for the back office tasks are semidetached from the scan displays and have a structure that allows "suspend-continue" behavior.

13.2.2 Scenarios and Cases

The significant new facility provided by the Swanwick system is the electronic mediation of coordination between sectors. In its simplest form, it works as follows:

- Team A have an aircraft coordinated into their sector(s) (via a prior invocation of the facility); at this point the aircraft may still be in a previous sector.
- The Team A Planner determines the flight level at which he or she wants the aircraft to leave the sector (along its flight plan route) and enters this (exit) level into the system.
- The system determines the sector and Team (Team B) to which the aircraft will be transferred and the time at which that Team should be notified of an "offer" (this sequencing is important as it helps the receiving Planner deal with offers in an efficient order).
- At the determined time, the system will offer the flight to Team B. The Team B Planner will assess the offer and (normally) tell the system that he or she accepts it.
- The system will inform Team A that the offer has been accepted. Once the aircraft is in Team A's sector, the Team A Tactical will climb or descend the aircraft to the agreed level (as necessary) and ultimately transfer the aircraft to Team B.

There are, of course, many variations and complications. Significant ones include these:

- The receiving Team may not accept the offered level, and the two Teams will have to negotiate a level.
- The neighboring sector might be another air traffic control unit (an "external" sector). There are four distinct types of external sector varying in the amount of data provided to them by the system and the manner in which that data is transfered.
- Coordination to a neighbor above or below the offering sector is different because the transfer does not relate to a simple boundary.
- There are occasions when the coordination process for an aircraft has to be suspended (e.g., if the aircraft has to be held in a sector).
- Coordinations sometimes have to be undone and remade, especially if a flight is diverted.
- When a single Team is responsible for several sectors, the boundaries between those sectors can be ignored. However, when the sectors are subsequently split apart, coordinations have to be introduced between the now separate sectors.

Our technique for analysing the coordination tasks and designing the system support was based on scenarios. Note that, unlike Carroll (2000; chap. 5), we use *scenario* to mean a specific, detailed story of use; that is, each scenario includes details of aircraft, flight plan, sectors, and levels. We disagree with Carroll's notion of using scenarios without such concrete data. Like Beyer and Holtzblatt (1998), we think that if a system is to be useful to real people performing real tasks, its designers must start with detailed concrete data.

Our technique is to start by capturing a "core set" of scenarios covering the key cases, i.e., the high-frequency and high-criticality cases that have the most impact on the effectiveness of the air traffic operation. We use this core set of scenarios to drive the development of a domain model (chap. 1) and a high-level definition of the user interface (the set of windows and parts, plus a definition of which domain objects are displayed and which domain actions can be invoked in each window/part). We then identify further candidate scenarios by using what-if variations of the sequences of events and/or the data (levels, routes, and sectors; see chap. 5). The candidate scenarios that are not readily accommodated by the design at that point are selected and used to refine and evolve it. There are two key features to our approach. First, we do not develop a comprehensive set of scenarios at the beginning (or otherwise attempt an exhaustive analysis). Rather, we allow the set to grow in parallel with the evolving design. Second, we are always prepared to backtrack in our design thinking when one of the later scenarios proves that we made a poor choice.

We also applied one other analysis technique in making design choices: We considered the training implications of our proposed design. The electronic coordination system has millions of potential states, and although these are comprehensible to someone familiar with the design models, we knew that we would not be training the controllers to understand that level of detail (air traffic controllers are not systems engineers). Thus, we took an approach of reviewing the evolving design from a training perspective, asking the question, When a controller (who understands the core concepts) sees this (new or modified) concept, what will he or she infer?

13.2.3 Fixing Problems: Task Analysis in Reverse

No computer system exists in isolation. Every system is always constrained by the outside world. In the case of Swanwick, we had a significant constraint in the form of the existing flight data–processing system, NAS. One of NAS's responsibilities is the exchange of flight data with adjacent air traffic control units—a limited form of electronic coordination. Because the rules for these exchanges were an imperfect match with the Swanwick electronic coordination system, there is a range of cases in which what NAS has done or will do for a coordination with an external sector is incompatible with what the Swanwick electronic coordination system would do for a coordination between two internal sectors.

It was not possible to remove these differences within the project constraints and timetable. Changing the interaction with the external units would require changing their systems (and the international agreements governing the exchanges). Restricting the internal coordination to the NAS model would have significantly limited its effectiveness. It is anticipated that the differences will be removed in the long term when the next generation of the international standards is brought in.

We addressed this problem by applying a reverse form of scenario analysis. First, we conducted an exhaustive analysis of the two systems to construct a set of theoretical problem scenarios in which the two systems diverged. We then looked for real-world situations in which these problems might actually occur. We could then ask the air traffic controllers to assess the consequences of the difference(s) and jointly look for a solution that eliminated any adverse effects.

13.2.4 Workload and Fallback Modes

Computer systems fail. The Swanwick system is designed to detect when parts of it fail and to degrade gracefully into one of a number of "fallback modes." One key aspect of this design is that the user interface allows the controllers to handle the immediate impacts of a failure, to continue to provide an effective service in a fallback mode, and to transition back to normal operation on recovery.

We used two analysis techniques in designing this part of the system. First, we analysed how the operation might work in the absence of parts of the system. From our high-level models of the system, we could easily classify parts of the user interface as being unaffected, partially affected but still useful, or left with no useful function. This gave us the basis to work with the project air traffic controllers to posit and assess a method of operation for each fallback mode. In some cases, we could identify additional backup facilities that were necessary, typically manual means to get data into a subsystem that would normally be supplied by the subsystem that had failed.

The other technique we used was to develop a coarse time line model, similar to a GOMS approach (chap. 4), to assess proposed ATC procedures for entering and leaving fallback modes. We established a set of "unit tasks" that would occur (e.g., "Pass an estimate by telephone to the next sector") and nominal times for each (18 seconds for passing an estimate). We then took a real traffic sample (from a busy day in July 1999) and looked at what would happen if there was a failure at a busy time. By applying a coarse-grained equation (work remaining = work outstanding + work arriving − work done), one minute at a time, for each member of the sector team, we could construct an approximate picture of how well they could keep up with the load. An illustration is provided in the next two subsections.

13.2.4.1 *Example Unit Tasks: Assistant Tasks A1, A2, and A3*

Assistant Tasks A1, A2, and A3 occur when the Assistant has access to the flight data and can print strips for the Planner (Table 13.1). A1 and A2 are of similar duration because the major part of the task is telephone based. The printing (A3) task is conducted after the telephone call has ended.

13.2.4.2 *Extract From Traffic Sample Walk-through*

This walk-through (Table 13.2) examines the Planner and Assistant workload with a failure happening at 1015. For each minute, the actions that arise, along with their estimated times,

TABLE 13.1
Example of Assistant Task Descriptions with Task Durations (s = seconds)

Assistant (A1)	*s*	*s*	*Assistant (A2)*	*Comments*
Locate exit strip	3			
Call next sector	1	1	Receive call	
		3	Answer call	
Inform of flight	10	10	Note call sign, fix, level, time	
End call	1	1	Call ends	
	18	15		

Assistant (A3)	*s*	*Comments*
Prints PFS, one for each fix	6	Using flight is Active Assume 3 fixes/3 strips
Collect PFS	15	
Strip up	9	
Place on board	6	
	36	

TABLE 13.2
Extract from Traffic Sample Walk-through

Time	*Planner*			*Assistant*		
1015	P3 BRT522			A11 EDD (DLH433)	60	
1016	P2 EIN652	23		A1 GIL100C	18	
				A2 CLX602	15	
				A2 HZSJP3	15	
				A2 MPH631	15	+3
1017	P1 GIL100C	33		A3 CLX602	36	
				A3 HZSJP3	36	+15
1018	P2 CLX602	23		A3 MPH631	36	
	P2 HZSJP3	23		A1 USA27	18	+9
1019	P1 USA27	33		A2 SAB542	18	
	P2 MPH631	23		A2 ELY832	15	
				A3 SAB542	36	+18
1020	P2 SAB542	23		A3 ELY832	36	
				A1 IWD3333	18	+12
1021	P1 IWD3333	33		A1 COA37	18	
				A1 MPH631	18	
				A1 EIN652	18	+6
1022	P1 COA37	33		A2 DLH440	18	
	P1 MPH631	33	+6	A3 DLH440	36	0

are listed for each person; each action is identified by the code for the action and the callsign of the subject flight. At the end of each minute, the time, if any, required to catch up is also listed for each person (unused time is "lost").

Such models are very crude: They do not allow for variations in the task and assume that there are no dependencies between team members or between teams. Nevertheless, they allowed us to identify potential problems and guided subsequent work. In particular, they allowed us to ensure that the procedures were basically viable prior to running simulation trials involving many people.

13.2.5 Details, Details, Details

The complexities of the real world and the subtle details of real tasks are crucial factors in successful systems development. How the details are handled is fundamental in determining how well the system supports its users but is also a major influence on the size and complexity of the system. Thus, it is vital that the system designers understand the task details and can use that knowledge to find system solutions that are effective in supporting task performance but do not require excessive complexity.

In the case of the Swanwick system, the unavoidable complexity came from two sources. First, there are the physical constraints of geography (the locations of airports and of national boundaries) and of aircraft performance. Second, over its long experience in air traffic control, National Air Traffic Services has developed procedures and control techniques that have been proven to be both safe and expeditious (chap. 1). These methods are context sensitive, especially

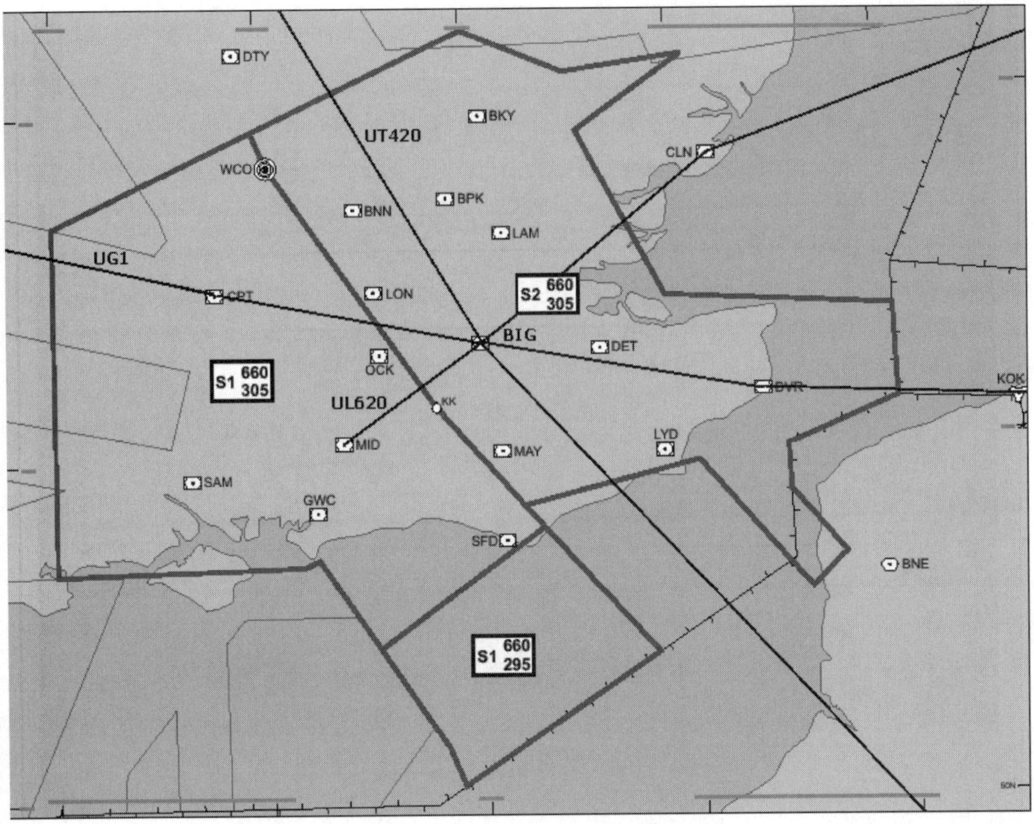

FIG. 13.2. Sector 1, Sector 2, and the Biggin Hill conflict region.

with respect to the geometry of the sectors and routes. Together, these factors significantly limit the division of the airspace into sectors and constrain both the data needed by a sector team and their likely actions for each aircraft. The following examples illustrate how this complexity affected the design of the new system.

The first example comes from the airspace over London: sectors 1 and 2. As can be seen from the simplified map in Fig. 13.2, there are two routes through sector 1 (UG1 and UL620, which starts at MID) that converge in sector 2 airspace at Biggin Hill (the BIG fix), where they also cross UT420. (The full map is more complex; there are five routes crossing at BIG, and many more in the vicinity.) The flying time from the sector 1 boundary to BIG is too short for the sector 2 Tactical controller to separate aircraft on the convergent routes, so the procedure requires the sector 1 Tactical controller to ensure that such aircraft are already separated before handing them over to sector 2. To do this, the sector 1 controller needs to know the times at which the relevant aircraft will reach BIG even though BIG is outside sector 1's volume. This means the system has to support a logical association between fixes and sectors that it can use when determining the fixes for which to provide a control team with times (which it does by printing paper strips). This association is separate from the similar data that the system needs in order to determine the fixes to use in support of electronic coordination between these sectors.

It would, of course, keep the system simpler if the sector boundaries could be chosen to avoid such cases. However, the number of routes and crossings makes this practically impossible. Moreover, using smaller sectors (and hence more controllers) has been a significant factor

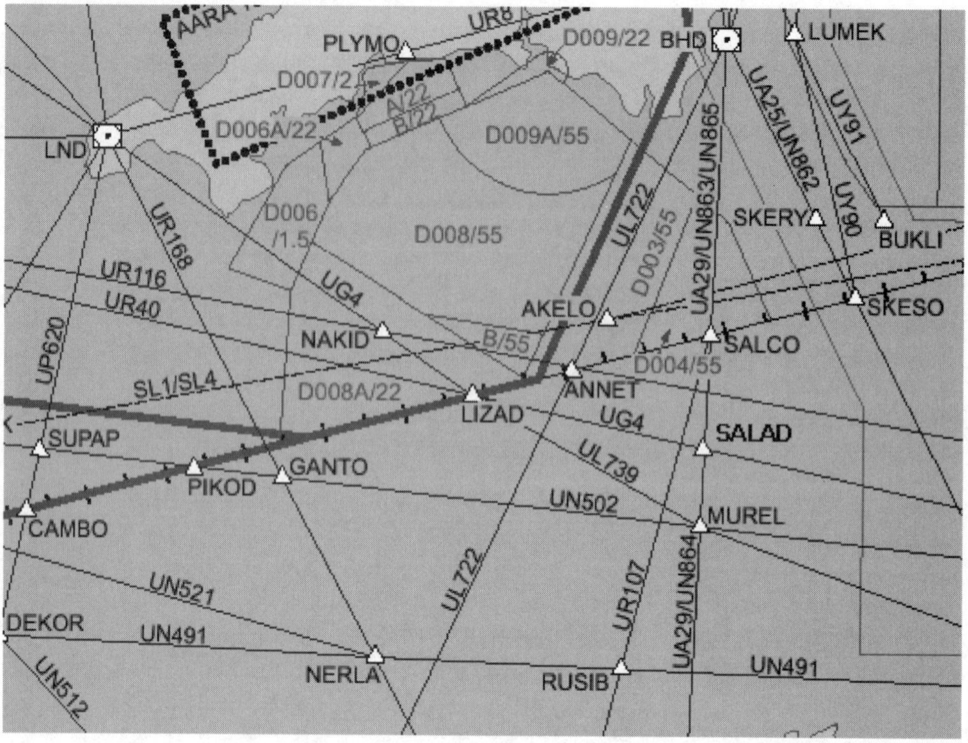

FIG. 13.3. SALAD convergence.

in increasing air traffic capacity (although we are approaching the limit at which additional overhead will outweigh gains), and such configurations are more likely to occur with more, smaller sectors. Thus, even if the situation could be avoided today, it would be likely to occur in the future, and so the system needed the ability to support it.

A similar example occurs in the Berry Head area (Fig. 13.3), but in this case the convergence occurs in French airspace and the two routes are in different London sectors (6, Berry Head (BHD); 9, Lands End (LND)). The air traffic procedure in this case is to make the sector 6 planner responsible for keeping aircraft on the two routes separated. In order for that controller to perform this task, he or she has to be involved in the coordination of aircraft on route UG4 with the French, even though those aircraft will not pass through sector 6 at all. The system solution in this case was to treat the aircraft as if they did pass through sector 6 (the association between routes and sectors is logical rather than geometric). This keeps the system relatively simple (it does not know that this is a special case), but one cost is that the sector 6 and sector 9 controllers must know that the system will behave differently for such aircraft in comparison with those on straightforward routes. (As each sector has a set of procedural rules that controllers must know in order to work on that sector, this cost is marginal.)

An example where the system does have to have separate rules for a range of cases occurs with electronic coordination. As noted above, the coordination procedure differs when the neighboring sector is above or below the offering sector rather than horizontally adjacent. The specific differences are these:

- There is no time at the boundary. The aircraft could climb or descend through it at any point, and thus the system cannot calculate an offer time (and cannot provide the optimal sequencing that it determines for horizontal cases).

- If there are two or more vertical neighbors along a route, the system cannot determine the target sector from the exit level (which it can for the horizontal case).
- By convention, the level for a vertical coordination is set to just inside the offering sector (i.e., just below the ceiling or just above the floor). However, the receiving sector Team will often amend that level to one within their airspace, and the aircraft will subsequently be transferred, having already been cleared to climb or descend into the receiving sector's volume.

Although most coordinations are horizontal, vertical coordinations make up a significant minority (or even a majority for some sectors). Thus, to provide effective support for the coordination tasks, both the user interface and the system's application logic have to treat vertical coordination cases differently from horizontal ones. In fact, the full story is more complex still, as there can be steps in the sector ceilings and floors that cause further complications.

Every piece of task analysis we undertook had to address such complexities, and understanding the details of the air traffic domain was one of the key challenges of the project. All the task analysis techniques we employed provided insights and assistance but only when employed as part of a systematic, thorough, and skilled approach to user interface design.

13.3 TASK ANALYSIS IN USER INTERFACE DESIGN

We take a pragmatic position on what kind of task analysis can and should be done when designing a user interface. First, we undertake analysis in order to support design; we do not consider analysis to be an end in itself. Our starting point is always the requirement, design issue, or other problem we are tackling, and we approach analysis with specific questions that we want to answer. The questions vary considerably in depth and scope, ranging from the broad "What information sources do air traffic controllers use?" to the very detailed "How does the sector 17 planner decide whether to hold an EGKK (Gatwick) inbound at TIGER?" We choose techniques and timescales to match.

Further, we do not treat analysis as a separate activity done before design starts. Our approach is to iterate the design, starting with the high-level concepts and approach and gradually adding details (in this the approach is similar to that described in chap. 8). At each stage, we do the analysis that allows us to make the design choices for the level of abstraction at which we are working (with exploratory forays into the details to validate the choices). As illustrated above, we apply different analysis techniques depending on the problem we are trying to solve.

In our experience, this design-led approach is both more efficient (because we do not expend effort analyzing everything in full detail) and more effective (because we know what information we need) than a large up-front analysis. Although we agree with Beyer and Holtzblatt (1998) on the importance of detail, we disagree with them when they suggest that the analysis should be largely complete before design starts. Our approach needs some care in its planning and management: Too little analysis is liable to lead to flawed designs, and having design-focused milestones and deliverables can cause the analysis activities to be underresourced. We do not advocate skimping on task analysis, but we do argue that it should be an integral part of a process designed to deliver designs and build systems.

REFERENCES

Carroll, J. M. (2000). *Making use: Scenario-based design of human-computer interactions.* Cambridge, MA: MIT Press.

Beyer, H., & Holtzblatt, K. (1998). *Contextual design.* San Francisco: Morgan Kaufmann.

III

Human Perspectives

A great deal of the research work on task analysis has been undertaken by psychologists because understanding the tasks a system carries out requires more than even a detailed listing of observable behavior. The chapters in this section primarily concentrate on human issues, and most rely on some particular psychological or ergonomic model, although chapter 17 argues that design should be driven by the functionality of the thing being designed and not by the interaction between it and its users.

The eight chapters in this part can be divided into three groups. First, chapter 14 introduces psychological issues in task analysis, and chapters 15 and 16 present very different examples of mental models used in cognitive task analysis. The next three chapters focus on how to analyse tasks using a mental model basis for task classification and integration (chap. 17), error identification (chap. 18), and task and subtask triggering (chap. 19). The final chapters address task analysis and human issues from less common perspectives: the physical requirements of a novel interaction device (chap. 20) and activity theory, based on a Russian social psychological model developed in the 1930s (chap. 21).

Chapter 14
Task Analysis Through Cognitive Archeology
Frank Spillers

Readers who do not have a background in psychology are advised to start part III with this chapter. Cognitive archeology involves the inference of unobservable psychological states and processes from evidence of how things are used in tasks. Chapter 14 provides a framework for the introduction of many of the psychological issues addressed in more detail in other chapters in this part and elsewhere in the handbook. Cognitive archaeology also focuses on the importance of artifacts in triggering, sequencing, and ending tasks.

Chapter 15

Cogntive Task Analysis in Interacting Cognitive Subsystems

Jon May and Phillip J. Barnard

Interacting cognitive subsystems (ICS), with a history of more than 30 years, is a proper cognitive psychologists' model of cognition. This chapter should disabuse psychologically naive readers from thinking that all this psychology stuff is vague, woolly, or "for the birds". ICS is a general model of cognition which is then applied to task analysis. The chapter has an extensive appendix that provides considerable detail on the operation of the cognitive subsystems and how they interact in tasks.

Chapter 16

Data Analysis for the Critical Decision Method

B. L. William Wong

Chapter 16 uses a model of cognition that is based on widely held experiential notions of human psychology and is opposite to the model used in the preceding chapter. The critical decision method (CDM) relies on task experts recalling critical incidents and being able to explain the reasons for their behavior. The CDM approach to cognitive task analysis, although lacking great cognitive detail, obviously still has value for investigating important instances of real-world complex task performance, particularly where the work achieved was not satisfactory. The chapter describes two different analysis methods suitable for use in different project circumstances and discusses the use of interviews as a source of task analysis data.

Chapter 17

Using Task Analysis for Information Requirements Specification: The Sub-Goal Template (SGT) Method

Thomas C. Ormerod and Andrew Shepherd

The Sub-Goal Template (SGT) method extends HTA's decomposition of goals to encompass the production of interface requirement specifications. The SGT method also provides a classification of cognitive activity that is simple but applicable to an extremely wide range of tasks. The core of the method is based on stereotypical plans that organize the sequence of subgoals. The analyst selects the appropriate template and then fills in the subgoals. The psychological model used by SGT is intermediate between those used in chapter 15 and chapter 16. This is well illustrated by the difference in the level of the detail provided by the sample analyses of similar transport-related tasks in this chapter and chapter 16. The SGT method should help make HTA more accessible and easier to apply.

Chapter 18

Task Analysis for Error Identification

Chris Baber and Neville A. Stanton

The task analysis for error identification (TAFEI) method described in chapter 18 extends HTA to cover the identification of potential user errors. TAFEI, which has been used in a wide range of application domains, has the twin virtues of methodological simplicity and applicability very early in the design process. Also of considerable interest is that the underlying cognitive model used by TAFAI differs from many other task analytic ones in the way it theorizes about how people have and use plans of their activities. In this model, users employ global, prototypical routines that are translated into local, transitory routines when they interact with things. The psychological model is close to that proposed in chapter 30.

Chapter 19
Trigger Analysis: Understanding Broken Tasks
Alan Dix, Devina Ramduny-Ellis, and Julie Wilkinson

Only after the event can task performance be described in a linear fashion as tasks or subtasks following each other in a neat causal chain. Chapter 19 investigates the critical issue of what causes tasks or subtasks to commence or recommence when they do. Although sometimes the ending of one task or subtask is what "triggers" the next, in many cases a sophisticated decision about what to do next needs to be made. The triggers that Trigger Analysis identifies may be mental or prompted by things in the task environment, and actual triggers are often subtle and some are even unconscious. The approach can be used to indicate where tasks can go wrong by assessing faulty triggers.

Trigger analysis is method designed to be used with and augment other task analysis methods, many of which inadequately model the specialized decision making involved in how people sequence their tasks and subtasks. This in itself makes chapter 19 important one within this Handbook's context.

Chapter 20
The Application of Human Ability Requirements to Virtual Environment Interface Design and Evaluation
William Cockayne and Rudolph P. Darken

Task analysis has nearly a century of history behind it (chap. 3), although for more than the first half of the 20th. century the focus was on people's physical rather than psychological behavior. The authors of chapter 20 convinced the editors that their Human Ability Requirements (HAR) approach had a rightful place in this handbook because it involves such a novel input device—an omnidirectional treadmill for use in virtual reality applications—that not even the physical behavioral requirements were known. It also discusses taxonomies and classification schemes, which, as chapter 1 argues, are essential (but not sufficient) for the construction of task analytic models of the world.

Chapter 21
Activity Theory: Another Perspective on Task Analysis
Phil Turner and Tom McEwan

Last century's global, political "East-West" separation into the "Communist" and "Free" worlds was also reflected in an intellectual separation between these worlds. In the last decade or so, a small band of HCI and software engineering researchers (e.g., Nardi, 1996) have revisited the rather different psychosociological work done in the USSR following Vygotski's pioneering research in the 1930s. Chapter 21 introduces activity theory and then shows how it can be applied as a form of task analysis to facilitate the design of a high-fidelity virtual reality training simulator. Activity theory formalizes the identification of contradictions between what is required and what is desired in the performance of a task. Resolving these contradictions is at the heart of good design of artifacts and tasks. Chapter 21 also provides an important demonstration of the claim in Chapter 1 that task analysis is at the core of virtually all HCI work even when the HCI approach adopted is not a traditional Western one.

REFERENCES

Nardi, B. A. (1996). *Context and consciousness: Activity theory and human-computer interaction.* Cambridge, MA: MIT Press.

14

Task Analysis Through Cognitive Archeology

Frank Spillers
Experience Dynamics

Analysis and observation of the user's task domain unveils a window of understanding into the behavioral patterns, contexts, and scenarios that are required and utilized by users to attain success in completing a task. Task analysis (TA) forms the foundation for interaction, behavior and usage since prioritisation of design elements is a serious issue for both users and designers.

Successful user interface (UI) designs often utilise insight into tasks by studying user interactions, intentions, and expectations. User interaction itself amounts to the interplay of cognition and information processing as embodied by task routines or sequences that are commonly captured in a TA. This chapter focuses on procedures for uncovering cognitive processes relative to user goals and tasks including decision-making systems, the impact of information overload on screen display, and the significance of user roles to tasks. The notion of a "cognitive archeology" as a means to investigating task cognition will be explored and explained as a novel best practice in TA.

"Cognitive archeology" or the capture of cognitive processes required and utilized by users for task completion offers a necessary insight into the interchange of cognitive and task generated needs as they unfold at the design level. Concentrated analysis of explicit and implicit needs, decision-making processes, procedural knowledge, and motivation strategies creates a means for prioritization thereby improving interface effectiveness.

14.1 INTRODUCTION

The aim of task analysis is to gain a detailed understanding of the tasks, goals, and expectations that the user brings to an existing or new system. Perhaps one of the fastest ways to understand the context of the task is to uncover the landscape and nuances of the physical and mental environments.

14.1.1 Cognitive Profiling

Analysis of the physical task environment (see chap. 1) sets the stage for understanding task activities, and cognitive profiling explains why users choose the behaviors that make tasks easier to execute. Typical task variables such as time, sequence, repetition, learning, and error handling have correlates in the cognitive domain, and these correlates can often affect how tasks are approached by users and measured by practitioners under usability-testing conditions.

Cognitive engineering plays a catalytic role in the task life cycle, because perceptual and motor skills involved in performing a task originate from neurological functions and cognitive activities. Cognition can often be invisible to the observer. Therefore, it is important to thoroughly understand the clues offered by a user's external behavior, because those needs will likely correlate to some cognitive need. In task analysis, the main concern regarding task completion is with the severity and complications of "cognitive overload" (Kirsh, 2000; Woods, Patterson, & Roth, in press). Successful completion of a task implies that the user is in control and able to manage workload while minimizing information overload. A supporting architecture of perceptual, experiential, semantic, and procedural knowledge compatible with the task environment always characterizes tasks that seem easy to complete.

The capture of cognitive activity exclusively as data points appears to be a byproduct of task analysis rather than a central, integral data source for understanding tasks (Beyer & Holtzblatt, 1998; Hackos & Redish, 1998; Preece, Sharp, Benyon, Holland, & Carey, 1994). For instance, the focus in a "contextual inquiry" is on gaining insight into the context of use levied by the physical environment. To date, little emphasis in this widely practiced technique has been placed on how cognitive processes influence environment, error handling (e.g., chap. 18), and problem solving. Kirsh (2000) shed light on this issue by viewing the environment more as an "activity space." He did not place emphasis on the environment itself as much as on how people alter their environments to increase ease of use and reliability.

Similar to analysis of physical environment constraints and opportunities, cognitive profiling offers explanations as to why users choose behaviors in task execution. Typical task variables such as speed, exploration, repetition, accuracy, learning, and error are impacted not only by how a user engages with a task but also by the user's decisions, roles, and interaction with artifacts and by the meaning the user imposes on the task problem. Hamilton (1996) and Albers (1997, 2000) both underscored the need to understand how users construct questions about their tasks and to understand task-specific decision making. According to Hamilton, neglect of these variables in task analysis constitutes a "major failure" in current design methodology.

14.1.2 Cognitive Activity as a Precursor to Task Performance

Cognitive needs that arise during task performance often represent the user's implicit need to cope with and adapt to environmental constraints. According to Kirsh (2000), users adapt to an environment not only by changing mental processes and behavior but by altering the very environment posing the adaptive challenge. The users cognitive patterns are not more important than the detection of external behaviors associated with a task, but rather it is the mental patterning itself that offers complete insight into what qualifies as success for the user when translated to the user interface design.

Consider this example of altering one's environment to solve a problem (see also Fig. 14.1):

> Currently no RV (recreational vehicle) TV antenna manufacturer offers a warning signal, light, or sound with their products. The result is that every year thousands of RV enthusiasts drive away from a campsite with the antenna raised and accidentally break it. To compensate, RVers in the

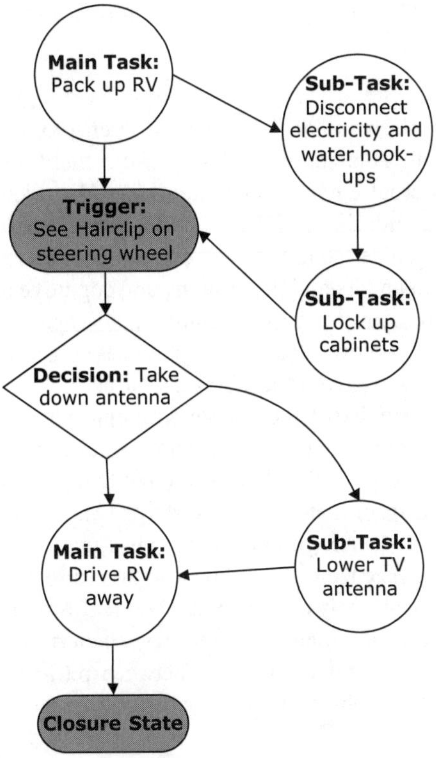

FIG. 14.1. Example of the RVers task model.

United States use a hair clip or similar object attached to the steering wheel to remind them that the TV antenna needs to come down upon departure. RVers report that on occasions where antenna damage did occur, they had forgotten to attach the reminder to the steering wheel.

Every task combines cognitive, physical, and perceptual actions (Drury, Paramore, Cott, Grey, & Corlett, 1987). These actions are characterized by a defining starting point (trigger; see chap. 19), a middle point (often involving feedback from the environment), and an ending or closure state. Tasks often contain nested subtasks or related main tasks and are typically impacted by either/or decisions within the task model (see Fig. 14.1). The process of identifying a task model is central to Hierarchical Task Analysis (HTA; chap. 3), which treats tasks as goals and subgoals rather than actions (Annett & Stanton, 2001). Preece et al. (1994) define the goal as the state of the system the user wants to achieve, the task as the activities required to achieve the goal, and the action as the steps required to complete the task (see also chap. 1). Thus, user goals overlap both physically and mentally in relation to tasks.

Task sequence is often influenced by fluctuations in decision making and in the environment. Performance can be degraded because of the need to accommodate a poor environment or by the physiological and cognitive resources of the user. For instance, task degradation is easily understood when the variable of stress or fatigue is introduced. Decline in task performance due to error, fatigue, or other reasons is well documented. Kilcarr (2002) reports significant impairment of driving ability and reaction time caused by fatigue and sleep deprivation and a diminishing of quality and quantity of work by 30% as estimated by employees at work. Task performance can clearly be hampered or enhanced by cognitive conditions and unpredicted nuances.

14.2 COGNITIVE ARCHEOLOGY

14.2.1 Origins

A new approach led by Renfrew (1994) in the field of archeology has led to an investigation of ancient human cognition. This approach departs from traditional archeology approaches that focus on study of external environments and material evidence left behind by human organization. In this new approach, the focus is on the belief systems and thought processes that enabled humans to organize, communicate, and engage in tasks. Since the early 1980s, archeologists have sought to understand the thinking and cognitive realm of primitive peoples. They have made efforts to reconstruct the cognitive landscape of past societies in order to answer questions about their knowledge, skills, purposes, practices, and thought processes and their symbolic behaviors relative to artifacts and structures found at archeological sites. Insights into the cognitive activity of the people who engaged with surviving physical and spiritual artifacts reveal important details about their social organization as well as individual and group behavior. Renfrew (1994) described cognitive archeology as "the study of past ways of thought as inferred from material remains" (p. 12). Segal (1994) argued that archeology is foundationally rooted in cognition in the sense that material objects only become archeological data if their existence can be shown to be a direct or indirect consequence of intelligent behavior. Unlike task analysis, which involves real-time analysis, cognitive profiling of past cultures looks at material remains and long-term patterns evident from the archeological record. Similarities can be drawn, however, between retrieval of past (static) cognitive activity and modern (real-time) observation, as both concern the study and interpretation of artifacts.

Cognitive archeology itself represents a shift toward detection of the internal mental portion of user behavior that is omnipresent in task activity. Cognitive archeology in task analysis involves elicitation, interception, and capture of the cognitive activities that a user finds beneficial for successful task completion.

14.2.2 The Importance of Artifacts

Task analysis involves the systematic analysis of workplace, workflow, and user interaction within the task environment. Task activities center around creation of and interaction with artifacts of both a physical and cognitive nature (see chap. 19). Hutchins (1995) defined cognitive artifacts as physical objects made by humans for the purpose of aiding, enhancing, or improving cognition. Norman (1991) defined a cognitive artifact as an "artificial device designed to maintain, display, or operate upon information in order to serve a representational function" (p. 19). Cognitive artifacts are those elements whose function is to aid or simplify task performance. *Examples of cognitive artifacts include the following*:

- a novel symbol system used on a calendar
- notes used to decipher a computer system
- a glance backward upon exiting
- a checkmark
- circle or color-coding scheme
- the sound of a warning timer
- a mental note
- a novel warning indicating an open or closed state
- changes in temperature or sound
- shifts in kinesthetic sensation or visual stimulus.

Like the artifacts gathered in archeological settings, contemporary artifacts include task-specific tools such as calculators, "cheat sheets," sticky notes, and handwritten notes or logs (Hackos & Redish, 1998, p. 139). Physical artifacts serve an important role in the sequencing, triggering, and closure of a task or set of tasks. Goel and Pirolli (1992; cited in Pearce, 1994) found that artifacts are designed in order to assist with problem structuring and problem solving. Artifacts thus act as the glue that binds user cognition, information processing, workload management, and task accomplishment.

14.3 COGNITIVE DATA IS REVEALED DURING TASK DECOMPOSITION

One of the aims of task analysis is to understand user needs and expectations in order to preserve the task integrity in the design requirements of a new system. The emphasis in cognitive archeology is on mental processes and interactions between task, environment, and cognitive resources. Cognitive archeology applied to task analysis procedures bridges the gap between primarily behavior based approaches (e.g., many of those discussed in part IV) and cognitive task analysis (CTA) methodologies (see chaps. 15 and 16). Unlike the behavior-focused methods, CTA is knowledge based, concentrating on internal representations, language, knowledge structures, and cognitive and perceptual filters (Wilt, 1999). Cognitive archeology would appear to approximate the emphasis that CTA gives to understanding a user's cognitive resources and learning challenges and the areas of a design where a user is likely to err.

In using the cognitive archeology approach, task analysis and ethnographic study methods are augmented for detection of the following issues.

- What motivation, expectations, and influences does the user bring to the task? (This question concerns the user's knowledge, values, and biases.)
- How does the user alter perception and compensate mentally so as to achieve task success? (This concerns the artifacts and the user's mental process.)
- How supportive or distracting is the environment? (This concerns context and sensory qualities.)
- What actions trigger decision making? (This concerns the user's role and task needs.)

The task analysis technique of "activity sampling" or the unstructured ethnographic interview yields the greatest amount of information for task decomposition. Additionally, HTA adds insight into task sequence, hierarchies, goals, and subgoals. HTA, however, becomes limited as a method when the aim is to determine wider issues related to the task, such as social influences, subjective experience, and distinctions between global and local task interaction. Hence, Hutchins (1995) introduced the notion that cognition is distributed across agents in an interconnected "stream of activity" and does not remain static within an individual's cognitive domain or task. In this regard, task data can be viewed as nonlocal, meaning that they also reside outside the user's cognitive process, "activity space," and environment. Instead, task data often reach beyond the user's task into the collective organization and the field where the work is conducted (i.e., the application domain; chap. 1). Note that usability techniques such as "card-sorting" are attempts to address the discrepancies in how users perceive, categorize, interpret, and apply meaning to the language of their particular domain.

The primary object of approaching task analysis through cognitive archeology or "cognitive workflow" (Kirsh, 2000) is to chart user motivation and needs, decision-making systems, influences, and socioenvironmental triggers (chap. 19).

An analysis of a supermarket shopping task (see chap. 12) might include the representation shown in Fig. 14.2, codified by a cognitive artifact—the familiar grocery list. The motivation

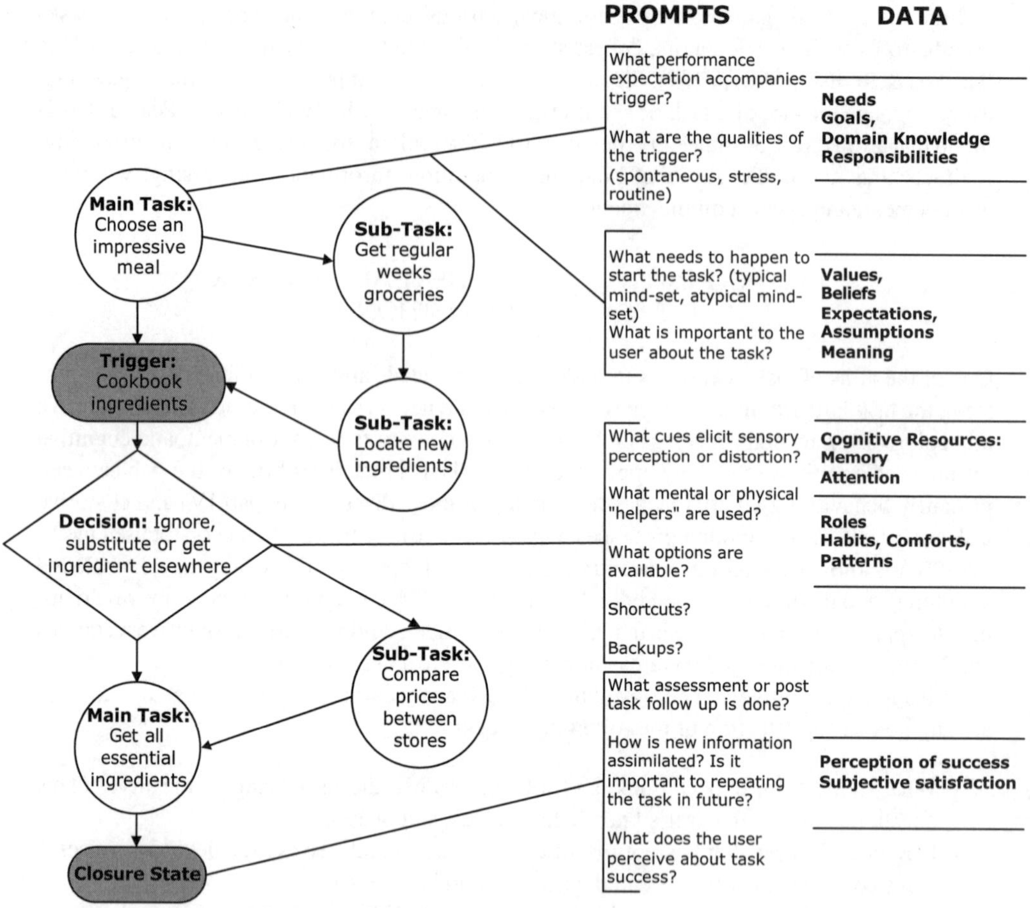

FIG. 14.2. Capturing cognitive processing at the task level.

may be to cook a special dinner to impress guests. The choice of brand-name items may be important in this context owing to the social influence and expression of brand loyalty. Outside of these attributes, the observer would want to know which occasions other than the "special dinner" triggered the use of the list and whether or not brand loyalties were consistent. Which items were regular purchases and which were irregular? Did irregular items appear on the list? What influence did mood have on the creation of the list? Was the list actually used while in the store? What items were ignored from the list? How many items that were not on the list were purchased? From this task analysis, the researcher might understand more about the social activity of supermarket shopping and how different variables and artifacts are utilized. By gaining a detailed understanding of the tasks, the researcher would also be able to identify opportunities to introduce new supermarket products and services. Profiling the cognitive archeology of the user group would allow business and marketing decisions to be made that were informed by an understanding of user behavior and the cultural and task dynamics of supermarket shopping.

The main issue in this example is that simply modeling a task does not constitute effective task analysis (e.g., chaps. 1 and 3). Task analysis can be used as a justification to force-fit requirements to a product. For instance, a supermarket Web site that has a shopping list feature online simply because the artifact exists in the real world might be missing the cognitive clues around how a shopping list is used and misused. For example, analysis of a shopping list may reveal that shoppers actually enjoy impulse buying in a supermarket and do not want to feel restricted by a shopping list. Such cognitive data offer greater insight into design and feature definition, or as Colbourn (1995) states, "Technology must be matched to the cognition required" (p. 13).

14.3.1 Error as an Opportunity Indicator

One of the aims of task analysis is to understand how tasks unfold, influence, and/or interfere with user success when applied to user interface design and screen display. Perhaps the most logical place to begin simplifying a user's task on screen is to understand what problems and issues the user is having under normal conditions. Error handling reveals the limitations and opportunities that demarcate what is suitable to present to the user in the user interface. Lewis and Norman (1995) noted that determining what information is most useful to the user is the biggest challenge involved in the design of error recovery systems. Understanding where users fail and how errors occur offers a strategic leverage point for exceeding user expectations with the system.

The following questions capture task specific criteria that ought to be reflected in the design in order to reduce error occurrence and increase the usefulness of error handling. (The questions are asked directly before the user begins a task):

- What does X mean to you? (Semantic construction.)
- What is important to you about X? (Values elicitation.)
- How do you know you need X? (Implicit or explicit knowledge.)
- Whom do you talk to about X? (Influence of other users.)

14.3.2 Prioritizing Tasks in Decision Making

The significance imposed on the task by the user has a direct relationship to the user's motivational state and perceived needs in the social and task environment. Whereas techniques such as "contextual inquiry" (Beyer & Holtzblatt, 1998) focus the researcher's observation on the context (location) of the task, and ethnography (see chap. 6) places emphasis on cultural and social influences, few methods explicitly encourage the profiling of decision making as a pivotal component of the task. Albers (2000) stated that current task analysis techniques are not designed to support complex decision-making. Albers (1997) suggested that serious attention be paid to decision-making systems because they are integral to complex problem-solving and mental-modeling. The designer, according to Albers, must understand the cognitive context or "cognitive workload" (Kirsh, 2000) or risk failing to support the user's mental model. Albers (1997) emphasized that cognitive context is very situation specific and that generalizations about user workload may be misleading: "Information about typical work situations is unreliable for designing information systems. No two work situations are identical since the context of a decision situation and the heuristics brought to bear depend on minute differences in situational and personal characteristics" (p. 8). According to McGrew (1998), 270 the characteristics that determine how a person makes a decision in the real world can easily be captured and modeled, and for some purposes this is so (e.g., in the analysis of recalled critical incidents; see chap. 16).

Good decision-making is facilitated by effective presentation and the display of information in a sequence and manner that is easy to understand (Albers, 1997; Hsee, 1995). Selecting the best display is critical, especially when uncertain or ambiguous information must be presented to and understood by the user (Coury & Strauss, 1998, p. 327).

ROLE:	Daily Finances Manager	Investment Strategist	Frequency:
TASK:			
• *Update Chequebook*	**X**		Low
• *Transfer Funds*	**X**	**X**	High
• *Make Deposits*	**X**	**X**	High
• *Write out cheque to pay bills*	**X**		Medium
• *Review statements*	**X**	**X**	Medium
• *Query a statement*	**X**		Low
• *Monitor Electronic Debits*	**X**		Low
• *Monitor Checking Balance*	**X**		High
• *Monitor Savings Balance*		**X**	Medium

FIG. 14.3. (Role-task) decision prioritization matrix with sample data for a banker user.

14.3.3 Looking for the Roles Users Play

Mapping user tasks to the roles users play (Beyer & Holtzblatt, 1998; Marine, 2001; Smith & O'Neil, 1996; chaps. 6, 7, 23, and 24) offers tremendous insight into decision making at the task level. In this regard, users are not viewed as fixed agents but as dynamic actors that utilize all cognitive resources available in order to solve their problems. During task activity, users engage in "role switching" (Beyer & Holtzblatt, 1998) in the sense that the decision making process requires a different set of behaviors and hence the need to play a new role (note the decision that triggers a subtask in Fig. 14.2). Role switching is an important element to identify in a task analysis because it often causes tasks to be interrupted and left open or suspended (chap. 19). Role detection also allows for prioritization of task and data presentation in the user interface based on frequency of user needs (see Fig. 14.3).

Tasks that are utilized across user roles are especially important to present to the user in the design. The number of roles satisfied should influence where to position features and functionality that support critical task activities, including complex decision-making.

14.3.4 Knowledge and Memory Management

One of the crucial factors in task prioritization is the amount of procedural or structural knowledge the user brings to the task. Users who are experts in their domain tend to handle errors and role switching more easily. The explanation for this probably includes the fact that experts have had multiple experiences to generalize from, causing them to remember to close open tasks and allowing them to troubleshoot through better decision-making. CTA methods have been utilized to elicit structural knowledge used in decision-making systems (Randel, Pugh, & Wyman, 1996; chap. 16). Adams (1993), in his study of expert decision makers in aviation systems, found that the main cognitive constraints on decision making include attention span and long-term and short-term memory. Memory management plays a critical role in role switching and error handling. Adams found that, compared with novices, experts utilize a highly organized body of procedural and conceptual knowledge and have faster access to this and prior experience scenarios.

Decision making data is essential in task analysis. Hsee (1995, cited in Hamilton, 1996) found that, when presented with a factor tempting to the decision maker but not relevant to the problem, the decision maker will often rationalize using the irrelevant factor. It seems that users, when solving problems, attempt to manipulate cognitive resources, including decision-making factors, in order to gain advantage in workload management. Designs that reflect poor capture of decision-making systems have a tendency to be less effective because they cause distraction and cognitive overload, which lead to errors. When user decision-making is supported in information design, the novice user, like the knowledge matter expert, gains enhanced cognitive resources, including attention, memory, creativity, and superior situational awareness.

14.3.5 How Environments Mediate Workload

The role of the physical environment and its artifacts (Beyer & Holtzblatt, 1998; Hackos & Redish, 1998; Hutchins, 1995; Norman, 1991; see also chap. 19). Physical environments define the context of use of a system and also the constraints and opportunities available to the user. Environments are not static but dynamic entities. Kirsh (2000) stated that people structure their workspaces in a way that promotes "scaffolding" or "cognitive safety nets" (Spillers, 2001). Cognitive safety nets are the mechanisms, processes, and artifacts that users utilize simultaneously in the task environment. As Kirsh explained, "Well designed environments ought to take note of the cognitive needs people have in performing their tasks and build in *scaffolding* that simplifies the way people make cognitive use of their environments to increase task reliability" (p. 6). The processes by which users create and utilize cognitive artifacts are an example of the scaffolding phenomenon.

In conducting the task analysis, the analyst pays particular attention to the impact of sensory stimulation at key points of the task life cycle. For instance, where the user might be influenced by visual information, the interface must accommodate the need, providing less or more information bandwidth through appropriate interaction techniques. In the financial investment community, a pie chart that shows the breakdown of amounts allocated to a portfolio can have tremendous value in monitoring financial conditions and in decision making. For Daimler-Chrysler Research and Technology North America, the design of choice for users of telematics systems incorporates the use of hard and soft keys in an interface, combined with message prioritization based on (a) operational relevance, (b) time criticality, and (c) safety relevance (C. Kirn & S. Wreggit, personal communication, February 18, 2002). What types of tasks are associated with buttons and knobs, scrolling devices, and navigation aids is of paramount importance for safety, ease of use, and error handling. The issues the designer is faced with include managing the presentation of sensory information with sensitivity to divided attention, situational awareness, memory, and data overload.

14.3.6 Cognitive Overload and Stimulus Screening

Sensory information forms the building blocks of experience. Part of our perceptual processes as humans involves the sensing, evaluating, and interpreting of cues and feedback from the environment. The most pervasive problem with human information processing involves "cognitive overload." Kirsh (2000) found that increases in the quantity of information reduce the meaningfulness and usefulness of the information. The level of arousal goes up in conjunction with environmental load, according to Mehrabian (1976), and after people exceed some arousal maximum, their performance degrades with further increases in their arousal. The body's physical response to high levels of arousal is fatigue, the stage after information overload. Fatigue is a natural reaction to what Mehrabian terms *general adaptation syndrome* (GAS).

Whether high levels of arousal play a part in task observation and cognitive profiling is of paramount importance. Mehrabian (1976) found that people differ in their capacity to handle cognitive load and can be categorized by their level of ability to screen out stimuli from the environment. Screeners are people who can cope with a higher level of environmental distraction, such as noise and extreme or unexpected sensory input. Nonscreeners have less tolerance for high arousal states and are distracted or tire easily. Based on Mehrabian's work, detection of stimulus screening questions can be added to the task analyst's investigation:

- Does the task require a high or low level of screening?
- What percentage of users profiled fit the task-screening requirement?
- How does data presentation need to adjust to arousal levels caused by the environmental load?
- Is this user able to handle increased cognitive loading?
- Is the ability to screen affected by roles and stages in the task life cycle?

Although stimulus screening certainly takes place in the automotive environment, designers must accommodate nonscreeners and base designs on the lowest common denominator. "Driver distraction" is a widely investigated area of interest in transportation research. In their human factors research, Kirn and Wreggit have considered several dependent measures as indicators of driver distraction. One of these key measures is subjective workload, as measured by a modified NASA TLX procedure; a method for evaluating the visual and cognitive distraction of a telematics system design has been pioneered at Daimler-Chrysler's Vehicle Systems Technology Center (VSTC). Utilizing a design technique similar to "forcing functions" (Lewis & Norman, 1995, p. 691), the VSTC has yielded positive results for driver workload management. In order for drivers to read e-mail inside a moving vehicle, Kirn and Wreggit limited the display to two lines (the subject and who the message is from). To retrieve the message, the driver must stop the vehicle or listen to the message through an audio channel. The team confirmed previous research that indicates that all subtasks must be able to be conducted in individual glance times of 2 seconds or less. In addition, tasks and subtasks were examined with regard to their effect, if any, on driving performance.

Similar studies have been conducted on the use of mobile phones when driving. A study by Cain and Burris (1999) revealed that receiving a call is the most hazardous activity associated with using a cell phone when driving. The authors attributed the element of danger to the fact that drivers cannot choose the time when a call is received. Mobile phone use while driving illustrates how unexpected stimuli interferes with the ability to screen stimuli specific to the operation of the vehicle.

14.4 CONCLUSION

Profiling the user's cognitive landscape offers a way to understand the emotional and intellectual factors that may affect memory, attention, confusion, situation awareness, interruption, and completion of a task. Future applications of task analysis, including the development of cognitive archeology, must take into account the "need to capture the importance of perceptual and cognitive processing in models of human performance" (p. 49; cited in Zachary, Le Mentec, & Ryder, 1996). Albers (2000) and Kirsh (2000) both point to improved work analyses that expose the details of how people exploit their environments to make the cognitive aspects of their work easier.

The key attributes of human cognition as related to the cognitive activity present in task analysis include attention, learning, knowledge, and memory. Variables that tax these cognitive elements may have repercussions on how easy the user finds the task when represented on a computer screen. Furthermore, immediate task influences such as time pressures, conditions of the task, habits, sequences, roles, and duties are constrained by the environmental load. The task environment provides data impacting design decisions by offering insight into the requirements necessary to bypass cognitive overload and the ability of users to screen out information irrelevant to their tasks.

Physical environments create the need for cognitive artifacts that trigger and facilitate interactions between the task environment and the user's information-processing needs (Norman, 1991). Artifact detection has paramount importance in user needs analysis for user interface design. Cognitive artifacts, when explored in detail, can offer insight into the challenges the user faces and how well he or she is able to cope with workload management issues.

Profiling of the user's mental processes leads to a clearer distinction between desired and required, perceived and actual needs relating to successful task completion. Identifying task attributes for interface design involves understanding the flux in user roles, motivation, goals, priorities, and decision-making systems. Cognitive archeology encourages detection of the cognitive expectations and needs that the user constructs and perceives to simplify his or her task activities. Cognitive archeology forms a basis from which to understand user needs and interface requirements while allowing for prediction of the overall effectiveness of an interface over time with greater accuracy.

REFERENCES

Adams, R. J. (1993). *How expert pilots think: Cognitive processes in expert decision making*. Fort Belvoir, VA: Defense Technical Information Center.

Albers, M. J. (1997). Decision making: A missing facet of effective documentation. In *Proceedings of the 14th Annual International Conference on Computer Documentation*. Available: http://www.people.memphis.edu/~malbers/texts/sigdoc96.htm.

Albers , M. J. (1998). Goal-driven task analysis: Improving situation awareness for complex problem-solving. In *Proceedings of the 16th annual International Conference on Computer Documentation*. Available: http://www.people.memphis.edu/~malbers/texts/sigdoc98.htm.

Albers, M. J. (2000). Information design for Web sites which support complex decision making. In *Proceedings of the STC 2000 Annual Conference*. Available: http://www.people.memphis.edu/~malbers/texts/STC2000.htm.

Annett, J., & Stanton, N. (2001). *Task analysis*. London: Taylor & Francis.

Beyer, H., & Holtzblatt, K. (1998). *Contextual design: Defining customer-centered systems*. San Francisco: Morgan Kaufmann.

Cain, A., & Burris, M. (1999). *Investigation of the use of mobile phones while driving*. University of South Florida, Center for Urban Transportation Research. Available: http://www.cutr.eng.usf.edu/its/mobile_ phone_text.htm.

Colbourn, C. J. (1995). *Constructing cognitive artefacts: The case of multimedia learning materials*. Available: http://wwwtools.cityu.edu.hk/ct1995/colbourn.htm.

Coury, B., & Strauss, R. (1998). Cognitive models in user interface design. *Proceedings of the Human Factors and Ergonomics Society, 1*, 325–329.

Drury, C. G., Paramore, B., Cott, H. P. V., Grey, S. M., & Corlett, E. N. (1987). Task analysis. In G. Salvendy (Ed.), *Handbook of human factors* (pp. 370–401). New York: Wiley.

Goel, V., & Pirolli, P. (1992). "The structure of design problem spaces." Cognitive Science, V. 16, pp. 395–429.

Hackos, J., & Redish, J. (1998). *User and task analysis for interface design*. New York: Wiley.

Hamilton, F. (1996). Predictive evaluation using task knowledge structures. Available: http://www.cs.bath.ac.uk/~pwild/papers/hamilton96predictive.pdf.

Hammond, K. R. (1998). Judgment and decision making in dynamic tasks: Information and decision technologies. In J. M. Flach, P. A. Hancock, J. K. Card, & K. J. Vincente (Eds.), *Global perspectives on the ecology of human-machine Systems* (pp. 3–14). Hillsdale, NJ: Lawrence Erlbaum Associates.

Hsee, C. (1995). Elastic justification: How tempting but task irrelevant factors influence decisions. *Organizational Behavior and Human Decision Processes, 62*, 330–337.

Hutchins, E. (1995). *Cognition in the wild*. Cambridge, MA: MIT Press.

Kilcarr, S. (2002, February). The silent killer. *Fleet Owner.* Available: http://fleetowner.com/ar/fleet_silent_killer/

Kirlik, A (1995). Requirements for psychological models to support design: Towards ecological task analysis. In J. M. Flach, P. A. Hancock, J. K. Card, & K. J. Vincente (Eds.), *Global perspectives on the ecology of human-machine systems* (pp. 68–120). Hillsdale, NJ: Lawrence Erlbaum Associates.

Kirsh, D. (2000). A few thoughts on cognitive overload. *Intellectica*. Available: http://icl-server.ucsd.edu/~kirsh/Articles/Overload/published.html.

Lewis, C., & Norman, D. (1995). Designing for error. In R. Baecker, J. Grudin, W. A. S. Buxton, & S. Greenberg (Eds.), *Readings in human computer interaction: Toward the year 2000* (pp. 686–698). San Francisco: Morgan Kaufmann.

Marine, L. (2001). *Does your design solve the right problem?* Paper presented at the Usability Professionals Association annual meeting, Las Vegas.

McGrew, J. F. (1998). Real world decision-making styles and their consequences. In *Proceedings of the Human Factors and Ergonomics Society* (pp. 268–271 vol. 1). Chicago.

Mehrabian, A. (1976). Public places and private spaces. New York: Basic Books.

Pearce, A. 1994. *Design of artifacts from a cognitive engineering perspective*. Available: http://maven.gtri.gatech.edu/sfi/resources/pdf/TR/TR004.PDF.

Norman, D. A. (1991). Cognitive artifacts. In J. M. Carroll, (Ed.), *Designing interaction: Psychology at the human-computer interface* (pp. 17–38). Cambridge: Cambridge University Press.

Preece, J., Sharp, H., Benyon, D., Holland, S., & Carey, T. (1994). Human-computer interaction. Harlow, England: Addison-Wesley; pp. 30–42.

Randel, J., Pugh, H., & Wyman, B. (1996). *Methods for conducting cognitive task analysis for a decision making task*. San Diego, CA: Naval Personnel Research and Development Center.

Renfrew, C. (1994). The Ancient Mind: Elements of Cognitive Archaeology. Cambridge: Cambridge University Press, pp. 6–38.

Segal, E. M. (1994). *Archaeology and cognitive science*. Available: http://cas-courses.buffalo.edu/classes/psy/segal/ARCHCHAP.htm.

Shepherd, A. (1989). Analysis and training in information tasks. In D. Diaper (Ed.), *Task analysis for human-computer interaction*. (pp. 15–55). Chichester, England: Ellis Horwood.

Smith, M., & O'Neill, E. J. (1996). Beyond task analysis: Exploiting task models in application implementation. Available: http://www.acm.org/sigchi/chi96/proceedings/shortpap/Smith_M/sm_txt.htm.

Spillers, F. (2001). *The cognitive "safety-net" and its implication on design*. Unpublished manuscript.

Wilt, G. (1999, February 10). *Task analysis*. Paper presented at George Mason University. Fairfax, VA.

Woods, D., Patterson, E., & Roth, E. (in press). Can we ever escape from data overload? A cognitive systems diagnosis. In E. Hollnagel & P. C. Cacciabue (Eds.), *Cognition, technology and work* (pp. 1–8). London: Springer-Verlag.

Woodson, W. E. (1981). *Human factors design handbook*. New York: McGraw-Hill.

Zachary, W., Le Mentec, J-C., & Ryder, J. (1996). Interface agents in complex systems. In C. Nutuen, & E. H. Park (Eds.), *Human interaction with complex systems: Conceptual principles and design practice* (pp. 35–52). Norwell, MA: Kluwer.

15

Cognitive Task Analysis in Interacting Cognitive Subsystems

Jon May
University of Sheffield

Philip J. Barnard
MRC-CBU

Cognitive task analysis (CTA) techniques seek to model the mental activity of a task operator. With the rise of computing artifacts, the focus of CTA has changed from supporting the tutoring of operators, to modeling knowledge application, to modeling cognitive processes. Descendants of knowledge-based approaches include GOMS, and produce quantitative temporal behavioral predictions for well defined interfaces. The increasing pace of design, and the dominance of small design teams has led to a demand for more flexible techniques. This chapter describes a particular approach to CTA using a cognitive theory called Interacting Cognitive Subsystems (ICS). A CTA in ICS requires a prior task analysis to have been conducted, but the analyst then identifies the configuration of cognitive processes necessary to transform information about the task, through the phases of goal formation, action specification, and action execution; for novices, occasional (normal), and expert operators. The availability of procedural knowledge, experiential, and abstracted memories influence the ease of processing, and the scope a design offers for their development informs ease of learning and skill acquisition. The location of a particular form of buffered processing predicts subjective awareness of different aspects of the task, and of task complexity. Two notations supporting analysis are described. The close coupling of the analytic approach and the underlying theory enables a CTA in ICS to provide supportive evaluation, allowing iterative redesign. It is also allowing further research linking ICS to formal models of systems analysis (Syndetics) and to other methods of TA, namely TKS, to extend both techniques to collaborative and multiple task performance.

15.1 A RATIONALE FOR COGNITIVE TASK ANALYSIS

The pace of technological change and the ubiquity of computer-based devices has brought about a significant change in the focus of task analysis over the past 20 years. As Annett describes in chapter 3 of this volume, the early time-and-motion focus on physical units of task performance has been superseded by an emphasis on the mental operations required of an operator of a device. Cognitive task analysis (CTA) began by recognizing that the key to skilled task performance lay in the possession and appropriate use by the operator of task-relevant knowledge and sought to develop models of the required knowledge to improve the tutoring of operators and speed the progression of an individual from task novice to task expert. Psychological theories of learning and knowledge structuring, such as Newell's SOAR architecture (Newell, 1990), gave rise to the realization that the task could itself be designed to facilitate knowledge acquisition, and so CTA began to be seen as a stage in task design, modifying the task device or structure, rather than as something to be done to modify the operators' knowledge.

The descendants of knowledge-based approaches to CTA are represented by the GOMS methods (see chap. 4). Initially, such methods were aimed at optimizing tasks for expert performance, as they were implicitly following a large-scale industrial model of both design and use. This required a structured methodology for CTA suitable for use in the design process and for dealing with end products that would be used repeatedly by highly skilled operators (products in which a fraction of a second of execution time saved could result in billions of dollars of costs saved).

Toward the end of the 1980s, dissatisfaction with the cost and effort of employing such methods, together with a recognition that not all task design fitted this mold, led to a search for analytical methods that could be used on a smaller scale. This was partly driven by the rise of small information technology (IT) firms and the in-house development of IT, where the design and implementation of computer applications was often carried out by one or two individuals. The pace of change in IT was another factor, since many IT applications were used now not by experts but only occasionally by people who were at best infrequent users of each of a large number of computer-supported tasks. This and the need for tasks to be user friendly and allow people to "walk up and use them" changed the focus away from designing for expert use toward designing tasks to be comprehensible by novices, supporting the recall of task steps during infrequent use.

In such a rapidly changing technological world, the rationale for CTA has become clear. Technology, devices, tasks, and even the knowledge of the human operator will vary widely, but one thing remains unchangeable: the human cognitive architecture. Even if the content of our memories and skills changes with time, the way that we process information, form and recall memories, and automate our skills does not change. Analysts ought to be able to make use of supportive methods based on cognitive psychology to fit the design of tasks to this single unchanging factor.

For a method to be applicable in the fast-paced world of small-scale IT development, it needs to be usable alongside a wide range of other approaches and at almost any stage of design, from initial task conception through to postdesign firefighting and usability evaluation. It has to be very tolerant in the amount of detailed design and domain information that it requires. It should also be able to guide designers explicitly toward improvements and changes in their initial ideas rather than just giving evaluative output. In other words, it must provide supportive evaluation for iterative redesign.

One such approach has been evolving steadily since the mid-1980s, largely with the support of the European Union, through the international basic research projects Amodeus and Amodeus2 (Esprit BRA 3066 and 7044; Barnard, 1993; Barnard et al., 1995). These projects brought together cognitive psychologists, computer scientists, IT designers, and design researchers to explore their interrelationships and involvement in human computer interaction (HCI). One result was the systematic application of a cognitive architecture called Interacting Cognitive Subsystems (ICS; Barnard, 1985; Barnard & May, 2000) to CTA, and a description of this approach forms the focus of this chapter.

ICS is not designed as a theory to support task analysis or design but is a general purpose representation of cognitive operations. It is a theory of cognitive function derived from empirical research in psychology and is open to empirical testing. In addition to the HCI applications of Amodeus, it has been used within clinical psychology and as a basic cognitive model within experimental psychology (e.g., Scott, Barnard, & May, 2001). For HCI designers, a handbook has been published for use in master's degree HCI courses (May, Scott, & Barnard, 1995) and tutorials at HCI conferences given to teach the approach to entrants to the IT industries in Europe (Scott & Barnard, 1997). A CTA in ICS attempts to identify the cognitive resources that an individual operator would need to employ in performing a task. Having identified these resources, the analysts can note conflicts where a single resource is required to do different things or is required by competing simultaneous tasks. The availability of prior knowledge to support processing can be estimated systematically, and the scope that processing allows for the progressive development of knowledge can be assessed. The behavioral consequences of the analysis are qualitative rather than the quantitative, time-based products of GOMS, since they highlight points in the task where processing demands are high and so where errors may occur (see also chap. 18). Support for redesign is implicit but guided by the analysts' understanding of the theoretical concepts that they have used in building the CTA. A close coupling between the underlying theory and the analytical process is seen by the proponents of ICS as a key strength (May & Barnard, 1994).

Because ICS addresses only the cognitive operations of a single user, it has not yet been applied to tasks in which several people collaborate, and because it takes as its starting point a description of the task to be designed, it does not include a conventional task analysis phase. Collaborative tasks can in principle be understood by modeling the processing of each individual, but owing to its form ICS allows a more ambitious scaling up of the approach. The ICS architecture has been influenced by ideas about information processing drawn from computer science, and the generic nature of the components of the overall system has allowed parallels to be drawn with formal models of systems analysis known as interactors. This has been used to develop a quasi-formal analytical method called *syndetics* (Greek for "bringing together") in which a computing system and the human cognitive architecture are modeled conjointly as interactors, no distinction being made within the model as to the human or technological origin of particular interactors (Duke, Barnard, Duce, & May, 1999).

In principle, a syndetic analysis could be applied to collaborative working, with individuals being classed as interactors within the work group. Research is currently investigating this possibility, which will be discussed further toward the end of this chapter, along with work that is linking CTA in ICS with Task Knowledge Structures (TKS; Johnson & Johnson, 1991). TKS is a knowledge-based method of task decomposition, and there are some interesting parallels with the way that knowledge is formed and abstracted in ICS. TKS has also conventionally addressed the individual task operator, and current research is exploring a combination of ICS and TKS to support their joint extension to collaborative and multiple task performance.

Before discussing this future work, the current state of CTA in ICS is presented, beginning with a brief description of the ICS architecture for human cognition. (A more detailed explanation of the processes and concepts involved in ICS is presented in Appendix A, and a glossary of terms is contained in an Appendix B). The general process by which a CTA is conducted is then elucidated by working through the steps an analyst would take in identifying the configuration of cognitive processes required by a task, the knowledge and memory use required to support processing, and the complexity of the exchanges between different processing resources. Two forms of notation that have been developed to support analysis will be described, one lexical, allowing the processes to be economically denoted, and the other graphical, allowing temporal and thematic changes in the content of processing to be represented. The description of ICS in this chapter is intentionally generic, and where examples are given, they are grounded in task performance in general. The terms *operator* and *device* are used in preference to *user* and *computer* to emphasize this. What follows is not a cookbook for the application of CTA but an explanation of the principles underlying CTA in ICS. Understanding the principles is a necessary component of learning to use ICS as an analytical technique, since it is this aspect that allows solutions to be inferred once problems have been identified.

15.2 INTERACTING COGNITIVE SUBSYSTEMS (ICS)

Models within cognitive psychology have usually been developed to describe the mental functions necessary for a particular phenomenon, such as face perception, reading, semantic memory, working memory, and so on. Typically, these models consist of box and arrow diagrams and in a sense are similar to a hierarchical task analysis conducted on aspects of human thought: The boxes represent things that logically have to be done, and each box can potentially be broken down into smaller units. Arrows drawn between boxes can represent either process relationships such as "this function enables that function" or "this function requires that function." The cognitive architecture represented by ICS is different and sometimes difficult for those experienced in cognitive psychology to grasp (perhaps for the same reason, it seems easier for computer scientists to follow the ideas). It starts from the assumption that mental operations can be thought of independently of the functional roles they support and that a limited number of general purpose operations can be used to perform all cognitive tasks. Phenomena such as short-term memory, semantic memory, attention, perception, and executive control arise from the activity of these operations, but the operations themselves are not specific to these phenomena. The boxes in ICS are not functions that need to be carried out but information-processing operations that carry out functions, and the arrows indicate a flow of information between operations.

The general purpose operation that ICS carries out is the transformation of a mental representation from one form, in which it describes a particular class of information about the world, into another form, in which it describes a different class of information. In performing this operation, ICS will store the incoming representation in a local form of long-term memory called the image record and will access proceduralized knowledge acquired from previous experiences with the incoming information to try to produce the appropriate output representation. In general within cognitive psychology, proceduralized knowledge is knowledge about how to do something that has become so well learnt that the action can be carried out without conscious effort. In ICS, proceduralized knowledge is embedded within the transformation processes, and it allows them to produce an output for a particular input without accessing the image record. If it is poorly developed, ICS may try to access previous experiences that are similar to the incoming representation from the image record, to elaborate and refine it. Finally, if the output representation that is produced is used successfully in task performance, the procedural knowledge relating the incoming and output representations is updated or strengthened, as appropriate. A key aspect of CTA in ICS is identifying the demands on proceduralized knowledge, the scope for its development (that is, skill acquisition), the support provided by the image record contents, and the scope for its development (that is, schematic abstraction in memory).

These three basic operations—storage, transformation, and revival of information by memory—are grouped together within a "subsystem" dedicated to the processing of one particular class of information. (Stored information is said to be "revived by memory" rather than "retrieved from memory" because memory is seen as an active entity rather than a passive store.) The overall system is built up of nine such subsystems, each one taking as its incoming representation either information from the sensory organs (the eyes and ears, which sense distal information sources, and bodily sensations, which sense proximal information sources such as taste, smell, temperature, and body states) or, crucially, cognitive information that has been output by another subsystem. Together these nine subsystems operate in parallel to continually process a flow of information. Figure 15.1 details the subsystems and shows how they are linked together. Two important features of the organization of the system should be noted: (a) Each subsystem

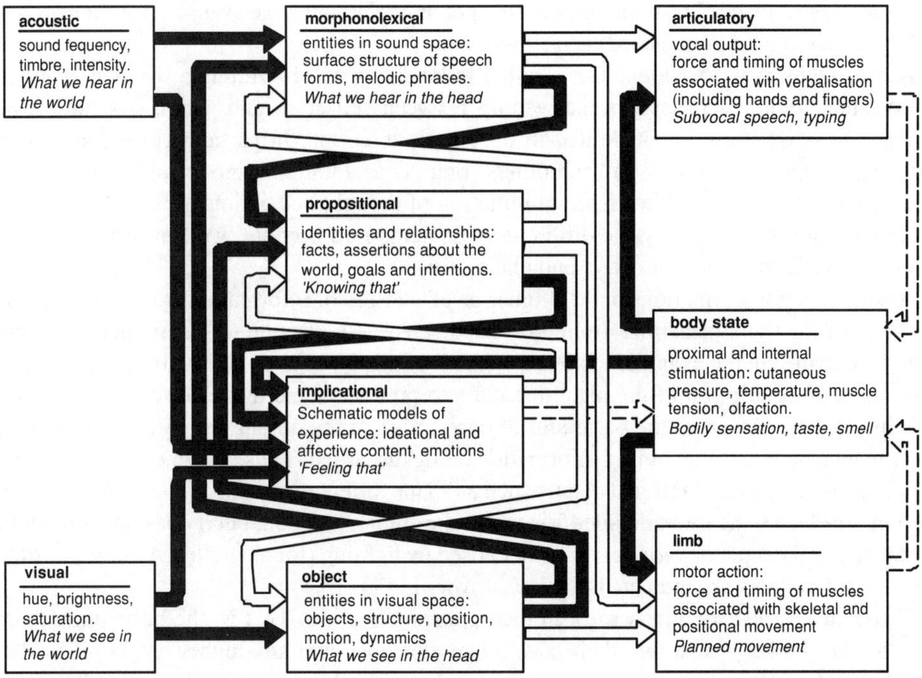

FIG. 15.1. The nine subsystem of ICS and the classes of information processed at each subsystem. The arrows indicate the potential flows of information, showing how the output of one subsystem can be used as the incoming representation of another subsystem. Black arrows represent "abstractive" flow, white arrows "elaborative" flow (see text). Dashed arrows are indirect, because they represent information exchange mediated by changes in the body.

produces more than one class of information but not all possible types (in fact, no more than three types), and (b) most subsystems have more than one source of incoming information.

Looking in more detail at the nine classes of information encompassed within the scope of ICS, we can see three broad categories. First, there are three sensory subsystems, two on the left of Fig. 15.1 operating on Acoustic and Visual sensations and one on the right operating on Body State sensations. Second, there are four central subsystems operating on different levels of interpreted meaning derived from the sensory information. These are the Morphonolexical and Object subsystems, which primarily process information from their respective sensory subsystems, and the Propositional and Implicational subsystems, which deal with increasingly abstract levels of meaning. Third, there are two effector subsystems, Articulatory and Limb, whose output is used to execute physical actions in the world.

No subsystem does much on its own. Just seeing something, for example, is a flow of information from the eyes through the Visual subsystem (which extracts both quantitative featural and qualitative affective information from the sensation and discards a lot of the sensory detail), the Object subsystem (which identifies objects and their spatial and activity relationships), the Propositional subsystem (which infers the meaning of the objects being seen and the actions they are performing and their relevance to the self and one's goals), and the Implicational subsystem (which detects conformance or violation of goals, revises current goals, and produces affective responses in the body, subsequently detected by the Body State subsystem). This is the simplest flow, leaving out any mental naming (Morphonolexical), vocalization (Articulatory), or motor activity (Limb) and thus any interaction with the seen object. Any cognitive task will inevitably require the involvement of processes within several subsystems, and so the CTA must consider the procedural knowledge available for each process and the image record contents of each subsystem.

At first sight, Fig. 15.1 looks like a rather redundant way of building a stimulus-response architecture, with several stages intervening between sensation and action. As the arrows indicate, however, flow is not limited to this single direction. In particular, the four central subsystems can all take some of each others' outputs as additional inputs, allowing loops of processing as well as top-down internal initiation of thought and action. Taking the nature of the information processed by the different subsystems into account, we can distinguish two types of transformation: abstractive and elaborative.

Abstractive transformations build "higher level" classes of information from more detailed classes, and the transformations from the Visual subsystem through to the Implicational subsystem described in the example of seeing were all of this type. They all involve combining the basic units of description of the incoming class to produce some higher level of description. This higher level becomes the basic unit of description in the output representation, with the constituent detail of the incoming information being discarded. Thus, an Object level of representation contains organizational details such as shape, orientation, and overall color that have been inferred from the more detailed Visual level of representation, but it does not contain the minute textural details or variation in color caused by lighting (for example). Abstractive flows between subsystems are indicated by black arrows in Fig. 15.1.

Elaborative transformations work in the opposite direction. That is, the basic units of description are broken down into their constituent elements, and any higher level organization is lost. An elaborative transformation works to add in detail that was not present at the higher level of description. The elaborative flows are shown by the white arrows in Fig. 15.1. Some of them correspond to what might be seen as the response side of a stimulus-response flow of processing, such as those from Morphonolexical to Articulatory and from Object to Limb, but others operate purely within the architecture, with information flowing from the Implicational to the Propositional subsystem and thence to the Morphonolexical and Object subsystems. The hierarchical, structured nature of information is a central concept in ICS, and it is discussed further in section 15.4.2.

Coupling these iterative and combinative flow patterns with each subsystem's independent access to memory means that ICS is able to encompass the rich variety of human cognitive experience. It also means that there is seldom just one way to do anything, which threatens to make CTA using the full model a frustratingly noncommittal exercise. Luckily, the simplest way of processing information through the architecture is associated with greater speed of execution, less ambiguity of behavioral output, and less subjective effort. Therefore, the more complex the alternate processing route required, the worse the results in terms of the conventional metrics of usability or task design.

A CTA in ICS is a more qualitative exercise than the quantitative assessments typically resulting from models such as those in the GOMS tradition. Every CTA results in a family of different cognitive task models (Fig. 15.2), and each model is a snapshot of the cognitive activity of a particular user who is at a specific phase in the task, has a given level of expertise at the task, and is working within a specific task context. The three levels of user expertise shown in Fig. 15.2 are simply nominal, representing people who have no previous experience of the task in question (the "novice" level), a great deal of experience (the "expert" level), or some experience (the "normal" level, which would be the level of a typical person). Although the very short term dynamics of cognitive activity are represented within a single model (which gives access to conventional interface design information such as "Will users notice this button?"), consideration of the relationships between models for different phases of activity within a task can address the short-term dynamics of task performance (e.g., "Will users do the elements of this task in an appropriate order?"). Finally, the relationships between models for different levels of user expertise can inform the analyst about the longer term dynamics,

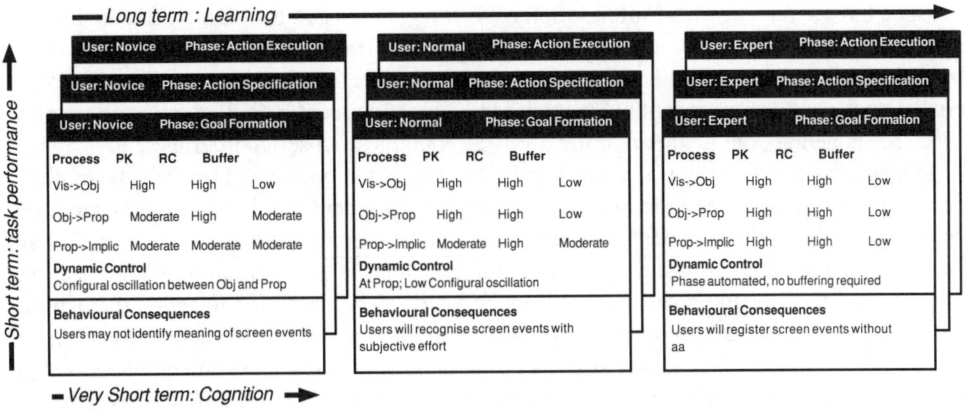

FIG. 15.2. A family of cognitive task models for novice, normal, and expert users at three phases of task performance. Each model details the configuration of processes, the status of their procedural knowledge, record contents, and requirements for memory access, together with the implications for dynamic control of cognition and likely behavioral consequences. A single model represents the very short-term dynamics of cognition; a stack of three phases represents the short-term dynamics of task performance; and changes across the three stacks represent long-term dynamics, or task learning.

such as likely patterns of learning, development of preferred patterns of use, and persistent errors.

15.3 THE PROCESS OF A CTA IN ICS

The family of cognitive task models illustrated in Fig. 15.2 indicate that conducting a CTA in ICS requires the analysts to identify the level of expertise of the individual they are concerned with as well as the phase of task performance. Within a phase, the analysts have to identify the configuration of subsystems that will be involved in the flow and hence the transformation processes operating within each subsystem (indicated in Fig. 15.2 by the "Process" column in each model). For each process, the analysts must consider the degree of existing proceduralized knowledge that is available (the "PK" column) and the availability of experiential image record contents that may be recruited to assist in strengthening the incoming representation (the "RC" column). In situations where there is little proceduralized knowledge and weak record contents to support the transformation, the process will attempt to operate in a buffered mode. In this mode, the process, instead of working directly on the incoming information supported by record contents, works solely on record contents.

The fourth column in the models (Fig. 15.2) describes the likelihood that each process will attempt to work in buffered mode. This is an important indication of the subjective effort that will be felt in task performance, because it operates like a bottleneck in the flow of information and gives rise to focal awareness of the information being processed. Subjectively, the individual feels that he or she is having to concentrate on whatever aspect of the task is being buffered. Tasks that have little proceduralized knowledge may require the buffer to move back and forth between two or even more subsystems (oscillation), making the individual feel that he or she is having to switch attention between several different aspects of the task. In contrast, a task for which proceduralized knowledge is available for the complete flow may not require buffering and may be able to be performed automatically, or at least without the individual needing to concentrate on it. Identifying the need for buffering allows this degree of dynamic control of cognition to be assessed.

Once these three classes of information have been established for each phase, the interrelationships of the phases of activity can be considered. The three phases of goal formation (realizing that something needs to be done), action specification (working out how to accomplish a goal), and action execution (carrying out actions to meet a goal) can, in the simplest tasks, be sequential, but neither real life nor ICS require a strict serial ordering.

In practice, an individual is usually doing several things at once, and there may be multiple cues in the world or in the individual's active cognitive content that indicate a new goal (see chaps. 14 and 19). While actions are specified to meet these goals, the task context may change, leading to a modification of the goal (i.e., a further round of goal formation). While actions are executed to meet the goal, the task context needs to be evaluated to refine the actions, modify plans (i.e., further action specification), and perhaps form subsidiary goals (i.e., yet more goal formation). Within any one task, it is more likely that these three phases will be richly interleaved than serial, and it is also conceivable that the phases of different tasks may be interleaved as the individual divides his or her effort between them. Interleaving of tasks and interleaving of phases of activity may be straightforward if the configurations of the processes that they require differ, and the tasks and phases may even be able to continue in parallel if they do not compete for the same resource (i.e., the same transformation processes or image records). However, as soon as competition occurs, then the architecture needs to be able to shift between different flows of information, adding the costs of disengagement and reengagement.

To summarize, CTA in ICS consists of identifying four aspects of cognitive resources:

1. The configuration of transformation processes in the flow of information between subsystems.
2. The proceduralized knowledge available for each process (and potential change).
3. The image record contents available within each subsystem (and potential change).
4. The dynamic control in terms of locus of buffer and interleaving of phases or tasks.

The following sections show how these four aspects can be determined in practice by giving examples of the situations that each one might be involved in. These sections assume some familiarity with ICS terminology and concepts (the general operation of an ICS subsystem is described in the appendix to this chapter).

15.4 CONSTRUCTING A CTA

Constructing configurations of ICS transformation processes is the first step in a CTA. Once each process has been noted, the next step is to consider the degree of proceduralized knowledge available to support the processes' actions and what record contents might be available to supplement it. Because knowledge acquisition is the development of record contents and then proceduralized knowledge, it generally results in one process learning to produce an output directly from its input without needing to engage in loops with other subsystems or accessing its own image record contents. The configurations for normal task operators and then experts will usually be compressions of the novice configuration. By starting with the novice pattern, as in the examples in the appendix, the configurations for the more practiced classes of operator can be inferred relatively easily.

15.4.1 Learning and Errors

The characteristic patterns of learning and skill acquisition also emerge directly from this process of modeling. Development of record contents and proceduralized knowledge will be greatest in the subsystems that have the most buffering, and reliance on memory access in these subsystems means that the task is effectively "controlled" by the class of information that they operate on. Errors and confusions between similar entities within the task will be most likely to occur at this level. Changes in the task that alter the information being used to control the task, perhaps by modifying the nature or grouping of the task entities, will have differing degrees of impact depending on the relationship between the changes and the level at which task control resides. If the changes are at the same level of information as the task control, then they will be most disruptive, since all that has been gained by buffered processing and the construction of new record contents will have to be relearnt. If the changes occur at a different level, then they may be easily accommodated, since all that will have to be relearnt will be mappings in what were presumably well-proceduralized processes.

For example, suppose that someone has learnt to associate several task steps by learning an Implicational relationship between them: They might all share some common functional attribute such as "deletion" or "coloring." This relationship would be represented in ICS by the development of proceduralized knowledge in the Implicational-to-Propositional process, so that forming the implicational representation of deletion or coloring would allow the Implicational-to-Propositional process to derive abstract propositional task steps. The Propositional subsystem would receive this information stream, subsequently using its Propositional-to-Morphonolexical process to output verbal names for the task steps. Redesigning the task to change the verbal forms required for task execution will have little detrimental effect, provided that the task attributes required for goal formation are not altered, because all that the individual will have to learn will be new Propositional-to-Morphonolexical mappings (which might already exist). Adding a new task step that fits the same implicational form as the existing steps will also not be problematic, although it might be occasionally omitted while learning is taking place. Changing the sequence in which the task steps have to be executed or interspersing new steps that do not fit the same implicational mold as the learnt steps will be very detrimental, since the controlling implicational representations will have to be relearnt. The operator will need to regress to the novice level for this task yet will also have existing but now incorrect record contents that must be overcome in learning rather than having nothing interfere, as a true novice would.

15.4.2 Controlling Attention

Tasks that require the operator to notice some event in the world and to carry out a single response are nice to model but rare in practice. It is more often the case that the operator must search the world for task-relevant information to form task goals; specify the actions with reference to other, perhaps dynamic attributes of task entities; and execute their actions in a changing world while monitoring the behavior of other task entities to ensure that their planned actions and inferred goals are accurate. We can do all this because we can direct our attention around the world, actively seeking information from different sensory streams and from different entities within those streams. Attention is task motivated and is task directed. The ICS architecture allows us to model this without requiring an attention module or having to postulate a central executive that controls and schedules actions. As has been described, the location of buffered processing accounts for the level of information that provides the sense of focal awareness, but the explanation for how we can control attention within this level relates to the abstraction and elaboration of information as it is transformed by the processes in the different subsystems.

The generic information-processing architecture of ICS subsystems allows us to use a generic representational notation for the structure of information regardless of its class or content (Fig. 15.3). Whatever the class, information consists of basic units of meaning. At the Object level, for example, the basic units are entities in visual space; at the Propositional level, they will be the abstract identities of entities and their interrelationships. In a conventional

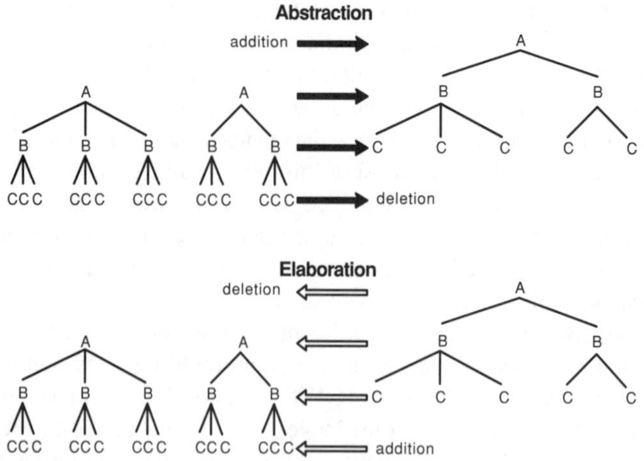

FIG. 15.3. Information at all levels in ICS is composed of basic units of meaning (B) that have constituent substructures (C) and can be clustered into higher order assemblies (A). Abstraction between subsystems uses the assemblies to derive the basic units of the output representation; elaboration uses the constituent substructures. Abstraction adds new higher order assemblies in the output representation while losing the detail of the input's constituent structures; elaboration adds new detailed constituent structures in the output while losing the higher order assemblies from the input.

computer interface window, the Object basic units might be graphical icons, checkboxes, text labels, or parts of the window frame itself (e.g., Word Icon–PDF Icon–Word Icon). At a Propositional level, they might be the identities recognized from these Objects and their interrelationships (e.g., {PDF document–version-of–Word document}–earlier-version-of–Word document), or they might be the steps of a task that the user is carrying out (e.g., {Select Icon–Change Window–Select Menu Option}). These basic units will each have a detailed constituent structure. Objects consist of parts that can include other smaller objects and ultimately parts such as edges, corners, surfaces (e.g., one form of Word icon consists of a blue italic *W* on a document icon that holds some small lines of text); Propositions can be subdivided into attributes and relationships (e.g., task steps can be broken down into subtasks and their sequencing down to and even beyond the keystroke level). Basic units also group together into higher order units, or assemblies (to give an A-B-C hierarchy of structure, as shown in Fig. 15.3). Objects group together to form regions of a scene (the icons are part of an array, which is part of a window, which is part of the display); Propositions assemble together to form sequences of related knowledge (the task steps are part of an overall goal).

Each transformation process changes the qualitative nature of the information being output, but the chunks of information that correspond to the basic units of the output representation relate to the input representation in a straightforward way. A process that is abstracting information (the black arrows in Fig. 15.1) is using the assemblies from its input representation to form the basic units of its output representation, whereas a process that is elaborating information (the white arrows) is using the constituent structures of its input to form the basic units of its output representation. Each process discards information. In particular, the output of an abstractive process does not include the constituent structure of its input representation, and the output representation of an elaborative process does not include the assembly units of its input. Each process also needs to add in new information that was not present in its input. An abstractive process needs to construct new, higher order assemblies to group together the basic units, and an elaborative process needs to construct new, more detailed constituent structures for each basic unit it is outputting. This "new" information is to some extent inherent in the input information, but it is effectively the contribution of memory, either through use of record contents or proceduralized knowledge. Cognitive processing is the repeated application of memory to refine information.

When a subsystem is receiving two inputs simultaneously, one might be from an elaborative process, the other from an abstractive process. One will be rich in the detailed constituent structure of the information (because it is derived from the basic units of information used in the prior process) but have weak assembly information (because it has been inferred from memory); the other will have detailed assembly information but be weak in the constituent structures. Together they may blend to complement each other. They can only do this, though, if their basic units are coherent; where they are, the detailed hierarchical information may overwrite the weaker information. Basic units that are in one stream but not the other may not be included in the blended representation or may be included but with either a weak degree of detail or a weak linking in to the organization of the other basic units.

This idea of hierarchical structure, abstraction, and elaboration allows the model to include an analysis of when streams of information are coherent and can blend and when they are incoherent and must be alternately processed or processed separately by different processes within a subsystem (Barnard & May, 1995). It also gives the model a way of representing thematic changes in processing such as those that reflect a change in attentional focus within a subsystem. Focal awareness (and hence attention) is determined by the location of the buffer, and the content of awareness is the information that is being buffered. Attention shifts between different classes of information reflect changes of the buffer between subsystems, as described earlier, but when attention shifts within a subsystem, the unit of information that is being buffered changes instead. Since the subsystem itself cannot alter its inputs it must follow from a change in the information arriving on the input array. There are consequently two types of attentional shift: an exogenous shift that is caused by a change in the content of an abstracted stream, driven ultimately by a change in the external world, and an endogenous shift that is caused by a change in the content of an elaborated stream, driven by internal processing.

Both exogenous and endogenous shifts in attention can also be modeled using the ideas of hierarchical structure. Although there can be several basic units at a level of representation, one will be the current topic of processing and hence the psychological subject of awareness in the subsystem (and if buffered, of the whole architecture). The other basic units are in awareness (but less focally), as predicate elements of the topic. These terms are taken from the systemic grammar of Halliday (e.g., Halliday, 1970) but can be interpreted generally to cover all classes of information, not just linguistic representations. Also in awareness are the topic's assembly structure and constituent structure. Changes in attention can be made to any of the other basic units (i.e., within the predicate structure), to the assembly of the topic (i.e., "up" the hierarchy in an abstraction), or to the constituent structure of the topic (i.e., "down" the hierarchy in an elaboration).

15.4.3 Graphical Notation to Represent Attentional Shifts

Shifts in attention can only occur when there is a change in the stream of information being processed. Endogenous streams can change when the output from a process that is feeding into the subsystem retopicalizes on a different element because of a change in the activity of the process (i.e., a different rule within the process gains control of the output). Endogenous shifts in attention are therefore are a consequence of activity in at least two subsystems. Figure 15.4 illustrates the nature of attentional shifts in the object level of representation when a computer

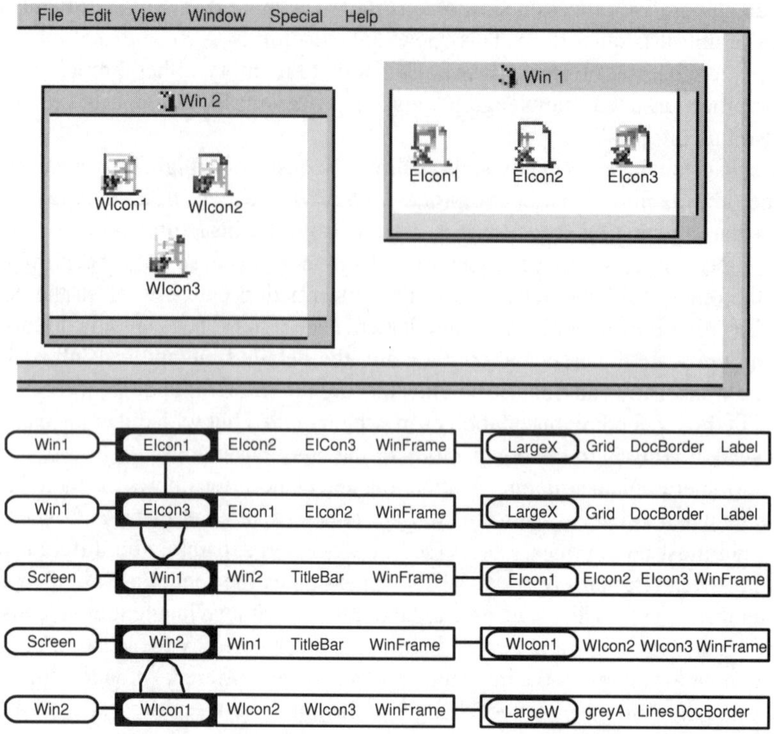

FIG. 15.4. A transition path diagram (TPD) representing successive shifts in attention around a simple window interface. Each row in the TPD represents a moment in time, with the assembly, basic units, and constituent structure that are in awareness at a particular level of representation (here the object level). Time is represented vertically, so reading the diagram from top to bottom reveals successive shifts in attention, first to an element in the predicate of the initial topic, then to their shared assembly unit. There is then a transition to a predicate of this higher order unit, followed by a transition down into its substructure.

user attends to various elements of a graphical user interface. Each row of the figure shows the elements of the representation that are currently in awareness: at the left, the assembly of the current topic; on its right, the topic and its predicate basic units; and at the right, the constituent structure of the current topic. Time is represented on the vertical axis, with each successive row representing a new state of attention in the subsystem.

The nature of the shift in attention between each row is indicated by the symbol joining the successive topics. A vertical bar represents a shift to an element of the predicate structure (that is, within a level). Note that the assembly unit remains the same because all basic units within the attentional focus are part of the same higher order unit, but also note that the constituent structure in awareness alters to reflect the detail of the new topic. An abstraction up the hierarchy to make the previous assembly unit the new topic is marked by a U-link. The previous basic units become the constituent structure, and new predicate elements enter awareness, together with a new assembly unit that groups them together with the topic. An abstraction down the hierarchy into the constituent structure of the topic results in the previous topic becoming the new assembly unit. One of the elements of the constituent structure becomes the new topic, with the others forming its predicate, and the new constituent structure depends on the topic.

Figure 15.4 represents shifts in visual attention around a scene. The figure represents the transitions that are made over time and so is called a Transition Path Diagram (TPD).

Although the basic units of the representation move up and down, the quality of the information represented does not change: It remains at the Object level. The process mapping this into a Propositional output, though, will be changing the quality of the information as well as its level—as will the process mapping it to the Limb subsystem. TPDs can be drawn for any of the nine levels of information processed in ICS, although the nature of the units represented will vary. In a Morphonolexical TPD, the units are phonemes or sound entities. In a Propositional TPD, the units are entities, their properties, and their relationships. In an Implicational TPD, the units are schematic patterns of affect and feeling.

TPDs are of most use in CTA in supporting reasoning about the need for oscillation between different configurations within task performance and in determining how well the structure of the information presented by the world conforms to the expectations of the operator, as represented by the task goals. Parallel TPDs drawn for Object and Propositional levels can allow the analyst to check that changes in the visual information available to the operator (or the operator's attentional shifts around an unchanging scene) match changes in the task goals (at the Propositional level), check that appropriate information is available just when the operator needs it, and determine whether this information might in fact force an exogenous shift in attention and so aid in goal formation. Parallel TPDs drawn for the Morphonolexical and Object subsystems can show how well sound and vision will support each other (that is, do sound events occur at appropriate times, given the influence of the POP loop on the Object structure?). Further details about the use of TPDs can be found in Barnard and May (1999).

15.4.4 Applying CTA in ICS

Drawing TPDs is helpful for an analyst who is becoming familiar with the ideas of CTA in ICS. With practice, though, the ideas about structure become second nature, and the notation plays a secondary role (namely, it helps the analyst explain the reasoning behind the analysis to others on the design team or to those who have commissioned the design). Similarly, the abbreviated lexical notation for representing configurations of processing can help an analyst spot duplicated resource requirements and possible points of blending or incoherence in the flow, but for a more experienced analyst, it will be used more for communicating and recording the CTA rather than executing it. The analytical steps of identifying the configuration of processing resources, and then their memory requirements, for the different phases of task performance and different classes of users, remains the core of the technique.

With any analytical technique, there will always be a degree of skill involved in its application. For instance, an experienced analyst will be able to draw on a history of similar analyses and jump to a conclusion without apparently engaging in any real analysis. The question of what role the theory or the analytic method play, compared with the skill of an experienced analyst, can be answered by the proof-of-concept expert system that was developed in the Amodeus projects. This was a collection of production rules that embodied a limited amount of knowledge about the domain of HCI and ICS and was thus able to produce CTAs for several scenarios of common HCI problems (May, Barnard, & Blandford, 1993). Three classes of rules were written: The first class asked questions about the scenario to aid in building an abstract task description; the second embodied ICS principles and used this task description to produce a family of cognitive task models (CTMs); the third deduced some behavioral consequences from the CTMs. No learning was possible (the rules were not modified automatically), and there was no skill in using the system (apart from learning how to describe the task when applying the first class of rules), and despite the limited nature of the third class of rules, it was found that the system could make interpretable and sensible predictions about novel scenarios other than those that its rules had been written to encompass.

15.5 FUTURE DEVELOPMENTS

The description of CTA in ICS in this chapter is based on its use in the domain of HCI, but the technique is not in principle limited to this. The descriptions used the terms *operator* and *task device* rather than *user* and *computer* or *interface* to reinforce the generality of the approach. The human cognitive architecture existed long before computers and graphical user interfaces, and a method that describes task performance in cognitive terms should be applicable just as much to the design of real artifacts and tools and to communicative tasks where several people exchange information without using any artifacts at all. Communicative tasks are often mediated by IT, though, and methods originally scoped to deal with individual HCI tasks are being examined to see how they can change their focus to be applied to collaborative situations in which group goals must be shared and other individuals' knowledge and goals must be recognized if task performance is to be successful.

In the introductory section, we noted two current areas of research that aim to extend the scope of CTA in ICS: Syndetics, which links the cognitive theory underlying ICS with ideas from formal methods of systems analysis, and the joint development of TKS and ICS to address collaborative and multiple task performance. Syndetics represents agents in an interaction as discrete, independent interactors that exchange information and attempt to coordinate their activity, whether they be humans, machines, or software components. It has been used to model conjoint human-machine-software systems such as intelligent whiteboards with gesture recognition, and it has been argued that it is potentially extensible to command-and-control situations as varied as a missile defence room on a naval warship and the routing of the Roman armies by the Carthaginian army at Cannae (Barnard, May, Duke, & Duce, 2000).

One of problems in the wider field of task analysis is relating cognitive task analysis to higher levels of analysis. For example, how do constraints that work in the individual mind and the technologies it has to support its work affect the behavior of a group or team? Going a step further, how do constraints at the level of the group or team affect the wider organization? The problem is one of communication between different levels of analysis. Traditional forms of task analysis embed task analysis for an individual within a team task analysis, which is in turn embedded in an analysis of the wider workflow within the overall organization. Attempts to build models that work across multiple levels of analysis have often foundered on the difficulty of relating the analytical approaches that apply at these different levels. Syndetic modeling (Duke et al., 1999) and the framework of systems analysis that supports it (Barnard et al., 2000) attempt to resolve this problem by couching the input and output of each level of analysis within the same notational form and by identifying two kinds of theory.

What we will call a type 1 theory concerns the behavior of a particular system at a particular level of analysis. All systems, in abstract terms, consist of a number of entities that interact to constrain the overall behavior of the system, and the task of a type 1 theory is to explain the way that these interactors behave. In task analysis, there will be different type 1 theories for explaining an individual's cognitive behavior (such as ICS), the behavior of a group or team (perhaps a task analysis model), and the workflow in an organization (perhaps a social psychology model). No system operates in isolation, however, for systems must themselves interact with other systems. A human interacts with other humans, with computers, with tools, and with organizations, among other entities. As well as consisting of interactors, then, each system is itself an interactor. To understand the workings of any system of interactors, it is necessary to know the external, observable behavior of other systems with which it interacts, but it is not necessary for a single type 1 theory to be able to model the internal behavior of those systems. That is the job of other type 1 theories scoped specifically for those interactors. A theory of the second kind (a type 2 theory) specifically relates these levels of analysis to one another. In task analysis, then, there needs to be a type 2 theory to explain how inferences about an individual task analysis can be taken up by the theory that addresses team tasks and another type 2 theory to explain how the results of team task analyses can be taken up by organization-level theories.

Recognizing the need for both types of theories is a crucial step, because it allows a theory that is restricted to a particular level of analysis to obtain information and insights from beyond its own analytic scope (by using type 2 theory to communicate with a type 1 theory of a different level of analysis). Equally, the inferences and conclusions of a type 1 theory at one level can be communicated to and have an effect on other levels of analysis. The current situation is that type 1 theories are far more common than type 2 theories. Indeed, there seems to be a reluctance by the theorists who have developed type 1 theories even to recognize that building type 2 theories to relate their work to that of other theorists is advantageous and strengthens their own approach.

The potential benefit that the syndetic approach offers is in understanding the way that these different levels relate together via type 2 theoretical interfaces. Using the same principles of abstraction and elaboration that ICS applies to the exchange of information within and between its cognitive subsystems, it is possible to argue that each level of theoretical analysis has a basic unit of analysis (whether this be a software entity, cognitive process, person, team, or organization) and a constituent structure and that these units can be assembled into higher order units. Successful communication between levels of analysis requires the recognition of mappings between the basic units of one level and another level's constituent structures (for an abstractive mapping) or assembly structures (for an elaborative mapping).

Just as a cognitive model such as ICS works with a configuration of mental processes, so a group or team can be regarded as configured to support particular interactions within the team and between that team and another. Figure 15.5 shows a purely hypothetical analysis of a manufacturing company, with its inputs and outputs and internal departmental structure. When the structure of one of the departments, which deals with incoming raw materials, is considered in more detail, the functions that it carries out are noted. Each of these functions is carried out by a human employee, whose cognitive processes can be modeled by ICS subsystems. The number and arrangement of the components of the manufacturing company and the raw materials department both mirror that of ICS. Realistic systems would not, of course, have this form. For figurative purposes only, the arrangement serves to make clear the analogy of the mappings between different levels and also the potential for applying, by extension, the same basic notations (e.g., configural definitions and TPDs) to the analysis of interactions within each level.

We can think of the parallels at the level of the more abstract concept of "interactors." At one level of analysis, the interactors are mental processes within the cognitive system; at the next level, they are people in a team; and at the highest level depicted, they are whole departments. The hierarchical decompositions, configural definitions used for ICS and the TPDs, could all potentially be applied at each level. In Fig. 15.5, each level contains what amounts to a control loop, where the behavior of the basic unit is partly determined by interactions among its constituents and partly constrained by the assembly of external interactors that the system engages with.

In this syndetic analysis, the manufacturing company is an interactor that exchanges information (and physical objects) with other interactors at the same level of granularity. These are basic units at an economic level of analysis. The constituent structure of the company consists of the major functions that need to be carried out, and these functions become the basic units of a lower level of organizational analysis. At this level, the company is the assembly, the departments are the basic units, and the roles carried out within each department are the

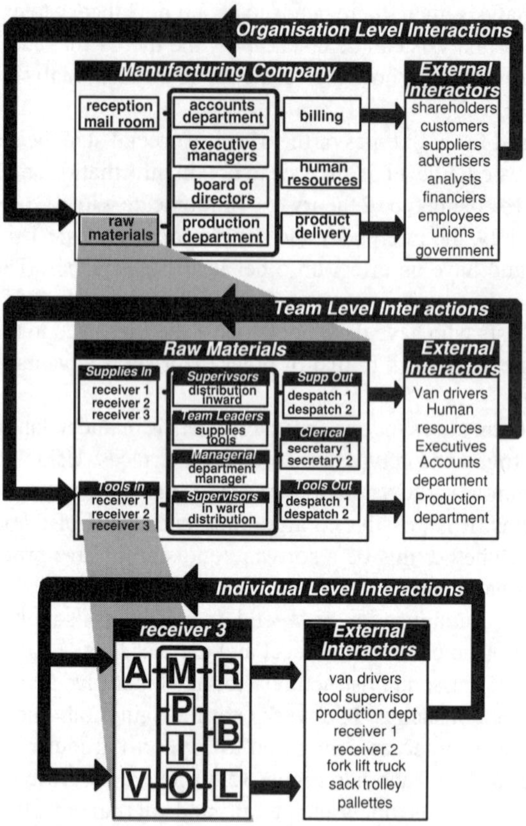

FIG. 15.5. In a syndetic model, a manufacturing company interacts with its world (consisting of other interactors at the same level of granularity), and it consists of a number of departments. Each department fulfills a number of roles, and each role is carried out by a human. The humans interact with their world (consisting of other interactors), and their cognitive behavior is modeled by postulating a number of subsystems.

constituents. Organizational theory provides the appropriate type 1 body of theory. At the next level of analysis down, the roles map onto humans, who are the basic units, and each unit interacts with other entities such as other people, computers, tools, and so on. Within each human, the constituent structure of cognitive subsystems can be seen, and so on. If desired, computational models of the activity within a process could be postulated, even neural models of their implementation within the brain. The key is that different theories and methods of analysis can operate at each level, but communication between them is only comprehensible once the mappings between units of analysis are made explicit. If mappings cannot sensibly be made, then it is likely that at least one level of analysis is missing, in which case the mappings are attempting to bridge too great an abstractive or elaborative gulf. Some concrete examples of parallels between system behaviors at each of these levels are outlined and discussed in Barnard et al. (2000).

This division of analysis into levels, each with its own body of type 1 theory and way of dividing the world up into analyzable entities, helps us to resolve the otherwise problematic absence of a task analysis phase in the CTA in ICS approach. Because ICS is a cognitive model of an individual, it does not make any theoretical sense to try to extend its scope to include task analysis, especially since there are already many well-explored methods and theories in this area. No one theory is grand enough to explain everything in a manner that is economical enough to be applicable. What is required is a coherent interface between levels. To link a cognitive task analysis with wider methods of task analysis, the interface can be an approach in which the basic unit of analysis is a human, the constituent structure is cognitive in nature, and mappings can be built between this cognitive structure and the information levels of ICS. The task analytic method can then be strengthened by abstraction from the outputs of ICS (being given a rationale for the cognitive behavior of its human), and ICS can be strengthened by elaboration from the outputs of the task analytic method (being given a rationale for the structure of the tasks being modeled).

TKS (Johnson & Johnson, 1991) is a task analytic method that provides exactly the sort of type 2 theoretical interface between system levels that we need. It produces a task description in terms of the structures of knowledge of task operations and task entities that an individual needs to possess to work with a well-defined subset of external interactors. TKS effectively specifies what is required to make the lowest level of control loop shown in Fig. 15.5 function, whereas ICS specifies only the internal interactors of the mental architecture. TKS includes notions of abstraction of task knowledge that is grounded in the external interactors, be they other members of the team or technological artifacts available for use or under design. These can then be related to the abstraction and elaboration of information within ICS and its re-representation at multiple levels of processing. In TKS, the controlling knowledge about an entity or a task step is contingent upon the task context. This itself originates in the properties of teams and artifacts. In ICS, the finer levels of "cognitive control" are similarly constrained by the location of the buffered process or the level of information that is in focal awareness. It is hoped that bringing these two approaches together will offer benefits to both, by offering ICS a link to a well-tested form of task analysis and providing TKS with analytical tools for reasoning about the abstraction and elaboration of task knowledge. These benefits are thought to be important in allowing the two approaches to be extended to situations where a person is performing multiple tasks at the same time or where several people are collaborating on a single task. In these situations, the analysis needs to take into account the conflicting demands of two task sequences and knowledge structures or the demands of the communication of task knowledge and of the assessment by one individual of the states of knowledge and intentions of others engaged in the task. This requires that both the prior task analysis feeding into ICS and the cognitive basis of task knowledge be given firmer foundations.

Developments along these lines show that, although CTA in ICS is a helpful problem-solving technique that can provide support to those who need to understand human cognitive processing in task performance, it is also an extensible technique that can deal with the changing demands of design. Its supportive nature and its extensibility are both consequences of the close relationship between the application representation (the methods presented in this chapter) and the underlying science base (the cognitive theory of ICS). These form a reciprocal pair, analogous to processing loops in ICS, with advances in one driving developments in the other. Research and application of task analytic techniques are not only of practical importance but have a crucial role to play in feeding back information to cognitive psychology and theories of systems analysis.

ACKNOWLEDGMENTS

The work reported in this chapter has been supported by the ESRC as part of the TICKS project and by the EU Training and Mobility of Researchers Network TACIT.

APPENDIX A: A SPECIFICATION OF ICS

15.A.1 COGNITIVE ACTIVITY WITHIN ICS

Although ICS subsystems operate on different classes of information, they each carry out the same general operations, as illustrated in Fig. 15.6. The architecture of each subsystem consists of an input array, an image record, and a number of transformation processes, one of which is a special copy process. The function of each of these is described below in general terms, and key concepts used in CTA in ICS are introduced as appropriate. These include blending of information streams; experiential, common, and active task records; buffering; and locus and oscillation of dynamic control.

15.A.1.1 The Input Array

All incoming information streams, whether from sensory organs or other subsystems, arrive at the *input array*. Streams originating from different subsystems will all contain the same class of information that the receiving subsystem specializes in, and so their origin is not really relevant from a processing point of view, and they can be considered as blending or combining at the input array. For analytical purposes, however, it is worth noting that the quality of the attributes contained within streams of different origin may well vary, and so the adequacy of the streams for further processing may differ.

15.A.1.2 Copying Information to the Image Record

Once information is present at the input array, the processes within the subsystem can operate on it. These processes are independent and can operate in parallel, each producing a different class of information as output. One process in all subsystems is a straightforward copy operation, which adds all incoming information to the subsystem's image record untransformed. The image record thus preserves a record of all information that has ever been received by the subsystem and so can be thought of as supporting *long-term memory* storage functions. The copy process is subjectively associated with a diffuse sense of awareness of information, the actual qualia of the experience depending on the class of information processed by the subsystem. (In Fig. 15.1, general summaries of the subjective nature of these qualia are given in italics.)

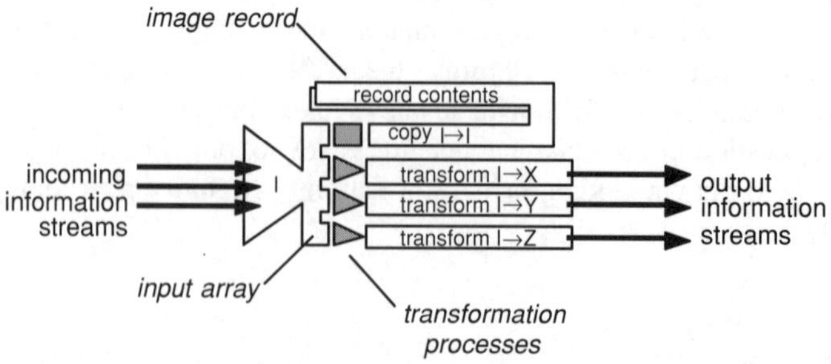

FIG. 15.6. The common architecture of each ICS subsystem comprises an input array where incoming streams of information blend; an image record; and several transformation, one of which is the copy process. Each transformation process produces a different class of output information.

15.A.1.3 Transformation Processes

In parallel with the copy process and with each other, the remaining transformation processes operate to produce their specific output representations. Each individual process can only produce one class of information as output, and there is only one of each class of transformation process within a subsystem. Each process can be thought of as bundles of mutually competing production rules, with "if" conditions contingent on the contents of the input array and "then" consequences that describe the output to be produced. The "if" clauses of several rules may match the incoming information to varying degrees, but in a winner-takes-all fashion, only the rule with the strongest match is able to express its "then" clause. Rules whose "then" parts describe different classes of information do not inhibit each other, however, and so exist within a different bundle (transformation process). In practice, it is not necessary to actually write out or speculate about the exact content of these rules. Unlike a GOMS model, an ICS model is not at a computational level and does not "run." All that is needed is to decide the degree to which the rules exist for a particular situation. In ICS terms, these rules are proceduralized knowledge: The existence of a rule whose "if" clause exactly matches the task-relevant attributes of the information on the input array and whose "then" clause is unambiguous and well defined corresponds to a high degree of procedural knowledge. The existence of one or perhaps several rules with close "if" matches and less well defined "then" clauses corresponds to progressively weaker proceduralized knowledge within that process.

15.A.1.4 Streams of Information

The mutual inhibition within transformation processes means that each one can only produce one stream of output information at a time and hence can only operate on one stream of information at a time. Different processes may be operating on different streams, however, and so a single subsystem can in principle produce several different outputs at once, from the same or from different incoming streams. When there is a single stream of information on the input array, the existence of independent processes means that it can be transformed into multiple forms for subsequent processing in other subsystems. Information thus becomes represented in different forms throughout the architecture and is recorded in these different forms in long-term memory.

15.A.1.5 Blending of Streams

Because there may be more than one stream of information arriving at the input array, processing may proceed on their blended product. Blending is only possible, however, if the streams really are coherent (i.e., they contain information about the same entities and one does not temporally lag behind the other). If they are not coherent, they will still be represented on the input array and copied to the image record, but the transformation processes will only be able to use one stream at a time. (Information-based constraints on blending are discussed in the section 15.4.2). The stream that is chosen depends on the strength of the proceduralized knowledge within each process, with the stronger "rule" choosing the stream that satisfies its "if" clause conditions.

15.A.1.6 Development of Proceduralized Knowledge

The outputs from these transformation processes are representations of information in forms that can be operated on by other subsystems. Figure 15.1 represents the relationships between subsystems as arrows, suggesting that information flows down channels between subsystems. In fact, it is probably better to think of the output parts of each transformation process as being physically identical to the input arrays of the receiving subsystem. This is because it is necessary for a process to know how "useful" its product has been, so that the rule that produced it can be strengthened relative to its competitors and procedural knowledge can consequently be acquired and refined. If the output side of a process can tell when its activity is being picked up by the processes operating within its partner subsystem, then it can carry out this learning in a back-propagating manner.

15.A.1.7 Accessing the Image Record Contents

The description so far covers situations in which there is adequate procedural knowledge within a process to produce a usable output. When this is not the case, there are two possible explanations: Either the incoming information is novel and no procedural knowledge exists to allow an appropriate output to be produced or the incoming information is impoverished or degraded in some way and the appropriate rule cannot fire strongly enough to beat its competitors. In these two situations, the process has to recruit additional information in order to elaborate or strengthen the incoming information so that it can be identified as similar to some other experience for which there is procedural knowledge or so that gaps or errors in the information can be filled in to enable the appropriate rule to fire. Both of these tactics require access to memory in the form of image record contents.

Whenever a process is unable to produce a strong or useful output representation, it will try to access the image record contents to supplement the information available from the input array. This can be done because the incoming information is simultaneously being copied into the image record. As it is copied, similar patterns of information that have been recorded in the past are activated, with the activation being stronger the more recent the experience and the more similar the match. The processes are able to latch on to this activation and to process this amplification of the incoming information instead of the information itself. In effect, the subsystem processes something that it is reminded of rather than the information itself. A constraint on the use of the image record, though, is that only one process can access it at a time, so if two or more processes are unable to use the incoming information directly, they will have to compete for image record access and use it sequentially. This has consequences both within the subsystem, slowing down operations while the processes wait for each other, and for the overall flow of information, as the two (or more) output representations can no longer be produced simultaneously.

15.A.1.8 Oscillation Within a Process

Although the incoming representation will be constantly changing, there may be occasions when a particular stream of information remains constant. To prevent the architecture locking up so that a single transformation process continually produces the same output, rules can be treated as susceptible to fatigue and unable to produce output indefinitely. As a rule declines in strength, a competitor may be able to overcome it and so gain control of the input stream. As different rules within a process sequentially gain control of the input to produce their own output representations, the process oscillates between different input streams, and its output also oscillates. If one of the outputs is not task relevant, then converging input streams from other subsystems may effectively act to suppress it by strengthening task-relevant information on the input array, reorienting the overall flow to the task in hand.

15.A.1.9 The Active Task Record

Switching between input streams (and hence between different flows of information) is inevitably time consuming, as the output of the subsystem may take some small duration to become stable or strong enough for receiving subsystems to operate on. It would thus be very inefficient for the architecture to oscillate or interleave different tasks. If it is assumed that a recently active rule has not become completely deactivated, though, but can reach its firing strength more quickly than one that has been used less recently, this cost can be reduced. Returning to a previously active rule is known as use of the active task record (ATR), and it is also assumed that the ATR can be reactivated using weaker input information than was needed to fire the rule in the first place and that it can be reactivated without necessarily reaccessing any image record contents. Oscillating between an ATR for a poorly proceduralized transformation and a stronger one provides a mechanism for the short-term rehearsal of information, especially when two subsystems reciprocally exchange information with each other.

15.A.1.10 Experiential and Common Task Records

Over time, as more and more experiences of a similar type are laid down in a subsystem's image record, the activations that occur come to represent abstractions of the generic features of those experiences. Although each individual experience may still be accessible given a close enough cue in the form of the incoming information, it is more likely that the abstracted form will be revived. This distinction between experiential task records (ETRs), which refer to specific episodes, and the abstracted common task records (CTRs) is important because, in the context of the overall cognitive system, it reflects access to particular instances that the individual remembers encountering as well as access to generic classes of knowledge about entities or tasks. Thus, ETRs may support knowing what to do with an object because one did this last time, whereas CTRs support knowing what to do because that is what one does with this object.

15.A.1.11 Buffered Processing

A special form of image record access is possible when the information that is being revived is that which has been copied in the immediately preceding moments (i.e., the proximal ETRs). When a process enters into buffered processing, it is in effect integrating incoming information over some short time window, giving it the ability to work on higher order chunks of information within the stream. This may be useful when the rate of change of the incoming information is simply too high for the transformation rules to be able to reach a stable firing level, producing a great deal of oscillation between rules without a stable output. It also allows a process to work on temporally dynamic patterns within the stream rather than the moment-by-moment values of the stream. For example, a process within the Acoustic subsystem can use buffered processing to identify the repetitive nature of a tone that warbles between two pitches as such instead of trying to identify the individual notes for the Morphonolexical subsystem to later integrate into pairs. While buffered processing is operating, of course, the rule that is firing is strengthening and learning to deal with the particular pattern of input information and so may be able to produce the output directly from the input array in the future (perhaps following a period in which it accesses the record contents for ETRs and CTRs).

15.A.1.12 Location of the Buffer (Locus of Dynamic Control)

Buffering can be thought of something that processes attempt as a last resort when unable to produce a usable task-relevant output stream. Subjectively, buffered processing is associated with focal awareness (over and above the diffuse awareness arising from the copy process described above). Because the existence of a buffer within a subsystem prevents other processes from accessing the image record, it prevents other processes in the subsystem from operating on anything other than proceduralized knowledge and ATRs. This tends to limit the number of different information flows that can be present within the architecture. The integration of incoming information across a time window that is suitable for the buffered process also entails that the rate of change of the output representation is now governed by that process and hence by activity within the subsystem. In practice, this means that only one buffer can be in operation at any moment (this is the reason why we only experience one point of focal awareness at a time, giving rise to an apparently unitary "stream of consciousness," or the ego). Identifying the subsystem in which the buffer is likely to be located is therefore an important part of any CTA in ICS, because not only does it define the subjective nature of the task phase for the individual concerned, but if there are competing demands for the buffer from processes within different subsystems, the overall complexity of the task performance will increase rapidly. The position of the buffer is called the *locus of dynamic control* because it defines the way in which the task is apparently being controlled by changes in attentional focus. In Fig. 15.2, the fourth column in each model lists the demand by each process for the buffer. The worse the procedural knowledge and the lower the support from record contents, the greater the likelihood that the process will attempt to move into buffered processing. The complexity of dynamic control is noted below each model.

15.A.1.13 Oscillation of Dynamic Control

It should be noted that there is no controlling subsystem in ICS determining where the buffer is to be positioned (and so no iterative problem of controlling the controller). The subsystems are each functionally independent, and they do not control each other in the sense of telling each other how to behave; operationally, however, they depend upon each other's outputs, and so the flow of information between them determines the nature of dynamic control. The architecture is thus data driven: If one process has the buffer, processing in another may falter and be unable to proceed without itself becoming buffered. Its output thus becomes unstable, and if this output was being used by the buffered process, it may no longer be able operate on even a buffered stream. Competition within its own rules may then lead to another process becoming active and the buffer failing. This would enable the process in the other subsystem to enter buffered processing, and the sequence may repeat, resulting in an oscillation of dynamic control between the subsystems. It is symptomatic of complex, attention-demanding tasks that the individual's focal awareness must shift between several sources of information.

15.A.2 THE NATURE OF TRANSFORMATION PROCESSES

Figure 15.1 includes 17 arrows linking the nine cognitive subsystems and three indirect arrows involving information exchange via the body. These (plus processes within the two effector subsystems, Articulatory and Limb, which result in overt behavioral expression) are the only routes by which information can be converted from one class into another. It is important to remember that a subsystem does not produce the class of information that its name indicates but operates on it. Thus the Propositional subsystem does not produce propositions, but the Morphonolexical, Object, and Implicational subsystems do, through one of their transformation processes. Understanding what each process can do, then, is essential for identifying the correct configuration for the class of user and the phase of task performance. The following sections describe the nature of each of the transformation processes and present examples of their application within a CTA. To avoid repetition and to draw out some of the similarities between processes in different subsystems, we describe the processes in groups. We also note opportunities for the development of proceduralized knowledge (PK) and record contents (ETRs and CTRs).

15.A.2.1 Notation for Process Configurations

Before continuing, a notation used to describe the processes will be introduced. Long sequences of subsystem names become unwieldy, and so a variety of systems of abbreviation have been developed. That employed here uses a single letter for each subsystem and hence a pair of letters for each process. In every case but one, the code letter is the initial letter of the class of information received by a subsystem. For example, the visual subsystem is coded as V and contains the VO and VI processes (Visual to Object and Visual to Implicational, respectively). The exception is the Articulatory subsystem, which is coded by R, since A is used for Acoustic.

A configuration of several processes can be written concisely by concatenating the letter pairs and using a colon as a divider (e.g., VO:OP:PI, with P standing for Propositional). This is suitable for linear information flow but does not work if two flows combine, one flow splits into two, or two subsystems are operating in a reciprocal loop. The former two situations can be dealt with by bracketing flows. For example, {AM}{VO:OM}:MP:PI would represent the blending of Morphonolexical (M) streams from the Acoustic and a Visual source, the latter via the Object subsystem. Reciprocal loops can be expressed using triplets, such as POP for an PO:OP loop, PIP for a PI:IP loop, and PMP for a PM:MP loop. Thus a visuo-motor flow that includes a loop to the Propositional subsystem would be VO:POP:OL (L for Limb). The Body State subsystem (B) can also be involved in a sort of loop with the Implication subsystem (i.e., BI:IB or BIB), but only through the indirect detection of bodily effects resulting from the IB process. The effector subsystems produce commands to direct speech and motor behavior, and although there may be many different efferent streams specialized for different musculatures, these are not usually of much individual importance for CTA and are coarsely grouped together as Rs (speech production), Lt (typing), and Lm (other general motor behavior).

15.A.2.2 Structural Transformations (AM, VO)

The two dominant human senses are hearing and seeing, and these are dealt with at the lowest cognitive level by the Acoustic (A) and Visual (V) subsystems. The AM process and the VO process have similar roles of recognizing patterns in the raw sensory data. AM picks out phonemes from speech and notes from music; VO picks out objects from the visual scene. These two processes build abstract structural representations of the world, discarding raw sensory data. By adulthood, we have all become somewhat "expert" at these two processes, in that we have a great deal of proceduralized knowledge for phonemes in our mother tongue and objects in our environment. Skilled musicians may develop "perfect pitch" whereby they can accurately differentiate notes in absolute terms rather than relatively, and they also become able to identify rapid triplets and larger groups of notes that less skilled listeners may well hear but be unable to decompose.

In designing a task environment, though, it is important to remember that the task-relevant stream of sound or visual information is unlikely to be the only one available. The question then becomes how strong the proceduralized knowledge for the intended stream is, in comparison with likely irrelevant streams (or those of other ongoing tasks). The quality of the stream is also important. A visual display that is difficult to see or of low resolution and a sound on a low-bandwidth line may not be adequate for the proceduralized knowledge to operate on without accessing record contents or, in extreme cases, buffering. Although we are aware of the qualia of (say) redness or a baby crying by virtue of the copy processes within the A and V subsystems, it is rare in other circumstances for the buffer to be used by these two processes—unless it is necessary, for example, to determine the precise color of a surface, transcribe a rapidly played piece of music, or learn the phonology of a new language. In tasks similar to these, we become focally aware of aspects of the sensory world.

These processes are very likely to be involved in the phase of goal formation, when, as is often the case, cues in the world are needed to alert the user to the need to carry out a task, but they will also be involved in action specification and action execution in dynamic environments where the nature of the task might vary from moment to moment or where the individual's behavior will need to be controlled and coordinated (e.g., a VO:OL:Lm flow).

15.A.2.3 Implicational Transformations (VI, AI, BI)

In parallel with the identification of structure within the world, the VI, AI, and BI processes are continually deriving highly schematic information about the implicational status of the world and the individual's bodily state. A repeating high-pitched note, a flashing red patch of color, or a sharp, hot, or buzzing cutaneous sensation will be interpreted by the relevant processes as signifying danger or threat. Other sensory patterns may be interpreted as pleasant or rewarding. These affective outputs are produced independently of the structural information and so do not contain information about the entity in the world that is threatening (or pleasant); however, they would include varying degrees of spatial information (VI, the most; AI, the least), allowing a subsequent search to be directed to the approximate location. These processes may underlie the apparently preattentional direction of processing toward self-relevant information, such as the well-known *cocktail party effect* whereby one can hear one's name in a noisy environment.

As with the VO and AM processes, these are of most importance in the goal formation phase, but they can also be used in controlling and coordinating action specification and execution in dynamic environments. Skilled operators may develop proceduralized knowledge that associates a particular pattern of sounds with correct functioning of their machinery and be sensitive to unusual sounds or to changes in the sound without being able to put their knowledge into words. They will just "feel" that something is right or wrong, "feeling" being the hallmark of Implicational information. A novice at driving a manual-transmission car has to attend to the sound of the car engine to know how fast it is revving whereas a normal or expert driver will change gear "without thinking" about it (a characteristic of a completely proceduralized information flow, here AM:ML:Lm, that does not require any buffering or image record access).

Note that the Body State subsystem produces Implicational information, but unlike the other two sensory subsystems it does not produce a "structural" level of information. This is because Implicational information is in fact its structural level; bodily sensations have very direct meaning for us and always convey affective information. Skilled use of olfactory sensation, as in wine tasting by an expert, is equivalent to the development of proceduralized knowledge for the BI process. Skilled use of tactile information in a case where the qualitative aspects of the information are crucial would also use the BI process, but a skill like reading Braille, where the exact nature of the sensations is crucial, is a planned process mediated by the feedback from Body State to the Articulatory and Limb subsystems (described below) and would not involve BI directly.

15.A.2.4 Effector Transformations (MR, ML, OL)

The two subsystems specialized for processing structural information are used to "drive" the effector subsystems by providing them with information streams via the MR, ML, and OL processes. Speech "in the head" at the Morphonolexical level is transformed into prevocal patterns at the Articulatory level by MR or, in the case of a skilled typist, into intended finger movements by ML. Visuospatial imagery of objects and one's body at the Object level is transformed into planned actions at the Limb level by OL. Because of their role in directing action, these processes are most often involved in the action execution phase of task performance.

As with the structural VO and AM processes, much effector activity is well proceduralized by adulthood, but novel devices (such as data gloves) can make us all into novices again while we learn how to offset or scale intended behaviors. Particular devices such as keyboards clearly require specialized procedural knowledge, which once developed allows a person to type a word simply by thinking it. Computer programs that involve key-bindings such as "Alt-F4" also require proceduralized knowledge if the individual is to execute the action directly upon thinking the command name "Quit." Unix users gradually develop the ability to associate two- or three-letter typing commands with quite complex and abstract operations, but the role of Morphonolexical representations in the process is made evident by the way that these commands gain pronounceable names, even when they are no more than consonant clusters, and can enter the users' everyday language.

As a novice develops into an expert, the phase of action specification in communicative tasks may well become so well proceduralized that it is difficult to identify any processes intervening between the formation of a goal and the execution of actions to achieve the goal. The whole phase may become a matter of the operation of proceduralized knowledge in one of these three processes, and so for experts it may be easier to model them as operating in action execution without the action specification phase.

15.A.2.5 Output and Bodily Feedback Transformations (Rs, Lm, BR, BL)

The Morphonolexical and Object subsystems produce planned and intended actions, but it is the Articulatory and Limb subsystems that turn the plans into actual commands for the musculature to execute. The Articulatory subsystem is predominantly concerned with speech actions, particularly those involving the tongue, mouth, larynx, and diaphragm (Rs). The Limb subsystem deals with motor actions that involve motion in space. Bodily feedback to control and coordinate these actions is provided by the Body State processes BR and BL, which operate on sensory representations of the external effects on the body (e.g., pressure and tactile sensations) and internal effects (muscle extension, limb and finger positions, and others). Together the Rs and BR processes form an indirect loop through the body and the world, as do the Lm and BL processes, in a similar way to the IB and BI processes.

Inevitably, given their location on the downstream side of the architecture, these processes are almost exclusively involved in the action execution phase. It is possible to envisage a goal that might be met by a physical response to a tactile stimulus without needing any higher level thought (no internal mental images of objects, mental thoughts, or propositions), but it is hard to see how this process would be the target of a cognitive task analysis.

15.A.2.6 Propositional Transformations (MP, OP)

The structural representations received by the Morphonolexical and Object subsystems can be thought of as mental models of the external world organized in space and time. As yet, though, they are essentially without meaning. It is the derivation of Propositional information from these representations by the MP and OP processes that allows us to identify objects and their relationships to one another and to hear the words represented by streams of phonemes. A task that requires a person to identify an object or a sound rather than just to make a motor action contingent upon its occurrence will involve the OP or MP processes. (This distinction is similar to that between choice responses and simple responses in experimental psychology). Note, however, that the identification derived at a propositional level is abstract and is not the same as naming (see the description of the PM process below).

The development of proceduralized knowledge in these processes represents the ability to recognize and categorize entities and their dynamics. Record contents would be used when the entity is not immediately recognizable or perhaps is a new exemplar of an existing category. An expert in a task may be able to directly derive a propositional meaning from a structural representation that a novice might need several interchanges of information between different subsystems to elaborate. This means that the availability of proceduralized knowledge here may allow shortcuts in the overall configuration of processes, different configurations being needed at different levels of task expertise.

15.A.2.7 Image Generation Transformations (PM, PO, OM)

Where the MP and OP processes allowed the generation of a Propositional idea from a mental image of sound or space, the PM and PO processes allow the reverse to occur. In effect, these allow the generation of mental imagery from abstract thoughts about the world. In picturing a hand holding a shiny red ball, one is using propositional knowledge to construct an object representation. The nature of that knowledge depends, of course, on one's own experience. People with a lot of experience of cricket will have developed a particular image that may contain detail about the finger positions relative to the stitching of the seam; others may have an image with less detail regarding these characteristics but more detail on aspects that reflect their own processing history. Constructing a stream of inner speech, perhaps as you plan what you are about to say or write, involves the PM process, for it entails converting abstract thoughts into more concrete representations with temporal order and intonation. Recalling the actual names of things is the formation of a Morphonolexical representation, either from Propositional information (by PM) or from a mental visual image at the Object level (by OM).

The OM process allows skilled reading in which a reader "hears" a phonological image in their head as soon as he or she sees a word shape or a morphemic fragment. This supports the reading of nonwords or new words for which no direct OP mappings have been acquired and hence supports the development of reading. Note that there is no direct MO processing making the reverse transformation possible, and so it is not possible to "see" mental scenes directly upon hearing sounds. Instead, a chain of MP:PO is required to first abstract a Propositional representation and then elaborate the Object imagery. This asymmetry in imagery gives visual sensation a more dominant role than auditory sensation, for the Morphonolexical subsystem is receiving information from three sources simultaneously (one of which is the OM process), whereas the Object subsystem only has two sources (VO and PO). The heard world is thus open to top-down influences of expectation and conformation with the seen world, but the seen world is not directly influenced by the heard world.

These three processes are involved in any thought that builds mental imagery from other internal processing. Novices at any task will almost certainly use these tasks, together with their reciprocal OP and MP processes, at all stages of task performance (as described in the section 15.A.2.9).

15.A.2.8 Central Engine (PI, IP, IB)

The Implicational level of representation is partly constructed directly from the three sensory subsystems but also from the Propositional information by the PI process. Our feelings and emotions are based not just on what we experience in sensory terms but also on our abstract thoughts. Repeatedly reading the words *sad, low, blue, glum, fail* uses the VO:OM:MP configuration to derive the identity level of meaning. The heart-sinking effect that this exercise has is a result of the PI process identifying the common implicational theme of depression and self-related threat that they all represent, followed by the IB process turning this representation into consequences in the body (for most people, a lowering of heart rate and blood pressure; for those prone to anxiety or in an anxiety-provoking situation, the opposite, perhaps along with an increase in breathing and a sense of panic).

The Implicational representations can also be used to derive a Propositional form of information by the IP process. Inevitably, the reciprocal processing that these two subsystems can indulge in dominates our internal mental lives. Most often, one of these two processes will be operating in buffered mode as we seek to understand the implications of the world around us and what the meaning of those implications is. These two subsystems are almost always going to be involved in novice goal formation and also generally in goal formation that is internally generated in response to a "wish" to get something done (such wishes result from a chain of implicational inferences about the world, our status within it, and our own bodily needs). Together the Propositional and Implicational subsystems are known as the Central Engine of cognition, in that they have a driving role in forming our patterns of thought and our sense of our selves.

Apart from goal formation, the implicational level of information is important in monitoring task performance. Following goal formation, a task description will be encoded as a propositional pattern of desired performance attributes (that is, the goal), and in parallel with constructing the output streams, the structural subsystems (M and O) will be generating an "upward" stream via MP and OP that represents the attributes of the planned actions (that is, the action specification). Mismatches between the active representation of the goal and the planned actions will be noted by the PI process and produce an implicational sense of impending error. The IP process may be able to use this to form an immediate subsidiary goal of checking the intended action specification in time to prevent the action execution.

In this description, it is clear that the interleaving of phases can become quite complex. Monitoring task performance is not a phase in itself but is an ongoing secondary task. When potential errors are noticed, the corrective action may be to interrupt the primary task, causing hesitation or long response intervals.

15.A.2.9 Processing Loops (PIP, POP, PMP)

A single subsystem cannot iterate on its own product, but as Fig. 15.1 shows, it may be able to engage in reciprocal exchanges of information with other subsystems (e.g., the Propositional and Implicational subsystems). In the descriptions of the individual processes above, it has been noted that the Propositional subsystem is reciprocally linked to the two structural subsystems and the Implicational subsystem. This gives it a special controlling role in keeping cognitive activity on track, in that Propositional representations derived from sensory streams (via MP and OP) are not taken for granted as veridical but must blend with each other and with the stream produced by IP. Since this latter stream will itself have been derived from earlier PI processing, there must be an internal feedback loop that serves to keep the active propositional representations coherent with the immediately preceding task context. This PI:IP loop (also called the PIP loop) is critically involved in all tasks that require judgment or assessment or higher level task monitoring.

The Propositional subsystem is also producing streams for the structural subsystems via PM and PO, and therefore, as well as blending their output with the IP stream, it is actually influencing their output by feeding back to them a time-lagged representation that is highly likely to have been derived from their own immediately previous processing. These MP:PM and OP:PO loops (also called the PMP and POP loops) can act to keep perceptual inferences from sensory information coherent over the short term despite inevitable changes in the quality and content of the sensory stream (changes in orientation, viewing position, lighting, auditory context, and so on). However, they can also influence the structural subsystems to interpret the sensory stream in a particular fashion, depending on the individual's expectations about the external world. Thus perception can be primed by the prior presentation of semantic associates, and events can be noticed more readily in particular task contexts where they are, effectively, anticipated. Ambiguous sensory patterns can be perceived unambiguously as one entity rather than another.

As expertise at a task develops, the proceduralized knowledge within the processes involved in these three loops may come to dominate activity such that the expert only perceives the task-relevant aspects of the world, in contrast to the novice, who has to engage in more active search for them and will be less certain about them. Although proceduralized knowledge will in most cases enhance task performance, it can also lead to errors in unfamiliar or rare situations, where the user seems to deny the evidence of his or her own eyes or ears. If in a thousand hours of operating some machinery, a certain warning light has never lit up, the expert may have more difficulty accepting that it is now lit than will a novice with ten hours' experience.

15.A.3 EXAMPLES OF CONFIGURATIONS OF COGNITIVE PROCESSES

Each process can only ever perform a small part of a task, and the real role that processes play can best be seen by seeing how they chain together to allow a flow of information to occur. In the following examples, the configurations for two small tasks are annotated to illustrate these roles and to give some idea of the configurations that might need to be constructed for other tasks. The three phases of Goal Formation (GF), Action Specification (AS), and Action Execution (AE) are set out one after the other, although of course they might not occur so sequentially.

The asterisks in the process notation in the example in Table 15.1 indicate exchanges of information with the external world via the sensory organs or with the body (other exchanges being within the cognitive architecture). This configuration would be that of a novice who knows what he or she is supposed to do but has not yet developed specialized procedural

TABLE 15.1
A Visual Event Alerts an Operator, Who Responds by Typing a Command

GF:	*VO	Visual sensory stream represented as perceptual objects
	OP	Identity of objects recognized as a scene
	PO	Feedback constraining recognition of scene
	PI	Inference of the meaning of scene
	IP	Goal formed from inferred meaning of scene
	IB*	Affective response to scene constructed
AS:	PI	Inference of meaning of the current task goal
	IP	Feedback constraining construction of current task goal
	*BI	Affective response to scene detected to blend with goal monitoring
	PM	Current goal transformed into verbal form of command name
	MP	Propositional meaning of command name fed back
AE:	ML	Finger and hand movements generated to type command name
	Lm*	Motor commands to hands and fingers executed
	*BL	Bodily feedback about hand and finger movements
	*BI	Inferred feedback of accuracy of hand and finger movements

TABLE 15.2
While Giving a Talk, A Speaker Decides to Point to a Diagram

GF:	MP	Meaning of currently planned speech fed back
	PI	Inference about lack of clarity of reference
	IP	Goal formed to provide additional referent to improve clarity
AS:	PI	Inference of the current task goal, i.e., point at referent in diagram
	IP	Feedback constraining construction of current task goal
	PO	Mental target image of current referent constructed
	OP	Propositional meaning of image fed back
AE:	OL	Finger and hand movements generated to point at referent
	Lm*	Motor commands to hands and fingers executed
	*BL	Bodily feedback about hand and finger movements
	*BI	Inferred feedback of accuracy of hand and finger movements

knowledge about the task. The activity on this phase arrives at a task goal, expressed as a proposition. Note that the propositional representation is being built from two different streams, from the Object and Implicational subsystems.

As the task becomes more practiced, the OP process would come to associate the appropriate task goal directly with the relevant Object information and so produce the task goal directly from the scene. This means that the PIP loop in the GF phase would become redundant; thus the Implicational meaning of the visual event may no longer be inferred, and the IB* process would not create an affective response. The GF phase would thus shorten to *VO:OP. In turn the AS phase would also no longer involve the Implicational level and would shorten to a :PMP: loop, which eventually might be completely proceduralized and merge with the AE phase configuration to make a single :PMP:ML:Lm* chain. Thus the acquisition of particular classes of knowledge over time can be identified, along with characteristic changes in behavior as skill develops.

The example in Table 15.2 shows a task goal being formed while the individual is engaged in another, related task, and hence the GF stream originates internally from a PMP loop that is part of the AS phase of another, higher order task (giving a talk). The use by the GF and the AS phases of the PIP loop indicates that there must be some sequencing in this configuration, and the three phases could be seen as one long chain. Skill acquisition in this task would allow the three phases to be smoothly integrated into a single flow, with gesture to physical referents being planned and executed alongside speech without interrupting the spoken stream (not represented in this example, but constructed by a flow of :PIP:PMP:MR:Rs*).

Gesture is mediated by imagery about the referent, which requires some existing knowledge of its presence in the diagram (it is assumed here that the speaker has his or her own slides). Without such knowledge, a visual search for the target might need to be performed (*VO:POP:OL:Lm*), with the POP loop evaluating objects in the scene until a match is located (May, Tweedie, & Barnard, 1993). In a situation where accuracy of reference is not paramount, the intention to make a physical gesture might be specified and executed without there actually being a real referent, and in fact this is what speakers normally do. The output stream involving the OL:Lm stream is in this situation a secondary stream, with a verbal output stream of :PIP:PMP:MR:Rs* achieving the primary task. Since buffering is highly likely to be within the PIP loop controlling the primary output stream, gestural activity in the secondary OL:Lm stream will not enter focal awareness. The speaker will gesture along with his or her speech without realizing.

APPENDIX B: GLOSSARY OF ICS TERMS

***** In the notation of configurations, indicates an exchange of information with the world (either as sensory input or effector output). For example, Rs* would be used for speech output from the articulatory subsystem.

: In the notation of configurations, divides the abbreviation of processes and indicates an exchange of information between subsystems (e.g., PO:OP).

A, AC Abbreviations for the Acoustic subsystem.

abstractive A transformation which produces a more abstract representation, by making the Assembly unit of the input into the Basic unit of the output

Acoustic A subsystem that receives acoustic sensory information and transforms it to Morphonolexical (by the AM process) and Implicational (by the AI process) levels. The acoustic representation includes the qualities of pitch, timbre, and intensity but not perceptual details such as speech forms or discrete sounds.

active task record A mapping that a transformation process has just been carried out and can be carried out again immediately without further input from the input array or can facilitate subsequent short-term activation of this mapping.

architecture The overall arrangement of the nine subsystems whose internal activity and interactions constitute human cognition.

ART Abbreviation for the Articulatory subsystem.

Articulatory A subsystem that receives information about the planned or actual state of the articulatory musculature and outputs effector commands to these muscles to produce speech (by the Rs processes).

assembly The "chunk" or superordinate grouping of the basic units within a representation that is currently being transformed by a process and can become the subject following a single thematic transition.

ATR Abbreviation for active task record.

basic unit The level of detail within a representation that is currently being transformed by a process; the nature of these units depends on the subsystem.

BIB A notional reciprocal loop involving the Implicational and Body State subsystems, using the BI process and the effects in the body brought about by the Implicational subsystem. This is involved in the perception of affective state, whereby both physical effects in the body (caused by activity or as noncognitive reflexes to stimuli) and cognitive appraisals of the state of the world combine to create a stable Implicational representation, which can then influence cognitive activity via the PIP loop.

blending The merging of two information streams into a single stream so that representations originating from two events in the world or from two different subsystems can be treated as a single representation and can act as input by a single transformation process.

Body State A subsystem that receives sensory information about proximal and internal stimulation of the body, including cutaneous pressure, temperature, muscle tension, visceral responses, and olfaction. It transforms this to the Implicational level (by the BI process), the Articulatory level (by the BR process), and the Limb level (by the BL process) to provide proprioceptive feedback.

BS Abbreviation for the Body State subsystem.

buffer See *Buffered processing*.

buffered processing A mode of image record access in which the revived representation is that which has just been copied to the image record. This allows processing to operate on an "extended present" that includes the immediate past and is used when the input stream is too degraded or is changing too fast for a stable output to be produced. This gives rise to a sense of focal awareness of the information being processed.

central subsystem One of the four subsystems (Morphonolexical, Propositional, Implicational, and Object) whose input and output is entirely cognitive (with the exception of the changes in the body brought about by the Implicational level) and hence which influence each other's processing closely.

cognitive task model The configuration, record contents, proceduralized knowledge, and dynamic control associated with a phase of task performance for a person with a particular level of expertise at a task.

common task record An abstract representation within the image record consisting of the commonalities of a large number of ETRs for similar instance of processing (i.e., a concept).

configuration A sequence of transformation processes to support the flow of information between subsystems during the execution of a phase of a task.

constituent structure The subcomponent elements of the current subject of the representation being transformed by a process, an element of which can become the subject following a single thematic transition.

copy process A process that copies the information from the input array into the image record without transforming it. This gives rise to a sense of diffuse awareness of the representations being copied.

CTM Abbreviation for Cognitive Task Model.

CTR Abbreviation for common task record.

DC Abbreviation for dynamic control.

dynamic control The degree of oscillation of buffered processing between different subsystems related to the need for different subsystems to facilitate the processing of streams of information at their own level.

effector subsystem The two subsystems (Articulatory and Limb) whose output consists of motor commands to the body's musculature, resulting in action in the world.

elaborative A transformation that produces a more detailed or elaborated representation by making the constituent structure of the input into the basic units of the output.

ETR Abbreviation for experiential task record.

experiential task record A single representation within the image record corresponding to a particular instance of the processing history of the subsystem (i.e., an episodic memory at that level of representation).

flow The exchange of information through a configuration of processes.

I, IMPLIC Abbreviations for the Implicational subsystem.

image record access Use by a transformation process of a revived representation from the image record as input rather than a representation from the input array.

Image record The long-term memory of a subsystem, which consists of a complete record of all information that it has ever received at its image record, provided that the copy process has been functional.

Implicational A subsystem that receives holistic, schematic information, transforms it to the Propositional level (by the IP process), and brings about affective changes in the body, which can be detected by the Body State subsystem (corresponding to an indirect IB process). The Implicational representation consists of high-level models of experience and includes ideational and affective content and the cognitive aspects of emotional states. Implications can be thought of as the meaning that is inferred from a set of propositions or the connotations of sensory stimuli.

input array The input side of a subsystem, where incoming streams of information arrive.

interleaving Changes between different configurations (i.e., the execution of two concurrent tasks) or between different phases of task performance.

L, LIM Abbreviation for the Limb subsystem.

Limb A subsystem that receives information about the planned or actual state of the skeletal musculature and outputs effector commands to these muscles to produce movement and positional change (by the Lm processes).

locus of dynamic control The subsystem within a configuration that is using buffered processing.

M, MPL Abbreviations for the Morphonolexical subsystem.

mapping The relationship between the input to a transformation and the output that is produced.

Morphonolexical A subsystem that receives structural sound-based information and transforms it to the Propositional level (by the MP process), the Articulatory level (by the MR process), and the Limb level (by the ML process). The Morphonolexical representation includes sound entities in space and time, the surface structure of speech, and musical phrases. It corresponds to what we "hear in the mind."

O, OBJ Abbreviations for the Object subsystem.

Object A subsystem that receives structural visual-based information and transforms it to the Propositional level (by the OP process), the Morphonolexical level (by the OM process), and the Limb level (by the OL process). The Object representation includes visuospatial entities in space and time and their position, structure, motion and dynamics. It corresponds to what we "see in the mind."

oscillation Changes of the location of buffered processing within a single configuration.

P, PROP Abbreviations for the Propositional subsystem.

phase A subdivision of cognitive activity during task performance: goal formation, action specification, and action execution. Although the phases are logically sequential, they may often be interleaved.

PIP A reciprocal loop involving the Propositional and Implicational subsystems and the PI:IP processes. This allows propositions about the state of the world and implications about the meaning of this state to influence each other, and it can lead to ambiguous facts being interpreted in a manner consistent with the current implicational state. This loop is involved in tasks that require some form of "central executive" in other models and is sometimes called the "central engine" of cognition.

PK Abbreviation for proceduralized knowledge.

PMP A reciprocal loop involving the Propositional and Morphonolexical subsystems and the PM:MP processes. This is involved in the perception and production of speech and allows expectations about the content of a speech stream to influence the perception of morphemes.

POP A reciprocal loop involving the Propositional and Object subsystems and the PO:OP processes. This allows propositions about the expected state of the visual world to influence its perception.

predicate Other basic units within the representation currently being transformed by a process, any of which can become the subject following a single thematic transition.

proceduralized knowledge Well-learnt mappings within a transformation process that allow an output representation to be produced rapidly from an input without requiring image record access.

Propositional A subsystem that receives factual, semantic information and transforms it to the Implicational level (by the PI process), the Morphonolexical level (by the PM process), and the Object level (by the PO process). The propositional representation consists of entities in the world and their relationships and includes facts, assertions about the world, goals, and intentions.

R Abbreviation for the Articulatory subsystem.

rate of change Each level of information is associated with a characteristic rate of change of the basic units within its representations derived from its processing history (i.e., the rate of change offered to it by its experience of the environment).

reciprocal loop A situation where two subsystems exchange representations directly because each can produce the other's level of representation as output. Reciprocal loops act to successively refine information toward a state consistent with both subsystems' record contents.

record contents The contents of the image record.

representation A pattern of information at a particular level within ICS.

revival The reactivation of one or more representations within a subsystem's image record caused by a match with the representation currently being copied.

rule Same as *mapping*.

sensory subsystem The subsystems (Visual, Acoustic, and Body State) that receive information from sensory transducers such as the eyes, ears, skin, and sympathetic nervous system rather than from other cognitive subsystems.

stream A sequence of representations within a single level of information derived over time from a single source (unless blended) and concerning a single aspect of the environment or of cognitive activity. Transformation processes can only operate on a single stream at a time.

structural subsystem The two subsystems (Morphonolexical and Object) that operate on perceptual structures rather than sensory information and are the immediate source of propositional knowledge about the current state of the world.

subject The basic unit within a representation that is the current focus of processing by a transformation process.

subsystem A collection of transformation processes and a copy process that all operate on the same level of representation and therefore all access the same input array and image record.

thematic transition A shift in processing by a transformation process from one element of a representation to another, resulting in a change in the theme of the information being processed (e.g., by the OP process, from one icon to another within the same array, from an icon to the array that it is part of, or from an array to one of its constituent icons).

topic Same as *subject*.

TPD Abbreviation for transition path diagram.

transformation process A process within a subsystem that transforms information into an output representation in another form.

transition path diagram A graphical notation that maps out successive thematic transitions within a level of information, with time on the vertical axis and the representational structures on the successive rows of the diagram.

V, VIS Abbreviations for the Visual subsystem.

Visual A subsystem that receives visual sensory information and transforms it to the Object level (by the VO process) and the Implicational level (by the VI process). The visual representations includes the qualities of hue, brightness, and saturation but not perceptual details such as edges or shapes.

REFERENCES

Barnard, P. J. (1985). Interacting cognitive subsystems: A psycholinguistic approach to short-term memory. In A. Ellis (Ed.), *Progress in the psychology of language* (Vol. 2, pp. 197–258.). London: Lawrence Erlbaum Associates.

Barnard, P. J. (1993). Modelling users, systems and design spaces (Esprit Basic Research Action 3066). In M. J. Smith and G. Salvendy (Eds.), Human-Computer Interaction: Applications and Case Studies. Advances in Human Factors/Ergonomics Vol. 19 A (Proc. of HCI International, 1993) pp. 331–336. Amsterdam: Elsevier.

Barnard, P. J., Bernsen, N. O., Coutaz, J., Darzentas, J., Faconti, G., Hammond, N. V., Harrison, M. D., Jørgensen, A., Löwgren, J., MacLean, A., May, J., & Young, R. (1995). *Assaying means of design expression for users and systems: Amodeus Project Deliverable 13* (Project final report). Brussels: Commission of the European Union.

Barnard, P. J., & May, J. (1995). Interactions with advanced graphical interfaces and the deployment of latent human knowledge. In F. Paternó (Ed.), *Proceedings of the Eurographics Workshop on the Design, Specification and Verification of Interactive Systems* (pp. 15–48). Berlin: Springer Verlag.

Barnard, P. J., & May, J. (1999). Representing cognitive activity in complex tasks. *Human-Computer Interaction, 14*, 93–158.

Barnard, P. J., & May, J. (2000). Towards a theory-based form of cognitive task analysis of broad scope and applicability. In J. M. C. Schraagen, S. F. Chipman, & V. L. Shalin (Eds.), *Cognitive task analysis* (pp. 147–163). Mahwah, NJ: Lawrence Erlbaum Associates.

Barnard, P. J., May, J., Duke, D., & Duce, D. (2000). Systems, interactions and macrotheory. *ACM Transactions on Computer Human Interaction, 7*, 222–262.

Duke, D. J., Barnard, P. J., Duce, D. A., & May, J. (1999). Syndetic modelling. *Human-Computer Interaction, 13*, 337–393.

Halliday, M. A. K. (1970). Language structure and language function. In J. M. Lyons (Ed.), *New horizons in linguistics*. Harmondsworth, England: Penguin.

Johnson, H., & Johnson, P. (1991). Task Knowledge Structures: Psychological basis and integration into system design. *Acta Psychologica, 78*, 3–26.

May, J., & Barnard, P. J. (1994). Supportive evaluation of interface design. In *Proceedings of the First Workshop on Cognitive Modelling and Interface Design* (pp. 185–203). Vienna: T. U. Wien.

May, J., Barnard, P. J., & Blandford, A. E. (1993). Using structural descriptions of interfaces to automate the modelling of user cognition. *User Modelling and User Adapted Interactions, 3*, 27–64.

May, J., Scott, S., & Barnard, P. J. (1995). *Structuring displays: A psychological guide*. Geneva: EACG.

May, J., Scott, S., & Barnard, P. J. (July 1997). *Modelling multimodal interaction: A theory-based technique for design analysis and support*. Tutorial presented at the Interact '97 conference, Sydney.

May, J., Tweedie, L. A., & Barnard, P. J. (1993). Modelling user performance in visually based interactions. In J. L. Alty, D. Diaper, & S. Guest (Eds.), *People and computers VIII* (pp. 95–110). Cambridge: Cambridge University Press.

Newell, A. (1990). *Unified theories of cognition*. Cambridge, MA: Harvard University Press.

Scott, S. K., Barnard, P. J., & May, J. (2001). Specifying executive representations and processes in number generation tasks. *Quarterly Journal of Experimental Psychology. A, Human Experimental Psychology, 54*, 641–664.

16

Critical Decision Method Data Analysis

B. L. William Wong
University of Otago

The critical decision method (CDM) is a retrospective cognitive task analysis technique (Klein et al., 1989) that involves systematically probing an interviewee about a particularly memorable incident to elicit information about his or her decision strategies and expertise. The purpose of this chapter is to describe how the qualitative data collected during CDM interviews can be analysed and interpreted. Two complementary analysis approaches will be described. The first approach is referred to as the structured approach and the second as the emergent themes approach. Each approach will be illustrated by examples from actual CDM case studies. This chapter will also discuss some important limitations in the analysis and interpretation of the data.

16.1 INTRODUCTION

The critical decision method (CDM) is a retrospective cognitive task analysis (CTA) method for investigating the decision processes invoked by people during major or significant incidents. By asking a systematic set of probes or question about these incidents, an investigator can identify the knowledge requirements, expertise, and goal structures involved in performing a decision maker's work. The insights gained from CDM investigations have been used to develop better user interface concepts for representing process and situational information.

There are two broad phases in the CDM: data collection and data analysis (see, e.g., chaps. 1 and 24). This chapter addresses the *data analysis* phase. A number of instructive descriptions of the interviewing procedure have been reported in the literature (Hoffman, Crandall, & Shadbolt, 1998; Klein, Calderwood, & McGregor, 1989; Wong, Sallis, & O'Hare, 1997). However, there are fewer descriptions of how the data should be analysed (Kaempf, Wolf, Thordsen, & Klein, 1992; Klein, 2000). Although it is as important to collect relevant data through the CDM interview, these data must be analyzed in a rigorous way to ensure that the findings can be usefully applied.

In this chapter, two complementary approaches to analyzing and making sense of CDM interview data will be described. The first approach is referred to as the structured approach, and the second as the emergent themes approach. The structured approach is very focused and uses an a priori classification framework to analyze the data. The use of such a framework, however, can lead to omissions of interesting concepts not previously anticipated. The emergent themes approach is exploratory and uses its own data, including unanticipated concepts, as a basis for its framework for analysis. Thus, the structured approach can be used in studies where the expected concepts are clear, whereas the emergent themes approach would be more appropriate in studies where the parameters of the domain are not known. The next section provides a brief background to the CDM. The remainder of the chapter describes the structured and emergent themes approaches, and discusses their limitations.

16.2 BRIEF OVERVIEW OF CDM

The CDM is a CTA method (Hoffman et al., 1998; Klein, 1993; chap. 15), and its main purpose is knowledge elicitation (Klein et al., 1989). CTA represents a class of methods for investigating the nature of cognitive work. These methods are used for identifying "information about the knowledge, thought processes and goal structures that underlie observable task performance" (Schraagen, Chipman, & Shalin, 2000, p. 3). In addition, they can be used to identify operators' mental models, their problem models, expert behavior, how novices become experts, difficulties in learning, how operators make critical and difficult judgements, and the demands that a given job places on the operator. CTA methods help us identify what knowledge, automatic and procedural skills, representation of processes, and understanding of decision making is needed to perform a task in order to match human capabilities with a system; how to design controls and displays; and how dynamic processes may be portrayed (Seamster, Redding, & Kaempf, 1997). A number of CTA methods exist for these purposes, including Concept Mapping (McFarren, 1987), Conceptual Graph Analysis (Gordon et al., 1993), Knowledge Analysis of Tasks (Johnson & Johnson, 1991), Cognitive Task Analysis-Based Instructional Design (Redding, 1995), Function-Based CTA (F-CTA; Roth & Mumaw, 1995), and Interacting Cognitive Subsystems (chap. 15). CDM falls within this group of methods.

The CDM is an extension of the critical incident method (Flanagan, 1954) that was used to investigate difficulties in aviation. The CDM was developed for research into how decisions were made in actual situations. This line of research led to the development of the recognition-primed decision (RPD) model (Klein, 1989). The RPD model differed from traditional utilitarian models of decision making that prescribe the generation of multiple options and the simultaneous evaluation of them (e.g., Simon, 1977). The RPD model was based on observations that decision makers in "naturalistic" environments—rapidly changing situations in which there is incomplete information, high stakes, lack of time, interrelated issues—do not practice decision making as prescribed by the classical utilitarian models. Instead, they engage in pattern recognition and storytelling strategies to gain an understanding of the situation. Once the problem is understood, they serially evaluate the options by mentally rehearsing and testing each course of action. The selected course of action is then implemented with or without modification, depending on its suitability for the problem and the availability of time. Readers are referred to Klein (1997) for a more detailed treatment of the RPD model.

The CDM has been used in a number of studies to identify decision strategies, expertise, and knowledge requirements. These studies have focused on the elicitation of specialized skills and cues in neonatal intensive care nursing (Crandall & Getchell-Reiter, 1993; Militello & Lim, 1995), the nature of decision making in naval command and control (Kaempf, Wolf, & Miller, 1993), situation awareness in emergency ambulance control and coordination (Wong & Blandford, 2001), and team decision making (Klein, 2001). Their findings have led to important insights for designing better decision aids (Klein, 1993). The CDM is good at exploring and characterizing decision needs rather than hypothesis testing. It uses specific and unusual incidents to gain a better understanding of the needs of abnormal and out-of-bounds situations and investigates incidents that are personally experienced and demanding, requiring the operators to perform at the limits of their ability.

16.3 CONDUCTING A CDM INTERVIEW

CDM interviews usually last between 1 hour and 1 and a half hours. Interviewees are asked to think back to a particularly demanding and memorable incident they experienced during the course of their work. These interviews are in-depth and are semistructured in nature. Open-ended questions are used to explore the breadth and depth of relevant issues associated with the decisions made during that incident.

The interviewee is first asked to describe the selected incident as a series of decision points on a timeline. This decision timeline is used as a framework for systematically probing the interviewee about how he or she made the decision at each decision point, the information cues attended to, his or her assessment of the situation, the options and information considered, the basis for the choice, and his or her goals when making the decision.

These interviews are usually recorded so that the investigators can concentrate on what is usually a demanding task of eliciting useful data while keeping the interviewees interested in sharing their experiences and knowledge. The recordings are then transcribed for subsequent analysis. The resulting transcripts are often voluminous (e.g., a 1-hour interview can easily amount to 15,000 words). These very rich data sets are also often messy (Milliken & Johnson, 1984): The data can be subjective and ambiguous; the meanings are context dependent; the data can lack obvious structure because of the open-ended nature of the interviews; they can be hard to reduce; and they can also be incomplete or irrelevant, especially if the interviewee is allowed to "rabbit on" (continue talking on the same point repetitiously or at too low a level of detail). In addition, similar concepts may be expressed differently by different interviewees, similar terms may be used to mean different things (ambiguity and context dependence), interviewees may jump from one issue to another and back again, descriptions of what they had done may be recalled at different stages during the interview, and aspects of a concept may be discussed in different parts of the interview (lack of structure). Consequently, there is a need to identify significant patterns, corroborate evidence across different incidents, and find an adequate structure grounded in the data to communicate the findings. Two methods of how this can be achieved are discussed next. Each approach is explained using examples. The structured approach uses an a priori framework of decision making for identifying cues, considerations, situation assessments, actions, and reasons for actions. The emergent themes approach, in constrast, allows concepts to emerge from the data and to subsequently structure their relationships so that they become meaningful.

16.4 THE STRUCTURED APPROACH

The main stages in the structured approach may be summarized as follows:

1. decision chart showing the decision process on a timeline with progressive deepening to illustrate how the decisions were made
2. incident summary
3. decision analysis tables based on the a priori decision-making framework
4. identification of items of interest in each incident
5. collation and comparing of common items of interest across incidents studied

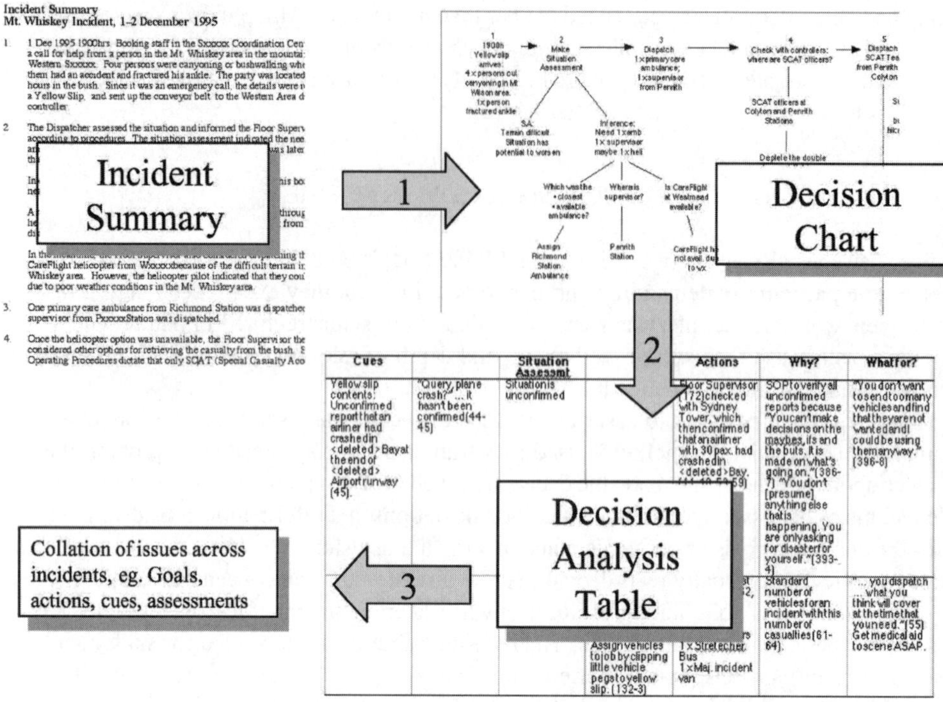

FIG. 16.1. Overview of structured approach for analyzing CDM data.

This process is shown in Fig. 16.1 and 16.2 Stages 1 to 4 are illustrated using one case example. This case concerns an emergency ambulance dispatcher in a major city who responded to several emergency calls after a high-speed commuter train crashed into the back of a slower historic train on the same track. Stage 5, which involves collating and comparing common items of interest across incidents, is illustrated using the findings from four similarly demanding cases at the same emergency ambulance control center.

16.4.1 Stage 1: The Decision Chart

A decision chart is a visual representation of the decisions made during an incident (Fig. 16.3). It employs a timeline to show the sequence in which the decisions occurred. It also shows, at key decision points, further considerations and deliberations and the outcomes at those decision points. Adding this information has been referred to as the *progressive deepening process*. Depending on how relevant other decision activities are to the study, additional timelines can be included to illustrate the parallel activities that often do occur during major incidents. Representing the progressive deepening analysis is helpful in explaining how the decisions were made and for indicating where further analysis in the next stage would be useful. The example described here will be referred to as the Cowan Train Station incident.

The decision chart is usually developed from a sketch made during the interview about the events and decisions of the incident. That sketch is typically used during the interview as a visual aid for identifying key points in the incident to explore using the CDM probes. See Wong et al. (1997) for a description.

Post interview, a study of the sketch and of the interview transcript is used to progressively describe the incident. In the Cowan Train Station incident (Fig. 16.3), Decision Point 1 (DP1) starts the sequence with two or three calls reporting a train crash 3 kilometers south of the Cowan

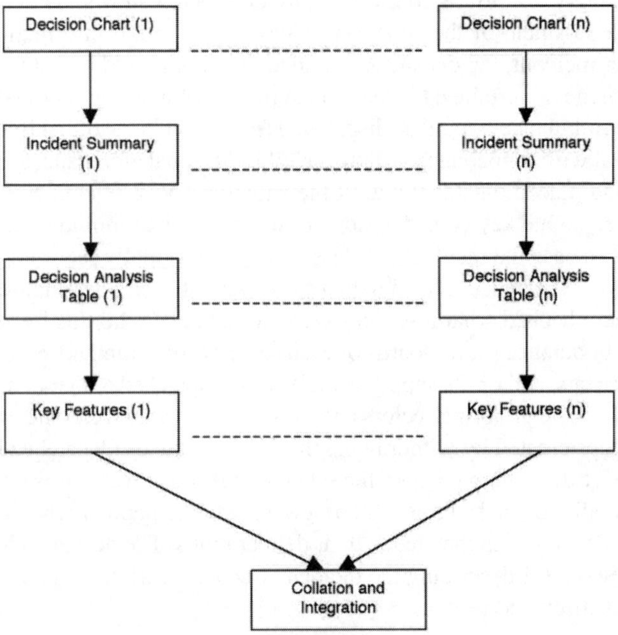

FIG. 16.2. Collating common items of interest across incidents.

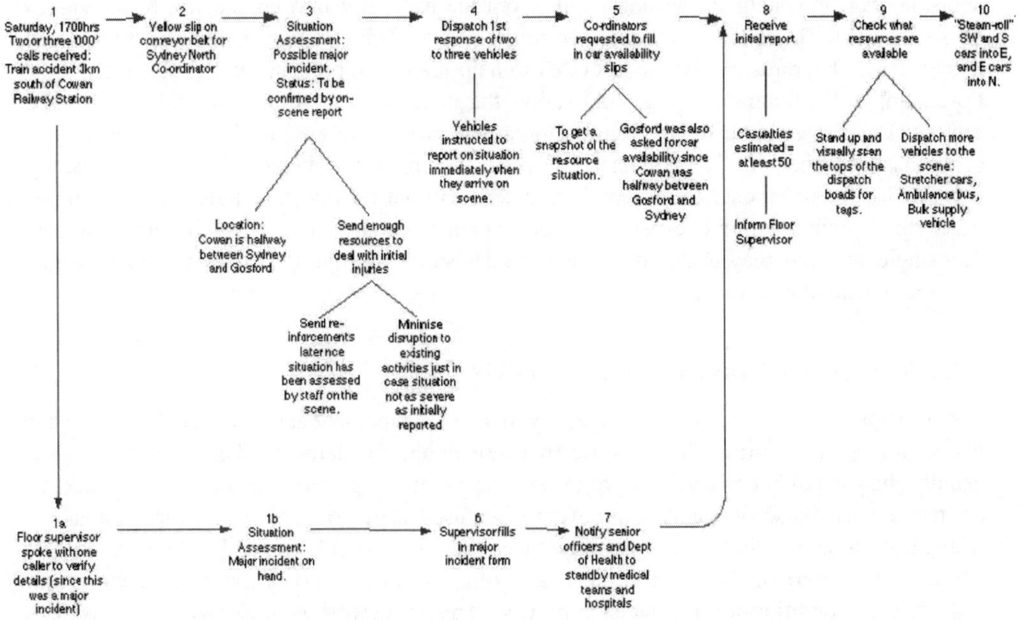

FIG. 16.3. Decision chart of the Cowan Train Station incident.

Train Station. For context purposes, DP1 branches off to a parallel set of activities (DP1a-1b-6-7): The dispatch supervisor, was informed and he assessed that this was a major incident and responded accordingly. Returning to the main timeline, DP3 shows the deliberations during the dispatcher's assessment of the situation: Although the initial information from the calls suggested a major incident, the dispatcher decided that this should be confirmed by on-scene reports. The dispatcher ascertained the location of the incident, as this would determine the stations that the first ambulances would be dispatched from. His plan was to initially dispatch three ambulances to deal with immediate casualties while additional ambulances from other sectors were identified and placed on standby until the situation could be confirmed by ambulances arriving at the scene. One key consideration at this time was to minimize disruption to other existing activities in case the incident was less severe than initially reported. At DP8, the first on-scene report was received, confirming that there were at least 50 casualties. DP9 describes how the dispatcher checked what resources were available. He did this by standing up in the room to scan the ambulance status boards of each sector. More ambulances as well as stretcher cars, an ambulance bus, and a bulk supply vehicle were dispatched to the scene. DP10 indicates that ambulances from neighboring sectors and those further away were "steamrolled"—spread out to cover the gaps created by withdrawing the 25–30 vehicles dispatched to the incident.

In the progressive deepening process, the specific time order of events is only approximated. This reflects the reality of such studies. Interviewees often remember what occurred but often not the specific order in which they made their deliberations. The decision chart, however, is a very useful framework for describing the incident. Such a description is known as an incident summary, and it is discussed next.

16.4.2 Stage 2: The Incident Summary

The incident summary provides a description of the incident. It uses the framework of the decision chart to organize relevant details from the transcript into an account of the stages of the incident. The incident summary from the Cowan Train Station incident is reproduced in Fig. 16.4. Diagrams can be incorporated to help readers appreciate spatial considerations. For example, the diagram in Fig. 16.4 shows the incident in relation to the five ambulance sectors. The arrows indicate how reinforcing ambulances were steamrolled across the city to compensate for the gaps in ambulance coverage. It summarizes the key information, making it easier for the researcher to review the incident without having to restudy the voluminous transcript. Summarization is especially useful when researchers investigate several incidents in a single study, as they often do. Having the decision chart and the incident summary aids when reviewing the incidents.

16.4.3 Stage 3: The Decision Analysis Table

The decision chart and the incident summary map out the incident and orientate the researcher to decision points during the incident. Key decision points are identified. These decision points usually show a number of levels of progressive deepening (e.g., more factors were considered or attended to). These key decision points are then used as the focal points for further analysis of the transcripts to identify relevant utterances (verbal protocols) made by the interviewee. These verbal protocols are then collated according to an a priori decision framework that reflects the recognition-primed decision process. This framework is illustrated in Fig. 16.5.

The decision analysis framework in Fig. 16.5 shows how a decision maker is presented with situational cues and information ("Cues and situational factors"). The information is processed, and sense is made of the information, resulting in an assessment of the situation ("Situation assessment"). Further mental processing leads to the generation of a course of

Incident Summary
Cowan Train Station Incident, Summer 1993

1. One Saturday afternoon in the summer of 1993, at about 1700hrs: Two or three triple '0' were received by the call-takers about a train crash in unpopulated areas about three km south of Cowan Railway Station. A south bound passenger inter-urban train from Sydney had caught up and hit the rear end of a 38-class historic train that was also heading south. The impact was severe enough to de-railed three carriages. Since this was a major incident, one of the calls was passed on to Floor Supervisor to verify the details.

2. The yellow slip was sent up the conveyor belt to the North Area Co-ordinator.

3. At this stage the incident was still unconfirmed by ambulance staff or other authorities. With 4 of the area's 12 ambulances already committed, there was a need to maintain an acceptable level of service to cover the level of emergency call outs expected on a Saturday, while responding to the immediate needs of the train incident.

4. Initially two or three vehicles were responded to the scene with instructions to provide more accurate information about the situation before dispatching more vehicles.

5. In the mean time, all area co-ordinators were asked to fill in the car availability slip. This form collated from each region, information about what vehicles are available and their probable ETA to the incident. The co-ordinators at Gadfly were also asked to provide this information, since Cowan was halfway between Sydney and Gadfly. This procedure takes about two minutes. Once the car availability slips have been collected from each co-ordinator, supervisor checks the slips and visually scans the boards to confirm the car availability.

6. The supervisor then fills in the yellow major incident form. It sets out the steps that the Co-ordination Centre should take for such major events, e.g. who to inform.

7. While waiting for the report from the scene, notify senior officers who may be sent to the scene as well were notified. The Department of Health was also notified so that they can put their medical teams and hospitals on stand-by.

8. Report from the scene was finally received and the situation confirmed. Number of patients was estimated as more than 50 people.

9. Quickly re-assess the need and the number of vehicles available in each area by visually scanning for tags sitting on top of the dispatch board for each area and referring to the cars availability slips. (Each tag represents one ambulance. Each type of ambulance is differentiated by the colour of the tag.) Co-ordinator refers to the availability slip only as a memory aid. Start dispatching. At the end of the accident, a total of about 25 - 30 vehicles were eventually dispatched to the incident.

10. Dispatch additional ambulances from the East Area to reinforce the North Area. Ambulances from the South-West and South were then "steam-rolled" in to the East to the incident or to cover gaps left by responding North Area ambulances.

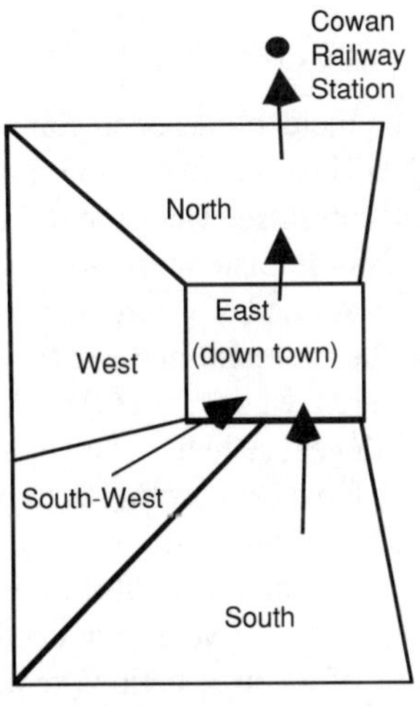

Schematic of the five ambulance areas of Sydney.
Arrows indicate where the ambulances were sent from and to.

FIG. 16.4. Sample incident summary.

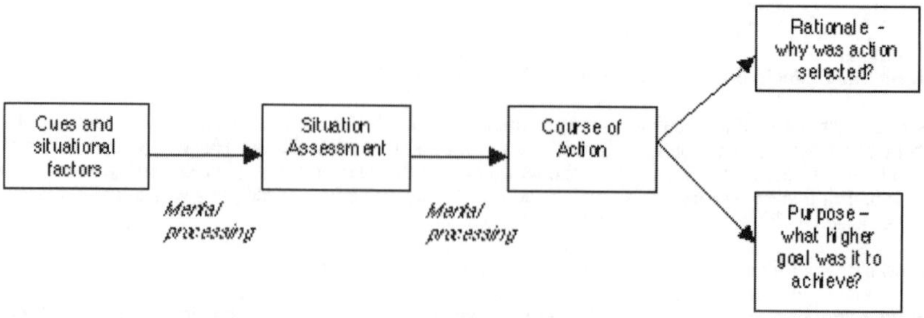

FIG. 16.5. Decision analysis framework.

action ("Course of action") and to the action itself. Decisions are not made independently but are based on reasons. The rationale ("Rationale") refers to the immediate reasons for the action. It answers the question why an action was carried out. For example, standard operating procedures dictate that notification of a major road traffic accident must be responded to by sending two ambulances immediately. The purpose ("Purpose") attempts to identify the higher goal that this action may be a part of. For example, the two ambulances dispatched to the road traffic accident may not be the nearest ambulances but were selected to balance the ambulance coverage in a region—a larger goal.

This framework may be implemented by a table for organizing the data for each decision point and collating the data across the various decision points. For example, all the cues that were needed or attended to during this incident have been collated in the "Cues" column in Table 16.1, which is taken from the decision analysis of the Cowan Train Station incident. Each row represents the analysis of a decision point. The data in the table cells may be actual excerpts from the transcripts or a summary of excerpts (the numbers indicate the lines in the source documents that the excerpts were taken from.) The recording of line numbers is important as it provides an audit trail so that conclusions drawn from the data are traceable to the source and could be independently verified.

16.4.4 Stage 4: Identification of Items of Interest

Once the decision analysis table for the incident has been completed, items of interest can be identified. Such items are identified based on the theoretical frame of reference applied to the study, which in this case is the Recognition-Prime Decision (RPD) model (Klein, 1989, 1997). According to the RPD model, the goals of human operators, together with their expectations, cues they perceive, and their action repertoire, play a significant role in the operators' recognizing the type of situation and assessing the situation. In the example of the Cowan Train Station incident, among the items of interest were the goal states that the dispatchers were attempting to achieve and the strategies they used (a goal state is a state of a process that someone tries to achieve through control of the process). Goals are useful starting points for directing analysis, although there might be many other items of interest.

The goal states were inferred from the statements of purpose reported by the dispatcher. Some of these statements were very clear and were contained in a single sentence. However, typically identifying the purpose of a set of actions requires several sentences or utterances. Consequently, the intended goal has to be carefully interpreted within the context of the discussion and collated under the "What for?" column of the decision analysis table (Fig. 16.1).

The next step is to consolidate all the relevant items of interest—in this example, the goal states identified—by grouping similar statements together and giving each grouping a

TABLE 16.1
Sample Decision Analysis Table

Cues		*Situation Assessment*		*Actions*	*Why?*	*What For?*
Yellow slip content: A commuter train had crashed into the rear of a historic train 3km south of Cowan Railway Station. (4, 50)	Saturday, 1700 hrs. 30% of all vehicles have already been committed (366)	Uncertain about the extent of injury. Initially dispatch minimum required to deal with immediate injuries. Reinforce as necessary later. (74, 75, 186)			Situation needs to be confirmed by on-scene staff before committing more resources. (57, 74, 81, 190, 234, 250, 254)	
Tags of vehicles on top of North Area board.				Dispatch 2–3 vehicles with instructions to report on arrival at scene. (57)	SOP. To confirm occurrence as a major incident and to determine more accurate estimate of number of vehicles needed.	To minimise disruption to on-going activities. To get a vehicle on the scene ASAP.
		The need to know what vehicles are available across all Areas.		Co-ordinators to fill in Car Availability Slip. (9–10, 57–60, 222–224) "And you also get somebody to sit next to the co-ordinator to write all these stuff down so you can replay it." (245)	"A snapshot picture so that you know what's available, know what to send." (115)	To enable re-deployment of vehicles to reinforce depleted areas.
Initial report from incident scene: Number of casualties = 50 patients (94, 280)		Will need 25–30 vehicles (ambulances, stretcher buses, bulk supply units) for an incident of this magnitude. (95, 254, 281, 292)	Assigning 25–30 vehicles to the incident will leave many gaps in coverage.	"You've already determined what's available, so then you allocate those that are available." (286–287)		"The goal would be to send the correct amount of ambulances to cope with the situation without shortfaling the areas that aren't involved." (361)

TABLE 16.2
Consolidating the Items of Interest

Goal States	*Statements of Purpose*
Maintain situation awareness	To determine the nature of the emergency. "A snapshot picture so that you know what's available, know what to send." (115)
Get resources moving	To get a vehicle to the scene ASAP.
Planning resource to task compatibility	To minimize disruption to ongoing activities. To enable redeployment of vehicles to reinforce depleted areas. (11–13) "The goal would be to send the correct amount of ambulances to cope with the situation without shortfaling the area that aren't involved." (361) "You don't want to send everything out of one area." (116) so that you will be able "to cover emergencies that are still going to happen . . . " (117, 123)

meaningful label (e.g. "Maintain situation awareness"). The consolidation is presented in Table 16.2 and discussed further in the next section.

16.4.5 Stage 5: Collation and Integration Across Cases

Once the items of interest have been identified and consolidated for each case, a specific item of interest can be compared across all cases being studied. It is through this process that sense can then be made of the data. Table 16.3 presents a summary of statements associated with the goal "planning resource to task compatibility" across five different incidents.

The table provides a useful method of showing how each concept or item of interest is supported by the available data. It is also possible to infer the representativeness of concepts identified. Reading across the cases shows that the goal "planning resource to task compatibility" was discussed by all the interviewees. Reading down the columns reveals the different aspects of the goal that were discussed.

Determining what resources are available and their locations in relation to the incident is another aspect of the cognitive work associated with the planning goal for such incidents. Subsequent analysis suggested that dispatchers first focused on the immediate area surrounding the incident to locate available ambulances and to compare the suitability of the identified ambulances for the job. This strategy was subsequently referred to as the "focus and compare" strategy. The need to focus around the immediate incident area and then to compare suitability thus forms the basis for how such information should be portrayed to facilitate the execution of the strategy.

In this section, a structured analysis of CDM data was illustrated using the Cowan Train Station incident. The structured approach showed (a) how an investigator can reliably reduce the quantity of data that one has to work with, (b) provide an audit trail to enable traceability of the findings and conclusions, and (c) systematically trawl through and make sense of the usually messy CDM interview data. Although effective, the structured analysis approach has one limitation. If strictly followed, it may not reveal interesting concepts that do not fit directly in the a priori decision analysis framework. This would negate the benefits of the exploratory and open-ended interview technique adopted. The next section presents an alternative approach

TABLE 16.3
Collation and Comparison of Items of Interest Across Cases

Mt. Whiskey	*King's Highway*	*Eastends*	*Brittany Bay*	*Cowan Railway*
To determine how many and what type of resources are needed by making an assessment of the situation.	To understand the nature of the accident in order to cover the amount of casualties.	Need to know what had happened so that you know what resources are going to have to start moving into the area.	"You don't want to send too many vehicles and find that they are not wanted and I could be using them anyway." (396–398)	To minimize disruption to ongoing activities.
To locate the nearest available ambulance.	Need to send the closest possible resource.	Prepare resources by identifying what vehicles will be available in the next few minutes.	"You dispatch . . . what you think will cover at the time that you need." (55)	To enable redeployment of vehicles to reinforce depleted areas. (11–13)
			To determine what resources are available and where these resources are.	
	To cover the holes made by bringing in your resources into the one area.		"At the same time you are calling in your other vehicles from the other areas to cover the situation [in the East area]." (188)	"The goal would be to send the correct amount of ambulances to cope with the situation without shortfaling the area that aren't involved." (361)
				"You don't want to send everything out of one area" (116) so that you will be able "to cover emergencies that are still going to happen." (117, 123)
Check if helicopter is available.				

that facilitates the identification of these concepts as they emerge from the data while still maintaining a systematic framework for analysing the data. This approach is referred to as the emergent themes approach.

16.5 THE EMERGENT THEMES APPROACH

The structured approach assumes a framework within which to collate relevant data. In contrast, the emergent themes approach does not impose such a framework but instead explores the data to identify ideas and their relationships. Based on an analysis of the transcripts, broad patterns—concepts and relationship between concepts—are identified. These broad patterns

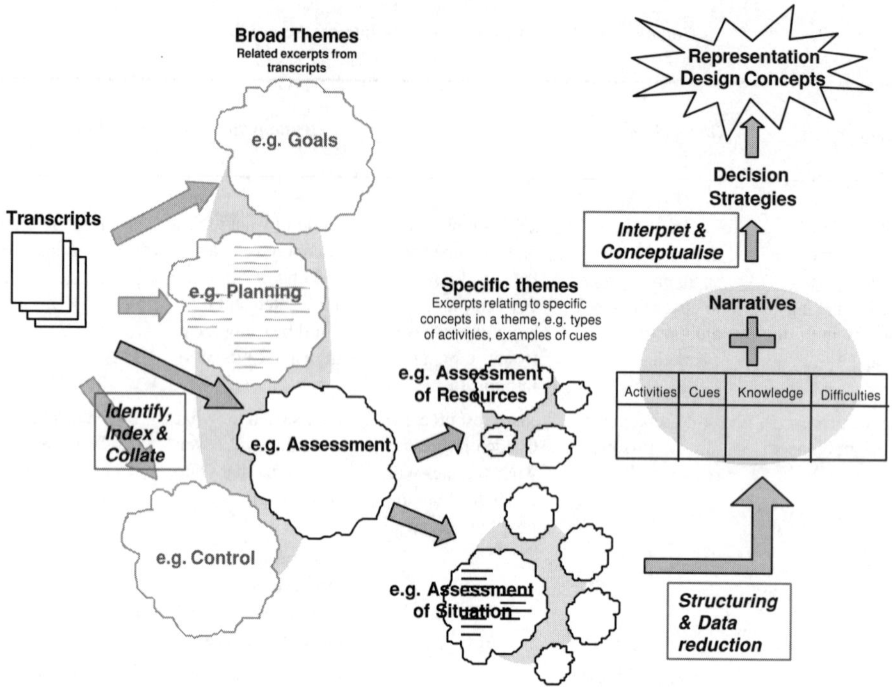

FIG. 16.6. Overview of the emergent themes approach.

form the initial structure that direct the next round of indexing the transcripts. The indexed data are then collated and further organized within each grouping to reveal the themes. Figure 16.6 provides an overview of the emergent themes approach.

In a separate study involving nine emergency ambulance controllers at a large metropolitan ambulance service control center, the emergent themes approach was used to identify key issues in ambulance control. Selected aspects of this study are used to explain the approach.

16.5.1 Index and Structure to Find Broad Patterns

One of the first tasks performed in this study was the identification of the goals that the dispatcher attempted to achieve. A review of the nine sets of transcripts identified a number of statements that were about the interviewees' goals. These statements were studied and organized into five groups of more closely related concepts (Table 16.4).

16.5.2 Theorize New Structure and Themes

Subsequent analysis led to a reclassification of the goal statements into five broad themes (Table 16.5). The first theme was the overall goal of "getting the right ambulance to the right place at the right time." The second theme identified the priorities and constraints that the dispatcher had to consider in all decisions, such as minimizing disruption to other sectors. The third and fourth themes dealt with decision processes assessing the problem, assessing the resource state, and planning and coordination and control. The last theme related to the need to be aware of happenings within and around the sector i.e., to have situation awareness. This framework then became the structure for the next stage of analysis.

TABLE 16.4

Initial Indexing and Collating to Draw Out Common Themes

To get the right number of ambulances to the right place as quickly as possible (ORCON) to ensure a good patient outcome (survivability).
To get the ambulances there as quick as possible and for a good result. (3/1266)
Maintain um . . . your [ORCON] standards. (3/1274)
Getting the right um . . . the right response to the um . . . incident. (4/1513)
To get an ambulance or ambulances to scene at the quickest possible way. . . . Get the vehicles rolling. (5/1488)
The main goal is to get the ambulance crews there and get . . . get patients . . . to get people treated as quickly as possible. (1/1175)
To ensure that the rest of the sector continues to perform.
The smooth running of the um . . . sector, um . . . while all this going on. (3/1273)
Smooth running of the service so that other people end up with a reasonable service. (6/762)
To keep critical incidents at the forefront of considerations.
P: And for this baby to be . . . it's always at the forefront of your mind. (3/1277)
P: The, um . . . thing about it is, when something like that happens, although you're still allocating for the rest of West London, it's always forefront in your mind, and it's always . . . it's always that sort of job, like a suspended baby, or a large road traffic accident will always, although you're just allocating, you're just allocating normally, but then they're always forefront in your mind. (3/49–52)
To understand what is happening in order to assess the need.
What the incident is, what response you need on it, get responses moving. (4/1514)
To find out exactly what was happening. (4/1531)
Until you find out what is happening, you can't mobilize at all. (4/1535)
To find out what was going on . . . find out what is happening. (5/1488)
To get the right vehicle to the right amount of patients is difficult. (6/754)
To communicate relevant information to the crews as quickly as possible (radio operator).
To give [the ambulance crews] the correct information clearly and concisely. (8/1666)
To keep them informed of the situations (8/1670)
Get updates from the police or the patient, condition might have deteriorated or police have no units to assign. (8/1674)
You're looking after the crews' safety as well, and what they're doing. (8/1688)

TABLE 16.5

Initial Framework for Further Analysis

Goal
To get the right ambulance to the right place at the right time
Priorities and constraints
Minimize disruption to rest or other sectors
Pay attention to critical incidents (while . . .)
Control other ambulances simultaneously
Decision processes
Assessment of problem
• Mental picture, understanding of what is going on
Assessment of resource status
• What is available to deal with the problem?
Planning
• What plans does the situation correspond with? Is situation novel and does it require construction of a new plan?
Co-ordinating and controlling
• Directing and communicating with crews
• Management of major incidents
Situation awareness
What is happening within and around sector?

TABLE 16.6
Sample of Statements Related to the Situation Awareness Theme

Aware of what's going on in and around their sectors so that they can start thinking ahead in case an incident escalates.
"For safety reasons, we need to know who's there . . . we keep a vehicle movement sheet . . ." "I know everything that is going on because all info comes through me . . ."
Other emergencies occur in a sector despite a major incident; hence still need to keep an eye on situation.
Knowledge of the area on which dynamic movements are tracked.
"We just try and picture everything . . . everything has to be a picture in your head of what's going on . . . [anticipating] what else they need . . . and what else we can be doing."
Use a map to help plan how to reach a major incident scene.
"You're the ones that are talking to the crew on the radio . . . and they are asking you and telling you the situation."
"Control ears"—time and effort minimization strategy for keeping track of what's going on around them.
"It's all up here, you know what's happening, you know what's going on, you know what's been done . . . it's difficult for someone to jump in, 'cause . . . you have not got a picture of it, you can't see what's going on, so you can only talk and you can only imagine what's going on."
Flick through the CAD displays, the tickets, review the box.

16.5.3 Explore Each Theme: Index and Structure Again

The initial framework can be used to explore the identified themes in more detail. If the transcripts have previously been indexed in qualitative research analysis software such as NUDIST (QSR, 1994), then all indexed occurrences of a theme can be easily retrieved. Otherwise, the researcher needs to review the transcripts again to identify, extract, and collate them. Table 16.6 containsis a selected sample of occurrences of the theme of situation awareness (this is the only theme that will be elaborated here).

To improve the usefulness of the statements on situation awareness, further structuring is necessary to make sense of the theme. The structure shown in Table 16.7 is used to explicitly show what activities are associated with situation awareness; the cues attended to and deliberations made; the knowledge and experience needed to be invoked in developing and maintaining situation awareness; the difficulties, problems, and likely mistakes (along with their consequences) associated with failing to maintain situation awareness; and the strategies used to maintain situation awareness. This structure links the analysis to the basic investigation framework used in the CDM. It provides a systematic way of describing what people do, what information they attend to and consider, what knowledge and experience they need, and what strategies they invoke to perform the decision-making task. Identifying the difficulties and likely mistakes provides an alternative perspective to triangulate the findings and a basis for reasoning about what a potential information system needs to support (in this example, how dispatchers maintain their situation awareness in the context of decision making).

16.5.4 Synthesis

In completing the analysis for this theme, the structured data are studied to develop conceptualizations about situation awareness. This is the process of synthesis—the bringing together of ideas in a meaningful manner. For example, synthesis of the data suggests that the dispatcher is the hub of all information about developments on the road. ("Everything that is going on comes through me") and that he or she uses a "control ears" strategy to keep track of what is going on in the control room. This synthesized strategy, which we will call the information hub strategy, requires a variety of functions to support the collation and integration of information from

TABLE 16.7
Structuring the Supporting Data for the Situation Awareness Theme

Activities
Aware of what's going on in and around their sectors so that they can start thinking ahead in case an incident escalates.
"For safety reasons, we need to know who's there . . . we keep a vehicle movement sheet . . ."
"I know everything that is going on because all info comes through me . . ."

Cues and considerations
Other emergencies occur in a sector despite a major incident; hence still need to keep an eye on situation.

Knowledge and experience
Knowledge of the area on which dynamic movements are tracked.

Strategies
"We just try and picture everything . . . everything has to be a picture in your head of what's going on . . . [anticipating] what else they need, . . . and what else we can be doing."
Use a map to help plan how to reach a major incident scene.
"You're the ones that are talking to the crew on the radio . . . and they are asking you and telling you the situation"
"Control ears"—time and effort minimization strategy for keeping track of what's going on around them.

Difficulties, problems, likely mistakes, and consequences
"It's all up here, you know what's happening, you know what's going on, you know what's been done . . . it's difficult for someone to jump in, 'cause . . . you have not got a picture of it, you can't see what's going on, so you can only talk and you can only imagine what's going on."
Flick through the CAD displays, the tickets, review the box.

different sources, across different modalities of communication, and over a period of time. The strategy then becomes an interface design objective that defines the manner in which the required information and functions are to be orchestrated to support the information hub strategy.

16.6 LIMITATIONS IN INTERPRETING AND REPORTING

Hoffman et al. (1998) identified six methodological issues associated with the conduct of CDM studies in general: method reliability, method adaptability, method efficiency, data quality, the manner in which interviewees relate an incident, and whether the incident has been personally experienced. Of relevance to this chapter is the issue of personal experience of the incident. The CDM is a retrospective method i.e., interviewees need to think back and recall an incident. If an interviewee has not personally experienced the incident, the interview will reveal little about the decision-making processes. For example, in an interview with a supervisor who had observed a dispatcher control ambulances during a major incident, the supervisor, although able to describe the overall development of the incident, was not able to describe the dispatcher's operational deliberations, nor identify the cues that the dispatcher had attended to. On another occasion, an opportunity arose to evaluate the accuracy of an interviewee's recall of an incident against computer and audio records of the incident. The interviewee's description of the incident correlated very closely with the sequence of events documented in the records despite the fact that the incident had occurred almost a month before. Furthermore, because of the critical nature of the incidents studied, interviewees have often been observed to recall what appear to be fine details associated with the incident. For example, one dispatcher recalled that, 7 minutes into an incident, he suddenly remembered that he had failed to notify the doctor in charge of the rural district. The 7-minute delay was significant and memorable, as the ambulance crew that day did not have a paramedic with them and therefore needed the services of a doctor.

High recall fidelity and appropriate access to cognitive processes, of course, cannot always be guaranteed, and some doubts have been expressed about the dependability of memory in chaps. 1 and 19.

Retrospection, as a technique, requires that the interviewees experienced the incidents themselves and that the incidents were so demanding or memorable that important details can be recalled with a reasonably high degree of accuracy. This suggests that care should be taken to select incidents that were personally experienced and particularly memorable. The iterative procedure of reconstructing an incident on a timeline and the careful application of specially designed probes, or questions, to address the recall of different aspects of the incident are necessary for ensuring the usefulness of the data collected. Although important, they do not directly address the issues associated with the analysis and interpretation of CDM data. The previous two sections described the mechanics of analyzing CDM interview data using the structured and the emergent themes approaches. The rest of this chapter discusses a number of limitations in the analysis and interpretation of qualitative data that CDM users should be aware of and what can be done about them.

16.6.1 Traceability

One common shortcoming in qualitative research is the "inadequate account of the process of data analysis" (Mays & Pope, 1996). In reporting quantitative research outcomes, summary tables and tests of statistical significance provide a simple means of explaining how the conclusions were reached. In qualitative studies such as those using the CDM, the reader should be given equivalent assistance—an audit trail of the analysis and interpretation process. This kind of traceability can be achieved is several ways:

- By providing the reader with adequate descriptive data (e.g., relevant excerpts from the transcripts) to set the context and to allow the reader to evaluate the significance of, for example, an experience or a string of events. This is referred to as a "thick description" (Patton, 1990).
- By making the format of the analysis transparent by showing how the data were summarized and collated.
- By reporting how the descriptions were interpreted. For example, what sensitizing concepts, including theories or frameworks in the literature, were used to interpret the data?
- By distinguishing between actual observations and the analyst's interpretation of those observations.

Collectively, these methods help the reader trace how various interpretations and conclusions were reached and therefore judge the validity of the findings.

16.6.2 Representativeness, Relevance, and Generalizability

Despite the rigorous approach to analyzing CDM data suggested in this chapter, the CDM still suffers from a limitation commonly experienced by case study methods: The number of respondents and settings is too small to make the results statistically representative. This has implications for the relevance and generalizability of the findings beyond the cases studied. It is not practical to conduct detailed case studies with a large number of respondents. Instead, stratified sampling techniques can improve the relevance of the coverage by allowing the analysts to be more selective in choosing the target groups and the kinds of incidents to study. For example, in the emergent themes approach discussed earlier, the study focused on the key dispatch controller's role rather than the assistants' roles. Furthermore, all the incidents studied were similar in terms of the level of difficulty in resource planning and coordination. The similarity in this regard provided a common focus and defines the utility of the findings in that context. Another approach for managing representativeness is to use triangulation techniques. For instance, by doing an observational study to identify skill-based behaviors that are often not explicitly articulated by interviewees. The analyst could use the observations to complement the information from the interviews to give a more complete representation of the work being performed.

16.6.3 Breadth of Issues

Representativeness is related to the breadth of issues identified in the incidents studied. A wide breadth of issues is characteristic of the situational variety inherent in naturalistic situations. Not all aspects of cognitive work may have been discussed by each interviewee. Should interesting issues reported by only a small proportion of the sample be reported? The incidents selected for CDM studies are usually demanding and often require the interviewees to perform at the limits of their ability. Identifying variety in the interviewees' cognitive tasks can provide very useful guidance for the design of information displays, especially for those catering for the unusual but having "potentially dangerous consequences if not well-supported" (Wickens, 2000). Again, the report should have enough details, such as the number of interviewees who discussed a particular concept, to allow the reader to judge whether the findings apply only narrowly or can they be used more broadly.

16.6.4 Reflexitivity and Reliability

Reflexivity refers to the bias that a researcher brings to the interview, the analysis, and the interpretation (Mays & Pope, 1996). Bias can occur during interviews, especially given the open-ended nature of CDM interviewing. Although bias cannot be totally eliminated, its effect can be reduced by acknowledging its possibility and by taking deliberate steps against it. For example, where a researcher is interested in identifying the existence of a concept, he or she should be careful not to use leading questions to establish its existence. Bias can also occur in the coding and interpretation of the data. In the CDM, this can be addressed by having multiple coders perform the analysis and comparing the extent of the correspondence in their coding of concepts. Such practices can help prevent the occurrence of the equivalent of Type I and Type II errors: reporting the occurrence of a concept when it is not really there and not reporting it when it is. Such errors can occur when individual analysts read too much into what was actually intended by the interviewee or by taking a statement out of context. Maintaining notes on the rationale for the coding and interpretation is helpful for tracing and explaining subsequent variability in results by different coders.

16.7 CONCLUSION

This chapter has described two complementary approaches for analyzing qualitative data collected from CDM interviews. The structured approach provides a mechanism for collating relevant information according to an a priori framework. The emergent themes approach allows common themes and concepts to be systematically identified and studied across different interviews and incidents. Both approaches are intended to reduce and organize data to make them more manageable for interpretation. As rigorous and systematic as both approaches can be, they are not without limitations. These limitations have been observed in the broader tradition of qualitative research. Although the relevance of conclusions depends very much on the validity of the data itself, how the data are analyzed and interpreted is equally important. The analyst needs to address issues of representativeness, breadth of issues, and reflexivity. By providing sufficient data, including thick descriptions, the analyst will allow readers to trace the process by which the conclusions were reached and judge for themselves the validity and generalizability of the conclusions. Although good interviewing instruments are vital for ensuring that the CDM collects useful data, being able to manage the resulting voluminous data set is also important. In fact, it is essential for gaining significant insights into the nature of human cognitive work and the manner in which decision aids can be designed to lead to even greater human performance benefits.

REFERENCES

Crandall, B., & Getchell-Reiter, K. (1993). Critical Decision Method: A technique for eliciting concrete assessment indicators from the intuition of NICU nurses. *Advances in Nursing Science, 16*(1), 42–51.

Flanagan, J. C. (1954). The critical incident technique. *Psychological Bulletin, 51*, 327–358.

Gordon, S. E., Schimerer, K. A., & Gill, R. T. (1993). Conceptual graph analysis: Knowledge acquisition for instructional system design. *Human Factors, 35*(3), 459–481.

Hoffman, R. R., Crandall, B., & Shadbolt, N. (1998). Use of the Critical Decision Method to elicit expert knowledge: A case study in the methodology of cognitive task analysis. *Human Factors, 40*, 254–276.

Johnson, P., & Johnson, H. (1991). Knowledge analysis of tasks: Task specification for human-computer systems. In A. Downton (ed.), *Engineering the Human-Computer Interface* (119–144). London: McGraw-Hill.

Kaempf, G. L., Wolf, S., & Miller, T. E. (1993). Decision making in the AEGIS Combat Information Center. In *Proceedings of Human Factors and Ergonomics Society 37th Annual Meeting*, (pp. 1107–1111). Santa Monica, CA: Human Factors and Ergonomics Society.

Kaempf, G. L., Wolf, S., Thordsen, M. L., & Klein, G. (1992). *Decision making in the AEGIS Combat Information Center: Task 1* (Report prepared for the Naval Command, Control and Ocean Surveillance Center, San Diego, CA; draft technical report). Fairborn, OH: Klein Associates Inc.

Klein, G. A. (1989). Recognition-primed decisions. In W. B. Rouse (Ed.), *Advances in man-machine systems research* (pp. 47–92). Greenwich, CN: JAI.

Klein, G. (1993). *Naturalistic decision making: Implications for design* (SOAR 93–1). Wright-Patterson Air Force Base, Crew Systems Ergonomics Information Analysis Centre.

Klein, G. (1997). The Recognition-Primed Decision (RPD) Model: Looking back, looking forward. In G. K. Caroline & E. Zsambok (Eds.), *Naturalistic decision making* (pp. 285–303). Mahwah, NJ: Lawrence Erlbaum Associates.

Klein, G. (2000). Analysis of situation awareness from critical incident reports. In D. J. Garland (Ed.), *Situation awareness: Analysis and measurement* (pp. 51–71). Mahwah, NJ: Lawrence Erlbaum Associates.

Klein, G. (2001). Features of team coordination. In E. Salas (Ed.), *New trends in cooperative activities: Understanding system dynamics in complex environments* (pp. 68–95). Santa Monica, CA: Human Factors and Ergonomics Society.

Klein, G. A., Calderwood, R., & Macgregor, D. (1989). Critical decision method for eliciting knowledge. *IEEE Transactions on Systems, Man and Cybernetics, 19*, 462–472.

Mays, N., & Pope, C. (1996). Quality in qualitative health research. In N. Mays (Ed.), *Qualitative research in health care*. London: BMJ Publishing Group.

McFarren, M. R. (1987). *Using Concept Mapping to define problems and identify key kernels during the development of a decision support system*. Unpublished Master's thesis in Operations Research, Air University (USAF), Wright-Patterson AFB, OH.

Militello, L., & Lim, L. (1995). Patient assessment skills: Assessing early cues of necrotizing enterocolitis. *Journal of Perinatal and Neonatal Nursing, 9*(2), 42–52.

Milliken, G. A., & Johnson, D. E. (1984). Analysis of messy data. Belmont, CA: Lifetime Learning Publications.

Patton, M. Q. (1990). *Qualitative evaluation and research methods* (2nd ed.). Newbury Park, CA: Sage.

QSR. (1994). QSR NUD•IST. Qualitative Solutions and Research Pty Ltd, La Trobe University, Melbourne, Australia.

Redding, R. E.(1995). Cognitive task analysis for instructional design: Applications in distance education. *Distance EDucation, 16*(1) 88–106.

Roth, E. M., & Mumaw, R. J. (1995). Using Cognitive Task Analysis to define human interface requirements for a first-of-a-kind systems. In *Proceedings of Human Factors and Ergonomics Society 39th Annual Meeting*, (pp. 520–524). Santa Monica, CA: Human Factors and Ergonomics Society.

Schraagen, J. M., Chipman, S. F., & Shalin, V. L. (Eds.). (2000). *Cognitive task analysis*. Mahwah, NJ: Lawrence Erlbaum Associates.

Seamster, T. L., Redding, R. E., & Kaempf, G. L. (1997). *Applied cognitive task analysis in aviation*. Aldershot, England: Avebury Ashgate Publishing Ltd.

Simon, H. A. (1977). *The new science of management decision*. Englewood Cliffs, NJ: Prentice Hall.

Wickens, C. D. (2000). The trade-off of design for routine and unexpected performance: Implications of situation awareness. In D. J. Garland (Ed.), *Situation awareness analysis and measurement* (pp. 211–225). Mahwah, NJ: Lawrence Erlbaum Associates.

Wong, B. L. W., & Blandford, A. (2001). Situation awareness and its implications for human-systems interaction. In M. Apperley (Ed.), *Proceedings of the Australian Conference on Computer-Human Interaction (OzCHI 2001)* (pp. 181–186). Perth, Australia: CHISIG, Ergonomics Society of Australia.

Wong, B. L. W., Sallis, P. J., & O'Hare, D. (1997). Eliciting information portrayal requirements: Experiences with the Critical Decision Method. In H. Thimbleby, B. O'Conaill, & P. Thomas (eds.), People and Computers XII, *Proceedings of HCI '97*, (pp. 397–415). London: Springer-Verlag.

17

Using Task Analysis for Information Requirements Specification: The Sub-Goal Template (SGT) Method

Thomas C. Ormerod
Lancaster University

Andrew Shepherd
Synergy Consultants

Designers increasingly recognise the need for user-centred approaches to system specification, yet the last thing they want is for yet another design method to be imposed upon them. To receive widespread use, task analysis methods must be an integrated and primary part of the design process. This chapter describes the sub-goal-template (SGT) method for producing requirements specifications from task analyses. The SGT method captures a set of information requirements for each user task and documents the commonalities and differences among tasks and task sequences under constraints of context and limits of performance. Critically, this information is specified in a form that can be used directly by clients and designers as a specification of information requirements, and even as tender and evaluation documents. A brief overview of the SGT approach to hierarchical task analysis is illustrated with an example showing the design of supervisory control systems for a simplified railway network.

17.1 INTRODUCTION

Articles on task analysis often begin with a plaintive cry that system developers fail to consider the human user sufficiently early in the design process, focusing instead on the specification of functional requirements. We began most of the papers we have written to date on task analysis in a similar vein. Systems design methods emphasize the configuration of elements to process materials and information, but these design methods do not take full account of the tasks that must be carried out to supervise or control processing elements. This argument still needs to be made: Designers do not always recognize the need for user-centered methods such as task analysis, believing that system design methods are what design support is all about. Without understanding the tasks that human beings are required to carry out, designers risk creating design errors. If people are involved in setting up equipment, it is important to know this, because a design can affect where equipment is sited and the environmental protection that must be provided for personnel. Equally, it is important to establish, early on, how many people will be involved in operating a system because the general manner of their control, where they work, and how they will communicate with each other constitute design constraints that will affect the choice of system elements, site, and access. Indeed, recognizing the contribution that an operating team will make can affect engineering considerations such as the use of automation as well as how the operating team will interact with the system.

Despite recognizing the continued importance of task analysis in the design process as a means of identifying design constraints, in this chapter we want to rehearse the converse argument: Designers are *right* to be suspicious of task analysis methods that concentrate upon human-computer interactions at the expense of the engineering of the system (see also chap. 1). In reality, system designers increasingly recognize the need for user-centered approaches to system specification, but they are also concerned to preserve control over specifying the functionality of systems. This is understandable and essential for ensuring that systems meet their objectives. Designers must constrain the manner in which systems are used in order to preserve security, safety, productivity, and other commercial and functional benefits. Consider the following contexts:

- In service-sector organizations (e.g., telephone sales), careful procedures must be followed to protect customer identity information and to ensure transactions are completed securely.
- In the supervision and control of a physical plant, it is necessary to preserve safe operating conditions while at the same time optimizing productivity and maintaining the plant and equipment in a serviceable condition.
- The supervision and control of transportation systems must allow safe yet commercially practicable operation. Aircraft movement is constrained by rules for maintaining aircraft separation as well as by principles governing payload and flight. The movement of trains is constrained by signaling rules as well as timetabling.
- The practices and procedures adopted in the health services to provide care and treatment for patients must be dictated by nursing and medical knowledge to ensure the well-being of the patient and by legal constraints to protect the interests of staff and their employers.
- Designers of computer-based user interfaces need to record the constraints on how tools and equipment are controlled via the interface—pumps should not be started before they are primed, missiles should not be fired before the target is located and fixed, and so on.

In each of these contexts, there are critical user-centered issues: Operators need appropriate training, expertise, instruction, feedback, supervision, workload, communication channels, and so forth. However, there are also physical, temporal, regulatory, and other properties of practical and safe system operation that must be captured and understood. These are not issues about the capabilities or otherwise of the user: They are issues about the use of the system, and they should remain the central focus of the designer. Thus, the systems designer must be clear about operational constraints as well as the human factors involved in interacting with systems. Discovering these constraints and factors should be the aim of task analysis in interactive systems design. The modest extent to which task analysis methods have been adopted by the systems design world is well-documented (e.g., Diaper, 2001; Dillon, Sweeney, & Maguire, 1993; Lansdale & Ormerod, 1994). In our view, this slow uptake can be explained, in part, by the failure of task analysis methods to deliver both operational and interactive requirements.

The widespread use of task analysis methods is further inhibited by the fact that designers frequently find them to be nonintuitive and jargon riddled and to carry heavy learning overheads (see also chap. 11). Many task analysis and other human-centered methods do not deliver a usable product to the designer. They may alert the designer to important issues, but their outputs are not clear signposts to implementation. Consequently, designers end up working with poor-quality outputs of poorly conducted task analyses (cf. Parker, 2001) while carefully crafted methods remain the rarefied tools of the researchers who developed them. Our philosophy behind developing the subgoal template (SGT) method was to provide a single approach to capturing system operation and user information requirements (for detailed descriptions of the SGT method, see Ormerod, 2000; Ormerod, Richardson, & Shepherd, 1998, Shepherd, 1993). At the same time, we set out to minimize the need for learning, jargon, and special tools while providing designers and clients with recognizably useful output.

In developing the SGT method, we were mindful of the need to leave designers in control of the design process. Task analysis has been criticized (e.g., Benyon, 1992) for overconstraining the solution options that a designer can consider (see also chap. 7). In part, this view stems from the (in our opinion, incorrect) assumption that task analyses should model only existing user-system interactions. Where this is the case, then the use of a task analysis method condemns the designer to work within the constraints of the existing interaction method. This need not happen if the analyst is alert to the fact that neutrally stated operations may, in principle, be redescribed using different methods that use different technologies. For example to "transfer luggage from one airplane to another" may be achieved by the baggage handler making lots of journeys on foot, by loading a truck, or by installing a conveyor belt. To explore this task at a lower level of detail, it is necessary to focus on one technology rather than another, for each has implications for what the baggage handler will be required to do. If the analyst can remain alert to the distinction between the functional requirements of the task (i.e. to transfer the luggage) and the method by which a transfer is accomplished, then it is possible to consider different ways of doing things.

Most task analysis methods stress the need for the analysis process to be iterative and reflexive, and we concur with this view. Task analysis should be a process of information gathering, organization, negotiation, and revision. However, the analyst needs stopping rules (see chaps. 1, 3, 4, and 22). On the one hand, stopping rules should provide an unambiguous decision procedure that enables the analyst to determine when a task has been analyzed to a point at which its structure has been decomposed, documented, and contextualized in sufficient depth that nothing important remains unreported to the designer. On the other hand, stopping rules should force an analysis to cease before it starts dictating *how* tasks should be implemented by the designer. Maintaining a clear separation between task analysis and design is extremely difficult, but we believe that this separation needs to be addressed. Our belief provided a key motivation for developing the SGT method.

In the remainder of this chapter, we describe the SGT method and how it provides designers with information requirements at an appropriate level of analysis to allow creative design. First, we examine how the scheme inherits the processes and products of Hierarchical Task Analysis (HTA; chap. 3), recognizing both the strengths of HTA and its perceived weakness in delivering to designers a directly usable product. We then outline the SGT method itself and present a brief example of how the SGT method can be applied to the design of supervisory control systems.

17.2 HIERARCHICAL TASK ANALYSIS

HTA was developed by John Annett and Keith Duncan in the 1960s as a general method for examining complex tasks (e.g., Annett & Duncan, 1967; Annett, Duncan, Stammers, & Gray, 1971; Duncan, 1972; Shepherd, 2001). A more complete account of HTA is provided in chapter 3 of this handbook, so here we focus on key concepts that are inherited by the SGT method.

17.2.1 Operations

HTA focuses the analyst's attention on the industrial or commercial goal to which the operator's effort in executing the task is directed. Its principal unit of description is the operation. An operation is a statement that briefly describes the operator's current objective, expressed as an imperative. For example, the operation to correct a fault in a washing machine might be expressed as "repair washing machine." This form of description is versatile and can be used to describe broader activities ("maintain domestic appliances") or more focused activities ("replace inlet filter from the cold water feed"). This facility to describe activities at different levels of detail enables the analyst to adjust the grain of description for different parts of the task. Thus a task can be decomposed into its components, which can in turn be further decomposed.

An operation is stated in a way that is neutral with respect to its solution (Duncan, 1972, used the phrase "psychologically celibate"). It is important to distinguish between an operation and the behavior that enables it to be carried out. As far as a system is concerned, the precise strategy by which an operator carries out an operation is immaterial provided it is successful and does not introduce unacceptable consequences. Performance strategy could be guided by a job aid, recalled as a consequence of repetition during training, recalled from a recently successful encounter in similar circumstances, developed through the operator's analysis of a current situation, or prompted by the affordances of a display.

Operational behavior, however it is executed, is goal directed. The operator selects actions from those available at the task interface in order to realize the goal. Actions are regulated by information gleaned from the system and can be described using Annett and Duncan's input-action-feedback (I-A-F) classification. This information will provide detail about the system (input), indicating when action is necessary and guiding choices between appropriate actions (action) as well as indicating when action may cease (feedback). Using this classification, the analyst identifies which aspects are likely to be at the root of performance problems and so guides the generation of hypotheses to overcome these problems.

17.2.2 Stopping Analysis

The analyst needs principles to determine when there is benefit in proceeding with further redescription. For this purpose, Annett and Duncan (1967) suggested a stopping rule that depends on analyst estimates of the probability of inadequate performance (P) and the cost of inadequate performance (C). Their stopping rule was that analysis should be taken no further than a level of redescription where the product of P and C was acceptable to the client. To estimate C, the analyst establishes what is of value in the system, including the money that might be lost (production costs) and the health and safety of personnel and the public. Costs can be offset by benefits. For example, anticipated long-term benefits might offset immediate investment for which a return has not yet been realized. Incurred costs could also provide opportunities for staff to obtain experience and gain expertise that will provide a return in the longer term. The cost factor is an amalgam derived from the analyst's enquiries and should reflect the system and not the analyst's values. To estimate P, the analyst can use reliability data or performance-shaping factors. The analyst would consider the task demands, the personnel employed to take responsibility for the task, the resources available to support the task, and the environment in which the task takes place. The $P \times C$ rule is a heuristic rather than a prescriptive arithmetic rule. Thus, if C is unacceptably high, further analysis is warranted no matter how low P is (assuming that it will not be a perfect zero). If C is quite high, it might still be worthwhile to explore the operation in greater detail, even where P is low. If P and C are both low, there would be little benefit in exploring the operation because it would not substantially affect overall performance.

17.2.3 Plans

Another important component of HTA is the plan. If an operation is decomposed into component operations, there needs to be an organizational system to indicate the conditions under which each component is carried out in order to complete the superordinate operation. For example, painting a window frame entails "prepare window-frame," "apply primer," "apply undercoat," "apply gloss-paint." These operations must be done in an appropriate order—sanding down and washing must precede applying the different coats of paint, each of which should be applied only when previous coats have dried properly. Plans are crucial to HTA. They capture many of the subtleties of tasks that cannot be explained merely by reference to operations. Shepherd (2001) illustrated numerous plans, showing how different types of plan combine to account for complex performance. Thus, some plans deal with sequences of activity, some with cycles of activity, and some with choices. Combining these in an HTA hierarchy results in apparently complex arrangements that are better understood by seeing how they result from the interaction of simpler plans.

17.2.4 Utility and Transparency of HTA

Annett and Duncan's basic ideas were informed by principles from human factors, systems thinking, and operational practices. The HTA method employs natural language to express operations and task conditions in plans. Consequently, system designers may engage their clients directly in agreeing and expressing what must be done. Equally, the client or the client's agent can validate an analysis to confirm that it properly reflects the tasks to be carried out.

By adopting an operation as the main unit of description, HTA maintains neutrality in task descriptions. An operational description indicates what must be achieved and not how it is achieved. The designer is at liberty to propose different technologies that might engage the human operator in physical work, mental work, or no work at all. In providing a human factors or resources solution to a performance problem, an analyst may evaluate and select many solutions on the basis of cost, consistency with other solutions, and utility. Thus, human performance problems may be solved by improving the environment, personnel selection, training, interface design, or a combination of these and other methods.

The stopping rule—or, more pertinently, the principle that the hierarchical description may be developed to different levels of detail in different places according to the importance of these places—means that the values of the system and not simply those of human factors are taken into proper account in the human factors contribution to the design process. It means that human factors cannot dictate the route that the human operator's contribution will take unless it can be first demonstrated that the issue addressed is important.

17.2.5 HTA and Interface Design

HTA has been shown to be capable of providing useful descriptions of a variety of tasks in many contexts. Equally, it has been shown to aid human factor and resource decision-making, including the design of teams and jobs, operating procedures, selection methods, interface design, and training, as well as reliability assessment and quantification. For many of these solutions, HTA provides a structure that facilitates the application of other human factor methods and principles (Shepherd, 2001). In addressing the issue of *interface design*, we are seeking a method that preserves the benefits of HTA with regard to thoroughness, flexibility of description, utility, and transparency. Interface technologies support communication between operators and systems. Typically, the operator elicits information about the system's status in order to make decisions about how best to exercise control over the system, then expresses these decisions in a way that will help effect the required changes. For this purpose, those aspects of HTA concerned with stopping rules need to be reviewed. We should distinguish between existing systems where a redesign of an interface is envisaged and new systems where an interfacing method has yet to be prescribed. In an existing system, the decision to revise an interfacing method may be taken following a completed task analysis, including an estimation of the risk of using the existing interface technology. The analyst must establish how the operator uses the existing technology. The decision to change may involve judgments about the suitability or otherwise of improved training, job aids, recruitment procedures, or other organizational factors (chap. 1). If redesign of an interface is judged essential, then it will be necessary to backtrack up the task hierarchy to a place where the functional requirement of interaction is expressed in broad detail, before the level at which the interfacing technology was assumed. This level is the starting point from which a new technology or approach can be devised. If interface design is considered for a new system, then the analysis should proceed as far as this point of functional specification and no further, since there would then be a need to make assumptions about how interfacing is to be accomplished. At this stage, a method for interface design can be considered, and it is at this point that the SGT method is utilized.

Neutrality regarding the design hypothesis is maintained within HTA by two features. First, the analyst preserves operational neutrality in task descriptions, thereby enabling different types of solution to be proffered. Second, HTA moves from general operational statements toward specific methods where devices for delivering solutions are assumed. This means that an analysis can, in principle, be developed in a number of ways so long as different technological solutions can be entertained. If one such development proves unsatisfactory, then the analyst may revert to the common neutral point and consider another.

17.3 THE SGT METHOD

The task of specifying the information requirements consists of specifying the reception and transmission of information (including exercising control) that the operator requires to carry out tasks in order to operate a system. The SGT scheme was devised to provide a nomenclature for classifying and organizing the stereotypical tasks that operators undertake so that the information necessary for each task could be collated across the analysis to give a requirements specification for system designers. Our aim was to retain the simplicity and utility of HTA by providing a scheme that captures the vast majority of operations in as few categories and hierarchical levels as possible (that is, to minimize the breadth and depth of task classes). In general, we adopt the view shared by others in the task analysis community (e.g., Payne & Green, 1986) that recognizing consistent structures in any task environment is good design practice. The resulting SGT scheme comprises a categorization of operator task by subgoal, given an overall goal to control a system in pursuit of a desired outcome.

The SGT method terminates in the assignment of subgoal templates to each task element in a task analysis, at which point information requirements can be assigned. To get to that stage requires the analyst and engineer or software developer to undertake a process of ongoing negotiation. Figure 17.1 shows a schematic description of this process, illustrated as a task analysis of a generic process-control operation (to maintain a system). The main stages of this generic process-control operation are explained in Table 17.1. This shows maintaining plant operations to be described in terms of two main activities, "monitoring plant" and "dealing with significant perturbations." The analysis consists of two kinds of activity: task decomposition, in which tasks are broken down into their constituent subtasks, and task redescription, in which tasks are reconceptualized using the concepts and notation of the SGT method.

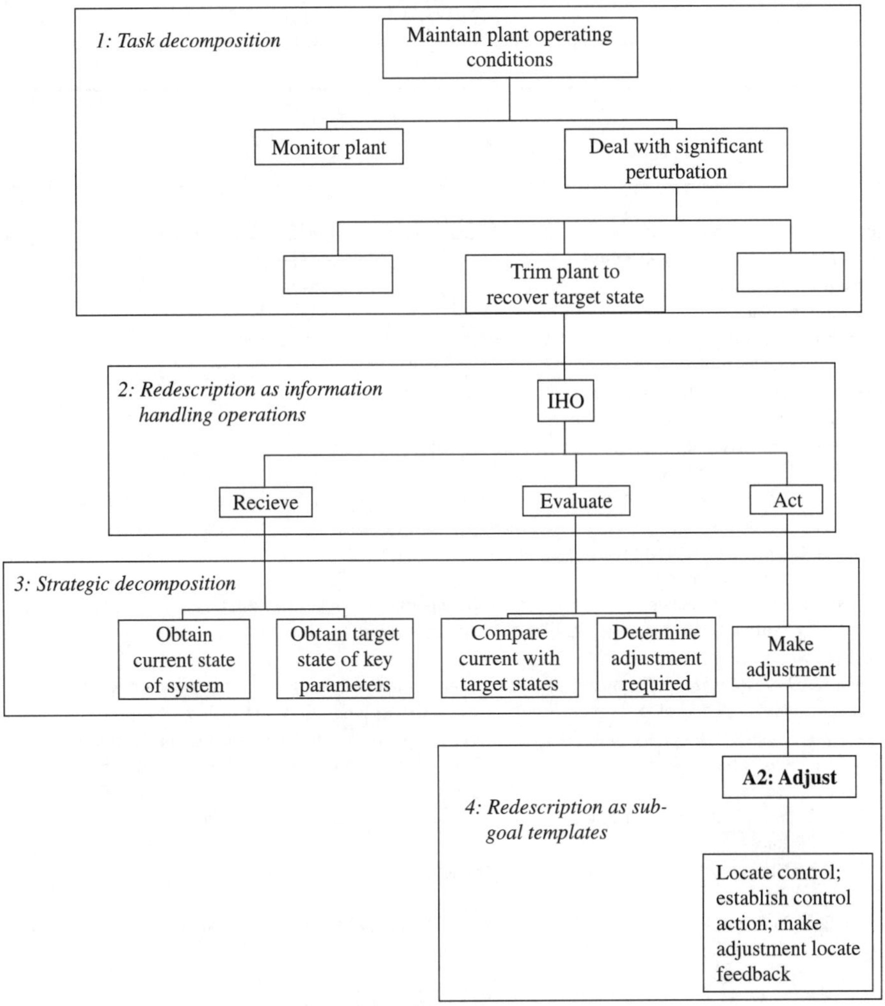

FIG. 17.1. The hierarchy of information-handling operations, showing (1) task decomposition, (2) redescription as information-handling operations (IHOs), (3) strategic decomposition, and (4) redescription as subgoal templates with associated information requirements.

17.3.1 Information-Handling Operations (IHOs)

When an analyst and an engineer or a systems developer work together to specify the control of a plant or software application, they deal with two types of operation: functional operations, which concern the states that have to be attained for the system to work, and informational operations, which specify information and control interventions necessary to attain these states. Our starting point for the SGT method is the recognition that all systems, whether they are process plants or PC applications software, will at some point require information to be handled outside the system being controlled (i.e., in the application domain; chap. 1). Indeed, that is generally the reason why people are employed. These are information-handling operations. A task can be redescribed as an information-handling operation if it provides the point where system activity requires an information-based intervention.

TABLE 17.1
The Relationship Between Monitoring and Fault-Handling Tasks

Tasks	*Explanations*
0 Maintain plant operating conditions	This is a common requirement in all system supervisory tasks.
plan 0: Throughout - 1. If perturbation is judged significant - 2.	
1 Monitor plant.	
plan 1: Do 1, then, if there is a deviation from target, do 2.	
1.1. Monitor specific parameters.	
plan 1.1: Do 1 & 2, then do 3.	
1.1.1. Obtain current state of key parameters.	
1.1.2. Obtain target state of key parameters.	These 3 activities are the essence of obtaining information in monitoring tasks.
1.1.3. Compare current with target states.	
1.2. Determine whether perturbations are significant.	Some perturbations result from temporary blips and will recover if left.
2 Deal with significant perturbation.	
plan 2: Do 1, then, according to plan, 2, 3, 4, 5, & 6.	Diagnosis and at least one other activity are required to deal with the problem, finishing with 8.
2.1. Diagnose and plan appropriate remedial action.	This may involve identifying a cause, but primarily it is concerned with determining how to proceed to maintain the system state.
2.2. Trim plant to recover target state.	It may be sufficient to "fine-tune" the system.
2.3. Put standby equipment on line.	It may be sufficient to switch to standby equipment.
2.4. Compensate for off-specification conditions.	Production may be optimized by adjusting running rates elsewhere.
2.5. Isolate problem.	The problem may be isolated such that other aspects may proceed in the best manner possible.
2.6. Shut down system.	A complete shutdown may be required to make the system safe for repair.
2.7. Rectify fault.	Usually repair or replacement.
2.8. Recover target operating state.	When a repair is complete, operating conditions must be recovered.

Information-handling operations involve the same three classes of activity: receiving information, evaluating information, and acting on information. For example, when a red wavy line appears under the text being typed in Microsoft Word, the system is engaging the user in an information-handling operation. The user has to receive information (that there is a problem with a word located above the wavy line), evaluate it (the word processor does not recognize the spelling of "SGT" but it is an acronym that we wish to retain), and act (to instruct the spell checker to learn or ignore the acronym). A "receive" component might involve an operation to obtain information about the status of the system (including categorical states, trends, prognoses), receive a communication (usually a command to execute a specified action, achieve a specified state, or adjust current operating goals), or retrieve data (about targets, previous states pertinent to current control, etc.). In effect, information can come from the system, from another operator, or from within the operator. An "evaluate" component might require the operator to monitor a system, diagnose a fault, plan a sequence of activities, or manage and oversee a change to a system. An "act" component might require an operator to carry out action to change the state of the system, issue a communication (usually to another person to execute a specified action, achieve a specified state, or adjust current operating goals), or record data for future use. Once the analyst has decomposed a task goal sufficiently to recognise that a subtask is an information-handling operation, the method requires that decomposition ceases, albeit temporarily. The reason for stopping the decomposition process here is that it is the first point at which the neutrality of task analysis is challenged.

There are potentially an infinite number of ways in which an information-handling operation might be conducted. For example, the designers of word-processing software might have decided to delay the indication of spelling errors until the user has invoked a spell check command or they might have decided to provide automatic correction of errors. The way in which an information-handling operation is carried out is a *strategic* decision. The task analyst cannot simply decompose it—the strategy must be negotiated with the client, engineer, software developer, and so on. Often, the strategy will involve elements that are not supported by interface design. For example, in a power plant, the subtask "rectify fault" might be dealt with by requiring the operator to diagnose and adjust the system or by triggering an automatic mandatory shutdown. The task analyst cannot decide what the appropriate strategy will be in advance of negotiation. Once the analyst has identified the designer's or client's (current) chosen strategy for achieving an information-handling operation, each of the "receive," "evaluate," and "act" components can be further decomposed into their constituent tasks.

17.3.2 Subgoals

To provide an unambiguous stopping point at which the analysis can be handed over to the interface designer, there needs to be a specification of the lowest level of operator task below which there should be no further decomposition. The initial SGT scheme (Shepherd, 1993) adapted the structure of process-control activities outlined by Annett and Duncan (1967). At any particular point in time, a process controller may have a subgoal to change the state of a system (by action), to observe the state of a system (by monitoring), to communicate information with or about a system (by communication), or to repair a system (by diagnosis). We subsequently extended the scheme to provide a nomenclature for classifying the stereotypical tasks that arise in the operation of other kinds of interactive system (Ormerod, 2000). This extension involved the addition of two subgoals: to exchange information with a system (by entering or extracting data) and to navigate through a system (by moving, locating, or browsing). These additional subgoals broadened the remit of the scheme to account for office information systems, ubiquitous computing devices, and so forth.

However, the extensions came at the cost of coherence and overlap in the scheme. One of our goals here is to present a revision to the scheme. In particular, the original Communication subgoal of Shepherd (1993) is merged within the Exchange subgoal. Furthermore, we recognized that the "diagnosis" set of subgoals included in the original scheme must be understood at a strategic level and have dropped this subgoal task from the scheme.

The revised classification scheme is shown in Table 17.2. At a lower level, each subgoal is differentiated by "task elements" (i.e., specific variants of the subgoal). The reason for this lower level of specification is to capture the point that operators may set up and execute the same subgoal for a number of different purposes, depending on the stage and conditions of the activity they are accomplishing. So, for example, the Monitoring subgoal is subdivided into three specific types of monitoring element. At any particular point, the purpose of monitoring

TABLE 17.2
Subgoal Templates (SGTs), Specific Task Elements, and Information Requirements (IRs)

SGTs	*Task Elements*	*Context for Assigning SGT and Task Element*	*Information Requirements*
Act		Perform as part of a procedure or subsequent to a decision made about changing the system.	Action points and order; Current, alternative, and target states; preconditions, outcomes, dependencies; halting, recovery indicators.
	A1 Activate	Make subunit operational: switch from off to on.	Temporal/stage progression, outcome activation level.
	A2 Adjust	Regulate the rate of operation of a unit maintaining "on" state.	Rate of state change.
	A3 Deactivate	Make subunit nonoperational: switch from on to off.	Cessation descriptor.
Exchange		To fulfil a recording requirement. To obtain or deliver operating value.	Indication of item to be exchanged; channel for confirmation.
	E1 Enter	Record a value in a specified location.	Information range (continuous, discrete).
	E2 Extract	Obtain a value of a specified parameter.	Location of record for storage and retrieval; prompt for operator.
Navigate		To move to an informational state for exchange, action, or monitoring.	System/state structure, current relative location.
	N1 Locate	Find the location of a target value or control.	Target information, end location relative to start.
	N2 Move	Go to a given location and search it.	Target location, directional descriptor.
	N3 Explore	Browse through a set of locations and values.	Current/next/previous item categories.
Monitor		To be aware of system states that determine need for navigation, exchange, and action.	Relevant items to monitor; record of when actions were taken; elapsed time from action to the present.
	M1 Monitor to detect deviance	Routinely compare system state against target state to determine need for action.	Normal parameters for comparison.
	M2 monitor to anticipate cue	Compare system state against target state to determine readiness for known action.	Anticipated level.
	M3 Monitor transition	Routinely compare rate of change during state transition.	Template against which to compare observed parameters.

may be to check that the system is functioning as the operator expects (to detect deviance), to check the system prior to further action (to anticipate change), or to inspect the rate of change from one state to another (to inspect transition).

The point of the task element level of categorization is to provide designers with a way of supporting the basic operations that an operator must accomplish. For example, traditional process-control displays have presented the operator with a bank of indicators, alarms, and controllers (or their computer-based equivalents), one set for each component (tank, valve, pipe, etc.) in the plant. The shift from pneumatic to computer-based control raised a problem regarding the monitoring of complex processes: How can an operator monitor a large system when he or she can only view a small fraction of it at any one time on a computer monitor? Relying on automated alarms is insufficient. When one device goes wrong in a heat exchange loop, for example, every alarm will go off on every device as the interdependencies among components work their way through the system. Supporting monitoring by allocating displays and alarms to every device is a recipe for information overload of the kind implicated in one of the most widely publicized accidents in recent years, namely, the Three Mile Island emergency (Rubinstein & Mason, 1979). Task elements specify the tasks that operators must carry out in order to ensure the proper functioning of a plant. They are intended to aid the designer in specifying precisely how to support the activities of monitoring by appropriate information without overloading the operator (chap. 14).

One obvious question is whether the subgoal templates and task elements that we have specified are complete or whether there remain further subgoal templates or task elements to be discovered. We are open to the possibility that others may be found, though we would seek strong grounds for expanding the scheme, since we aim to keep it as simple as possible. The only justification that we would accept for expanding the scheme would be if new subgoal templates or task elements were found that had unique *information requirements*, a topic to which we now turn.

17.3.3 Information Requirements

As well as providing a taxonomy of task types, the SGT method captures the basic information requirements of each subgoal template element, and it is here that we believe the method provides the designer with precisely the information needed to develop a detailed systems design specification. Determining the information requirements of a specific subgoal template element seems at first problematic because an operator needs to conduct a number of descrete activities in order to execute a SGT-level operation. Take, for example, the subgoal template "A2 Adjust." This task element might itself be decomposed into the following sequence of activities:

- Locate adjustment device.
- Review operation of adjustment device.
- Review adjustment to be made.
- Execute adjustment.
- Monitor system response.
- Deal with deficiencies.

This set of activities is prototypically the "template" of operations needed to execute the "Adjust" subgoal task (hence the name "subgoal templates"). However, given that each of these operations is itself a subgoal template element, one could end up recursively decomposing operations ad infinitum (indeed, the "Deal with deficiencies" operation will itself entail an adjustment operation). To provide an absolute boundary to the process of operational decomposition, we have converted each of the stereotypical sequences of operations that are necessary to execute an subgoal template into a list of the minimum information that would be needed if these operations were to be undertaken. So, for example, the "Adjust" task element requires information on the following:

- Preconditions, action points, and order (to locate device, review operation, and execute adjustment).
- Current, alternative, and target states (to review adjustment to be made).

- Rate of state change, plus outcomes and dependencies (to monitor system response).
- Halting and recovery indicators (to deal with deficiencies).

In developing the SGT method, we adopted the following rule: Task elements are added to the scheme only where they possess one or more unique information requirements. This rule follows from the belief that we should provide the minimum necessary task nomenclature. Information requirements, broadly defined, comprise the data that an agent (be it human or automated, single or multiple) needs in order to accomplish a task safely and efficiently. Some information requirements are common to the task elements of a subgoal template. So, for example, all the task elements of the Action subgoal template require a specification of action points and the order in which they must be sequenced; the current, alternative, and target states; necessary preconditions, outcomes, and dependencies; and halting, reversal, and recovery information. The "Adjust" task element, however, necessarily requires information on the rate at which a system changes in response to the ongoing adjustment. It is an essential information requirement that distinguishes it from actions for which such information is optional (and might, therefore, be unnecessary or even distracting).

The information requirements of each task element are neutral as to the chosen method of implementation. For example, one way of supporting the information requirement to provide halting and recovery indicators would be some kind of "undo" facility. Alternative methods might include some kind of shutoff or "panic button," an automatic limiter, and so on. The nature of the solutions provided for each information requirement must, in our view, remain the preserve of the designer, but the information requirement offers a cue to their provision. The resultant design should then be subject to task analysis to confirm that it is appropriate and will support reliable performance.

17.3.4 Sequencing Elements

The inclusion of plans in HTA recognizes the importance of indicating the conditions under which component operations are carried out in order to satisfactorily complete a task goal. Including sequencing elements in the SGT method was, to some design researchers, controversial, because it mixes data and control flow in the same specification (e.g., Benyon, 1992). However, as Ormerod (2000) argued, sequences of operations dictate information requirements as much as operations themselves. For example, if the designer is developing an interface to start up a plant, then it is essential that the same interface support the parallel task of monitoring the state of the plant during startup.

The SGT method capitalizes on the fact that there are, in principle, only four ways in which operations can be sequenced relative to each other:

1. Operations may follow each other in a strict and mandatory sequence, the outcomes of earlier operations providing the operating conditions for later operations. For example, "prepare," "startup," "maintain," and "shutdown" operations will always appear in a strict linear order regardless of intervening operations or conditions.
2. Operations can be contingent on the outcomes of earlier operations or contextual factors and so arise in a sequence only if the necessary preconditions exist. For example, operations involving adjustment will typically be contingent on other operational or contextual outcomes.
3. Operations may require parallel performance. For example, system control actions generally must be performed alongside monitoring activities.
4. Operations may not have a predetermined order but may instead be undertaken as and when the operator wishes to do so. For example, routine documentation updates (e.g., personnel

TABLE 17.3
Sequencing Elements Used to Construct Plans Containing Subgoal Template Elements

Code	*Type*	*Syntax*
S1	Fixed sequence	S1 Do X . . .
S2	Contingent sequence	S2 If (c) then do X . . . If not (c) then do Y . . .
S3	Parallel sequence	S3 Do together X . . . Y . . .
S4	Free sequence	S4 In any order do X . . . Y . . .

logs) may take place whenever the operator has time to do them. Their inclusion in a sequence of operations may be mandatory, but the precise point at which they occur may be determined online rather than in advance.

Table 17.3 shows the notation used to capture these sequencing elements. There are two points to observe about the notation. First, the "S2 Contingent sequence" element makes explicit the (usually implicit) sequencing constraint that there are always two states and/or outcomes that must be considered by the designer. What the operator tasks are if the preconditions do arise and what the operator tasks are if the preconditions do not arise. For example, a common contingency might be "If the alarm sounds, then call the supervisor." An "if not" condition may simply give rise to the operational statement "continue" (e.g., "If an alarm does NOT sound, then continue to monitor the process"). Nonetheless, failure to make such equilibrium-maintaining information explicit can lead to problematic gaps in the information requirements specification.

Second, the notation requires that each sequence be nested within a procedural flow of control. Nesting of sequence elements may, in some instances, reverse the order of tasks as they are naturally specified in a task decomposition. Table 17.7 (part of the railway network case study described below) contains an example of such a reversal. The need to "deal with problems" may arise at any time during the operations of "ensuring railway conditions" and "Power up." Thus, the plan for sequencing these three operations begins with an "S3 Do in parallel" sequencing element that makes explicit the need to deal with problems throughout the operational sequence.

17.4 EXAMPLE: SUPERVISORY CONTROL OF A RAIL NETWORK

System supervisory tasks arise when human operators are employed to maintain conditions in a complex system. There are many such tasks, including the supervision of transport systems, such as railway networks (chap. 16) and air traffic control (chaps. 1, 13, 19, and 25), and automated production units, such as flexible manufacturing systems and process plants concerned with chemical manufacture (chap. 3), petroleum refining, and power generation. In each of these cases, we encounter staff responsible for controlling operations, usually through symbolic representation of the elements under control. Control staff must oversee automated systems to make sure that they are operating according to design and that no problems have arisen. Staff may be required to supplement automatic control, alter operating rates and conditions, control intermediate stages of system startup and shutdown, or deal with problems. There are several reasons why such remote, symbolic control is required: The control staff is in a position of safety away from risky physical environments, can oversee a large number of elements distributed remotely, and can collate information from different sources in one place.

Supervisory systems have a wide range of technologies but share many characteristics, such as the less than full control that operators have over events. For example, the movement of aircraft is a consequence of pilot decisions. The flight of an aircraft must be monitored by the air traffic controller through radar and voice communication with the flight crew and controlled by issuing instructions, which may or may not be wholly complied with. Chemical reactions generally abide by the laws of chemistry and physics, but their precise effects may vary according to processing conditions. A doctor may prescribe a course of treatment, but the patient's response needs careful monitoring to establish the precise effects of treatment. Generally in such systems, controllers must obtain information, make judgments about how target conditions can best be reached or reestablished, act on these judgments, then monitor the effects of their actions and adjust where necessary. Having appropriate information, action controls, and feedback is paramount.

A simplified railway control task illustrates the application of the SGT method to supervisory tasks. The railway example shows how the SGT method presents explicit challenges for a designer engaged in providing dialogue features to support a task. Table 17.4 describes the main operations involved in this task. In HTA, the task is described as "Supervise railway

TABLE 17.4
Decomposition of Main Railway Control Tasks

Task	*Explanation*
0 Supervise railway system *plan 0: S3 Do together 1 and 2.* *S2: If problems are identified through system monitoring or if incident is notified, do 3, else continue.*	Most railways operate on a daily cycle, with services commencing in the early hours and then increasing in frequency to offer a full timetable. Toward the end of the day, the services are reduced according to the timetable. As track clears, engineering hours are scheduled.
1 Provide services	Controllers bring trains into service at the start of the day and introduce further trains according to the timetable. As trains are no longer required, they are removed from service.
2 Maintain best service	When trains are introduced into service, they must be controlled to comply with the timetable, without risk to services, equipment, staff, or the public.
3 Deal with system problems	If problems arise, they must be dealt with promptly and effectively in order to maintain safety and service standards.

Note. This table shows the tasks as falling into two parallel activities (S3 sequence) and one nested contingent activity (S2 sequence).

TABLE 17.5
Decomposition of Activities Involved in "Provide Services" Task and Their Redescription as Information-Handling Operations (IHOs)

Task			*Explanation*
1		Provide services	
	plan 1:	*S1 At start of day/shift - 1.1.* *S2 If start of day - 1.2.* *If not - continue.* *S1 Throughout day - 1.3.* *S2 If timetable alerts -1.4.* *If not - continue.* *S1 At end of day - 1.5.*	Service provision entails some activities at the start of the day, then a series of activities according to the timetable.
	Subtask:	*IHOs:*	
	1.1.	Revise timetables and rotas according to information	Various documents must be read to establish any last-minute changes to the timetable.
	1.2.	Carry out overall system startup checks	Ensure that all systems are operational and that there are no significant resource deficiencies that would prevent services from starting.
	1.3.	Monitor timetable	The timetable must be monitored throughout the day, because it will prompt many of the necessary control interventions.
	1.4.	Make section of track available	When prompted by the timetable, steps are taken to ensure that track sections may be safely used.
	1.5.	Clear line section at end of day	When the last train has passed through a section of track at the end of the day, ensure that the track can be released for safe engineering work.

system," a global task that can be decomposed into three main operations governed by a plan. Note that each of the resulting subtasks can be further decomposed by the analyst. None of them necessarily require an information-based intervention for their execution.

The activities involved in the subtask "1 Provide services" are decomposed further in Table 17.5. In many system supervisory tasks, there are initial stages where the controller makes adjustments to operating targets: Here these adjustments are concerned with amending the timetable. The controller then makes system checks to establish any impediments to progressing with the startup. The decomposition of the "Provide services" subtask reveals a series of activities that are unambiguously information-handling operations. It is inconceivable that any of these supervisory activities could be conducted without information being received, evaluated, and acted upon. It might be an automatic system that conducts these activities (e.g., some kind of intelligent agent software that automatically adjusts the timetable according to received information regarding deviations from normal running), although in the context of rail network supervision, the involvement of a human operator at this level is likely if not mandatory, on grounds of ensuring safe operation. Regardless of the involvement of a human operator, the activities are information-handling activities, and the client, engineer, or software developer needs to work with the analyst to further understand the strategies that might be considered to execute each of these activities.

Table 17.6 shows a strategic decomposition of the activity "Revise timetables and rotas according to information." This task is undertaken to review the timetable according to more recent events. The assignment of "Monitor to detect deviance" subgoal templates to each

TABLE 17.6
Strategic Decomposition and Assignment of Subgoal Templates (SGTs) to "Revise Timetables and Rotas According to Information" Task

Task Element	*Explanation*	*SGT*
1.1. Revise timetables and rotas according to pertinent information	Timetables and rotas are typically published in advance but should be revised at the start of each day.	
plan 1.1: S3 Do together 1.1.1 and 1.1.2. S2 If necessary, do 1.1.3. If not, continue.		
1.1.1. Monitor sources of information		
plan S4: In any order 1.1.1: do 1.1.1.1–>1.1.1.3.		
1.1.1.1. Inspect log book	Logged information from previous shifts may affect current activities.	M1
1.1.1.2. Inspect traffic circulars and line notices	Special timetabling modifications may be prompted and communicated through traffic circulars.	M1
1.1.1.3. Examine nightly engineering notices and safety arrangements	Work during engineering hours may affect the availability of sections of track, thereby interfering with the timetable.	M1
1.1.2. Evaluate information and plan revisions		
1.1.3. Make adjustments to control instructions		
plan S4: In any order 1.1.3: do 1.1.3.1–>1.1.3.3.		
1.1.3.1. Make changes to timetable	The timetable must be revised according to planned revisions.	A2
1.1.3.2. Make changes to staff rota	Staff rotas must be updated.	A2
1.1.3.3. Make changes to working practices	Special arrangements may need to be communicated regarding safe operating.	A2

of the monitoring subgoals and their associated information requirements (see Table 17.2) draw the designer's attention to the types of information that must be provided to support this strategy. However, treating these task elements as monitoring activities is only one possible strategy for revising timetables and rotas. An alternative strategy might be to train each operator to undertake regular sampling of information sources and to communicate these to another operator or device. This strategic approach would then be coded as a sequence of information exchange activities ("E2 Extract" followed by "E1 Enter" subgoal template elements).

The point of the strategic decomposition is to make explicit the point at which the emerging design becomes committed to specific methods for implementing information-handling operations and to encourage a review of alternative methods. It also provides the designer with the impetus to consider more than one strategy, thereby avoiding the problems of satisficing (Simon, 1981) associated with poor-quality design processes. Interface design may be affected by many aspects of work design, such as assigning communication responsibilities to other personnel. Often a controller's decision making can be substantially simplified by requiring a colleague to be more explicit in his or her encoding of events. This is particularly the case in industrial environments where people must interpret the implications of information recorded in logs without any clear format. For example, the person entering the information could be made responsible for recording the clear implications for operating so that less ambiguity would ensue. All of these contextual factors require explication, and a high-quality HTA should capture them. However, the stage of strategic decomposition provides a semiformal point at which the significance of such contextual constraints is evaluated for every strategic proposal.

Table 17.7 shows a similar examination of the strategic decomposition of "Make section of track available." The operations associated with ensuring acceptable conditions, under the strategy embodied in this analysis, require confirmation from engineers and other staff that satisfactory states obtain. It would be possible to install a signal to be sent by the engineers concerned, but, for reasons of safety, it is advisable always to supplement this with a direct communication and for the controller to enter this value on the control screen, together with the

TABLE 17.7
Strategic Decomposition and Assignment of Subgoal Templates (SGTs) to The Subtask "Make Section of Track Available"

Task Element	*Explanation*	*SGT*
1.3. Make section of track available *plan 1.3: S3 Do together 1.3.3 and S1 On timetable prompts, do 1.3.1. S1 When conditions are acceptable for starting up the section, do 1.3.2.*		
1.3.1. Ensure railway conditions are acceptable for section startup *plan 1.3.1: S4 In any order do 1.3.1.1–>1.3.1.3.*		
1.3.1.1. Establish whether engineers are clear from overnight possession	Safety practices require controllers to confirm that engineering work has been completed.	E2
1.3.1.2. Establish whether sub-station equipment is available	There must be confirmation that the section of track can be powered up.	E2
1.3.1.3. Establish whether overnight signals maintenance has been cleared and tested	Signalling systems must be shown to be fully operational. This may mean that a formal test procedure is implemented.	E2
1.3.2. Power up section of line *plan 1.3.2:* Do 1.3.2.1, then 1.3.2.2. then 1.3.2.3.		
1.3.2.1. Receive line clear/line safe message	A communication will be sent from the track to indicate that the section of line is clear. The controller must await this cue for further action.	M2
1.3.2.2. Reset tunnel telephone lines		A1
1.3.2.3. Recharge current rail section		A1
1.3.3. Deal with problems	Problems arising from operations will be dealt with by direct communication with engineering staff.	

authority given. Thus, even when conditions allow for automation of an information-handling operation, there are often sound operating grounds for retaining human intervention.

17.5 CONCLUSION

In this chapter, we have presented a revision of the SGT method, an extension of HTA devised for the purpose of supporting the development of interfaces. The method involves the decomposition of tasks to a point where information-handling operations are recognized, that is, operations in which information-based intervention becomes mandatory for continued system operation. At this point in the analysis, the method necessitates the elicitation, through negotiation with clients, designers, and so forth, of one or more strategies for implementing each information-handling operation. These strategies are then further decomposed until the operations can be redescribed as subgoal templates, that is, stereotypical operation categories that capture the vast majority of operator activities regardless of overall goal, technology, or performance context. Each subgoal template (and at a lower level, each task element, that is, specific type of subgoal template) has associated with it a set of unique information requirements (the needed information is that, which, regardless of chosen implementation method, will allow the activity to be performed at the interface).

One of the purposes of this chapter is to update the SGT method so that it is no longer geared specifically to process control domains. There are important differences in the development of process-control and other interactive systems that need to be recognized by designers. Perhaps most significant among these is the apparent nondeterministic nature of many process-control systems. Systems that embody complex processes such as heat exchangers have a nasty tendency to behave in an unpredictable and unstable fashion, making the role of the operator in monitoring and adjusting the system a critical one. Despite this important distinction between process-control and other interactive systems, we believe there is much to be gained by delivering subgoal templates as a single coherent method of task analysis. Moreover, the embedding of hidden but proactive devices within information systems such as Web browsers (e.g., avatars and other forms of interface agent) means that interactive systems are becoming, at least in perceived interaction style, increasingly nondeterministic. Thus, the concerns of process-control domains increasingly apply to other domains of interactive systems development.

We have endeavoured to provide a method that has the qualities of simplicity, flexibility, explicitness, and utility. Good design provides individual operations support through interface design as well as addressing collaboration, training, job design, personnel selection, fault diagnosis, maintenance, and health and safety issues. Separating these perspectives into independent design issues is a bad thing. Yet, the complexity of an integrated understanding is very challenging. Good design processes involve many perspectives and contributors, all of whom have to be able to share an understanding of methods and tools and their outputs. Therefore, good design support is also minimal design support. The beauty of HTA is its simplicity and naturalness, and we have endeavoured to continue this tradition in the SGT method. Task analysis will always be a skilled activity, but the SGT method endeavours to facilitate the acquisition and execution of analytic skills by providing a clear modus operandi for pursuit of task analysis within an ongoing design process with minimal learning overheads and an explicit statement of outcomes for designers.

At the same time, good design is creative—designers should not be told what to do by task analysts. Good task analyses, despite what some have argued, do not condemn designers to the limitations of previous approaches to implementing operations, but good task analyses are not always easy to conduct. The SGT method endeavours to ensure good practice in task analysis, recognizing that task analysis is a process of negotiation between analyst and engineer (or client, designer, or user group), not a process of capturing an absolute standard for implementing a set of operations.

We believe that there is a currently unrecognized danger of trying too hard to drive the design of systems by user-centered considerations. One risk of focusing too much on user operations is that it may undermine the need to drive design by a clear functional specification of objectives and constraints (see Vicente, 1999, for a similar argument regarding the importance of focusing on the functional status of plant in designing support for operator tasks). The SGT method is explicit about the points at which it is neutral to implementation and the points at which it must lose its neutrality. The key to successful application of the method is the recognition that, regardless of the domain of application, the engineering of the system should drive the design of human-system interactions, not the reverse. Only then will designers take on board a complete assessment of the needs of human users in the specification of complex interactive systems.

REFERENCES

Annett, J., & Duncan. K. D. (1967). Task analysis and training design. *Occupational Psychology, 41*, 211–221.

Annett, J., Duncan, K. D., Stammers R. B., & Gray. M. J. (1971). *Task analysis*. London: Her Majesty's Stationery Office.

Benyon, D. (1992). The role of task analysis in systems design. *Interacting With Computers, 4*, 102–123.

Diaper, D. (2001). Task analysis for knowledge descriptions: A requiem for a method. *behavior and Information Technology, 20*, 199–212.

Dillon, A., Sweeney, M., & Maguire, M. (1993). A survey of usability engineering with the European IT industry: Current practice and needs. In J. L. Alty, D. Diaper, & S. Guest (Eds.), *People and computers VIII* (pp. 81–94). Cambridge: Cambridge University Press.

Duncan, K. D. (1972). Strategies for analysis of the task. In J. Hartley (Ed.), *Strategies for programmed instruction: An educational technology* (pp. 19–81). London: Butterworth.

Duncan, K. D. (1974). Analytical techniques in training design. In E. Edwards, & F. P. Leeds, (Eds.), *The human operator and process control* (pp. 283–320). London: Taylor & Francis.

Lansdale, M. W., & Ormerod, T. C. (1994). *Understanding interfaces: A handbook of human-computer interaction*. London: Academic Press.

Ormerod, T. C. (2000). Using task analysis as a primary design method: The SGT approach. In J. M. C. Schraagen, S. F. Chipman, & V. L. Shalin (Eds.), *Cognitive task analysis* (pp. 181–200). Mahwah, NJ: Lawrence Erlbaum Associates.

Ormerod, T. C., Richardson, J., & Shepherd, A. (1998). Enhancing the usability of a task analysis method: A notation and environment for requirements specification. *Ergonomics, 41*, 1642–1663.

Parker, K. (2001). An approach to requirements analysis for decision support systems. *International Journal of Human-Computer Studies, 55*, 423–433.

Payne, S. J., & Green, T. R. G. (1986). Task-action grammars: A model of the mental representation of task languages. *Human-Computer Interaction, 2*, 93–133.

Rubinstein, E., & Mason, J. F. (1979, November). An analysis of Three Mile Island: The accident that shouldn't have happened. *IEEE Spectrum*, pp. 33–42.

Shepherd, A. (1993). An approach to information requirements specification for process control tasks. *Ergonomics, 36*, 807–819.

Shepherd, A. (2001). *Hierarchical task analysis*. London: Taylor & Francis.

Simon, H. A. (1981). *The sciences of the artificial* (2nd ed.). Cambridge, MA: MIT Press.

Vicente, K. J. (1999). Wanted: Psychologically relevant, device- and event-dependent work analysis techniques. *Interacting With Computers, 11*, 237–254.

18

Task Analysis for Error Identification

Chris Baber
The University of Birmingham

Neville A. Stanton
Brunel University

Human error is a significant contributor to product failure. However, it is uncommon for designers to explicitly consider the potential for human error in the design of products. In this chapter, it is proposed that "human error" arises as a consequence of the interaction between user and product, and that modeling this interaction can allow insight into possible error paths. Using a simple representation of product functioning, based on state-space diagrams, task analysis for error identification indicates paths between states that are open to the user but which do not support the achievement of the user's goal; such paths are considered to be erroneous. From this perspective, one of the aims of product design is to minimize paths to error.

18.1 INTRODUCTION

Product evaluation often involves user trials. However, such trials require a prototype to be available. Admittedly, the prototypes need not be fully functioning products, but there may well be a need to have made most of the major design decisions prior to building such prototypes. This means that prototypes might already reflect significant design decisions and assumptions concerning *how* the user will interact with the product. An alternative approach is to perform an analytic evaluation of product concepts—in other words, to develop predictive models of user performance (e.g., in terms of the time it takes to perform a sequence of actions or in terms of possible user errors) and to use these models as a means of evaluating design ideas. In this manner, designs can be compared and evaluated prior to committing to a specific concept. We term such an approach *analytical prototyping* (Baber & Stanton, 2001).

Traditionally, human-computer interaction (HCI) has focused on models of user performance that are concerned either with performance time or with information processing, as in GOMS (Card, Moran, & Newell, 1983; chap. 4). Although such approaches have proved useful, they tend to concentrate on "expert" (i.e., error-free) performance. Furthermore, the focus of such models has tended to be on the activity that the user performs on the product rather than on the *interaction* between user and product. In other words, many of these models treat the product as a neutral object of user action and do not consider the effects of changing product state on either user activity or user decision-making. Consequently, we set ourselves the challenge of developing a method that could describe user-product activity, could focus on user error, and could support analytical prototyping. Before presenting a worked example of the method in action, the chapter begins with a discussion of the assumptions underlying the method. These assumptions form part of a theory of human action, although, as the reader will see, the theory remains very much "under development."

18.2 ASSUMPTIONS

The method that we have developed is based on a theoretical underpinning that reflects, we feel, current thinking about HCI. Significantly, we are struck by the commonly reported occurrence of people failing to read instructions or manuals prior to attempting to use a product. To us, this indicates people are approaching the products as if they were experts, that is, as if they had a complete and appropriate repertoire of routines to use on the product. Thus, on receiving a new digital camera as a present, the prospective user might take the camera out its packaging, check that the batteries are inserted, and then immediately try to take a picture; only when the interaction breaks down will our fictitious user turn to the user manual.[1] This implies that people possess either all the relevant knowledge needed to use the product (which might seem a little odd given that this is their first encounter with the specific product) or sufficient knowledge to imbue them with confidence when using the product. The first assumption is that (paradoxically, perhaps) users behave like experts; in other words, they approach the product with a repertoire of appropriate actions that, if successful, will make the product work. Only when users do not feel that they possess this repertoire (whether through lack of experience or confidence) do they seek help from another source, such as a manual or a "local expert."

The question, therefore, is what happens in the user-product interaction to support such behavior. In any interaction, a product can be assumed to present a set of functions to the user (via its features, its display, its controls, etc.), in the form of a *system image* (Norman, 1986). The user conceptualizes a *representation* of how the product works, possibly in terms of pairing features with functions (e.g., the user might assume that pressing a specific button would turn the product on) or possibly in the form of *routines* based on previous experience. When there is a mismatch between the user's representation and the system image, the user could make errors or become frustrated.

The notions of system image, representation, and routines, taken together, have led to development of the notion of rewritable routines. To present this notion, it is necessary to take a slight detour into the realm of mental models. It is well known that there are many different and competing views on the nature of mental models. For example, Johnson-Laird (1983) presents mental models as temporary frameworks developed during the task of reading (or other tasks involving information extraction) in order to allow the reader to maintain the "gist" of a story and to develop hypotheses of plot development or character motivation and action. A model of this type is developed on an ad hoc basis and requires the recruitment of knowledge from

[1] This, of course, presents one view of how people interact with new products. An alternative view is that the prospective user reads the manual from cover to cover, fully digesting any instructions and guidelines, in order to acquire sufficient knowledge to use the product, and only then begins to take pictures. The reality probably lies between these extremes. However, it is suggested that both versions present the user as expert (one by dint of experience, the other by dint of reading).

long-term memory and its combination with information derived from the text. Consequently, the mental model represents a specific state of affairs in the text (or the world).

Alternatively, the notion of mental models as representations of the workings of products appears to be popular in the HCI literature. This view can be traced at least as far back Craik's (1947) work on tracking control, which suggested that operators need to represent the system they are controlling in order to anticipate its dynamics. Researchers might accept that mental models allow users to make inferences about the system with which they are interacting and that a mental model provides a "problem space" (in Newell & Simon's [1972] terms). However, it is generally accepted that such mental models are imprecise, incomplete, and inconsistent. Consequently, the problem space might itself be messy and problematic.

O'Malley and Draper (1992) have suggested that, rather than the mental model containing a representation of the product, what is required is some means of filling the gaps in the system image and interpreting any information provided by the product. The proposals of O'Malley and Draper have more than a passing resemblance to the suggestions of Johnson-Laird (1983), and we take this as the starting point for our concept of rewritable routines.

To conclude this brief discussion, users' rely on highly fragmented knowledge of the product, rely on a variety of metaphors to contrast the product with other products, and can be heavily influenced by the system image. This suggests that "mental models" need not be complete, coherent internal representations of a product (i.e., they do not have to fully describe the physical and functional aspects of the product) and that they are often insufficiently detailed to predict the consequences of user actions.

By way of introduction to the notion of rewritable routines, consider the mobile telephone shown in Fig. 18.1. The system image presents several types of information to user, in a variety of formats. Some information is permanently accessible (e.g., the static visual display, the technical and communications interfaces, and the manual), whereas other information requires user action to be accessed (e.g., the dynamic visual display and audio display). User actions can be performed using manual controls or speech.

Based on the previous discussion, it is proposed that the system image *implies* a set of routines the person can use and that the selected routine will depend on the user's goal and previous experience. In order to explore this point, assume that a person who has not previously encountered this model of mobile telephone is set the simple goal of turning it on. This will require the user to identify a button or key that can be expected to turn on the telephone. The user might examine the side or top of the telephone to look for an "on" button, might see the green lifted-handset icon and assume this means make a call (and by analogy, turn on), might recognize the Ⓘ icon as an ISO symbol for "on," might assume that "OK" meant turn on, or might press any of the other keys on the basis of some other assumption. The point

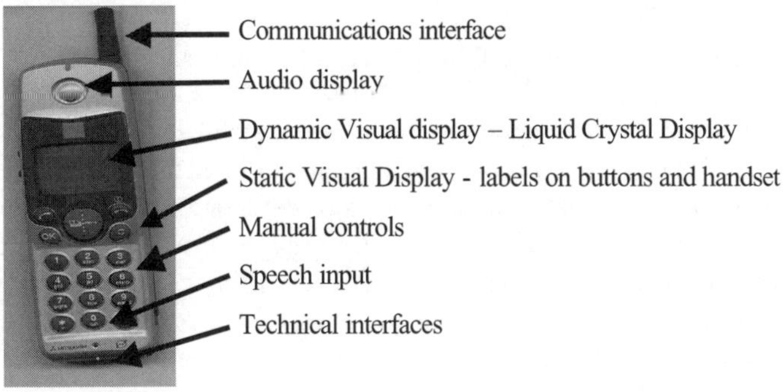

FIG. 18.1. Part of the system image of a Mitsubishi mt401 mobile phone.

is that, for the first-time user, there are multiple possible routines that can be activated in pursuing the goal of turning on the telephone. The selection of a routine will be influenced by prior experience and by the system image. One could imagine that a given user might assume that the green lifted-handset icon indicates the initial action that one makes using a conventional telephone. In other words, the user brings a routine to the use of the telephone that incorporates the following sequence of tasks: "Pick up handset, dial number, hold conversation, replace handset." The system image provides cues to the first step in this sequence of tasks (i.e., "pick up handset"). Of course, this is the wrong action, and the user should press the key labeled with a red lowered-handset icon (i.e., on this model of mobile telephone, one begins by "hanging up"). This particular model of mobile telephone has to be switched on using an "on" key on the handset. Notice that the ISO symbol for "on" is positioned above the lowered handset on the righthand side below the LCD. In effect, the user is being asked to the press the key showing a lowered handset to turn on the telephone. The user's choice of action will be based on the appearance of the telephone (its system image) and the user's prior experience, which leads him or her to assign meanings to the keys, icons and, other objects in the system image.

We assume that people require very little knowledge of the internal workings of a product in order to use it. For instance, one does need to know how a car engine works in order to drive the car. We also assume that all interaction with a product will be goal based and purposeful—that people have a reason for using the product and that they seek to match their goals with the product's functions. We further assume that interaction between user and product proceeds through a series of states, as hypothesized by Card et al. (1983). At each state, the user interrogates the system image for a correspondence between goal and function. Ideally, the goal and function would match and the action would be obvious. In the ergonomics literature, such matching is known as *stimulus-response compatibility* (e.g., turning a control knob clockwise causes a pointer to move to the right on a linear scale). People appear to have well developed stereotypes for some forms of stimulus-response compatibility and to base their actions on these stereotypes. Thus, at one level of behavior, users can simply match the system image with a stereotyped response. Such stereotyped responses can be thought of as *global prototypical routines*. Errors can arise when users mistakenly match an object in the system image with their goal or when a strong stereotype overrides the correct action. For example, in work on ticket-vending machines, Baber and Stanton (1996) found that users often mistakenly attempt to insert their money as the first action (that this error is common is supported by the prevalence of labels on these machines that read, "Do not put your money in first"). It is proposed that this error is the result of a conflict in two global prototypical routines. In one routine, users follow a "vending machine routine" (i.e., insert money, make selection, retrieve item; see chap. 26), and in another routine, users follow a "ticket kiosk routine" (i.e., request ticket, pay money, collect ticket). It would appear that the designers had assumed the latter routine but users often follow the former.

One rarely achieves the overall goal with a single action, and users need to keep track of their position in a sequence of goals (as proposed by the GOMS of Card et al., 1983). Further, sometimes there does not appear to be a clear match between goal and function. At this stage, users need to engage in problem solving and to infer the appropriate action on the basis of the system image. Rather than carrying a representation of the product throughout the interaction, users only require information when they are unsure of the appropriate action. The routine represents a set of actions (or a single action) that is deemed appropriate for a given state. Such routines can be thought of as *state-specific routines.* Interpretation of the system image in terms of the current goal state might draw on knowledge related to other products (i.e., through analogy or metaphor) in order to infer an appropriate action. Once the action has been performed, then the knowledge is no longer required. In this way, the routines are "rewritable" in that they can be overwritten by subsequent information. Thus, users might invoke one or two pieces of information from long-term memory (using metaphor or analogy) in order to determine an appropriate action. However, in order to minimize working memory load, the users will rarely need to maintain this information throughout the interaction and will concentrate on monitoring their progress toward the goal.

It is assumed that the majority of routines that one uses will be local. A local routine will relate solely to the immediate demands of interacting with the product (e.g., matching the actions that could be performed on the basis of the system image with the actions that should be performed to work toward the user's current goal). This is because movement through states in human-machine interaction is punctuated by brief periods in which the machine responds to user actions. From this perspective, users will employ multiple routines to determine relevant action as and when required. This suggests that users neither need nor use mental models, as opposed to routines, of the product being used. Rather, users seek to draw on previous experience to infer actions only when necessary. This further suggests that basing a design on a single metaphor (or even on a collection of metaphors) might not be useful; users will have different experiences and so might fail to appreciate the relevance of the supplied metaphor. Consequently, a general design proposal is that the "exits" from each state in a transaction need to be clearly marked, need to be clearly related to potential user goals, and need to be kept to a minimum (so as not to overload or challenge problem-solving abilities). In this way, the description would seek to consider knowledge-in-the-world, knowledge-in-the-user's-head, and knowledge-in-the-context.

Although the ideas presented in this section are by no means radical, they have led us to propose that there is a need to represent the interaction between user and product in a manner that makes it easy to consider these ideas during initial design activity. Furthermore, we want our approach to produce quantitative data (i.e., time and error data) that will support early evaluation of products in terms of user performance.

18.3 TASK ANALYSIS FOR ERROR IDENTIFICATION

The basis of Task Analysis for Error Identification (TAFEI) is the assumption that user-product activity proceeds through a goal-oriented series of *states*; that is, that each user action modifies that state of the product until the user has reached a specific goal. This means that the interaction can be easily represented in terms of a simple finite-state machine. However, it is important to note that the progression from state to state is dependent on the user's goal. This reduces problems of combinatorial explosion that are often associated with finite-state descriptions, as will become clear in the worked example below. At each state, the user needs to select an appropriate action in order to progress toward the goal. However, it is assumed that more than one action will be possible in each state. Thus, selection will depend on routines (section 18.6 describes how this notion is being incorporated into TAFEI). It is assumed that there are a small number of "legal" actions—that is, actions that will progress the user to the goal (typically in the region of 1 or 2)—and that all other possible actions are "illegal."

Procedurally, TAFEI comprises three main stages. In using TAFEI, we first describe the human side of the interaction. We tend to employ Hierarchical Task Analysis (HTA; Chapter 3), but any technique for describing human activity could be used. HTA suits our purposes for the following reasons: (a) It is related to specific tasks, (b) it is directed at a specific goal, and (c) it allows consideration of task sequences (through "plans"). Second, we construct state-space diagrams (SSDs) to represent the behavior of the artifact. We map plans from the HTA onto the SSD to form the TAFEI diagram. Finally, we devise a transition matrix to display state transitions during device use. TAFEI aims to assist the design of artifacts by illustrating when a state transition is possible but undesirable (i.e., illegal). Making all illegal transitions impossible should facilitate the cooperative endeavour of device use.

The first step in a TAFEI analysis is to obtain an appropriate HTA for the device. As TAFEI is best applied to scenario analyses, it is wise to consider just one specific goal as described by the HTA (e.g., a specific, closed-loop task of interest) rather than the whole design. Once this goal has been selected, the analysis proceeds to the construction of SSDs for device operation.

A SSD essentially consists of a series of states that the device passes through, up to and including the goal state. For each series of states there will be a current state and a set of possible exits to other states. At a basic level, the current state might be "off," with the exit condition "switch on" taking the device to the state "on." Thus, when the device is "off," it is "waiting for" an action (or set of actions) that will take it to the state "on." It is important to have, after completing the SSD, an exhaustive set of states for the device under analysis *in terms of the specific goal being considered*. Numbered plans from the HTA are then mapped onto the SSD, indicating which human actions take the device from one state to another. Thus the plans are mapped onto the state transitions (if a transition is activated by the machine, this is also indicated on the SSD, using the letter *M* on the TAFEI diagram). This results in a TAFEI diagram.

As mentioned above, in each state there will be a set of actions available to the user; we describe these as actions that the product is "waiting for." Thus, when you insert your card into an automatic teller machine ("cashpoint"), the machine will be waiting for <enter PIN>. When the PIN is correctly entered, the machine will be waiting for <select cash, select cash + receipt, select balance, etc.>; selecting any one of these options will move the product to a new state. If the user selects an option that does not fit with the overall goal (e.g., if the user selects <balance> when the goal is <cash>), then the action is illegal. In other words, illegal actions lead the user away from the goal. Clearly, the definition of an illegal action is dependent on the user's goals rather than on any characteristic of the product. The TAFEI approach seeks to identify illegal actions in pursuit of specific goals. Once illegal actions are identified, the next step is to propose ways of redesigning the product in order to minimize such actions.

The most important part of the analysis (from the point of view of improving usability) is the transition matrix. All possible states are entered as headers on a matrix. The cells represent state transitions (e.g., the cell at row 1, column 2 represents the transition between state 1 and state 2), and these are then filled in one of three ways. If a transition is deemed impossible (i.e., you cannot go from state X to state Y), "-" is entered into the cell. If a transition is deemed possible and desirable (i.e., it progresses the user toward the goal state; a correct action), it is a legal transition, and "L" is entered into the cell. If, however, a transition is possible but undesirable (a deviation from the intended path; an error), it is an illegal action, and the cell is filled with an "I." The idea behind TAFEI is that usability may be improved by making all illegal transitions (errors) impossible and limiting the user to only performing desirable actions. It is up to the analyst to conceive of design solutions to achieve this.

Examples of applications of TAFEI include prediction of errors in using kettles (Baber & Stanton, 1994; Stanton & Baber, 1998), comparison of word-processing packages (Baber & Stanton, 1999), withdrawing cash from automatic teller machines (Burford & Baber, 1994), medical applications (Baber & Stanton, 1999; Yamaoka & Baber, 2000), recording on tape-to-tape machines (Baber & Stanton, 1994), programming a menu on cookers (Crawford, Taylor, & Po, 2001), programming videocassette recorders (Baber & Stanton, 1994; Stanton & Baber, 1998), operating radiocassette machines (Stanton & Young, 1999), recalling a phone number on mobile phones (Baber & Stanton, 2001), buying a rail ticket on the ticket machines in the London Underground (Baber & Stanton, 1996), and operating high-voltage switchgear in substations (Glendon & McKenna, 1994).

All of these examples of applying TAFEI share common features, which define the operational parameters of the technique. First, they are all applied to purposeful use of the device drawn from the goals of a task analysis. Second, each of the devices offers a clear and logical sequence of activity. The tasks are discrete and step by step rather than continuous and concurrent. Third, there are clear system boundaries between the device and human elements. Given these requirements, it is no wonder that most of the analyses have tended toward single-user, single-device systems. Theoretically it should be possible to take a nested systems approach to analyse systems of greater complexity by addressing different levels and different scenarios. Some movement in this direction has begun with the analysis of operating high-voltage switchgear in substations (Glendon & McKenna, 1994). Before this is taken any further, it is sensible to assess the performance of the TAFEI technique on relatively simple systems first.

18.4 WORKED EXAMPLE

This section presents a worked example to illustrate the method. The product selected for this example is a digital watch. Although the number of states that the watch can enter is limited, the product is sufficient to explain how the method is applied.

We begin by considering the system image of the product. The system image presents the controls, display, labels, and instructions that the product presents to the user. Figure 18.2 contains a simple labeled schematic of the digital watch. It is often useful to begin the analysis with a simple sketch to help focus on the key features of the product. This step can be applied even if the product is at the concept stage (i.e., in the absence of a working product).

The next stage is to determine specific user goals. The selection of appropriate goals is obviously a matter of the analyst's judgment. However, it makes sense to pick a couple of goals that can be reached with only one or two user actions, a couple that might require between three to five actions, and a couple that require more than five actions. In this way, the analysis can address levels of complexity in using the product. Definition of a goal is essential to focus the analysis, constrain the number of states that need to be considered, and constrain the path through the states. The user goal also defines the concept of illegal action. In this example, two user goals will be presented. In general, we would advise using a set of between two and five user goals that are representative of typical product use and reflect different functions.

$USER\_GOAL_1$: Change date to March 1st
$USER\_GOAL_2$: Start stopwatch

Each user goal can be decomposed in two ways: (a) in terms of the states through which the product passes in order to realize the goal, and (b) in terms of the user actions required to achieve the goal. It is important to recognize these as two separate processes rather than to attempt to combine them into one process. Here we focus on just one of the goals.

18.4.1 State-Space Diagram

Figure 18.3 shows the states that our watch will pass through in order to support the actions required to change the date. By way of explanation, the watch needs to be in the mode selection

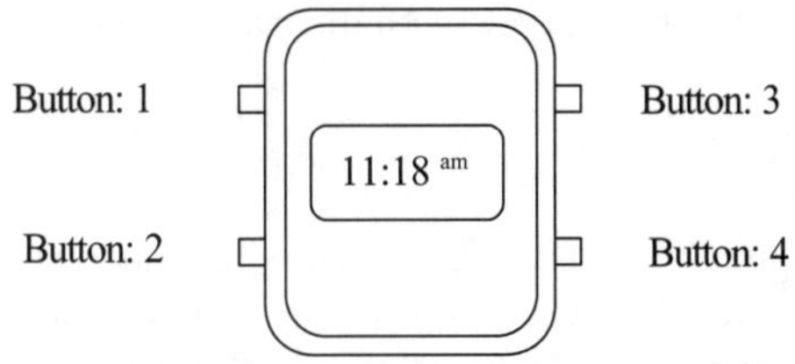

FIG. 18.2. A simple labeled schematic of product.

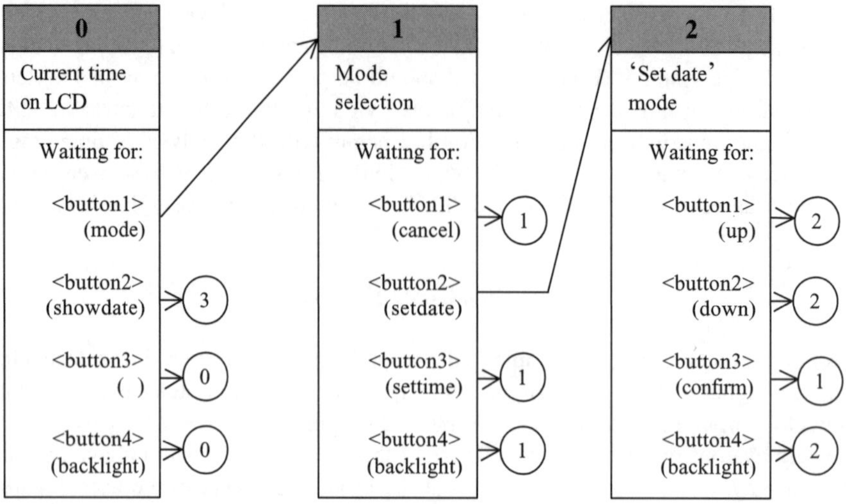

FIG. 18.3. State-space diagram for USER GOAL$_1$: Change date to March 1st.

state 1 before the user can change any setting. The watch allows one parameter to be changed, such as month or day or hour or minute. When a change has been made, the watch returns to the mode selection state 1 and then can either accept a new setting or can return to the initial state (0). So, in a change of date, the watch's states will be: 0 (initial) > 1 (mode selection) > 2 (change month) > 1 (mode selection) > 2 (change day) > 1 (mode selection) > 0 (initial). In Fig. 18.3, the numbers in circles refer to the state that the product will move to following a given transition.

From the figure, it is apparent that the watch exhibits nonlinear task sequencing; in other words, many of the user's actions will result in the watch either remaining in its current state or returning to a previous state. One consequence of such operation, we think, is that the user does not have a sense of a task progressing toward a goal but rather feels that actions are random and poorly structured. Note also that the definition of each of the buttons (with the exception of 4) changes with each state of the watch. Such inconsistency might be misleading and confusing to users, particularly in the absence of clear labeling or the instruction manual.

The numbering of the SSD is specific to the goal being described; in other words, if a different sequence of states in pursuit of a different goal is constructed, then the state <Mode Selection> might be assigned a different number. Note also that the numbering of the "waiting for" states first uses the numbers in the diagram and then increments numbers of "undrawn" states. Thus, the last state in Fig. 18.3 is numbered 2, but the first "waiting for" state for State 0 is "press button 2<showdate>," which is not part of the task sequence for this goal and is assigned to be state 3.

Notice also that the "waiting for" states all involve the use of four buttons but that the function of the buttons changes according to the mode of the watch. The SSD indicates the resulting function for each button, dependent on the watch's current state.

18.4.2 User Actions

Having established the manner in which the watch will change states as the user pursues a specific goal, the analyst's next task is to consider the actions that the user will perform. The analyst could construct a hierarchical task analysis for this process or could simply reflect on the sequence in which buttons are being pressed. We prefer to use the HTA notation for two

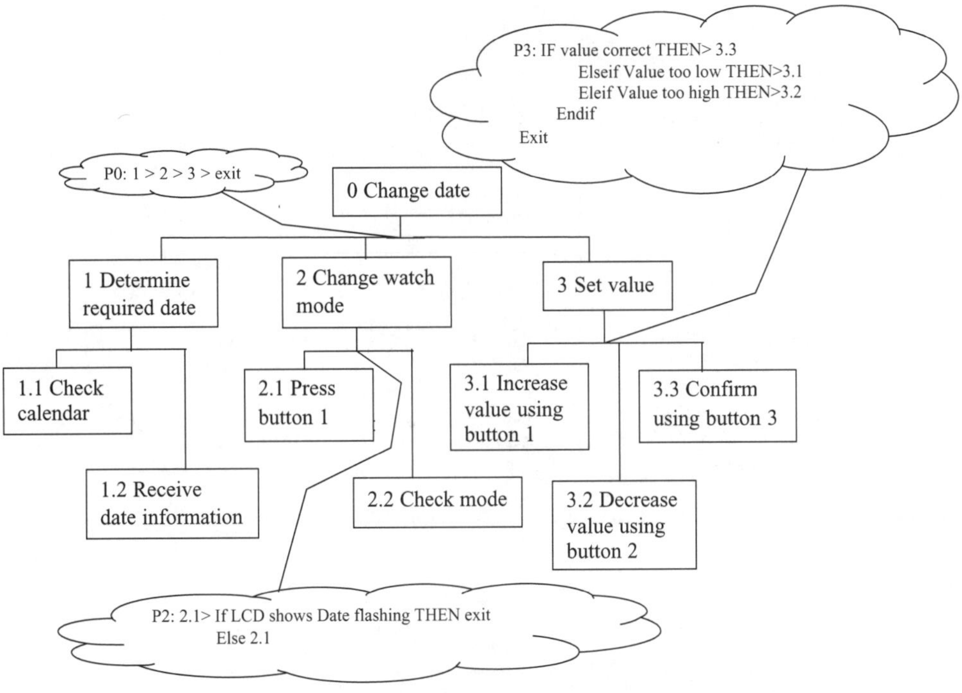

FIG. 18.4. HTA for USER GOAL$_1$: Change date to March 1st.

reasons: (a) it is an agreed standard form of analysis, at least amongst UK ergonomists, and (b) it allows consideration of factors outside the simple button-pressing sequence. Figure 18.4 shows a hierarchical task analysis for the goal of changing the date.

18.4.3 TAFEI Diagram

The plans and actions on the hierarchical task analysis can now be added to Fig. 18.3. The numbers for plans or actions are simply written on the arrows that correspond to the appropriate user activity. Thus, the arrow from State 0 to State 1 can be labeled "p2" and the arrow from State 1 to State 2 can be labeled "p3" (Fig. 18.5).

In this manner, we have a simple representation linking user actions to product states. Labeling the transitions in terms of the hierarchical task analysis provides a convenient means of indexing the two diagrams. Furthermore, at this stage, it is possible to use TAFEI for creating scenarios. For instance, the design team could consider what user action would lead the user to fail to perform p2 (e.g., pressing any other button). The answer might simply be a *slip* (i.e., the user intended to press button 1 but inadvertently pressed another button) or it might be a *mistake* (i.e., the user intended to press one of the other buttons because he or she was using an inappropriate routine or misinterpreting the system image).

18.4.4 Transition Matrix

The identification of user error is initially performed using a transition matrix. We begin by asking whether it is possible to move from a given state to another state. It is important to realize that we are not considering *why* a user might make such a transition, only whether there is anything in the design to *prevent* such a transition. Given that all of the buttons on the watch can be pressed at all times, the transition matrix for this product will be as shown in Fig. 18.6.

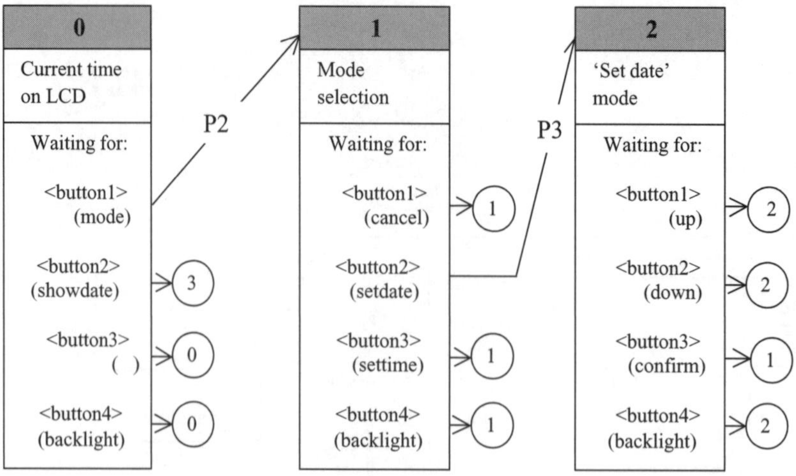

FIG. 18.5. TAFEI diagram.

		To States			
		0	1	2	3
From states	0	I	L	I	I
	1	L	I	L	I
	2	I	L	L	I
	3	I	I	I	I

FIG. 18.6. Transition matrix for USER $GOAL_1$: Change date to March 1st.

In Fig. 18.6 moving from State 0 to State 1 is the only course of action that will result in the achievement of the goal. However, the design of the watch allows the user to press any other button (i.e., effectively to move to one of the other states), and any such action will not lead to the goal; hence, it is illegal.

From Fig. 18.6, it is clear that even in so simple a product there is potential for user error, with some 11 illegal actions (compared with 5 legal actions). The issue for the design team is how to redesign the watch in order to reduce the number of illegal actions. Although here is not the place to debate ways in which the design team might approach this task, the reader can imagine discussions involving changing of labeling, clear indication of mode, highlighting of "available" keys, and so on.

18.5 VALIDATION

In order to demonstrate that the technique is usable, we have conducted a series of experiments into the reliability and validity of TAFEI. Empirical evidence of a method's worth should be one of the first requirements for its acceptance by the ergonomics and human factors community. Stanton and Stevenage (1998) suggested that ergonomists should adopt criteria similar to the standards set by the psychometric community (i.e., research evidence of reliability and validity prior to wide use of the method). It may be that the ergonomics community is largely unaware of the lack of data (Stanton & Young, 1997) or assumes that the methods provide their own validity (Stanton & Young, 1999).

In an earlier study, Baber and Stanton (1996) aimed to provide a more rigorous test of the predictive validity of TAFEI. Predictive validity was tested by comparing the errors identified by an expert analyst with those observed during 300 transactions with a ticket machine in the London Underground. Baber and Stanton (1996) suggested that TAFEI provides an acceptable level of sensitivity based on the data from two expert analysts ($r = 0.8$). A major claim for heuristics approaches is their speed of use. This study suggests that, given equal time, an heuristics approach is less productive than the structured human error identification (HEI) approach of TAFEI. HEI asks analysts to consider human activity rather than device characteristics. This means that the analysts tend to uncover potential problems that users might experience with the device.

Participants also showed performance improvement over the first two experiments, increasing the hits and reducing the misses without compromising the false alarms. Both reliability and validity achieved acceptable levels, and the data compare well with those of previous studies (Stanton & Stevenage, 1998). The signal detection paradigm is a useful way of coding error prediction data, and the present study reinforces this approach. In this instance, signal detection refers to the ability of an error identification to "hit" (correctly predict) errors without incurring too high a cost of "false alarms" (predicting errors that cannot occur) and without too many "misses" (failures to predict an error). Our approach is to develop the TAFEI model and use it to predict errors, then conduct user trials that allow us to collect observations on errors that people make during their interactions. One might argue that a hit rate of 5.4 out of 11 (49%) is relatively poor, but, as argued above, any technique that can predict human error reliably can prove useful in product design and evaluation.

18.6 CONCLUSION

TAFEI has been designed as a method with a specific focus. The aim is to support designers in developing products with minimal potential for user error. It is proposed that by explicitly exploring the potential for error in the early stages of design, one can reduce the error rate in later stages, allowing user trials to focus on other aspects of use. The method pairs a simple description of device operation (using state-space diagrams) with a description of user activity in order to propose likely illegal operations.

We feel that TAFEI is of benefit for two reasons: (a) It forces analysts to consider human activity and to consider problems in the light of this activity, and (b) it forces the focus of attention away from device features. This means that it is possible to suggest radical revisions to a design (in order to reduce predicted errors) rather than seeking to modify specific features. Finally, TAFEI does not need real users performing real tasks with real products. Thus, the method is applicable to very early stages of design (see Baber & Stanton, 1999, for a discussion of how TAFEI can be applied as a design tool).

Currently, the method is being extended to provide quantitative predictions of user performance (Baber & Stanton, 2001). Each user action can be assigned a time using, for example, the times proposed by Card et al. (1983) to estimate transaction times for interacting with a product. Such an approach provides data similar to those generated by keystroke-level models. However, assuming that each state offers more than one exit (i.e., has more than one "waiting for" component), there is a possibility that interactions can branch between states. Consequently, we have been using MicroSAINT to model user interactions. In MicroSAINT, transitions between states are described in terms of time and probability. Thus, for Fig. 18.4, we might assume a time of 200 ms to press Button 1 and a probability of 0.8 to press Button 1 (and 0.1 for Button 2, 0.05 for Buttons 3, and 0.05 for Button 4). Obviously a full set of probabilities are required prior to developing the method further, and we are currently working

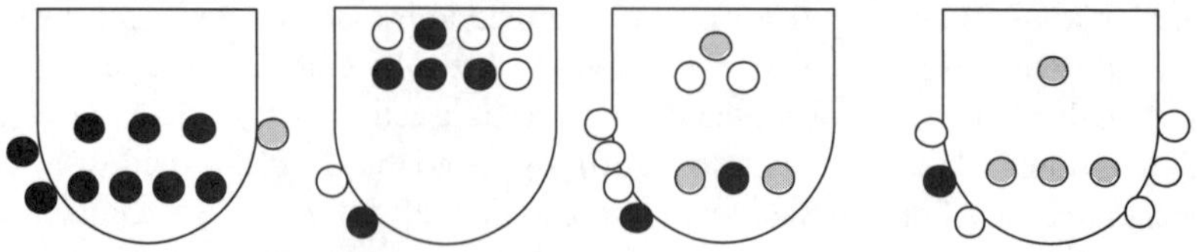

FIG. 18.7. Responses to CD player task.

with the HEART method proposed by Williams (1986). The point is that, from Fig. 18.4, we can imagine a situation in which a user erroneously presses Button 3, is taken back to State 0, and then needs to recognize the mistake before pressing Button 1. In this manner, it is possible to develop quite sophisticated models of user activity. Of course, such models do not explain why the user would make an error, which is why we are also developing the rewritable routines theory.

Although the notions of global prototypical routines and state-specific routines appear plausible, there is a need to both validate these concepts and propose an architecture for cognition that supports them. This forms part of an ongoing project. For instance, Fig. 18.6 shows the results of an initial study into people's response to the system images of a compact disc (CD) player. In the study, nine people were given blank templates of a CD player. They were also given a set of 10 functions. Their task was to indicate which of the rectangles on the system images they thought were associated with the functions. If response to a blank system image was solely guesswork, then the patterns of response would be random (i.e., we would not expect any response to be in the majority). If, on the other hand, people had global prototypical views of how products were laid out, then we would anticipate that some responses would occur much more often than others.

In Fig. 18.7, the filled (black) circles indicate majority responses (i.e., where seven or more people made the same response, the shaded circles indicate where five to seven people agreed, and the blank circles indicate where less than five people agreed. Notice that there are fewer black circles on the righthand CD player. Because participants were given the same set of functions for each CD player and the only difference was the position of the circles on the templates provided to them, the implication is that people were more consistent in their identification of controls on the two lefthand players than on the other two. One explanation is that some system images (e.g., the two on the left) evoke stronger global prototypical routines than others. Consequently, when faced with a new product, people will look for specific controls in the area where they expect to the find the controls. There is only one black circle on the far right CD player. This identifies the volume control and is positioned in a similar location on all four diagrams, suggesting that the participants "expect" this control to be in this location.

This experiment has been repeated with other participants and for other types of product (e.g., mobile telephones and personal digital assistants) and has achieved similar patterns of results, suggesting that people do hold expectations about product layout and that, as in the case of display-control compatibility (e.g., Sanders & McCormick, 1992), people anticipate that certain features will be in certain locations and that certain functions will be operated from certain locations. Given people's tendency to behave as if they were expects, many of the slips arising from interaction with products occur simply because the user is tries to operate a control that is not in its expected place. TAFEI provides a method that allows designers to explore design concepts from the perspective of a theory of human-product interaction.

REFERENCES

Baber, C., & Stanton, N.A. (1994). Task analysis for error identification. *Ergonomics, 37*, 1923–1942.

Baber, C., & Stanton, N. A. (1996). Human error identification techniques applied to public technology: Predictions compared with observed use, *Applied Ergonomics, 27*, 119–131.

Baber, C., & and Stanton, N. (1999). Analytical prototyping. In J. M. Noyes & M. Cook (Eds.), *Interface technology* (pp. 175–194). Baldock UK: Research Studies Press.

Baber, C., & Stanton, N. (2001). Analytical prototyping of personal technologies: Using predictions of time and error to evaluate user interfaces, In M. Hirose (Ed.), *Interact IFIP TC 13 International Conference on Human: Computer Interaction '01* (pp. 585–592). Amsterdam: IOS Press.

Burford, B., & Baber, C., (1994). A user-centred evaluation of a simulated adaptive autoteller, In S. A. Robertson (Ed.). *Contemporary Ergonomics* 1994. London: Taylor and Francis, 64–69.

Card, S. K., Moran, T. P., & Newell, A. (1983). *The psychology of human-computer interaction*. Hillsdale, NJ: Lawrence Erlbaum Associates.

Craik, K. J. W. (1947). The theory of the human operator in control systems (Pt. 1). *British Journal of Psychology, 38*, 56–61.

Crawford, J. O., Taylor, C., & Po, N. L. W. (2001). A case study of on-screen prototypes and usability evaluation of electronic timers and food menu systems. *International Journal of Human Computer Studies, 13*, 187–202.

Glendon, A. I., & McKenna, E. F. (1994). *Human safety and risk management*. London: Chapman & Hall.

Johnson-Laird, P. N. (1983). *Mental models*. Cambridge: Cambridge University Press.

Newell, A., & Simon, H. A. (1972). *Human information processing*. Englewood Cliffs, NJ: Prentice Hall.

Norman, D. (1986). Cognitive engineering. In D. Norman & S. Draper (Eds.), *User-centered system design: New perspectives on human-computer interaction* (pp. 31–61). Hillsdale, NJ: Lawrence Erlbaum Associates.

O'Malley, C., & Draper, S. (1992). Representation and interaction: Are mental models all in the mind? In Y. Rogers, A. Rutherford, & P. A. Bibby (Eds.), *Models in the mind: Theory, perspective and application* (pp. 73–91). London: Academic Press.

Sanders, M. S., & McCormick, E. J. (1992). *Human factors in engineering and design*. New York: McGraw-Hill.

Stanton, N.A., & Baber, C. (1998). A systems analysis of consumer products. In N. A. Stanton (Ed.), *Human factors in consumer products* (pp. 75–90). London: Taylor & Francis.

Stanton, N. A., & Stevenage, S. V. (1998). Learning to predict human error: Issues of acceptance, reliability and validity. *Ergonomics, 41*, 1737–1756.

Stanton, N. A., & Young, N. A. (1997). Ergonomics methods in consumer product design and evaluation. In N. A. Stanton (Ed.), *Human factors in consumer products* (pp. 21–53). London: Taylor & Francis.

Stanton, N. A., & Young, M. S. (1999). *A guide to methodology in ergonomics*. London: Taylor & Francis.

Williams, J. C. (1986). HEART: A proposed method for assessing and reducing human error. In *Proceedings of the 9th Advances in Reliability Technology Symposium*. Bradford, England: University of Bradford.

Yamaoka, T., & Baber, C. (2000). Three point task analysis and human error estimation. In *Proceedings of the Human Interface Symposium* (pp. 395–398).

19

Trigger Analysis: Understanding Broken Tasks

Alan Dix
Lancaster University

Devina Ramduny-Ellis
Lancaster University

Julie Wilkinson
University of Huddersfield

Why do things happen when they happen? Trigger analysis exposes the triggers that prompt subtasks to occur in the right order at the right time, and assess whether tasks are robust to interruptions, delays, and shared responsibility. Trigger analysis starts with task decomposition via any suitable method, proceeding to uncover what trigger prompts each subtask. The obvious answer: "because the previous subtask is complete" is often the precondition, not the actual trigger. Previous analysis has uncovered certain primary triggers and potential failure points. The complete analysis produces an ecologically richer picture, with tasks interacting with and prompted by their environment.

19.1 INTRODUCTION

Many task analysis and workflow techniques decompose the overall task into smaller subtasks, processes, or activities. The order of these subtasks is then typically specified or described (e.g., plans in hierarchical task analysis [Shepard, 1995; chap. 3], links between processes in a workflow, and temporal connectives in ConcurTaskTrees [Paternò, 1999; chap. 24]). For short-term tasks performed by one person without interruption, this may be the end of the story:

> *To photocopy a document*: (a) open copier lid; (b) put original on glass; (c) close lid; (d) select number of copies; (e) press copy button; (f) when copying complete, remove copies; (g) remove original.

But what happens if the number of photocopies is large and so you go for a cup of tea between (e) and (f)? Someone comes in and interrupts between (f) and (g)? Instead of a small photocopier, this is a large print machine, and different people are responsible for different stages?

Trigger analysis deals with exactly these issues. Why do things happen when they happen and do they happen at all? By exposing the triggers that prompt activities and subtasks to occur in the right order at the right time, trigger analysis allows us to decide whether tasks are robust to interruptions, delays, and shared responsibility (even across organizational boundaries).

Trigger analysis starts with a task decomposition obtained by any suitable method (and can therefore be used in combination with many task analysis and workflow methods). It then proceeds to uncover what trigger causes each subtask to occur. The initial answer is, "because the previous subtask is complete," but completion of that subtask is often merely a precondition, not the actual trigger.

Previous empirical and theoretical analysis has uncovered a small set of primary triggers, including the simple "previous task complete," timed events ("every hour I check the mail"), and environmental cues ("the document is in the in-tray"). For each class of trigger, there is a set of subsequent questions, for example, "What happens if you are interrupted between tasks?" "How do you know when it is the right time?" "Are there several tasks with the same environmental cues?"

Triggers are what make activities happen when they do. A closely related issue is knowing where in the task sequence you are. Environmental cues often act in both roles, as triggers saying that something should happen and as placeholders saying what should happen. Typically, the complete analysis produces a highly ecologically rich picture, and rather than cognitively driven tasks acting on the environment, we see tasks interacting with and prompted by their environment. Note that trigger analysis is not an alternative task analysis technique or notation but instead an additional concern that should be grafted on to existing analysis methods.

This chapter examines the trigger analysis method in depth, with particular reference to its grounding in empirical work. We start with a brief discussion of the theoretical underpinning for the fieldwork undertaken. We then suggest some reasons why prolonged interactions tend to break down in organizational contexts and provide an explanation of the five trigger types that have emerged during our investigations. We also present a second, related pattern, the 4Rs framework, in detail, as we believe the 4Rs form an important and fundamental unit of work. Finally, we describe a comprehensive application of the 4Rs and the benefits they provided for the analysis of work in a systems development project.

19.2 THEORETICAL BACKGROUND

The roots of trigger analysis lie in two principal theoretical foundations: the study of the pace of interaction (Dix, 1992, 1994) and status–event analysis (Abowd & Dix, 1994; Dix, 1991; Dix, Finlay, Abowd, & Beale, 1998). The primary basis consists of the issues surrounding pace—that is, the rate at which users interact with computer systems, the physical world, and one another. Thinking about pace makes one concentrate on the timescale over which interaction occurs, both the similarities between interactions of widely different pace and also the differences.

19.2.1 Status–Event Analysis

Status–event analysis is a collection of formal and semiformal techniques all focused on the differences between events (things that happen) and status (things that always have a value). Applications of status–event analysis have included auditory interfaces (Brewster, Wright, & Edwards, 1994), formal analysis of shared scrollbars (Abowd and Dix, 1994), and software architectures for distributed agent interfaces (Wood, Dey, & Abowd, 1997).

Status–event analysis allows the distinction between an actual event (some objective thing that occurs) and a perceived event (when an agent, human or machine, notices that an event has occurred). Sometimes this is virtually instantaneous, but more often there is a lag between the two. Many formal and informal analyses of events assume simultaneity between cause and effect. However, accepting that there is often a gap allows us to investigate what actually causes secondary events to occur when they do. The lag between the actual event and the perceived event (e.g., the time it takes to notice that e-mail has arrived) can in turn influence the pace of interaction.

Status–event analysis also looks at the effect of events on agents. An event may simply be intended to inform, but more often an event is intended to initiate actions, which in turn may cause further events. The actions of agents may change the status of the agent or the world, but changes in status are themselves events that may trigger further action.

Furthermore, most notations in computing focus primarily on events with little, if any, description of status phenomena. However, we shall see that environmental cues—things that are around us, such as piles of papers or notice boards—play an essential role in acting as triggers and placeholders.

The analysis of status and events has allowed us to see some common features of human-human, human-computer, and internal computer interactions. For example, it is common to see status mediation whereby one agent communicates an event to another by manipulating a status that will eventually be observed by the second agent. Moreover, polling, the periodic observation of a status phenomenon to detect change, is not just a low-level computational device but something people do as well, as described in more detail in chapter 13. The rich interplay of status and event phenomena is reflected in the ecological perspective that colors the analytic stance of our current study.

Because status–event analysis gives equal weight to status and event phenomena, it is able to address interstitial behavior—the continuous relationships between status phenomena in the interstices between events. This interstitial behavior is often what gives the "feel" to computer systems—you move the mouse and the window is dragged around the screen. Part of the power of the human motor system is its capability for near continuous control and response, as in sports, driving, manipulating objects, and so on, and paradigms such as direct manipulation and virtual reality use this power.

This continuous interaction is important in many nondesktop technologies, including aspects of ubiquitous computing (see Dix, 2002b). However, in this chapter we concentrate more on office style interactions and the discrete activities commonly dealt with in task analysis. Often the lowest level subtask or activity may involve continuous interaction, but here we look principally at the events that drive the transitions between these.

19.2.2 Pace of Interaction

The pace of interaction with other people or with a computer system is the rate at which one sends messages or commands and then receives a response. It varies from tens of milliseconds in a video game to hours or days when interacting by post.

The pace of interaction is influenced by and influences three principal factors (see Fig. 19.1):

1. the intrinsic pace of the channels through which the users communicate
2. the pace of the collaborative tasks
3. the users' own natural pace for different forms of mental activity.

Problems may occur when there is any significant mismatch between any of these and the resulting pace of interaction (Dix, 1992; Dix, Ramduny, & Wilkinson, 1998).

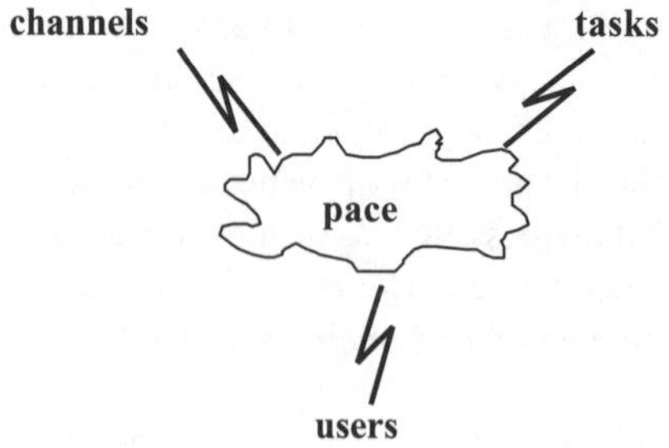

FIG. 19.1. Factors affecting pace.

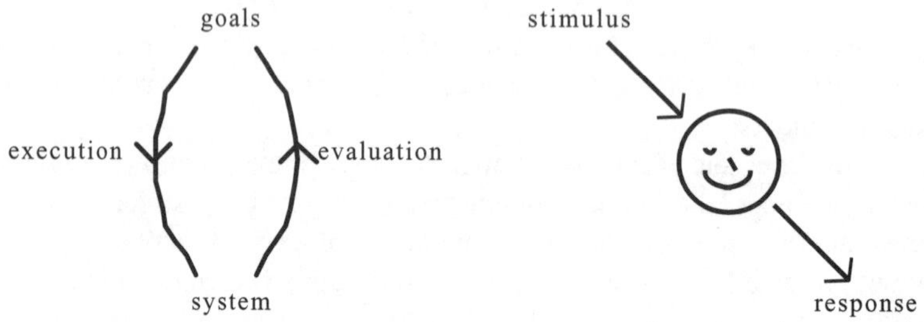

FIG. 19.2. (a) Norman's execution–evaluation cycle and (b) the stimulus–response cycle.

19.2.3 Prolonged Interaction

The study of pace helps us to understand interactivity in a wider context. A system or collaborative process is not interactive because it is fast or has instant feedback. Instead, interactivity is about the appropriate pace of interaction in relation to the task at hand. Indeed, in many collaborative situations the pace of communication may be over days or weeks.

The reason for the prolonged nature of these interactions varies: It may be because of the communication medium (e.g., normal postal delays) or the nature of the task (e.g., a doctor requiring an X-ray has to wait while the patient is wheeled to radiology and back). One of the key points is that models of interaction that concentrate on a tight cycle between action and feedback break down (Dix, 1992). This is typified by Norman's (1986, 1988) execution–evaluation cycle: A user has a goal, formulates actions that will further that goal, executes the actions, and then evaluates the results of those actions against the expected outcome and the goal (see Fig. 19.2a). This model effectively assumes that the results of the user's actions are immediately available. If the delay between executing actions and observing the results is greater than short-term memory times, then the evaluation becomes far more difficult. This problem has been called the "broken loop of interaction."

Another model of interaction, used in industrial settings, is the stimulus–response model (see Fig. 19.2b). Commands and alarms act as stimuli, and the effective worker responds to these in the appropriate manner. However, in a pure form, this model does not allow for any long-term plans or goals on the part of the worker; the worker is treated in a mechanistic manner, as a cog in the machine.

To incorporate both these perspectives, we need to stretch out the interaction and consider the interplay between the user and the environment over a protracted timescale. We use the term *environment* to include interaction with other users, computer systems, or the physical environment. Such interaction is typically of a turn-taking nature: The user acts on the environment, the environment "responds," the user sees the effects and then acts again.

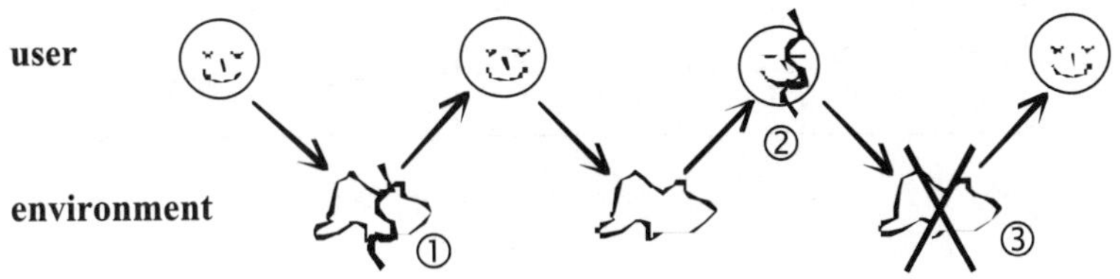

FIG. 19.3. Problems for prolonged interaction.

This process is illustrated in Fig. 19.3. Notice how the Norman loop concentrates on the user–environment–user part of the interaction whereas the stimulus–response model concentrates on the environment–user–environment part. We can see various ways in which long-term interaction affects this picture:

① *Action–effect gap*. The user performs an action, but there is a long delay before the effects of that action occur or become apparent to the user. For example, you send an e-mail and some days later get a reply. The problem here is loss of context: How do you recall the context when you eventually receive the feedback? When the reply comes, you have to remember the reason why the original message was sent and what your expectations of the reply were. The way in which e-mail systems include the sender's message in the reply is an attempt to address this problem. In paper communication, the use of "my ref./your ref." fulfills a similar purpose.

② *Stimulus–response gap*. Something happens to which the user must respond but for some reason cannot do so immediately. For example, someone asks you to do something when you meet in the corridor. The problem here is that you may forget. Hence the need for to-do lists or other forms of reminder (see chap. 14). In the psychological literature, this has been called *prospective memory* (Payne, 1993).

③ *Missing stimulus*. The user performs an action, but something goes wrong and there is never a response. For example, you send someone a letter but never get a reply. For short-term interactions, this is immediately obvious. You are waiting for the response, and when nothing happens, you know something is wrong. However, for long-term interactions, you cannot afford to do nothing for several days waiting for a reply to a letter. In this case you need a reminder that someone else needs to do something—a to-be-done-to list!

Trigger analysis focuses particularly on the second problem, the stimulus–response gap, but all three problems need to be considered when analysing the robustness of a process.

19.3 FUNDAMENTAL CONCEPTS OF TRIGGER ANALYSIS

19.3.1 Triggers: Why Things Happen When They Happen

Workflows and process diagrams decompose processes into smaller activities and then give the order between them. Similarly in HTA, plans give some specification of the order of subtasks, and in CTT, these temporal orders are made more specific using operators derived from LOTOS (Paternò, 1999; chap. 24).

Figure 19.4 shows a simple example, perhaps the normal pattern of activity for an office worker dealing with daily post. Notice the simple dependency: The post must be collected from the pigeonhole before it can be brought to the desk and before it can be opened. However, look again at the activity "open post"—when does it actually happen? The work process says it

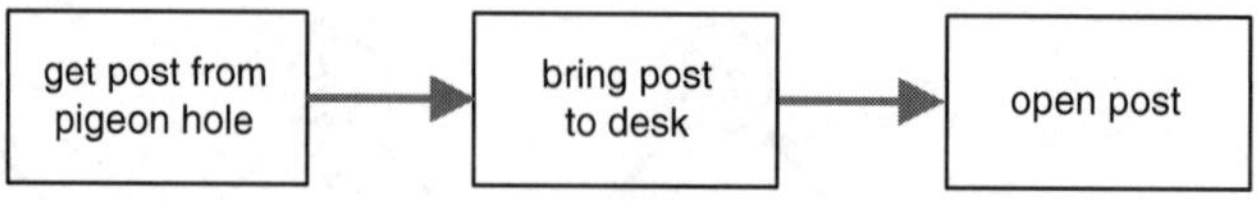

FIG. 19.4. Simple work process.

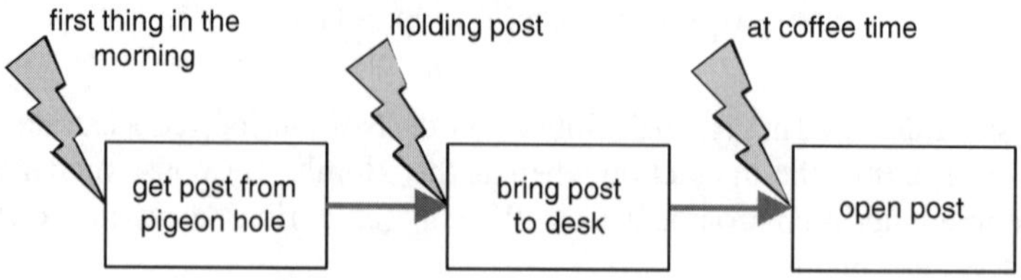

FIG. 19.5. Triggers for activities.

doesn't happen before the "bring post to desk" activity is complete, but does it happen straight away after this or some time later?

In previous work (Dix, Ramduny, & Wilkinson, 1996, 1998), we looked in detail at the triggers that cause activities to happen when they happen. In the case of opening post, this could easily be something like "at coffee time" rather than straight away. In our work, we identified a number of common triggers:

- *immediate:* straight after previous task
- *temporal:* at a particular time or after a particular delay
- *sporadic:* when someone thinks about it
- *external event:* some event occurs such as a phone call or the receipt of a message
- *environmental cue:* something in the environment prompts action.

We will look at a more detailed breakdown later.

Now we can augment the work process with triggers for each activity (Fig. 19.5). Notice how we have examples of several types of trigger, two temporal and one environmental (letters in the office worker's hand prompting her to carry them to her desk).

Triggers are important not only because they help us understand the temporal behavior of the task but because they tell us about potential failure modes. If two environmental triggers are similar, one might do parts of the task out of sequence. If a trigger may not occur or be missed (likely for sporadic triggers), activities may be omitted entirely. Triggers also help us assess the likelihood of problems due to interruptions. For example, immediate "just after" sequences are disrupted badly by interruptions but environmental cues tend to be robust (because they are still there).

Sometimes triggers are seen in the plans of HTAs and sometimes "waiting" subtasks are included for external events, but such cases are exceptional. The normal assumption is that tasks are uninterrupted. However, it is straightforward to add a trigger analysis stage to most task analysis methods.

In terms of the ecology of interaction, triggers remind us that tasks are not always performed to completion. In the television quiz program *Mastermind*, the time limit would sometimes buzz when the quizmaster, Magnus Magnusson, was in the middle of a question. He would calmly announce, "I've started, so I'll finish," and then continue the question. In the heat of the real world, it is rarely so easy to ignore interruptions and distractions (chap. 14). In practice, tasks are interleaved with other unrelated tasks or, potentially more confusing, different instances of the same tasks, and may be interrupted and disrupted by other activities and events. Furthermore, the performance of the tasks is dependent on a host of sometimes fragile interactions with the environment and apparently unconnected events.

19.3.2 Artifacts: Things We Act on and Act With

Notice that one of the trigger types is an environmental cue—a thing in the environment that prompts us to action. Some years ago one of the authors received a telephone call reminding him to respond to a letter. He could not recall receiving it at all, but searching through a pile on his desk he found it and other letters, some several weeks old, unopened and unread. What had happened? His practice was to bring the post upstairs to his desk but not always read it straightaway. Not being a coffee drinker, it was not coffee time that prompted him to open the post but just the fact that there was unopened post lying on his desk. This process had worked perfectly well until there was a new office cleaner. The new cleaner did not move things around on the desk but did straighten higgledy-piggledy piles of paper. However, he had unconsciously been aware that unopened post was not tidy on the pile. This had effectively been acting as a reminder that post needed dealing with—this is a trigger. But with the new cleaner, post that for some reason got missed one day would then look as if it was tidily "filed" in a pile on the desk.

This story is not unique. The ethnographic literature is full of accounts of artifacts being used to manage personal work and mediating collaborative work. In some cases the effect has to do with the content of the artifacts, what is written on the paper, but in many cases it has to do with the physical disposition—by orienting a piece of paper toward you, I say, "Please read it." In the example above, the cue that said, "Post needs to be opened," was purely in the letters' physical orientation (not even their position).

Of course, artifacts do carry information and are often the inputs or products of intellectual work. Furthermore, in physical processes, the transformation of artifacts is the purpose of work.

One example that has been studied in detail in the ethnographic literature is air traffic control (see also chaps. 1, 13, and 25), and all these uses of artifacts are apparent in air traffic control (Hughes et al., 1995). Of central importance are flight strips (Fig. 19.6), small slips of card, one for each aircraft, recording information, such as flight numbers, current height, and headings. An aircraft's flight strip is essential for the controller managing the aircraft, but the strips taken together constitute an at-a-glance representation of the state of the airspace for other controllers. In addition, the strips are in a rack, and the controllers slightly pull out the strips corresponding to aircraft that have some issue or problem. This acts partly as a reminder and partly as an implicit communication with nearby controllers. Finally, the strips in some way represent the aircraft for the controllers, but of course the real purpose of the process is the movement of the aircraft themselves.

Task models often talk about objects, either implicitly in the description of subtasks or explicitly in the task model. However, the objects are always "second class"—users act on them, but they are not part of the task. CTT and most work process notations do talk about automated tasks but not about the artifacts, whether electronic or physical, included in the interaction.

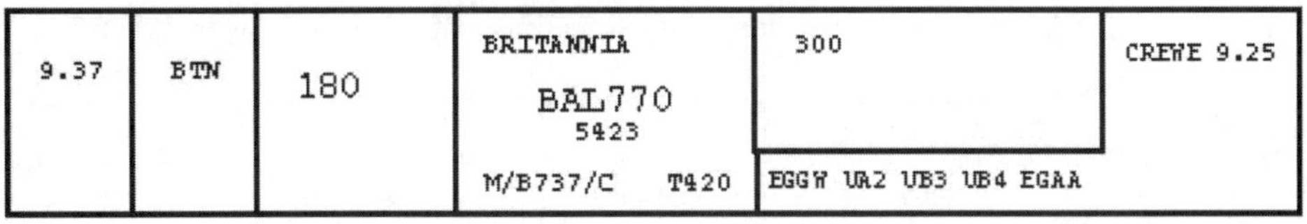

FIG. 19.6. Air traffic control flight strip.

In Unified Modelling Language (e.g., Bennett, Skelton, & Lunn, 2001) and other object-oriented design methods (e.g., Jacobsen, 1992; see also chaps. 7, 11, 12, 22, 24, and 27), it is common to give a life cycle description of "objects." However, this is usually because the intent is to store and automate the object electronically. Also, workflow analysts study document life cycles, again largely because of the intention to automate.

The task analysis chapter of Dix, Finlay, et al. (1998) treats physical objects as "first class" in an example of entity-relationship style task analysis. The method used was based largely on the ATOM method (Walsh, 1989), but to our knowledge this type of method has not gained widespread acceptance.

There is no reason why most task analysis methods should not adopt some form of artifact tracking. The tracking may be as simple as recording which artifacts are triggers for, used by, modified by, or produced by any particular subtask. For tasks where artifacts are particularly central, more sophisticated artifact life cycles could sit alongside the task description. These life cycles may be mundane (letter closed → letter open), but this is the point: Users recruit their everyday knowledge and physical properties of the world to coordinate their activity.

19.3.3 Placeholders: Knowing What Happens Next

It is half past five in the evening. The busy office building is beginning to become quiet as people pack up to go home. One or two employees work late in their offices, but as the evening wears on, they too go home. Soon there is only the hum of vacuum cleaners and the clatter of waste bins as the office cleaners do their work, until eventually the last light goes out and the building sleeps. A few employees have taken papers and laptops home and continue to work, but eventually they too put aside their work and sleep.

It is three o'clock in the morning, and in the darkness and silence of the office and the deep sleep of all the employees, where is the *memory* of the organization? The next morning at nine o'clock the office is a flurry of activity. It has not forgotten and has restarted its activities, but how?

We have already discussed two aspects of this memory: information required to perform tasks, and triggers that remind us that something needs to happen. However, there is one last piece of this puzzle that we have only hinted at. As well as knowing *that* we need to do something, we need to know *what* to do next. In the complex web of tasks and subtasks that make up our job, *where* are we?

In fact, when looking at triggers, we have already seen examples of this. The post being untidy on the desk said, "Something needs to happen," but the fact that it was unopened said, "It needs to be opened." We have already noted that similar triggers could cause subtasks to be performed out of sequence. If we only have a small number of dissimilar tasks, moving out of sequence is unlikely to happen, as we can remember where we are in each task. However, as the number of tasks increases, especially if we are performing the same task on different things, we find it harder to remember where we are.

Let us look again at air traffic control. One of the controller's tasks is to manage the flight level of aircraft. A much simplified model of this activity is shown in Fig. 19.7. Because this is a shared task between the controller and the pilot, each box is labelled with the main actor (although Tasks 2 and 3 are both communications). Recalling earlier sections, we might ask what

FIG. 19.7. Flight-level management task.

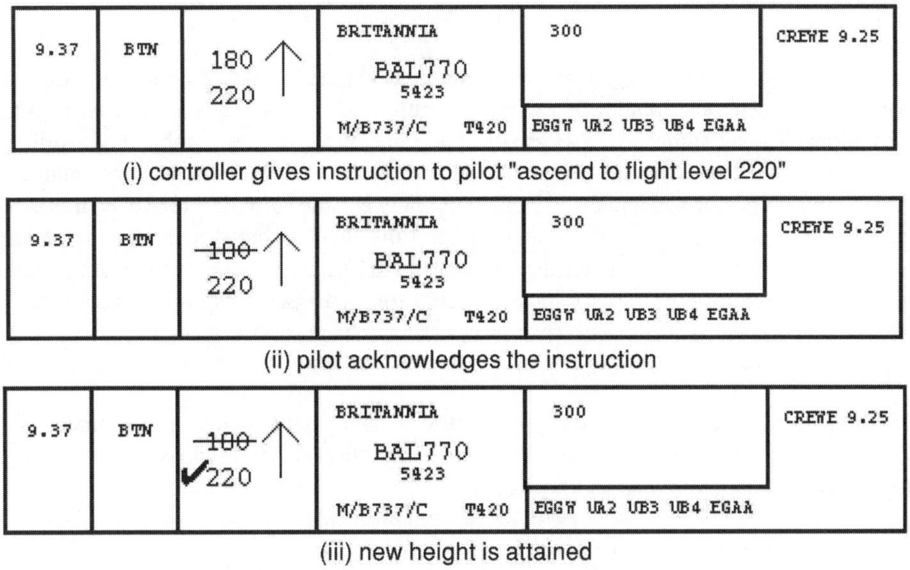
9.37
BTN
180
220
BRITANNIA
BAL770
5423
M/B737/C
T420
300
CREWE 9.25
EGGW UA2 UB3 UB4 EGAA
(i) controller gives instruction to pilot "ascend to flight level 220"
(ii) pilot acknowledges the instruction
(iii) new height is attained

FIG. 19.8. Flight strip annotated during task.

information is required at each stage. For example, Task 1 would depend on radar, the locations of other planes, planned takeoffs and landings, and new planes expected to enter the airspace.

Note that Box 5 is not really a task but more a "state of the world" that signifies task completion. However, it is important, as the controller will need to take alternative action if it does not happen. Of course, without appropriate placeholders, the controller might forget that a plane has not achieved its target level. This may either cause trouble later, as the old level will not be clear, or even provoke potential conflicts between aircraft.

In fact, the flight strips do encode just such a placeholder (see Fig. 19.8). When the controller informs the pilot of the new height, he writes the new level on the flight strip (i). When the pilot confirms she has understood the request, the controller crosses out the old level (ii). Finally, when the new level has actually been reached, the new level is ticked (iii).

Virtually all task-modeling notations treat the placeholder as implicit. The sequence of actions is recorded, but not why the user should do things in the way proposed. Of course, one purpose of task analysis has been to produce training—that is, to help people learn what appropriate processes are—but learning this does not help them to actually remember where they are in the processes.

As with other forms of information, placeholders may be stored in different ways:

- in peoples' heads (remembering what to do next)
- explicitly in the environment (to-do-lists, planning charts, flight strips, workflow system)
- implicitly in the environment (Is the letter open yet?).

Although often forgotten, placeholders are crucial for ensuring that tasks are carried out effectively and in full. At a fine scale, it is rare to find explicit records, as the overhead would be too high. Instead, memory and implicit cues predominate. As users' memory may be unreliable when they are faced with multiple tasks and interruptions, it is not surprising to find that various forms of environmental cues are common in the workplace. However, electronic environments do not have the same affordances to allow informal annotations or fine tweaking of artifacts' disposition.

19.4 FINDING TRIGGERS

This chapter is focused on triggers and associated placeholders. However, these are important because of the general problem of missing stimuli, and they are linked to issues such as interruptions. The techniques we use are designed to expose all these problems as well. Part of the data we collect is on *what* is done. In traditional workflow fashion, we catalog the various activities performed and the dependencies between the activities. However, this is only intended as the superstructure of the analysis, not the focus. Instead, our focus is on *when* activities are performed and *whether* they happen at all. The central and distinguishing feature of our work is therefore the way we look explicitly for the triggers that initiate activities.

19.4.1 Standard Data Collection

Because of the similarity of our study to traditional task analysis, we can use many of the same sources for data collection: documentation, observation, and interviews. However, trigger analysis gives us an additional set of concerns.

19.4.1.1 Documentation

Documentation of long-term processes is likely to be relatively accurate, although it may omit the activities beyond organizational boundaries and also most of the triggers. However, we can use it as an initial framework that can be filled out by observation or during subsequent interviews.

19.4.1.2 Observation

Direct observation is a very effective technique, widely used in ethnographic studies or similar sorts of analysis. In many office situations, there are several instances of the same process at different stages of completion. For example, in an insurance office, many claims are processed, and each might be at a different stage. In these cases, using “a-day-in-the-life” observation may be sufficient. So long as we can see each activity during the study period, we can piece the activities together afterwards. Even if we never see a process run from end to end, we can reconstruct it from its parts. This is similar to observing a natural forest. The complete life cycle of a tree might be hundreds of years long, but by looking at trees at different stages of growth, we can build up a full picture over a much shorter period.

However, direct observation poses special problems when many of the processes of interest extend beyond the time frame of observation and are geographically dispersed. In particular, it may miss rare but potentially significant events. Observation usually fails in the following situations:

- Activities are sporadic and long term.
- Processes are in lockstep and long term.
- Unusual events occur.

19.4.1.3 Interviews

Where direct observation is impractical, interviewing can be used. Interviewing allows both prospective investigation (asking about what is currently going on) and retrospective investigation (asking about what has happened). However, interviewing is often regarded as problematic because the accounts people give of their actions are frequently at odds with what they actually do, although chapter 16 argues otherwise.

We are, however, in a strong position as we approach such interviews. Our analytic focus—the structure imposed by task analysis and the specific interest in triggers—allows us to trace omissions and inconsistencies and obtain reliable results from interviews. This is important as, although we would normally expect some additional direct observation, practical design must rely principally on more directed and less intrusive techniques.

19.4.2 Types of Triggers

Based principally on our theoretical analysis and refined by the results of our previous case studies (Dix, Ramduny, & Wilkinson, 1996, 1998), we have classified the different kinds of triggers that occur. These include the following five broad categories:

1. *Immediate triggers* occur when one activity begins immediately after the previous activity reaches completion.
2. *Temporal triggers* include periodic actions that happen at regular intervals or actions that occur after a particular delay (e.g., the expectation of receiving a response by a certain date or the generic task of reminding people based on some time interval).
3. *Sporadic triggers* arise when an individual responsible for some action remembers that it must be done.
4. *External events* include alarms and other signals (e.g., a wristwatch or automatic calendar set to give a reminder at a specific time) and specific events (e.g., a telephone call, a face-to-face request, the receipt of a message, the completion of an automatic activity, or even an event in the world).
5. *Environmental cues* are things in our environment that remind us that something ought to be done, whether explicitly recorded (e.g., a diary entry or to-do list) or implicit (e.g., unanswered e-mails or a half-written letter in the typewriter).

Some of these triggers may be evident from observation, such as when a telephone rings and someone does something (external event). However, others are less clear, and this is where direct questioning helps ("Why did you do that then?"). In particular, it may be difficult to tell whether someone just remembers something (sporadic) or something acted as a reminder (environmental cue). The reason that these cases are difficult to determine is that both types of triggers relate to the user's perception and psychological state rather than observable actions.

After having identified the primary triggers, we need to assess their robustness by asking follow-up questions. Sometimes if the primary trigger fails, there is a secondary trigger that stops the process from failing. For example, if someone says that he or she always does two things one after the other (immediate), we can ask what would happen if there was an interruption. The answer may be that the person would need to remember (sporadic) or that the half-finished task would act as a reminder (environmental cue).

Sometimes the primary trigger is sufficiently complex to need its own secondary trigger. So, depending on the answers, we may have several levels of questions:

1. "Why did you send the reminder letter?" "Because it is 2 weeks since I sent the original letter" (temporal).
2. "How do you know it is two weeks?" "Because I wrote in my diary when I sent the letter" (environmental cue).
3. "What makes you check your diary?" "I always check it first thing each morning" (temporal).
4. "When you send the first letter, what happens if you are interrupted before entering the expected date in the diary?" "I always remember" (sporadic secondary to immediate primary).

Notice that both Questions 3 and 4 are suggested by the answer to Question 2. Question 3 is about chasing the chain of events that led to a particular trigger. In contrast, Question 4 arises

TABLE 19.1
Triggers, Failure Modes, and Follow-up Questions

Trigger Type	*Failure Mode*	*Follow-up Questions*
Immediate	Interruption	Does the second activity always proceed immediately? If there is any possibility of a gap, then look for secondary or backup trigger(s).
Temporal		
Periodic action	May forget	How do you remember to perform the action in the relevant period?
Delay	Poor memory	How do you remember the delay? This may be part of a routine, like consulting a diary every morning, but if it is an hourly activity, then how do you know when it is the hour? Perhaps the clock strikes, an external event—another trigger.
Sporadic		
Memory	May never happen	How do you remember to do something? If a request is made verbally, the recipient has to remember that the request is outstanding until either it can be performed or some record is made of the commitment, a reminder—another trigger.
External event		
Automatic	Reliability	How reliable is the medium of communication?
Communication	Delay	Are there likely to be any communication delays?
Environmental cue		
Explicit	Is moved	Can the cue be disrupted?
Implicit	Failure to notice	How do you remember to examine the cue? Perhaps a periodic activity or a note in a diary—a temporal trigger. Can the cue be missed?
	Ambiguity	Can the cue be confused with others, leading to skipped or repeated activities?

because the answer to Question 2 was an environmental cue and things in the environment need to have been put there by a previous activity. Note that this should also lead to a reexamination of the task associated with sending the original letter, as this should include the "write expected date in diary" activity.

We could continue asking follow-up questions indefinitely, but at some point we must stop. We can either believe that a trigger does always occur as specified or, if not, look at the whole process and assess the consequences should the activity fail to trigger at all and perhaps look at any delays associated with noticing it.

Table 19.1. lists some of the main ways in which different types of trigger can fail and the sort of follow-up questions that can be asked. It is not complete and is being amended as we learn more. An evolving and more detailed table with examples can be found on the trigger analysis web pages: http://www.hcibook.com/alan/topics/trigger-analysis/

19.4.3 Studying Artifacts: Transect Analysis

As we have seen, artifacts, both physical and electronic, are an inseparable part of an ecologically valid understanding of work and leisure. Tasks may be initiated to create an artifact (write a chapter abstract), tasks may occur because of artifacts (the memo requesting an action, the broken machine), artifacts may mediate tasks split between several people (patient records, whiteboard), artifacts may record where you are in a task (document in in-tray, office planner), and electronic artifacts may even control tasks (workflow systems; e.g., Winograd's [1988] Coordinator).

Studying artifacts can therefore give us a rich understanding of the tasks that they are part of, especially when the tasks are complex, are long lived, or involve different people.

One artifact-centered method that is particularly useful for uncovering triggers and placeholders is *transect analysis*. This method takes an ecologically rich approach to looking at artifacts in their physical context, and it views physical disposition as being as important as the artifact itself. It focuses on uncovering the task in praxis as performed in the actual work environment. Transect analysis looks at a snapshot of a work environment (desk, office, or potentially organization), either at a particular time (noon on Tuesday) or over a relatively short period (a day in the life).

Transect analysis is inspired by use of the transect in environmental studies to investigate a cross-section of an environment on a particular day. If the ecology is diverse and nonseasonal, it is possible to build up a picture of the life cycles of particular organisms and the interrelations between them even though they are not being studied over time. Similarly we can look at each artifact in the work environment and ask, "Who is using it?" "What for?" "Why is it here?" "What would happen if it were somewhere else?" "How does it relate to other artifacts?" In particular, we look for instances of the same kind of artifact in different places (e.g., several invoices in different stages of processing) within one person's immediate environment (in-tray, center desk, out-tray, at an angle on a pile) or between people. By piecing together these snippets of human–artifact interaction, we can create models of task processes and artifact life cycles.

This analysis can be done initially by the analyst but at some stage should be used as a prompt for the actual users. If asked questions away from the actual environment, they may not volunteer many of the unofficial procedures that make the workplace work. However, when shown a document in location and asked, "Why is it there?" they will usually be able not only to knit together the scattered threads of processes but also add details beyond the immediate workplace. If interviews are carried out in the workplace, then transect analysis can be woven seamlessly into the normal interviewing process.

19.5 LARGER STRUCTURES

Trigger analysis can be used in the context of virtually any task analysis. In the case studies that we have done, we have used a simple process diagram as a method of recording the task analysis and asked specific questions about the overall robustness of the process. However, we believe that the lessons and the broad techniques would be applicable to any type of process-oriented task analysis, such as HTA.

19.5.1 Processes and Activities

We record the processes as a series of circles or bubbles, one for each activity. Each bubble names the activity and the person or persons who perform it. Lines between the bubbles record dependencies, and arrows at the beginning of each bubble record the trigger for the activity (see Fig. 19.9).

There are plenty of methods for recording processes, and this is not the focus of our work, so we take a minimalist approach. We do not attempt to record all the complexities of real processes in a single diagram. Instead, we use many separate diagrams, often concentrating on specific scenarios. The crucial thing is to look for the trigger of each activity.

The level of analysis is also governed by this focus. In general, we place activity boundaries wherever there is the likelihood of a delay or gap. The most obvious such break occurs when people at different sites perform subsequent activities in a process. However, there are often distinct activities performed sequentially by an individual, as in the letter-reading example above. In principle, such analysis could go down to the full detail found in HTA. This would be reasonable if, for example, interruptions were possible in the middle of typing a letter.

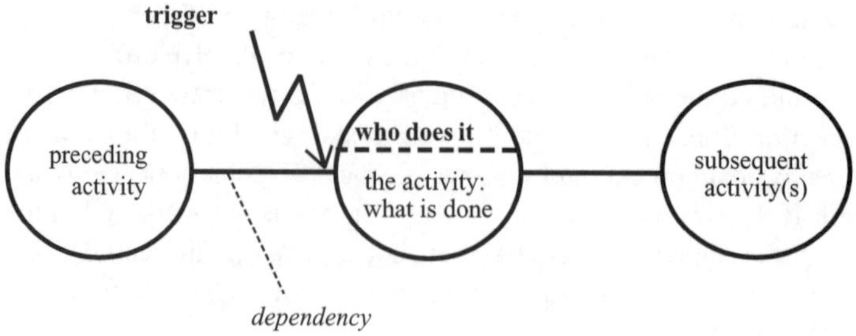

FIG. 19.9. Recording processes.

We deliberately use the term *activity* rather than *action* to emphasize that the lowest level of our analysis is far from atomic. Activities may be shared between individuals. For example, having a meeting or dictating a letter would be regarded as a single activity involving several people. Again, one could dissect such an interaction, but this would be the remit of conversational analysis. We may also ignore details of an activity because they are uninteresting or we do not have sufficient knowledge about them. For example, if we issue an order to an external organization and then wait for the goods to arrive, we may not be interested in the internal processes of that firm. Finally, we include some activities that would normally be omitted in a traditional process model. In particular, we often include the receipt of a message as a distinct activity. This is done to emphasize the gap that may occur between receipt and response.

19.5.2 Process Integrity

We can assess the reliability of the work process by asking questions about the triggers for activities. However, nothing is ever 100% correct, and it is inevitable that triggers will fail for some reason, activities will be missed, and perhaps the whole process will stop because something goes wrong. A process model combined with a well-founded assessment of the reliability of each activity can allow us to assess the robustness of the whole process. If someone fails to complete some activity and the next activity is never triggered, what happens? Does the whole process seize up or will the failure eventually be noticed?

Note that this is not an ad hoc procedure. We can *systematically* go to each trigger and ask what happens to the entire process if the trigger fails. Furthermore, by looking at the process as a whole, we can improve our assessment of the reliability of any trigger. For example, if the trigger for an activity is that a report is in someone's in-tray, we can examine the wider context and assess the likelihood that the report will indeed be there when required.

19.5.3 The 4Rs

Although our initial focus was on the individual triggers, we began to notice an emerging pattern as we recorded the processes during our case studies. We call this pattern the 4Rs: request, receipt, response, and release.

Figure 19.10 shows a simplified version of a case study on the operation of a conference office (Dix et al., 1996). We can see a general structure emerging:

Request: Someone sends a message (or implicitly passes an object) requiring your action.
Receipt: You receive it via a communication channel.
Response: You perform some necessary action.
Release: You file or dispose of the things used during the process.

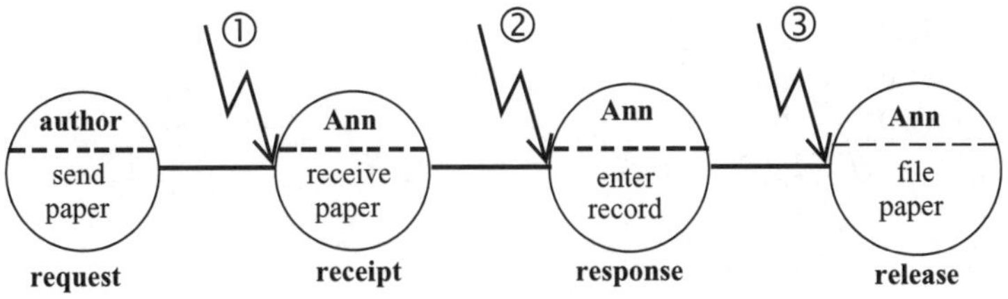

FIG. 19.10. The 4Rs.

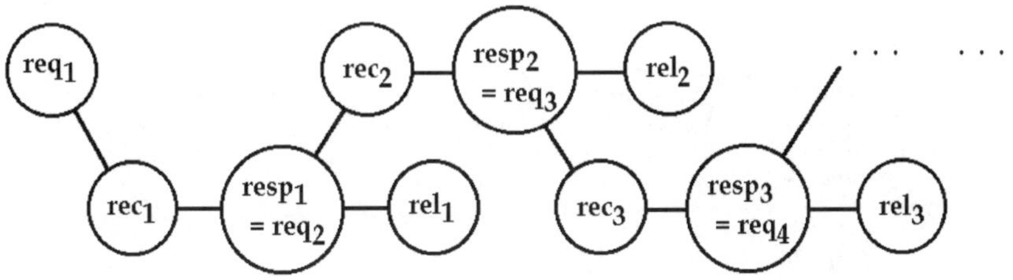

FIG. 19.11. A 4Rs chain.

At this point, if the functional goal has been achieved, the process can be considered to have reached completion.

The process in Fig. 19.10 is very similar to the process that one of the authors follows when dealing with e-mail. When the mail arrives, he reads it (or at least notes its arrival) but does not deal with it immediately. It stays in his in-tray until he has replied or otherwise dealt with it. Only at that stage does he file it in a folder or discard it. If he is interrupted after replying, the original message is still in the in-tray (secondary trigger). Once, whilst he was in the middle of replying to a message, the machine crashed (interruption). When some time later he again read his e-mail, he mistakenly (and unconsciously!) took the continued presence of the e-mail in the in-tray as signifying an interruption before filing (secondary trigger), and hence he filed the message without replying.

We believe that the 4Rs constitute a fundamental unit of long-term work. Not only does the pattern of activities occur among different processes, but we also see a similar pattern of triggers. ① is always simply some sort of communication mode and can be assessed for reliability and timeliness. The response activity is typically triggered by ②, the presence of a document or other object. The release activity triggered by ③, which is of the "immediately follows" kind, removes that cue but also relies on its existence as a secondary trigger. The problems with the author's e-mail will occur elsewhere! The existence of generic patterns makes it easier to uncover problem situations quickly and to take solutions found in one situation and adapt them to another.

Our case studies show that the 4Rs are normal: The same pattern recurs with similar triggers and similar failure modes. We have also seen that it is normative: If the 4Rs pattern is nearly followed but with some deviation, the deviation is an indication of possible problems. It is also frequently the case that the response of one 4Rs pattern forms the request activity initiating a new 4Rs pattern. A chain of such patterns constitutes a sort of long-term conversation (see Fig. 19.11). The 4Rs appear to make up a pervasive, generic pattern at a lower level than the patterns identified in Searle's speech-act theory (e.g., Winograd & Flores, 1986) and are perhaps the equivalent of adjacency pairs found in conversational analysis.

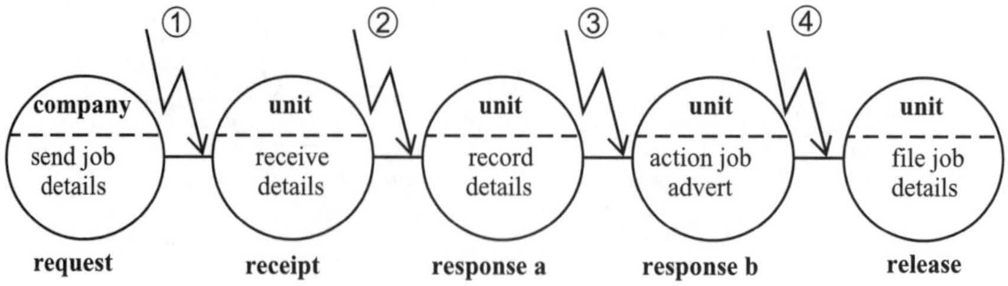

FIG. 19.12. Initial job advert.

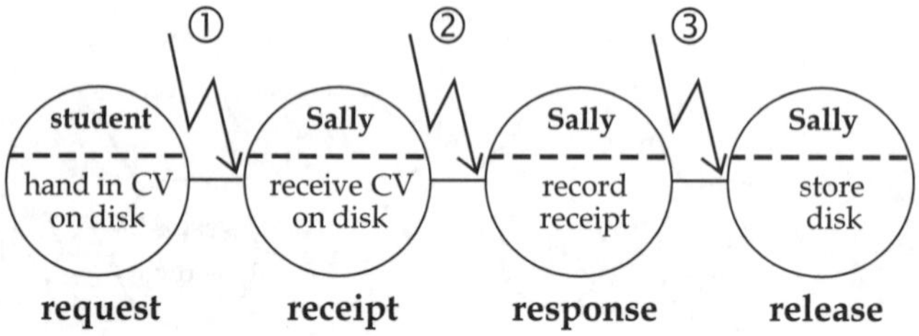

FIG. 19.13. Students submit CVs.

19.6 EXAMPLE: THE STUDENT PLACEMENT UNIT

A key aspect of validating the trigger analysis was to apply the framework to prolonged interactions in an extremely busy office environment—the Placement Unit at the School of Computing and Mathematics at the University of Huddersfield. The unit is staffed on a full- and part-time basis by administrative and academic staff respectively and is responsible for helping some 200 sandwich course students secure 1-year placements in industry every year. Besides dealing with students seeking placements (predominantly recruitment and skills-building processes), the unit also supports those already on placement (including processes in assessment and monitoring).

Also, the MaPPiT project[1] had just been launched at the university, and part of the project remit was to develop a process support system in Lotus Notes (e.g., Lloyd & Whitefield, 1996) for the placement unit, thereby automating many placement activities.

Contact with companies occurs via all media, the telephone being the most common and the most interruptive. Outright winners in the interruption stakes, however, are the students, for whom the unit has an "open door" policy between 10 a.m. and 4 p.m.

Figures 19.12 to 19.14 show that the activity triggers had the potential to be seriously delayed, sometimes indefinitely, if they had been left to be resolved by outside companies.

19.6.1 Case 1: Job Adverts

The establishment of a new placement starts with the initial request from a company for a placement student (see Fig. 19.12). This is a 4Rs pattern but with a two-stage response. One of the dangers of multistage responses is that the triggers at ② and ③ are often similar, which can lead to problems.

---

[1]MaPPiT stands for Mapping the Placement Process with Information Technology, a HEFCE-supported 2-year project funded by the Fund for the Development of Teaching and Learning. Details available at http://www.hud.ac.uk/scom/mappit/

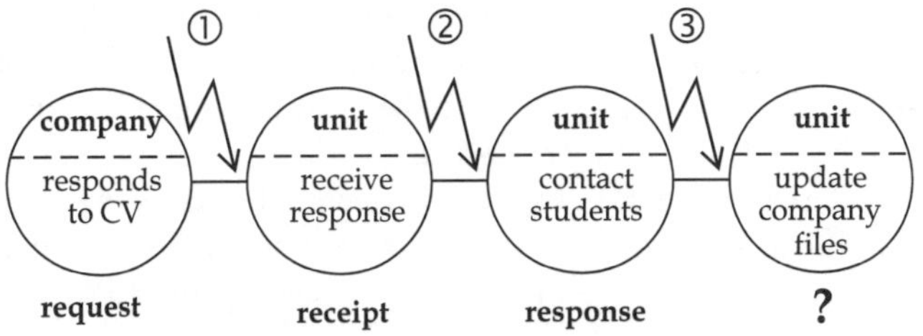

FIG. 19.14. Company decision.

In the advert process the company initiates the request. Many placement providers recruit annually, diarizing to send the unit their requirements. They drive the process by setting closing dates for applications. Another set of companies are triggered into providing placement details by a standard letter from the unit sent out fortnightly via a diary system when it is deemed most appropriate to make the contact. Looking at ①, we asked the administrative staff how they would know if details failed to arrive, thus breaking the chain of activities. At present, the only backup is the diary, so a time delay occurs between the failure occurring and the next fortnightly (sometimes monthly) check for responses from the previous month's companies. A follow-up question would be, "How do you remember to look in the diary?" The answer is that the paper-based diary remains highly visible on the placement officer's desk. However, as the year progressed, we noticed the diary being checked less and less.

The next activity, "record details," should ideally follow on directly from the first activity. This class of trigger is insecure and liable to interruption—a common occurrence in the unit. The staff member then has to remember what to do next. Usually, there is the environmental cue on the desk, such as a note jotted down from a phone call, a fax copy, a letter, or an open e-mail message. The follow-up question was, "What if you do not record the receipt of details in the diary?" Another staff member could check the diary, see the assumed nonreceipt of details, and annoy a company by chasing for details already sent in unless he or she checked the job adverts log (response b) first or checked the company file for the ad. All this checking seemed unnecessarily complicated.

With Lotus Notes in mind, the project team accepted these current problems as needing resolution. The diary could easily become electronic with built-in navigators (agents) that automatically trigger reminders to execute activities. Gone is the need to remember to check the diary, as reminders appear in individuals' to-do lists. Even to-be-done-to lists can be constructed. Receipt of a job description need only be recorded in one place, necessitating only one checking activity before chasing a company. Furthermore, the electronic record means that any inconsistency between recorded details can be displayed and thus act as a trigger at ③.

19.6.2 Case 2: Submission of CVs

A large proportion of the unit's placement providers happily accept standard CVs from students, so CVs are lodged with the unit and checked by placement tutors very early in the year. Figure 19.13 shows this process, and again we see a straightforward 4Rs pattern.

The first two activities usually happen face to face, so there is little risk of breakdown at ①. We noticed that some students, unfortunately, ignored the office hours of the unit and so "posted" CV disks under the door after staff had left for the day, creating a risk that the disk would be disposed of, damaged, lost, or misplaced. Similarly, the unit's activities here are all exposed to interruptions and therefore incompletion. Also, staff changes in the unit brought some new faces unaccustomed to the subprocess of receiving and recording students' disks.

In one case, a student made alterations to the CV and returned the disk, but a new member of staff promptly lost the update while being interrupted to do other more complex tasks. In other cases, a disk was found to be corrupt when the CV was required, leading to another set of interactions with the student and further delays.

The planned Notes implementation for this completely bypasses the current error-prone process. Students would fill in a CV template using a Web browser. The CV is then automatically submitted to a Notes database that logs the receipt and sends an e-mail to the unit to confirm that the CV has been submitted on time. Students can update the CV at any time without bothering the administrative staff and in the full knowledge that the latest version will be sent to potential placement employers.

19.6.3 Case 3: Company Decisions

After seeing the students' CVs, companies decide on students to shortlist for interviewing. Figure 19.14 demonstrates how the pace of interaction can really slow down when pursuing students to arrange interviews or to provide feedback if rejected.

At the response stage, the pace slows considerably once the students are on vacation and hard to track down. Much time can be spent trying to contact students, and assuming contact is made, the next activity can be stalled by a phone call or face-to-face enquiry. We rely on an individual's good short-term memory and/or an environmental cue to ensure the sequence is fulfilled.

Although the process appears at first to be a simple 4Rs pattern, we have put a question mark against the last activity. The release usually consumes or destroys the environmental cues that have prompted previous activity. It is not clear that this is the case for this process—what are the environmental cues? The company's decision will arrive in a letter or be recorded on paper, but the slow pace of the response means that the cue may be lost or grow stale, ceasing to be salient because it has been around too long.

Redesigning this sequence in Notes, we decided that it would be better to record the receipt of the company's contact electronically when the contact actually happens, invariably by means of a phone call, fax, or letter. All the information can therefore be displayed on screen, and so if incomplete because of an interruption, it cannot be discarded without a prompt. Note how this action has established an environmental cue within the electronic environment. In the revised process, this cue is removed when the contact details are complete, thus making the pattern a true 4Rs, with robust triggers throughout. Moreover, if the response stage becomes drawn out and relies on chasing student responses, there is the possibility of automatically signaling if the expected reply is not forthcoming, thus supplying a to-be-done-to facility.

From the three cases discussed above, we can see how different levels of automation have been suggested by the trigger analysis and the 4Rs pattern. At one extreme, the strategy involved the complete bypassing of the human process, but in the others only parts were automated. More importantly, the analysis has ensured that the Notes implementation does not hide existing triggers, as often is the case with electronic filing, but instead enhances the triggers with automatic reminders and electronic environmental cues.

19.7 CONCLUDING REMARKS

This chapter has considered the analysis of work processes with a view to understanding more clearly why things happen when they do. There is a strong indication that work activity is more prone to breakdowns and failures when the interaction is prolonged, when the pace of interaction is slow, or when the activity is of a collaborative nature and crosses organizational boundaries. When all three conditions are met, failure becomes almost a certainty.

We recommend trigger analysis as a useful method for uncovering what prompts a task to happen. One particular advantage of trigger analysis is that weaker triggers can be identified swiftly and action can be taken to render the tasks more robust, as we saw with the student placement example. Moreover, the classification of the triggers by type allows a simpler analysis. A further benefit of this method is that individuals may conduct investigations with little or no expertise in knowledge elicitation techniques. The follow-up questions keep it simple, and the analysis can be used with virtually any other forms of task analysis.

Trigger analysis does not only stop at triggers; artifacts and placeholders also have their roles to play. Any activity in a workplace involves artifacts of work, and we recommend that current task analysis methods should accommodate some form of tracking for artifacts, placing them on a more equal footing with the user in the task context than is now common. Just as triggers remind us that *something* needs to be done, placeholders must be robust so that we know clearly *what* needs to be done.

The importance of environmental cues means there is another rich source of information: the work environment itself. We look at an office and see papers and files on the desk, Post-it notes, an in-tray, a wall calendar. Why is that file on the desk? What will happen to it? What would happen if it were not there? We know that environmental cues can be triggers for activities, and so we take each item in the environment and look for the activity it triggers or the coordinating role it fulfils. At the very least, a piece of paper left on the desk is saying, "File me please."

The 4Rs framework was found to be a fundamental unit of interaction and work—a handy "template" for relatively straightforward work analysis. A key feature of the persistent pattern of 4Rs is its ability to highlight weaknesses in prolonged interaction. We can establish whether the 4Rs are complete, as incomplete 4Rs have always revealed a weak trigger, environmental cue, and/or placeholder.

Finally, the study of work situations should not be isolated from the real context in which the work is conducted. Trigger analysis fits easily into such an environment, as it needs to examine and understand that environment. Most importantly, it can complement other task analysis methods that are also suitable for field investigations.

In summary, trigger analysis can augment many task decomposition methods. In addition to the description of *what* should happen found in most such methods, it captures *why* they occur *when* they do and *whether* they are likely to be missed or misordered. The strong emphasis on all types of triggers, including environmental cues, leads to an ecologically rich analytic approach.

19.8 FURTHER INFORMATION

The most complete theoretically rooted account of triggers and the 4Rs is the Dix et al. (1998) article in *Interacting with Computers*. For an analysis of how rich ecological features, including triggers and placeholders, can augment more formal task analysis models and notations see Dix (2002a). Finally, for new information and hyperlinked material, consult the trigger analysis web pages: http://www.hcibook.com/alan/topics/trigger-analysis/

REFERENCES

Abowd, G., & Dix, A. 1994. *Integrating status and event phenomena in formal specifications of interactive systems.* In *Proceedings of the 2nd ACM SIGSOFT symposium on foundations of software engineering* (pp. 44–52). New York: ACM Press.

Bennett, S., Skelton, J., & Lunn, K. (2001). *Schaum's outline of UML.* New York: McGraw-Hill.

Brewster, S. A., Wright, P. C., & Edwards, A. D. N. (1994). *The design and evaluation of an auditory-enhanced scrollbar.* In *Proceedings of the SIGCHI conference on Human factors in computing systems: celebrating interdependence* (pp. 173–179). New York: ACM Press; Reading, MA: Addison-Wesley.

Dix, A. J. (1991). *Formal methods for interactive systems.* London: Academic Press.

Dix, A. J. (1992). Pace and interaction. In *Proceedings of HCI'92: People and computers VII* (pp. 193–207). Cambridge: Cambridge University Press.

Dix, A. (1994). Que sera sera: The problem of the future perfect in open and cooperative systems. In *Proceedings of HCI'94: People and computers IX* (pp. 397–408). Cambridge: Cambridge University Press.

Dix, A. (2002a). Managing the ecology of interaction. In *Proceedings of First International Workshop on Task Models and User Interface Design (Tamodia 2002)* (pp. 1–9). Bucharest: INFOREC Publishing House.

Dix, A. (2002b). Towards a ubiquitous semantics of interaction: Phenomenology, scenarios and traces. In *Proceedings of DSV-IS 2002* (pp. 238–252). Berlin: Springer-Verlag.

Dix, A., Finlay, J., Abowd, G., & Beale, R. (1998). *Human-computer interaction* (2nd ed.). Hemel Hempstead: Prentice-Hall.

Dix, A., Ramduny, D., & Wilkinson, J. (1996). Long-term interaction: Learning the 4Rs. In *CHI '96 Conference Companion* (pp. 169–170). New York: ACM Press.

Dix, A., Ramduny, D., & Wilkinson, J. (1998). Interaction in the large. *Interacting with Computers, 11*(1), 9–32.

Hughes, J., O'Brien, J., Rouncefield, M., Sommerville, I., & Rodden, T. (1995). Presenting ethnography in the requirements process. In *Proceedings of the IEEE conference on Requirements Engineering* (pp. 27–34). New Jersey: IEEE Press.

Jacobsen, I. (1992). *Object-oriented software engineering.* Wokingham: Addison-Wesley.

Lloyd, P., & Whitefield, R. (1996) *Transforming organisations through groupware: Lotus Notes in Action.* Berlin: Springer-Verlag.

Norman, D. A. (1986). New views of information processing: Implications for intelligent decision support systems. In E. Hollnagel, G. Manuni, & D. D. Woods (Eds.), *Intelligent decision support in process environments* (pp. 123–136). Berlin: Springer-Verlag.

Norman, D. A. (1988). *The psychology of everyday things.* New York: Basic Books.

Paternò, F. (1999). *Model-based design and evaluation of interactive applications.* Berlin: Springer-Verlag.

Payne, S. J. (1993). *Understanding calendar use. Human-Computer Interaction, 8,* 83–100.

Shepherd, A. (1995). Task analysis as a framework for examining HCI tasks. In A. Monk, & N. Gilbert (Eds.), *Perspectives on HCI: Diverse approaches* (pp. 145–174). London: Academic Press.

Walsh, P. (1989). Analysis for task object modelling (ATOM): Towards a method of integrating task analysis with Jackson system development for user interface software design. In D. Diaper (Ed.), *Task analysis for human-computer interaction* (pp. 186–209). Chichester, England: Ellis Horwood.

Winograd, T. (1988, December). Where the action is. *Byte*, pp. 256–258.

Winograd, T., & Flores, F. (1986). *Understanding computers and cognition: A new foundation for design.* Reading, MA: Addison-Wesley.

Wood, A., Dey, A. K., & Abowd, G. D. (1997). *CyberDesk: Automated integration of desktop and network services.* In *Proceedings of the 1997 Conference on Human Factors in Computing Systems (CHI '97)* (pp. 552–553). New York: ACM Press.

20

The Application of Human Ability Requirements to Virtual Environment Interface Design and Evaluation

William Cockayne
Stanford University

Rudolph P. Darken
Naval Postgraduate School

This chapter investigates the application of human ability requirements (HARs) to task analyses of physical user interfaces, typical of virtual environments. The components of these tasks typically consist of whole or partial body interaction using the hands, arms, or legs in ways that are similar to, but not identical to task performance in the real world. The chapter describes a method of human performance evaluation that is applied to locomotion tasks in virtual environments. The chapter makes recommendations for how this methodology can be used to develop taxonomies for other areas of human interaction of general interest to the research community. We also discuss how this taxonomy can be used by researchers and practitioners to both better understand the human ability requirements of tasks in their virtual environment system and also to develop better interaction techniques and devices based on the HARs for a given task or task component.

20.1 INTRODUCTION

The purpose of this chapter is to describe a classification of human ability requirements and its application to task analysis in human-computer interaction (HCI) for virtual environments (VEs). The objective of this classification is to structure existing and current research into a coherent framework supporting the development and analysis of interaction techniques and devices. This classification methodology, or taxonomy, can be used to better understand interaction techniques already seen in the literature and also to develop new techniques that are related to existing ones in terms of human abilities—the only components of the man-machine system not likely to change substantially over time (chap. 15 also argues that human cognitive abilities are unchanging). The methods and classifications presented here have not been developed in a vacuum but have instead been derived from a well-defined taxonomic methodology based on human ability requirements (Fleishman & Quaintance, 1984). The chapter shows how this taxonomy can be used by comparing task components performed in the real world to these same task components performed on a new device for interaction in VEs. The comparison is based solely on human performance characteristics. The classification elucidates the strengths and weaknesses of the device in terms of a user's ability to perform the task component in the VE.

To relate the approach described here to traditional task analysis methodologies, we can view aggregate tasks in the context of VEs as tasks involving purposeful behavior, such as "find a specific object" or "move to the road intersection." Components of these tasks might include knowing where to go (*cognitive*), physical movement (*motor*) from place to place (even if abstracted through a device), and visual perception of the environment to guide movement and to know when to stop (*perceptual*). Our focus here is on the physical components of these tasks that we can directly relate to natural human abilities. Consequently, we use the term *task components* to refer to the low-level physical subtasks that aggregate to form high-level tasks more suitable to traditional task analysis methodologies. We also discuss how these new techniques fit into the overall process of system design that includes traditional task analysis.

A comprehensive and structured approach to this work is required to overcome the limitations of the present body of research on human-computer interaction, which consists mostly of single cases. Single cases do not substantially increase our understanding of the overall problem of human-computer interaction or how future solutions might be devised in other than an ad hoc, intuition-driven fashion. To date, there has been no concerted effort to analyze and compare VE interaction techniques in terms of human performance. Additionally, there has been no comprehensive analysis of the abilities required to perform task components in a VE in such a way that their performance in this environment can be compared with their performance in the real world, which may be of vital concern.

Some would argue that because VEs enable experiences that cannot exist in the real world, linking VE task components to real-world task components via human abilities is unnecessary. We contend that humans acquire skills in the real world, our abilities have evolved in such a way as to optimize performance in the real world, and it is these skills that we bring with us into a VE for the execution of tasks. Consequently, whether in a virtual or real environment, our abilities are the same, and therefore they constitute a solid foundation for the evaluation of task components, techniques, and devices. HCI has long made use of existing skills and knowledge in interface design. This is the basis for the use of metaphors and mental models (e.g., Cooper, 1995; Dix, Finlay, Abowd, & Beale, 1998; Preece et al., 1994; and most of the chapters in part II of this Handbook). VE's also make use of metaphors and mental models, to leverage existing skills and knowledge of users. Although we do not contend that all tasks in VEs must be performed in a manner similar to their real-world counterparts (e.g., that virtual walking must involve physical walking), we do believe that execution of any task in a VE must involve physical action on the part of the user and therefore can be analyzed on the basis of natural human abilities.

One approach that engineering psychology and human performance evaluation have used to solve this problem is to develop and apply classifications and taxonomies of human performance. The development of taxonomic systems for classifying tasks that people perform and of the dimensions that are used to classify them has helped to bridge the gap between basic research and application in three ways. In particular, these taxonomies have

> (1) eliminated redundant terms, (2) disclosed similarities and differences between "operations" in the laboratory and the applied world as well as between various subject matter areas within research, and (3) alerted behavioral scientists to possible sources of variance that may contaminate or even negate their research findings in the operational setting. (Fleishman & Quaintance, 1984, p. 4)

The chapter begins with a discussion of taxonomic science and classification as related to the development of the Human Ability Requirements (HARs) taxonomy for human performance evaluation. It discusses the extension of real-world taxonomy method and tools into VEs and how these can be used to extend and complement conventional task analyses. It is the linking of human abilities as required by task components to interaction techniques and devices that is of concern. Our research was based on the need to understand how humans perform physical tasks in the real world in order to guide the design and implementation of interaction techniques and devices to support these tasks in VEs.

We applied these techniques to two problem domains: bipedal human locomotion (Darken, Cockayne, & Carmein, 1997) and two-handed whole-hand interaction (Cockayne, 1998). Two classifications were developed for the real world and for VEs. The two classifications are used to analyze a large number of real-world task components, externally validate the classification, analyze human bipedal locomotion on the omni-directional treadmill, and analyze the two-handed interaction tasks of a field medic. We present only the bipedal locomotion analysis here.

The chapter makes recommendations for how this methodology can be used to develop taxonomies for other areas of human interaction of interest to our community. It also discusses how this taxonomy can be used by researchers and practitioners to better understand the human abilities requirements of task components in their VE system and to develop better interaction techniques and devices based on the HARs for a given task component.

20.2 CLASSIFICATIONS AND TAXONOMIES

Most of the taxonomies in the HCI and VE literature are *Linnaean* taxonomies. These taxonomies attempt to classify entities and groups in terms of their "essence." There are no set rules or procedures for how an entity is classified. The method involves significant subjective judgment as to the fundamental characteristics of an entity or group of entities. More importantly, the context in which an entity is to be classified has everything to do with the language used to describe it. This is a dominant and recurring issue in the VE literature. An engineer might describe a glove device in terms of its components (e.g., fiber optics, stress sensors, and so on) whereas a psychologist might describe it in terms of the tasks for which it can be used (e.g., pointing, grasping, and so on). What is needed is a method that maintains consistency and repeatability in an objective manner across researchers, organizations, studies, and systems.

Such a method is called a *numerical taxonomy* and is the basis for the classification presented in this chapter. The primary advantage of a numerical versus a Linnaean taxonomy is that, once an item is placed in a numerical taxonomy, something is immediately known about it in the language of the taxonomy—in this case, human performance. This is not necessarily the case in a Linnaean taxonomy, as it does not have a consistent set of rules for inserting new items.

The distinction between a taxonomy and a classification is often blurred in present HCI and VE literature. The methods underlying the development and extensions of classifications are the domain of *taxonomic theory*. The process and method used in the development and later in the extension of lists of grouped knowledge is a *taxonomy*. One of the initial results of a taxonomy is the production of a *classification*, or listing, of the knowledge being studied.

Although a classification is useful to practitioners and researchers in allowing for the grouping of knowledge, it is limited in usefulness without an underlying taxonomy. The taxonomy directs involved parties to a more detailed understanding of the assumptions inherent in the classification. In addition, the taxonomy is crucial in allowing researchers to extend or enhance the classification based on new knowledge or research. Most importantly, a classification is *heuristic*; it guides future investigations by stating hypotheses. The delivery of an underlying taxonomy for any VE classification being developed is crucial in an immature field such as ours. The terminology presented above is well understood in the field of human performance and can be stated as follows:

> A *taxonomy* is the theoretical study of systematic classifications including their bases, principles, procedures, and rules. The science of how to classify. A *classificatory system* is the end result of the process of classification, generally, a set of categories. A *classification* is the ordering or arrangement of entities into groups or sets on the basis of their relationships, based on observable or inferred properties. (Fleishman & Quaintance, 1984, p. 21)

Taxonomic research and the development and application of classifications typically occurs in areas of intensive research activity, where there is a strong desire on the part of participants and more importantly practitioners to address the growing development of unorganized facts. The application of classification development to fields of research allows the fields to more rapidly develop by enhancing communication using a common language. The field becomes more comprehensive in nature, allowing participants and observers to understand the present state of the field as well as developments that occur and use this knowledge to drive their own contributions. Classifications allow participants to better communicate the results of studies, resolve confusion between research institutions and programs, and point out areas of neglected study. These assumptions underlie the work presented here.

At the same time, classifications and taxonomies do not show researchers where to find the next set of answers. What they do is help researchers understand the knowledge that they possess. By extension, the researchers are then able to describe areas of knowledge or research that are not complete.

20.2.1 Human Ability Requirements

With this general understanding of taxonomies, classifications, and their uses, this chapter focuses on one well-developed taxonomy, that of human ability requirements (HARs). This method for classifying human abilities has been in development since the mid-1960s and has been used in a variety of applications (e.g., Rose, Fingerman, Wheaton, Eisner, & Kramer, 1974; Wilson, Barnard, Green, & MacLean, 1988). The taxonomy was initially intended to provide a classification of human abilities for use by the military for training and job placement of skilled personnel. The resulting classification has continued to be refined and is presently codified in the Fleishman–Job Analysis Survey (F-JAS; Fleishman, 1995).

An example of a human ability described in the F-JAS is *static strength*, which is defined as "a human's ability to use continuous muscle force to lift, push, pull, or carry objects." The human abilities defined in the F-JAS are grouped into four metaclasses: cognitive, psychomotor, physical, and sensory/perceptual. Two other metaclasses of human abilities are being defined by Fleishman: interactive/social and knowledge/skills. The basis for using the F-JAS is that a number of human abilities can be used to describe the way that a human solves a problem (that is, executes a task or task component). The resulting analysis of task performance, in terms of required human abilities, can be used to evaluate interaction techniques and devices across disparate application domains because the basis, human abilities, remains constant and universal. It should also be noted that the use of this taxonomy does not preclude the use of any other that may be more domain specific. For example, the analysis of locomotion described here can be combined with the "taxonomy of virtual travel techniques" described by Bowman, Koller, and Hodges (1997), resulting in a comparison of interaction techniques for travel and their associated human ability requirements for execution.

Each of the human abilities defined in F-JAS (and, by extension, this classification) is presented with a representative name and definition. The definition allows for the analysis

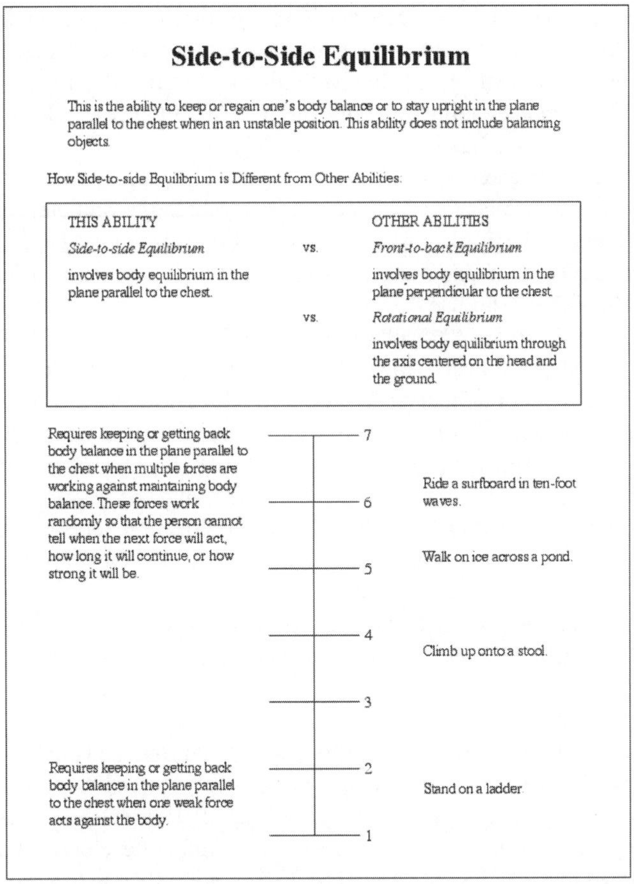

FIG. 20.1. Human Ability Requirement definition for side-to-side equilibrium.

of human task components using an absence/presence evaluation and reporting scheme. Absence/presence evaluation is simply the use of a standard definition to decide whether the idea or task component presented in the definition is absent or present in the thing being studied. In addition to the definition, each of the human abilities is also represented by a seven-point scale using a behaviorally anchored rating technique that anchors both the high and low ends of the scale with additional definitions and task examples. The use of a scaled analysis allows the application of the taxonomy to move from being largely qualitative to quantitative. Each of the task components can be further analyzed, beyond absence/presence, using the scales to derive ordinal ratings. The scales for abilities included in F-JAS can be found in Fleishman (1995). The scales developed or extended in reviewing users of VEs can be found in Cockayne (1998) and Darken et al. (1997). An example of one of the scales developed in our work is presented in Fig. 20.1. Fleishman's work provides a basis for developing of a new view of humans in VEs through the relationship between task components and human abilities.

20.2.2 The Need for Classification in Virtual Environments

The critical problem that this system of classification addresses is that researchers and practitioners in virtual environments do not speak the same language. The relatively recent engineering breakthroughs that have enabled the creation of VEs have brought together an eclectic group of professionals, and the professionals within each discipline apply their field's methods to issues associated with humans in computer-generated, synthetic spaces. In attempting to address these issues, these same people are beginning to discover that the study of humans in VEs requires a broad understanding and application of multiple research methods that, until now, have only been used to investigate real-world tasks.

Human performance research in VEs is being explored by professionals from myriad fields of research: human factors, behavioral and cognitive psychology, computer science, industrial engineering, and biomechanics, to name a few. Each of these disciplines has its own methods of addressing issues related to human performance. Each field's supporting research on humans in VEs has brought with it expertise that was developed over years of study as well as its own terminology, methods, histories, and acknowledged leaders.

Vast amounts of data have been collected from a large number of studies in an attempt to understand various aspects of human performance in VEs. However, it is often difficult, if not impossible, for new research to build on previous research because there is no common ground on which to construct a useful dialogue. Consequently, the research being performed is not providing a general understanding of the issues nor allowing the development of better VE systems by practitioners. In situations such as this, a classification helps in framing a problem area and providing that common ground on which to discuss its implications:

> The need for classification is based on the ability to generalize across events, an important goal of science, and for establishing and enhancing communication among scientists. (Melton, 1964, p. 325)

The development of uniform, transferable concepts, definitions, and models of user behavior is all too uncommon in areas of research such as ours. There is often little time for retrospection in research; even this fact is commonly unacknowledged. The lack of general methods and rules hinders the potential impact of present research on future development and products.

It should be noted that the classification presented here is only one of many possible classifications utilizing a Linnaean taxonomy. Our community is interested in human performance, and therefore we chose human abilities as the basis of our work. The methodology could be used as the basis for the classification of problem sets unrelated to human performance.

Before going further, it is important to state one assumption of the applications discussed throughout the remainder of this chapter. We assume VEs are a class of computer simulations in which a human plays an integral role. If a human is not in the real-time loop, then the system being discussed is not a VE but simply a set of algorithms. This distinction is not meant to diminish the importance of pure computer simulations but to help define a virtual environment. However, it would not be appropriate to restrict the term *virtual environment* to only include systems involving head-mounted displays (HMDs) or projection-based systems such as CAVEs (Cruz-Neira, Sandin, & DeFanti, 1993). We generalize the term to include any computer-based, real-time simulation involving a representation of space and a human user interacting in that space. This could mean anything from a computer game to a scientific visualization application, for example.

At the heart of any VE system are the tasks that a user can and will perform in that system. The performance characteristics of these tasks are clearly not identical to their real-world counterparts using today's technology. What is needed is a way to cross barriers from the real world to the virtual world by extending taxonomic methods of human performance to include VEs. The universal objective of building effective VE systems is to capture the full extent of human skills and to facilitate their usage. This requires a VE system to provide a comprehensive model of human performance in order to be compatible with human capacities and limitations.

This claim is especially true for VE training systems that seek to replicate aspects of the real world as well as for nonrealistic or abstract VEs, which seek to explore experiences not presently possible in the real world. As stated earlier, involving the human in either type of system automatically requires an acceptance of the strengths and limitations of human skills developed in the real world. If a human is to be involved, then interaction techniques must be based on something the user is familiar with because they will be executed using natural human abilities. This is not to suggest that VE interfaces must always mimic the real world to the highest degree. This is certainly not our point. Simply, user performance in VEs is based on expectations and abilities developed and grounded in the physical world.

The application of classifications and taxonomies is not new to VE and HCI research, but they have not yet been applied in a broad manner to the full array of problems being addressed. In most areas of VE human interaction, there are presently no taxonomies or classifications of human abilities. As we attempt to answer new questions, this approach can help generalize, communicate, and apply the research findings outside of the applications and situations in which the research was developed.

20.3 APPLICATION TO VIRTUAL ENVIRONMENT RESEARCH

The taxonomic method presented here can be used to classify any aspect of human abilities. The F-JAS system covers a broad range of human abilities that can be used as a basis for any interaction in VEs. However, there is a limitation resulting from assumptions made by the original researchers. The initial research and resulting classifications treat the real world as a constant. For instance, humans in the real world can feel things that they pick up with their hands. They depend on the fact that the ground will not move out from under their feet except in rare circumstances (e.g., unstable parts of California). These are all reasonable assumptions about the real world that may be invalid when applied to a VE. Understanding the repercussions that such assumptions have on research in VEs and also how to remove them has been the preoccupation of a great deal of the initial work in this project.

This section presents an overview of the development of specific classifications for VEs, the analysis inherent in their development, the methods used, and the resulting knowledge and tools. We refer throughout the section to Fig. 20.3, which presents a flowchart of how the methods are used in VE research and development to build VE systems and extend the classification for future research. We proceed by presenting a series of examples based on research recently completed, research which centers on a classificatory system developed and utilized to study active human locomotion in virtual environments (Darken et al., 1997). This system is based on F-JAS human abilities and has been extended to remove assumptions that are invalid in VEs. A second classification of significance that will not be presented here is that for two-handed whole-hand input (Cockayne, 1998).

The present analysis shows that when devices or techniques are used in a VE to perform task components common to the real world, the mapping between abilities and requirements can be quite different. This is typical of VE interfaces. The analysis then brings the virtual and the real together to show what can be gained from this process.

20.3.1 Locomotion in Virtual Environments

Locomotion is an essential part of everyday life. So too must it be an essential part of interaction in VEs. There have been a number of techniques proposed and implemented to facilitate the function of viewpoint movement from one virtual location to another (see Hollerbach,

2002). These techniques include active methods such as pedaling devices (Brogan, Metoyer, & Hodgins, 1998), walking in place (e.g., Iwata & Fujii, 1996; Templeman, Denbrook, & Sibert, 1999), foot platforms (e.g., Iwata & Yoshida, 1999), and treadmills (e.g., Darken et al., 1997; Hollerbach, Xu, Christensen, & Jacobsen, 2000; Iwata & Yoshida, 1999) as well as a variety of passive (or indirect) methods such as logarithmic rate control (Mackinlay, Card, & Robertson, 1990), scaling and zooming (Stoakley, Conway, & Pausch, 1995), and viewpoint metaphors (Ware & Osborne, 1990).

In the real world, each of us possesses certain abilities that we use to perform locomotion tasks. Each of us knows how to walk from one location to another. However, when we enter the virtual world, constraints inherent to the technology do not allow us to use the same abilities we developed for the real world in an exact and identical fashion. Although each of us can walk hundreds of yards at a time in any typical day, we could not do so in a virtual world due to space and cabling limitations. These limitations can only be overcome with the use of specialized devices or interaction techniques (e.g., triggered wands or point-and-fly).

20.3.1.1 Active and Passive Locomotion

The class of human locomotion we are concerned with here is called *active locomotion*, and when we use the term *locomotion*, we mean active locomotion. Active locomotion relies on a person's physical exertion and on the person's body for transport—it is not an abstraction. This is in contrast to passive (or indirect) locomotion, which is transport via a vehicle or vehicular metaphor. In passive locomotion, a person's motion or exertion indirectly results in transport. This difference separates walking and running from riding a bicycle, driving a car, or being a passenger on an airplane. These two kinds of locomotion use distinctly different human abilities. Active locomotion relies heavily on the human's use of limb movement, balance, coordination, and numerous other physical traits. Passive locomotion tends to bypass these traits. Passive motion instead relies on more cognitive or abstracted models of control or on physical abilities that are unrelated to active locomotion. Abstracting the task component in order to use special interaction techniques transfers the task from the realm of active locomotion to the realm of passive locomotion because body movements will not be directly used to move. A possible exception to this might be a walking-in-place technique such as Gaiter (Templeman et al., 1999).

There are strong arguments for both classes of locomotion, most of which are dependent on other task demands on the user. In instances where there is a need to simulate the exertion associated with locomotion or where the user's hands are busy, the most obvious solution, if it could be proven feasible, would be to use active locomotion to move through the virtual world. This has only recently become practical with the development of specialized locomotion devices. One such device is the omni-directional treadmill (ODT), which is the focus of discussion here.

There are two primary criteria for a locomotion device or technique: (a) accuracy and control and (b) cognitive demand (attention). The ideal locomotion device or technique would facilitate rapid movement over vast distances without sacrificing accuracy or control (Mackinlay et al., 1990) and would be so transparent to the user that using it would become automatic (Posner & Snyder, 1975). We should not equate the notion of naturalness with automaticity. It is clear that unnatural (or learned) actions can become automatic with practice (e.g., playing the piano or riding a bike). However, within the context of locomotion devices for virtual worlds, particularly for training applications, we suggest that locomotion must be transparent even for novice users. We assume that users have to be trained for some primary task other than locomotion and that expending time and effort training to use the system is unacceptable. Therefore, a direct comparison of locomotion on the ODT and real-world locomotion is warranted and appropriate.

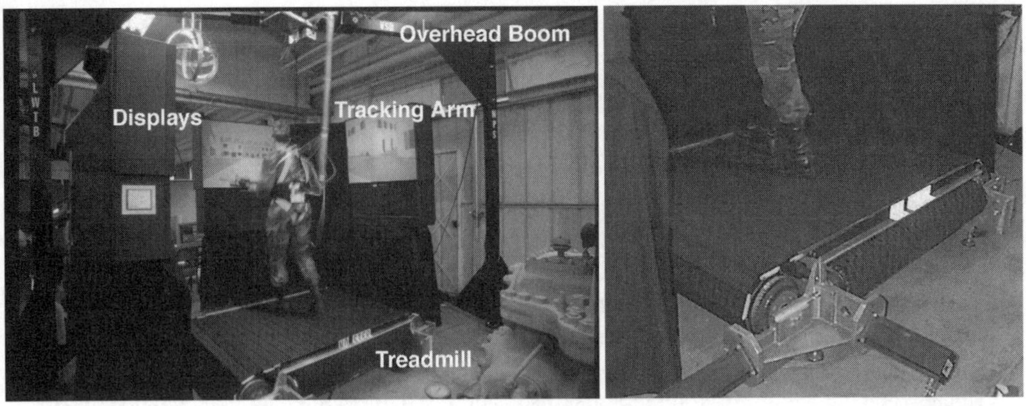

FIG. 20.2. The Omni-Directional Treadmill.

20.3.1.2 The Omni-Directional Treadmill (ODT)

The ODT (Fig. 20.2) allows for bipedal locomotion in all directions of travel in the plane parallel to the ground. From an engineering perspective, this device is a major breakthrough. However, our purpose is to investigate this device from a human factors perspective, comparing it to real-world locomotion to determine if it achieves the performance levels necessary for widespread use.

A discussion of the engineering principles behind the ODT can be found in Darken et al. (1997). For our purposes, it is important to understand that motion on the ODT is produced by a vector sum from two orthogonal belts that operate simultaneously and allow movement in any direction desired. When the user is moving, the entire active surface area is in motion. The mechanical tracking arm determines the relative position of the user to the center of the top belt.

As the user moves on the surface of the ODT, he or she must be returned to the center of the surface via a reactive force applied to the feet through the treadmill. The ODT is not passive but actively applies forces to the user. Herein lies the most important aspect of the ODT in terms of human locomotion and usability.

There are two fundamental types of movement associated with the use of the ODT:

User-initiated movement: The user attempts to walk from the ODT's center to some position.
System-initiated movement: The ODT attempts to return the user to its center.

As the user moves off center and the ODT responds with a centering reaction, the ground under the user (the platform) must move accordingly. While standing at its center, a user has approximately .6 meters of active surface to move on in any direction. As the system tracks the user's position on its surface, the information is passed to an algorithm that determines how to adjust the treads in order to recenter the user. The lag inherent in this loop must be extremely low but can never be eliminated entirely. If the user accelerates quickly from a rest state, the ODT has very little time to recenter in order to keep the user from running off the platform. Furthermore, if the user should change direction during this response time, the ODT must then determine what the best vector of return is to bring the user to center. That vector may not line up with the user's center of mass, possibly causing a loss of balance.

This problem is what makes the issue of bipedal locomotion on the ODT so complex. Not only is precise tracking required, but the related communications, filtering, calculated response, and actual response must occur correctly with essentially no lag in order to adequately simulate the real world. If the centering action is noticeable, it may interfere with locomotion tasks and associated higher level tasks, possibly degrading or negating training value.

Having made our case that locomotion on the ODT has significant differences (and also similarities) with real-world locomotion, the next step is to compare locomotion on the ODT to real-world locomotion via the HARs classification.

20.3.2 Definition of the Problem

The development process begins with a desire to build a virtual environment system (Fig. 20.3). In our case, we intend to build a training system involving bipedal locomotion, but this is not a requirement. The first step is to conduct an appropriate task analysis (chaps. 1 and 3). The purpose of the analysis is to identify the necessary task components (later, interaction techniques and devices will be chosen to support these). To assist in developing a training system, the task analysis specifically must address *what is to be trained*. This is especially critical when the intent is to train motor skills (as opposed to cognitive or perceptual skills) or when motor skills are necessary for training some other skill.

The application of HARs to task components in no way should be considered a replacement for traditional task analysis techniques. They are purely complementary. HARs are not useful, in and of themselves, for analyzing patterns of behavior or for understanding cognitive processes. Similarly, traditional task analysis methods typically are poorly suited to describe task performance at the task component level of human abilities. Furthermore, not all task analyses need to reach this level of granularity, but for those that do (particularly where detailed comparisons need to be made between performance in a computing system like a VE and real-world performance), the use of HARs can be an effective technique. In our work, we have found the use of HARs in VE training systems to be particularly useful when combined with a conventional task analysis.

At the Naval Postgraduate School, we have conducted a number of cognitive task analyses (chaps. 15 and 16) of this type in support of our work in developing VE training systems for the Office of Naval Research. We completed an analysis of ship handling for underway replenishment (when a Navy ship refuels and resupplies while at sea) (Norris, 1998) and later extended this to include pier-side ship handling (Grassi, 2000). Ship handling involves a three-person team on the bridge of the ship who must coordinate their actions. We used an augmented GOMS notation (chap. 4) to describe the tasks in order to capture the parallel activities of each team member. This analysis was used by the Naval Air Warfare Center Training System Division to build a VE ship-handling trainer that is now in use at the Surface Warfare Officer School in Newport, Rhode Island. We then did an analysis of tactical land navigation (Stine, 2000) using the applied cognitive task analysis (ACTA) methodology (Militello, Hutton, Pliske, Knight, & Klein, 1997). ACTA worked well for this analysis because we could study navigation expertise outside the scope of the entire team. This was not true for our most recent cognitive task analysis, which was of close-quarters combat (squad-level building clearing) (Aronson, 2002). In this case, a team of four must quickly, efficiently, and safely clear a building of potential threats. We used parts of the ACTA methodology here but resorted back to our augmented GOMS notation for representing the components of the overall task.

In the case of the ODT, the training task involves coordinated actions between multiple people on foot (e.g., tactical infantry navigation, close-quarters combat). It is necessary to include both locomotion (translation from place to place) and maneuvering (any other motion that does not translate the body in space). We can describe active locomotion tasks in terms of four primary factors:

1. *Relative velocity*: Rest, walk, or jog. This defines the approximate relative velocity of the user when not accelerating or decelerating. Running is not possible on the ODT, nor is crawling or kneeling.

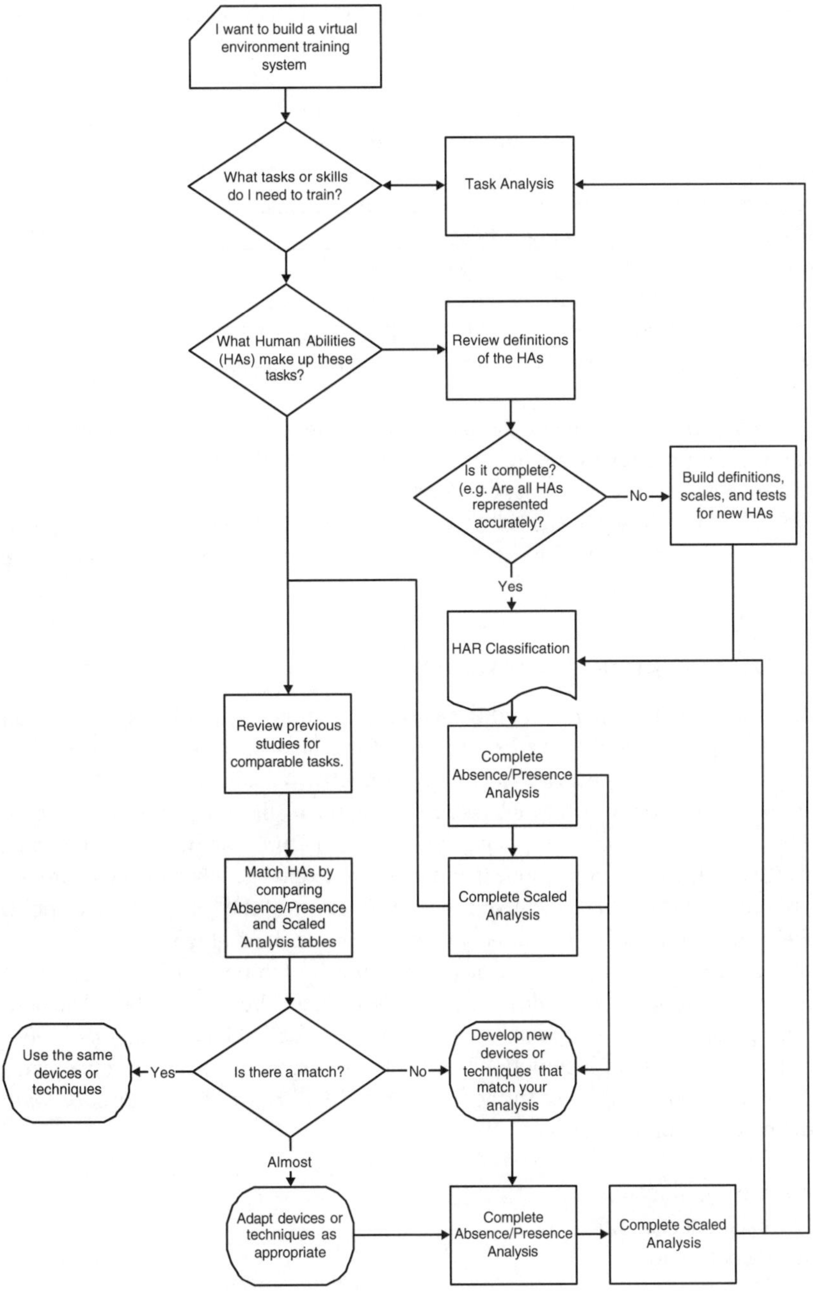

FIG. 20.3. A flowchart of the HAR process as applied to VE system development.

2. *Transition*: Accelerate or decelerate. As will become evident in the scaled analysis, the rate of acceleration or deceleration is a critical factor. It may or may not imply a change in gait.
3. *Movement direction*: Forward or backward. Sidestepping is considered a maneuvering task component.
4. *Direction change*: Straight or turn. This describes whether a direction change takes place during a transition or at constant velocity.

We can then construct a task component list for our analysis based on these factors. The list includes the following:

- *walk*: At least one foot is touching the ground at all times
- *jog*: Neither foot may be touching the ground at any time
- *acceleration from rest to a walk or jog*: Change of state
- *deceleration to rest from a walk or jog*: Change of state
- *acceleration from walk to jog*: Change of gait
- *deceleration to walk from jog*: Change of gait
- *turning in place* (no forward or lateral movement): Maneuvering action
- *sidestepping* (purely lateral movement): Maneuvering action
- *tilting upper body without foot movement*: Maneuvering action.

The next step is to decompose the task components from the task analysis into human abilities. It is clear from reviewing the full list of human abilities in the F-JAS that many can be ruled out immediately. For example, our locomotion task allowed us to immediately discard human abilities such as *finger dexterity*. We now need to analyze the task components in the real world so that we will have a basis for comparison when we apply these procedures to the VE.

20.3.3 Analysis of the Real World

If it is determined that the human abilities currently in the working HARs classification (the obvious starting point would be the F-JAS) are sufficient and also are represented in a manner fitting the task component, then we can proceed directly to the analysis phase. We perform an initial absence/presence review of each task component with the intention of looking for aspects of the task component that are not being covered by the present HARs. Special attention is paid to aspects of the task components that are known to change when the tasks are performed in VEs. We determine that a refinement is needed because of the forces that are applied to the user when walking on the ODT but that are not present in the real world.

Aspects of the task components that not covered by the present HARs are collected in order to look for similarities and differences in these "new" human abilities. The new human abilities are grouped, and a small set of new HARs are defined as extensions of the present classification. In the case of locomotion, the HARs developed are actually refinements of two HARs that existed previously. We further refine the F-JAS HAR *gross body coordination* into the following three components:

side-to-side coordination
front-to-back coordination
rotational coordination.

Similarly, we refine *gross body equilibrium* into the following three components:

1. side-to-side equilibrium (see Fig. 20.1)
2. front-to-back equilibrium
3. rotational equilibrium.

We now need to complete the classification by developing definitions of the new HARs, developing the behaviorally anchored seven-point scales that each would need, and reviewing the whole classification for completeness. The definitions are built based on the wording, terminology, and approach of the definitions for the preexisting HARs. Special attention is paid to defining the exact nature of the ability covered by each HAR, and the definition is used to build quick charts on how the HAR under review differs from similar HARs (see the top of the scale in Fig. 20.1).

With the classification definitions completed, we next need to create the scales for each of the abilities. As mentioned before, each of the definitions utilizes a behaviorally anchored seven-point scale. The creation or utilization of each of the scales is a more iterative task than the creation of each particular definition, as these scales are expected to change over time as more task components are analyzed. The reader will note that the scale presents two more pieces of data for use in analysis: extended definitions and task examples. The extremes of the scale, 1 and 7, are anchored with extended definitions. The complete presence of the ability, at 7, and the complete absence of the ability, at 1, are presented in the definitions. These definitions appear on the left side of the scale in Fig. 20.1. Along the right side are task examples anchored at points equivalent to their presence to give the analyst real-world grounding.

At this point, the classification is usable for the analysis of task components in the real world or the VE. Note that the development of taxonomies is an iterative process and is subject to revision as new systems are developed. This first revision of the taxonomy and classification does not purport to be complete or immutable.

One of the first applications of the classification is in the reanalysis of the real-world task components using an absence/presence analysis. The results of applying an absence/presence analysis to a series of human active locomotion tasks in the real world are shown in Fig. 20.4. The analysis, which is of a very small subset of locomotion tasks, is presented primarily so that the reader can see the level of the task components. The absence/presence analysis of the task components is useful for reviewing the complexity of the tasks from a multimodal viewpoint. One thing that we have discovered in these types of analyses is that sometimes the most innocuous-seeming real-world task components are the most complex problems from a

	1. Multilimb Coordination	2. Rate Control	3. Response Orientation	4. Reaction Time	5. Speed of Limb Movement	6. Static Strength	7. Explosive Strength	8. Dynamic Strength	9. Trunk Strength	10. Extent Flexibility	11. Dynamic Flexibility	12. Gross Body Coordination	12(A). Side-to-Side Coordination	12(B). Front-to-Back Coordination	12(C). Rotational Coordination	13. Gross Body Equilibrium	13(A). Side-to-Side Equilibrium	13(B). Front-to-Back Equilibrium	13(C). Rotational Equilibrium	14. Stamina	15. Speed of Transport
Accelerate from Rest to Walk or Jog	●				●	●	●	●	●		●	●	–	–	–	●	–	–	–		●
Decelerate from Walk or Jog to Rest	●				●	●		●	●			●	–	–	–	●	–	–	–		●
Accelerate from Walk to Jog	●				●			●	●		●	●	–	–	–	●	–	–	–	●	●
Decelerate to Walk from Jog	●				●			●	●		●	●	–	–	–	●	–	–	–		●
Walk	●				●			●	●		●	●	–	–	–	●	–	–	–	●	●
Jog	●				●			●	●		●	●	–	–	–	●	–	–	–	●	●
Turn in Place (no forward or side movement)	●					●			●	●		●	–	–	–	●	–	–	–		
Side-Step	●					●			●	●		●	–	–	–	●	–	–	–		
Tilt Upper-Body without Foot Movement	●					●			●			●	–	–	–	●	–	–	–		

● Present
– Not Applicable

FIG. 20.4. An absence/presence analysis of real-world locomotion.

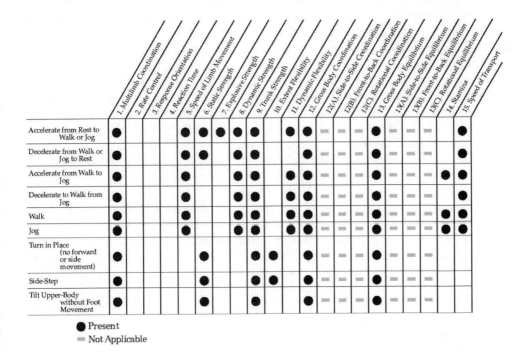

	1. Multilimb Coordination	2. Rate Control	3. Response Orientation	4. Reaction Time	5. Speed of Limb Movement	6. Static Strength	7. Explosive Strength	8. Dynamic Strength	9. Trunk Strength	10. Extent Flexibility	11. Dynamic Flexibility	12. Gross Body Coordination	12(A). Side-to-Side Coordination	12(B). Front-to-Back Coordination	12(C). Rotational Coordination	13. Gross Body Equilibrium	13(A). Side-to-Side Equilibrium	13(B). Front-to-Back Equilibrium	13(C). Rotational Equilibrium	14. Stamina	15. Speed of Transport
Accelerate from Rest to Walk or Jog	5	1	1	1	5	2	6	6	5	1	4	5	—	—	—	5	—	—	—	2	5
Decelerate from Walk or Jog to Rest	5	1	1	1	5	2	1	5	5	1	4	4	—	—	—	4	—	—	—	1	5
Accelerate from Walk to Jog	5	1	1	1	5	1	2	6	5	1	4	5	—	—	—	5	—	—	—	2	4
Decelerate to Walk from Jog	5	1	1	1	5	1	1	5	5	1	4	4	—	—	—	4	—	—	—	1	4
Walk	5	1	1	1	4	1	1	4	3	1	3	3	—	—	—	3	—	—	—	2	2
Jog	6	1	1	1	5	1	1	4	4	1	4	4	—	—	—	4	—	—	—	4	3
Turn in Place (no forward or side movement)	4	1	1	1	1	2	1	1	3	2	1	2	—	—	—	3	—	—	—	1	1
Side-Step	5	1	1	1	1	3	1	1	3	2	1	4	—	—	—	4	—	—	—	1	1
Tilt Upper-Body without Foot Movement	2	1	1	1	1	3	1	1	5	1	1	3	—	—	—	3	—	—	—	1	1

1, 2, 3, 4, 5, 6, or 7 Scaled Score
— Not Applicable

FIG. 20.5. A scaled analysis of real-world locomotion.

HAR standpoint. It is also important to note that occasionally a human ability that would seem to be required for real-world locomotion (e.g., *rate control*) actually fails the absence/presence test on the basis of the precise definition of the human ability and its relationship to the task component in question. (See the appendix at the end of this chapter for definitions of the human abilities required for human locomotion.)

The next step in the application of the classification is the analysis of the task components using the scaled analysis. The application of the scaled analysis to the series of human active locomotion tasks in the real world can be seen in Fig. 20.5. The task components being analyzed, as well as the classification, are the same as in Fig. 20.4. The use of the scaled analysis is crucial when the analyst seeks to understand the weight of each HAR in the task component being performed. Whereas the absence/presence analysis allows for quick and simple comparison of task components based on the abilities that they require, the scaled analysis allows for a comparison of task components that utilize the same abilities.

Our work is only half done at this point. We now need to cross the boundary from the real world to the virtual world in order to understand how locomotion is performed under the constraints imposed by the ODT. This will lead to an evaluation of the ODT as well as a substantive analysis of where it excels and fails in terms of human performance.

The flowchart in Fig. 20.3 shows that at this point we would normally compare the results of these analyses to previous studies that used the same measures and methods. We would then determine if we found a match between what our analyses determined the HARs of our task components to be and what HARs were supported by existing devices and interaction techniques suitable to our needs. If we were to find a direct match (e.g., someone else had already developed a technique that directly matches what we need), then we would choose to use that technique or device. If a close match was found (e.g., the technique almost fits our needs but not quite), we would likely adapt that technique or device for our purposes and then reevaluate using both the absence/presence and scaled evaluation methods. And if no match if found, then we have to break new ground, developing our own technique or device. We would also perform absence/presence and scaled analysis evaluations so that future developers could use our results.

20.3.4 Analysis of the Virtual Environment

With the real-world analysis completed, we can now study the same task components implemented in a VE. This analysis can be performed to build a realistic VE (in which case the verisimilitude is meant to be exact) or an abstract environment that utilizes inherent human skills in an optimal manner. In either case, the approach is similar to that of the real world.

The first step is to again perform an absence/presence analysis of the task components but this time while in a VE using the ODT. An obvious question here is, Which virtual environment? When discussing the real world, assumptions can be made about the common experiences of all humans. When discussing a VE, what are the metrics? This is the point where the analysis could devolve into a hardware and software analysis. The answer that we offer to this question is, It doesn't matter what VE is being used—a human is still a human. The idea behind using HARs is that the task components are analyzed based on human abilities. Human abilities must be considered the very basis of the capacity of a human to execute a task. Increase in performance over time has to do with optimizing the utilization of these abilities and possibly changing which abilities are used.

For example, over time, a person adapts to walking on ice and does not fall as much. He has not acquired new human abilities but has adapted to the environment and the requirements of the task. Expert performers, as compared to novices, may execute a task differently and may even use different task components (see chap. 15). Although the natural abilities of the novice and the expert may be the same, the way they are employed in the execution of a task may differ. Therefore, it is important to analyze the right user population for the same reasons this is important in traditional task analyses.

We can say that the abilities of a human, regardless of their skill level, do not change when the human moves from one environment to the other. When a human walks around on the ground, on a train, or on a boat, the abilities utilized by the human may be different, but the human's natural abilities have not changed. This implies that the analysis of a VE is predicated on the use of equipment with which the VE is constructed because this is the physical interface between the user and the virtual environment. The VE analysis is apparatus dependent. We happen to be using the ODT in our analysis, but any locomotion device or technique could have been used.

Referring back to the real-world absence/presence analysis, Fig. 20.6 reviews the same task components analyzed in Fig. 20.4 but now as performed on the ODT. The task components in Fig. 20.4 were completed in the real world and analyzed independent of any thoughts of placing the subject in a simulated environment. The task components in Fig. 20.6 were analyzed in the context of the user performing the tasks on the ODT. When reviewing the information contained in the two charts, the reader can see that there is a distinct difference in the HARs in each scenario. To understand the actual quantities of each HAR used, we perform a scaled analysis.

Using the same procedures as in Fig. 20.5, we analyze the same task components using the scales (Fig. 20.7). As with the absence/presence analysis, the task components analyzed in Fig. 20.5 were completed in the real world whereas the task components in Fig. 20.7 were analyzed in the context of the user performing the tasks on the ODT. Although comparison of the absence/presence charts for both scenarios indicate the different HARs used, these charts provide a more quantitative measure of the HARs for the task components studied.

Now that the analysis has been completed, we are in a position to state definitively where the ODT excels and fails. The "perfect" VE locomotion device (assuming that our metric is

	1. Multilimb Coordination	2. Rate Control	3. Response Orientation	4. Reaction Time	5. Speed of Limb Movement	6. Static Strength	7. Explosive Strength	8. Dynamic Strength	9. Trunk Strength	10. Extent Flexibility	11. Dynamic Flexibility	12. Gross Body Coordination	12(A). Side-to-Side Coordination	12(B). Front-to-Back Coordination	12(C). Rotational Coordination	13. Gross Body Equilibrium	13(A). Side-to-Side Equilibrium	13(B). Front-to-Back Equilibrium	13(C). Rotational Equilibrium	14. Stamina	15. Speed of Transport
Accelerate from Rest to Walk or Jog	●	●	●	●	●	●	●	●	●		●	—	●	●		—	●	●			●
Decelerate from Walk or Jog to Rest	●	●	●	●	●	●		●	●			—	●	●		—	●	●			●
Accelerate from Walk to Jog	●	●	●	●	●			●	●		●	—	●	●		—	●	●		●	●
Decelerate to Walk from Jog	●	●	●	●	●			●	●		●	—	●	●		—	●	●			●
Walk	●	●	●	●	●			●	●		●	—	●	●		—	●	●		●	●
Jog	●	●	●	●	●			●	●		●	—	●	●		—	●	●		●	●
Turn in Place (no forward or side movement)	●	●	●	●		●			●	●		—	●	●	●	—	●	●	●		
Side-Step	●	●	●	●		●			●	●		—	●	●	●	—	●	●	●		
Tilt Upper-Body without Foot Movement	●	●	●	●		●			●			—	●	●		—	●	●			

● Present
— Not Applicable

FIG. 20.6. An absence/presence analysis of VE locomotion on the ODT.

	1. Multilimb Coordination	2. Rate Control	3. Response Orientation	4. Reaction Time	5. Speed of Limb Movement	6. Static Strength	7. Explosive Strength	8. Dynamic Strength	9. Trunk Strength	10. Extent Flexibility	11. Dynamic Flexibility	12. Gross Body Coordination	12(A). Side-to-Side Coordination	12(B). Front-to-Back Coordination	12(C). Rotational Coordination	13. Gross Body Equilibrium	13(A). Side-to-Side Equilibrium	13(B). Front-to-Back Equilibrium	13(C). Rotational Equilibrium	14. Stamina	15. Speed of Transport
Accelerate from Rest to Walk or Jog	6	4	7	7	5	3	6	7	6	1	4	—	6	6	1	—	5	6	1	4	6
Decelerate from Walk or Jog to Rest	6	4	7	7	5	3	1	6	6	1	5	—	5	7	1	—	4	7	1	1	7
Accelerate from Walk to Jog	6	4	7	7	5	1	3	7	6	1	4	—	6	5	1	—	5	5	1	4	6
Decelerate to Walk from Jog	6	4	7	7	5	1	1	6	6	1	5	—	5	6	1	—	4	6	1	1	6
Walk	6	3	3	4	4	1	1	5	4	1	3	—	5	3	1	—	4	3	1	4	4
Jog	7	2	4	5	5	1	1	5	5	1	3	—	6	3	1	—	4	3	1	6	4
Turn in Place (no forward or side movement)	5	3	3	3	1	3	1	1	4	5	1	—	5	4	6	—	4	4	6	1	1
Side-Step	6	3	4	7	1	4	1	1	4	5	1	—	7	6	5	—	7	6	5	1	1
Tilt Upper-Body without Foot Movement	3	2	3	6	1	4	1	1	6	1	1	—	4	7	1	—	4	7	1	1	1

1, 2, 3, 4, 5, 6, or 7 Scaled Score
— Not Applicable

FIG. 20.7. A scaled analysis of VE locomotion on the ODT.

replicating real-world bipedal locomotion) would result in the identical HARs shown in Fig. 20.4 and 20.5. That is, locomotion on the device would be exactly like locomotion in the real world. However, this is obviously not the case.

Reviewing the task component *accelerate from rest to walk or jog* can elucidate this. First, in only 4 of the 15 abilities is the amount of skill required to perform the task component similar, shown by the sameness of the values in the two charts (e.g., *speed of limb movement*). Five of the 15 abilities are within one point of each other, but in all cases the VE requires a greater amount of skill (e.g., *static strength*). The remaining 6 skills display a two-point or greater difference (e.g., *rate control*). Finally, in this task component, as with all of the task components presented in the charts, the abilities of *gross body coordination* and *gross body equilibrium* must be extended based on orientation in the VE. Simply, the task in the VE is highly dependent on the position and orientation of the user on the ODT platform whereas the real-world version of the task has no such requirement. The user has to exert more physical and more mental energy within these two abilities in the VE as compared with the real world.

At this time, the ODT is usable as a locomotion device. Approximately half of the abilities that a human uses in the real world to complete an active locomotion task are used similarly on the ODT. The corollary is that approximately half of the tasks are more complex in the VE, which leads to the conclusion that, though usable, the ODT does not accurately replicate active locomotion tasks as performed in the real world. These results are being utilized in the development of the next-generation ODT.

20.4 CONCLUSION

This chapter focuses on the analysis of human performance in the real world and human performance using a locomotion device in a VE. This example shows how the analysis methods discussed can address a major problem facing practitioners and researchers today: What devices and techniques should be used in VE design in order to solve a particular problem? From our analysis, a practitioner could make reasonably accurate assumptions as to how people will perform in a VE using the ODT for locomotion. Also, researchers interested in locomotion devices can use the analysis to determine how important certain factors are in their design criteria. Since it is unlikely that we will see a device that perfectly replicates human bipedal locomotion in the near future, it would be useful to know what corners to cut in order to get close without making wholesale sacrifices in usability and performance.

The approach presented in this chapter can be used for many additional purposes (see Fig. 20.8). It can help in the comparison of interaction techniques and devices, the development of new interaction techniques and devices, the analysis of training requirements, and the general sharing of knowledge in a common language.

It has been used in our research and at many other institutions to address aspects of all of the above areas. The initial intent was to use this approach to develop better training systems and analyze associated learning (Fleishman & Quaintance, 1984). Once it became developed, it proved to have the underlying advantage of creating a common language and method and enhancing communication between parties. Also, extending the application of the classifications to new devices and methods of human-computer interaction has made it easier to understand, structure, and communicate related research. Finally, with the creation of new taxonomies, and new classifications, Fleishman's work has been applied to completely new classes of interaction and devices that were not possible when the work was first started.

There is a strong desire in the HCI and VE research and application communities for a cookbook solution to the problems that are being studied. Although the methodology presented here does not present a simple solution to these problems, it furthers the development of

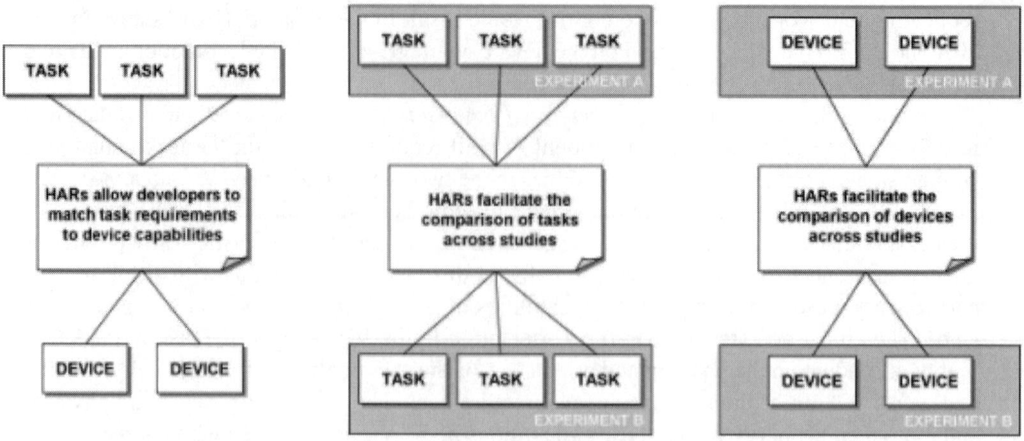

FIG. 20.8. Utilizing Human Ability Requirements classifications.

comprehensive classifications and taxonomies that will greatly help us achieve this lofty goal. These tools provide a better understanding of the underlying problems, give structure to the questions and the answers, and allow a quantitative approach to human interaction studies. With understanding, structure, and quantification, researchers are given the ability to better communicate their work and thus facilitate breakthroughs and the solution of problems.

APPENDIX: DEFINITION OF HUMAN ABILITIES REQUIRED FOR ACTIVE LOCOMOTION

1. *Multilimb coordination*. This is the ability to coordinate the movements of two or more limbs (e.g., two legs, two hands, one leg and one hand). The ability does not apply to tasks in which trunk movements must be integrated with limb movements. It is most common to tasks where the body is at rest (e.g., seated or standing) while two or more limbs are in motion.
2. *Rate control*. This is the ability to make timed, anticipatory motor adjustments relative to changes in the speed and/or direction of a continuously moving object. The purpose of the motor adjustments is to intercept or follow a continuously moving stimulus whose speed and/or direction vary in an unpredictable fashion. This ability does not extend to situations in which both the speed and direction of the object are perfectly predictable.
3. *Response orientation*. This is the ability to select and initiate the appropriate response relative to a given stimulus in a situation where two or more stimuli are possible and where the appropriate response is selected from two or more alternatives. The ability is concerned with the speed with which the appropriate response can be initiated and does not extend to the speed with which the response is carried out. This ability is independent of the mode of stimulus presentation (auditory or visual) and also of the type of response required.
4. *Reaction time*. This ability involves the speed with which a single motor response can be initiated after the onset of a single stimulus. It does not include the speed with which the response or movement is carried out. This ability is independent of the mode of stimulus presentation (auditory or visual) and also of the type of motor response required.
5. *Speed of limb movement*. This ability involves the speed with which discrete movements of the arms or legs can be made. The ability deals with the speed with which the movement can be carried out after it has been initiated; it is not concerned with the speed of initiation of

the movement. In addition, the precision, accuracy, and coordination of the movement are not considered under this ability.

6. *Static strength.* This is ability to use continuous muscle force to lift, push, pull, or carry objects. This ability can involve the hands, arms, back, shoulders, or legs. It is the maximum force that one can exert for a brief period of time.

7. *Explosive strength.* This is ability to use short bursts of muscle force to propel oneself, as in jumping or sprinting, or to throw objects. It requires gathering energy for bursts of muscular effort.

8. *Dynamic strength.* This is ability of the muscles to exert force repeatedly or continuously over time. The ability involves the degree to which the muscles do not "give out," or fatigue. The ability is involved in supporting, holding up, or moving objects or the body's own weight repeatedly over time.

9. *Trunk strength.* This ability involves the degree to which one's stomach and lower back muscles can support part of the body or the position of the legs, repeatedly or continuously over time. The ability involves the degree to which these trunk muscles do not "give out," or fatigue, when they are put under repeated or continuous strain.

10. *Extent flexibility.* This is the ability to extend, flex, or stretch muscle groups. It concerns the degree of flexibility of muscle groups but does not include repeated or speed flexing.

11. *Dynamic flexibility.* This is the ability to make repeated trunk and/or limb flexing movements where both speed and flexibility of movement are required. It includes the ability of these muscles to recover from the strain and distortion of repeated flexing.

12. *Gross body coordination.* This is the ability to coordinate movements of the trunk and limbs. This ability is most commonly found in situations where the entire body is in motion or being propelled.

12(A). *Side-to-side coordination.* This is the ability to coordinate movements of the trunk and limbs along the axis passing through both of the user's shoulders. This ability is most commonly found in situations where the entire body is in motion or being propelled across the plane of the user's chest.

12(B). *Front-to-back coordination.* This is the ability to coordinate movements of the trunk and limbs along the axis passing through the user's chest. This ability is most commonly found in situations where the entire body is in motion or being propelled in the plane perpendicular to the user's chest.

12(C). *Rotational coordination.* This is the ability to coordinate movements of the trunk and limbs in rotation about the axis passing through the user's head and the ground. This ability is most commonly found in situations where the entire body is in motion or being propelled around the axis perpendicular to the ground.

13. *Gross body equilibrium.* This is the ability to maintain the body in an upright position or to regain body balance, especially in situations where equilibrium is threatened or temporarily lost. This ability involves only body balance; it does not extend to the balancing of objects.

13(A). *Side to side equilibrium.* This is the ability to maintain the body in an upright position or to regain body balance, especially in situations where equilibrium is threatened or temporarily lost. This ability involves only body balance across the plane of the user's chest; it does not extend to the balancing of objects.

13(B). *Front-to-back equilibrium.* This is the ability to maintain the body in an upright position or to regain body balance, especially in situations where equilibrium is threatened or temporarily lost. This ability involves only body balance in the plane perpendicular to the user's chest; it does not extend to the balancing of objects.

13(C). *Rotational equilibrium.* This is the ability to maintain the body in an upright position or to regain body balance, especially in situations where equilibrium is threatened or temporarily lost. This ability involves only body balance around the axis perpendicular to the ground; it does not extend to the balancing of objects.

14. *Stamina*. This ability involves the capacity to maintain physical activity over prolonged periods of time. It is concerned with resistance of the cardiovascular system (heart and blood vessels) to breakdown.

15. *Speed of transport*. This ability involves the speed with which the human propels the whole body through space. The ability deals with the speed with which the movement can be carried out after it has been initiated; it is not concerned with the speed of initiation of the movement. In addition, the precision, accuracy, and coordination of the movement are not considered under this ability.

REFERENCES

Aronson, W. (2002). *A cognitive task analysis of close quarters battle.* Unpublished master's thesis, Naval Postgraduate School, Monterey, CA.

Bowman, D., Koller, D., & Hodges, L. (1997). *Travel in immersive virtual environments: An evaluation of viewpoint motion control techniques*. Paper presented at the Virtual Reality Annual International Symposium (VRAIS '97), Albuquerque, NM.

Brogan, D. C., Metoyer, R. A., & Hodgins, J. K. (1998). Dynamically simulated characters in virtual environments. *IEEE Computer Graphics and Applications, 15*, 58–69.

Cockayne, W. (1998). *Two-handed, whole-hand interaction.* Unpublished master's thesis. Naval Postgraduate School, Monterey, CA.

Cooper, A. (1995). *About face: The essentials of user interface design*. Foster City, CA: IDG Books.

Cruz-Neira, C., Sandin, D., & DeFanti, T. (1993). Surround-screen projection-based virtual reality: The design and implementation of the CAVE. In *Computer graphics: Proceedings of Special Interest Group on Computer Graphic '93* (pp. 135–142). New York: ACM Press.

Darken, R., Cockayne, W., & Carmein, D. (1997). The omni-directional treadmill: A locomotion device for virtual worlds. In *Proceedings of User Interface Software & Technology '97* (pp. 213–222). New York: ACM Press.

Dix, A., Finlay, J., Abowd, G., & Beale, R. (1998). *Human-computer interaction*. London: Prentice Hall.

Fleishman, E. (1995). *Fleishman job analysis survey: Rating scale booklet*. Bethesda, MD: Management Research Institute.

Fleishman, E., & Quaintance, M. (1984). *Taxonomies of human performance; The description of human tasks*. Orlando, FL: Academic Press.

Grassi, C. (2000). *A task analysis of pier side ship-handling for virtual environment ship-handling simulator scenario development.* Unpublished master's thesis, Naval Postgraduate School, Monterey, CA.

Hollerbach, J. M. (2002). Locomotion interfaces. In K. Stanney (Ed.), *Handbook of virtual environments: Design implementation and applications* (pp. 239–254). Mahwah, NJ: Lawrence Erlbaum Associates.

Hollerbach, J. M., Xu, Y., Christensen, R. R., & Jacobsen, S. C. (2000). *Design specifications for the second generation Sarcos Treadport Locomotion Interface.* Paper presented at the International Mechanical Engineering Congress and Exposition, Orlando, FL.

Iwata, H., & Fujii, T. (1996). *Virtual perambulator: A novel interface device for locomotion in virtual environments.* Paper presented at the Virtual Reality Annual International Symposium (VRAIS '96), Santa Clara, CA.

Iwata, H., & Yoshida, Y. (1999). Path reproduction tests using a Torus Treadmill. *Presence: Teleoperators and Virtual Environments, 8*, 587–597.

Mackinlay, J. D., Card, S. K., & Robertson, G. G. (1990). Rapid controlled movement through a virtual 3D workspace. In *Computer graphics: Proceedings of Special Interest Group on Computer Graphics '90* (pp. 171–176). New York: ACM Press.

Melton, A. (1964). *Categories of human learning*. New York: Academic Press.

Militello, L. G., Hutton, R. J. B., Pliske, R. M., Knight, B. J., & Klein, G. (1997). *Applied cognitive task analysis (ACTA) methodology*. Fairborn, OH: Klein Associates.

Norris, S. D. (1998). *A task analysis of underway replenishment for virtual environment ship-handling simulator scenario development.* Unpublished master's thesis, Naval Postgraduate School, Monterey, CA.

Posner, M., & Snyder, C. (1975). Attention and cognitive control in information processing and cognition. In R. L. Solso (Ed.), *The Loyola Symposium* (pp. 55–85). New York: Lawrence Erlbaum Associates.

Preece, J., Rogers, Y., Sharp, H., Benyon, D., Holland, S., & Carey, T. (1994). *Human-computer interaction*. Menlo Park, CA: Addison-Wesley.

Rose, A. M., Fingerman, P. W., Wheaton, G. R., Eisner, E., & Kramer, G. (1974). *Methods for predicting job-ability requirements: 2. Ability requirements as a function of changes in the characteristics of an electronic fault-finding task (AIR Tech Report 74–6).* Washington, DC: American Institutes for Research.

Stine, J. (2000). *Representing tactical land navigation expertise.* Unpublished master's thesis, Naval Postgraduate School, Monterey, CA.

Stoakley, R., Conway, M. J., & Pausch, R. (1995). *Virtual reality on a WIM: Interactive worlds in miniature. Paper presented at ACM Special Interest Group on Human-Computer Interaction '95*, Denver, CO.

Templeman, J. N., Denbrook, P. S., & Sibert, L. E. (1999). Maintaining spatial orientation during travel in an immersive virtual environment. *Presence: Teleoperators and Virtual Environments, 8*, 598–617.

Ware, C., & Osborne, S. (1990). Exploration and virtual camera control in virtual three dimensional environments. In *Computer Graphics: Proceedings of Special Interest Group on Computer Graphics '90* (pp. 175–183). New York: ACM Press.

Wilson, M. D., Barnard, P. J., Green, T. R. G., & MacLean, A. (1988). Knowledge based task analysis for human-computer systems. In G. C. van der Veer, T. R. G. Green, J. Hoc, & D. M. Murray (Eds.), *Working with computers: Theory verses outcome* (pp. 47–87). London: Academic Press.

21

Activity Theory: Another Perspective on Task Analysis

Phil Turner
Napier University

Tom McEwan
Napier University

In this chapter, we introduce activity theory, describing its origins and its principal researchers and thinkers. We stress its descriptive power and its usefulness in defining uniquely a unit of analysis for work. We contrast its development with that of task analysis. We conclude with a brief demonstration of this descriptive power by using it as an organizing framework of the evaluation of a collaborative virtual environment.

21.1 INTRODUCTION

In contrast to the Western tradition of task analysis stands a parallel historical development, namely that of *activity theory* (AT). For the purposes of this chapter, we introduce only one strand of this work, which is more fully described as *cultural-historical activity theory* (CHAT, although the terms AT and CHAT are generally used interchangeably). Our reasons for introducing the reader to this are several-fold. First, there is something intrinsically interesting about considering an alternate aetiology and subsequent lineage of an independent line of research and reasoning concerning task analytic approaches to understanding the dynamics of work. Second, the collective nature of CHAT, derived from its Marxist roots, potentially offers a means of answering the now well established criticisms of task analysis and of human-computer interaction (HCI) as a whole (e.g., Bannon, 1991). Third, CHAT finds a place for uncomfortable issues (at least to Western thinking), such as the roles of consciousness and motivation in human purposive activity. Clearly we have set ourselves a challenge in making a case for CHAT, but we also address the greater challenge for demonstrating the utility of CHAT in action. Here we use an illustration drawn from a collaborative virtual reality (CVE) development project, DISCOVER, and show how CHAT can define and organize the evaluation of the resulting CVE.

21.2 ORIGINS

The aetiology of CHAT is complex, drawing upon, as it does, a number of different continental philosophical traditions, the most important being Marxism and Hegelian thought. Add to this its birth during the early days of the Russian revolution and we have a system of thought that is necessarily collectivist and socialist. A system, or body, of thought is preferable to theory as a description of *CHAT*, for it is not falsifiable or predictive in character, two of the hallmarks of a scientific theory. Instead *CHAT* provides a strongly descriptive conceptual framework and vocabulary directly attributable to the work of Lev Vygotski. Vygotski was a typical all-round genius who lived in the early 20th century and contributed to pedagogical thought, learning theory, the psychology of language and thought, and developmental psychology, then promptly died of tuberculosis in his late thirties. His works on CHAT, particularly in the domains of pedagogy, child psychology, and the psychology of language and thought, remain standard texts and are still taught to undergraduates. A reasonably thorough treatment of his thinking could run to many volumes, so instead we consider two or three of his key observations before turning to the contribution of one of his students, Leont'ev.

21.2.1 Vygotski

There is no definitive biography of Vygotski, but Alex Kozulin's introduction to Vygotski's (1986) *Thought and Language* is generally regarded as the closest there is to one, these brief notes draw heavily on it. A major theme of Vygotski's theoretical framework is that social interaction plays a fundamental role in the development of cognition. Vygotski (1986) stated,

> Every function in the child's cultural development appears twice: first, on the social level, and later, on the individual level; first, between people (inter-psychological) and then inside the child (intra-psychological). This applies equally to voluntary attention, to logical memory, and to the formation of concepts. All the higher functions originate as actual relationships between individuals. (p. 57).

A further aspect of Vygotski's thinking is the idea that the potential for cognitive development is limited to a certain time span that he calls the "zone of proximal development" (ZPD), which he defines thus: "The distance between the actual development level as determined by independent problem solving and the level of potential development as determined through problem solving under adult guidance or in collaboration with more capable peers" (p. 187).

The spatial metaphor in the above quotation is mirrored in the real world as a zone or field to which an individual belongs and which may be populated by experts, tools, and other cultural artifacts. Vygotski's theory was an attempt to explain consciousness as the end product of socialization. For example, in the learning of language, our first utterances with peers or adults are for the purpose of communication, but once mastered they become internalized and allow "inner speech."

21.2.2 Leont'ev

Leont'ev was one of Vygotski's students who, among other things, went on to develop a number of key concepts for what was to become activity theory. Unlike traditional task analysis, Leont'ev argued for the study of activity to be based on an understanding of the individual's *object,* which is usually interpreted as *objectified motive*—motive made visible or tangible. It is important to see this concept as richer than the concept of a goal as traditionally defined by task analysts, and motivations may be viewed as precursors to goals. Goals might be

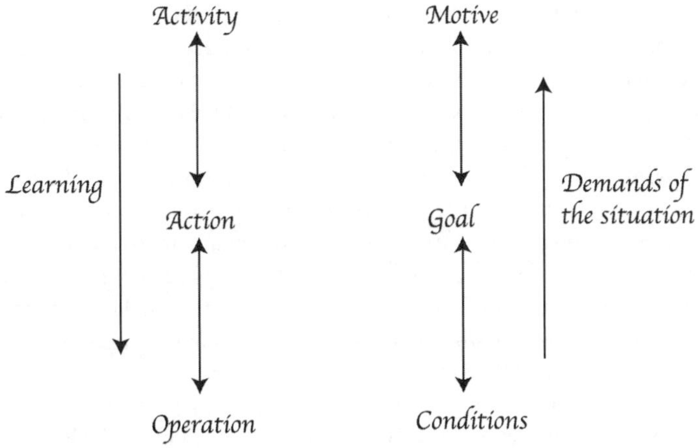

FIG. 21.1. The activity hierarchy.

thought of as either fixed or having been changed, thus disrupting the model of the system. Motivations might better be regarded as ever changing and with implications that are also fluid and based on the context. This way of looking at things has three important consequences: First, it allows us to identify uniquely a unit of analysis—the activity—by distinguishing between motivations. An activity is uniquely identified by its motive, which is collective—the motivation of a (small) group. Second, it exposes the psychological nature of activity theory. Both Vygotski and Leont'ev were psychologists and pedagogues and not designers or work study specialists. (Their orientation does introduce a serious problem, namely, that motivation cannot be observed directly but can only be inferred.) Third, we can introduce the concept of alienation (drawn from Marxism) to cover cases in which an individual's motivations are at odds with the collective's motivation. Activities are realized by an aggregation of *mediated actions*, which in turn are achieved by a series of low-level operations. This structure, however, is flexible and may change as a consequence of learning, context, or both.

The definition of a goal in task analysis varies, as is clear from other chapters of this handbook (e.g., chaps. 1, 3, 28, and 30). In Fig. 21.1, *goal* has the narrow definition used by Annett et al. (1971). Note that it is the *action* that must be learnt in order to accomplish the *goal*, whereas the *activity* is primarily concerned with the *motive*. These three levels are repeated in many other domains, not as discrete entities but as concentric layers, such as in the unit, integration, and acceptance layers of software testing. We might think of the *operation* and *conditions* level as being the how, the *action- goal* level as being the what, and the *activity-motive* level as being the why. This in turn is equivalent to the layers of data, information, and knowledge used in the field of knowledge management or to Smithson and Hirschheim's (1998) layers of efficiency, effectiveness, and understanding. Recently we have also seen that Gibson's (1977) notion of affordance can usefully be considered as three layers: usability, embodiment, and fitness for purpose (Turner & Turner, 2002).

By way of example, consider the process of learning to use a complex interactive device such as a motorcar. The object of the activity is likely to be quite complex, encompassing perhaps the need to be able to drive because of work commitments, the desire to attract the opposite sex, peer pressure, the desire to satisfy the indulgent parent who purchased the car; or the need to participate in a robbery. These could change from day to day or even minute to minute (see Chap. 30). The activity is realized by means of an aggregation of actions (e.g., obtain driving license, insure car, take driving lessons, learn the highway code, get a job to pay for the petrol, etc.). These individual actions in their turn are realized by a set of operations (e.g., get driving license application form, complete form, write out check for the license, send off license, etc.). This, of course, is an incomplete, *static* description of the activity, whereas humans are constantly learning with practice. For example, when a novice driver is first presented with the intricacies of a gear-stick (manual gear shift), the process of disengaging the engine, shifting gear, and reengaging the engine is likely to be under conscious control. The focus of attention is at the operations level, but with practice attention will tend to slide down the hierarchy as the action becomes automatic. Over time actions become automatic, and the activity itself is effectively demoted to level of an action—unless circumstances change. Such changes might include driving on the opposite of the road, using a different make of motorcar, driving a lorry, or being faced with the possibility of a collision. In such circumstances, consciousness becomes refocused at the level demanded by the context.

This alternate formulation of the nature and structure of an activity is of interest for a number of reasons: First, this theory of activity basically has a hierarchical structure. Second, it introduces the ideas of consciousness and motivation at the heart of the activity. Leont'ev offers a mechanism by which the focus (and locus) of consciousness moves between these various levels of abstraction—up and down the hierarchy depending on the demands of the context. (Note that chap. 15 discusses such skill acquisition from a traditional, Western cognitive perspective.)

21.2.3 The Role of Engeström in Modern CHAT

> To be able to analyze such complex interactions and relationships, a theoretical account of the constituent elements of the system under investigation is needed. ... Activity theory has a strong candidate for such a unit of analysis in the concept of object-oriented, collective and culturally mediated human activity. (Engeström & Miettinen, 1999, p. 9)

In the late 1980s Engeström, a Finnish academic, extended activity theory by incorporating a model of human activity and methods for analysing activity and bringing about change in organizations in a manner reminiscent of participatory design (e.g., Cole & Engeström, 1993; Engeström, 1987, 1995, 1999). Engeström's work has been adopted and elaborated by many researchers in Scandinavia (e.g., Bardram, 1998a, 1998b), the United States (e.g., Nardi, 1996), Australia (e.g., Hassan, 1998), and Great Britain (e.g. Blackler, 1993, 1994; Turner & Turner, 2001a, 2001b; Turner, Turner, & Horton, 1999). Engeström's account of activity theory is probably the dominant formulation of AT in use in the study of information systems, HCI, and CSCW. In such research, there is perhaps a greater focus on the role of activity per se than on history and culture.

As we have already discussed, central to activity theory is the concept that all purposive human activity can be characterized by a triadic interaction between a *subject* (one or more people) and the group's *object* (or purpose) mediated by *artifacts* or tools. In activity theory terms, the subject is the individual or individuals carrying out the activity; the artifact is any tool or representation used in that activity, whether external or internal to the subject; and the object encompasses both the purpose of the activity and its product or output. Subsequent development of activity theory by Engeström and others has added more elements to the original formulation. These are *community* (all other groups with a stake in the activity), *division of labor* (the horizontal and vertical divisions of responsibility and power within the activity), and *praxis* (the formal and informal rules and norms governing the relationships between the subjects and the wider community for the activity). Thus, activities are social and collective in nature.

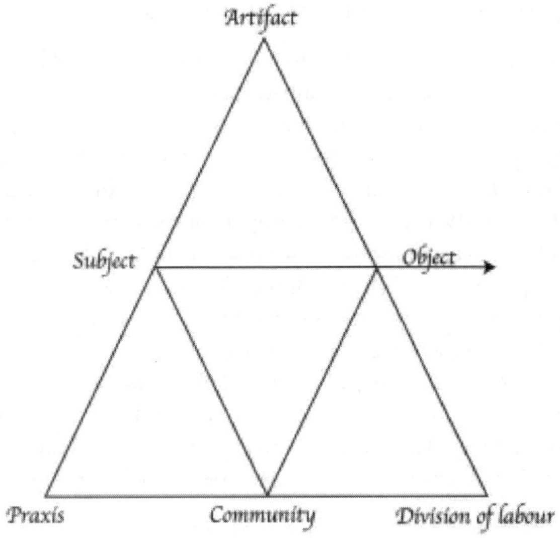

FIG. 21.2. An activity triangle (schema).

The relationships between the elements of an activity are often represented by an activity triangle (Fig. 21.2).

The use of activity triangles is widespread in the activity theory literature, but it must be remembered that this is only a partial representation of an activity. The triangle should be regarded as a nexus, existing as it does in a continuum of development and learning and in turn masking its internal structure (that is, the individual *actions* by which it is carried out). It would be a mistake to view activity triangles in the same way that we might view a Universal Modeling Language (UML) diagram. The UML specification (Unified Modeling Language Specification v 1.4, 2001) is very prescriptive in its symbolism of text, nodes, lines, and shapes. The activity triangle represents a holistic model in which all the components are in a state of tension. The lines do not have equivalent meaning, and the nodes themselves are a mix of entities and concepts (some represent *praxis* and *division of labor*, not as nodes, but as a greater context). Selecting any three points gives 120 potential triads, relatively few of which are (yet) significant. What is of interest is how certain pairs of nodes are mediated by a third node, essentially the notion described in the previous paragraph. Appreciating the effects of such mediation is vital if analysts are to stop analyzing and start understanding the system that they wish to change.

Activity theory is perhaps unique among accounts of work in placing such a strong emphasis on the role of collective learning. Vygotski's work on developmental learning has been a major influence on the thinking of Engeström, who has extended the idea to encompass collective learning, which he has termed *expansive learning* (Engeström, 1987). Engeström has demonstrated the usefulness of expansive learning, with its cycles of internalization, questioning, reflection, and externalization in the development of activities in a variety of domains (Engeström, 1990, 1999). The drivers for these expansive cycles of learning and development are *contradictions* within and between activities. Although this is something of a departure from Vygotski, it has proved particularly valuable to HCI and CSCW researchers (e.g., Holt & Morris, 1993; Nardi, 1996; Turner & Turner, 2001a). We now consider contradictions in more detail.

21.2.4 Engeström's Description of Contradictions

Activities are dynamic entities that have their roots in earlier activities and bear the seeds of their own successors. They are subject to transformation in the light of *contradictions*. Figure 21.3 is an illustration of an activity system (that is, a group of related activities) complete with potential contradictions. Those contradictions found within a single node of an activity are called *primary contradictions*. Primary contradictions might manifest as a faulty mediating artifact (e.g., bug-ridden software) or as a heterogeneous subject group with, say, ill-matched training and skills. That these two problems have self-evident solutions (fix the bug, train the people) for the here and now is of little comfort. For example, fixed bugs allow new uses, which reveal other previously unknown bugs, and trained staff move on to other employers or become resistant to retraining when the context changes.

Contradictions that occur between nodes are called *secondary contradictions*. In practice, contradictions of this kind can be understood in terms of breakdowns between actions or sets of actions that realize the activity. These actions are typically polymotivated (that is, executed by different people for different reasons or by the same person as a part of two separate activities),

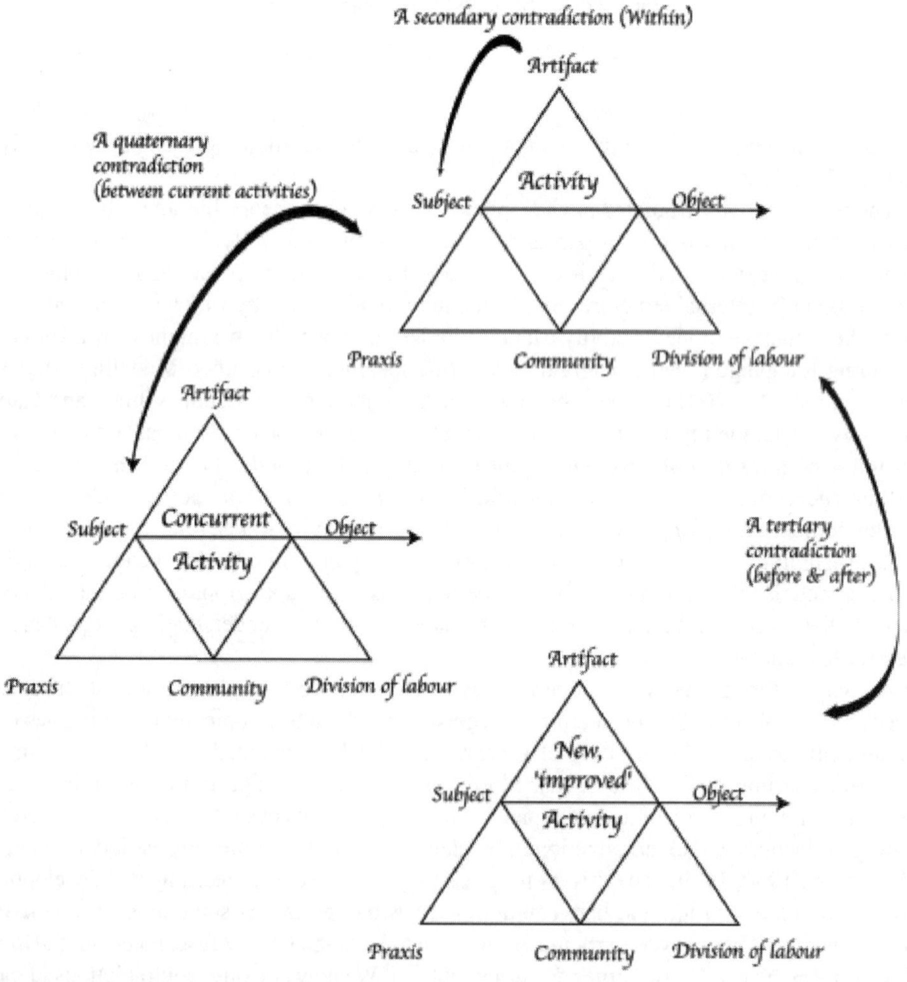

FIG. 21.3. An activity system and potential contradictions.

and this polymotivation may be at the root of subsequent contradictions. (See chaps. 1 and 30 for further discussion of multiple goals.)

Tertiary contradictions may occur when an activity is remodeled to take account of new motives or ways of working. Thus, they occur between an existing activity and what is described as a "culturally more advanced form" of that activity. A culturally more advanced activity is one that has arisen from the resolution of contradictions within an existing activity and may involve the creation of new working practices (praxis) or artifacts or a new division of responsibilities. Finally, contradictions between different co-existing or concurrent activities are called as *quaternary contradictions*.

From this, it can be seen that a complex and continuously evolving web of contradictions may emerge. Primary and secondary contradictions in an activity may give rise to a new activity, which in turn spawns a set of tertiary contradictions between it and the original activity, and this may be compounded by quaternary contradictions between coexisting activities.

21.2.5 Identifying Contradictions

Having described the role of contradictions and how they may be classified, we now turn to how they are identified. Engeström (1999) described contradictions as being identified by disturbances in the free running of an activity. Thus at the local level (e.g., an office or specific organizational division), contradictions might include bottlenecks, various folktales as to why a procedure is the way it is, and differences of opinions as to the what, why, and when of an activity. Engeström (1999) gave examples of such disturbances from a medical case study, including mismatches between administrator forms, uncertainty about the division of responsibilities between doctors and nurses, and uncertainty about the sequencing of procedures. A further example is Holt and Morris's (1993) retrospective analysis of the contradictions in the activity systems operating in the events leading up to the Challenger shuttle disaster. They concluded that fundamental contradictions in and between the activities underpinning the development, construction, and deployment of the shuttle were ultimately responsible for the loss of the vehicle and crew.

Contradictions are distinguished from disturbances in that many disturbances may map onto a single contradiction. Thus disturbances are the visible manifestations of underlying contradictions. In practice this means that understanding the dynamics of the current work, making visible its nuances, and identifying any disturbances therein are the necessary precursors to identifying contradictions.

21.3 FROM DESCRIPTION TO EVALUATION

We now turn to a demonstration of activity, theory in action. As we have said, activity theory offers a structured and structural account of an activity, complete with its context and indications of internal dynamics and contradictions. These strengths can now be used to order the evaluation of a novel application in situ, namely, a collaborative virtual environment (CVE). We begin by describing the work the CVE is intended to support—in this instance, training—and then consider evaluation of the CVE.

21.3.1 The Case Study

Safety-critical training in the maritime and offshore domains is recognized as extremely important by all stakeholders in these industries but is almost prohibitively expensive. Current methods require trainees to be located at a specialist training site that is typically equipped with costly physical simulators. The DISCOVER project aimed to provide a CVE-based series of team training simulations that would dramatically reduce the need for senior mariners and oil rig workers to have to attend courses at specialist centers. Although the system would be made available at such institutions, it could also be used over the Internet from offshore or on board ships. The consortium comprised four marine and offshore training organizations based in the United Kingdom, Norway, Denmark, and Germany; virtual reality technology specialists; training standards bodies; a number of interested employers; and Napier University.

21.3.1.1 Current Training

Space prevents a full treatment of safety-critical training at all of the training organizations, so we confine ourselves to one example, which we shall call the Center. At the Center, the training scenarios are played out in a room adapted from a conventional lecture room. The "bridge" area is found behind a screen in one corner of the room and contains the ship's blueprints laid out on a table, alarm and control panels, communication devices, various reference manuals, and a crew list. The other piece of simulation equipment is in the main body of the room (Fig. 21.4). This rather resembles a large domestic tea-trolley with four shelves, each representing one deck in a four-deck section of the ship. Each shelf has a blueprint of the relevant deck. These blueprints, and those on the bridge, can be annotated with schematic depictions of hazards such as smoke and are populated by miniature models of crew members who can be moved around, knocked over to simulate injury or death, and so on. The simulation is completed by an "engine room" located in one of the tutor's offices down the corridor from the lecture room. The engine room is equipped with a pair of walkie-talkies and more blueprints.

A typical scenario at the Center concerns a badly maintained ship taken over by the current crew at short notice and carrying a hazardous cargo, which subsequently catches fire. A fire team is sent to investigate, and the situation is exacerbated by crew members' being overcome

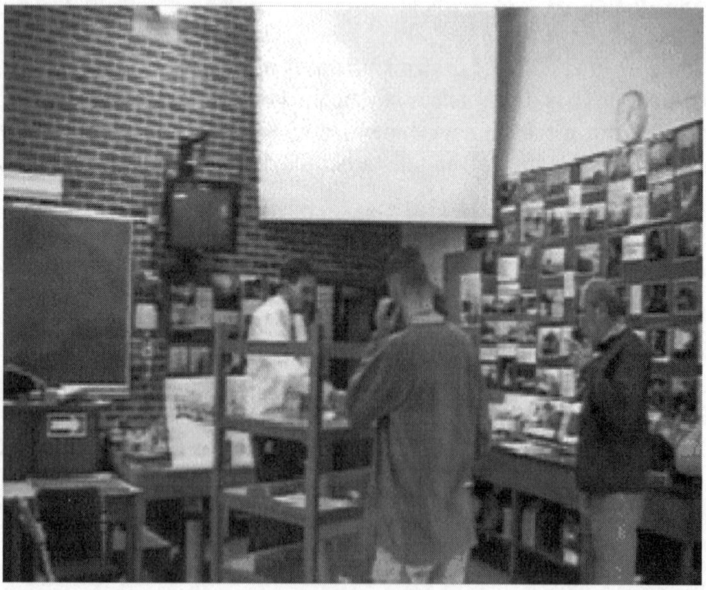

FIG. 21.4. Around the "trolley" at the center. The man in white is a trainer, the other two are trainees (both ship masters). The tea trolley is the wooden structure between the men. Each shelf represents a ship's deck. The decks are populated by Subbuteo football figures that represent members of the crew.

by smoke, power failures, engine stoppages, and sundry other hazards. Trainees form teams that are assigned to the bridge, to fire fighting duty (working around the trolley), and to the engine room. Other trainees act as nonparticipant observers. Tutor-trainee interaction, which is intense, relates to the plot of the scenario and the teams' handling of it. The tutors point out aspects that a team might have overlooked, hint at possible actions, and generally keep the action running smoothly. As problems escalate, the teams become very evidently engaged in the action. Figure 20.4, which is a still image from a training session, shows a trainer (the figure in white) and two trainees, all seen using communication devices (cell phones and walkie-talkies). At the center is the trolley mock-up of the ship's section. The trainer is in the act of moving some of the figures representing the remainder of the crew into the casualty position, a development that will be reported back to the bridge by the incident team leader. Once the action has run its course, a full debriefing takes place, comprising discussion and feedback about the teams' actions, the real-life scenario, and alternative approaches. Tutors take pains to ensure this is trainee led, and discussions are amplified by the tutors' recall of particular incidents together with comments by the observers.

The staff at the Centre have varying models of how the DISCOVER CVE might support their work. From the organizational point of view, it is hoped that the system will enable training to be delivered in a more flexible and economical manner, allowing skills to be acquired, practiced and even assessed without the need for mariners to attend in person. This model requires an environment that is self-contained, supports all the different types of interaction described above, runs over the Internet, and has the added value of simulating some conditions more realistically than current methods. Tutors would need the facility to modify events in the environment, as in current practice.

In this view, trainees interact with the environment, each other, and any other role players inside the CVE, with the possible addition of videoconferencing for discussions, debriefing, and tutor-trainee interaction. Another view is that the CVE would be a more or less direct substitute for the "tea trolley" embodying the ship's section but have the advantages of increased realism. The trainees would remain physically located at the center, and most interpersonal interaction would be outside the environment. It will be appreciated that the first of these alternatives is much more demanding, both in technical terms and in its implications for organizational change. Interviews with training staff indicate that they see the concept of remote delivery as an interesting development with added potential for an enhanced degree of realism, the acquisition of new skills for themselves, and additional business.

21.3.1.2 The Envisaged DISCOVER Training Solution

The CVE itself is designed to run on standard, high-end, networked desktop PCs, the only special purpose equipment being high-specification soundcards and audio headsets. The environment represents the interior of a ship (maritime version) or an offshore platform (offshore version). The users are present as avatars in the environment. Trainee avatars have abilities designed to mimic real action and interaction as closely as possible and have access to a small number of interactive objects such as fire extinguishers, alarm panels, and, indeed, bodies. Communication again imitates that in the real world, being mediated through voice when avatars are in the same room and telephone, walkie-talkie, or public announcement (PA) when they are not. Tutors are not embodied as avatars but have the ability to teleport to any part of the environment to see the location of trainees on a map or through a bird's-eye view, track particular trainees, and modify the environment in limited ways, for example, by setting fires or locking doors. It should be stressed that the environment is intended to support the training of emergency management and team coordination skills rather than lower level skills. Figures 21.5 and 21.6 are screenshots taken from the maritime version of the DISCOVER CVE.

FIG. 21.5. Avatars on the bridge.

FIG. 21.6. The tutor's view of the bridge.

21.3.2 The Tasks of Evaluation

As is usual in projects such as DISCOVER, we had defined the purpose of the evaluation in the project proposal in open and fairly vague terms (e.g., does the CVE support the learning of key skills?). What emerged from actually engaging in the project proper was a need to address a series of wider issues. The first of these issues to emerge was acceptability, for both trainers and trainees. A second issue was the need to partition the evaluation itself logically by adopting a suitable framework. We now discuss these in turn.

21.3.3 Is the Software Acceptable?

We undertook a contradictions analysis of the current training activity to investigate the acceptability of the CVE to the trainers, trainees, training organization, and training standards bodies. The first step in this process was to identify disturbances in the current activity and

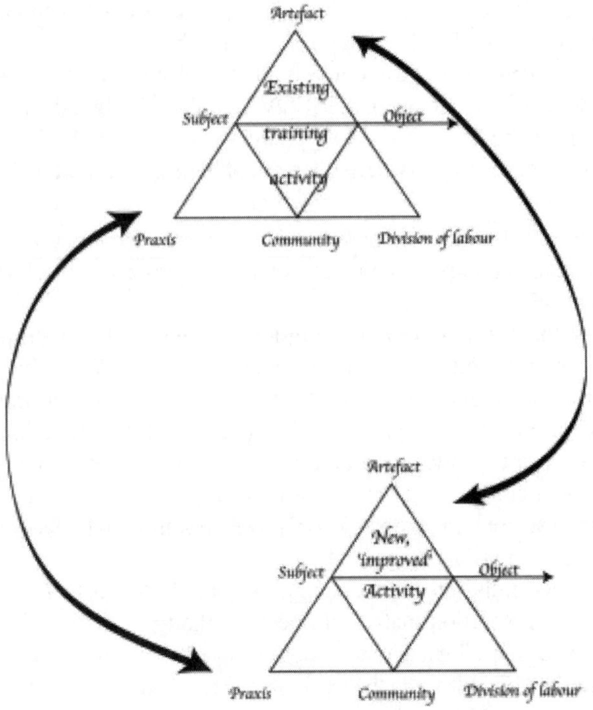

FIG. 21.7. Potential contradictions between the existing and the new (improved) training activities.

contradictions between the current and the culturally more advanced (new, improved) activity (Fig. 21.7). It should be stressed that the following discussion of disturbances and potential contradictions is intended to be indicative only, not canonical or definitive.

Disturbances and potential contradictions within the current activity included these:

1. *Concerning the tea trolley.* This prop had been proved to be very effective in supporting collaborative decision making but remained, well, just a tea trolley. There is a matter of image for the Center. Resolving this contradiction—effective but *low-tech*—became one of the major thrusts of the DISCOVER project. The challenge was to build a CVE with all of the flexibility, ease of use, and effectiveness of the trolley. No easy task.

2. *The problem with travel.* Senior mariners are valuable people and had to travel to the Center to be trained. Their absence was expensive for their employers, travel to and from the Center was expensive, as was their accommodation, and so forth. Yet getting all of these people together in one place was the most effective way of training them. This problem generated a set of requirements for the DISCOVER system and was resolved by designing the CVE to operate in a distributed fashion over the Internet.

Potential contradictions between the current and new activities included the following three:

3. *The Doom problem.* This problem was identified by the trainers themselves. They posed the question, "How can we seriously deliver safety-critical training using something that looks like a video game?"

4. *Training the trainers problem.* The trainers are experienced senior mariners and expressly not computer people (their words), and they collectively voiced concerns about the need to train

them in the use of the DISCOVER CVE. They also openly speculated about the consequences of introducing the technology for the structure of their jobs.

5. *The identity problem.* This problem was identified by one of the maritime standards bodies. They raised the issue of using DISCOVER in a distributed manner in places like Southeast Asia and on the high seas. How could we guarantee the identity of the person engaged in training given that the participants are collaborating remotely?

Each of these potential contradictions and many more were, in the main, worked through in a series of what-if scenarios (chap. 5). Others were largely glossed over with the excuse of "We will have to wait and see."

In the same way that Norman (1999) modified his view of the appropriateness of using Gibson's affordances in design as a result of the rather crass way in which designers would analyze a need for an affordance here, an affordance there, we do not propose to categorize definitively each of the above five potential disturbances or contradictions. The point is to understand that each can be understood and resolved as one or more types of contradiction but that later implementations may yet reveal other types. Perhaps intuitive ability is what separates a good analyst from one who ticks off boxes. For those who feel uncomfortable with this uncertainty, here is a partial categorization.

The tea trolley issue falls into the category of potential disturbances (leaving aside the possible quaternary contradiction that would occur if the tea lady needed to take the trolley away). The key word is "prop." All actors recognize that, by suspending disbelief, an audience will attribute a number of culturally based capabilities to such a prop motivated by their participation in a spectacle. Moving from an improvised theatrical experience to a Hollywood movie involves some of the same challenges as turning a training prop suitable in a given situation into a believable CVE. Perhaps in stating that "there is a matter of image for the Center," we hint at a tertiary contradiction, in that the trolley works but that in the new improved system it fails to meet the expectations.

The second problem is essentially a quaternary contradiction between the activity of establishing the requirements for the CVE and other activities in the manners' working lives, and the solution is evident.

The other three are potential tertiary contradictions.

21.3.4 A Structured Evaluation of the CVE

We now turn to the structured evaluation of the CVE, which we based on the three-level model of an activity shown in Fig. 21.1.

21.3.4.1 Evaluating the Fitness for Purpose

The purpose of the CVE was to support the teaching and learning of emergency management skills for use in offshore and maritime contexts. It was essential that stakeholders should have confidence in the software as affording a means for such training and trust that the skills learnt in the environment would be effective in real emergencies. Clearly, the evaluation of the fitness for purpose can only be undertaken with the participation of individuals from the community concerned. In one of the trials of early versions of the software, we had access to several maritime officers (including the captain of a well-known passenger ship), who completed custom-designed questionnaire items about their confidence in the future use of the system as well as taking part in debriefing sessions. More substantive evaluation for perceived fitness for purpose focused firstly on data from the tutor sessions discussed below. Here data were collected through custom-designed questionnaire items, posttrial discussions, and analysis of verbalizations and behavior from the video record.

As for pedagogic effectiveness, trials are planned that incorporate realistic training scenarios with inbuilt checkpoints for the display of specific management behaviors at appropriate times. These trials will be complemented by observation employing criteria based on the measures of team effectiveness derived by the TADMUS project (Cannon-Bowers & Salas, 1998) in its research into training for decision making under stress and on the deeper aspects of pedagogy in the Laurillard (1993) model. Finally, DISCOVER must receive the seal of approval from industry validating bodies.

Despite this detailed planning, it has turned out to be impractical to run rigorous comparative trials of DISCOVER against conventional training (because of the restricted availability of trainees and the related difficulty of ensuring matched groups). Still less will it be possible to "prove" the effectiveness of DISCOVER training in genuine emergencies. It remains the case that at the current state of knowledge the verification of the transfer of virtual reality–based training to the real world is very much an active issue for research (Caird, 1996, explores this issue in some detail).

21.3.4.2 Evaluating the Action Level

Here the evaluation focus is on how effectively are actors embodied *in* the environment and how effectively they can collaborate *through* the environment. An additional concern is evaluating the related issues of perceptions of fidelity, presence, and engagement.

Trainees in the DISCOVER environment needed to be able to find each other (Fig. 21.8), communicate by appropriate means with fellow trainees and tutors, monitor what others were doing, and interact with various items in the environment (e.g., to pick up an avatar overcome by smoke). Tutors had to be able to gather sufficient information from monitoring activity

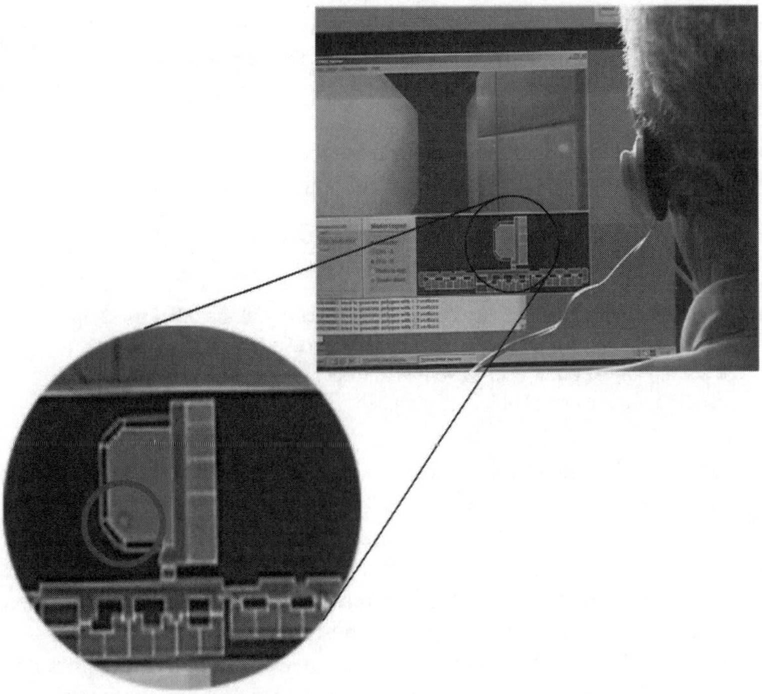

FIG. 21.8. A trainee trying to figure out which way he is facing (his location is represented by a flashing circle).

in the CVE to provide guidance and posttraining feedback, communicate with trainees, and modify interactive objects in the CVE (e.g., change the location of fires). It had also been stressed by all stakeholders from the training and employer organizations that the CVE must be extremely realistic and imbue a strong sense of presence if it was to be considered fit for its purpose of providing training. Realism was necessary for two reasons. First, existing physical ship simulators mimic their seagoing equivalents closely, so closely that officers undergoing simulator training can be dismissed should they run the simulator aground. Second, one of the key elements in emergency management training is engagement in the emergency scenario and consequently the experience of a suitable degree of stress.

Here the choice of techniques was constrained by the limited range of ready-made tools for evaluating aspects of collaboration in virtual environments and again by the availability of subjects. Aspects of communication and coordination (primarily, being able to see, hear, and address other users) were evaluated in parallel with the ergonomic elements in the very early trials mentioned above. Once the software was reasonably stable and more coworking features had been added, more complex trials were carried out.

As before, proxy subjects were largely used to identify the most immediate issues concerning affordances for embodiment and communication. These subjects undertook structured tasks, including the range of collaborative activities employed in a realistic training situation and the underlying collaborative actions identified by the COVEN hierarchical task analysis approach (ACTS Project COVEN, 1998), such as being able to identify each other and so forth. Short posttrial questionnaires were administered using items derived from the task analysis.

COVEN (COllaborative Virtual ENvironments) was a European project that addressed "the technical and design-level requirements of VR-based multi-participant collaborative activities in professional and citizen-oriented domains" (Normand & Tromp, 1996, pp. 76–78). The COVEN consortium comprised 12 partners drawn from industry and academe. COVEN created a vast number of outputs, among which was a cognitive walk-through evaluation technique for collaborative virtual environments. The use of a cognitive walk-through as a relatively low-cost early usability evaluation technique is well established. Prior to the walk-through, user tasks and characteristics are identified, typical scenarios of use are constructed, and then the expert "walks through" the task using the technology, reviewing the actions necessary to accomplish tasks, and estimating where target users would encounter problems. A generic checklist of questions based on psychological theory structures the process. COVEN provides a cognitive walk-through specifically designed for the evaluation of CVEs. A sample of the checklist is shown below (ACTS Project COVEN, 1998):

1. Can the user locate the other user(s)?
2. Can the user recognise the identity of the other user(s), tell the other users apart?
3. Are the communication channels between the users effective?
4. Are the actions of the other user(s) visible and recognisable?
5. Can the user act on a shared object while keeping the other user(s) in view?

For users adopting the role of tutor, an additional set of tasks and questionnaire items was derived from Laurillard's (1993) model of teaching and learning. At this level, we sought to address, not the efficacy of any teaching or learning, but rather the support of the environment for such pedagogic actions as setting or modifying task goals, monitoring trainees, and giving feedback. Again, observers monitored the progress (or, occasionally, lack of progress) of the scenario, supported by checklists mirroring the questionnaire content.

Finally, issues of fidelity and presence were also covered, initially through a short series of items, adapted from VRUSE, in the posttrial questionnaire and the observers' checklist. The Human Factors Toolkit for VR evaluation (Kalawsky, 1999) comprises an overall framework incorporating suitably adapted elements of the MUSiC evaluation tools (such as usability context analysis), performance measures, and two specially designed questionnaire instruments, VRUSE and VSART. VRUSE treats usability as a multifactorial dimension, the factors comprising functionality, user input, system output, user guidance and help, consistency, flexibility, simulation fidelity, error correction/handling and robustness, sense of immersion/presence, and overall system usability. VSART, which focuses on presence, is similarly multifactorial. The instrument probes demand for and supply of attentional resources, situation awareness, information quality and quantity, and "technological factors," including, for example, field of view. VRUSE has been extensively validated and used by authors other than its originators, whereas VSART appears to be at an earlier stage of development. Neither instrument addresses collaborative issues.

The final version of the software was evaluated through trials with experienced tutors from one of the training organizations involved in the project. (Evaluation techniques had been planned for trials with "real" trainees, but in the event personnel could not be made available. This work continues outside the scope of the project at one of the training organizations.) Tutors undertook a realistic training scenario authored by one of the training organizations. They took turns playing tutor and trainee roles. NASA's questionnaire measures of presence (PQ) and immersive tendencies (ITQ) were developed as part of an investigation into the potential of virtual environments for military training. The immersive tendencies instrument investigates an individual's tendency to become immersed in experiences in general, and presence is defined as the subjective experience of being in one environment while physically being in another. As with Kalawsky's (1999) work, both aspects are treated as multifactorial, the PQ treating presence as comprising involvement/control, naturalness, and interface quality. This time the NASA ITQ questionnaire (which measures immersive tendencies; Witmer & Singer, 1998) was administered before the trial started, followed up by a questionnaire instrument incorporating the collaborative and pedagogic aspects as before, coupled with the NASA PQ—the ITQ counterpart that measures presence. These trials were videotaped to allow further analysis of the evaluation data.

21.3.4.3 Evaluating the Operational Level

The aspects of the CVE to be evaluated at this level are those concerned with the ergonomics and usability of the means provided to interact with the CVE, including the now standard range of GUI controls as well as input and output devices. Aspects to be considered are their perceptibility, ease of operation, provision of feedback, and, in general, the list of low-level usability heuristics to be found in any textbook. Here we were concerned, inter alia, with the affordances of such features as the push buttons provided to activate virtual communication devices such as the phone and walkie-talkie and the use of the mouse click as a means of opening doors, setting off fire extinguishers, and generally activating objects. The design of these had been a subject of much debate. Issues included whether, for realism, a phone should have the usual set of buttons reproduced virtually or whether users would find a dialogue box more convenient. We also needed to evaluate the basic input devices, such as the mouse, which was for moving through the environment (employers were keen that the system should run on a standard worktop PC and peripherals), and the headsets, which were used for verbal communication.

The overall emphasis in the choice and construction of techniques for level 1 was to obtain basic usability data with minimal consumption of analyst and user resources. The affordances were primarily investigated through user trials, starting from the earliest versions of the software. Early trials employed mostly proxy subjects who represented the eventual user population as closely as possible in terms of relevant background skills and experience. Use of proxy

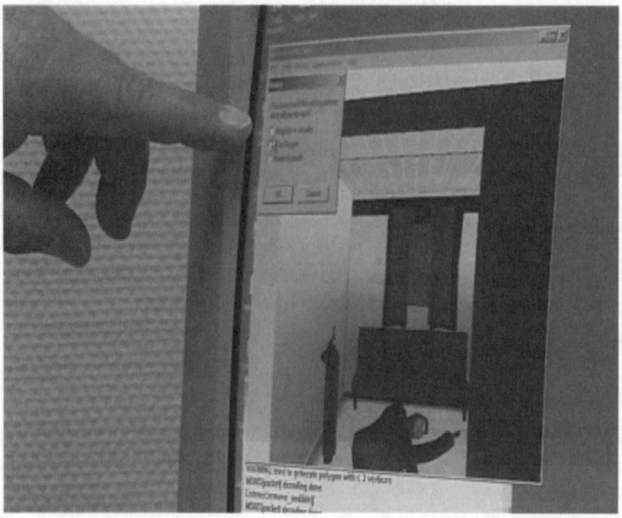

FIG. 21.9. A trainee examining a dialogue box.

subjects allowed us to conserve the scarce resource of real users for more polished versions of the software and for fitness-for-purpose issues. Subjects undertook realistic single-user and collaborative tasks matched to the functionality of the software version under review while monitored by observers. Figure 21.9 shows a user contemplating a dialogue box.

In later trials, the main evaluation focus shifted to the action and operation levels, but usability continued to receive some attention. Posttrial questionnaires were compiled by adapting usability items from standard usability instruments and VRUSE (Kalawsky, 1999) and with guidance from Kaur, Maiden, and Sutcliffe (1997). Although the custom-built questionnaires did not now have the strong validation of their parents, the questions could be tailored to the particular context of the DISCOVER CVE while keeping the overall instruments to a manageable length. Observers augmented the self-report data. The trials were supplemented by usability inspections structured by standard heuristics.

In the event, most usability problems were identified by an initial quick expert check of the interface, but the other techniques adopted were able to provide substantive data to back up these observations.

21.4 CONCLUSION

This chapter began with a brief introduction to the dominant formulation of activity theory. From there we built on the structural and dynamic descriptions offered by the theory to organize the evaluation of a CVE. Although effective enough in practice, activity theory is primarily a psychological theory of human development, learning, and (to some extent) motivation, which, and it will require a reorientation of thinking about these topics.

Activity theory should not be seen as a slightly odd variant of mainstream task analysis. Western task analysis is considerably better developed, operationalized, and tested. Instead, as more of the subjects of task analysis become defined as knowledge workers, we should take account of activity theory's emphasis on the centrality of learning and development in using such things as interactive devices, information appliances, and even collaborative virtual environments.

We have offered only a taste of how one aspect of activity theory's rich conceptual language might be applied to what is a complex and taxing problem. Rather than know the "cost of everything and the value of nothing," the analyst has to learn to understand and work with the real-world messy things that human beings do with and in systems.

ACKNOWLEDGMENTS

Special thanks to Dr. Susan Turner for leading the evaluation work package and for letting us rifle her evaluation deliverable for bons mots. Also thanks to Oli Mival for capturing the evaluation footage (or is it meterage?) and for the screenshots. We gratefully acknowledge the contributions of our colleagues on the DISCOVER project in providing the sites and subjects for the fieldwork herein described and in developing the DISCOVER software. The project was financially supported by the EU ESPRIT program.

REFERENCES

ACTS Project COVEN. (1998). N. AC040 D3.5, Usage Evaluation of the Online Applications Part B: Collaborative Actions in CVEs 2.

Annett, J., Duncan, K. D., Stammers, R. B., & Gray, M. J. (1971). *Task analysis*. London: Her Majesty's Stationery Office, London, UK.

Bannon, L. J. (1991). From human factors to human actors. In J. Greenbaum & M. Kyng (Eds.), *Design at work: Cooperative design of computer systems* (pp. 25–44). Hillsdale, NJ: Lawrence Erlbaum Associates.

Bardram, J. E. (1997). Plans as situated action: An activity theory approach to workflow systems. In J. A. Hughes, W. Prinz, T. Rodden, & K. Schmidt (Eds.), *The Proceedings of the Fifth European Conference on Cosmputer Supported Cooperative Work* (pp. 17–32). Kluwer Academic Publishers.

Bardram, J. E. (1998a). Scenario-based design of cooperative systems. In F. Darses and P. Zaraté (Eds.), *The Proceedings of Second International Conference on the Design of Cooperative Systems '98*, (pp. 20–31).

Bardram, J. E. (1998b). Designing for the dynamics of cooperative activities. In *The Proceedings of the Computer Supported Cooperative Working, 98* (pp. 89–98). Seattle, Washington: ACM Press.

Blackler, F. (1993). Knowledge and the theory of organizations: Organizations as activity systems and the reframing of management. *Journal of Management Studies, 30*, 863–884, Seattle, Washington.

Blackler, F. (1994). Post(-)modern organizations: Understanding how CSCW affects organizations. *Journal of Information Technology, 9*, 129–136

Caird, J. K. (1996). Persistent issues in the application of virtual environment systems to training. *IEEE: Human Interaction with Complex Systems, 3*, 124–132.

Cannon-Bowers, J. A., & Salas, E. (1998) *Decision making under stress*, APA, Washington, DC: American Psychological Association.

Cole, M., & Engeström, Y. (1993). A cultural-historical approach to distributed cognition. In G. Salomon (Ed.), *Distributed cognitions: Psychological and educational considerations* (pp. 3–45). Cambridge: Cambridge University Press.

Engeström, Y. (1987). *Learning by expanding: An activity-theoretical approach to developmental research*. Helsinki: Orienta-Konsultit.

Engeström, Y. (1990). *Learning, working and imagining. Twelve studies in activity theory*. Orienta-Konsultit, Helsinki.

Engeström, Y. (1993). Developmental studies of work as a testbench of activity theory: The case of primary care medical practice. In S. Chaiklen & J. Lave (Eds.), *Understanding practice: Perspectives on activity and context* (2nd ed.), (pp. 64–103). Cambridge: Cambridge University Press.

Engeström, Y. (1995). Objects, contradictions and collaboration in medical cognition: An activity-theoretical perspective. *Artificial Intelligence in Medicine, 7*, 395–412.

Engeström, Y. (1998). The tensions of judging: Handling cases of driving under the influence of alcohol in Finland and California. In Y. Engeström & D. Middleton (Eds.), *Cognition and communication at work* (pp. 199–232). Cambridge: Cambridge University Press.

Engeström, Y. (1999). Expansive visibilization of work: An activity theoretic perspective. *CSCW, 8*(1-2), 63–93.

Engeström, Y., & Miettinen, R. (1999). Introduction. In Y. Engestrom, R. Miettinen, & R.-L. Punamäki (Eds.), *Perspectives on Activity Theory*. Cambridge: Cambridge University Press, 1–16.

Gibson, J. J. (1977). The Theory of affordances. In R. Shaw & J. Bransford (Eds.), *Perceiving, acting and knowing: Towards ecological psychology* (pp. 67–82). Hillsdale, NJ: Lawrence Erlbaum Associates.

Hassan, H. (1998). Activity theory: A basis for the contextual study of information systems on organisations. In H. Hasan, E. Gould, & P. Hyland (Eds.), *Information systems and activity theory: Tools in context* (pp. 19–38). Wollongong, Australia: University of Wollongong Press.

Holt, G. R., & Morris, A. W. (1993). Activity theory and the analysis of organizations. *Human Organization, 52*(1), 97–109.

Johnson, C. (1999). Evaluating the contribution of desktop VR for safety-critical applications. In *Proceedings of* 18th *International Conference, SAFECOMP '99*, (pp. 67–78). Toulouse, France: Springer Verlag.

Kalawsky, R. S. (1999). VRUSE: A computerised diagnostic tool for usability evaluation of virtual/synthetic environment systems. *Applied Ergonomics, 30*, 11–25.

Kaptelinin, V., Nardi., B., & Macaulay, C. (1999). The activity checklist: A tool for representing the "Space" of context. *Interactions, 6*(4), 27–39.

Kaur, K., Maiden, N., & Sutcliffe, A. (1997). Interacting with virtual environments: An evaluation of a model of interaction. *Interacting With Computers, 11*, 403–426.

Kuutti, K. (1991). Activity theory and its applications in information systems research and design. In H.-E. Nissen, H. K. Klein, & R. Hirschheim, (Eds.), *Information Systems Research Arena of the 90's* (pp. 529–550). Amsterdam: Elsevier.

Kuutti, K., & Arvonen, T. (1992). Identifying potential CSCW applications by means of activity theory concepts: A case example. In J. Turner & R. Krauts (Eds.), *The Proceedings of Computer Supported Cooperative Working'92* (pp. 233–240). Toronto: ACM Press.

Laurillard, D. (1993). *Rethinking university teaching*. London: Routledge.

Nardi, B. (1996). Some reflections on the application of activity theory. In B. Nardi (Ed.), *Context and consciousness* (pp. 18–44). Cambridge, MA: MIT Press.

Norman, D. A. (1999). Affordance, conventions and design, *Interactions, 6*(3), 38–43.

Normand, V., & Tromp, J. (1996). Collaborative virtual environments: The COVEN project. In *Proceedings of the Framework for Immersive Virtual Environments Conference, Five 96, December 1996, 76-78* (pp. 67–82). Hillsdale: NJ.

Smithson, S., & Hirschheim. R. (1998). Analysing information systems evaluation: Another look at an old problem. *European Journal of Information Systems, 7*, 158–174.

Suchman, L. A. (1987). *Plans and situated actions*. Cambridge: Cambridge University Press.

Turner, P., & Turner, S. (2001a). A web of contradictions. *Interacting With Computers, 14*(1), 10–14.

Turner, P., & Turner, S. (2001b). Describing team work with activity theory. *Cognition, Technology and Work, 3*(3), 127–139.

Turner, P., & Turner, S. (2002). An affordance-based framework for CVE evaluation. In X. Faulkner, J. Finlay, & F. Détienne (Eds.), *People and computers XVI: Memorable yet invisible* (pp. 89–103). London: Springer-Verlag.

Turner, P., Turner, S., & Horton, J. (1999). From description to requirements: An activity theoretic perspective. In *Proceedings of the International ACM SIGGROUP confernece on Suppporting Group Work '99* (pp. 286–295). Phoenix: ACM Press.

Unified Modeling Language Specification v 1.4 (2001). Available: http://www.omg.org/technology/documents/formal/uml.htm

Vygotski, L. (1986). *Thought and language*. (Al. Kozulin, Trans. & Ed.). Cambridge, MA: MIT Press.

Witmer, B. G., & Singer, M. J. (1998). Measuring presence in virtual environments: A presence questionnaire. *Presence, 7*(3), 225–240.

IV

Computing Perspectives

The editors admit that "Computing Perspectives" is not an ideal title for this part. Indeed, Fabio Paternò thought it "limiting" to include his chapter under such a heading because "the ConcurTaskTrees (CTT) approach captures some important psychological concepts and integrates them with other elements in a rigorous framework." The editors do not disagree with Fabio and even admit there is a strong case for having his chapter (chap. 24) in part I, which covers the fundamentals of task analysis. Actually, all of the main methods described in part IV can capture some psychological concepts and provide some form of rigorous framework.

The editors, however, believe that the six chapters in part IV do naturally belong together, and they defend their decision to group them together with two arguments. First, the chapters are all directly concerned with the use of task analysis in the software life cycle, and they nearly all propose serious computer assisted software engineering (CASE) tools to support task analysis.

Second, most of the chapters exhibit a similarity of approach, and most of the authors know of each other's work—and probably each other. Indeed, in the early days of editing this handbook the editors rather flippantly thought of part IV as presenting the "mainland European perspective." In any case, within part IV the chapter references and the editorially imposed cross-referencing form a discernible cluster. There is also considerable overlap of content between the taxonomies offered in chapters 22–25.

Part IV is important because the main task analysis methods discussed are targeted at software engineering development, and these methods, particularly where supported by CASE tools, are therefore practical for use in commercial projects (see part II). The methods do differ in regard to the type of project for which they are designed. For example, TOOD (chap. 25) is only suitable for large application domains, especially those that are safety critical, whereas DIANE+ (chap. 26) is explicitly designed for less complex systems. All the chapters that deal with specific methods take an object-oriented approach.

Chapter 22
Choosing the Right Task Modeling Notation: A Taxonomy
Sandrine Balbo, Nadine Ozkan, and Cécile Paris

Chapter 22 can be seen as an introduction to part IV. Its taxonomy of a representative set of task analysis methods, emphasizing their notational differences, provides a tabular classification with the aim of helping readers to appreciate the differences between task analysis approaches and help them select an appropriate one. Chapter 22 is thus complementary to chapter 6 (which also presents a taxonomy) and offers a solution to the rather gloomy view expressed in chapter 1 that people often choose a task analysis method just because they are familiar with it.

Chapter 23
Environments for the Construction and Use of Task Models
Cécile Paris, Shijian Lu, and Keith Vander Linden

Almost as a continuation of the preceding chapter, chapter 23 begins by presenting a classification of CASE tools to support task analysis. The chapter then describes the ISOLDE Environment, which provides tools for extracting task information from a variety of sources, undertaking task analyses based on the DIANE+ method (chap. 26), and linking these to an object-oriented, UML-based software engineering approach.

Chapter 24
ConcurTaskTrees: An Engineering Approach for Task Models
Fabio Paternò

ConcurTaskTrees (CTT) have been developed by Fabio Paternò over a considerable period of time, and CTT and the LOTOS notation (e.g., Paternò & Faconti, 1992) constitute one of best established and most widely cited approaches to providing task analysis with structured and formal representations. As already noted, there is a strong case for including this chapter in part I, on the fundamentals of task analysis. Up-to-date descriptions of CTT and its CASE tool CTTE are provided, and they are illustrated by means of an example. The chapter also discusses their use with alternative devices, such a palm-tops and WAP phones (see chap. 26), and in Computer-Supported Cooperative Work (CSCW) applications (chap. 7).

Chapter 25
Using Formal Specification Techniques for Modeling Tasks and Generating HCI Specifications
Mourad Abed and Dimitri Tabary

Traditionally, software engineering has been about building large software systems (e.g., Sommerville, 1989), and the TOOD method and its large TOOD-IDE CASE tool are very much in this tradition. For the foreseeable future and beyond, there will continue to be application domains requiring work systems that involve very large amounts of software, much of it bespoke. This chapter ends with a detailed example from air traffic control (see also Chapter 13 and also Chapters 1 and 19).

Chapter 26

One Goal, Many Tasks, Many Devices: From Abstract User Task Specification to User Interfaces

Jean-Claude Tarby

DIANE+ is another widely cited task analysis method that relates task analysis and software engineering but for more modest applications than that of TOOD (described in chapter 25). Chapter 26 introduces DIANE+ and then provides a highly topical example of how the method can support the design of different interfaces on different platforms (PC, WAP phone, and regular telephone).

Chapter 27

Linking Task and Dialogue Modeling: Toward an Integrated Software Engineering Method

Chris Scogings and Chris Phillips

Lean Cuisine+ extends the Lean Cuisine approach to logical menu design to include the modeling of tasks and dialogues. As this chapter describes and demonstrates in its example, Lean Cuisine+ can be combined with an object-oriented, UML-based software engineering approach at a logical, abstract level (i.e., without specifying the appearance of menu and dialogue items).

REFERENCES

Paternò, F., & Faconti, G. (1992). On the use of LOTOS to describe graphical interaction. In A. Monk, D. Diaper, & M. D. Harrison (eds.), *People and computers VII* (pp. 155–173). Cambridge: Cambridge University Press.

Sommerville, I. (1989). *Software engineering* (3rd Ed.). Reading, MA: Addison-Wesley.

22

Choosing the Right Task-Modeling Notation: A Taxonomy

Sandrine Balbo
The University of Melbourne

Nadine Ozkan
Lanterna Magica

Cécile Paris
CSIRO MIS

A number of task modeling notations have been developed in the human-computer interaction (HCI) communities, often with different goals and thus different strengths. Usually, each model is designed to be employed at a specific phase of the software development life cycle (SDLC). Without an explicit understanding of the different attributes of these models, it is difficult to select a specific one to achieve one's goals. To address this issue, we propose a taxonomy consisting of several axes. The elements of the taxonomy have been defined according to various criteria we have found useful to consider when having to decide on a specific methodology or task model.

In this chapter, we first present our taxonomy. We then illustrate it with a few examples. We choose these models either because we used them ourselves or because they are the most popular in the literature.

22.1 INTRODUCTION

A number of task-modeling notations have been developed in the human-computer interaction (HCI) community, often with different goals and thus different strengths. For example, each notation is usually designed to be employed at a specific phase of the software development life cycle (SDLC) or for the design of a specific type of system.

Without an explicit understanding of the different attributes of these notations, it is difficult for analysts to select a specific one to achieve their goals. To address this issue, we have tried to organize the various notations in a taxonomy consisting of several axes, building on the work of Balbo (1994), Brun and Baudoin-Lafon (1995), and van Welie, van der Veer, and Eliëns (1998). The elements of the taxonomy have been defined according to various criteria we have found useful to consider when deciding on a specific task-modeling notation. (see chap. 20 for a discussion of other taxonomies and classifications.)

We first present our taxonomy, then illustrate the main axis in the taxonomy using sample notations. We chose these notations either because we have used them ourselves or because they are representative of the literature. Together they cover a wide range of issues that must be addressed.

The aim of our taxonomy is twofold. On the one hand, it is intended to help system designers (in a wide sense) choose an appropriate notation. The process of choosing a notation includes determining what attributes are required of a notation to suit specific purposes, considering different available notations that possess the appropriate attributes, and finally deciding among them. We believe that a clear understanding of the various attributes of different notations and what they can cover can help ensure an appropriate choice of notation. Our goal is to improve and facilitate the practice of task modeling, but the taxonomy is also meant to provide a state-of-the-art global view of task modeling practice. The research community needs to be able to assess the usefulness of task modeling, understand what is covered by current notations, determine what types of new notations might be needed, and so on.

22.2 TAXONOMY DEFINITION

We combine and build on factors from the taxonomies introduced by Balbo (1994), Brun and Baudoin-Lafon (1995), and van Welie et al. (1998). We found, however, that these taxonomies should be augmented and updated to reflect our specific aims and to incorporate details of recent types of systems or recent constraints imposed on systems, including groupware, Web applications, applications for mobile devices, and multiplatform applications (i.e., applications deployed on several platforms, such as PCs and hand held devices).

In this section, we present the various axes of our taxonomy. Each axis is described, along with its potential values. Note that, whereas the first two axes of our taxonomy refer to when and how a notation is used, the other axes refer to the various characteristics a notation may have.

22.2.1 Axis 1: Goal of Using a Notation

The first axis of the taxonomy refers to the desired goal in using task modeling. Task modeling can be used for a variety of purposes: for system design, for system evaluation, for the automatic generation of parts of a system, and so on. Moreover, task models can have different levels of abstraction, ranging from high-level user-oriented models to detailed system-oriented ones (Paris, Tarby, & Vander Linden, 2001), and each level serves a different purpose throughout the software development life cycle (SDLC). Hence, in order to systematize our approach, we refer to the various phases of the SDLC as potential values for the "goal" axis. We borrow the phases defined by IBM in Bardon, Berry, Bjerke, & Roberts (2001) because they seem sufficiently general to accommodate other methodologies. These phases are described below.

22.2.1.1 First Value: Discovery and Definition

This first phase consists of an exploration of the various concerns that will impact system design. Its purpose is to create an understanding of the requirements of the various stakeholders, the environment in which the system will be used, the type of functionality required, the various constraints that must be taken into account, and so forth. When task modeling is used in this phase, the intent is to generate high-level rather than fine-grained descriptions. The purpose here is to "identify the main tasks and the main objects that are relevant for the user. Here, the level of abstraction is very high; the tasks are "black boxes" and the objects are objects that the user employs" (Paris et al., 2001, p. 317). A similar approach is developed in the Scenario

Based Design (SBD) methodology (Rosson & Carroll, 2001; chap. 5), where *problem* scenarios are used to analyze requirements.

We briefly illustrate the use of task modeling here by examining the design of "innovative" systems (Ozkan, Paris, & Simpson-Young, 1998). Innovative systems are those that either create new possibilities and new tasks or have a major impact on existing tasks, virtually transforming them into new ones. Task modeling, even before the real design of a system starts, is useful for innovative systems, as it allows the designers to map out the environment in which the future system will be embedded, including its place in that environment and its potential impact on that environment. Performing a task analysis facilitates projecting the future environment and understanding the system's place within it. It also highlights some of the requirements that the future environment will place on the system. Although this is important for any application, it is crucial for the design of innovative systems. For example, the introduction of an automatic online help generator can dramatically change the tasks and functions of technical writers (Paris, Ozkan, & Bonifacio, 1998a). A task modeling analysis will help understand the impact of the technology, and the possibilities it may create. (Refer to Ozkan, Paris, & Balbo, 1998, and Paris, Balbo, & Ozkan, 2000, for further discussion of this use of task modeling.)

22.2.1.2 Second Value: Design

By *design*, we mean system design, user interface design, and documentation design, all of which are activities ideally taking place in parallel based on common specifications. This is the phase when task modeling is traditionally mostly used because it provides an understanding and an explicit representation of how the system will function in relationship to the user. Thus the main use of task modeling is for user interface design. Task modeling here also usually impacts the system architecture through the identification of common functions across an application, which correspond to common architectural modules.

A specialized use of task modeling at this stage consists in the definition of generic functionalities, that is, functionalities that should function consistently across a suite of applications (e.g., the "File Open" functionality). Here we touch on the idea of a "task model library" that is valid across several applications and would eventually be implemented in a software library, ensuring consistency and learnability over several applications. We have, for example, defined a task library for the UI-Card application suite (UICard Project, 2002). "The UI-Card is a novel use of smart card technology that brings one-touch direct access to the Internet. It enables access to a range of applications for communication, entertainment, education and information" (http://www.uicard.com/).

A similar specialized use of task modeling at this stage is in defining the user interface of an application that must be deployed on several platforms and that must function as consistently as possible across them (Thévenin & Coutaz, 1999).

Finally, there is increasing evidence (e.g., Balbo & Lindley, 1997) that task modeling can be used for another type of design often overlooked in software development, namely, documentation design. Task modeling can help organize the functional system documentation (as opposed to the other types of system documentation; Paris et al., 1998a).

22.2.1.3 Third Value: Development and Deployment

For the implementation phase, we follow Brun and Baudouin-Lafon (1995) in claiming that a notation can have two purposes:

It can be used to derive the generic functions of the system. Some task modeling notations may highlight the fact that some functions of the application are generic throughout all its modules and should be implemented as such. (This is akin to the architectural concerns

discussed in the previous section; here, however, we refer specifically to concerns of implementation efficiency.)
It can be used to automatically generate all or parts of the final system. This could include automatic generation of part of the system itself, portions of the user interface, or even portions of the end-user documentation. (Refer to Vanderdonckt & Berquin, 1999, for an example of the automatic generation of portions of the user interface, and Paris, Ozkan, & Bonifacio, 1998b, for an example of the automatic generation of portions of the end-user documentation.)

The appropriate degree of formality of a notation depends on what parts of the final system should be generated: By definition, formal notations (expressed in mathematical or logical terms) are those used for the automatic generation of code whereas semi formal notations can be used for the automatic generation of end-user documentation.

22.2.1.4 *Fourth Value: Evaluation and Prediction*

Some task-modeling notations were designed specifically for user interface evaluation. Of special interest are predictive notations, such as the extended family of GOMS models (John & Kieras, 1994; chap. 4). These can be used to predict time on task, successful task completion, and the effect of errors on time performance. Other notations, whether specifically designed for evaluation or not, are commonly used to design empirical usability tests, either experimental or more naturalistic. (See Paternò, 1999, for a description of various model-based usability evaluation techniques.)

22.2.2 Axis 2: Usability for Communication

Although task-modeling notations are usually not designed explicitly for this purpose, our own experience (e.g., Paris et al., 2000) as well as that of other sources (e.g., van Welie, van der Veer, & Koster, 2000) shows that the communicative function of task models is one of their major values: In practice, task models often act as a lingua franca among all or a subset of the parties involved in and impacted by the development and introduction of an application. In general, communication throughout the SDLC is a crucial and frequent activity that involves people from widely different backgrounds: software architects, user interface designers and evaluators, graphic artists, implementers and testers, technical writers, end users, and various stakeholders (see also chaps. 8, 9, & 11). The main communication activities in which communication is essential are as follows:

- illustration and validation of user requirements
- definition of the vocabulary to be used in the user interface and in the system's documentation
- validation of feasibility
- design and validation of the system's documentation
- integration of the usability evaluation into the system development life cycle.

The usability of a notation covers two distinct notions, and hence two axes in our taxonomy are related to it. The first notion concerns the ease with which a task model can be read and understood for the purposes of communicating the design to participants of the SDLC other than the task-modeling team. Here, the notion of usability integrates the notions of learnability and readability. The second notion concerns the ease with which task models are generated, maintained, and manipulated (modified) by the task-modeling team (c/f axis 3).

To support effectively one or several of these, a notation must typically be easy to read and quick to be learned by neophytes (that is, people with no background in task modeling).

Obviously, graphical, semiformal notations are more suited to this purpose than are formal notations. (See Ozkan, Paris, & Balbo, 1998, for an experimental study of a graphical notation's usability.)

22.2.3 Axis 3: Usability for Modeling Tasks

Another related but less frequent concern regarding usability is the degree of ease with which task models are generated and manipulated. This concern is relevant where a notation is to be used strictly by the design and development team (e.g., for automatic generation purposes), when there are tight deadlines, or when there is high staff turnover. In these cases, efficient communication is crucial, but the notation's usability must be high for *professionals*. In addition, the notation should, ideally, be commonly used and abundantly documented and illustrated.

22.2.4 Axis 4: Adaptability

Adaptability is the characteristic of being easily modifiable to fit a new situation when necessary. It is obviously important when applying task models in novel and varied situations or using them in innovative ways. We distinguish three types:

1. *adaptability to new aims* (e.g., can a task notation originally intended for the design phase be employed for evaluation?)
2. *adaptability to new types of systems.* (This applies to notations that are specialized. Notations have been developed for specific types of systems, such as real-time and simulation systems. Can such notations be successfully and easily applied to other types of systems?)
3. *adaptability to new systems requirements*, (e.g., a new requirement for some types of applications is that they must be deployed on several platforms—for instance, hand held device—and PC with similar or at least compatible user interfaces [see chaps. 24 and 26]. Can the notation accommodate multiple platforms?)

Adaptability is obviously linked to "expressive power": one may decide to choose a notation that is limited in expressive power but flexible enough to be easily adapted.

22.2.5 Axis 5: Coverage or Expressive Power

Coverage is the breadth of what the notation can express. Specific task attributes that may be important to represent (depending on the application) are these:

- *optionality*. Whether a task is mandatory or optional.
- *task agency*. Who must perform the task: the user, the system, or both?
- *levels of decomposition*. Is decomposition allowed and is there a limit to the permitted number of levels of decomposition?
- *temporal relationships among tasks*. These include synchronization of actions (sequencing, alternatives, composition, and iteration), parallelism (interleaved or true parallelism), and delays.
- *nonstandard actions*, such as actions involved in the management of errors of interface feedback to the end user. For example, a notation used for the automatic generation of online help should allow the representation of system feedback upon a user action.

Coverage is obviously linked to "extensibility." A notation that does not meet some particular coverage requirement can still be adequate provided its coverage can be extended.

TABLE 22.1
The Taxonomy, Its Axes, and Their Values

Axis	*Values*
1. Goal	Discover and define, design, Develop and Deploy, evaluate and predict
2. Usability for communication	High, medium, low
3. Usability for modeling	High, medium, low
4. Adaptability	High, medium, low
5. Coverage	High, Medium, low
Optionality	*Yes, no*
Agency	*Yes, no*
Decomposition	*Yes, no*
Temporal	*Yes, no*
Nonstandard	*Yes, no*
6. Extensibility	High, low

Note. The elements that may help in setting a value for the coverage axis are in italics.

22.2.6 Axis 6: Extensibility

Linked to coverage and to adaptability is the ease with which a notation's coverage can be extended. Often, a notation used in a situation or for a purpose other than that originally intended will require some adaptations to extend its coverage. The choice of a notation, then, depends on whether extensions can be made easily or not. For example, Diane+ (chap. 26), when used for the automatic generation of system documentation, was easily augmented in order to represent notions specific to this domain (Paris et al., 2000). In general, it is a safe rule of thumb that graphical, semiformal notations can easily be extended to express relationships or concepts not originally included. Other notations, such as the more formal ones, are more difficult to extend.

22.2.7 Synthesis

We have presented our taxonomy by discussing its various axes. Table 22.1 summarizes the six axes, and the values for each. As we have emphasized, these axes are not independent. In fact, there are many interdependencies between them. The taxonomy cannot be used to systematically reason about notations. Formal reasoning is not our purpose here. As stated above, the aim of the taxonomy is to serve as a repository of current task-modeling practices. In a sense, it is a "living" entity that should be updated regularly so that it remains a useful resource for both researchers and practitioners.

22.3 EXAMPLES OF TASK-MODELING NOTATIONS

In the previous section, we described the various characteristics that one could seek when choosing a task-modeling notation and various attributes that a task-modeling notation could cover. In this section, we discuss six well-known and widely used notations, examining which characteristics they exhibit and which attributes they cover. This section thus illuminates the meaning of each attribute by means of concrete examples and enables the reader to understand how well-known notations fit into the taxonomy.

It is important to realize that, in some cases, the way we "rate" a notation is subjective and reflects our experience. For example, although it is possible to state whether a notation allows for the representation of task decomposition, it would be hard to state that a notation universally has the value "high" for adaptability. This is a value judgment, and our value judgments will be based on our experience.

We review the following notations:

- Use Cases and scenarios (Constantine & Lockwood, 1999; Jacobson, Christerson, Jonsson, & Övergaard, 1992; chap. 5).
- MAD (Rodriguez & Scapin, 1997; Scapin & Bastien, 2001; Sebillotte, 1995; chaps. 6, 22, and 24).
- Diane+ (Tarby & Barthet, 1996, 2001; chaps. 6, 25, and 26).
- UAN (Hix & Hartson, 1993; chaps. 23, 24, and 27).
- Site maps and functional flows as used for the development of Web sites at Modem Media (www.modernmedia.com), an international online advertising and web agency.
- KLM and CPM-GOMS (Card, Moran, & Newell, 1983; John & Kieras, 1994; chap. 4).

We have personally employed these notations, either in industry or with industrial partners, with the exception of KLM and CPM-GOMS.

Each notation is illustrated with the same example: an extract of the task "booking a flight" as defined on the www.Go-fly.com Web site. For the sake of simplicity—and to be able to go straight to the points we wish to illustrate—we only keep a subset of all the actions that are required. In particular, for the step corresponding to the selection of the flight, we only represent the selection of the departure and arrival airports. Although this example is simple, the task still requires the representation of the following:

- a hierarchy of tasks
- both user and system actions
- sequences of tasks and actions
- parallel actions (e.g., all the parameters for the flight besides the departure and arrival airports can be entered in parallel).

Note that the sample task is a simple and short subtask. Consequently, the task-modeling notations will be used only to model behavior, not the associated psychological processes, although all the notations illustrated here can be used for this purpose as well.

In each of the subsections below, we briefly introduce the notation, then describe the task of booking a flight as represented by the notation, and finally discuss the attributes covered by the notation.

22.3.1 Use Cases and Scenarios

A *use case* is a narrative description of the interaction between a user (in some user role) and a system. The case describes falls within the specific intended use of the system and is directed toward a goal that is meaningful to the user. Although Jacobson et al., (1992) define use cases as tools to specify all the functionalities of a system, they are meant to be informal and imprecise. Consequently, their main use is in the discovery and definition phase of the SDLC. In the later versions of the Unified Modeling Language (UML), use cases take a nonnarrative form

(Rumbaugh, Jacobson, & Booch, 1999). In UML, uses cases are used primarily to capture the high-level user-functional requirements of a system (Kenworthy, 2002), and this process again corresponds to the first phase of the SDLC. Another nontextual type of use case has been developed by Constantine and Lockwood (1999): the *essential use case*. Use cases of this type are designed to highlight more explicitly the interaction itself. We use essential use cases to illustrate our example below.

First, however, we briefly describe *scenarios* in order to establish their similarities to use cases. A scenario is a particular instantiation of a use case, and many scenarios will be extracted from one particular use case. Scenarios are also mainly represented using narrative descriptions. They describe the setting, the actor(s), and the events of a user-computer interaction. In addition, they include information about the user's mental activities (e.g., goals, plans, and reactions). There is generally no notation associated with them and no formal approach to deal with them, although the approach proposed by Rosson and Carroll (2001; chap. 5) restricts the type of narratives one can write by requesting a specific set of stories corresponding to the elements one is trying to elucidate. For example, to analyze requirements, one should have *problem scenarios*, which tell a story of current practice; to envision new designs, *activity scenarios*, which specify functionalities, should be written. *Information scenarios* provide details about the information that the system will provide to the user. Scenarios of yet another type, *interaction scenarios*, describe user actions and system feedback. Generally, scenarios are realistic and rich in details.

In contrast, use cases, in particular, essential use cases, extract the basics of the interaction capturing the essence of a subset of scenarios. An essential use case, as introduced by Constantine and Lockwood (1999), represents a streamlined version of a use case. The notion of a *use case map* is also defined. This map interrelates the set of use cases for one development. We illustrate this concept in the example below. Fig. 22.1 presents the use case map, and Tables 22.2 and 22.3 show the essential use cases for the task of booking a flight.

In a use case map, use cases can be related in various ways. For example, in Fig. 22.1 there are two types of relations between use cases: extensions (links labeled *extends*) and composition (links labeled *uses*). The "select other parameters" use case extends "find the right flight," as it expresses optional tasks that are to be run in parallel to the mandatory task of "find the right flight." The composition link indicates that a use case is a component of another use case, so "choose a flight" is a component, or sub-use case, of "find the right flight." In Tables 22.2 and 22.3, composition is indicated by a ">" symbol in front of the sub-use case.

Use cases, scenarios, and even essential use cases are fairly informal ways to perform a task analysis. Consequently, they are mainly used in the discovery and definition phase of the

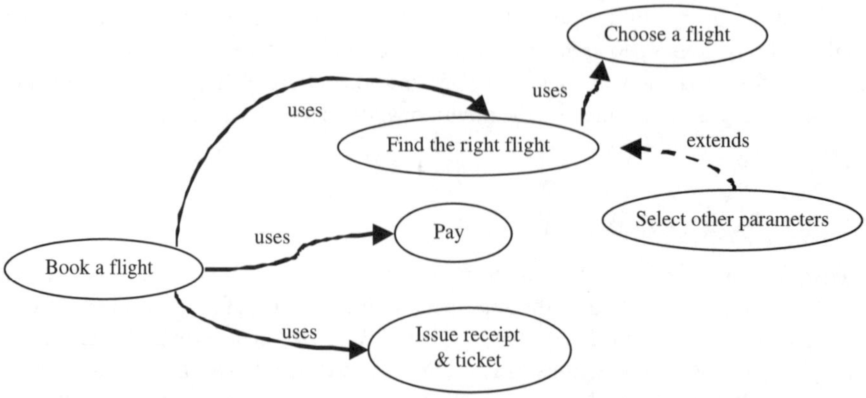

FIG. 22.1. Use Case map extract for Booking a flight.

TABLE 22.2
Use Case for Booking a Flight

User Intention	*System Responsibility*
> Find the right flight > Pay	> Issue receipt & ticket

TABLE 22.3
Use Case for Finding the Right Flight

User Intention	*System Responsibility*
Select departure airport Select return airport > Choose a flight	Update return airport list Flight selected

SDLC to provide a rich setting for the later stages of the design process. Essential use cases may be also employed in designing the interface, as they provide a streamlined specification of the interaction between the user and the system. Although they are useful for facilitating communication among all the parties involved (everyone understands a narrative), care must be taken: Because language is imprecise, it is hard to be sure that all parties interpret the same textual narrative the same way.

Because uses cases and scenarios are textual narratives, they are adaptable and can describe what needs to be described. Similarly, their expressive power is quite large, once again because they are written in a rich representation (that is, language). On the other hand, their rich means of expression decreases their formality and ability to be exact and unambiguous. For example, it may actually be quite awkward to express unambiguously various formal relationships between actions (e.g., whether a task may be interrupted or not). Clearly, use cases and scenarios have no "generative power"—no elements of the final system can be generated automatically.

22.3.2 Méthode Analytique de Description (MAD)

MAD (Rodriguez & Scapin, 1997; Scapin & Bastien, 2001; Sebillotte, 1995; see also chaps. 6 and 24) is both a task notation and a methodology for the acquisition of task knowledge based on extensive interviews with end users.

In MAD, tasks are decomposed hierarchically and linked by temporal and logical relations in an associated graphical representation. A task or subtask is represented as a tree whose nodes are tasks, subtasks, or elementary actions. Each task, subtask, or elementary action is characterized by its name (which appears in the tree) along with additional information recorded either on the tree itself or in separate text tables. The additional information states the task's goal, its initial and final states, pre- and postconditions, and some attributes. These attributes indicate whether the task is iterative, whether it is optional or mandatory, whether it is a priority task (that is, a task allowed to interrupt other tasks), and whether it is interruptible.

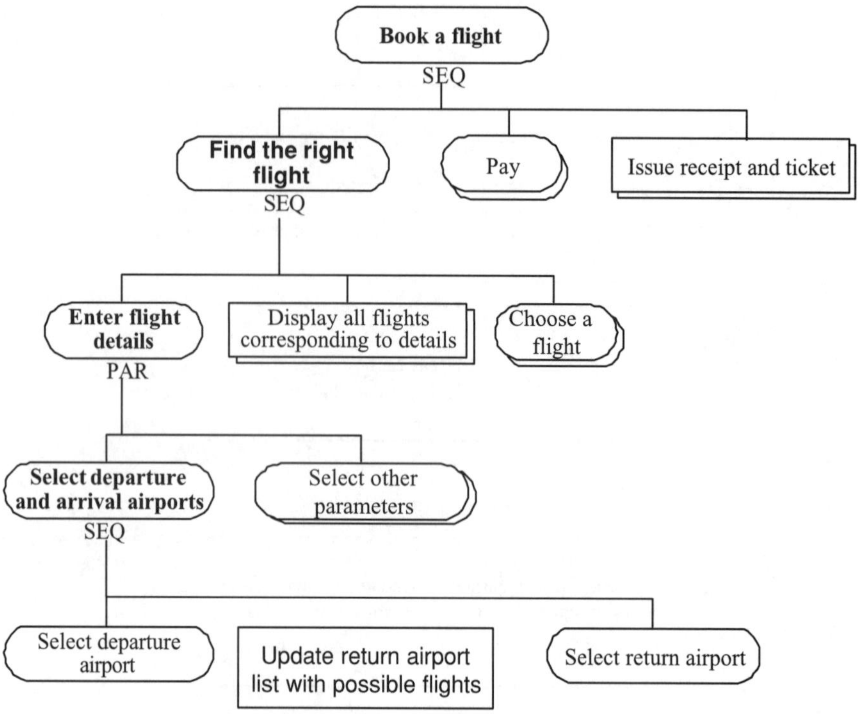

FIG. 22.2. Extract of the MAD description.

A representation of our sample task using the MAD notation is shown in Fig. 22.2. To "book a flight," the user must first "find the right flight," then "pay." The system will then "issue receipt and ticket." System actions are indicated by a rectangular box. These three actions are sequential, as indicated by the annotation "SEQ." Figure 22.2 shows several levels of decomposition, reflecting the hierarchy of tasks and subtasks. Some of the tasks are decomposed elsewhere (that is, outside this diagram), indicated by three-dimensionality. In Fig. 22.2, "pay," "issue receipt and ticket," and "choose a flight" are decomposable tasks whose decomposition is not shown here. Only task names are included for the sake of simplicity, not their associated notes, which typically would appear in the separate text tables.

MAD stipulates that end user task analysis and description contributes to user requirements analysis and is thus intended to be used at the beginning of the SDLC, that is, in the discovery, and definition phase. However, MAD task descriptions can also be exploited in later phases, such as in the design of the user interface and of the help content, as described for the APLCRATES project in Balbo and Lindley (1997).

With respect to our taxonomy, then, MAD can be used for the first two values of "goal." Its appropriateness for the "development and deployment" and "evaluation" phases is not as clear. In the development and deployment" phase, MAD could be used to identify and specify the generic functions of the system. It could not be used to generate automatically any part of the final system. Finally, although a MAD description can be manually used in evaluation to compare actual usage to intended usage (as specified in the model), its use in evaluation is not formal or systematic.

Our experience shows MAD to be a very good tool for communication, partly because of its semiformality. The mixture of graphical notation and textual description is easy to understand and sufficiently precise to ensure common understanding. In fact, the notation played a major communication role in a project in which the development team had to address decision support system development, database definition, online help, and user manual definition in conjunction with the end users and the clients (Balbo & Lindley, 1997; Paris et al., 2000).

In the context of the APLCRATES project, the MAD notation was used for two purposes: to help in the design of the user interface and the documentation, and to represent the expertise extracted from the domain experts. Because MAD had not originally been designed for this letter purpose, we had to extend and adapt the notation. We encountered no major difficulties in doing so. We thus found the notation adaptable to this new situation and extensible to fit the new requirements. On the other hand, in the Isolde project, where we wanted to exploit task models for the automatic generation of online help, we preferred using Diane+ instead of MAD because of the MAD's low level of formality, which limits its generative power (as a lot of information is still represented in free text).

In terms of coverage, MAD allows for the representation of a hierarchy without any restrictions on the number of levels allowed. The attribute "agency" (that is, whether an action is to be performed by the user or the system) can be represented in a limited way: MAD cannot represent the notion that an action may need to be performed by both agents. MAD can represent some of the temporal relationships among actions. It cannot, however, represent formally, as part of its diagrammatic notation, the number of iterations (i.e., this task is to be repeated *n* number of times). This information can be expressed, but it has to be done through the textual descriptions attached to each task. Because MAD is a semiformal notation that employs both graphics and text, it combines formal coverage (through the graphical notation) and informal coverage (through text), and its extensibility is high, but as more information is represented in textual form, its generative ability diminishes, and there is an increased risk of ambiguity.

22.3.3 Diane+

Diane+ (Tarby & Barthet, 1996, 2001; chaps. 6, 25, and 26) is a descendent of Diane (Barthet, 1988). Diane+ similar to MAD in that it is also to be used during the user requirement analysis phase and can guide the design of the user interface. Like MAD, it uses a graphical notation to represent task decomposition as well as temporal and logical relationships among tasks. In Diane+, though, more information is *explicitly* and *formally* represented. For example, a Diane+ diagram explicitly indicates whether a task is to be accomplished by the end user, the system, or a combination of both. Diane+ also allows for the formal representation within the diagram of concepts such as iteration and optionality. Figures 22.3 and 22.4 show the Diane+ version of our flight booking example. The top-level functionalities are shown, as well as the "select departure and arrival airports" function.

In Diane+ diagrams, a task is represented by a box containing the name of the task (usually quite short) and, when appropriate, the constraints on the number of times the task can be

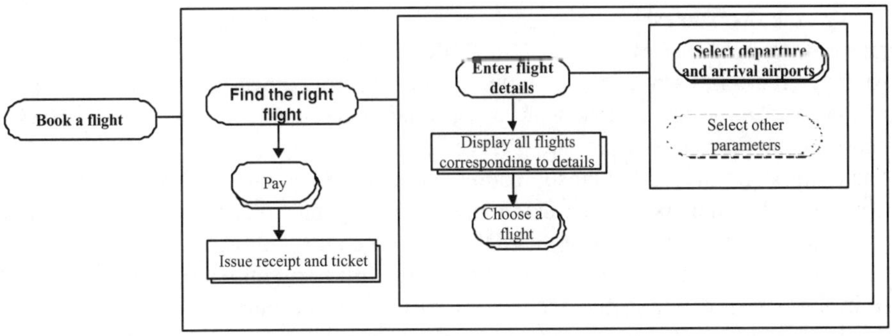

FIG. 22.3. Description of Flight Booking example using DIANE+.

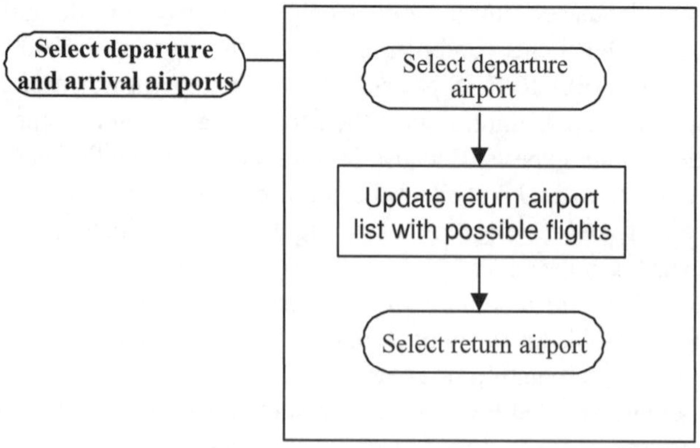

FIG. 22.4. The task "Select dep. & ar. airports" described in DIANE+.

executed (to represent iteration). These constraints are expressed by two integers (i, j) where i is the minimum number of times the task can be executed and j is the maximum. Default values are (0, n) for optional tasks and (1, n) for mandatory tasks. In the original Diane+, another 2-tuple of integers is associated with the task to express the number of iterations on a task's subtasks. In our work with Diane+, although we kept the iteration constraint on a task, we discarded the ability to express the number of iterations on a task's subtasks, as it seemed confusing and subtasks iterations could be expressed in alternate ways. The decomposition of a task is expressed in the big rectangle linked to that task. Like MAD, Diane+ does not impose any restrictions on the number of levels that can exist in a decomposition hierarchy.

Diane+ represents optionality explicitly: A box drawn with a full line indicates a mandatory task, and a dashed line indicates an optional task. The shape of the box represents agency: An elliptical box indicates that the task is purely manual (i.e., performed by the end user); a rectangular box indicates an automatic task (i.e., performed by the system); and, finally, a box with rounded ends indicates an interactive task (i.e., performed by the end user together with the system). In Figs. 22.3 and 22.4, all tasks shown are either interactive or automatic. As in MAD, three-dimensionality indicates that the task is to be decomposed elsewhere. For example, "select departure and arrival airport" is shown as decomposable in Fig. 22.3, and its decomposition is given in Fig. 22.4.

Task preconditions, system feedback to a task, and task termination can also be indicated explicitly in a diagram, although none are shown in our example.

Diane+ allows for the representation of temporal relations among actions such as sequencing and parallelism. The sequence is represented using the arrow symbol. When tasks can be done in parallel, they are shown in the same decomposition box but without any arrow between them (e.g., the two subtasks composing "enter flight details" in Fig. 22.3). Diane+ can also represent Boolean connectors such as OR, XOR, and AND.

Although it too is a semiformal notation, Diane+ is more powerful in its formal representation than MAD. In some situations, its augmented representational power can be crucial, as demonstrated in the Isolde case study (Paris et al., 2000).

The discussion thus far has mostly addressed the coverage provided by Diane+. We now turn briefly to the other axes of our taxonomy. Originally, the notation was developed mostly to be used in the first two phases of the SDLC. We have ourselves used it extensively in these phases. We note in particular our use of the notation (a) during the discovery and definition stage in the design of innovative technology to identify the place of new technology in a working environment (Paris et al., 1998a) and (b) during the design phase of system design to develop a set of functionalities to be consistent across applications (UICard Project, 2001).

It has since been advocated for use during the remaining phases of the design cycle (Paris et al., 2001). We have illustrated its use during the development and deployment phase of the SDLC by automatically producing from task models written in the notation excerpts of the online help to be associated with a system (a major aspect of the Isolde project), thus showing that it could help achieve one of the aims of that phase of the SDLC. We also found that the models used in the UICard project indeed helped the software engineers derive generic functions of the systems to be implemented and helped them structure their code.

Our experience shows that Diane+, like MAD, can be used as a communication tool. We exploited the notation in the UICard project as a tool to convey to software designers the functionalities required. We also performed an experiment to test the usability of the notation with technical writers (Ozkan, Paris, & Balbo, 1998). This experiment tested the usability of Diane+ for communication (can the technical writers understand the models?) and for modeling (can the technical writers produce Diane+ models?). Finally, we found that we were able to both extend and adapt the notation for our purposes, namely, the automatic generation of online help, without problems.

22.3.4 User Action Notation (UAN)

Like MAD, Diane+, and use cases to some extent, User Action Notation (UAN; Hix & Hartson, 1993; chaps. 23, 24, and 27) is based on hierarchical task decomposition. UAN was created to answer the need to formalize the communication between the user interface designers and the development team. To this end, UAN concentrates on providing a way to represent the user interface specifications of an interactive system—more precisely, to represent what users will see and how they will interact with the system. UAN was designed to be used during user requirement analysis as well as during user interface design.

The notation for representing that decomposition is not graphical but tabular, similar to essential use cases. At its highest levels (see Tables 22.4 to 22.7), the notation allows for the representation of task decomposition and the temporal relationships that may exist among subtasks (sequence, iteration, repeating choice, concurrence, etc.). Sequencing is represented vertically, in the absence of any other symbol, while the "||" symbol refers to the fact that the two subtasks can be executed in parallel. Table 22.4 describes the sequence of two tasks composing the high-level task "book a flight."

TABLE 22.4
Decomposition of "Book a Flight"

Task: Book a flight (Find the right flight   Pay)

TABLE 22.5
Decomposition of "Find the Right Flight"

Task: Find the right flight (Enter flight details   Choose a flight)

TABLE 22.6
Decomposition of "Enter Flight Details"

Task: Enter flight details (Select Dep.&Arr. airports ||Select other)

TABLE 22.7
Decomposition of "Select Departure and Arrival Airports"

Task: Select Dep.&Arr. airports (Select departure airport Select arrival airport)

TABLE 22.8
Decomposition of "Select Departure Airport"

Task: Select departure airport		
User Actions	*Interface Feedback*	*Interface State*
~(Pulldown departure airports)	Airports pulldown appears	
Mv		
~(Dep. airport pulldown item)	Airport name is highlighted	
M^	Departure airport remains in pulldown field	Departure airport is selected and arrival airport list is updated

At the elementary action level, the notation uses a three-column table. Each column represents an aspect of the task to be modeled. In the left column, we find the (temporal) sequence of physical actions to be performed by the end user in order to execute the task (using a specific syntax). The middle column describes the feedback from the system on the execution of the end user's actions. Finally, the right column records how the corresponding system variables change as a result of the actions described in the first column.

Table 22.8 shows an elementary action–level decomposition of the task "select departure airport." To accomplish this task, the end user must move the mouse onto the pull-down menu "departure airports" ("~(Pulldown departure airports)"), and then click on the mouse ("*Mv*"). As a result of this user action, the system displays the departure airport list ("*Airports pulldown appears*"). Moving the mouse on top of an item in the list (*~(dep. airport pulldown item)*") triggers the system to highlight that particular item, as shown in the interface feedback column. Then the end user must stop the cursor on top of the desired airport and release the mouse "*M^*". At this point, the system variable is set to the selected departure airport, and the arrival airport list is updated. These are system internal operations.

This level of description can be seen as very low level, as it represents the sequence of primitive actions that the end user performs and how the system reacts. It is, however, a detailed, systematic, and nonambiguous description of the interface, of how the user is to interact with the system, and of the effects of the interactions on the underlying system. It thus links user actions with internal system actions.

UAN is closer to specific software specification (for the user interface, at least) than MAD or Diane+ are. It was designed as a tool to communicate the interface requirements to software specialists, not as a means of communicating with end users. As a consequence, its communication and modeling usability capacity is lower than notations like Diane+ and MAD. The formalism used to describe the interaction makes it less readily understandable to the novice. UAN enables, via its table format, decomposition and the representation of agency and temporal relationship; to a certain extent, allows nonstandard coverage; but does not provide for formal representation of the optional character of a task. Analysts need some experience in using UAN to master the notation, although, like MAD and Diane+, because of its semiformality, the notation is easily extendable to meet new needs.

22.3.5 Functional Flows (FFlows)

Functional Flows (FFlows) are used in the design of Web sites. As a general rule, FFlows are developed by HCI experts in collaboration with the other members of the team. Once the high-level interaction between pages is defined by the site map (which shows the organization of the main pages of a site), FFlows are used to kick-start and share the detailed understanding and representation of the specific behavior of pages and pop-ups. An individual FFlow is a fine-grained representation of the interaction between a well-defined subset of pages and pop-ups. Figure 22.5 shows the various components of a FFlow, and Fig. 22.6 shows an extract from the FFlow for booking a flight.

All pages and pop-ups are present in the FFlows, but only some of the user actions and system actions are represented. The user and system actions that are represented are actions that trigger the loading of another page or pop-up.

Like the elementary action level of UAN, FFlows are designed once the interaction choices have been made. They present the detail of the behavior of a well-defined subset of pages within a Web site. With regard to our taxonomy, FFlows are of most use in the "design" and "develop and deploy" phases, especially the latter. FFlows are used to define and organize the fine-grained interaction between the various functionalities provided on a Web site. They can also assist in validating and testing a Web site while it is under construction, in the same way as use cases and scenarios, although these are less helpful for validation and testing because, as textual narratives, they are less formal and more likely to be interpreted in different ways.

Because FFlows specify in detail the behavior of the pages within a site, they support communication between all parties involved in the design of the site, from the stakeholders to the developers. The graphical nature of the representation makes FFlows robust communication tools.

We use FFlows is to represent Web interaction, and for this purpose we employ a limited number of graphical elements suited for such interaction. The adaptability of the notation for

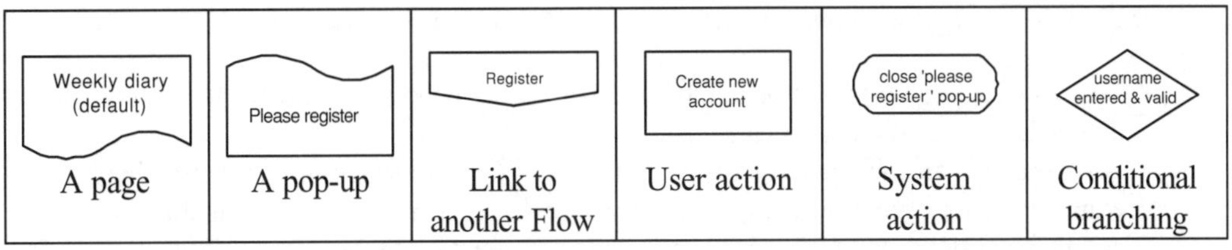

FIG. 22.5. Graphical elements of a FFlow.

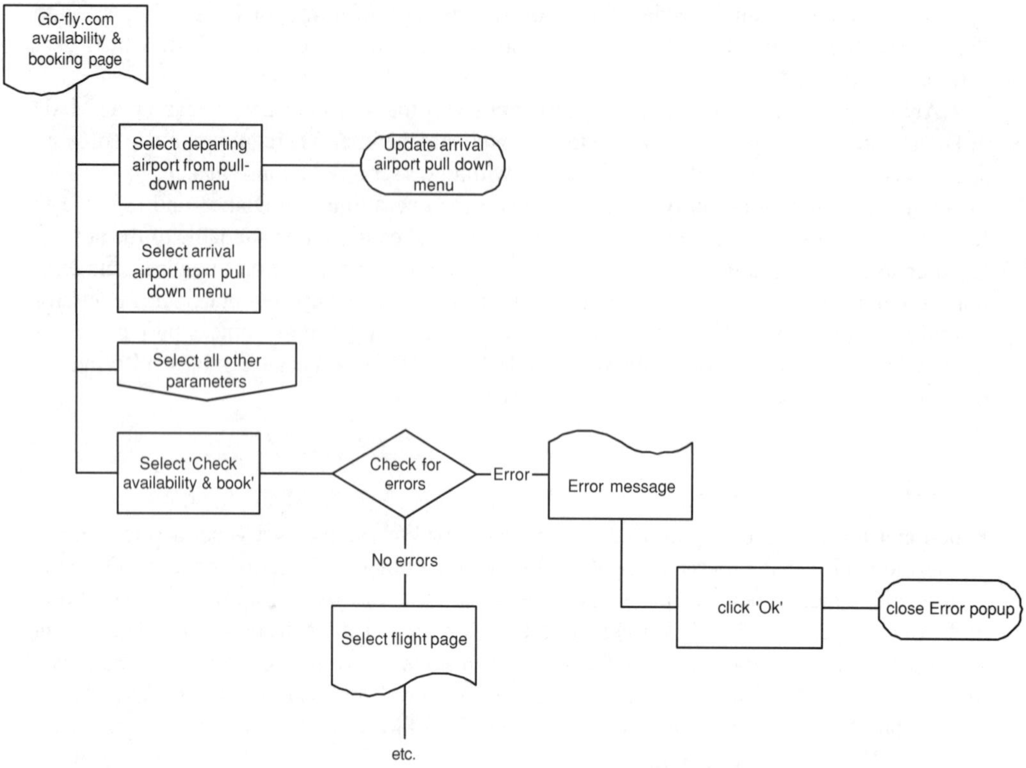

FIG. 22.6. An extract of the FFlow for Booking a Flight.

other applications is unclear. In this environment, the coverage is low, as there is no need to represent optionality or decomposition.

22.3.6 The KLM and CPM Models of the GOMS Family

We present below brief descriptions of two of the GOMS (Goal, Operator, Method, Selection) family of models (chap. 4). We also briefly illustrate CPM-GOMS (John & Kieras, 1994), which is based on the Model Human Processor. The "CPM" indicates this notation models cognitive, perceptive, and motor tasks.

At the keystroke level, GOMS is used as a quantitative tool for predicting performance. In contrast, all the notations discussed previously are used mostly during the initial stages of the SDLC (e.g., MAD and the use cases) or to support development (e.g., UAN and FFlows). GOMS is designed to assist in creating models of the expected behavior of expert users. It provides four levels of abstractions, depending on the designer's goal: *tasks, functions, arguments*, and *physical actions*. At the keystroke level, there are only physical actions. The GOMS basic elements are *goals, operators, methods*, and the *selection rules*:

- A goal defines a state to be reached. It is realized through the execution of a set of operators and is decomposed hierarchically.
- An operator is an elementary action (not decomposed further) within the abstraction level considered. Its execution triggers a change both in the mental model of the end user and in the system. All operators have associated execution times inferred from empirical experiments, and it is through them that predictive evaluation is performed. Given a sequence of operators, it is possible to predict how long the goal will take to achieve.

Execution times for the operators at the keystroke level are listed in Table 22.9. All operators belonging to the same abstraction level have a similar execution time.
- A method describes the process to reach a goal. It represents user know-how and is expressed through decomposition of subgoals and operators.
- A selection rule is used when more than one method can be used to reach a goal. It provides a way to determine which method should be chosen.

Going back to our flight booking example, the selection of a departure airport from the pull-down list would be expressed as shown in Table 22.10.

The keystroke level model can be employed to guide the user interface design process by predicting the time required for a user to perform a task or by helping in comparing analyses to detect problems in software already designed.

Figure 22.7 shows the representation of our sample task using GOMS-CPM (we removed any mention of execution time). Unlike the other GOMS models, GOMS-CPM allows cognitive and perceptual tasks to be operators and recognizes that these operators do not need to be performed serially but can be done in parallel (as required by the task). Figure 22.7 shows the operators presented along a time axis, and the width of each box represents the time needed to perform the action. Several distinct processes may be performed in parallel, each contributing to the accomplishment of an individual goal. This GOMS modeling method is often called the *critical path method* because of its ability to show the precise subtasks that determine the duration of a task. In Fig. 22.7 the critical path is shown by means of the grayed boxes. Given such a model, GOMS-CPM predicts execution time based on an analysis of component activities.

TABLE 22.9
Keystroke Operators and Associate Execution Times

Operator	*Execution Time*
K (Keystroking): on the keyboard, for an expert	0.35 sec.
P (Pointing): using the mouse	1.03 sec. + variable
H (Homing): movement of the hand when leaving the mouse or keyboard	0.4 sec.
D (Drawing)	0.161 sec. + variable
M (Mental activity)	1.35 sec.
R (Response time): system response time	varies

Note: From *The GOMS Family of Analysis Techniques: Tools for Design and Evaluation* (p. 12), by B. John and D. Kieras, 1994. Reprinted with permission.

TABLE 22.10
KLM Model

Move cursor on top of departure airport pull-down list:	M + P
Click mouse down:	K
Response time to display pull-down menu:	R
Move cursor on top of desired airport:	M + P
Release mouse:	K

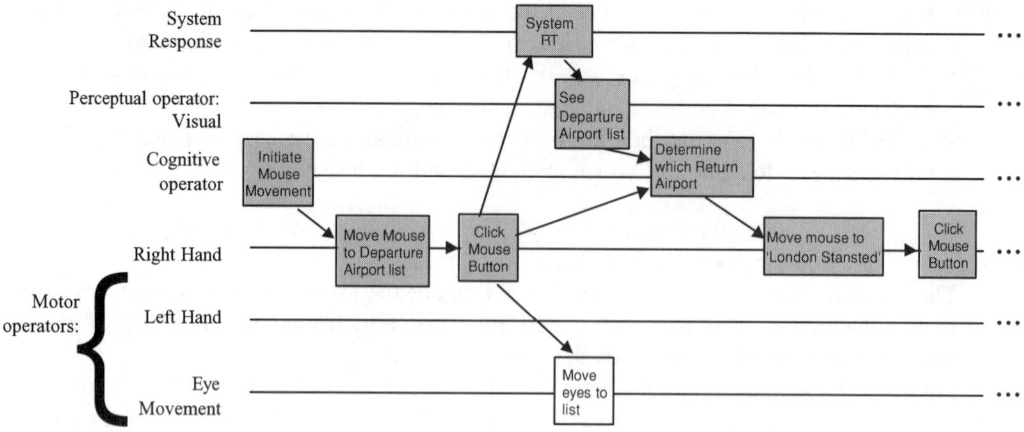

FIG. 22.7. GOMS-CPM model.

22.3.7 Using the Taxonomy to Compare Notations

The purpose of our taxonomy is to help compare and contrast task-modeling notations and identify their strengths. This taxonomy provides a basis on which to choose a task-modeling notation that supports a specific set of attributes and is best suited for a specific purpose. Table 22.11 shows how the notations we have discussed rate on the six axes of our taxonomy.

22.4 CONCLUSION

In this chapter we have proposed a taxonomy of task-modeling notations. This taxonomy helps in understanding and comparing the attributes of different notations in a systematic way—their strengths and limitations and appropriateness for use. The potential impact of the taxonomy is twofold. First, it can assist system designers and evaluators in choosing a notation suited to their purposes and in applying task modeling effectively. There is currently such a large number of task-modeling notations available that choosing among them can be time consuming and difficult to do. Second, the taxonomy can assist in understanding the global picture of task-modeling notations today and in determining where there is a need for new notations (e.g., to cover attributes currently unaddressed or only partly addressed).

We hope this taxonomy will continue to be a living entity and will evolve with time. As the taxonomy stands now, it is built from our experience and is hence subjective. It will obviously undergo changes as our experience widens and as others contribute to it. The taxonomy should be kept up to date and reflect current practice. That means it will need to accommodate new types of systems and evolving approaches to system design and evaluation. Our hope is that the taxonomy will eventually act as a repository of best practices that belongs to and evolves within the task-modeling community and contribute to its shaping and strengthening.

22.5 ACKNOWLEDGMENT

Cécile Paris gratefully acknowledges the support of the Office of Naval Research. Through Grant N00014-96-0465, it subsidized the Isolde project, which contributed to the analysis required to investigate the issues presented in this chapter, and also subsidized the writing of this chapter.

TABLE 22.11
Rating of Task-Modeling Notations

Axis	*Use Cases*	*MAD*	*Diane+*	*UAN*	*FFlows*	*GOMS*
1. Goal	**Discovery & Definition,** Design, Evaluation & Prediction	Discovery & Definition, Design	Discovery & Definition, Design, Development & Deployment	Discovery & Definition, **Design,** Development & Deployment	Design, **Development & Deployment,** Evaluation & Prediction	Discovery & Definition, **Evaluation & Prediction**
2. Usability for communication	High	High	High	Medium	High	Medium
3. Usability for modeling	High	High	High	Low	Medium	Low
4. Adaptability	High	High	High	Medium	(unclear)	Low
5. Coverage	High	Medium	High	Medium	Low	Medium
Optionality	Yes	No	Yes	No	No	No
Agency	Yes	Yes (partially)	Yes	Yes	Yes (partially)	No
Decomposition	Yes (variable)	Yes (variable)	Yes (variable)	Yes (variable)	No	Yes (fixed)
Temporal	Yes	Yes (partially)	Yes (partially)	Yes	Yes (partially)	Yes (partially)
Nonstandard	Yes	No	Yes	Yes	Yes	No
6. Extensibility	High	High	High	High	High	Medium

Note. The SDLC stage for which the notation was designed is set in boldface.

REFERENCES

Balbo, S. (1994). *Évaluation ergonomique des interfaces utilisateur: Un pas vers l'automatisation.* PhD doctoral dissertation, University of Grenoble.

Balbo, S., & Lindley, C. (1997). *Adaptation of a task analysis methodology to the design of a decision support system.* In *Proceedings of Interact'97.* Human-computer interaction, Interact'97 (pp. 355–361). London: Chapman & Hall.

Barthet, M-F. (1988). *Logiciels interactifs et ergonomie, modèles et méthodes de conception.* Editions Dunod Informatique. Paris.

Bardon, D., Berry, D., Bjerke, C. J., & Roberts, D. (2001). *Crafting the compelling user experience: Using a methodical software engineering approach to model users and design. Make IT Easy.* Available: *www-3.ibm.com/ibm/easy/eou_ext.nsf /Publish/1650.*

Brun, P., & Beaudouin-Lafon, M. (1995). A taxonomy and evaluation of formalisms for the specification of interactive systems. In the *Proceedings of HCI'95* conference on people & computers X. (pp. 197–212). Cambridge University Press.

Card, S. K., Moran, T. P., & Newell, A. (1983). *The psychology of human-computer interaction.* Hillsdale, NJ: Lawrence Erlbaum Associates.

Constantine, L., & Lockwood, L. (1999). *Software for use: A practical guide to the models and methods of user-centered design.* Reading, MA: Addison-Wesley; New York: ACM Press.

Hix, D., & Hartson, R. (1993). *Developing user interfaces: Ensuring usability through product and process.* John Wiley & Sons. New York.

Jacobson, I., Christerson, W., Jonsson, P., & Övergaard, G. (1992). *Object-oriented software engineering: A use case driven approach.* Reading, MA: Addison-Wesley.

John, B., & Kieras, D. (1994). *The GOMS family of analysis techniques: Tools for design and evaluation* (Carnegie Mellon University School of Computer Science Technical Report No. CMU-CS-94-181). CMU University: Pittsburgh.

Kenworthy, E. (2002). *The object practitioner's guide.* Available: http://www.zoo.co.uk/~z0001039/PracGuides/

Ozkan, N., Paris, C., & Balbo, S. (1998). *Understanding a task model: An experiment.* In *People and computers XIII* (pp. 123–137). Sheffield, UK: Springer-Verlag.

Ozkan, N, Paris, C, & Simpson-Young, B. (1998). Towards an approach for novel design. In *Proceedings of OzCHI'98* (pp. 186–191).

Paris, C., Balbo, S., & Ozkan, N. (2000). Novel uses of task models: Two case studies. In JMC Schraagen, S. E. Chipman, & V. L. Shalin (Eds.). *Cognitive task analysis* (pp. 261–274). Mahwah, NJ: Lawrence Erlbaum Associates.

Paris, C., Ozkan, N., & Bonifacio, F. (1998a). The design of new technology for writing on-line help. In *People and computers XIII* (pp. 189–205). Sheffield, UK: Springer-Verlag.

Paris, C., Ozkan, N., & Bonifacio, F. (1998b). Novel help for on-Line help. In *Proceedings of the 16th Annual International Conference on Computer Documentation (SIGDOC '98)* (pp. 70–79). Association for Computing Machinery. ISBN: 158113004X.

Paris, C., Tarby, J.-C., & Vander Linden, K. (2001). A flexible environment for building task models. In People and Computer XV (pp. 313–330). Sheffield UK: Springer-Verlag.

Paternò, F. (1999). *Model-based design and evaluation of interactive applications.* Springer-Verlag. ISBN: 1-85233-155-0.

Rodriguez, F. G., & Scapin, D. L. (1997). Editing MAD* task descriptions for specifying user interfaces at both semantic and presentation levels. In M. D. Harrison & J. C. Torres (Eds.), *Proceedings of the Eurographics Workshop on Design, Specification, and Verification of Interactive Systems '97 (DSV-IS '97)* (pp. 193–208). Vienna: Springer-Verlag.

Rosson, M. B., & Carroll, J. M. (2001). *Usability engineering: Scenario-based development of human-computer interaction.* San Francisco: Morgan Kaufmann.

Rumbaugh, J., Jacobson, I., & Booch, G. (1999). *The Unified Modeling Language reference manual.* Reading, MA: Addison-Wesley.

Scapin, D., & Bastien, C. (2001). Analyse des tâches et aide ergonomique à la conception: L'approche MAD*. In C. Kolski (Ed.), *Interactions homme-machine pour les systèmes d'information: Tome 1. Analyse et conception de l'IHM* (pp. 85–112). Hermès. Toulouse-France.

Sebillotte, S. (1995). *Methodology guide to task analysis with the goal of extracting relevant characteristics for interfaces* (Technical report for Esprit project P6593). Rocquencourt, France: INRIA.

Tarby, J.-C., and Barthet, M.-F. (1996). The Diane+ method. In *Proceedings of the Second International Workshop on Computer-Aided Design of User Interfaces* (pp. 95–119). Namur-Belgium. Presse. Universitaire de Namur.

Tarby, J.-C., & Barthet, M-F. (2001). Analyse et modélisation des tâches dans la conception des systèmes d'information: Mise en œuvre avec la méthode Diane+. In C. Kolski (Ed.), *Interaction homme-machine pour les systèmes d'information: Tome 1. Analyse et conception de l'IHM* (pp. 117–141). Hermès. Toulouse-France.

Thévenin, D., & Coutaz, J. (1999). *Plasticity of user Interfaces: Framework and research agenda.* In M. A. Sasse & C. Johnson (Eds.), *Proceedings of the Interact '99* (pp. 110–117). ISO Press. Edinburgh. UK.

UICard project. (2002). A CISRA online project description. Available: www.research.canon.com.au/uicard/index.html.

van Welie, M., van der Veer, G., & Eliëns, A. (1998). An ontology for task world models. In *Proceedings of the 5th International Eurographics Workshop on Design Specification and Verification of Interactive Systems (DSV-IS '98)* (pp. 57–70). Available: http://www.cs.vu.nl/~gerrit/gta/.

van Welie, M., van der Veer, G., & Koster, A. (2000). Integrated representations for task modelling. In *Proceedings of the Tenth European Conference on Cognitive Ergonomics* (pp. 129–138). ISBN: 91-7219-828-1 EACE-Paris.

Vanderdonckt, J., & Berquin, P. (1999). Towards a very large model-based approach for user interface development. In N. W. Paton & T. Griffiths (Eds.), *Proceedings of the First International Workshop on User Interfaces to Data Intensive Systems* (UIDIS '99). (pp. 76–85). Los Alamitos, CA: IEEE Computer Society Press.

23

Environments for the Construction and Use of Task Models

Cécile Paris
CSIRO

Shijian Lu
CSIRO

Keith Vander Linden
Calvin College

Task models are useful because they support the design and evaluation of interactive systems from the point of view of the intended users. Unfortunately, they are also difficult to build. To help alleviate this problem, tools have been built that aid in the construction of task models in a variety of ways in the hope that designers will be encouraged to build explicit task models and reap the benefits of their use and reuse. This chapter will review the range of tools that have been built, and, as an example, will discuss ISOLDE, an environment that integrates an heterogeneous set of such tools and supports the construction, use, and reuse of task models.

23.1 INTRODUCTION

Task models are explicit representations of user tasks that can help support certain rigorous forms of task analysis. They are recognized as useful constructs that can be exploited throughout the software development life cycle (SDLC; chap. 22), either on their own or as components of full interface design models. For example, designers and implementers of interactive systems can use them to assess the complexity of the tasks that the users are expected to perform, anticipate the amount of time required to perform the tasks (Card, Moran, & Newell, 1983; chap. 4), simulate the behavior of the resulting system (e.g., Bomsdorf & Szwillus, 1996), automatically build interfaces based on the model (e.g., Puerta, 1996), and generate user-oriented instructions (e.g., Paris, Ozkan, & Bonifacio, 1998). In addition, task models can be used to facilitate communication between the various stakeholders in the design process (e.g., Balbo & Lindley, 1997; see also chaps. 8, 9, 11, & 22). Although there are a variety of task-modeling languages (chaps. 6 and 22), they can all support each of these goals to some degree.

Known to be useful, task models are also known to be difficult to build and maintain. This is, to some degree, true of informal, textual models but is particularly true of the formal or semiformal models required to support the benefits mentioned above. To support these benefits throughout the full SDLC, task models must be built at different levels of abstraction (chap. 22). Although it might not be difficult to model a set of anticipated user goals at a high level, it becomes progressively more difficult as more levels of detail are included. Representing the task model for a large interactive system all the way from the highest level user goals down to the keystroke level could be a daunting task indeed. In addition, there may be separate task models for different views of the interaction—one, say, from the user's point of view and another from the system's point of view. Finally, in those cases where models are actually built, it is likely that different notations will be used to represent different aspects of the interaction, thus rendering the resulting model incoherent and difficult to use fully. For example, high-level views of users and their goals may be recorded in one notation (e.g., use case diagrams), and low-level views of the explicit user interface actions may be recorded in another (e.g., UAN; chaps. 22, 24, and 27). Taken together, these issues tend to prevent explicit and coherent task models from being built and used extensively in practice (Paris, Tarby, & Vander Linden, 2001).

Given that task models are useful for analysis but difficult to build, it is desirable to provide tools that aid in their construction and consolidation. Task model editors, for example, can help a task analyst build, manipulate, and save explicit task models in some coherent modeling language. Such editors include features specifically designed to support task modeling and are thus easier to use than more general purpose graphics or text editors. Furthermore, they can produce a coherent representation of the various aspects of the task model, which could then be manipulated by other systems (e.g., an XML-based representation). As we will see in the next section, task model editors have indeed been built for a number of task-modeling languages.

It is clear, however, that although specialized task model editors might be easier to use than general-purpose editors, they still do not render the construction of formal or semiformal task models easy. The task analyst still has to construct the model from scratch. It would thus be helpful to provide additional tools that facilitate the extraction of task model information from other sources, say, from models or artifacts already built for other purposes during the SDLC or even from existing prototypes if they are available. Because no single source can provide all the input for the full task model, a set of extraction tools could be developed, each one focusing on a different knowledge format. An environment could then be built in which the output of the extraction tools could be consolidated and represented using a coherent task-modeling language.

This chapter argues that the construction and consolidation of task models capable of supporting the benefits listed above require the use of an extensible task-modeling environment, that this environment must be built on top of a uniform modeling language, and that it must integrate a task model editor with an expandable set of heterogeneous task extraction and construction tools. Such an environment would thus provide task analysts with a higher level interface to a task model editor while allowing them to leverage information sources already built for other purposes. The more powerful, flexible, and yet integrated the tools, the more likely designers are to build models and reap the benefits of using them. Having such an environment may also encourage a greater exploitation of task models, including for unforeseen innovative uses.

The chapter begins with a review of the range of tools that have been built to date, identifying the modeling language they support and their primary source of task knowledge. It then presents an extended discussion of Isolde, which is intended to serve as an example of the sort of environment being advocated.

23.2 A REVIEW OF TASK MODEL CONSTRUCTION AND EXTRACTION TOOLS

A number of task-modeling tools have been developed that facilitate the construction of task models, including dedicated editing tools and tools that extract task knowledge from other sources. Each of these tools is designed to represent the task model in a particular formal or semiformal modeling language. This section reviews the range of available tools, starting with task model editors. Table 23.1 summarizes this review.

The most common task-modeling tools are the specialized task model editors. These tools support the task analyst, in consultation with other stakeholders in the SDLC, in building task models with features tailored to a specific modeling language. Though different in many ways, these editors all include special features for entering and representing user tasks, rendering them more supportive for task modeling than more general purpose editors like MS Powerpoint, VISIO, or standard text editors. Such editors exist for several modeling languages, including CTT (Paternò, Mancini, & Manconi, 1997; Paternò, Mori, & Galimberti, 2001), Diane+ (Tarby & Barthet, 1996; chaps. 6, 22, 25, and 26), GTA (van der Veer, Lenting, & Bergevoet, 1996; chaps. 6, 7, and 24), GOMS (Baumeister, John, & Byrne, 2000; chap. 4), MIMIC (Puerta, 1996), and UAN (Hix & Hartson, 1994; chaps. 22, 24, and 27). The special features that they provide tend to center around a particular task-modeling language (Bentley & Johnston, 2001). For example, CAT-HCI, GOMSED, QGOMS, and GLEAN3 (see Table 23.1) support task models written in the GOMS language and thus provide facilities to enter GOMS attributes. Editors have also been designed for other related modeling languages, such as UML (Rumbaugh, Jacobson, & Booch, 1999), as well as Statecharts (Harel, 1987) and PetriNets (Navarre, Palanque, Bastide, & Sy, 2001). Table 23.1 lists of current task model editors together with the task-modeling language they support. This table also indicates larger environments

TABLE 23.1
A Summary of Task Model Editors and Extraction Tools

Construction Tool	*Description*	*Language*	*Environment*	*Reference*
CAT-HCI	Task editor	GOMS		Williams, 2000
CRITIQUE	Evaluation tool	GOMS		Hudson et al., 1999
CTTE	Task editor	CTT	CTTE	Paternò et al., 1997; 2001
EL-TM	Text analysis	CTT	CTTE	Paternò & Mancini, 1999
RemUSINE	Evaluation tool	CTT	CTTE	Paternò & Ballardin, 1999
Euterpe	Task editor	GTA		van Welie et al., 1998
GLEAN3	Task editor	GOMS		Kieras et al., 1995
GOMSED	Task editor	GOMS		Wandmacher, 1997
IMAD	Task editor	MAD	Alacie	Gamboa et al., 1997
Mobi-D Editors	Editor(s)	MIMIC	Mobi-D	Puerta, 1998
U-Tel	Text analysis	MIMIC	Mobi-D	Tam et al., 1998
QGOMS	Task editor	GOMS		Beard et al., 1996
Quantum	Task editor	UAN	Ideal	Hix & Hartson, 1994
TAMOT	Task editor	Diane+	Isolde	Paris et al., 2001
Isolde:U2T	UML to task	Diane+	Isolde	Lu et al., 1999
Isolde:UIR	Event recorder	Diane+	Isolde	Paris et al., 2001
Isolde:T2T	Text analysis	Diane+	Isolde	Brasser & Vander Linden, 2002
VTMB	Task editor			Biere et al., 1999

in which a task editor may be a component. As we will see below, these environments may contain additional tools to aid in the construction of task models.

With the aid of these task model editors, task analysts can build the models that support the range of practical uses in the SDLC discussed above, but they must do so manually. This is good in that it most likely forces the designer to wrestle with key issues in the design of an interface, but it is also clear, unfortunately, that building a complete task model for any reasonably complex device, even using lite-methods such as those described in chapters 8 and 9, is a daunting enough prospect to prevent designers from doing it in practice, supported by a task model editor or not. This is particularly the case with larger or more complex interactive devices. Researchers have, therefore, sought to develop tools that take advantage of other sources of information in the construction of a model. These sources, which include written task descriptions, other models of various sorts, and prototype demonstrations, will now be discussed in turn.

One commonly available knowledge source is a written task description, the most informal way of representing a task. These descriptions can come in a number of forms, including task scenarios and use cases (e.g., Constantine & Lockwood, 1999; Jacobson, Christerson, Jonsson, & Övergaard, 1992; Lewis & Rieman, 1994; chap. 5), instructional texts (Carroll, 1990), and other more informal descriptions. These descriptions provide the objects and actions of the domain of the application (knowledge that we term *domain knowledge*; see also chap. 1) and also specify the series of interactions between the intended user and the application. When present, therefore, they represent an important source of information from which to build a task model. The advantage of this source is that it is relatively easy to produce. The primary disadvantage is that natural language is too informal to be used rigorously. A task-modeling environment can, however, provide support for extracting more formal task information from written task descriptions. Table 23.1 calls these tools "text analysis" tools. Two of these tools, U-Tel (Tam, Maulsby, & Puerta, 1998) and EL-TM (Paternò & Mancini, 1999), support the task analysts as they *manually* identify the objects and actions in a task description and assemble them into tasks and task models in the task model editor. In this approach, the task analysts perform the text analysis. The third text analysis tool, Isolde:T2T (Brasser & Vander Linden, 2002), attempts to extract task model knowledge *automatically* from written text using a natural language parser to perform the text analysis.

Another viable source of task knowledge is a model represented in any one of a number of other modeling languages. For example, Universal Modeling Language (UML; Rumbaugh et al., 1999; chaps. 7, 11, 12, 19, 22, 24, and 27) is primarily a system-modeling language, but it also includes some task-oriented information, primarily in the use case diagrams and potentially in the scenario diagrams and state diagrams. It also includes some domain knowledge in its class diagrams. It is thus possible to extract domain and task-modeling knowledge from these diagrams. Table 23.1 labels such tools "UML to task." Isolde:U2T is an example of such a tool (Lu, Paris, & Vander Linden, 1999). These tools can support the construction of task models in situations where other models are created as part of the SDLC. Similarly, a task analyst may be able to consolidate task knowledge from models represented in other task-modeling languages (Limbourg, Pribeanu, & Vanderdonckt, 2001).

A final source of task knowledge is a system prototype, when one exists. Prototypes are especially useful in capturing low-level user actions. To extract these actions, the task analyst can hook an existing prototype up to any one of a number of user interface event recorders, and the recorder will then collect GUI events as potential users or designers use the prototype (Hilbert & Redmiles, 2000). Because this approach requires the existence of a running system, it has been more commonly applied to the evaluation of existing user interfaces. For example, RemUSINE (Paternò & Ballardin, 1999) uses an event recorder to support evaluation (see Table 23.1). It can, however, also be applied to the construction of task models provided that a prototype is available early enough in the design process. Isolde:UIR (Paris et al., 2001) is an example of such a tool (see Table 23.1). One key issue with these tools is linking the low-level interface events collected by the recorder with higher level user goals (Hoppe, 1988). This typically requires resorting to other knowledge sources and linking them together using a task editor.

Each of the potential sources of knowledge mentioned above provides different aspects of the required task knowledge and is appropriate in different contexts. For example, objects, actions, and high-level user goals may be obtained from a written task description or a UML use case diagram but not from event recorders. Conversely, low-level interface actions can be obtained more easily by event recorders. It is thus clear that no single source or tool will support the construction of all the elements of a complete task model at all levels of abstraction. Therefore, it is more appropriate to provide a task-modeling *environment* that integrates an extensible set of heterogeneous tools within a unified framework. This framework would most likely designate a selected task-modeling language that would represent information collected from any of the tools supported by the environment. Examples of this sort of environment include CTTE (Paternò et al., 1997, 2001; chap. 24), Mobi-D (Puerta, 1997), and Isolde (Paris et al., 2001) (see Table 23.1).

23.3 THE ISOLDE ENVIRONMENT

The previous sections advocated the use of an extensible, heterogeneous environment that provides specialized tools for extracting task information from a variety of sources. The environment should also provide a task model editor that allows a task analyst to integrate, refine, and extend the extracted knowledge within a common modeling format or still build the model from scratch if desired. As an example of this concept, this section presents the Isolde environment. Isolde's architecture, shown in Fig. 23.1, is centered around the Tamot task model editor for the Diane+ modeling language (Tarby & Barthet, 1996; chaps. 6, 22, 25 and 26) and its underlying XML-based representation language. Isolde includes tools to facilitate the construction of task models (shown on top of the figure as "Knowledge Acquisition Tools") and tools that exploit task models (shown at the bottom of the figure as "Analysis or Application Tools").

This section presents the various modules of this environment, concentrating on the tools that facilitate the construction of a task model: The task model editor (Tamot) and three acquisition

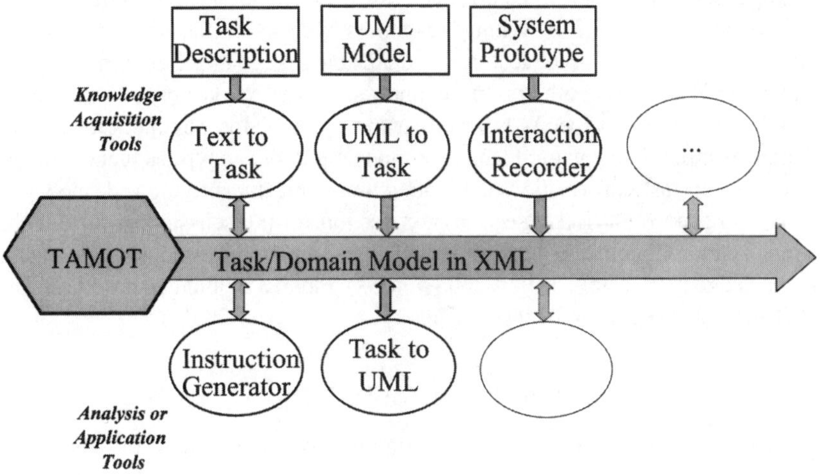

FIG. 23.1. The Isolde environment.

tools specific to the three knowledge sources mentioned above—written text (Text-to-Task, T2T), other models (UML-to-Task, U2T), and system prototypes (User Interaction Recorder, UIR). The tools are presented in the context of a running example, one showing the use of Tamot itself. Isolde currently also includes two application tools: the Instruction Generator and the Task-to-UML tool. These are described briefly, only as an illustration of how a task model can be exploited in a variety of ways once it has been built. Importantly, as shown in Fig. 23.1, the environment is extensible, both in terms of the acquisition tools and the application tools. The only requirement on an extraction tool is to produce a Diane+ model in Isolde's open, standardized XML-based representation. Similarly, an analysis or application tool must be able to take such a representation as input.

23.3.1 The Task Model Editor: Tamot

Tamot is a general-purpose task and domain model editor (Lu, Paris, & Vander Linden, 2002). Its task representation is based on the Diane+ modeling formalism, which supports a number of task attributes, such as task annotations (e.g., repetition and optionality) and procedural relationships (e.g., goals and sequences). Diane+ was originally designed to support the early phases of the SDLC (e.g., requirement analysis and design) but has recently been proposed for use in all stages of the SDLC (Paris et al., 2001).

Tamot is similar to other graphically based task editors (e.g., Beard, Smith, & Danelsbeck, 1996; Paternò et al., 2001; van Welie, van der Veer, & Eliëns, 1998). It enables users to construct a task model manually. Tamot has been used by task analysts at CSIRO, in commercial projects, and it has also been shown to be usable by technical writers (Ozkan, Paris, & Balbo, 1998). Within the Isolde environment, however, Tamot's primary role is to allow users to consolidate, modify, and extend the task knowledge produced by the extraction tools. Consequently, Tamot reads in and outputs files in Isolde's open, standardized XML-based representation.

Figure 23.2 shows Tamot editing a task model. As mentioned earlier, this chapter uses a task model for Tamot itself as an example. On the upper left, Tamot displays a hierarchy of user tasks. On the upper right, it displays a set of windows showing the graphical representation of the tasks at various points in the hierarchy. In this example, we see the high-level task of "design a file with Tamot," which is decomposed into several subtasks, one of which is creating a model. This subtask is further decomposed, and the figure also shows the sequence of primitive actions that occur when performing the task "Open an existing file": The user must choose the open menu item, select the file, and click the OK button. (User actions are shown as boxes with rounded corners.) The system's responses to these actions are shown as rectangles.

In addition to supporting all the features of the Diane+ language, Tamot includes a number of other features. It supports both bottom-up and top-down approaches to task modeling (see chap. 1) and allows users to specify tasks either through a dialog box mechanism or through natural language input. In the natural language mode, the user can type sequences of sentences that are parsed into task information and loaded automatically into the task model as tasks, annotations, and procedural links (see the next section for a discussion of this mechanism). Tamot's representation includes both task knowledge (tasks and procedural relationships) and domain knowledge (e.g., actors, actions, and objects). Finally, Tamot can produce a customizable HTML-based report from the task model.

23.3.2 The Text-to-Task Extraction Tool (T2T)

As mentioned earlier, written texts are a commonly available source of task and domain knowledge. The HCI community has produced tools that allow users to *manually* acquire knowledge from such descriptions (e.g., Paternò & Mancini, 1999; Tam et al., 1998). T2T,

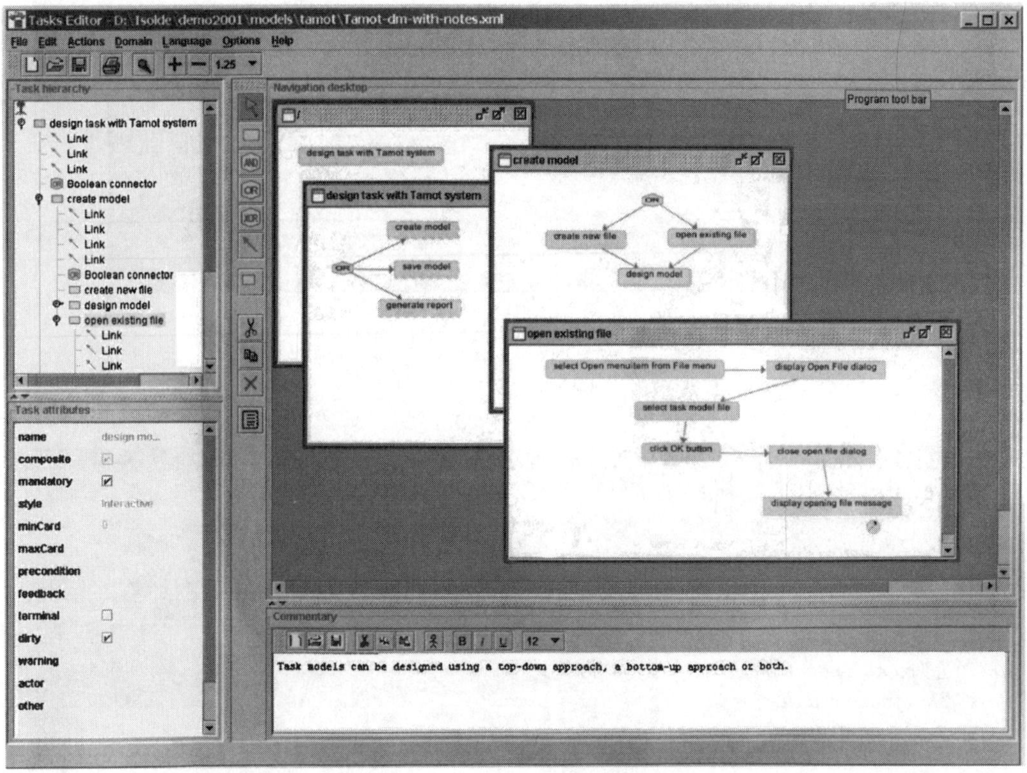

FIG. 23.2. The Tarmot Task Model Editor.

our text-to-task extraction tool, uses information extraction technology and natural language parsing to *automatically* acquire from the text knowledge that can then be reused to create a task model task (Brasser & Vander Linden, 2002). This is also the facility used in Tamot to acquire objects and actions from the task names.

The core of T2T is a finite-state grammar built as a 30-state augmented transition network (cf. Jurafsky & Martin, 2000). At the sentence level, this grammar identifies the basic elements mentioned in a text—the objects and actions of tasks and their relationships; in particular, *who* is doing the *action* to *what*—and thus provides information about the process (action), the actor (who), and the actee (what). The roles of source (*from* where), destination (*to* where), instrument (*with* what), and location (*where)* are identified as well. Any remaining elements of the sentence are left as canned strings (e.g., complements or adverbs). T2T distinguishes between nouns and verbs based on the classifications of WordNet (Fellbaum, 1998). Although it can work automatically, it also allows the user to modify any incorrectly assigned constituent boundaries or head-word assignments using an *interactive parsing tool*, shown in Figure 23.3. Here, T2T has identified the sentence's action ("select"), its object ("menu item"), and its source ("menu").

When there are expressions of multiple tasks, as will be the case for a written description (as opposed to a single task name), the grammar extracts the procedural relationships between them (e.g., sequence, preconditions, task or subtask relations). It then creates the appropriate domain knowledge, which includes representations of the actions and of their related objects. Figure 23.4 shows a sample instructional text and the task model that T2T derives from it. The system has extracted the purpose of the procedure ("open an existing task"), the sequence of

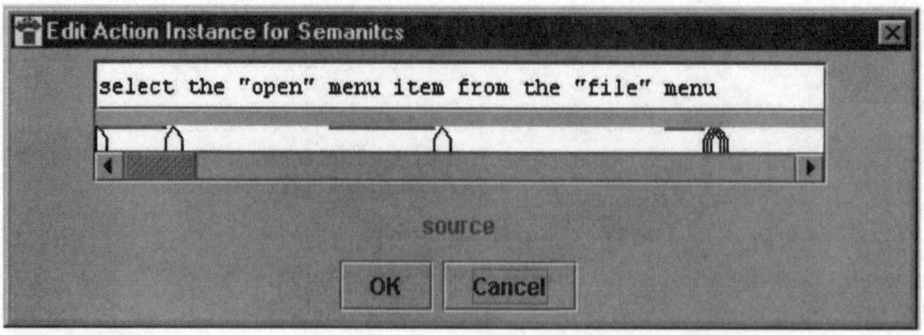

FIG. 23.3. The interactive T2T interface.

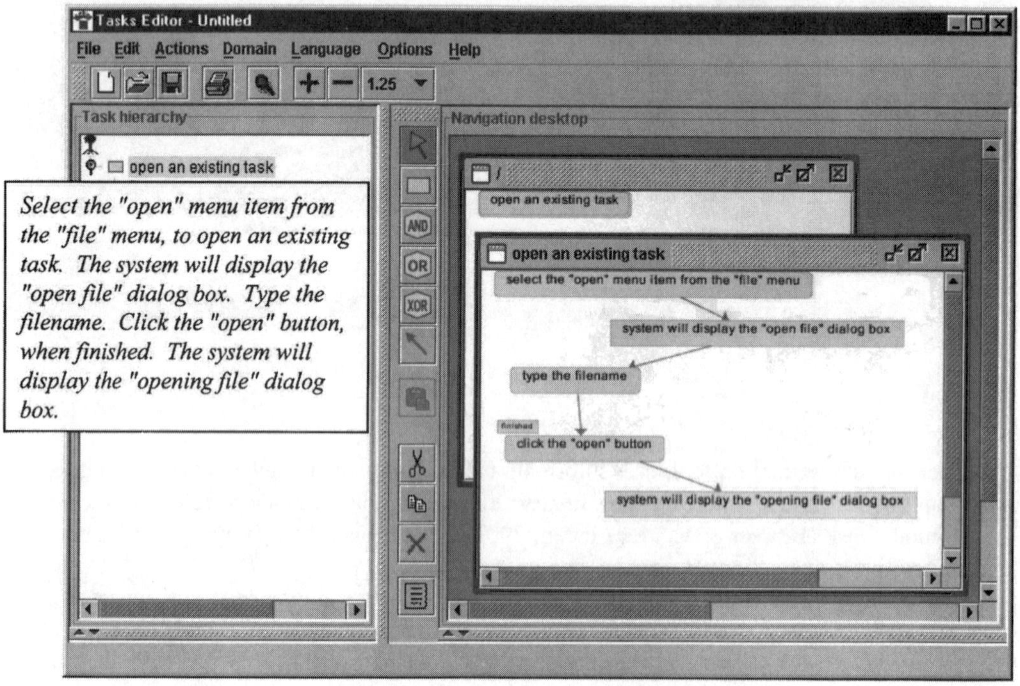

FIG. 23.4. A task model extracted by T2T.

actions ("select", "type" and "click"), the condition ("when finished"), and the system actions ("display" and "display").

The T2T grammar was based on a corpus of 28 task descriptions taken from documentation written for a variety of software engineering courses, and it was evaluated on an additional 9 descriptions using the traditional *recall* and *precision* measures. These indicate how many of the tasks the system was able to recognize (recall) and the correctness of the tasks it recognized (precision) (cf. Jurafsky & Martin, 2000). The results of this preliminary evaluation are shown in Table 23.2, together with the problems encountered. Some of these problems are due to the small syntactic coverage of the parser at this point. Others are due to the fact that the parser currently does not attempt to resolve references (e.g., pronouns and other anaphora).

TABLE 23.2
Evaluation of T2T

	Recall	*Precision*	*Problems*
Elementary task units	90.1% (118/131)	68.2% (118/173)	Infinitives, "and" and punctuation
Purposes	17.8% (5/28)	100% (5/5)	"If you want to . . ."
Preconditions	42.8% (3/7)	75% (3/4)	"First, you must . . ."
Domain Entities	47% (155/324)	47.9% (155/323)	Phrasal verbs, anaphora

23.3.3 The UML-to-Task Extraction Tool (U2T)

Design models represented in UML are common in software engineering (Rumbaugh et al., 1999), particularly in Object-oriented analysis and design. If they already exist, these models can serve as another source of task knowledge. The UML-to-task tool U2T performs this extraction using the standard Rational Rose scripting language (see also chap. 24). It has been discussed in some detail elsewhere (Lu et al., 1999; Vander Linden, Paris, & Lu, 2000). We will thus only briefly describe it here.

UML design models typically include both system structure models, such as class diagrams, and system behavior models, such as use case diagrams and scenario/interaction diagrams. These three types of diagrams are the key sources of task knowledge in UML. Examples of them for our Tamot example are shown in Figs. 23.5 and 23.6. A use case diagram (Fig. 23.5(a)) identifies the users and their goals with respect to an application. A class diagram (Fig. 23.5(b)) identifies the basic classes of domain objects and the actions that can be performed on them. An interaction diagram (Fig. 23.6, lower right) is associated with basic use-cases and represents sequences of interactions between objects instantiated from the classes in the class diagram.

Although UML models are system oriented, they contain knowledge relevant for the task model. U2T extracts the task-oriented information from UML models based on a set of heuristics derived from an analysis of the common semantic ground between system behavior models and task models (Lu et al., 1999). In brief, the relationship between the two types of models is as follows:

- Use cases can be seen as equivalent to composite tasks in a task model.
- The containment of use cases by other use case diagrams can be seen as equivalent to the hierarchical structure in a task model.
- Classes and methods in class diagrams can be seen as objects and actions common to the domain of application.
- Messages in scenario diagrams originating from a user object can be seen as user tasks in sequence.
- System-originated messages that operate on a visible system object and follow a user message can be seen as a system response to the user action.

U2T, which is implemented in Rose script and can be seen as an extension to Rational Rose, extracts task information from UML and exports it in Diane+ format to the Tamot task model editor. Figure 23.7 shows the result of U2T extracting the task model from the scenario diagram for "add level" of Fig. 23.6.

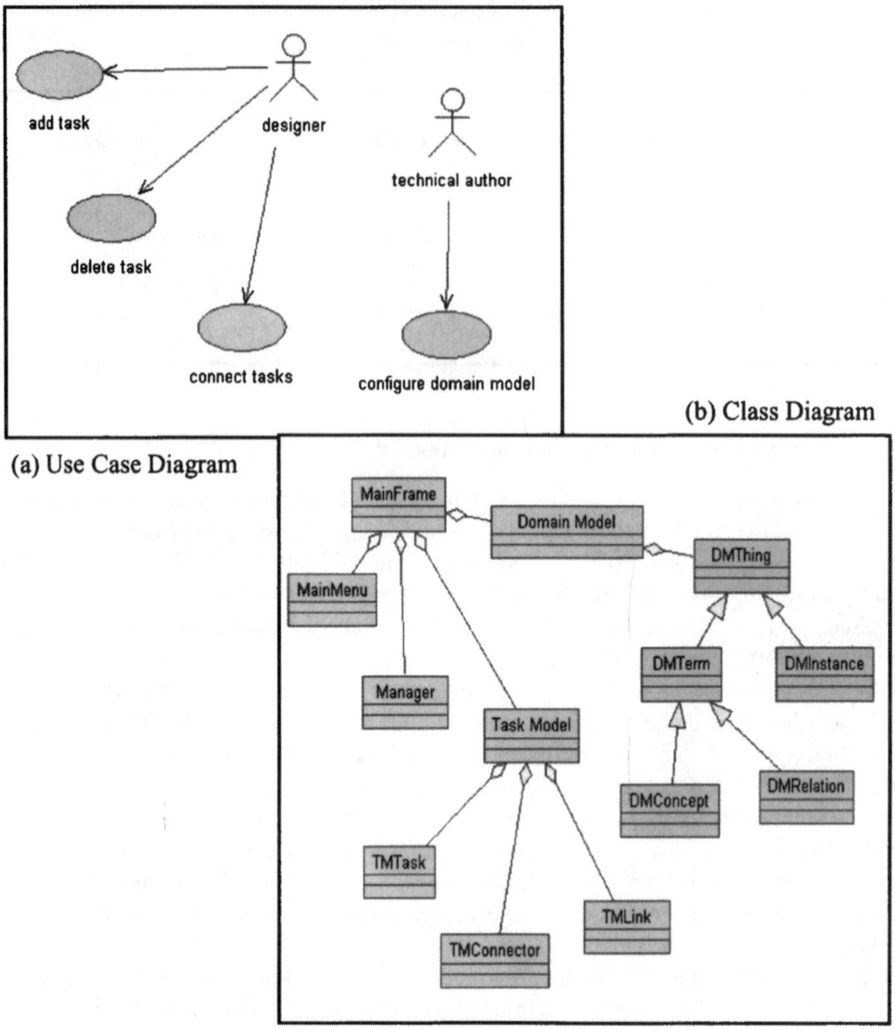

FIG. 23.5. Sources of task information in UML models.

23.3.4 The Event Recorder: User Interaction Recorder (UIR)

System prototypes, if they exist, implement the GUI objects used in their interface and the interaction dialogue used to interact with them. Because tools exist that can extract the objects and record the dialogue as the prototype runs (Hilbert & Redmiles, 2000), the prototype can also serve as a knowledge source. UIR, our user interaction recorder, uses an object extraction tool and an event-recording tool, both developed at Sun Microsystems for the Java platform (Sun Microsystems, 2002), to extract task and domain knowledge.

Figure 23.8 shows both parts of UIR in operation. The window in the lower right is the user event listener window, and the one in the upper left is the object extractor window. The user event listener shows a hierarchy of user tasks (in the Tasks pane). Because it cannot infer these high-level goals, it allows the user to build them manually or to import them from another source (e.g., an existing task model or UML model).

In this example, the user goals include opening, saving, and building a task model. The user can record the low-level steps required for each goal by selecting the goal, pressing the record button, and demonstrating the task in the real application (in this case, in Tamot itself).

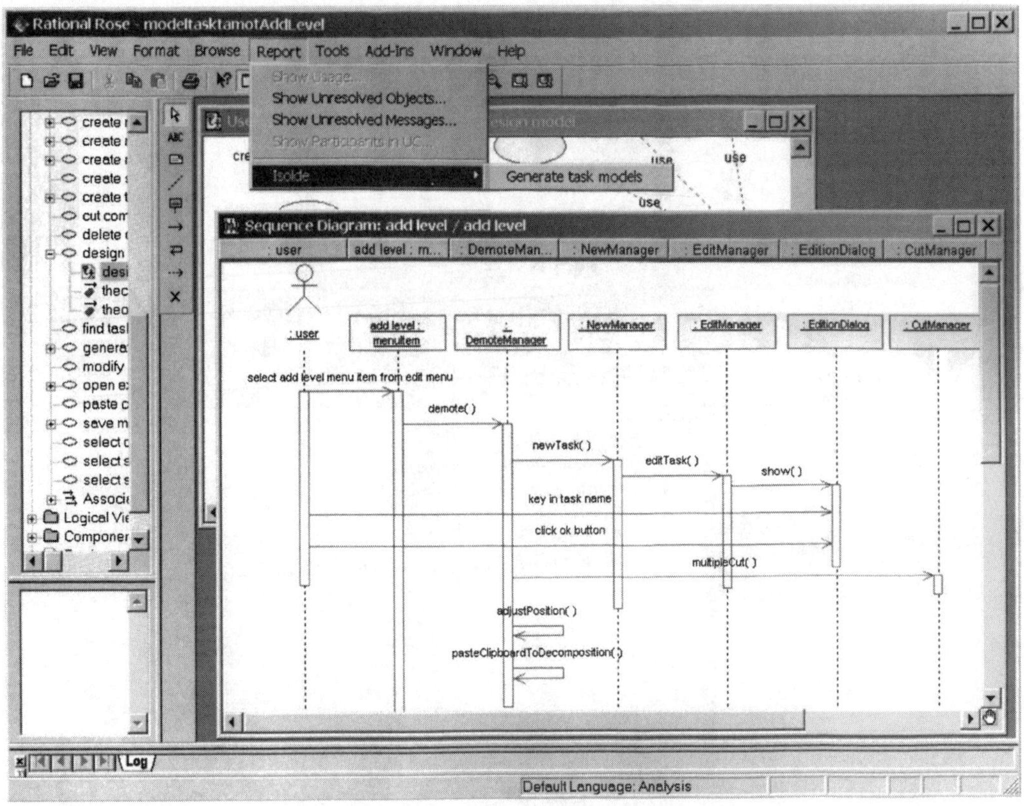

FIG. 23.6. Rational Rose with the U2T extraction tool.

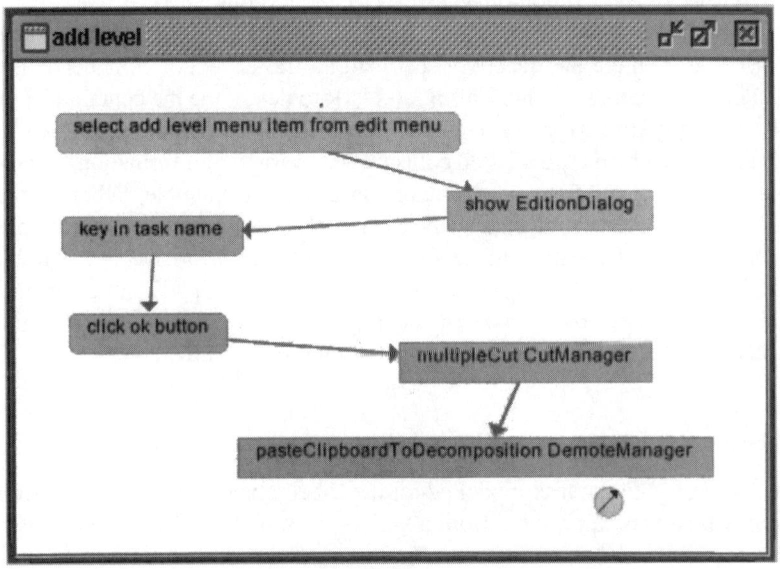

FIG. 23.7. A generated task for "add level."

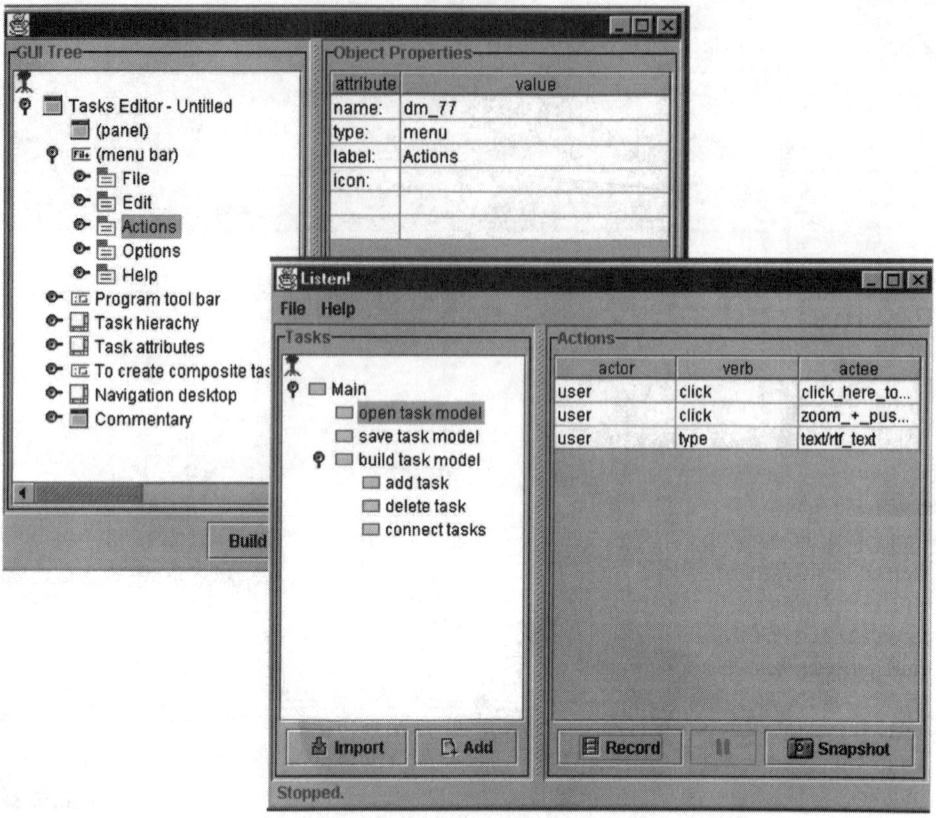

FIG. 23.8. The user interaction recorder.

The sequence of events (for opening a task model, in this example) is recorded and shown as a vertical sequence of actor-action-object triples (in the Actions pane).

The UIR object extractor, the window in the upper left of Fig. 23.8, extracts the hierarchy of GUI objects used in the interface. We see that the "Tasks" editor contains a "menu bar," which contains the "Actions" menu. This allows UIR to determine the containment hierarchy of GUI objects, supporting expressions like "Select the task icon *from the tool bar*."

The Java tools on which UIR is based collect a wide variety of information, some of which is too low level to be useful for task or domain modeling. For example, rather than recording an event like "the user types a filename," the event listener records a separate key-press event for every letter typed into the filename text field. UIR includes heuristics that help it translate from the low-level events and raw widgets it extracts to the higher level, more domain oriented actions and objects required by the task and domain models. Any knowledge that is still missing or incorrectly recorded in any way can be added or modified within Tamot.

23.3.5 Isolde: Integrating and Enabling

The extraction tools and the task model editor described above, taken together, form an environment in which task knowledge from a variety of sources and in a variety of notations can be constructed, extended, refined, and finally consolidated. The result is a set of unified models all represented in one modeling language that is stored in an XML-based notation. The environment is flexible in that it caters to a variety of situations, depending on what is most appropriate. The task modeler may build the task model by hand, extract task information from any of the sources discussed above, or use any combination of these approaches. Importantly, the environment is also open in that new extraction tools can be added to the environment provided that they support the XML-based representation.

Once task models are defined in the environment, they are ready to be exploited for a variety of purposes. Their use and potential reuse is enabled in the Isolde environment by a set of application or analysis tools. The Isolde environment currently includes two such tools (see Fig. 23.1), which are discussed briefly here.

The Instructional Text Generator (ITG) uses task models to drive the generation of user-oriented instructions for the task being modeled (Paris et al., 1998). It takes as input a task model with its associated domain knowledge and generates a draft of a hierarchical HTML-based set of instructions. The task model must include both the high-level goals of the user and the low-level actions that achieve those goals. The domain model must include, for each task, a specification of the actor, the action being performed, the object being acted upon (if any), and other relevant information (e.g., location or manner). ITG also uses a number of linguistic resources (e.g., discourse, lexical, and grammatical), some in deep representations and others in shallow form, and a natural language generation engine, which includes text planning, sentence planning, and realization. ITG provides one control over the linguistic resources via some *style parameters*, which allow the production of the text to adhere to different styles, at both text and sentence levels. Automating parts of this task is useful because writing and maintaining user-oriented documentation is crucial to the usability of a product but is also labor intensive and tedious. A recent evaluation of the text produced, aimed at assessing its effectiveness compared with human-authored text, showed no significant differences between the two (Colineau, Paris, & Vander Linden, 2002).

T2U, a task-to-UML tool, can be seen as the inverse of the U2T tool discussed above. It converts task models into UML system models (Lu, Paris, Vander Linden, & Colineau, 2003). The system models that are produced, however, are necessarily limited by the user-oriented nature of the task models they take as input. T2U can, nevertheless, extract some basic system-oriented information on which a system designer can base a full UML model. This can provide some flexibility in the SDLC in that developers can build models using whatever tool or notation they prefer and then port the relevant information one way or the other.

23.4 CONCLUSION

As this chapter has explained, task models are useful in the development of interactive systems but also difficult to build and maintain. The difficulty is due to the fact that in order to support a variety of task applications and analyses, task models should include representations of various levels of information, from the highest level user goals down to the lowest level GUI events, and they should be represented in a single, coherent representation scheme. To help address this difficulty, analysts are advised to use an extensible, heterogeneous task-modeling environment that integrates tools that extract task information from a variety of existing sources along with tools that support the creation, manipulation, and maintenance of task models, all within a coherent representation scheme. They may find it useful to consider the range of existing environments and tools reviewed in this chapter.

23.5 ACKNOWLEDGMENTS

The authors thank the past and present members of the ISOLDE team, including Sandrine Balbo, Todd Bentley, Michael Brasser, Nathalie Colineau, Thomas Lo, Nadine Ozkan, Maryline Specht, Robert Tot, and Jean-Claude Tarby. This work has been supported by the Office of Naval Research (grant N00014-96-0465), CSIRO, and Calvin College.

REFERENCES

Balbo, S., & Lindley, C. (1997). *Adaptation of a task analysis methodology to the design of a decision support system.* In *Proceedings of Interact'97.* IFIP Transactions Series, Boston, MA Kluwer Academic Publishers (pp. 355–361).

Baumeister, L. K., John, B. E., & Byrne, M. D. (2000). A comparison of tools for building GOMS models. In *Proceedings of the ACM Conference on Human Factors in Computer Systems* (CHI 2000) (pp. 502–509). New York: ACM Press.

Beard, D., Smith, D., & Danelsbeck, K. (1996). QGOMS: A direct-manipulation tool for simple GOMS models. In *Proceedings of the ACM Conference on Human Factors in Computer Systems (CHI 1996)*, New York, NY, ACM Press (pp. 25–26).

Bentley, T., & Johnston, L. (2001, November). *Analysis of task model requirements.* In *Proceedings of OZCHI'01,* the Conference for the Computer-Human Interaction Special Interest Group of the Ergonomics Society of Australia, Fremantle, Western Australia: Australia (pp. 1–6).

Biere, M., Bomsdorf, B., & Szwillus, G. (1999). Specification and simulation of task models with VTMB. In *Proceedings of the ACM Conference on Human Factors in Computer Systems (CHI '99)* (pp. 502–509). New York: ACM Press.

Bomsdorf, B., & Szwillus, G. (1996). Early prototyping based on executable task models. In *Proceedings of the ACM Conference on Human Factors in Computer Systems (CHI '96)* (Vol. 2; pp. 254–255). New York: ACM Press.

Brasser, M., & Vander Linden, K. (2002). Automatically eliciting task models from written task descriptions. In *Proceedings of the fourth International Conference on Computer-Aided Design of User Interfaces (CADUI 2002)*, Dordrecht, the Netherlands, Kluwer Academic Publishers, (pp. 83–91).

Card, S. K., Moran, T. P., & Newell, A. (1983). *The psychology of human-computer interaction.* Hillsdale, NJ: Lawrence Erlbaum Associates.

Carroll, J. (1990). *The Nurnberg funnel: Designing minimalist instruction for practical computer skill.* Cambridge, MA: MIT Press.

Colineau, N., Paris, C., & Vander Linden, K. (2002, July). An evaluation of procedural instructional text. In *Proceedings of the International Conference on Natural Language Generation (INLG 2002)* (pp. 128–135). New Brunswick, NJ, ACL Press.

Constantine, L., & Lockwood, L. (1999). *Software for use: A practical guide to the models and methods of user-centered design.* Reading, MA: Addison-Wesley; New York: ACM Press.

Fellbaum, C. (Ed.). (1998). *WordNet: An electronic lexical database.* Cambridge, MA: MIT Press.

Gamboa Rodríguez, F., & Scapin, D. (1997). Editing MAD* task descriptions for modifying user interfaces at both semantic and presentation levels. In M. Harrison & J. Torres (Eds.), in *Proceedings of the Fourth International Eurographics Workshop on Design Specification and Verification of Interactive Systems (SDV-IS '97)* (pp. 193–208). Vienna, Austria, Springer-Verlag.

Harel, D. (1987). Statecharts: A visual formalism for complex systems. *Science of Computer Programming, (8),* 231–274.

Hilbert, D., & Redmiles, D. (2000). Extracting Usability information from user interface events. *Computing Surveys, 32,* 384–421.

Hix, D., & Hartson, R. (1994). *Ideal: An environment to support usability engineering,* in *Proceedings of the 1994 East-West International Conference on HCI,* Lecture Notes in Computer Science 876, Vienna, Austria, Springer, (pp. 95–106).

Hoppe, H. U. (1988). Task-oriented parsing: A diagnostic method to be used by adaptive systems. In *Proceedings of the ACM Conference on Human Factors in Computer Systems (CHI '88)* (pp. 241–247). New York: ACM Press.

Hudson, S., John, B., Knudsen, K., & Byrne, M. (1999). A tool for creating predictive performance models from user interface demonstrations. In *Proceedings of the 12th Annual ACM Symposium on User Interface Software and Technology* (UIST '99). (pp. 93–102). New York, NY: ACM Press.

Jacobson, I., Christerson, W., Jonsson, P., & Övergaard, G. (1992). *Object-oriented software engineering: A use case driven approach.* Reading, MA: Addison-Wesley.

Jurafsky, D., & Martin, J. (2000). *Speech and language processing.* Murray Hill, NJ: Prentice Hall.

Kieras, D. E., Wood, S. D., Abotel, K., & Hornof, A. (1995). GLEAN. In *Proceedings of the eighth ACM Symposium on User Interface and Software Technology* (*UIST 1995*) (pp. 91–100). New York, NY: ACM Press.

Lewis, C., & Rieman, J. (1994). *Task-centered user interface design: A practical introduction.* Shareware book available: ftp://ftp.cs.colorado.edu/pub/cs/distribs/clewis/HCI-Design-Book/

Limbourg, Q., Pribeanu, C., & Vanderdonckt, J. (2001). Towards uniformed task models in a model-based approach. In C. Johnson (Ed.), *Interactive Systems: Design, Specification, and Verification: Proceedings of the 8th International Workshop (DSV-IS 2001)* (pp. 164–182). Berlin: Springer-Verlag.

Lu, S., Paris, C., & Vander Linden, K. (1999). Towards the automatic generation of task models from object-oriented diagrams. In S. Chatty & P. Dewan (Eds.), *Engineering for human-computer interaction* (pp. 169–190). Boston: Kluwer.

Lu, S., Paris, C., & Vander Linden, K. (2002). Tamot: Towards a flexible task Modeling tool. In the *Proceedings of Human Factors 2002,* Melbourne Australia, November 2002.

Lu, S., Paris, C., Vander Linden, K., & Colineau, N. (2003). *Generating UML diagrams from task models.* In the *proceedings of CHINZ'03,* the 4th Annual International Conference of the New Zealand chapter of the ACM's SIGCHI, July 3-4, 2003, Dunedin, New Zealand, 2003. Also available as CSIRO/MIS Technical Report 2002/101.

Navarre, D., Palanque, P., Bastide, R., & Sy, O. (2001). A model-based tool for interactive prototyping of highly interactive applications. In *Proceedings of the 12th IEEE International Workshop on Rapid System Prototyping* (pp. 136–141). New York, NY: IEEE Press.

Ozkan, N., Paris, C., & Balbo, S. (1998). Understanding a task model: An experiment. In H. Johnson, K. Nigay, & C. Roast (Eds.), *People and computers XII* (pp. 123–138). London: Springer.

Paris, C., Ozkan, N., & Bonifacio, F. (1998). Novel help for on-line help. In *Proceedings of 16th Annual International Conference on Computer Documentation (SIGDOC '98).* (pp. 70–79). New York, NY: ACM Press.

Paris, C., Tarby, J., & Vander Linden, K. (2001). A flexible environment for building task models. In *People and Computers XV-Interaction without Frontiers ICM-HCI 2001*) (pp. 313–330), London: Springer-Verlag.

Paternò, F., & Ballardin, G. (1999). Model-aided remote usability evaluation. In A. Susse, & C. Johnson (Eds.), *Proceedings of the Seventh IFIP Conference on Human-Computer Interaction* (Interact '99) (pp. 431–442). Amsterdam, the Netherlands: IOS Press.

Paternò, F., & Mancini, C. (1999). Developing task models from informal scenarios. In M. W. Altona, & M. G. Williams (Eds.), *Companion Proceedings of the Conference on Human Factors in Computing (CHI '99),* (pp. 228–229). New York NY: ACM Press.

Paternò, F., Mancini, C., & Meniconi, S. (1997). ConcurTaskTrees: A diagrammatic notation for Specifying Task Models In *Proceedings of Interact'97* (pp. 362–369). IFIP Transactions Series, Boston, MA Kluwer Academic Publishers.

Paternò, F., Mori, G., & Galimberti, R. (2001). CTTE: An environment for analysis and development of task models of cooperative applications. In *Proceedings of the ACM conference on Human Factors in Computer Systems (CHI 2001).* (Extended abstracts; pp. 21–22). New York, NY: ACM Press.

Puerta, A. R. (1996). The Mecano Project: Enabling user-task automation during interface development. In *Spring Symposium on Acquisition, Learning and Demonstration: Automating Tasks for Users* (pp. 117–121). Menlo Park, CA: AAAI Press.

Puerta, A. R. (1997). A model-based interface development environment. *IEEE Software, 14*(4), 41–47.

Puerta, A. R. (1998, January). *Supporting user-centered design of adaptive interfaces via interface models.* Paper Presented at the First Annual Workshop On Real-Time Intelligent User Interfaces For Decision Support And Information Visualization, San Francisco.

Rumbaugh, J., Jacobson, I., & Booch, G. (1999). *The Unified Modeling Language reference manual,* Reading, MA: Addison-Wesley.

Sun Microsystems. (2002) *Java Event Listener and Java Monkey.* Available: http:java.sun.com

Tam, R. C., Maulsby, D., & Puerta, A. R. (1998). U-TEL: A tool for eliciting user task models from domain experts. In *Proceedings of IUI '98: International Conference on Intelligent User Interfaces* (pp. 77–80). New York, NY: ACM Press.

Tarby, J.-C., & Barthet, M.-F. (1996). The Diane+ method. In J. Vanderdonckt (Ed.), *Computer-Aided design of user interfaces* (pp. 95–119). Namur Belgium. Presses Universitaires de Namur.

Vander Linden, K., Paris, C., & Lu, S. (2000). Where do instructions come from? Knowledge acquisition and specification for instructional text. In T. Becker & S. Busemann (Eds.), *IMPACTS in natural language generation: NLG between technology and application* (DFKI report D-00-01. pp. 1–10). Schloss Dagstuhl, Germany.

van der Veer, G. C., Lenting, B. F., & Bergevoet, B. A. J. (1996). GTA: Groupware task analysis: Modelling complexity. *Acta Psychologica, 91*, 297–322.

van Welie, M., van der Veer, G. C., & Eliëns, A. (1998). EUTERPE: Tool support for analyzing co-operative environments. In *Proceedings of the Ninth European Conference on Cognitive Ergonomics,* (pp. 25–30). Rocquencourt, France EACE Press.

Wandmacher, J. (1997). Ein werkzeug fur GOMS-analysen für simulation und bewertung von prototypen beim entwurf. *Tagungsband PB97: Prototypen für Benutzunsschenittstellen, 19,* 35–42.

Williams, K. (2000). An automated aid for modeling human-computer interaction. In J. M. Schraagen, S. Chipman, & V. Shalin (Eds.), *Cognitive task analysis.* (pp. 165–180). Mahwah, NJ: Lawrence Erlbaum Associates.

24

ConcurTaskTrees: An Engineered Notation for Task Models

Fabio Paternò
ISTI-C.N.R.

Task models represent the intersection between user interface design and more engineering approaches by providing designers with a means of representing and manipulating a formal abstraction of activities that should be performed to reach user goals. For this purpose they need to be represented with notations able to capture all the relevant aspects. In this Chapter, the ConcurTaskTrees notation is described and discussed with particular attention to recent evolutions stimulated by experiences with its use and the need to address new challenges.

24.1 INTRODUCTION

One universally agreed-on design principle for obtaining usable interactive systems is this: *Focus on the users and their tasks*. Indeed, of the relevant models in the human-computer interaction (HCI) field, task models play an important role because they represent the logical activities that should support users in reaching their goals (Paternò, 1999). Thus, knowing the tasks necessary to goal attainment is fundamental to the design process.

Although task models have long been considered in HCI, only recently have user interface developers and designers realized their importance and the need for engineering approaches to task models to better obtain effective and consistent solutions. The need for modeling is most acutely felt when the design aims to support system implementation as well. If we gave developers only informal representations such as scenarios (chap. 5) or paper mock-ups (chap. 1), they would have to make many design decisions on their own, likely without the necessary background, to obtain a complete interactive system.

In order to be meaningful, the task model of a new application should be developed through an interdisciplinary collaborative effort involving the various relevant viewpoints. If experts from the various fields contribute appropriately, the end result will be a user interface that can effectively support the desired activities. This means that the user task model (how users think that the activities should be performed) shows a close correspondence to the system task model (how the application assumes that activities are performed).

There are many reasons for developing task models. In some cases, the task model of an existing system is created in order to better understand the underlying design and analyze its potential limitations and how to overcome them. In other cases, designers create the task model of a new application yet to be developed. In this case, the purpose is to indicate how activities should be performed in order to obtain a new, usable system that is supported by some new technology.

Task models can be represented at various levels of abstraction (chap. 1). When designers want to specify only requirements regarding how activities should be performed, they consider only the main high-level tasks. On the other hand, when designers aim to provide precise design indications, then the activities are represented at a smaller granularity, including aspects related to the dialogue model of a user interface (which defines how system and user actions can be sequenced).

The subject of a task model can be either an entire application or one of its parts. The application can be either a complete, running interactive system or a prototype under development. The larger the set of functionalities considered, the more difficult the modeling work. Tools such as CTTE (publicly available at http://giove.cnuce.cnr.it/ctte.html) open up the possibility of modeling entire applications, but in the majority of cases what designers wish to do is to model some subsets in order to analyze them and identify potential design options and better solutions.

Although task models have long been considered in human-computer interaction, only recently have user interface developers and designers realized their importance and the need for engineering approaches to task models to better obtain effective and consistent solutions. An engineering approach should address at least four main issues (these are also discussed in chap. 22):

- The availability of flexible and expressive notations able to describe clearly the possible activities. It is important that these notations are sufficiently powerful to describe interactive and dynamic behaviors. Such notations should be readable so that they can also be interpreted by people with little formal background.
- The need for systematic methods to support the specification, analysis, and use of task models in order to facilitate their development and support designers in using them to design and evaluate user interfaces. Often even designers who do task analysis and develop a model do not use it for the detailed design of the user interface because of the lack of structured methods that provide rules and suggestions for applying the information in the task model to the concrete design. Structured mothods can also be incorporated in tools aiming to support interactive designers.
- Support for the reuse of good design solutions to problems that occur across many applications. This is especially relevant in an industrial context, where developers often have to design applications that address similar problems and thus could benefit from design solutions structured and documented in such a way as to support easy reuse and tailoring in different applications.
- The availability of automatic tools to support the various phases of the design cycle. The developers of these tools should pay attention to user interface aspects so that the tools have intuitive representations and provide information useful for the logical activities of designers.

Task models can be useful both in the construction and the use of an interactive system. They can aid designers and developers by supporting high-level, structured approaches that allow integration of both functional and interactional aspects. They can aid end users by supporting the generation of more understandable systems.

This chapter first presents the basic concepts on which ConcurTaskTrees (Paternò, 1999) has been developed. The chapter not only provides an introduction of the notation but also discusses some issues that have been revealed to be important during its development and application: the relationships between task models and more informal representations, how to use task models to support the design of applications accessible through multiple devices, how to integrate them with standard software engineering methods such as UML (see chaps. 7, 11, 12, 19, 22, 23 and 27), how to represent task models of multiuser applications, and what tool support they need in order to ease and broaden their analysis by designers. The chapter also touches upon the experience of using ConcurTaskTrees (CTT) in a number of projects.

24.2 BASIC CONCEPTS

Of the relevant models, task models play a particularly important role because they indicate the logical activities that an application should support to reach user goals.

A goal is either a desired modification of a state or an inquiry to obtain information on the current state. Each task can be associated with a goal, which is the state change caused by its performance. Each goal can be associated with one or multiple tasks. That is, there can be different tasks that achieve the same goal.

Task descriptions can range from a very high level of abstraction (e.g., deciding a strategy for solving a problem) to a concrete, action-oriented level (e.g., selecting a printer). Basic tasks are elementary tasks that cannot be further decomposed because they do not contain any control element.

For example, making a flight reservation is a task that requires a state modification (adding a new reservation in the flight database) whereas querying the available flights from Pisa to London is a task that just requires an inquiry of the current state of the application. Making a reservation is a task that can be decomposed into lower level tasks such as specifying departure and arrival airports, departure and arrival times, seat preferences, and so on.

It is better to distinguish between task analysis and task modeling. Whereas the purpose of task analysis is to understand what tasks should be supported and what their related attributes are, the aim of task modeling is to identify more precisely the relationships among such tasks. The need for modeling is most acutely felt when the design aims to support system implementation as well. If developers were given only a set of scenarios, they would have to make many design decisions on their own in order to obtain a complete system. This is why UML has widely been adopted in software engineering: Its purpose is to provide a set of related representations (ranging from use cases to class diagrams) to support designers and developers across the various phases of the design cycle.

There are various types of task models that can be considered:

- *The task model of an existing system.* The purpose of this model is to describe how activities should be performed in an existing, concrete system in order to understand its limitations, problems, features, and so on.
- *The task model of an envisioned system.* In this case the goal is to define how a new system, often supported by some new technology, should aid in the achievement of users' goals or improvement of some aspect of existing systems.
- *The user task model.* This model indicates how the user actually thinks that tasks should be performed. Many usability problems derive from a lack of direct correspondence between the user and the system task model (Norman, 1986; chap. 18 and 19).

The design cycle in Fig. 24.1 shows one way of viewing the various task models. Often people start with the task model of the existing system and have requirements for designing a

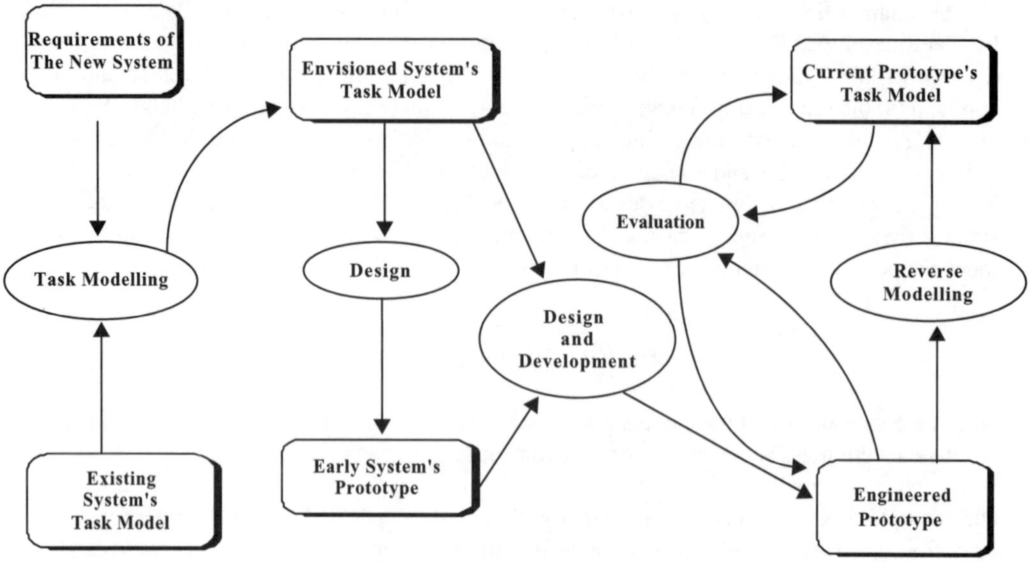

FIG. 24.1. Use of task models in the design cycle.

new system (e.g., the use of some new technology). The modeling activity is used to create a task model of the envisioned system—a model that can act as the starting point for the design of the concrete user interface. In order to have a first discussion regarding the preliminary design ideas, designers often use early, rapid prototypes produced using paper, mock-ups, and similar things. The task model and the early prototypes are the input for the actual design and development of the engineered prototype that will then become the delivered system. Constraints in the development of the system can result in a system whose task model is different from the initial model. So, if task models are used to support the evaluation of the system, then a kind of reverse engineering has to be done in order to identify the actual system task model that needs to be evaluated.

More generally, task models can be used for many purposes:

- *To improve understanding of the application domain.* The modeling exercise requires designers to clarify what tasks should be supported, their mutual relationships, their constraints, and so on. In an air traffic control project, we developed task models together with designers who had been working for a long time in this domain without this technique (Paternò, Santoro, & Sabbatino, 2000). The designers admitted that this exercise raised a number of issues that they never considered before.
- *To record the results of interdisciplinary discussion.* Because task models are logical descriptions of the activities to support, they should take into account the viewpoint of all the relevant stakeholders (designers, developers, end users, domain experts, the client, etc.).
- *To support effective design.* If a direct correspondence is created between the tasks and the user interface, users can accomplish their goals effectively and satisfactorily.
- *To support usability evaluation.* Task models can be used in evaluation in various ways. For example, they can be used to predict time performance, as in the GOMS approach (John & Kieras, 1996; chap. 4) and to analyse the user's actual behavior, as in WebRemUSINE (Paganelli & Paternò, 2002).

- *To support the user during a session.* They can be used at run time to automatically generate task-oriented help, including explanations of how to enable disabled actions and perform them.
- *To provide documentation on how to use a system to accomplish the desired tasks.*

24.3 INFORMAL DESCRIPTIONS VERSUS TASK MODELING

One issue often discussed is whether to use structured abstractions or more informal descriptions during the design cycle. This section explains similarities and differences between these approaches.

Scenarios (Carroll, 2000; chap. 5) are the best known of the informal types of description. The importance of scenarios is generally recognized, and their adoption in current practice is impressive. Often scenarios are used even by people who have no knowledge of the HCI literature. When people want to discuss design decisions or requirements for a new system, they find proposing scenarios (i.e., examples of possible use) to be very intuitive. A lively debate on the relationship between task analysis and scenarios can be found in the journal *Interacting with Computers* (2002, issues 4 and 5).

One question is, What is a scenario? Sometimes it seems that any type of description of system use can be a scenario. This brings to mind a discussion on use cases (Cockburn, 1997) in which 18 definitions of use cases were classified based on four issues (purpose, contents, plurality, and structure). Likewise, distinguishing between various types of informal descriptions—stories, scenarios, use cases, and task analyses—can be useful. Stories include some emotional aspects (Imaz & Benyon, 1999); scenarios provide a detailed description of a single, concrete use of a system; use cases are a bit more general and describe groups of use; and task analyses aim to be even more general.

In particular, scenarios, on the one hand, and task analysis and modeling, on the other, should not be confused. Actually, each can be applied without the other, even though it is advisable to use both of them. It is telling that in software engineering scenarios are mainly applied in the requirements phase. In fact, their strong point is their ability to highlight issues and stimulate discussion while requiring limited effort to develop, at least as compared with more formal techniques. However, because a scenario describes a single system use, there is a need for other techniques that may provide less detail but broader scope. Here is where task analysis comes in.

It is often difficult to create a model from scratch, and various approaches have been explored to make it easier. CRITIQUE (Hudson et al., 1999) is a tool that is designed to create KLM/GOMS models (Card, Moran et al., 1983) from the analysis of use session logs. The model is created following two types of rules: The types of KLM operators are identified according to the type of event, and new levels in the hierarchical structure are built when users begin working with a new object or when they change the input to the current object. One limitation of this approach is that the task model only reflects the past use of the system and not other potential uses. U-Tel (Tam, Maulsby, & Puerta, 1998) analyzes textual descriptions of the activities to support and then automatically associates tasks with verbs and objects with nouns. This approach can be useful, but it is too simple to obtain general results. The developers of Isolde (Paris, Tarby, Vander Linden, 2001; chap. 23) have considered the success of UML and provide some support to import use cases created by Rational Rose in their tool for task modeling.

The ConcurTaskTrees Environment (CTTE; Mori, Paternò et al., 2002) supports the initial modeling work by offering the possibility of loading an informal textual description of a scenario or use case and interactively selecting the information of interest for the modeling. In this way, the designer can first identify tasks, then create a logical hierarchical structure, and finally complete the task model. The use of these features is optional: Designers can start to create the model directly using the task model editor, but such features can ease the work of modeling.

To develop a task model from an informal textual description, designers first have to identify the different roles. Then they can start to analyze the description of the scenario, trying to identify the main tasks that occur in the description and refer each task to a particular role. It is possible to specify the category of the task in terms of performance allocation. In addition, a description of the task can be specified, along with the logical objects used and handled. Reviewing the scenario description, the designer can identify the different tasks and add them to the task list. This must be performed for each role in the application considered.

When each role's main tasks in the scenario have been identified, it might be necessary to make some slight modifications to the newly defined task list. The designers can thus avoid task repetition, refine the task names so as to make them more meaningful, and so on. Once the designers have their list of activities to consider, they can start to create the hierarchical structure that describes the various levels of abstraction among tasks. The final hierarchical structure obtained will be the input for the main editor, allowing specification of the temporal relationships and the tasks' attributes and objects.

24.4 HOW TO REPRESENT THE TASK MODEL

Many proposals have been put forward for representing task models. Hierarchical Task Analysis (Shepherd, 1989; chap. 3) has a long history and is still sometimes used. More generally, such notations can vary according to various dimensions:

Syntax (textual vs. graphical). Some notations are mainly textual, such as UAN (Hartson & Gray, 1992; chaps. 22, 23, and 27), in which there is a textual composition of tasks enriched with tables associated with the basic tasks. GOMS is also mainly textual, even if CPM-GOMS has a more graphical structure because it has been enhanced with PERT charts that highlight parallel activities. ConcurTaskTrees and GTA (van der Veer, Lenting et al., 1996; chaps. 6 and 7) are mainly graphical representations aimed at better highlighting the hierarchical structure. In ConcurTaskTrees the hierarchical structure is represented from the top down, whereas in GTA the representation is from left to right.

Set of operators for task composition. There are substantial differences in regard to these operators among the proposed notations. As Table 24.1 shows, UAN and CTT provide the richest set of temporal relationships and thus allow designers to describe more flexible ways of performing tasks.

Level of formality. Occasionally notations have been proposed without sufficient attention paid to the definition of the operators. When task models are created using such a notation, it is often unclear what is being described because the meaning of many instances of the composition operators is unclear.

The main features of ConcurTaskTrees are as follows:

Focus on activities. ConcurTaskTrees allows designers to concentrate on the activities that users aim to perform. These are what designers should focus on when designing interactive applications that encompass both user- and system-related aspects, and they should avoid

TABLE 24.1
Comparison of Task Model Notation Operators

	GOMS	*UAN*	*CTT*	*MAD*	*GTA*
Sequence	X	X	X	X	X
Order independence		X	X		X
Interruption		X	X	X	
Concurrency	Only CPM-GOMS	X	X	X	X
Optionality			X	X	
Iteration		X	X		X
			X		X
Objects			X		X
Performance	X		X		X
Pre-post condition	X	X	X	X	X

low-level implementation details that, at the design stage, would only obscure the decisions they need to make.

Hierarchical structure. A hierarchical structure is very intuitive (chap. 1). In fact, when people have to solve a problem, they tend to decompose it into smaller problems, still maintaining the relationships among the various parts. The hierarchical structure of this specification has two advantages: It provides a wide range of granularity, allowing large and small task structures to be reused, and it enables reusable task structures to be defined at both low and high semantic levels.

Graphical syntax. A graphical syntax often, though not always, is easier to interpret, because the syntax of ConcurTaskTrees reflects the logical structure of tasks, it has a treelike form.

Rich set of temporal operators. A rich set of possible temporal relationships between the tasks can be defined. This set provides more possibilities than those offered by concurrent notations such as LOTOS. The range of possibilities is usually implicit, expressed informally in the output of task analysis. Making the analyst use these operators is a substantial change from normal practice. The reason for this innovation is to get the designers, after doing an informal task analysis, to express clearly the logical temporal relationships. Such an ordering should be taken into account in the user interface implementation to allow the user to perform at any time the tasks that should be enabled from a semantic point of view.

Task allocation. How the performance of the task is allocated is indicated by the related category, and each category has a unique icon to represent it. There are four categories: tasks allocated to the system, tasks allocated to the user, tasks allocated to an interaction between the system and user, and abstract tasks. Abstract tasks, represented by a cloud icon, include tasks whose subtasks are allocated differently (e.g., one subtask is allocated to the user and one to the system) and tasks that the designer has not yet decided how to allocate (see also chaps. 6 and 22).

Objects and task attributes. Once the tasks are identified, it is important to indicate the objects that have to be manipulated to support their performance. Two broad types of objects can be considered: user interface objects and application domain objects. Multiple user interface objects can be associated with a domain object (e.g., temperature can be represented by a bar- chart or a textual value).

For each single task, it is possible to directly specify a number of attributes and related information. They are classified into three sections:

General information. General information includes the identifier and extended name of the task; its category and type; its frequency of use; informal annotations that the designer may want to store; indication of possible preconditions; and indication of whether it is an iterative, optional, or connection task. Although the category of a task indicates the allocation of its performance, task types allow designers to group tasks based on their semantics. Each category has its own types of task. In the interaction category, the task types include selection (the task allows the user to select some information), control (the task allows the user to trigger a control event that can activate a functionality; chap. 19), editing (the task allows the user to enter a value), monitoring, responding to alerts, and so forth. This classification can help in choosing the most suitable interaction or presentation techniques for supporting the task performance. Frequency of use is another useful type of information because the interaction techniques associated with more frequent tasks need to be better highlighted to obtain an efficient user interface. The platform attribute (e.g., desktop, PDA, cellular, etc.; see also chaps. 22 and 26) allows the designer to indicate for what type of devices the task is suitable (Paternò & Santoro, 2002). This information is particularly useful in the design of nomadic applications (applications that can be accessed through multiple types of platforms). It is also possible to specify if there is any specific precondition that should be satisfied before performing the task.

Objects. For each task, it is possible to indicate the objects (name and class) that have to be manipulated to perform it. Objects can be either user interface or domain application objects. It is also possible to indicate the right access of the user to manipulate the objects when performing the task. Since the performance of the same task in different platforms can require the manipulation of different sets of objects, it is possible to indicate for each platform what objects should be considered. In multi-user applications, different users may have different right accesses.

Time performance. It is also possible to indicate the estimated performance time (including minimal, maximal, and average performance times).

Figure 24.2 presents the temporal operators provided by CTT in greater detail.

It is important to remember that in CTT hierarchy does *not* represent sequence. In Fig. 24.3, the goal of the designer is to specify that "in order to book a flight, I have to select a route, then select a flight, and lastly perform the payment." However the specification is wrong because the sequential temporal evolution should be represented linearly from left to right instead of from the top down.

When there is a need to get back to some point in the specification, it is possible to use the structure of the model and the iterative operator. For example, in Fig. 24.4, once the task *CloseCurrentNavigation* is performed, the task *SelectMuseum* is enabled again because the parent task is iterative. This means that it can be performed multiple times, and so its first subtask is enabled once the last one has been completed.

Optional tasks have a subtle semantics in CTT. They can be used only with concurrent and sequential operators. Their names are enclosed in square brackets. For example, in Fig. 24.5, *Specify type of seat* and *Specify smoking seat* are optional tasks. This means that once the mandatory sibling tasks (*Specify departure* and *Specify arrival*) are performed, then both the optional tasks and the task after the enabling operator (*Send request*) are enabled. If the task after the enabling operator is performed, then the optional tasks are no longer available.

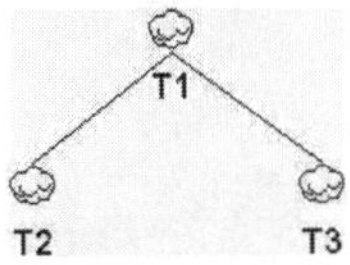	**Hierarchy** Tasks at same level represent different options or different tasks at the same abstraction level that have to be performed. Read levels as "In order to do T1, I need to do T2 and T3", or "In order to do T1, I need to do T2 or T3"
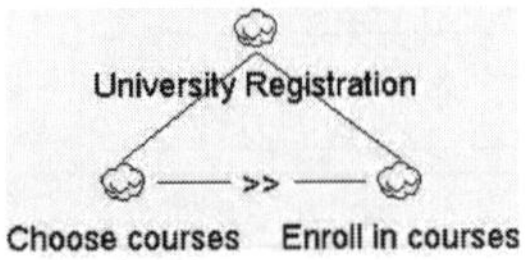	**Enabling** Specifies second task cannot begin until first task performed. Example: I cannot enroll at university before I have chosen which courses to take.
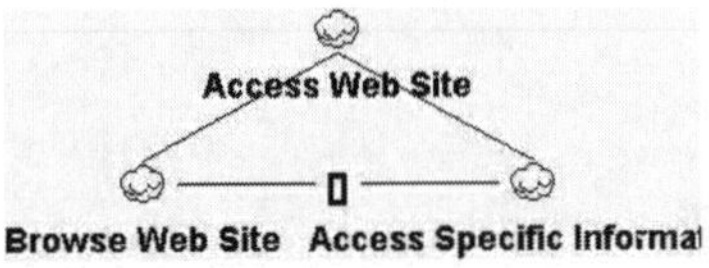	**Choice** Specifies two tasks enabled, then once one has started the other one is no longer enabled. Example: When accessing a web site it is possible either to browse it or to access some detailed information.
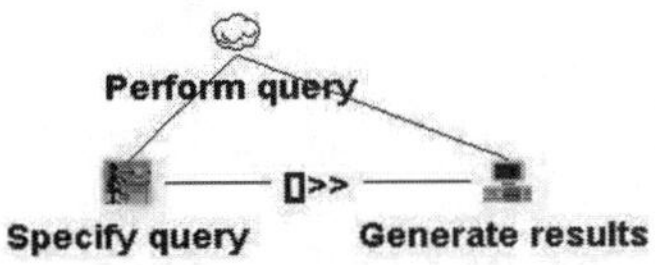	**Enabling with information passing** Specifies second task cannot be performed until first task is performed, and that information produced in first task is used as input for the second one. Example: The system generates results only after that the user specifies a query and the results will depend on the query specified.
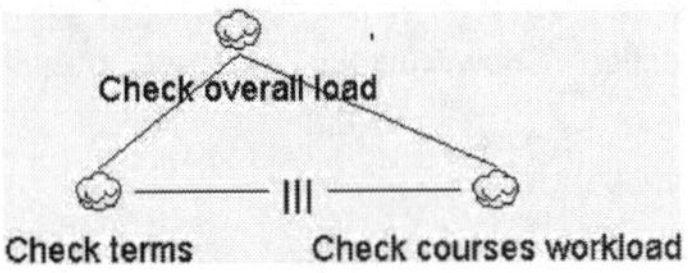	**Concurrent tasks** Tasks can be performed in any order, or at same time, including the possibility of starting a task before the other one has been completed. Example: In order to check the load of a set of courses, I need to consider what terms they fall in and to consider how much work each course represents
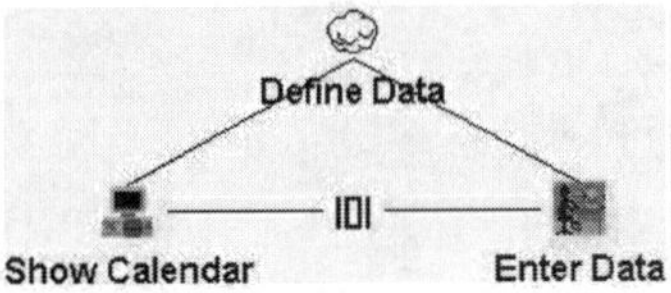	**Concurrent Communicating Tasks** Tasks that can exchange information while performed concurrently Example: An application where the system displays a calendar where it is highlighted the data that is entered in the meantime by the user.
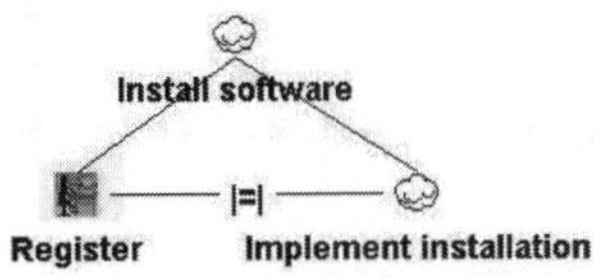	**Task independence** Tasks can be performed in any order, but when one starts then it has to finish before the other one can start. Example: When people install new software they can start by either registering or implementing the installation but if they start one task they have to finish it before moving to the other one.
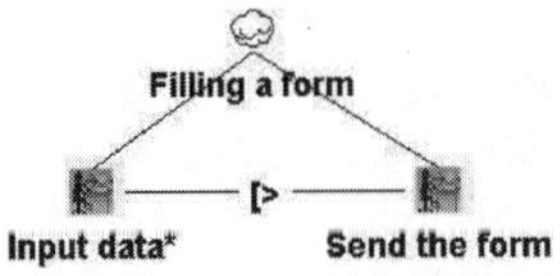	**Disabling** The first task (usually an iterative task) is completely interrupted by the second task. Example: A user can iteratively input data in a form until the form is sent.
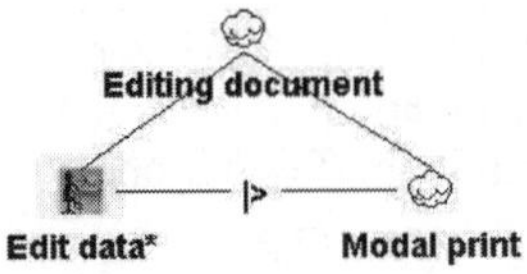	**Suspend-Resume** First task can be interrupted by the second one. When the second terminates then the first one can be reactivated from the state reached before Example: Editing some data and then enabling the possibility of printing them in an environment where when printing is performed then it is no possible to edit.

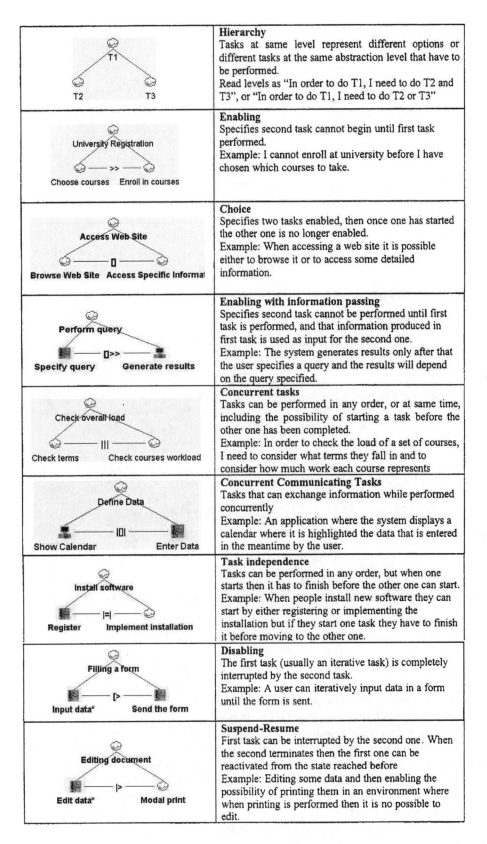

FIG. 24.2. Temporal operators in ConcurTaskTrees.

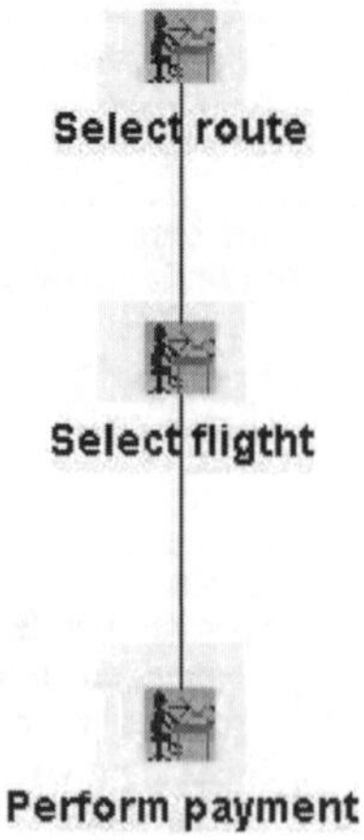

FIG. 24.3. Example of wrong specification of sequential activities.

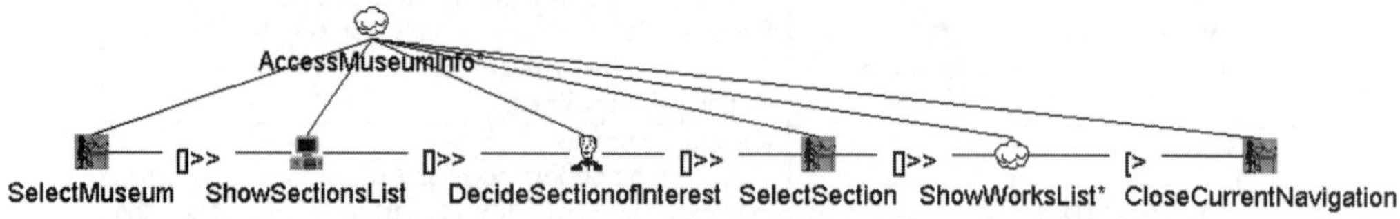

FIG. 24.4. Example of parent iterative task.

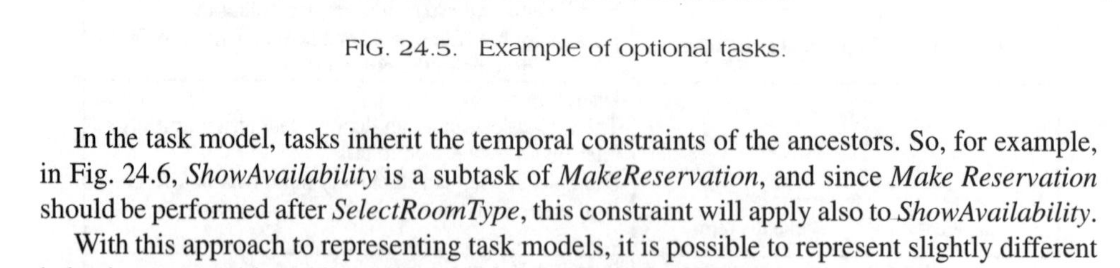

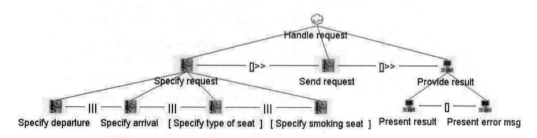

FIG. 24.5. Example of optional tasks.

In the task model, tasks inherit the temporal constraints of the ancestors. So, for example, in Fig. 24.6, *ShowAvailability* is a subtask of *MakeReservation*, and since *Make Reservation* should be performed after *SelectRoomType*, this constraint will apply also to *ShowAvailability*.

With this approach to representing task models, it is possible to represent slightly different behaviors even with small modifications in the specification. The two examples in Fig. 24.7 have the same structure, but the iteration in one case is in the parent task, in the other it is in the subtasks. In the first example, therefore, the parent a task (*Booking flights*) can be performed multiple times, and each time the two subtasks should each be performed once (i.e., overall they will be performed the same number of times). In the other example, the parent task has two iterative subtasks, and thus each of them can be performed an undefined number of times (i.e., the number of times each is performed is independent of the number of times the other is performed).

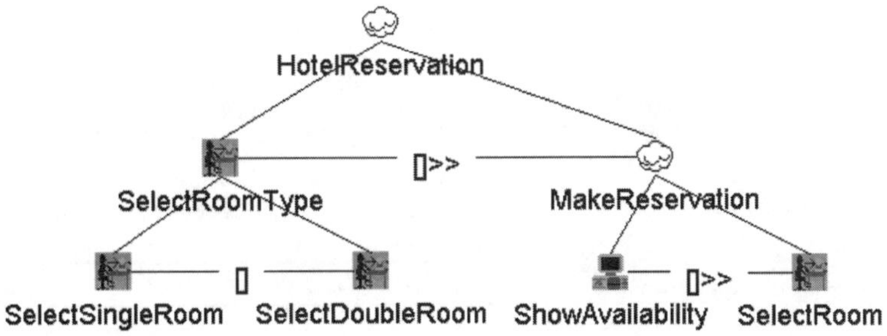

FIG. 24.6. Example of inheritance of temporal constraints.

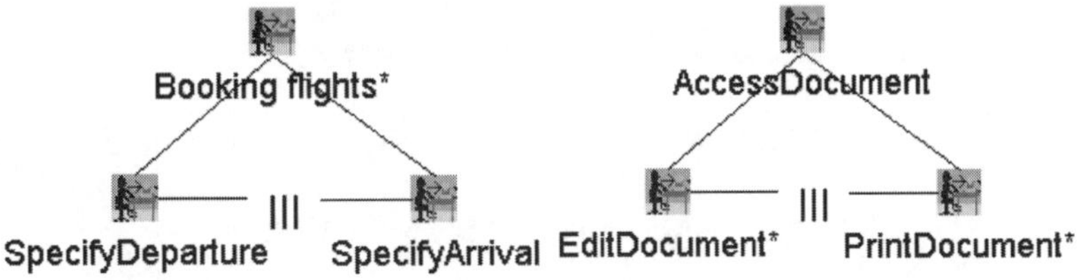

FIG. 24.7. Example of different structures in task model specifications.

Some of the features described were not included in the original version of CTT. The order independence operator has been added to provide an immediate description of a specific type of constraint. In the case of order independence, when a task is started, it has to be completed before another task is begun. With concurrent tasks, it is possible to start the performance of another task before the completion of the first task started.

Also recently added is the platform attribute. It allows designers to specify what platforms the task is suitable for. In addition, it is possible to specify, for each object manipulated by the task, the platforms suitable for supporting the task. This type of information allows designers to better address the design of nomadic applications—those that can be accessed through a wide variety of platforms. The basic idea is that designers start by specifying the task model for the entire nomadic application and can then derive the system task model for each platform by filtering tasks through the platform attribute.

24.4.1 An Example of Specification

Figure 24.8 represents the task model of a mobile phone. The task model is presented in an incomplete form for the sake of brevity. It is important to note that the task model depends on the features of one specific type of cellular phone. It shows nonetheless that task models can be useful for analyzing the design of more recent mobile systems.

The model is structured into two sequential parts: One dedicated to connecting the phone and the other one to handling the communication. At any time, it is possible to switch off the phone. The first part is a sequence of interactive and system tasks. The second part is dedicated to the actual use of the phone, which is iterative because multiple uses are possible in a session. First of all, the user has to decide what he or she wants to do. Next, there are two main choices: either make a call (the most frequent function) or use other functions. In case of a call, the user can either select the number from a list or recall and enter it. The other possible uses are not detailed for the sake of brevity.

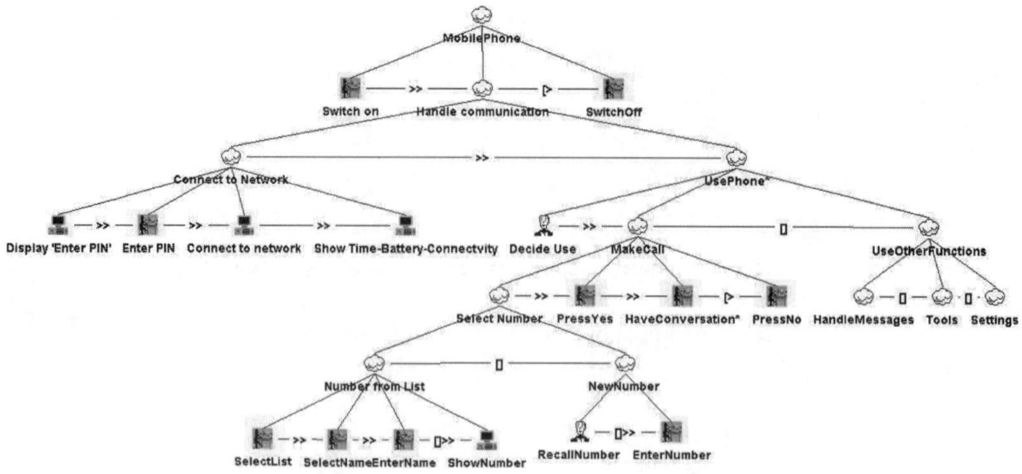

FIG. 24.8. Example of task model specification with ConcurTaskTrees.

24.5 UML AND TASK MODELS

Model-based approaches have often been used in software engineering. If we consider UML, one of the most successful model-based approach for the design of software systems, we notice a considerable effort is usually made to provide models and representations to support the various phases and parts of the design and the development of software applications. However, of the nine representations provided by UML, none is particularly oriented to supporting the design of user interfaces. Of course, it is possible to use some of them to represent aspects related to the user interface, but it is clear that this is not their main purpose. To integrate the two approaches (task models represented in ConcurTaskTrees and UML), various basic strategies can be used (see below). These can exploit, to different extents, the extensibility mechanisms built into UML itself (constraints, stereotypes, and tagged values), which can extend UML without requiring changes in the basic UML metamodel. The basic strategies are as follows:

Representing elements and operators of a task model by an existing UML notation. For example, if you consider a ConcurTaskTrees model as a forest of task trees, where ConcurTaskTrees operands are nodes and operators are horizontally directed arcs between sibling nodes, you can represent the model as UML class diagrams. Specific UML class and association stereotypes, tagged values, and constraints can be defined to factor out and represent properties of and constraints on CTT elements.

Developing automatic converters from UML to task models. For example, it is possible to use the information contained in system-behavior models supported by UML (i.e., use cases, use case diagrams, and interaction diagrams) to develop task models.

Building a new UML for interactive systems. A new UML can be obtained by explicitly inserting ConcurTaskTrees in the set of available notations while still creating semantic mapping of ConcurTaskTrees concepts into a UML metamodel. This encompasses identifying correspondences, at both the conceptual and structural levels, between ConcurTaskTrees elements and concepts and UML ones and exploiting UML extensibility mechanisms to support this solution.

Of course, there are advantages and disadvantages in each approach. In the first, it would be possible to have a solution compliant with a standard that is already the result of many long discussions involving many people. This solution is surely feasible. An example is given in Nunes and Falcao (2000): CTT diagrams are represented as stereotyped class diagrams. Furthermore, constraints associated with UML class and association stereotypes can be defined so as to enforce the structural correctness of ConcurTaskTrees models. However, two key issues arise: whether the notation has enough expressive power and whether the representations are effective and support designers in their work rather than complicate it. The usability aspect is important not only for the final application but also for the representations used in the design process. For example, activity diagrams are general and provide sufficient expressive power to describe activities. However, they tend to provide lower level descriptions than those in task models, and they require rather complicated expressions to represent task models describing flexible behaviors.

Thorny problems also emerge for the second approach. In particular, it is difficult to first model a system in terms of object behaviors and then derive a meaningful task model from such models. The reason is that object-oriented approaches are usually effective for modeling internal system aspects but less adequate for capturing users' activities and their interactions with the system.

The third approach seems to offer more promise as a way to capture the requirements for an environment supporting the design of interactive systems. However, care should be taken to ensure that software engineers who are familiar with traditional UML can make the transition to this new method easily and that the degree of extension from the current UML standard remains limited. More specifically, use cases could be useful in identifying tasks to perform and related requirements, but then there is no notation suitable for representing task models, although there are various ways to represent the objects of the system under design. This means that there is a wide gap that needs to be filled in order to support models able to assist in the design of user interfaces.

In defining a UML for interactive systems (Paternò, 2001), the designers can explicitly introduce the use of task models represented in CTT. However, not all UML notations are equally relevant to the design of interactive systems; the most important in this respect appear to be use cases, class diagrams, and sequence diagrams.

In the initial part of the design process, during the requirement elicitation phase, use cases supported by related diagrams should be used. Use cases are defined as coherent units of externally visible functionality provided by a system unit. Their purpose is to define a piece of coherent behavior without revealing the internal structure of the system. They have shown to be successful in industrial practice.

Next, there is the task-modeling phase, which allows designers to obtain an integrated view of functional and interactional aspects. In particular, interactional aspects (aspects related to the ways of accessing system functionality) cannot be captured well in use cases. In order to overcome this limitation, use cases can be enriched with scenarios (i.e., informal descriptions of specific uses of the system). More user-related aspects can emerge during task modeling. In this phase, tasks should be refined, along with their temporal relationships and attributes. The support of graphically represented hierarchical structures, enriched by a powerful set of temporal operators, is particularly important. It reflects the logical approach of most designers, allows the description of a rich set of possibilities, is declarative, and generates compact descriptions.

In parallel with the task-modeling work, the domain modelling is also refined. The goal is to achieve a complete identification of the objects belonging to the domain considered and the relationships among them. At some point there is a need for integrating the information between the two models. Designers need to associate tasks with objects in order to indicate what objects should be manipulated to perform each task. This information can be directly introduced in the task model. In CTT it is possible to specify the relationships between tasks and objects. For each task, it is possible to indicate the related objects, including their classes and identifiers. However, in the domain model more elaborate relationships among the objects are identified (e.g., association, dependency, flow, generalization, etc.), and they can be easily supported by UML class diagrams.

There are two general kinds of objects that should be considered: presentation objects, those composing the user interface, and application objects, which are derived from the domain analysis and responsible for the representation of persistent information, typically within a database or repository. These two kinds of objects interact with each other: Presentation objects are responsible for creating, modifying, and rendering application objects. The refinement of tasks and objects can be performed in parallel so that first the more abstract tasks and objects are identified and then the more concrete tasks and objects. At some point, the task and domain models should be integrated in order to clearly specify the tasks that access each object and, vice versa, the objects that are manipulated by each task.

24.6 TASK MODELS FOR MULTIPLATFORM APPLICATIONS

The availability of an increasing number of device types ranging from small cellular phones to large screens is changing the structure of many applications (see chap. 26). It is often the case that within the same application users can access its functionality through different devices. In a nomadic application, users can access the application from different locations through different devices. This means that context of use is becoming increasingly important, and it is essential to understand its relationship with tasks.

In general, it is important to understand what type of tasks can actually be performed in each platform. In a multiplatform application, there are various possibilities:

1. Same task on multiple platforms in the same manner (there could be only a change of attributes of the user interface objects).
2. Same task on multiple platforms with performance differences:
   - Different domain objects. For example, presenting information on works of arts can show different objects depending on the capability of the current device.
   - Different user interface objects. For example, in a desktop application it is possible to support a choice among elements graphically represented, whereas the same choice in a Wap phone is supported through a textual choice.
   - Different task decomposition. For example, accessing a work of art through a desktop can support the possibility of accessing related information and further details not supported through a Wap phone.
   - Different temporal relationships. For example, a desktop system can support concurrent access to different pieces of information that could be accessed only sequentially through a platform with limited capabilities.
3. Tasks meaningful only on a single platform. For example, browsing detailed information makes sense only with a desktop system.
4. Dependencies among tasks performed on different platforms. For example, during a physical visit to a museum users can annotate works of art that should be shown first when they access the museum Web site.

All these cases can be specified in ConcurTaskTrees through the use of the platform attribute that can be associated with both tasks and objects. Such an attribute indicates what platforms are suitable to support the various tasks and objects, thus allowing designers to represent the task model even in the case of nomadic applications.

24.6.1 Task Models for Cooperative Applications

Providing support for cooperative applications is important because the increasing spread and improvement of Internet connections makes it possible to use many types of cooperative applications. Consequently, tools supporting the design of applications where multiple users can interactively cooperate are more and more required.

In our approach, when there are multi-user applications, the task model is composed of various parts. A role is identified by a specific set of tasks and relationships among them. Thus, there is one task model for each role involved. In addition, there is a cooperative part whose purpose is to indicate the relationships among tasks performed by different users.

The cooperative part is described in a manner similar to the single user parts: It is a hierarchical structure with indications of the temporal operators. The main difference is that it includes cooperative tasks—those tasks that imply actions by two or more users in order to be performed. For example, negotiating a price is a cooperative task because it requires actions from both a customer and a salesperson. Cooperative tasks are represented by a specific icon with two persons interacting with each other.

In the cooperative part, cooperative tasks are decomposed until tasks performed by a single user are reached (these are represented with the icons used in the single-user parts). These single-user tasks will also appear in the task model of the associated role. They are defined as *connection tasks* between the single-user parts and the cooperative part. In the task specification of a role (see, e.g., the top part of Fig. 24.9), connection tasks are identified by a double arrow under their names.

With this structure of representation, in order to understand whether a task performance is enabled, it is necessary to check both the constraints in the relative single-user part and the constraints in the cooperative part. It may happen that a task without constraints in the

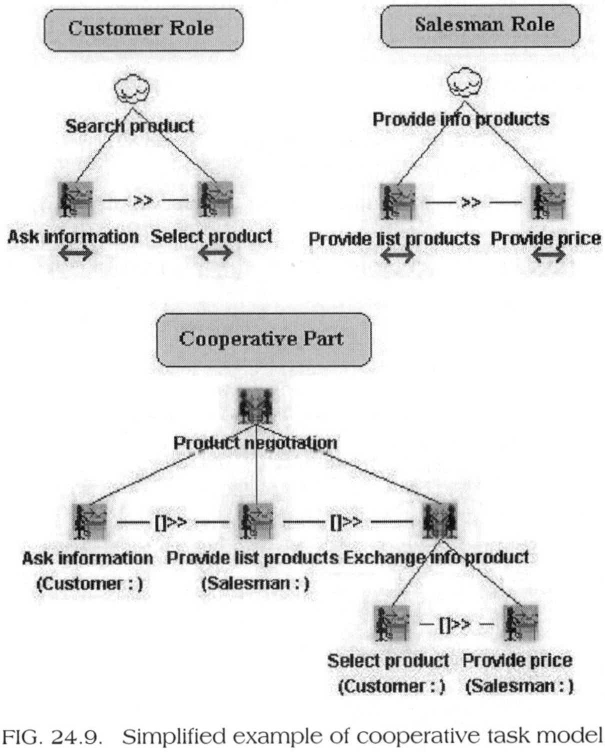

FIG. 24.9. Simplified example of cooperative task model.

single-user parts is not enabled because there is a constraint in the cooperative part, indicating that another user must first perform another task. Figure 24.8 shows a *Search Product* task performed by a customer and a *Provide Info Product* task performed by a salesperson. If we consider each part of the task model in isolation, these two tasks can be started immediately. However, if we consider the additional constraint indicated in the part below, we can see that the performance of the *Provide list products* task (by the salesperson) needs to wait for the performance of the *Ask information* task (by the customer) in order to be enabled.

We are aware that in cooperative applications, users interact not only while performing their routine tasks but also when some external event blocks the work flow. Moreover, such events can create situations where previously defined tasks need to be changed and/or repeated. The task model associated with a certain role enables the distinguishing of individual tasks and of tasks that involve cooperation with other users. Then, moving on to the cooperative part, it is possible to see what the common goals are and what cooperation among multiple users is required in order to achieve them. The set of temporal operators also allows the description of flexible situations where, when something occurs, activities need to be performed in a different manner. Of course, specifying flexible behaviors requires an increase in the complexity of the specification. Trying to anticipate everything that could go wrong in complex operations would render the specification exceedingly complex and large. The extent to which designers want to increase the complexity of the specification to address these issues depends on many factors (e.g., the importance of the specification for the current project, the resources available, and their experience in task modeling).

The CTTE tool allows different ways to browse the task model of cooperative applications. The designers can interactively select the pane associated with the part of the task model of interest. (In Fig. 24.9 the task model is composed of a cooperative part and two roles, customer and sales representative.) In addition, when connection tasks are selected in the task model of a role (they are highlighted by a double arrow below their name), it is possible to automatically visualize where they appear in the cooperative part and vice versa.

24.6.2 Tool Support

Table 24.2 summarizes some features deemed important for analysis tools and shows whether task models are supported by some of the currently available tools. The features provided by CTTE (see Fig. 24.10) highlight its contribution to understanding the possibilities in terms of the analyses that can be performed with such tools. The first row indicates whether the tools are able to consider concurrent tasks, (that is, their analysis is not limited to sequential

TABLE 24.2
Comparison of Task-Modeling Tools

	CTTE	*EUTERPE*	*VTMB*	*QDGOMS*	*IMAD**	*Adept*
Concurrent task	X	X	X		X	X
Cooperative tasks	X	X			X	
Metrics	X	X				
Reachability	X					
Performance evaluation	X			X		
Simulator	X		X		X	

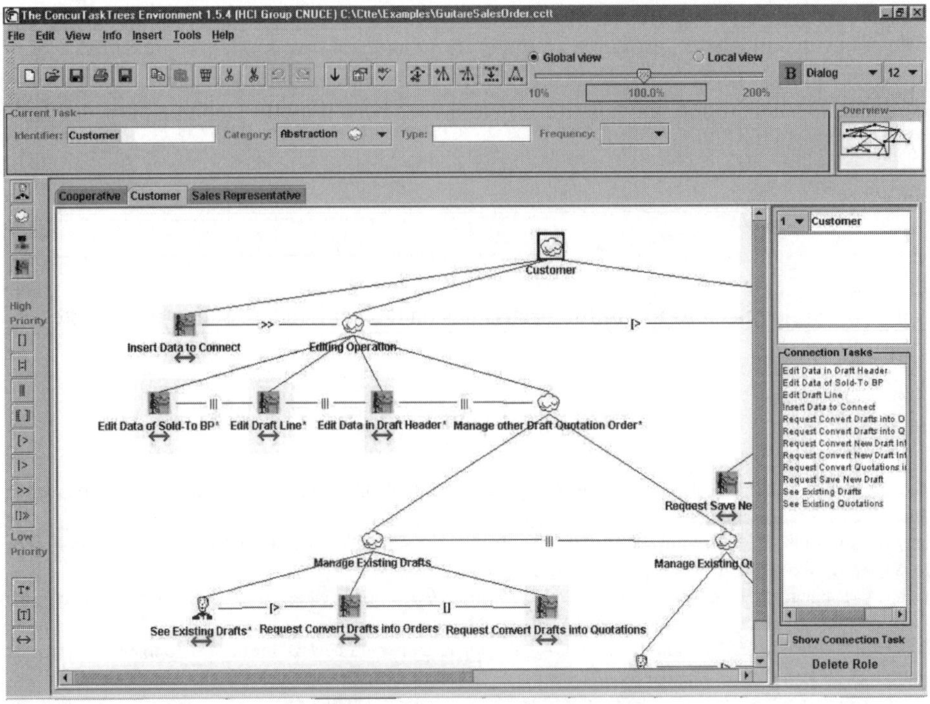

FIG. 24.10. The layout of CTTE.

tasks). The next row indicates their ability to analyze cooperative tasks involving multiple users. The third row indicates whether they are able to calculate some metrics. CTTE is able to compare two models with respect to a set of metrics, and EUTERPE (van Welie, van der Veer, & Eliëns, 1998; chaps. 6 and 7) also supports the calculation of some metrics to analyze a specification and find inconsistencies or problems. The remaining features considered concern the ability to predict task performance. This is usually supported by tools for GOMS such as QDGOMS (Beard, Smith, et al., 1996) that interactively simulate the task model's dynamic behavior. Interactive simulation is also another important feature supported by a few tools, such as VTMB (Biere, Bomsdorf, & Szwillus, 1999).

24.7 CONCLUSION

CTT and its related environment represent a contribution to engineering the use of task models in the design, development, and evaluation of interactive systems. In this chapter, the main features of the notation have been discussed, and indications how to address important related issues (e.g., integration with UML, design of multiplatform applications, and creation of task models for cooperative applications) have also been provided.

The approach has also stimulated a good deal of research work: Nunes and Falcao (2000) proposed a transformation from UML to CTT, Sage and Johnson (1997) proposed integrating CTT and a functional language to obtain a user interface prototyping environment, and Pribeanu, Limbourg et al., (2001) explored ways to extend CTT to include context-of-use information. Furthermore, the reference framework for designing user interfaces able to adapt to multiple platforms while preserving usability (Calvary, Coutaz, & Thevenin, 2001) considers task models (in particular, those represented by CTT) as a starting point for the design work, and the MEFISTO project, which aimed to identify methods supporting the design of air traffic control applications, provided an integration between task models represented by CTT and system models represented by ICO Petri Nets (Navaree, Palanque, Paternò, Santoro, & Bastide, 2001). In addition, the notation and tool have been used for teaching purposes and in various projects all over the world.

ACKNOWLEDGMENT

Thanks to Carmen Santoro for useful discussions on the topics addressed in this chapter.

REFERENCES

Beard, D., Smith, D., & Denelsbeck, K. (1996). Quick and dirty GOMS: A case study of computed tomography. *Human-Computer Interaction, 11*, 157–180.

Biere, M., Bomsdorf, B., & Szwillus G. (1999). Specification and simulation of task models with VTMB. In *Proceedings CHI'99* (Extended abstracts; pp. 1–2). New York: ACM Press.

Booch, G., Rumbaugh, J., & Jacobson, I. (1999). *Unified Modeling Language reference manual.* Reading, MA: Addison-Wesley.

Calvary, G., Coutaz, J., & Thèvenin, D. (2001). A Unifying reference framework for the development of plastic user interfaces. In *Proceedings Engineering HCI '01* (pp. 218–238). Springer-Verlag.

Card, S., Moran, T., & Newell, A. (1983). *The psychology of human-computer interaction.* Hillsdale, NJ: Lawrence Erlbaum Associates.

Carroll, J. (2000). *Making use: Scenario-based design of human-computer interactions.* Cambridge, MA: MIT Press.

Cockburn, A. (1997). *Structuring use cases with goals.* Available: http://members.aol.com/acockburn

Gamboa-Rodriguez, F., & Scapin, D. (1997). Editing MAD* task description for specifying user interfaces at both semantic and presentation Levels. In *Proceedings DSV-IS'97* (pp. 193–208). Springer-Verlag.

Gamma, E., Helm, R., Johnson, R., & Vlissides J. (1995). *Design patterns: Elements of reusable object-oriented software.* Reading, MA: Addison-Wesley.

Hartson, R., & Gray, P. (1992). Temporal aspects of tasks in the user action notation. *Human-Computer Interaction, 7*, 1–45.

Hudson, S., John, B., Knudsen, K., & Byrne, M. (1999). A tool for creating predictive performance models from user interface demonstrations. In *Proceedings UIST'99* (pp. xxx). New York: ACM Press.

Imaz, M., & Benyon, D. (1999). How stories capture interactions. In *Proceedings Interact'99* (pp. 321–328).

John, B., & Kieras, D. (1996). The GOMS family of analysis techniques: Comparison and contrast. *ACM Transactions on Computer-Human Interaction, 3*, 320–351.

Limbourg, Q., Ait El Hadj, B., Vanderdonckt, J., Keymolen, G., & Mbaki, E. (2000). Towards derivation of presentation and dialogue from models: Preliminary results. In *Proceedings DSV-IS 2000* (Lecture Notes in Computer Science 1946; pp. 227–248). Springer-Verlag.

Mori G., Paternò F., & Santoro, C. (2002). CTTE: Support for Developing and analyzing task models for interactive systems design. *IEEE Transactions on Software Engineering,*

Navarre, D., Palanque, P., Paternò, F., Santoro, C., & Bastide, R. (2001). A tool suite for integrating task and system models through scenarios. (pp. 88–113). Springer-Verlag.

Norman, D. (1986). Cognitive engineering. In D. Norman & S. Draper (Eds.), *User centered system design: New perspectives on human-computer interaction* (pp. 31–61). Hillsdale, NJ: Lawrence Erlbaum Associates.

Numes, N., & Falcao, J. (2000). Towards a UML profile for user interface development: The Wisdom approach. In *Proceedings UML'2000* (Lecture Notes in Computer Science; (pp. 50–58). Springer-Verlag.

Paganelli, L., & Paternò, F. (2002). Intelligent analysis of user interactions with Web applications. In *Proceedings ACM IUI 2002* (pp. 111–118). New York: ACM Press.

Paris, C., Tarby, J., & Vander Linden, K. (2001). A flexible environment for building task models. In *Proceedings of the HCI 2001,* Lille, France. (pp. xxx).

Paternò, F. (1999). *Model-based design and evaluation of interactive application.* Springer-Verlag.

Paternò. (2001). Towards a UML for interactive systems. In *Proceedings Engineering HCI '01* (pp. 175–185). Toronto: Springer-Verlag.

Paternò, F., & Santoro, C. (2002). One model, Many interfaces. In *Proceedings CADUI 2002* (pp. 143–154). Kluwer.

Paternò, F., Santoro, C., & Sabbatino, V. (2000). Using information in task models to support design of interactive safety-critical applications. In *Proceedings AVI'2000* (pp. 120–127). New York: ACM Press.

Pribeanu, C., Limbourg Q., & Vanderdonckt, J. (2001). Task modelling for context-sensitive user interfaces. In *Proceedings DSV-IS 2001* (pp. 49–68). Heidelberg: Springer-Verlag.

Sage, M., & Johnson, C. (1997). Haggis: Executable specifications for interactive systems. In *Proceedings DSV-IS'97* (pp. 93–109). Heidelberg: Springer-Verlag.

Scapin, D., & Pierret-Golbreich, C. (1989). Towards a method for task descrption: MAD. In *Proceedings Work with Display Unit* (pp. 78–92). The Netherlands: Elsevier.

Shepherd, A. (1989). Analysis and training in information technology tasks. In D. Diaper (Ed.), *Task analysis for human-computer interaction* (pp. 15–55). Chichester, England: Ellis Horwood.

Tam, R.C.-M., Maulsby, D., & Puerta, A. (1998). U-TEL: A tool for eliciting user task models from domain experts. In *Proceedings IUI'98* (pp. 77–80). New York: ACM Press.

van der Veer, G., Lenting, B., & Bergevoet, B. (1996). GTA: Groupware task analysis-modelling complexity. *Acta Psychologica, 91*, 297–322.

van Welie, M., van der Veer, G. C., & Eliëns, A. (1998). An Ontology for task world models. In *Proceedings DSV-IS'98* (pp. 57–70). Heidelberg: Springer-Verlag.

Wilson, S., Johnson, P., Kelly, C., Cunningham, J., & Markopoulos, P. (1993). Beyond hacking: A model-based approach to user Interface design. In *Proceedings HCI'93* (pp. 40–48). New York: Cambridge University Press.

25

Using Formal Specification Techniques for the Modeling of Tasks and the Generation of Human-Computer User Interface Specifications

Mourad Abed
University of Valenciennes

Dimitri Tabary
University of Valenciennes

Christophe Kolski
University of Valenciennes

The increasing presence and democratisation of interactive systems brings about an even greater need for automated or semi-automated generation of human-computer user interfaces (UIs). With the aim of generating the presentation and the dialogue parts of interactive systems, we suggest a method, based on the model-based user interface design (MBD) paradigm, named TOOD (task object oriented design). It defines a task model, which makes it possible to identify the objects handled as well as the distribution of tasks between the user and the machine. It then formalizes the dynamic behavior of the tasks and the objects in order to lead to the generation of a HMI prototype derived directly from the task model. A software environment named TOOD-IDE supports the method. A case study concerning HMI design for Air Traffic Control illustrates the approach.

25.1 INTRODUCTION

The current breakthrough in communication and information technologies places the emphasis on the domination of interactive systems at every level: interactive terminals for the general public, multisite organization of companies, migration of complex system control rooms (air traffic control, nuclear power plant control, etc.), development of interactive and personalized Internet sites, and so on. The increasing presence and democratization of interactive systems brings about an even greater need for the automatic generation of human-machine interfaces that integrate the needs and habits of the users.

One of the solutions created through research is the generation of a human-computer user interface prototype based on the model-based user interface design (MBD) paradigm. This paradigm refers to an explicit, mostly declarative, description that captures the semantics of the application and the knowledge necessary for the specification of the appearance as well as for the behavior of the interactive system (Sukaviriya, Kovacevic, Foley, Myers, Olsen, & Schneider-Hufschmidt, 1994; Szekely, 1996; Wilson, Johnson, Kelly, Cunningham, & Markopoulos, 1993); it is part of the line of UIMSs (user interface management systems) and provides a true alternative for the construction of interfaces. It recommends a top-down approach and requires the drawing up of formal specifications by the designers (Palanque & Paternò, 1997), rather than by a computer program. This knowledge, described in a high-level specialized language, is translated or interpreted for the total or partial generation of the application code.

Current research trends therefore increasingly suggest design methods employing various models: user task models, models of the data handled by the system, user models, and presentation models. Interface construction tools use such models to generate and check the dialogue and the interface of an application.

The central component of MBD is the model. It represents the application characteristics from the point of view of the interface in the widest sense, including the notions of dialogue and presentation. Szekely (1996) indicated that models can be classified based on their level of abstraction:

- At the highest level, we find the task and domain models. A domain model represents the objects handled by the application, their relationships, and the tasks that the application can perform on these objects. A task model represents the tasks that the user must be able to perform. They are generally broken down into a hierarchy of tasks and subtasks. Very often, the terminal tasks correspond to operations that can be performed directly using the application.
- The second level of abstraction represents the general structure of the application interface. This structure is described in terms of low-level interactions (e.g., selection of an element in a set), presentation units (corresponding to windows), and elements of information (a value or set of values, labels, constants, and so on). This structure provides an abstract definition of the information to be presented to the user and the dialogues used in order to interact with this information.
- The third level forms the concrete specification of the interface. It specifies the return of information from the second level in terms of toolbox elements (menu, check box, and so on).

The importance of MBDs, specifically task-based user interface designs, in the design, development, and assessment of an interactive application is generally recognized. However, the use of such task models has been greatly limited by the lack of automatic tools that support them (chaps. 1 and 23); the need for such support was shown during recent workshops (Bomsdorf & Szwillus, 1999). Indeed, when designers tackle real applications, they immediately feel the necessity of having tools that allow them, first, to analyze the results of their modeling work, then to modify that work in order to show it to other people or use it to implement the software artifacts that support the tasks identified.

With the aim of generating a prototype for the presentation and dialogue of an interactive system, we suggest a user task object oriented design method named TOOD (task object oriented design). It defines a task model that makes it possible to identify the objects handled as well as the distribution of tasks between the user and the machine. It then formalizes the dynamic behavior of the tasks and the objects to aid in the generation of a user interface prototype derived directly from the task model. A software environment named TOOD-IDE, created using the Java programming language, supports the method.

25.2 RELATED WORK

In this section, we give a nonexhaustive review of existing representative approaches based on the MBD paradigm. Table 25.1 presents a certain number of methods based on this paradigm and specifies the various models used to cover the development cycle. The table lists the tools used by these methods; the part of the interactive system generated, along with the generation mode; and the type of task model used in order to generate the user interface.

The comparison shows that all of these methods are capable of generating a user interface presentation and dynamics more or less automatically. They all define a dialogue model, which defines the links between the objects of the domain and their representation in the interface. The major difference between these approaches involves the effective integration of a user model in order to define the dialogue. The TADEUS, MECANO, and TRIDENT methods are concerned with a computing task model that precludes any integration of user knowledge in the model. However, these methods define a user model that includes the major characteristics associated with the type of user and the level of experience required. On the other hand, the GLADIS++, DIANE+ (chaps. 6, 22, 23, and 26), and ADEPT (chaps. 6 and 24) methods base their generation on a user task model. They belong to the current trend toward task-based user interface design and generate an interface based on user tasks.

These methods remain insufficient, from our point of view, and show the tendency of computer design specialists to favor formal models (e.g., entity-relation models, object-oriented design) that allow them to automate the design process to the greatest degree possible. Unfortunately, automation is not always possible with the user task models (declarative). Our work attempts to contribute to this research trend by developing a method based on the use of formal models that favor iterative design with assessments. The purpose of TOOD is to formalize the user task by jointly using the object-oriented approach and Object Petri Nets (OPNs). The object-oriented approach owes part of its success to the fact that it favors structuring because of its focus on classes, objects, systems, and subsystems. The OPNs combine user-friendly graphics with a mathematical formalism and thereby provides design possibilities for tools for the verification, validation, and assessment of performance and for prototyping. The semantics of object-oriented design are therefore sufficiently complete and rich for the representation of the static part of the task. In TOOD, the dynamics part is taken over by the OPNs.

In this chapter, we begin by giving details concerning the TOOD development cycle. We continue by defining its task model. We then describe its environment support by means of a simple case study. Finally, we give a succinct presentation of the operational model.

25.3 TOOD AND THE USER INTERFACE DEVELOPMENT CYCLE

The TOOD design process can be divided into four major stages (Fig. 25.1):

1. The analysis of the existing system and user needs is based on the human activity; it forms the entry point and the basis for any new designs.
2. The structural task model is concerned with the description of the user tasks in the system to be designed. Two models are created at this level: the static structural task model and the dynamic structural task model.

TABLE 25.1
Model-Based Design Method

MBD/Tool	*Specification of Need*	*Global Design*	*Detailed Design*	*Generation (Type of Generation)*	*Type of Task Model*
ADEPT/ADEPT (Markopoulos et al., 1996)	Task model (TKS; Johnson, 1999) User model	Abstract interface model	Concrete interface model	Presentation and dynamics (assisted)	User centered
DIANE+ (Tarby & Barthet, 1996)	Task model Domain model (OPAC)	Specification of goals, procedures, and actions	Dialogue model	Presentation, dynamics, and assistance (automatic)	User centered
GLADIS++/ALADIN++ (Ricard & Buisine, 1996)	User task model	Service model	Dialogue model	Presentation and dynamics (assisted)	User centered
TRIDENT/TRIDENT (Bodart et al., 1995)	Task model Entities/relations	Activity Chaining graph	Definition of presentation units	Presentation and dynamics (assisted)	Computing tasks
MECANO/MOBI-D (Puerta, 1996; Puerta & Maulsby, 1997)	Task model User model Domain model	Dialogue and interface model	Presentation and dynamics (automatic)		Computing tasks
TADEUS/TADEUS (Schlungbaum, 1998)	Task model User model Domain model	Design of the dialogue presentation and the dynamics according to two types of inter-window and intra-window dialogue		Presentation and dynamics (assisted)	Computing tasks

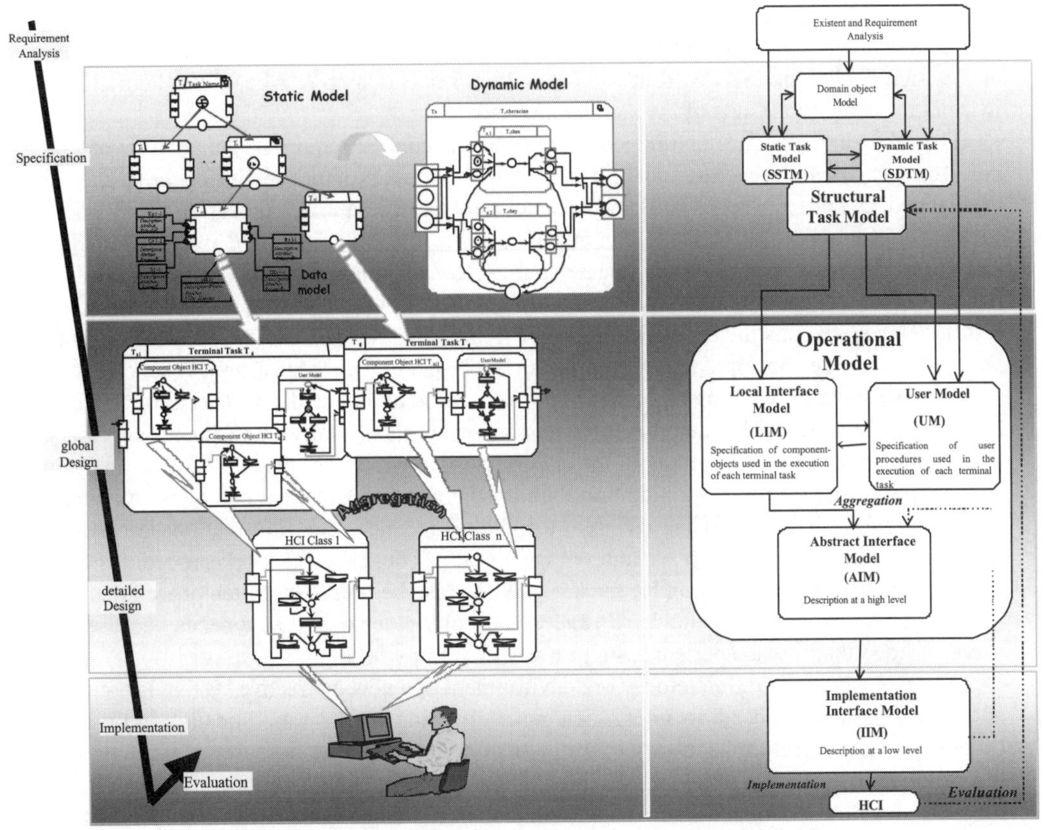

FIG. 25.1. TOOD and the user interface development cycle.

3. The operational model makes it possible to specify the user interface objects in a local interface model as well as the user procedures in a user model of the system to be designed. It uses the needs and characteristics of the structural task model to create an abstract interface model, which describes the user's objectives and procedures.
4. The creation of the user interface is concerned with the computerized implementation of the specifications resulting from the previous stage, supported by the software architecture defined in the interface implementation model (IIM).

The TOOD method is supported by an environment (TOOD-IDE) developed using Java. It makes model capture and syntactic checking easier. In addition, it supports the test and simulation activities of the dynamic task model. Examples of screen pages are given later, illustrated with models used to describe tasks related to air traffic control.

25.4 ANALYSIS OF THE EXISTING SYSTEM

In order to know what the user is supposed to do using the new system, we must find out what is achieved in real work situations (the activity analysis) by analyzing the existing version of the system or a similar one.

During this stage, we suggest describing the user interface of the existing system and user behavior (cognitive and physical) in a single system model. We demonstrate how the user's knowledge of the system can influence its resulting use. The construction of this model, described in our research projects (Abed, 1990; Bernard, 1994), associates Structured Analysis and Design Technique (SADT; Ross, 1977) and Petri nets (David & Alla, 1994; Petri, 1962) in order to model the results of the activity analysis.

When the characteristics and the complexity of the context require it, the analysis can be based on eye movement capture and the description of behaviors of the user during the task using specialized instruments (e.g., oculometer, electronic "spy"; Mullin, Anderson, Smallwood, Jackson, & Katsavras, 2001). This technique of gathering information is used together with pre- and postexperiment interviews and questionnaires. Readers may want to consult the detailed description of this environment and representation given in Abed and Angué (1994) and Abed (2001). This work constitutes a way of integrating the user, the task, and the environment into an user interface design approach. The approach is then centred on the task and is user oriented.

The analysis of the existing task takes in not only the activity but also its execution context. It describes the activities by identifying the human, material, and software resources necessary to accomplish the activity, along with the environment in which it is performed. The analysis concentrates more on the work situation than on the existing technical solutions. As stated previously, it is carried out using different techniques for gathering information; in cases in which it is impossible to carry out interviews, the users describe the work procedures they follow (see chap. 2). Information for the analysis can also be found in written procedures and by questioning experts in the field. The analysis usually gives rise to new needs because the users, through their own experience, suggest new ways of working.

The design and integration into a work environment of a new interactive system is not only a technical problem. Indeed, its integration generally brings about a modification in the way of working. It is therefore necessary to distinguish between the existing way of working and the new way of working following the integration of the interactive system. The results of the analysis of the existing system and needs are then described in three basic models: a task model, a user model, and a domain model. The information taken from the analysis is varied and, in our opinion, should be transposed into a single formal model to reduce ambiguity in interpretation. This model can then be used for specification as well as for design purposes.

25.5 STRUCTURAL TASK MODEL

After the previous stage, which concerns the analysis of the existing system and its user's activity, the structural task model makes it possible to establish a coherent and relatively complete description of the tasks to be achieved using the future system while avoiding the drawbacks of the existing system. Two models are developed for this: a structural static model and a structural dynamic model.

As in MAD* (Gamboa-Rodriguez & Scapin, 1997; chaps. 6, 22, and 24), DIANE+ (Tarby & Barthet, 1996; chaps. 6, 22, 23, and 26), and GLADIS++ (Buisine, 1999), the structural task model is designed to take both the user and his or her task into account in the development cycle. The objective is to provide a methodological tool representation for development teams, enabling them to abstract the information about the user task necessary for the formal design of the user interface and to permit its integration into the development cycle. It is a working language that facilitates dialogue and information exchange between the members of a development team.

The construction of the structural model involves four iterative stages:

- hierarchical decomposition of tasks
- identification of the data, named domain objects, necessary for the tasks' execution

- definition of the intertask relations and their dynamics
- continuation of the interactive process until all the terminal tasks (that is, nondecomposable tasks) have been identified.

25.5.1 Static Structural Task Model

The structural model enables the breakdown of the user's stipulated work with the interactive system into significant elements, called tasks. Each task is considered an *autonomous entity* that corresponds to a goal or subgoal and can be situated at various hierarchical levels. The goal remains the same in the various work situations. In order to perfect the definition, TOOD formalizes the concept of tasks using an object representation model in which each task can be seen as an *object*, an instance of the *task class*. This representation consequently attempts to model the task class by means a generic structure of coherent and robust data, making it possible to describe and organize the information necessary for the identification and performance of each task.

This parallel with software engineering guarantees a strong link between a user-centered specification based on ergonomic models (chap. 26 provides a list of ergonomic guidelines) and the software design based on the object model.

Two types of graphical and textual document define each task class (inspired by HOOD [Delatte, Heitz, & Muller, 1993] and SADT-Extended [Feller & Rucker, 1989]), as shown in Fig. 25.2. The task class is studied as an entity using four components: the input interface (II), the output interface (OI), the resources or means (M), and the body (B).

We also associate a certain number of identifiers with these describers, making it possible to distinguish the task class from the others: name, goal, index, type, and hierarchy (Table 25.2).

The *body (B)* is the central unit of the task class. For intermediate or hierarchical tasks, it gives the task procedure diagram, that is to say, the logical and temporal relations of the subtasks. In some way, these relations reflect the user's work organization. On the other hand, for terminal tasks, it defines the action procedures for the user interface–user couple. The specification for these procedures is produced in the task operational model. A task that belongs to a (decomposable) hierarchical level is called a *control task*.

Resources are human users and/or interactive system entities involved in the performance of the task. Four types of task are defined: manual, automatic, interactive, and Cooperative (Table 25.3; Tabary, Abed, & Kolski, 2000). A manual task is accomplished by one and only one user. An automatic task can be done by one or several system resources. An interactive task is accomplished by a user's interaction with a set of system resources. Finally, a cooperative task requires the activity of two or more users who interact either among themselves (human-human cooperation, as in the ConcurTaskTrees CTT model; chap. 24; Paternò, 2000) or with a collection of system resources (interactive cooperation).

The *input interface (II)* specifies the initial state of the task. It defines the data necessary for the execution of the task. These data are considered as the initial conditions to be satisfied at the beginning of the task. They include three categories of information (Table 25.4).

The *output interface (OI)* specifies the final state of the task. It is composed of two types of data (Table 25.5).

25.5.1.1 Domain Objects and Describer Objects

The input and output interface data of the task represents the domain objects. Using these objects, the task dynamics and the specification of the interface objects are defined, as will be described later. In a computing sense, these domain objects are entities that are independent of the dynamics and of the description of the task itself; they are created in order to model the

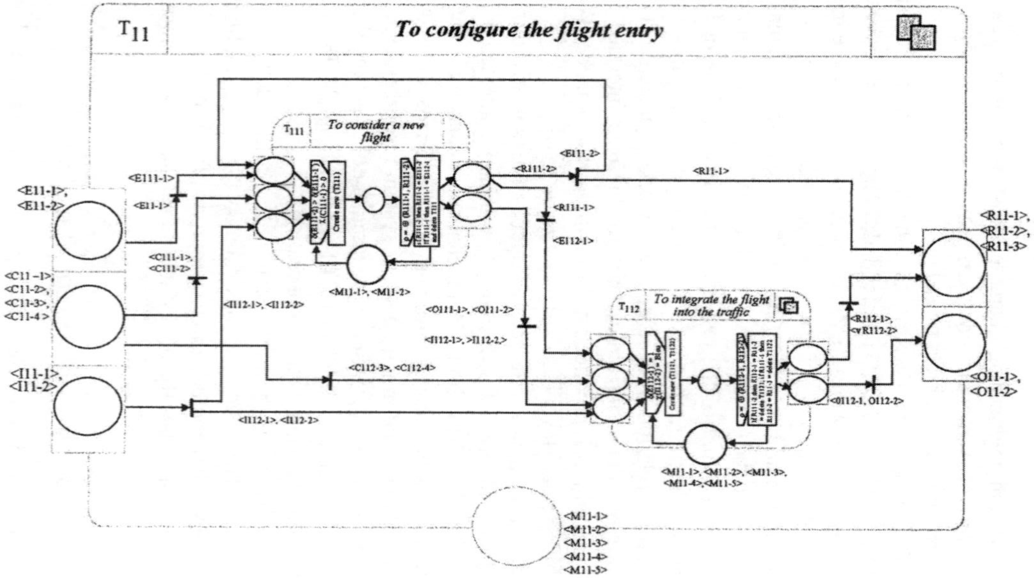

Task Class <*Name of the Task*>
{
Attributes:

// Index: Tx_x
// Goal:
// Hierarchy
Super Class < *Name of the Task* >
Sub-Classes <*List of the task*>
// Input Interface (II)
<*Objects Triggers List* >
<*Objects Input Data List*>
<*Objects Control Data List*>
//Output Interface (OI)
<*Reactions Objects List*>
<*Objects Output Data List*>

Implementation

Attributes:
- *att1* : --
- *att2* : --

Operations:

- *Op1* : -- //Description of the service... //.
- ...

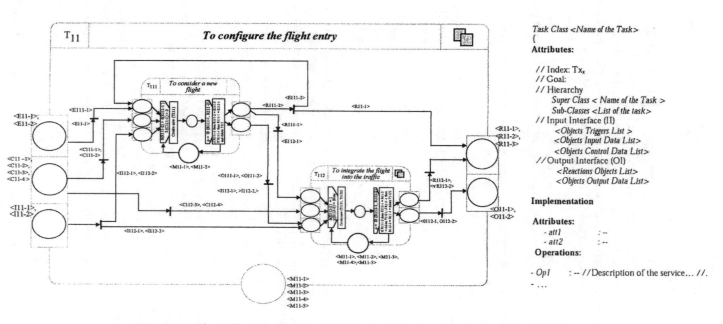

FIG. 25.2. Class diagram of the task class and domain objects (describer objects).

TABLE 25.2
Task Class Identifiers

Identifiers	*Description*
Name	Action verb followed by a complement (object treated by the task), reflecting the treatment to be performed by the task. It is preferable for the name to include vocabulary used by the users in order to respect their terminology during the development of the interface.
Goal	Explanation in natural language of the goal that the users or application wishes to achieve via the task.
Index (Tij)	Formal identifier of the task formed using the number of the master task, to which the sequential number corresponding to the said task is added.
Type	Nature of the task, designating its category: manual, automatic, interactive, or cooperative.
Hierarchy (■)	Number of task classes composing it; this is represented by a series of small squares.
Super Task	Parent task controlling the subtasks; it is also called the *control task*.
Subtask	Child task contributing to the performance of the parent task.

TABLE 25.3
Task Types

Types of Task	*Human Resource*	*System Resource*
Manual	1..1	0
Automatic	0	1..N
Interactive	1..1	1..N
Cooperative	2..N	0..N

Note. 1..1 = one and only one resource; 1..N = one or more resources; 2..N = at least two resources; 0..N = no or several resources 0 no resource.

data to be handled by the interactive application. A domain object can have several states (a life cycle notion); the value of its state directly affects the treatment associated with the object.

In the TOOD model, each domain object is referenced by one or several tasks via the *describer objects* of the task input and output interfaces. These references have a dual purpose: to resolve the problem of data sharing and to specify the nature of the domain object as concerns its role as data for the task. There are five types of object in accordance with the definition of the input and output interfaces of the task class: triggers, controls, input data, reactions, and output data. The abstract describer class represents the root of this structure. It has a name that defines each class and a unique index. It also defines an attribute called "ref," which references an element of data that is exchanged (see Fig. 25.2). The trigger class has an additional attribute named "priority," which is used to manage any conflicts arising when several trigger events are present simultaneously.

25.5.1.2 *Case Study: Air Traffic Control*

This example is an extract from the HEGIAS project (*H*ost for *E*xperimental *G*raphical *I*nterfaces of *A*dvanced Automation *S*ystem). HEGIAS concerns the reorganization and

TABLE 25.4
Input Interface

Attributes	*Description*
Events of interlocking (triggers)	Trigger events bring about the performance of the task. There are two types: explicit trigger events, which are set off by the system without user involvement, and implicit trigger events, which are set off by the user in order to perform the task.
Controls	Information that must be checked during the performance of the task. These controls affect the way in which the task is performed.
Input data	Information necessary during the performance of the task.

TABLE 25.5
Output Interface

Attributes	*Description*
Reactions	Results produced by the performance of the task. Their content indicates the following types of modification: physical (in this case it indicates the modification of the environment, such as an application call, change of state, etc.) and mental (indicates the modification or a new representation of the situation by the user). The reactions thus determine whether the aims are attained. If they are, the task will be repeated after a possible development of the situation.
Output data	Data transformed or created by the performance of the task.

computerization of an air traffic control position (chaps. 1, 13, and 19 also discuss air traffic control). The HEGIAS position mainly includes interfaces for displaying radar images and strip tables for new flights and integrated flights (a strip is grouping of information concerning a flight presented in a specific form). It also includes direct on-screen designation means, tactile screens, and air-ground communications (Abed & Ezzedine, 1998; Mahfoudhi, Abed, & Angue, 1995).

Each position involves two controllers, an *organic controller* and a *radar controller*. The *organic controller* deals with the strategic management of air traffic. This controller is responsible for traffic assessment and forecast and helps the radar controller. The organic controller is also responsible for coordination with adjacent sectors (a sector is a section of airspace managed by a control position). The *radar controller* is responsible for the tactical management of the traffic. He or she sees to the surveillance of the traffic in the sector, contacts the pilots, detects conflicts between flights, and resolves them. The radar controller is in charge of the sector and of the control decisions.

The elements of information (domain objects) about the flights used by the controllers are these:

- the path followed (that is the succession of beacons flown over by the aircraft)
- the times at which the beacons are passed (effective passing time and forecast or "estimated" time)
- the flight level(s) used (cruise or "stable," current, climbing, or descending).

The controllers also have other nonmodifiable information:

- the type of aircraft and its capacities (essentially the cruise speed and the climbing and descending rates)
- the departure airfield and destination airfield
- the aircraft reference (its name).

As we said at the beginning of this chapter, TOOD-IDE is an accompanying tool that refines the results as the analysis progresses. Therefore, the construction of the static task model begins with the identification and organization of the application tasks using diagrams. This organization is based on a hierarchical breakdown technique. It makes it possible to introduce increasingly fine levels of detail gradually through the structure of diagrams including the model. The root task, which is found at the highest level of abstraction and represents the overall objective of the system, is increasingly detailed in another diagram in which the tasks are connected through their interfaces (II) and (OI). This connection is shown by arrows indicating the domain objects processed by the tasks and also the intertask relations (sequence, choice, parallel, and iteration). These relations are expressed in a formal way in the dynamic model using the OPN formalism. Each of the control tasks (decomposable tasks) is made up of a set of subtasks, themselves made up of subtasks, going to the level at which the subtasks can no longer be broken down (then called *elementary tasks*).

The diagrams in Fig. 25.3 show a three-level breakdown for the example dealt with here. The root task "T_1: To plan the air traffic" gives rise to three subtasks, T_{11}, T_{12}, and T_{13}, which represent the three main functions of air traffic control respectively: management of incoming flights, management of outgoing flights, and assistance for the radar controller. It can be seen that task T_{11} can be activated by the event "$E_{11\text{-}1}$: Arrival of a new flight." This event calls upon two types of treatment corresponding to the two interactive tasks: "T_{111}: To consider a new flight" and "T_{112}: To integrate the flight into the traffic."

The terminal task T_{111} concerns the consideration of the parameters of the new flight (I_{11-1}: entry level, $I_{11\text{-}2}$: speed, $I_{11\text{-}3}$: route, $C_{11\text{-}1}$: type, $C_{11\text{-}2}$: reference, etc.). According to the context, this task finishes with either one of these two reactions:

- reaction "$R1_{11\text{-}2}$: Delaying of flight integration." This reaction reactivates the same task $E_{111\text{-}2} = R_{111\text{-}2}$ loop as long as the time left available to the controller to integrate is not exhausted. The time gap allows the controller to assess the flight entry conditions in the sector according to the evolution of traffic.
- reaction "$R_{111\text{-}1} = E_{112\text{-}1}$: Integration of new flight," which enables the trigger sequence of task T_{112}. This task is then broken down into two terminal subtasks, "T_{1121}: Classify the flight in traffic load" and "T_{1122}: Transfer flight to radar controller."

It is obvious that the organization of tasks at the level of each breakdown diagram is not done haphazardly but truly reflects a work strategy that could be adopted by the controller. The diagrams also specify for each task the system resources (interactive objects) and human resources (roles) necessary for it to be performed. For example, task T_{111} is carried out on the basis of an interaction between its two resources, "$M_{111\text{-}1} = M_{11\text{-}1}$: Organic controller" and "$M_{111\text{-}2} = M_{11\text{-}2}$: Table of new strips." The behavior and interaction of these resources are specified in the operational model (see section 25.6).

The TOOD-IDE task model editor generates task class description files in text form and also Java code corresponding to class specifications. Figure 25.6 shows an example of the textual specifications for task T_{111}.

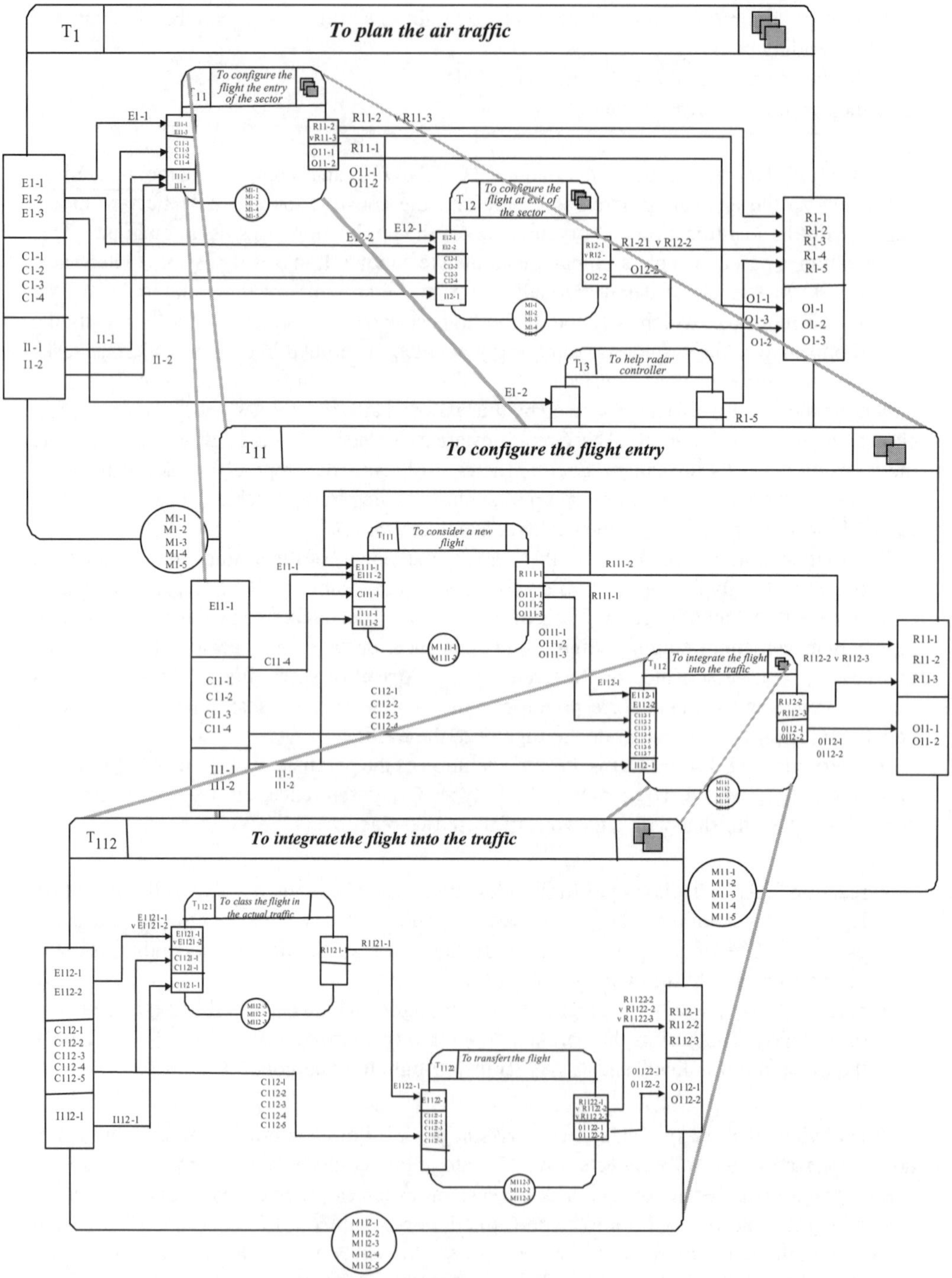

FIG. 25.3. Graphical specification of the task class "T_1: Schedule the traffic" (structural model).

Although the diagrams specify the intertask relations and the data handled, they do not provide a formal definition of the task dynamics and the influence of the data on the performance of the task. Indeed, one of the main problems that arise when using the notations available in the domain is that there are several possible interpretations of the models represented. To avoid this kind of drawback, the second phase consists in transforming the static task model into a dynamic task model. This transformation uses a formal specification technique based on OPNs that makes it possible to represent the data and the tasks in an integrated form. The advantage of this is that a greater degree of understanding of the dynamics of the system is obtained.

25.5.2 Dynamic Structural Task Model

The dynamics of the task are defined by the dynamics of its body. The dynamic task model aims at integrating the temporal dimension (synchronization, concurrency, and interruption) by completing the static model. The dynamic behavior of tasks is defined by a control structure called TCS (task control structure) based on an OPN (Sibertin-Blanc, 1985, 2001; see Fig. 25.4). It is merely a transformation of the static structure. The TCS describes the use of the input interface describer objects, the task activity, the release of describer objects from the output interface, as well as the resource occupation (Tabary & Abed, 2001).

The choice of OPNs is justified by the qualities of formalism they provide to describe complex, reactive, or concurrent behaviors. Among these extremely important qualities are the following:

- The graphic representation provides a tangible support for the comprehension of a model by a human actor.
- The formal definition leads to the production of specifications that are as free from false interpretations and ambiguity as possible.
- The performable character is ensured by the formal definition of notation. This makes it possible to interpret the OPN using a program and to simulate the functioning of the system while the specification is being worked on.

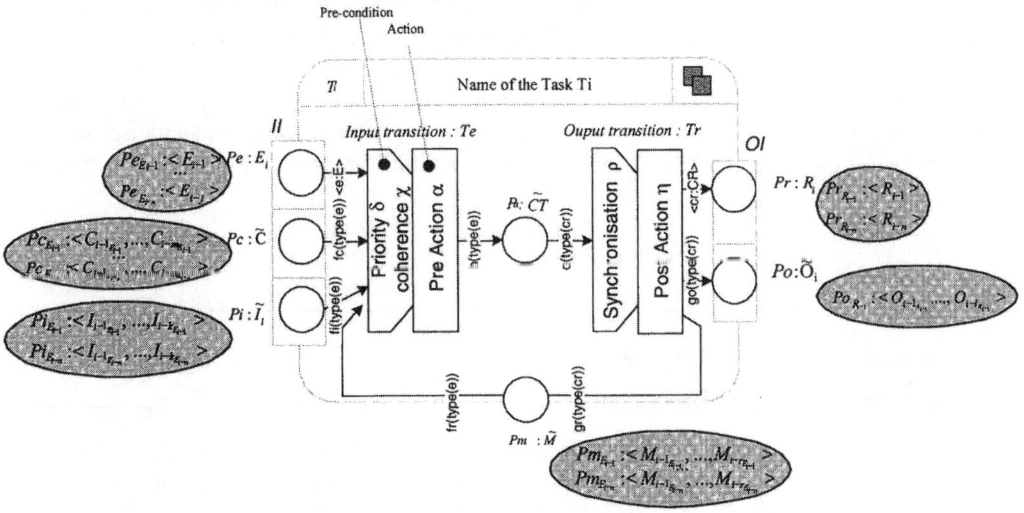

FIG. 25.4. Task control structure (TCS).

25.5.2.1 Definition of the Task Control Structure

The task control structure is a generic OPN made up of seven places and two transitions: Te (input transition) and Tr (output transition) (see Fig. 25.4). The tokens in the task control structure are the describer objects, the resources, and the subtasks identified in the task model. The net arcs are labeled by variables that play the role of formal transition parameters and make it possible to specify the flow of objects in the TCS, at the same time allowing the production and consumption of tokens. The transitions contain an action that can create or destroy objects or even call on the methods of the objects associated with their formal parameters. The crossability of a transition is conditioned by the presence of tokens in its input places and also by the value of these tokens, which is expressed using a condition called a *precondition.*

25.5.2.2 Temporal Constraints of the Task

Each TCS has an input transition Te and an output transition Tr made up of a precondition part and an action part. The functions associated with each transition make it possible to select describer objects and to define their distribution in relation to the task activity.

Precondition. The precondition part of transition Te is made up of two functions, δ and χ:

- *Priority function* δ makes it possible to select the highest priority trigger for the task. This function is at the basis of the interruption system. It allows the performance of a task to be initiated even if another, lower priority task is being carried out. However, the performance of the task in relation to this trigger remains subject to the verification of the completeness and coherence functions.
- *Coherence function* χ assesses the admissibility of these describers in relation to the conditions envisaged for the task. This function is a set of verification rules using simple logical or mathematical type operators and a unique syntax that makes it possible to formulate them.

Postcondition. The precondition of the output transition (called a *postcondition*) defines a function of *synchronization* ρ. This function specifies the state of the subtasks of the control task; this is necessary so that the control task can be finished. The definition of structural decomposition is ensured using AND, OR, and alternative operators (Table 25.6).

Preaction and Postaction. In the task control structure, the input and output transitions have actions called preaction α and postaction η respectively. For the input transition, the action creates the set of child tasks by giving them an initial context; these are data common

TABLE 25.6
Structural Decomposition

Decomposition	ρ	*Explanation*
AND	(Π(Ti) = T1.finished & ... & Tn.finished)	All subtasks must be finished.
OR	(Σ(Ti) = T1.finished + ... + Tn. finished)	A subset of tasks must be finished.
Alternative	($\oplus$(Ti) = T1.finished $\oplus$ ... $\oplus$ Tn.finished)	One and only one subtask must be finished.

to the control task and the child tasks. It also shows the interactive resources of the task. The output transition action guarantees the copying of the data shared between the child tasks and the control task, along with the possible creation of new reactions.

Activity Place. The activity place defines the temporal links between the tasks. These relations and their representations are described by the formalism provided by the OPNs. The reader will find more details on the dynamic model, its simulation, and the checking of its properties in Tabary and Abed (2001) and Tabary (2001).

The example given in Fig. 25.5 shows a formal description of the dynamics of tasks T_{111} and T_{112}. It can be seen that task T_{111} should be activated first because of presence of event $E_{111\text{-}2}$ ($\delta(E_{111\text{-}2}) > \delta(E_{111\text{-}1})$); this means that the organic controller may not take a new flight until the flight he or she is dealing with has been integrated. This must be done within the time limit given: control data "$C_{111\text{-}1}$: Temporization time" ($C_{111\text{-}1} > 0$). Indeed, this hierarchical organization of tasks triggers events according to their importance (alarm, temporal constraint, interruption, and so on), and the test of data values makes it possible, at the moment of specification, to determine communications between the various application windows and the user interaction, along with the values managed by the functional core. For instance, the activation of task T_{111} by the event "$E_{111\text{-}1}$: Arrival of a new flight" requires the presence of input data $I_{111\text{-}1}$ and $I_{111\text{-}2}$ and control data $C_{111\text{-}2}$, which can be used by the controller to assess the flight entry situation in the sector.

Each task control structure makes it possible to simulate the obtained model both graphically and mathematically. The graphical simulation shows the task dynamics, that is, the evolution of the network marks according to the presence of data in the places and the crossing of transitions (assessment of the preconditions). The mathematical formalism, based on algebraic techniques for the static verification of the OPNs, enables the designer to check certain properties such as the reinitialization, blocking, and repetitivity of the model (cf. Tabary & Abed, 2001, and Palanque & Bastide, 1997).

Figure 25.6 shows the file containing textual static and dynamic specifications. It corresponds to an instantiation of the task class structure visible on the right part of the Fig. 25.2 (attributes and operations). For instance, as seen in Fig. 25.6, the activation of Task T_{111} by event $E_{111\text{-}1}$ uses input data $I_{111\text{-}1}$ and $I_{111\text{-}2}$ ($fi(E_{111\text{-}1}) = < I_{111\text{-}1}, I_{111\text{-}2} >$).

25.6 OPERATIONAL MODEL

The objective of this stage is the automatic passage of the user task descriptions to the user interface specification. This passage completes the structural model describing the body of terminal task objects in order to answer the question how to execute the task (in terms of objects, actions, states, and control structure). In this way, the body of the terminal task no longer describes the subtasks and their relations but rather the behavior of its resources.

At this level we integrate the resources of each terminal task object into its body (Fig. 25.7). There are two types of resource: user and interface. The user is represented by the user model. This model describes the cognitive and physical behavior of the user according to the strategies used to apprehend the interactive system and perform the tasks, with reference to the analysis of the existing situation (Fig. 25.1). The interface or interactive objects describe the services provided (the functions available) and the states of the system, according to the dialogue.

The operational model has two stages for the specification of the interface. The first stage involves the specification of interactive objects for each terminal task based on a user model, according to a preliminary model called the local interface model. The second stage is intended

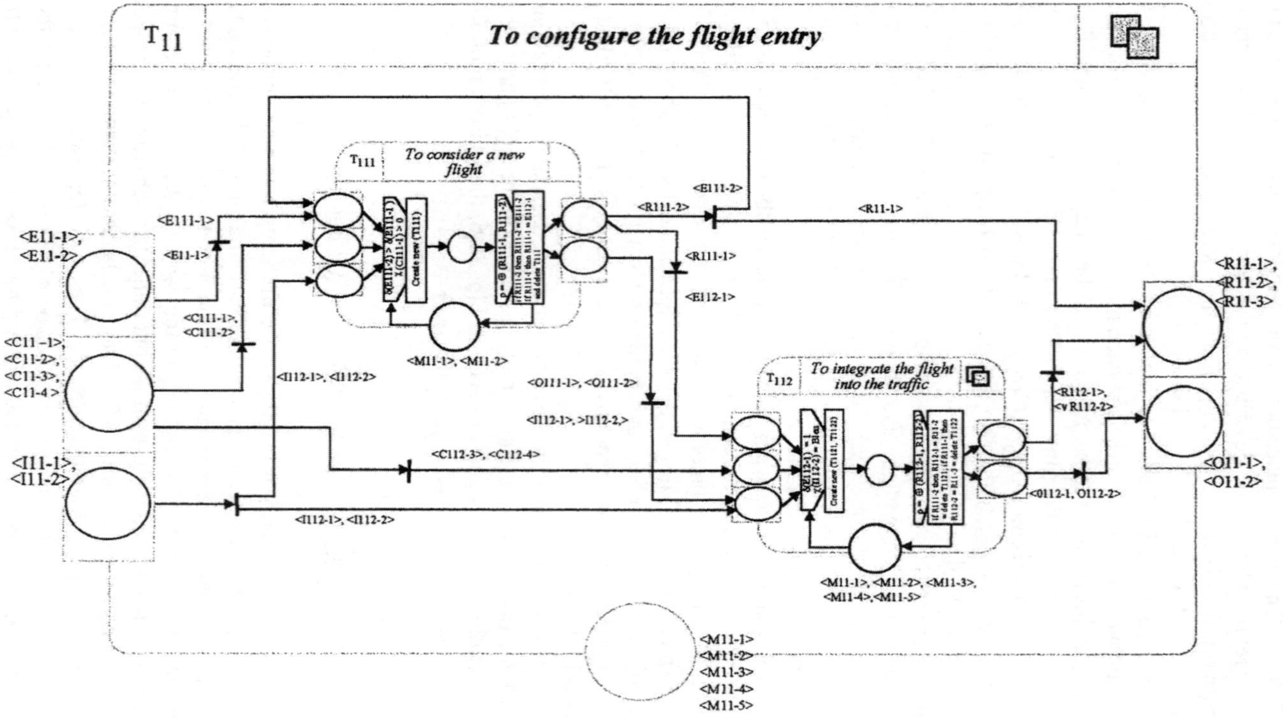

FIG. 25.5. Example of a dynamic model of the task "Configure the flight entry."

```
Task Class 'To consider a new flight (NF)'
{
// Static Specification
// Goal
the Organic Controller (OC) takes knowledge of information of the NF.
// Index
T111
// Type
type: interactive
// Hierarchy
        Parent Class:          'To configure the flight entry'
        Sub-Class:             'no'
// Triggers
E111-1  : = E11-1              : Arrival of a new flight
E111-2  : = R111-2             : Delaying of flight integration
// Control data
C111-1  : = C11-4              : Time of temporization (Tempo)
// Input data
I111-1  :=  I11-1              : Entry Flight Level (EFL)
I111-2      I11-2              : Route (BAL)
// Output data
O111-1                         : Identifier
O111-2                         : Entry level
O111-3                         : Route(points of passage and time)
// Reactions
R111-1  : = R11-1              : Knowledge and integration NF
R111-2                         : Delaying of flight integration
// Resources
M111-1 : = M11-1               : Organic Controller (OC)
M111-2 : = M11-3               : New Strips Table (NST)

// Dynamics Specification
// Incidence function
f_i (E111-1)= f_i(E111-2)= <I111-1, I111-2>
f_c (E111-1)= f_i (E111-2)= <C111-1>
// Priority Function
δ(E111-1) > δ(E111-2)
// Coherence function
χ(C111-1) > 0
// Synchronisation function
ρ= ⊕ (R111-1, R111-2)
```

FIG. 25.6. Example of textual specifications concerning the task "To consider a new flight."

to specify the global interface through the aggregation of the local interface models, following a second model called the abstract interface model (see Fig. 25.1).

25.6.1 User Interface Specification

The static specification of an interactive object describes all the information necessary for its correct utilization by the other objects. It therefore defines all its input and output interfaces in the same way as the task object, but it is *reduced* in the resource area (Fig. 25.7). Here the events represent calls for service, whereas a reaction represents the report on a service or the offer of a service to other objects in the system.

The dynamics specification describes the way in which the object evolves (spontaneously or in reply to service events or invocations); it performs operations when services are called for. Here again the behavior is described using an OPN called an ObCS (object control structure), which is similar to the ICO (interactive cooperative object) (Palanque & Bastide, 1997). The ObCS has a simple and easily understandable graphical representation, which makes it possible to model the object in terms of states, operations, and series or behaviors. The places on the network model the various possible states of the object. For example, the initial marking of the object forming the "new strips table" is represented by place P_1, a state indicating that the new strip is not displayed. The crossing of the transition "T_1: Display NS" following the service request "$E_{1\text{-}1}$: Display new strip," leads to a new state "P2 display of new strip" (i.e., all the data concerning the new flight).

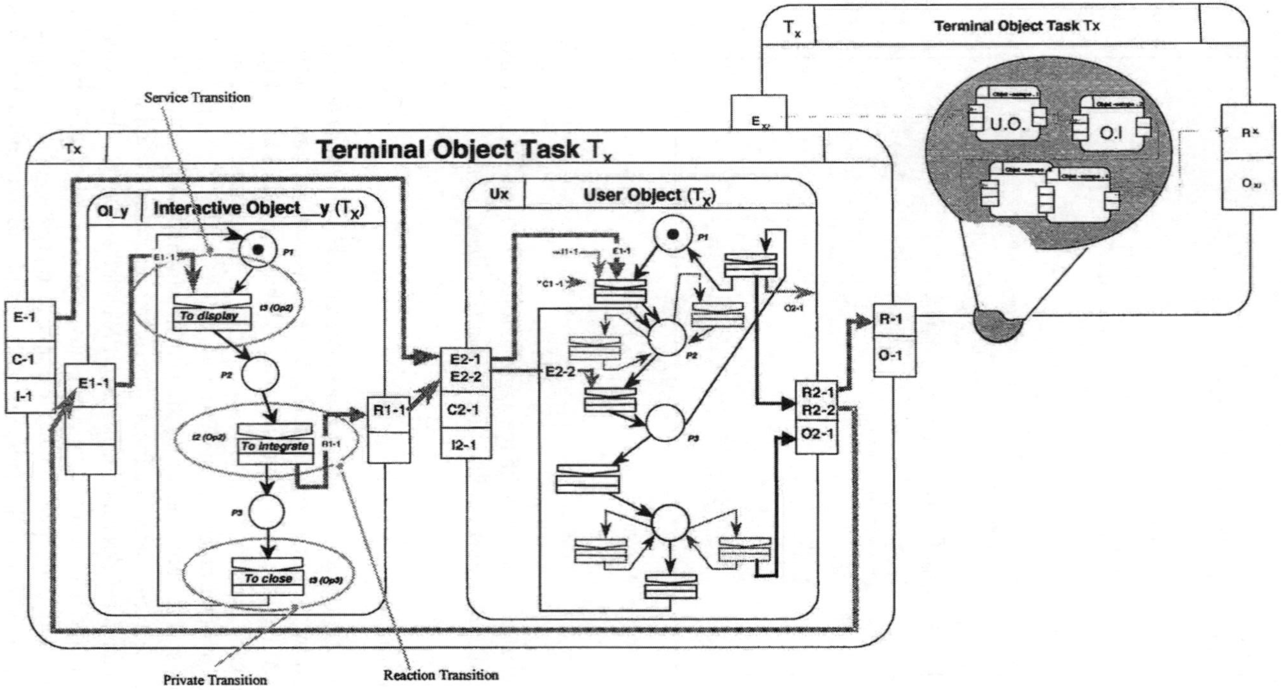

FIG. 25.7. Component objects of the body "terminal task."

The object operations are represented by the transitions of its ObCS. These operations make it possible to evolve from one state to another. Two types of transition are considered service requests and offers, as shown in Fig. 25.7:

- The service transition is associated with a trigger event from the input interface of the component object. Such events represent the services intended to be triggered by the user or another application component.
- Each reaction transition is associated with a report from the component object interface representing the offer of a service to another application object.

The other "private" transitions allow spontaneous evolution.

A transition includes a *precondition* part, which checks the presence and value of tokens, and an *action* part, which calls up the methods included in the objects associated with its formal parameters.

The user model is modeled using the same formalism (ObCS) as and in a similar way to the interactive object. It is therefore considered a user-type object.

For example, starting from the P_3 state (strip selected) of the interactive object "New Strips Table," the operator has the possibility of carrying out two actions: T_3 (open a road zoom) or T_5 (temporize the new strip), as shown in Fig. 25.8. On the other hand, the set of states and operations of an operator object represents the various procedures possible for the execution of the terminal task.

All the objects cooperate in a precisely defined manner to fulfill the aim of the terminal task object in response to a given functional context. An object is defined using its class (interface or operator) and provided with a set of states and a set of operations (or actions) that allow the change of these states.

Thus, the ObCS of a user object describes the probable behavior of a user (action on the tools, reading of information, mental activity, etc.). The places materialize the different operator states, which can be *perception, cognition*, or *action* (Fig. 25.8; see chap. 1 on the 3Ps, perceive, process, and perform). The transitions make it possible to model the passage between states.

On the other hand, the ObCS of an interactive object formally defines its human-machine dialogue for the task. According to the state of the dialogue, certain operations are no longer authorized. For example, operation "T_3: Opening of indication zoom" may only be performed if the route zoom is closed (Fig. 25.8). Thus the role of the ObCS is precisely to formalize this type of sequencing constraint between actions or the services offered. This availability is defined by the positioning of the places and the service transitions.

An example from air traffic control, corresponding to the terminal object task "To consider a new flight," requires the use of two component objects: a "New Strips Table: NST" and the "Organic Controller: OC" (Fig. 25.8). The behavior of the interactive object "New Strips Table: NST" is defined by five states: P_1, P_2, P_3, P_4, and P_5. Starting from each state, the organic controller can carry out a group of actions (transitions). From the P_3 state (strip selected), for example, the controller has the option of performing three actions: T_3 (open a road-zoom), T_5 (temporize the new strip), or T_6 (open indiactif-zoom). For the component-object "Organic Controller," the set of states and operations represents the different procedures possible in order to execute the terminal task "To consider of the new flight" according to a given context of use.

Therefore, the display of a new strip in the interactive object "New Strips Table" invokes, by event $E_{1\text{-}1}$, the operation service "Consult the NS" of the user model or object "Organic controller OC." According to his or her selection "Ch=," the organic controller carries out a preliminary reading of the new strip information ("Consult the road" or "Consult the level"). After this reading, the controller changes his or her state to "cognition" in order to evaluate

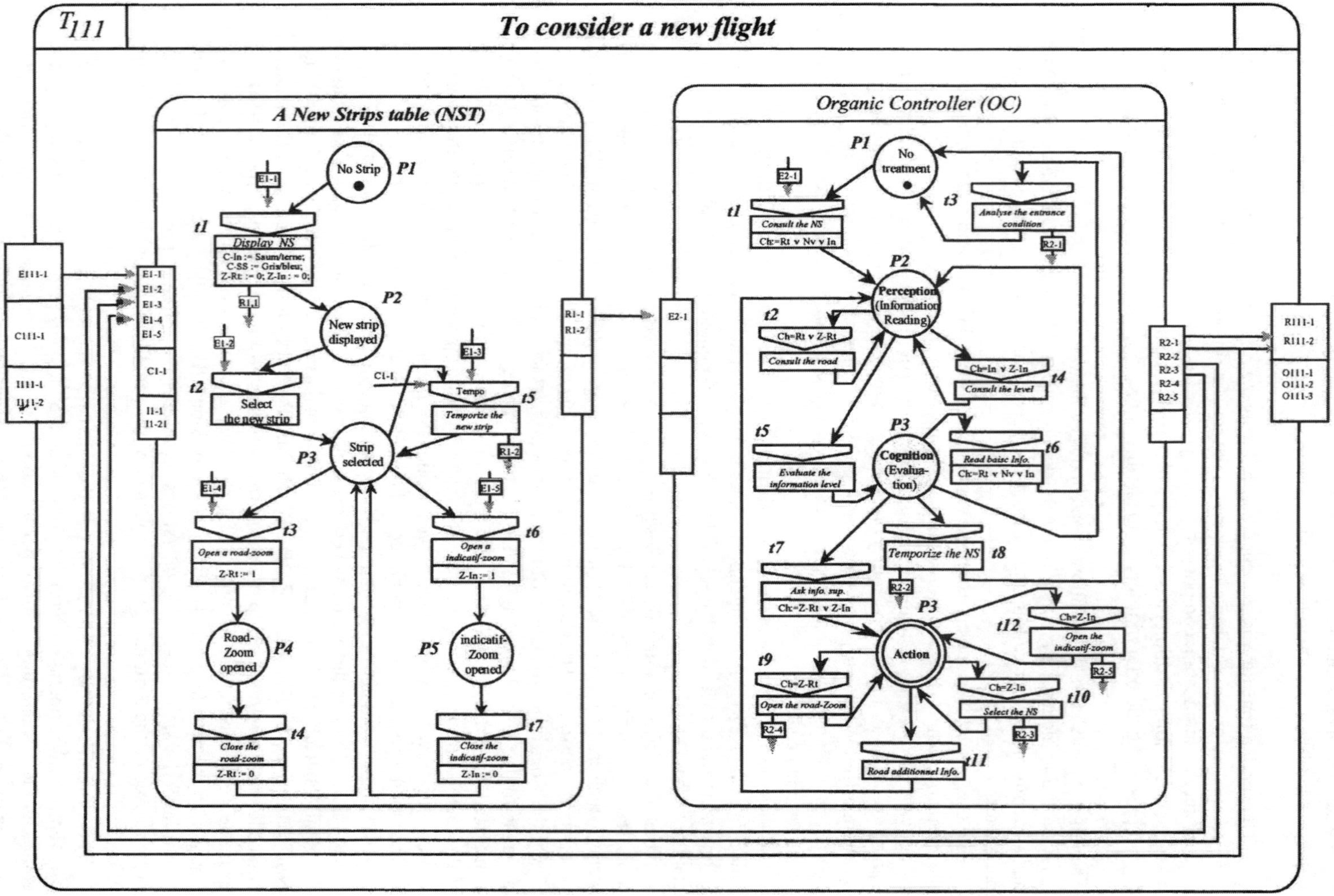

FIG. 25.8. A graphic specification of the MLI interactive object "New Strips Table" and UM "Organic Controller."

the information level. Then the controller decides to "read the basic information again" or "ask for additional information." The request for additional information expresses itself by a change of state to "Action" in order to "Select the NS," "Open the Road-Zoom," and "Select the Indicatif-Zoom." Both actions transmit $R_{2\text{-}3}$, $R_{2\text{-}4}$, and $R_{2\text{-}5}$ reactions to the object "New Strips Table." It should be noted that the organic controller carries out the action "Open a Road-Zoom" only after receiving the event $E_{1\text{-}4}$, confirming that the action "Select the NS" has been carried out. Once the road-zoom has been opened, the organic controller changes his or her state to "information reading" in order to read the additional information and then to the "situation evaluation" state to decide whether to read the information again, "temporize the NS" ($R_{2\text{-}2}$), or invoke the terminal task object "T_{112} ($R_{2\text{-}1}$): To analyse the entrance conditions."

25.6.2 Aggregation Mechanism

The local interface model is constructed using the specification of the interactive objects in relation to the terminal tasks that are independent of each other. In fact, the user interface, and consequently each interactive object, is not limited to a specific task or particular situation. In order to do this, we define an abstract interface model that describes the user interface object classes. The construction of a user interface object class implies the aggregation of all the interactive objects bearing the same operations or services in the local interface model. The aggregation mechanism is performed on the basis of composing OPNs. Several publications and research projects are to be found on this subject, including Paludetto (1991) and Li and Woodside (1995). The construction of the ObCS of a user interface class is done through the merging of places and/or transitions. The reader will find more details in Abed (2001) and Tabary (2001).

To illustrate this, the example in Fig. 25.9 corresponds to the class "New Strips Table" constructed using the aggregation of the component objects "New Strips Table" of (1) the terminal task object "T_{111}: To consider a new strip" and (2) the terminal task object "T_{1122}: To take decision on conditions of entrance."

25.6.3 User Interface Implementation

The user interface implementation model in the TOOD methodology is the presentation specification of the final interface as it will be seen by the user. It corresponds to the specification of the presentation components. The construction of this model takes place using the translation of objects, states, actions and ObCS into screens, menus, windows, and icons. This translation depends on a collection of ergonomic criteria and guidelines (Bastien & Scapin, 1995; Smith & Mosier, 1986; Vanderdonckt, 1999) and usability heuristics (Molich & Nielsen, 1990). Figure 25.9 shows the prototype of a future object-oriented interface, usable in air traffic control, that corresponds to the development of the implementation model. This development, made by CENA (Centre d'Etudes de la Navigation Aérienne), concerns the organic controller (OC) workstation. It includes three objects:

- *a radar picture* that displays the limits of the controlled sector, the plane tracks, and labels associated with the plane tracks
- *a new strips table* situated in the upper left part of the screen. Strips are presented according to an automatic ordering by geographical flow
- *a built-in strips table* situated in the lower left part of the screen

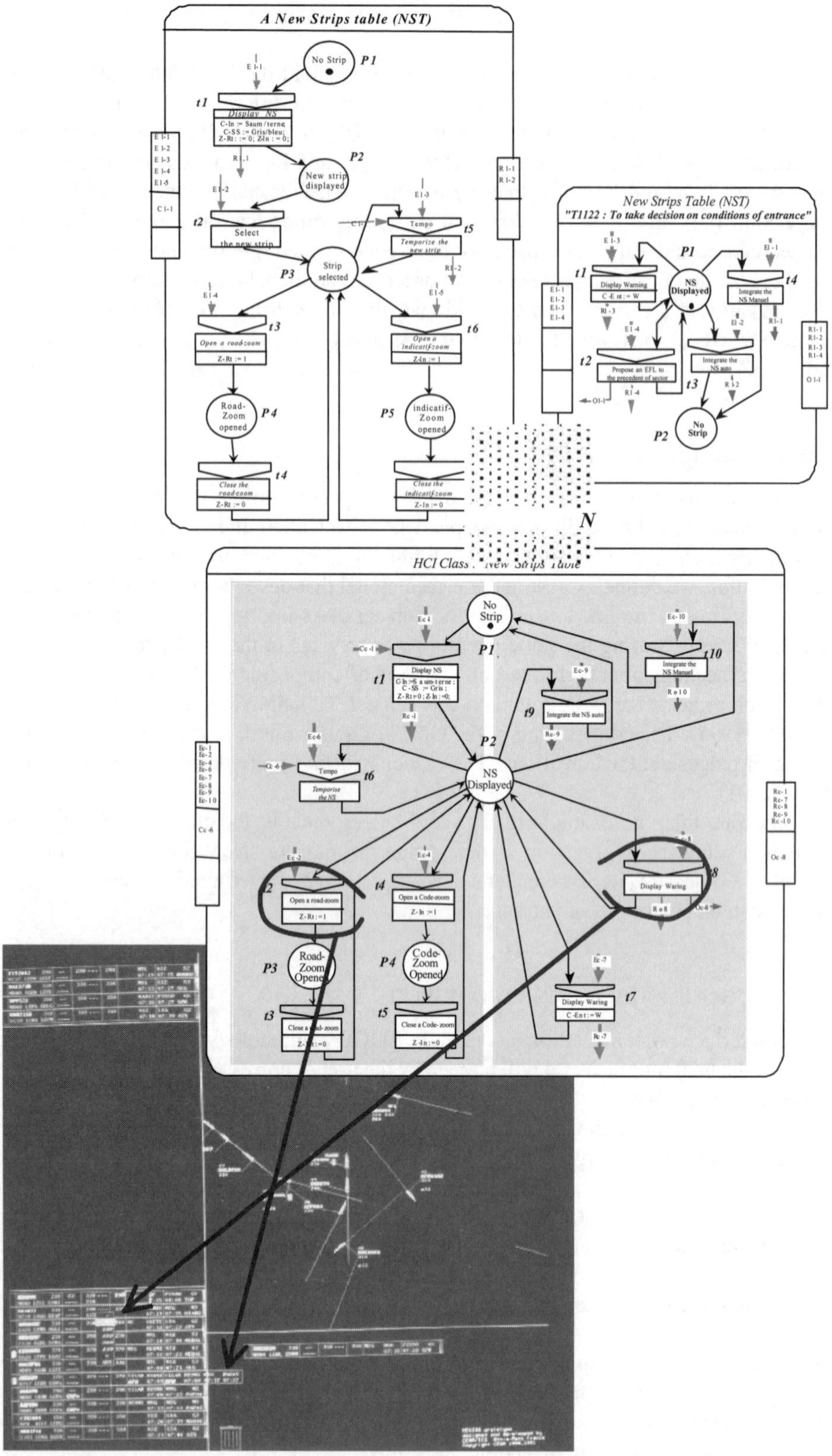

FIG. 25.9. Example of the aggregation of component objects of the local model in the user interface class.

25.7 THE TOOD-IDE ENVIRONMENT

Following in the footsteps of the model-based environments presented in section 25.2, the TOOD-IDE environment aims at continuing the structured and integrated development of interactive systems. It is based on several edition and modeling tools. The main ones are these:

- a task model editor for the input of the task model (static and dynamic).
- an object editor for the input of the domain object model, the user model, and the local interface model (static and dynamic).
- an operational model editor intended to link the user model and the local interface model.
- Automated design tools for helping the user in the creation of the local interface model, the user model, and the abstract interface model. There are three tools of this type:
  1. an abstract design tool that (a) helps the designer with the identification of interactive objects and users from the task model and (b) groups together the interactive objects in the abstract interface model
  2. a concrete design tool that ensures the passage from the abstract interface model to the implementation model (i.e., migration toward a PAC-type software architecture [Coutaz, 1997])
  3. an implementation tool that transforms the interface implementation model into usable and modifiable Java code. This tool has not yet been developed and is replaced by Jbuilder.

One of the advantages of the OPNs is their executable character. This functionality is put to use in the TOOD-IDE environment. After the code generation, it is possible to simulate the behavior of the ObCS. Thus, like in the case of most well-known OPN edition environments such as PetSchop (Navarre, Palanque, & Bastide, 2002), Great SPN (Chiola, Franceschinis, Gaeta, & Ribaudo, 1995), and ReNew (Kummer, Moldt, & Weinberg, 1999), the user alternates between different modes: network edition, compilation, generation, and simulation.

The simulator makes it possible to check the dynamic behavior of the task model according to the domain objects and the availability of resources. Three windows are displayed during the simulation (Fig. 25.10): "TOOD Editor," "Tree," and "Simulator." The "TOOD Editor" and "Tree" windows make it possible to display tasks during simulation.

The implementation of the TOOD-IDE environment was developed using Java language based on the JDK 1.3 libraries.

The perspectives of this environment concern the design support tools, the advice and criticism tools, the implementation tool, the description of task and object libraries, and the migration of the environment toward a cooperative distributed environment.

The TOOD method associated with the TOOD-IDE environment helps the analysts and designers by guaranteeing that task modeling will lead to a user interface object-oriented design in which the completeness and the coherence of most of the used objects (with their dynamic behavior) are automatically verified.

25.8 CONCLUSION

The interest of TOOD lies in its formalizing of users' tasks with a view to user interface design. In other words, it transforms information resulting from task descriptions into a high-level human-computer user interface specification with the aim of generating a concrete interface. In the TOOD method, this stage is divided into two major stages: Creation of the task model

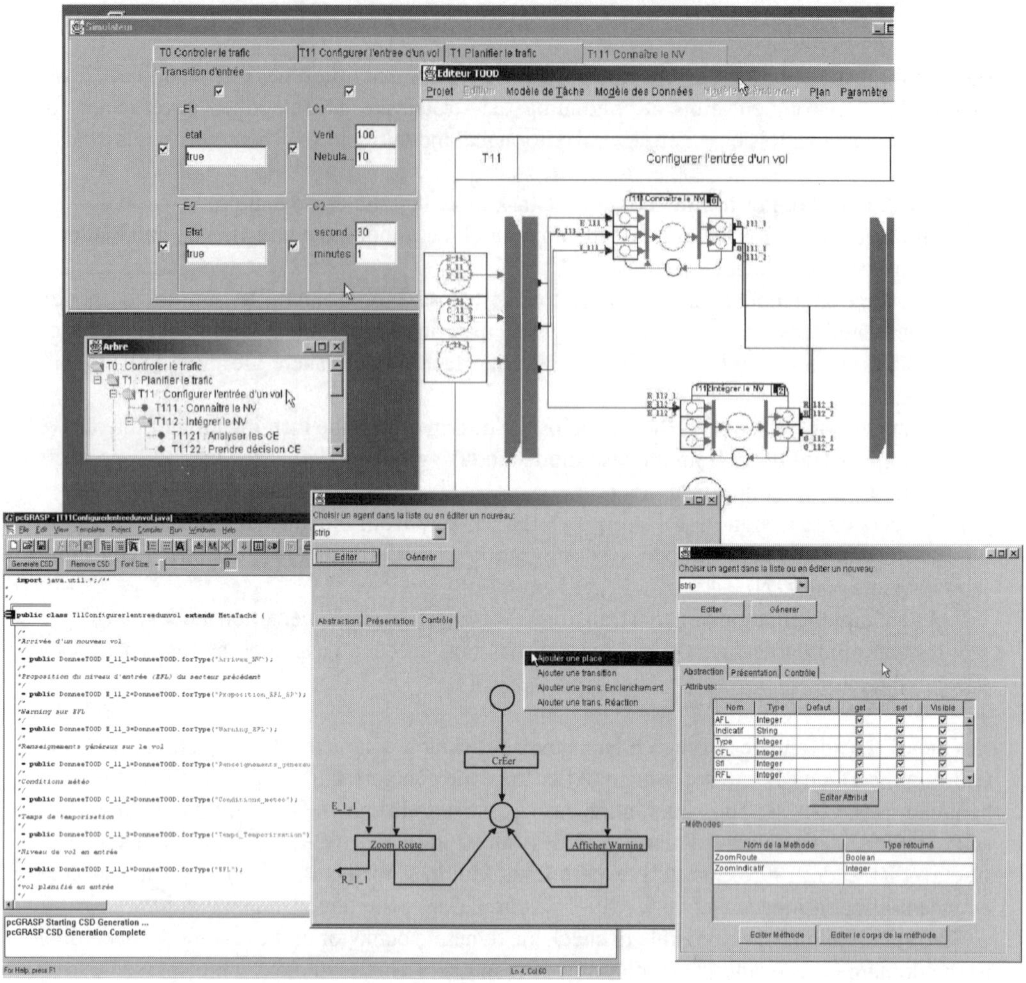

FIG. 25.10. Examples of the screens using the TOOD-IDE environment.

and creation of the operational model. These models provide support for the formalization of information necessary to the design.

The task model is based on the essential unit of the model: the task object. It can be described according to two aspects: static and dynamic. The static aspect describes the attributes of the object task and the information necessary for them (domain objects), along with their composition in several object tasks. The dynamic aspect, which is based on OPNs, describes the behavior of the task object—the conditions of interlocking and termination, the conditions of the domain objects used and their various states, their coordination, priority and synchronization—without dealing with the way in which this information will be presented in its final version, on the level of the physical interface.

The operational model consists in establishing the link between the task and the specification of the interface on a high level of abstraction. Two levels of specification are carried out: a local interface model specific to each terminal task object of the task model and an abstract interface model that is more complete, enabling the user interface specification through the intermediary of the first model. The local interface model, although based on the same detailed formalism (combining object and OPN), handles two types of objects: The interactive objects specify the dialogue in term of states, functionalities, and control structure, whereas the user type component objects describe the behavior of the user in terms of the procedures and actions on the interactive objects. The specifications of the interface are deduced starting from the user model.

By the aggregation of the local interface model interactive objects, TOOD makes it possible to specify the abstract model of the interface that contains the interactive objects. In fact, this abstract model represents the base of the model of the interface implementation. The latter corresponds to the low abstraction level specification of the presentation of the final interface such as it will be seen by the user. The construction of this model is carried out by the translation of the model's objects, states, actions, and sequences, abstracted in the form of screens, menus, windows, icons, and so on. At present, our idea is to migrate this specification to a multi-agent architecture of the PAC type in order to integrate the presentation aspects.

TOOD has been used in variety of projects (e.g., office automation applications, air traffic control, railway simulation). It can be used in a multi-user context. Given the difficulty of learning to use the TOOD method, its use is only justified within the framework of a complex application.

TOOD uses a single formalism for all its models, unlike the majority of the MB-IDE approaches.

In order to improve the portability of the interactive applications and particularly the portability of their human-machine interface, our work must ensure adaptability of the interfaces, the resources of interaction, and a specification that is independent of the interaction devices. For this, we plan to define the specifications in an XML-type language (Gerard, 2001; Harold & Means, 2001) to establish a generation of multiple codes (C++, Java, etc.) and multiple platforms (Palm-Pilot, mobile telephone, and all the new communication tools to come) in line with the work of Calvary, Coutaz, and Thévenin (2001) and Paternò and Santoro (2002; see Chapters 7, 24, and 26).

REFERENCES

Abed, M. (1990). *Contribution à la modélisation de la tâche par outils de spécification exploitant les mouvements oculaires: Application à la conception et à l'évaluation des interfaces homme-machine.* Unpublished doctoral dissertation, Université de Valenciennes et du Hainaut-Cambrésis.

Abed, M. (2001). Méthodes et modèles formels et semi-formels pour la conception et l'évaluation des systèmes homme-machine. Habilitation à Diriger des Recherches, Université de Valenciennes et du Hainaut-Cambrésis.

Abed, M., & Angué., J.-C. (1990, October). A new method for the conception, realization and evaluation of human-machine systems. Paper presented at the International Conference on System, Man and Cybernetics (SMC '94), San Antonio, TX.

Abed, M., & Ezzedine, H. (1998). Towards an integrated approach to human-machine design-evaluation of systems. *Journal of Decision Systems, 7*, 147–175.

Bastien, J. M. C., & Scapin, D. L. (1995). Evaluating a user interface with ergonomic criteria. *International Journal of Human-Computer Interaction, 7*, 105–121.

Bernard, J. M. (1994). Exploitation des mesures oculométriques dans la modélisation de la tâche prescrite et de l'activité réelle des opérateurs par réseaux de Petri. Doctoral dissertation, Université de Valenciennes et du Hainaut-Cambrésis.

Bodart, F., Hennebert, A.-M., Leheureux, J.-M., Provot, I., Vanderdonckt, J., & Zucchinetti, G. (1995). Key activities for a development methodology of interactive applications. In *Critical issues in user interface systems engineering* (pp. 109–134). Berlin: Springer-Verlag.

Bomsdorf, B., & Szwillus, G. (1999). Tool support for task-based user interface design. In *proceedings CHI '99 ACM SIGCHI Conference on Human Factors in Computing Systems* (pp. 169–170). New York: ACM Press.

Buisine, A. (1999). *Vers une démarche industrielle pour le développement d'interfaces homme-machine.* Unpublished doctoral dissertation Université de Rouen.

Calvary, G., Coutaz, J., & Thévenin, D. (2001). Supporting context changes for plastic user interfaces: A process and a mechanism. In A., Blandford, J. Vanderdonckt, & P. Gray (Eds.), *People and computers XV: Interaction without frontiers* (pp. 349–363). London: Springer.

Chiola, G., Franceschinis, G., Gaeta, R., & Ribaudo, M. (1995). GreatSPN 1.7: GRaphical Editor and Analyzer for Timed and Stochastic Petri Nets. *Performance Evaluation, 24*(1–2), 47–68.

Coutaz, J. (1997). PAC-ing the software architecture of your user interface. In *DSV-IS '97: 4th Eurographics Workshop on Design, Specification and Verification of Interactive Systems* (pp. 15–32). Vienna: Springer-Verlag.

David, R., & Alla, H. (1994). Petri nets for modeling dynamic systems: A survey. *Automatica, 30* 175–202.

Delatte, B., Heitz, M., Muller, J.-F. (1993). HOOD Reference Manual 3.1. Prentice Hall, London.

Feller, A., & Rucker, R. (1989). In: Shunk, D. L.: *Optimization of Manufacturing Systems Design. Proceedings of the IFIP WG 5.3 International Conference on Modeling and Simulation for Optimization of Manufacturing Systems Design and Application, 1989, Tempe, AZ, USA* (pp. 171–194). Amsterdam, Netherlands: North-Holland, 1990.

Gamboa-Rodrìguez, F. & Scapin D. (1997). Editing MAD* task descriptions for specifying interfaces at both semantic and presentation levels. In *DSV-IS'97* (pp. 193–208). Vienna, Austria: Springer-Verlag.

Gérard, F. (2001). *XML*. Paris: OEM.

Harold, E. R., & Means, W. S. (2001). XML in a nutshell. Paris: O'Reilly.

Johnson, P. (1999). Theory based design: From individual users and tasks to collaborative systems. In J. Vanderdonckt & A. Puerta (Eds.), *Computer-aided design of user interfaces II* (pp. 21–32). Dordrecht: Kluwer Academic Publishers.

Kummer, O., Moldt, D., & Wienberg, F. (1999). Symmetric communication between coloured Petri Net simulations and Java-Processes. In *Proceedings of the 20th International Conference on Application and Theory of Petri Nets* (Lecture Notes in Computer Science; Vol. 1639, pp. 86–105). Heidelberg: Springer-Verlag.

Li, Y., & Woodside, C. M. (1995). Complete decomposition of stochastic Petri Nets representing generalized service networks. *IEEE Transactions on Computers, 44* 1031–1046.

Mahfoudhi, A., Abed, M., & Angue, J. C. (1995). An object oriented methodology for man-machine systems analysis and design. In Y., Anzai, K., Ogawa, & H. Mori (Eds.), *Symbiosis of human and artifact: HCI International '95* (pp. 965–970). Amsterdam: Elsevier.

Markopoulos, P., Rowson, J., & Johnson, P. (1996). On the composition of interactor specifications. In *Formal Aspects of the Human Computer Interface* (pp. 1–6). London: Springer-Verlag.

Molich, R., & Nielsen J. (1990). Improving a human computer dialogue. *Communications of the ACM, 33* 338–¯348.

Mullin J., Anderson, A. H., Smallwood, L., Jackson, M., & Katsavras, E. (2001). Eye-tracking explorations in multimedia communications. In A. Blanford, J. Vanderdonckt, & P. Gray (Eds.), *People and computer XV: Interaction without frontiers* (pp. 367–382). London: Springer.

Navarre, D., Palanque, P., & Bastide, R. (2002). Model-based interactive prototyping of highly interactive applications. In C. Kolski & J. Vanderdonckt (Eds.), *Computer-aided design of user interfaces III* (pp. 205–216). Dordrecht, The Netherlands: Kluwer Academic.

Palanque, P., & Bastide, R. (1997). Synergistic modeling of tasks, systems and users using formal specification techniques. *Interacting With Computers, 9* 129–153.

Palanque, P., & Paternò, F. (Eds.). (1997). *Formal methods in human-computer interaction*. Springer-Verlag.

Paludetto, M. (1991). *Sur la commande de procédés industriels: Une méthodologie basée objets et réseaux de Petri*. Unpublished doctoral dissertation. Université Paul Sabatier de Toulouse.

Paternò, F. (2000). *Model-based design and evaluation of interactive applications*. Milan, Italy: Springer.

Paternò, F., & Santoro, C. (2002). One model, many interfaces. In C. Kolski & J. Vanderdonckt (Eds.), *Computer-aided design of user interfaces III* (pp. 143–154). Dordrecht, The Netherlands: Kluwer Academic.

Petri, C. (1962). *Kommunication mit Automaten*. Unpublished doctoral dissertation, University of Bonn.

Puerta, A. R. (1996). The Mecano Project: Enabling user-task automation during interface development. In *AAAI '96 Spring Symposium on Acquisition, Learning and Demonstration: Automating Tasks for Users* (pp. 117–121). Stanford, CA: AAAI Press.

Puerta, A. R., & Maulsby, D. (1997). Management of interface design knowledge with MOBI-D. In *Proceedings of ACM SIGCHI Conference on Human Factors in Computing Systems (CHI '97)* (pp. 249–252). New York: ACM Press.

Ricard, E., & Buisine, A. (1996). Des tâches utilisateur au dialogue homme–machine: GLADIS++, une démarche industrielle. In *Proceedings of IHM '96* (pp. 71—76). Paper presented at the Huitièmes Journes sur l'interaction Homme. Machine (IHM '96), Grenoble, France, September.

Ross, D. T. (1977, January). Structured analysis (SA): A language for communicating ideas. *IEEE Transactions on Software Engineering* pp. 16–34.

Schlungbaum, E. (1998). Knowledge-based support of task-based user interface design in TADEUS. In *Proceedings of ACM SIGCHI Conference on Human Factors in Computing Systems (CHI '98)*, Paper presented at the CHI '98 Workshop Handout From task to dialogue: Task based user interface design, Los Angeles, April, ACM Press.

Sibertin-Blanc, C. (1985). High-level Petri nets with data structure. In *Proceedings of 6th EWPNA*, paper presented at the 6th European Workshop on Application and Theory of Petri Nets, Espoo, Finland, June 1985.

Sibertin-Blanc, C. (2001). Cooperative objects: Principles, use and implementation: Concurrent object-oriented programming and Petri nets. In G. Agha, F. De Cindio, & G. Rozenberg, (Eds.), *Advances in Petri nets* (Lecture Notes in Computer Science, 2001; pp. 16–46). Berlin: Springer.

Smith, S. L., & Mosier, J. N. (1986). Guidelines for designing user interface software (Report EDS-TR-86-278). Bedford, MA: The MITRE Corporation.

Sukaviriya, N., Kovacevic, S., Foley, J. D., Myers, B. A., Olsen, D. R., & Schneider-Hufschmidt, M. (1994). Model-based user interfaces: What are they and why should we care? In *Proceedings of Seventh Annual Symposium on User Interface Software and Technology (UIST '94)* (pp. 133–135). New York: ACM Press.

Szekely, P. (1996). Retrospective and challenge for model based interface development. In J. Vanderdonckt (Ed.), *Proceedings of the second International Workshop on Computer-aided Design of User Interfaces (CADUI '96)* (pp. xxi–xliv). Namur, Belgium: Presses Universitaires de Namur.

Tabary, D. (2001). *Contribution à TOOD, une méthode à base de modèles pour la spécification et la conception des systèmes interactifs*. Unpublished doctoral dissertation Université de Valenciennes et du Hainaut-Cambrésis.

Tabary, D., & Abed, M. (2001). A software Environment Task Object Oriented Design (ETOOD). *Journal of Systems and Software, 60* 129–141.

Tabary, D., Abed, M., & Kolski, C. (2000). Object-oriented modelling of manual, automatic, interactive task in mono or multi-user contexts using the TOOD method. In Z. Binder (Ed.), *IFAC 2nd Conference on Management and Control of Production and Logistics (MCPL 2000)* (pp.101–111). Amsterdam: Elsevier.

Tarby, J.-C., & Barthet, M.-F. (1996). The Diane+ method. In J. Vanderdonckt (Ed.), *Proceedings of the Second International Workshop on Computer-aided Design of User Interfaces (CADUI '96)* (pp. 95–119). Namur, Belgium: Presses Universitaires de Namur.

Vanderdonckt, J. (1999). Development milestones towards a tool for working with guidelines. *Interacting With Computers, 72* 81–118.

Wilson, S., Johnson, P., Kelly, C., Cunningham, J., & Markopoulos, P. (1993). Beyond hacking: A model based approach to user interface design. In J. Alty, D. Diaper, & S. Guest (Eds.), People and Computers VIII (pp. 217–231). Cambridge: Cambridge University Press.

26

One Goal, Many Tasks, Many Devices: From Abstract User Task Specification to User Interfaces

Jean-Claude Tarby
Université des Sciences et Technologies de Lille

The permanent evolution and multiplicity of devices require an increasing time to produce interfaces adapted to these devices. Recent user interface modeling languages like UIML or XIML allow to be more and more independent from the constraints of these platforms. But the major problem remains: how to efficiently produce user interfaces? This chapter explains how abstract user tasks can be useful to quickly produce applications which can be represented similarly on different devices or differently on the same device.

26.1 INTRODUCTION

The permanent evolution and multiplicity of devices (workstation, laptop, PDA, Wap phone and so on; see chap. 24) means that more time is needed to produce user interfaces adapted to these devices. Recent user interface–modeling languages like UIML, XForms and XIML allow increasing independence from the constraints of these platforms, but the major problem remains: how to efficiently produce user interfaces. Most of the time, these new languages are contrained to give an abstract representation of the widgets used in the interface, for instance, by describing a button with a "Button" abstract entity that will be translated into a JButton in Java Swing or <input type="submit" name="..." value="..."> in HTML. The level of abstraction is rarely higher in these new languages, except for XIML, which provides different abstraction levels through formal models describing, for instance, the user model or the task model.

The actual generation of tools allows prototypes to be constructed very rapidly and efficiently (using a rapid application development approach), and companies are making more and more use of them. Consequently, they prefer to spend time prototyping rather than precisely specifying data, tasks, or the user interface. In many safety critical domains like air traffic control (chaps. 1, 13, 19, and 25), there is a need to provide highly detailed models of data, the user interface, cooperation between objects, and so on, so as to be as certain as possible that the system will meet its requirements. In such domains, formal approaches like Petri Nets, Statecharts, or Z schema are often used. This chapter addresses, not this category of applications, but applications in which the expense of formal methods is inappropriate.

The chapter explains how a task-based approach can be useful for quickly producing applications (or prototypes) that take into account the user requirements. The basis of this approach is the modeling of abstract user tasks that can be represented similarly on different devices (plasticity; Calvary, Coutaz, & Thévenin, 2001a, 2001b; chap. 24) or differently on the same device (customization). Working with tasks is well appreciated among users and human factors experts. Users explain their work by means of concrete words, and the designers model concrete tasks by skillful observing and interviewing of the users. The next step for the designers is to translate the tasks into an abstract level in order to apply the tasks on new or different devices. This approach can be summarized either as "one service, lots of user interfaces" or "one goal, many tasks, many devices."

Participative design is well practiced these days, but it is very hard for users and human factors experts to model their tasks using formal languages or representations like Petri Nets or Z languages. The problem is not one of formalism but of methodology. The software development life cycle (e.g., chaps. 1 and 22) should start in a soft manner with the user requirements by writing abstract tasks that will be refined and translated later into more concrete tasks, eventually ending in a more formal representation. The first part of the chapter explains the problem of writing abstract tasks with the Diane+ method. The second part lists several ergonomics criteria that are used to choose alternative user interfaces for implementation. The last part details an example to show how abstract tasks can be mapped onto different devices (Web site, Wap phone, and regular phone).

26.2 THE DIANE+ METHOD

Diane+ (Paris, Tarby, & Vander Linden, 2001; Tarby & Barthet, 1996, 2001; chaps. 6, 22, and 23), a descendent of Diane (Barthet, 1988), is used during the user requirement analysis phase and can guide the design of the user interface. Initially Diane+ was very structured. To develop a new application from an existing system, Diane+ required at least five steps based on procedures (Fig. 26.1). A procedure is a structured and detailed description of how to realize a goal. It is composed of a set of user and system tasks. The five steps required by Diane+ were as follows:

1. Collect the planned and the real procedures.
2. Extract from Step 1 the minimal procedure that includes the constraints related to the application domain, the user practice, and so on. Planned and real procedures must comply with the minimal procedure.

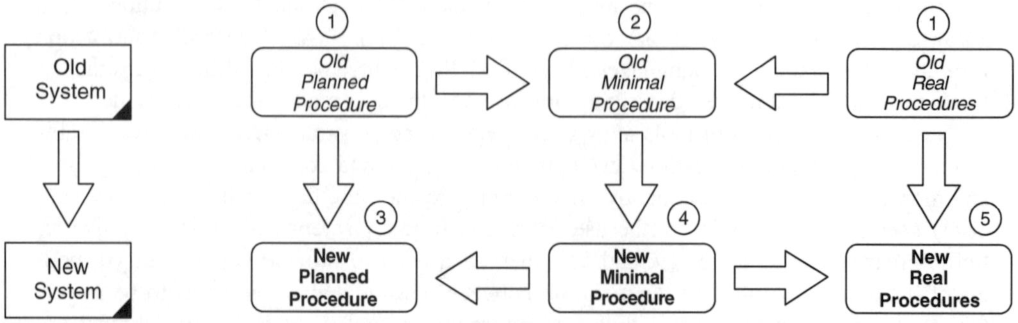

FIG. 26.1. The five steps of the Diane+ method in its early days.

3. Build the new planned procedures.
4. Build the new minimal procedures with respect to the planned procedures.
5. Capture the new real procedures and compare them to the new minimal and planned procedures in order to eventually modify the requirements.

The three last steps can produce real applications, mock-ups, or prototypes.

Diane+ is user centered and focuses on user tasks and data. It can be used by real users, stakeholders, human factors experts, and so on. From the early stages, Diane+ was appreciated by various participants in the design process, although its graphical representation is semi-formal. The practice of Diane+ has, however, changed since 1988 because people who apply Diane+ today usually design only the tasks of the new planned procedures and sometimes of the old planned procedures. This evolution seems to have been caused by the parallel evolution of software tools, which now allow very easy and efficient prototyping without modeling lots of information previously. Consequently, the methodology evolved into a two-level task-based requirements specification framework.

26.2.1 A Two-Level Task-Based Requirements Specification Framework

To be useful, task models must have two key characteristics (Paris et al., 2001; chap. 22): (a) they should be complete in the sense that all levels of user interaction should be included (i.e., there should be a collection of task models at different levels of abstraction), and (b) they should be coherent in that a uniform representation language should be used. Chapter 23 describes a flexible environment in which a coherent task model can be built without imposing unrealistic constraints on the design process. This chapter describes the requirements specification framework that is associated with this environment.

The "drink machine" example can be used to illustrate the framework. In this example, the user is thirsty and wants to drink something. The user does not mind how the system works to furnish a drink (e.g., how the bottle is delivered or how the coffee is heated). From the user's point of view, what is primordial is to drink. On the opposite end, it is important for the system to know how to deliver the bottle or how to heat the coffee but not who the user is, nor what the reason is for the user's visit to the drink machine. Both of these two active actors need to interact to reach the user's goal. The user knows his or her goal, and the system must do its best possible to satisfy the user.

The framework presented here describes the two points of view: the user's point of view (UPV) and the system's point of view (SPV).

26.2.2 The User's Point of View (UPV)

The objective of the UPV is to identify the main tasks and the main objects that are relevant for the user. Here the level of abstraction is very high; the tasks are considered to be "black boxes," and the objects are objects that the user really uses. In the UPV level, the system is hidden and the focus is on the user. This level is independent of the user interface and the platform. It must not integrate constraints like the operating system or the hardware.

The UPV level is close to the "essential use cases" (Constantine & Lockwood, 1999). It specifies tasks at a high level of abstraction, and it differentiates user tasks and system tasks. It differs from the "essential use cases" in that (a) it integrates the "use case maps" and the "use cases" and (b) the level of detail is greater because the UPV has no restriction on the number of task decompositions.

To obtain the UPV, the designers can ask users, "What do you want the system to do for you?" From the corpus of information gathered, the designers keep all information that refers to the actions and user's objects, eliminating references to specific interaction events (e.g., "click," "cut/copy/paste," "open file," "select menu," and so on). These are kept, as they may be of use for the specification of the SPV level, but they do not belong to this high level of abstraction. If a system already exists and the prospective system is to replace it, information for this level can also be gathered from all stakeholders and the user's manuals and other documentation.

The UPV can be refined into more details to identify more precisely what the user participation is to be like, independently of the platform and a user interface. Information about the system must remain hidden, but the feedback from the system may be present. For example, the UPV might indicate that the user will receive feedback such as "The price of the drink is displayed" but not "A 4-digit LED screen displays the price as XX.YY," which identifies a particular kind of display.

To obtain all the information needed at this level, the designer refines the tasks by interviewing users or writing and exploiting scenarios (Rosson & Carroll, 2001; chap. 5) using tools like U-Tel (Tam, Maulsby, & Puerta, 1998). To build the UPV level, the analysts should use words that describe actions and imply the participation of the user (e.g., "The user chooses a drink"). If necessary, words that describe global actions by the system can also be employed (e.g., "The system prints the bill" or "The system displays the price of the drink"). Note that, at this stage, paper and computerized mock-ups can be elaborated to verify whether the task model conforms to the user's behavior.

Continuing with drink machine example, the UPV level provides the specification shown in Fig. 26.2. "Find the drink machine" is modeled as a manual task and is passive in that the drink machine relies on its static properties (e.g., color, location, and so on) to be found (see also chaps. 1 and 18). In a design where the machine becomes active in some way when approached (e.g., a voice chip is activated), this subtask would be an interactive one.

The UPV may also include "undesirable" tasks. For instance, "Pay for the drink" is a mandatory task due to the social context, but it is not a desirable task for the user. Undesirable tasks usually come from social, economic, and legal contexts and from the context of use (a company's habits, for example). Recognizing undesirable tasks is very important because they often influence the design of the application and the user interface.

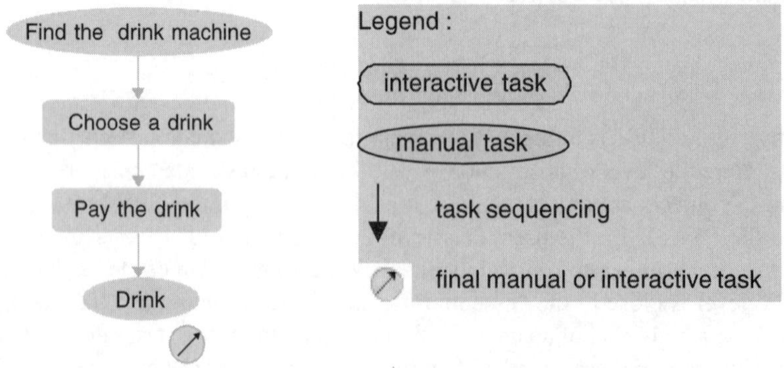

FIG. 26.2. The UPV level for the drink machine.

26.2.3 The System's Point of View (SPV)

The objective of this level is to identify *how the system satisfies the user's goal*. At this level, the system is finally introduced in several ways:

- Through the use of automatic tasks (an automatic task is executed only by the system).
- By specifying more precisely interaction actions (e.g., "Enter the password," "Validate," "cancel").
- By specifying all normal and abnormal exits for each task (e.g., what to do if the password is not correct, if the user cancels, if the data link is broken, etc.).
- By introducing the constraints due to the platform, such as (a) user interface guidelines pertaining to standard behaviors (e.g., "cut/copy/paste," "open/save") or (b) hardware with touch screen, voice recognition, and so on.

In the UPV, the system was present but not taken into consideration because the system was assumed to work as expected. The SPV level reflects how the system satisfies the user's goals from the system's perspective. That means the system itself cannot do any more to satisfy the user; for instance, in the drink machine example, the system has finished its work when it has delivered the drink, given the change, and said that the drink is ready (Fig. 26.3). Why the user wants to drink does not concern the system. So, when the system says that the drink is ready, its goal is reached.

An important point, made in chapter 1, is that system's goal can be and usually is different from the user's goal. This is the case in the drink machine example. The terminal event is not the same for the system and for the user; consequently, we differentiate them. The difference

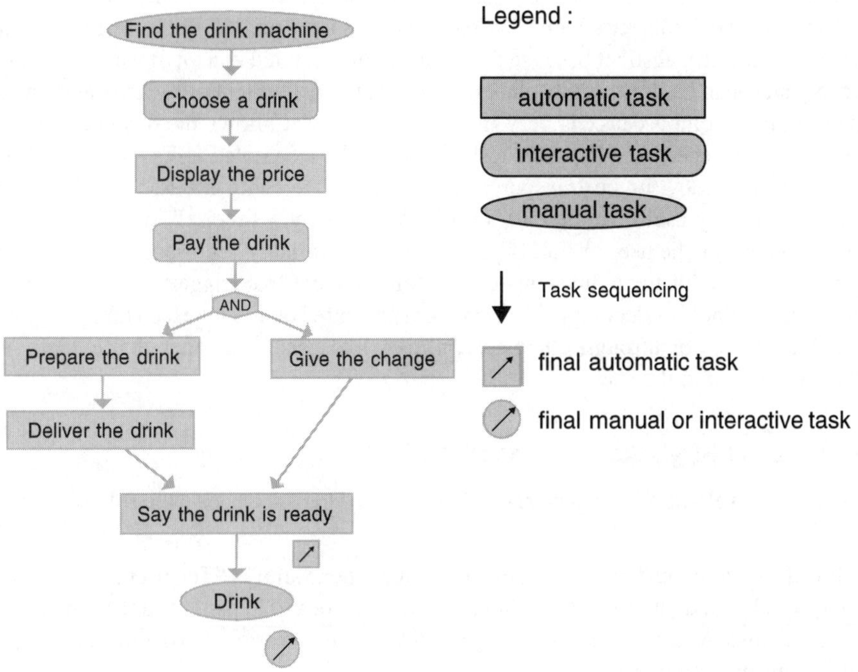

FIG. 26.3. Example of the SPV level for the drink machine (only the first level of decomposition of the planned procedure is represented).

between the goals influences the design of the prototypes. For instance, when the drink machine has reached its goal (even if a "Wait while the user takes his/her drink" automatic task was the last task before the system's goal), it can accept a new request from another user even if the previous user has not taken his or her drink and consequently has not reached his or her goal. That means that the prototype will accept that a new user chooses a drink after the "Say the drink is ready" automatic task. This itself identifies a potential prototype design error, for if drinks are not collected, the delivery system may jam or cause a mess. Thus, identifying the differences between human and machine goals is often vital.

Sometimes designers find it helpful to build prototypes to test the interaction specifications. In such cases, a GUI event recorder can be exploited to fill in the SPV automatically (Paris et al., 2001; chap. 23). Specifying user interface options (discussed later in this chapter) and exploiting an event recorder when the prototype is built can drastically help to consolidate the SPV level.

26.2.4 Discussion

26.2.4.1 Abstract Task Versus Complex Task

All the tasks in the UPV and the SPV are "abstract," meaning that they do not specify how they will be realised concretely in the final system. Abstract tasks exist in almost all task notations but not always with the same signification. For example in CTT (Paternò, Mancini, & Meniconi, 1997; chap. 24), an abstract task is a task that includes subtasks; this kind of task is called a complex task in the Diane+ notation. The notion of abstract task as presented in the UPV and SPV can be compared to the notion of a conceptual task. A conceptual (abstract) task may be translated into one or more elementary or complex tasks. For example, "Make a phone call" is an abstract task that can be decomposed into "Select a number from the list" or "Recall the last number" or "Enter a number." "Select a number from the list" can be decomposed, for instance, into "Select the list," "Select a name," "Select the number," and "Call the selected number." All these tasks are still abstract because they can be implemented in a lot of ways; for example, "Select a name" can be done with a click on a scrollable list widget or by a vocal command.

The notion of complex task is very important because it allows the designer to produce different user interfaces from the same task; for example, a complex task can be implemented separately in a window or be transparent on the interface (with no associated window or widget). So it is very useful, at the UPV or SPV level, to know precisely whether the abstract task is important for the user or not. If yes, it must be explicitly represented on the interface either by a window (or the equivalent) or through a widget that triggers the execution of its decomposition. If not, its decomposition can be represented on the interface through graphical artifices like frames or through other possibilities, like vocal messages that explain all the possibilities offered to the user.

26.2.4.2 Is SPV Always Necessary?

As in the UPV level, the SPV can be refined into more details to specify the following exactly:

- How the system must be used. In the drink machine example, "The user chooses a drink" must be changed into "The user chooses between Coca®, Pepsi®, Vittel®, and Evian®," and the ways the user interacts with the system must be specified in detail (selecting menu item, pushing buttons, etc.).
- How the system reacts to the user's actions with the system. This means that all feedback (e.g., icon or cursor modifications, dialog boxes, and so on) and all system decisions (e.g., normal/abnormal end of task, suspend a task, and so on) must be indicated.

Elaborate detailed SPV specifications can be very long and very complex. It is utopian to believe that software designers will write SPV specifications for large systems (see also chap. 22). The SPV level is, however, useful (and more practical) for subsystems or for systems with smaller task models (e.g., mobile phone, smart card) in at least three ways. First, it can inform the designer of the interface quality. Second, it can guide the production of instructions (manually or automatically). Finally, it can support usability evaluation directed toward checking whether the implemented system embodied the early design and comparing actual and desired user activities.

Writing detailed SPV specifications can be avoided partly by providing alternative interface implementations for the interactions and feedback associated with the abstract tasks. This option is discussed in the next subsection.

26.2.5 How to Obtain Alternative Interface Implementations

Once the UPV and SPV are built, mock-ups and prototypes are elaborated by coupling ergonomic rules with criteria. The criteria are used as guides to improve the application of ergonomic rules and to choose the right widgets, spatial disposition, interaction mode, and so on. The following is a list of sample criteria:

1. *Target equipment/platform*. What are the target platforms? What equipment is provided with them? If necessary, the SPV can be refined for each target platform (one SPV for PC, one SPV for Wap phone, etc.)
   - *Output device*. This criterion is important for all tasks that display information:
     - *Screen*. What is the resolution in pixels? How many lines and how many characters per line? Is it a graphic display or not? What is the size of free space on the screen? These questions are essential when the platform is as specific as a PC screen or a Wap phone screen.
     - *Vocal output*. Is the output mono, stereo, or quadraphonic? May the type of voice be chosen? How many languages are managed? "Display" tasks can be mapped to voice outputs. Current sound cards can manage more than 100 different sonorous sources, and this ability can be very useful when lots of information must be displayed or when the user is not familiar with computers.
     - *Printer*. Is the printer graphic? Is it a color printer? What kind of paper is used? What size, color, and so on? An interactive kiosk may print information, but usually the printers are dedicated and print on very small paper sheets. The printed information must be concise and readable.
   - *Underlying operating system*. What are the common functions? Is cut/copy/paste available and used by users? Is it a multiple or a full-screen mono window environment?
   - *Interaction mode*. Keyboard, mouse, touchpad, trackpad, voice recognition, gesture recognition, and so on? Is the application monomodal or multimodal? With a regular phone, voice recognition and keypad are usually used to enter data, but other media are available (fingerprint for the identification, gesture for navigation, and so on).
2. *Context*. It is essential to know the context of use to perfectly map tasks to the context:
   - *Physical environment*. Is the application used in a public place (e.g., train station) or a private place (e.g., at home)? Is it noisy or silent? Dark or bright? Closed (e.g., in a room) or open (e.g., in the street)? For example, in a public place, identifying the user by means of voice recognition is certainly not a good solution.
   - *Social*. Notion of privacy? For instance, a password must not be entered by a vocal order! Respect of social laws? Buying a product on the Internet entails legal procedures that must be integrated into the application.

- *Temporal.* Is the application used during day or night? Is the application time (and battery) consuming? Is the user in a hurry or not? If the application can detect whether the user is in a hurry, it should reduce the length of messages.
- *Individual.* Does the user have habits? Is it possible for the user to customize the application? Is the application expensive to use? For instance, if the application requires a permanent connection with a mobile phone, it could be very expensive.

3. *Kinds of tasks.* Three basic categories of tasks exist in Diane+—manual, interactive, and automatic—and their properties greatly influence the user interface:
   - Is the task a *"display"* task or an *"enter"* task? If so, what to display, what to enter? How to enter? Where to display?
   - Is the task a *"mandatory choice"* task? The system must not authorize other tasks until the user chooses or enters something.
   - Is the task an *"exclusive choice"* task? The user interface must put forward only the possible solutions.
   - Is the task an *interactive* task? The user interface must reflect the possibility of interacting (text field, push button, and so on).
   - Is the task an *automatic* task? Remember to display a message if the execution time is greater than one second. Give permanent feedback to the user.
   - Is the task a *mandatory* task? Use a message to strengthen the obligation or when the user avoids this task.
   - Is the task a *facultative* task? How does the user know that this task exists? How can the user trigger it?
   - Is the task a *repeated* task? How many times can the user execute the task? Does the user know the number of possible repetitions?
   - *Abstract or concrete* task? Is the task a mandatory step for the user? Is it important for the user to explicitly trigger or perform the task? For instance, "Search a film" (see the case study below) is too broad to be interesting, but "Search a film by title" is meaningful.
4. *Data related to the task.* The data model is elaborated at the same time as the UPV and SPV are built (Tarby, 1997, 1999; Tarby & Barthet, 1998).
   - What is the *volume of data to be sent*? *To be received*? One character, 10 pages, 1 Mb? The volume of data influences the time of execution, the cost of the connection, and so on.
   - How are *data linked together*? Are there similarities with other applications?
   - *Precise or imprecise information*? A price for a product is precise, but a comment given by the user about a product is not.
   - *Known or unknown information*? For instance, if the application asks for the color of eyes, this information can be evaluated precisely, but the number of people on Earth who have blue eyes cannot be known precisely.
   - *Formatted or not*? For instance, a date is usually formatted, but what is the format to respect?
5. *User's level.* The application may be adapted to it during the design or must be customizable during use.
   - What is the user's level *for the interaction mode*? Novice, occasional, expert? For instance, if voice recognition is imposed, are the users familiar with this interaction mode?
   - What is the user's level *for the application domain*? Novice, well-informed, expert? If the application is about movies, are the users cinema lovers or people who go to the cinema less than five times a year?

All the questions and criteria influence the offered solutions for alternative interfaces. These interfaces can be produced with "new" languages like UIML or XML/XSLT, but it is not guaranteed that the interfaces will offer exactly the same functionality. These new languages can only transform a conceptual representation into a concrete one, for example, a "button" into a JButton, or "emphasis" into a <b> HTML tag. Respecting the above-mentioned criteria improves the consistence of the interfaces, but today this verification can not be done automatically. The designer must check whether each task in the task tree can be applied for each device. Consequently, applying the above criteria for each device can entail modifying the task tree with regard to the underlying hardware or software. New or more detailed decompositions of tasks may appear, some tasks may disappear, and so on. The next section presents an example showing how some of the criteria are applied.

26.3 CASE STUDY: BOOKING CINEMA TICKETS WITH WEB, WAP, AND PHONE DEVICES

The New Millennium Company distributes 35 films every week. Most of them are recent and are scheduled all day long; the others are very old and famous films and are scheduled at special times, depending on the film (e.g., *Vertigo* only at 11:00 a.m.). The company aims to increase its popularity by allowing its clients to book tickets online. One constraint is that the booking should be done through its Web site, by a Wap phone, or by a regular phone. Each option involves a different platform, but the company requires that booking using the three platforms must be coherent. The database is stored on one server, and the different user interfaces will be produced with UIML or XML/XSLT. (Note that this example, although simple, covers the major notions of the methodology. The example could be changed to involve a company that manages entertainment in a big city; the users could search films, plays, and exhibitions; and so on.)

Following are two possible scenarios:

1. John is at home and wants to go to the cinema this afternoon because he has 3 free hours. Once connected to the system via his PC, he selects "search by session." John asks for playing times between 1:00 p.m. and 4:00 p.m., and the system displays the list of films that correspond to this time interval. After selecting the second film, John asks for more information about it (e.g., a summary of the film, the name of the director, and so on). John books a ticket for this film, and the system provides him with a confirmation of the booking.

2. It is 10:00 p.m. Mary and Philip are leaving a restaurant. While walking in the street, Mary says she wants to go to cinema to see the new film with Tom Cruise, but Mary and Philip do not remember the title and do not know if it is playing tonight. Mary gets her Wap phone and calls the New Millennium Company. She asks for a search by actor, and the system alphabetically lists all actors in current films. She selects Tom Cruise, and the system displays *Vanilla Sky*, along with all next sessions with free seats (the 10:30 p.m. session is full, so the system displays only the 11:30 p.m. performance). Philip wants a summary, so Mary selects "abstract" and a synthetic voice reads aloud the film's abstract.

Many UPV and SPV task models can be written for such scenarios. It is not possible to determine a priori which one is better, but they can be evaluated by real users and/or through prototypes based on the UPV and SPV task models.

One possible UPV is represented on Fig. 26.4 (bold lines represent decomposition of the task). Once the UPV is done, the SPV can be specified. In the example, only two tasks are

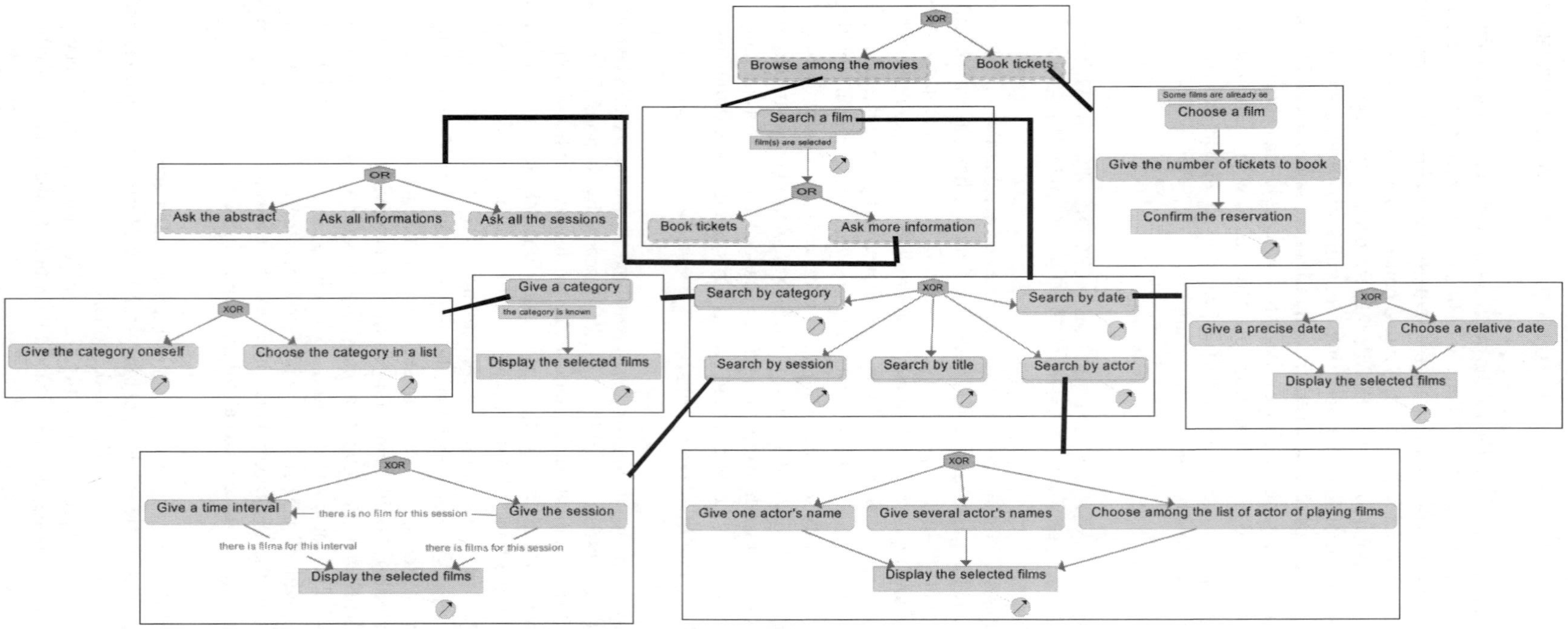

FIG. 26.4. Extract of the "New Millennium" UPV task model.

different between the UPV and SPV: "search by session" and "search by date." They will be presented later. Each task of the UPV and SPV can be specified precisely by splitting it into elementary subtasks or by constructing a formal model of interaction, with a Statechart (Horrocks, 1999), for instance. The next section show that the UPV and SPV provide enough information to build coherent user interface prototypes for a Web site, a Wap phone, and a regular phone without modeling precisely tasks with a formal notation. (Note: we anticipate the future of Wap phones by assuming that the next generation of mobile phones will be multimodal, meaning that the user could use keypad, touch screen, voice, and gesture and that the phone will answer at least vocally and graphically.)

26.3.1 Task 1: Choose Among the List of Actors in Current Films

Some people know the names of most actors because they love the cinema. So, it must be possible to enter a name or names ("Give one actor's name" and "Give several actors' names") and for the spelling to be rectified (the system should rectify the name transparently). Sometimes users do not remember the name of an actor, and it is necessary to provide a list of the main actors playing in the films of the week A search may be also made using the names of characters, e.g., Obiwan Kenobi in *Star Wars*. The question then is how to implement the task "Choose among the list of actors in current films" whether the platform is a Web site, a Wap phone, or a regular phone.

26.3.1.1 Web Site

Using a Web site is the easiest way to implement this task. Widgets like a simple scrollable list (Fig. 26.5) or a list with pictures (Fig. 26.6) could be used. A scrollable list respects mainly the screen space criteria, the list with pictures respects the application domain's user-level criteria (on the assumption that some users rarely go to the cinema). If the designer specified a multiple selection of actors in the UPV, both lists must respect this functionality.

26.3.1.2 Wap Phone

The screen criteria are very important for this device. Wap phone screens are very limited in size and resolution. Consequently, the system must not display too much information. One easy solution is to display only the names of actors in alphabetical order, eventually with a multiple selection (Fig. 26.7). This solution saves time and battery power because the volume of data is very small (temporal context criteria and data criteria).

The display of the list can be completed by a reading aloud by the system. With such vocal assistance, a user who is shortsighted can select an actor by clicking when the name is read, and a vocal confirmation can be made by the system (individual context criteria).

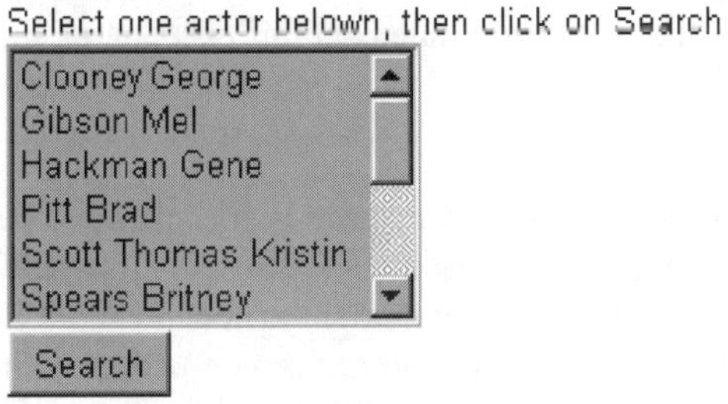

FIG. 26.5. Multiple selection of actors for "expert."

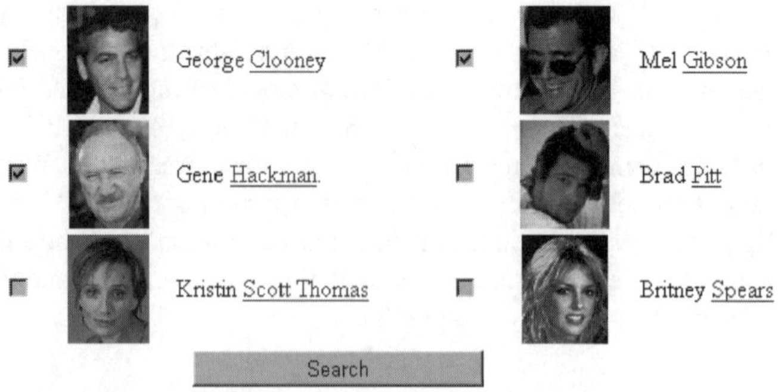

FIG. 26.6. Multiple selection of actors for "novice."

FIG. 26.7. Multiple selection of actors on a Wap phone.

26.3.1.3 *Regular Phone*

This device is very special because there is no screen (output device criteria). Therefore, to search by actor, vocal assistance could say, "Press 1 to give the name of one actor, 2 to give the name of several actors, or 3 to select from a list of actors in current films." If the user presses 3, the vocal assistance could say, "Select one actor by saying 'search' when his or her name is read" for a simple selection, or "Select each actor you want by saying 'yes' during the reading of the list and say 'search' to start the search" for a multiple selection.

26.3.2 Task 2: Search by Date

In the UPV, the user is supposed to give either a precise date (March 23) or a relative date (Monday), then the system displays the search result. In the SPV (Fig. 26.8), the distinction is not necessary because it is the role of the system to support these two possibilities. So, in the SPV, there is only one interactive task: "Give a date." The UPV is therefore important because it makes clear that the system must be able to understand what kind of date is given by the user, but it is not interesting for the user to have two different interactors on screen to perform this task. Again, the question is how to implement this task among the three platforms.

26.3.2.1 *Web Site*

Concerning the screen criteria, if the screen space is not limited, a calendar ActiveX control (Fig. 26.9) or an HTML table can be used (Fig. 26.10); if the screen space is limited, an HTML line can be used (Fig. 26.11). It is important to note that both solutions provide relative and precise dates (temporal criteria). The user can see immediately the current date (which is selected and/or in bold) and the current week for the films. The user just has to click on the required date.

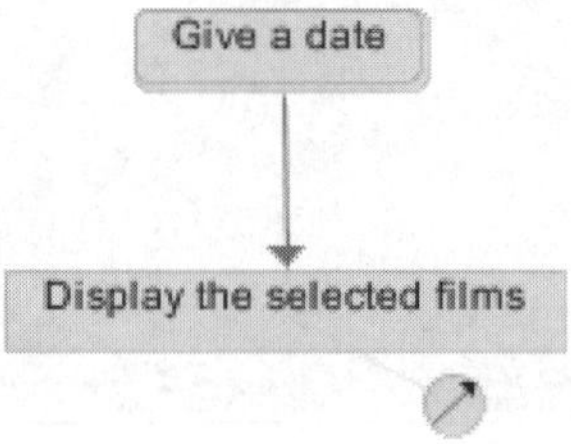

FIG. 26.8. "Search by date" in SPV.

Février 2002
Dim Lun Mar Mer Jeu Ven Sam
1 2
3 4 5 6 7 8 9
10 11 12 13 14 15 16
17 18 19 20 21 22 23
24 25 26 27 28

FIG. 26.9. Choose a date with an activeX control.

February 2002
Sun Mon Tue Wed Thu Fri Sat
1 2
3 4 5 6 7 8 9
10 11 12 13 14 15 16
17 18 19 20 21 22 23
24 25 26 27 28

FIG. 26.10. Choose a date with a HTML table.

Wednesday 13rd | **Thursday 14th** | Friday 15th | Saturday 16th | Sunday 17th | Monday 18th | Tuesday 19th

FIG. 26.11. Choose a day with a single HTML line.

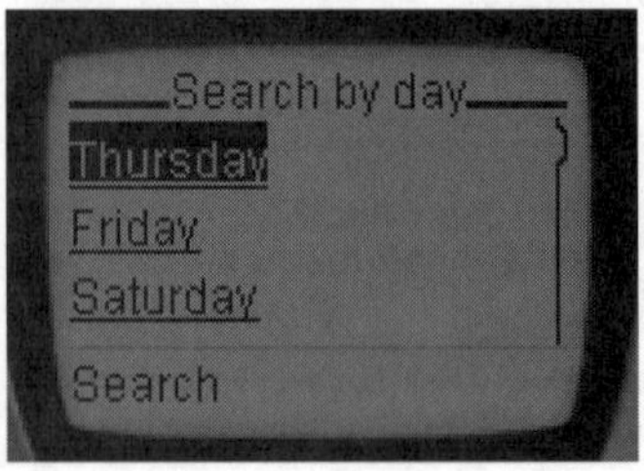

FIG. 26.12. Choose a day with a Wap phone.

26.3.2.2 Wap Phone

Displaying a calendar is not useful on a Wap phone because of the small size of the screen (output device criteria). Therefore, the HTML single line solution is reused here, and a simple list of the next few days is provided. In Fig. 26.12, the temporal context criteria are applied to provide more useful information to the user. The current day is displayed on top, and yesterday's date is not displayed because it is useless for the user.

26.3.2.3 Regular Phone

The SPV task model can be used to design the "interface." Therefore, the system can say, "What date do you want?" and it must then be able to understand whether the user answered "Friday," "Next Friday," "Tomorrow," "the 15th," and so forth. The social context criteria are applied here because the user can say "Friday," "Tomorrow," or "the 15th" for the same date.

26.3.3 Task 3: Search by Session

In the UPV, the user either gives a precise time (at 8:00 p.m.) or a time interval (after 8:00 p.m., between 8:00 p.m. and 10:00 p.m., and so on). If there is no film for a requested session (the "imprecise data" criteria are applied here so that the search is made within a number of minutes around the interval, the actual duration being determined by some form of usability assessment), the system proposes to the user a time interval. As for the search by date, in the SPV (Fig. 26.13) the user does not have to choose between these two possibilities. The system must be able to understand the request of the user (the exact time or time interval). Therefore, there is no more XOR choice for the system; the XOR operator is for the cognitive aspect.

26.3.3.1 Web Site

Depending on the free space, scrollable or dropdown lists can be used (Fig. 26.14). Whatever the solution is, an intelligent preselection must be done. In Fig. 26.14, the supposed time is 12:45 p.m.; therefore, the system selects by default the next session for the "from" session and the session at "from" +2 hours for the "to" session (temporal context criteria). More intuitive solutions can be used, such as hyperlinks with specific colors if the session is almost full (social context criteria).

26.3.3.2 Wap Phone

Similarly to "Select by date," simple lists can be employed, but the interface will be more intuitive if it displays only the next (free) sessions (Fig. 26.15). The temporal and social context criteria are applied. Another possibility is that the user says what time interval he or

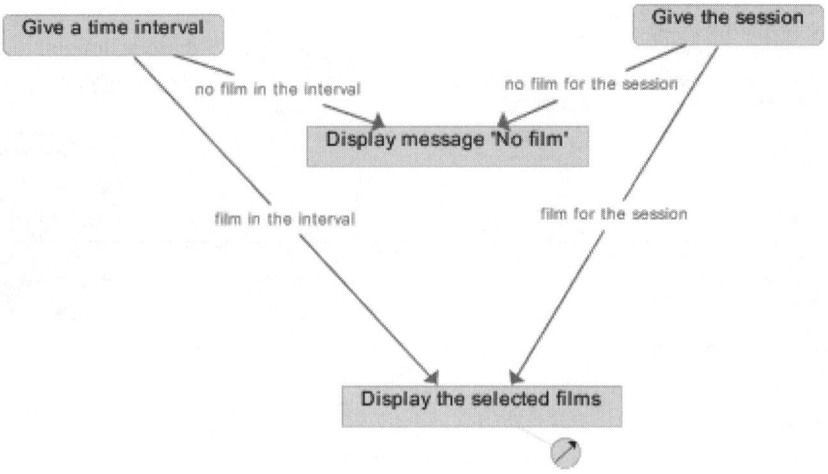

FIG. 26.13. Search by session in SPV.

FIG. 26.14. Select a session with a HTML page.

FIG. 26.15. List of the next sessions on a Wap phone.

she wants (e.g., "I want the sessions between 8:00 p.m. and 10:00 p.m."). Future multimodal Wap phones should be able to process this request.

26.3.3.3 *Regular Phone*

The SPV requirement can be applied here. The user may ask for films "at 8:00 p.m.," "after 8:00 p.m." "before 8:00 p.m.," "between 8:00 p.m. and 10:00 p.m.," "starting at 8:00 p.m.," "finishing before 8.00 p.m." and so on. The system must be able to recognize the user's request. Another solution would be for the system to say, "Here are the next sessions; say 'stop' to select one," then start to read the sessions. The social context and temporal context criteria are applied here; for instance, if the user is in a hurry, he or she does not want to spend time in menus selecting start time and end time.

26.3.4 Task 4: Display the Selected Films

This task appears after each search. The displayed information is supposed to be the same for each search result, but it can be adapted to the target platform.

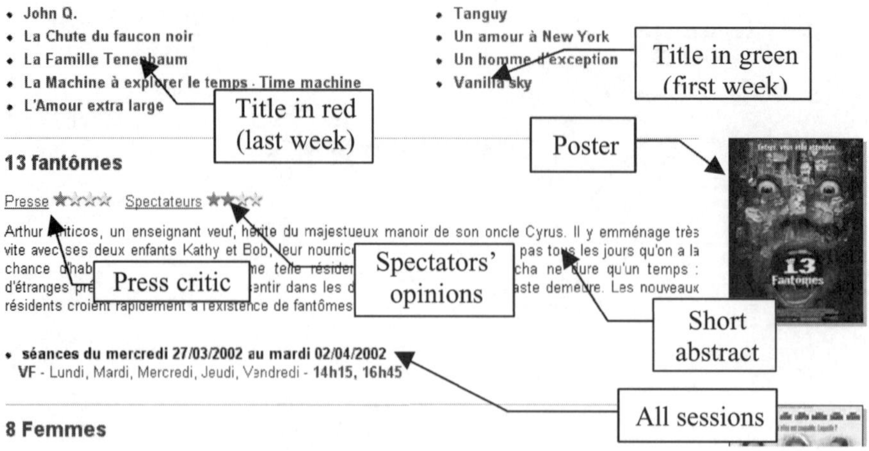

FIG. 26.16. Selected films with a distinction for first and last week.

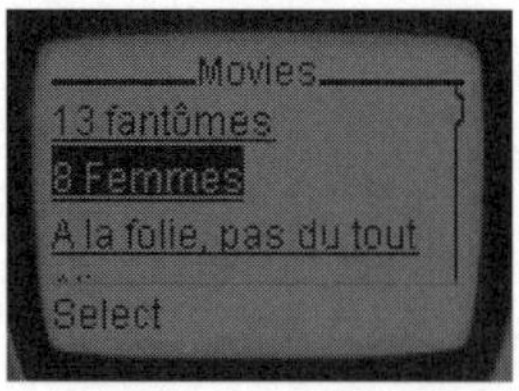

FIG. 26.17. Select films on a Wap phone.

26.3.4.1 Web Site

Current workstations can display lots of information on screen. This ability allows the display of more than just film titles. For instance, movies can be differentiated by colors (e.g., to specify whether it is the first or last week) and be associated with an abstract, the sessions, spectators' opinions, and so forth (Fig. 26.16). Social context criteria, data criteria, and output device criteria are applied here.

26.3.4.2 Wap Phone

Because of the volume of data criteria and the temporal criteria, the selected films should be displayed as a textual list (Fig. 26.17) while vocal assistance gives a small amount of information for each film (sessions, short abstract, etc.).

26.3.4.3 Regular Phone

In the case of a selection of a session by the user, vocal assistance could say, "There are six films at 8:00 p.m. You can select one film by saying 'stop' or pressing key 1 after the title. The titles are *13 Fantômes, 8 Femmes, A la folie, pas du tout,...*" In this solution, the temporal criteria and data volume criteria are applied first.

26.3.5 Task 5: Ask for the Abstract

"Ask more information" and "Ask for the abstract" are optional (dotted line) for the user. It is assumed that "Browse among the movies" is finished when the user has finished searching for

OCEAN'S ELEVEN
FAITES VOS JEUX !
Ocean's eleven
Film américain (2001). Action, Comédie. Durée : 1h 57mn.
Date de sortie : 06 Février 2002
Avec George Clooney, Brad Pitt, Julia Roberts, Matt Damon, Andy Garcia Plus...
Réalisé par Steven Soderbergh
Presse
Spectateurs
Donnez aussi votre avis !
BANDE ANNONCE
PHOTOS
Synopsis
Après deux ans passés dans la prison du New Jersey, Danny Ocean retrouve la liberté et s'apprête à monter un coup qui semble impossible à réaliser : cambrioler dans le même temps les casinos Bellagio, Mirage et MGM Grand, avec une jolie somme de 150 millions de dollars à la clé. Il souhaite également récupérer Tess, sa bien-aimée que lui a volée Terry Benedict, le propriétaire de ces trois somptueux établissements de jeux de Las Vegas.
Pour ce faire, Danny et son ami Rusty Ryan composent une équipe de dix malfrats maîtres dans leur spécialité. Parmi eux figurent Linus Caldwell, le pickpocket le plus agile qui soit ; Roscoe Means, un expert en explosifs ; Ruben Tishkoff, qui connaît les systèmes de sécurité des casinos sur le bout des doigts ; les frères Virgil et Turk Malloy, capables de revêtir plusieurs identités ; ou encore Yen, véritable contorsionniste et acrobate.

FIG. 26.18. Ask for an abstract on a web page.

FIG. 26.19. Ask for an abstract on a Wap phone.

a film, and the user can book tickets or ask more information after this step. The kinds of tasks and output device criteria are important here.

26.3.5.1 Web Site

A web page can contain a long abstract (Fig. 26.18) with information on the origin of the film, the names of the principal actors, spectators' remarks, and so on. This type of abstract is possible because there is enough place on the screen and the time needed to download the information is very short (screen space and temporal context criteria).

26.3.5.2 Wap Phone

In contrast, on a Wap phone the time required to download information can be long and expensive (temporal context criteria). Supposing that future Wap phones will be multimodal, a short synopsis can be displayed on the screen, one or two sentences, for instance (Fig. 26.19), while the long version is read aloud (when the user requests this).

26.3.5.3 Regular Phone

When the user has selected a film, vocal assistance confirms the action by repeating the title of the film, then it proposes to book tickets for this film or to obtain more information: "You have selected *Ocean's Eleven*. Press 1 to book tickets, 2 to obtain all sessions, 3 to obtain a synopsis, 4 to receive all information on this film." If the user presses 3, vocal assistance reads aloud the long version of the abstract (output device and volume of information criteria).

26.3.6 Summary of the Three Devices

Each device has limits as well as advantages that can help the designer produce the best interface. Here is a short summary.

With a workstation, the interface can be very busy, and multimedia information may be used. Lots of multimedia information can be presented simultaneously; this allows the designer to reduce the depth of the task tree by regrouping some tasks in a few windows instead of creating one window per task. If done rigorously, the efficiency of the application will be improved for the user.

The Wap phone has two hard constraints: the screen size and the cost of a call. These constraints limit the amount of the information. The navigation is very limited and not user-friendly, and a simple task can result in multiple interactions. For instance, the task "Enter a name" will be implemented by means of four interactions: go to the name field with the up/down Wap phone buttons, press "Enter" to open the field, enter the name with the phone keyboard (dealing with the well-known difficulties of using phone keyboards), and validate the entered name. Therefore, the Wap phone increases dramatically the depth of the task tree. The designer must limit the input tasks (or replace them with voice recognition if possible) and the amount of information and may rearrange the task tree to optimize the navigation. Note that with PDAs, because they offer larger screens, multimedia information, and so on, the limits are not as restrictive.

The regular phone has a very limited capacity to display information on screen, but it allows efficient navigation if there is good voice recognition. To implement the task tree, the designer must either eliminate all display information or replace it by vocal information. There is no visible context for the user because the navigation must be the designer's main preoccupation. The user must always know where he or she is in the task tree and should be able to navigate very quickly through the tasks. Consequently the user's vocal orders need to be recognized before the end of each running task.

26.4 DISCUSSION

Coupling the UPV and SPV with prototypes is a very efficient way to quickly build user interfaces, but its success will depend on the quality of the task modeling. This phase can be really improved by integrating the final users in the modeling through participative design and by using scenarios. Current prototyping tools like Visual Basic® or Dreamweaver® save time and allow added time for the task modeling and/or the scenarios, but it is difficult today to know a priori how much time is necessary for task modeling, writing scenarios, and prototyping. One solution is to precisely define words and syntax for the task modeling and the scenarios with regards to the users' habits, the application domain, and so on. For instance, "Select" or "Choose" implies an interactive selection among a limited choice (e.g., a list of available printers), "Give" or "Enter" assumes that the user must enter data because the choice is unknown or not limited, and so on.

The UPV-SPV method is very useful for information system applications (e.g., e-business or virtual campus Web sites). For example, we have modeled the major tasks for all categories of e-business Web sites (e.g., search a product, exchange a product, build a product, put a product up for auction, add a product to the caddie, order a product, request a vendor, and so on). That allows us to be sure that no functionality is forgotten when a new site is prototyped.

by title	13 fantômes ▾
by category	Horror ▾
by actor	Clooney George ▴ Gibson Mel Hackman Gene Pitt Brad Scott Thomas Kristin Spears Britney ▾

☐ ☐ ☐ ☐ ☐ *others...*

| by session | From 01.00 PM ▾ To 03.00 PM ▾ |
| by date | Wednesday 13rd \| Thursday 14th \| Friday 15th \| Saturday 16th \| Sunday 17th \| Monday 18th \| Tuesday 19th |

| Search | More information | Book tickets |

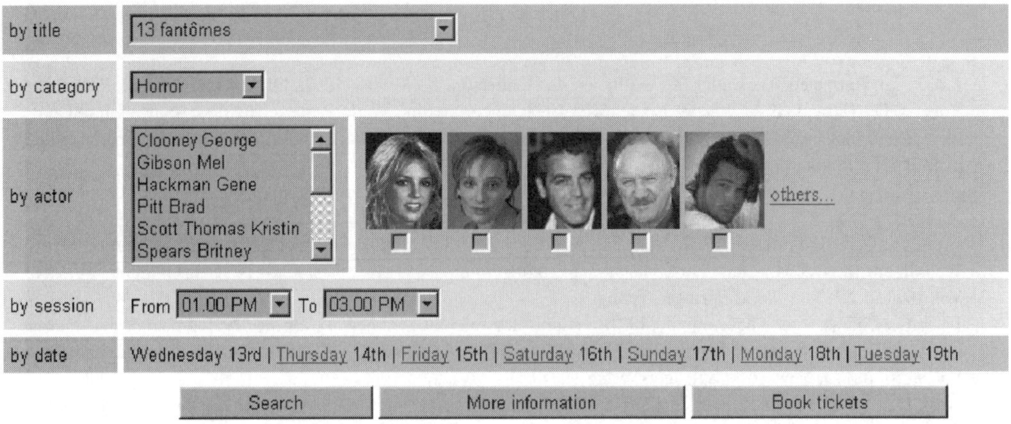

FIG. 26.20. Search a film, book tickets, and ask for more information all on the same page.

New Millennium
Book tickets
Films by title
Films by session
Go

FIG. 26.21. Another organization on the tasks for the Wap phone.

The UPV-SPV method is also useful for systems that do not require too much formal description, such as a photocopy machine. Some companies prefer this approach over spending time creating formal models with Statecharts (Horrocks, 1999).

Another advantage of the UPV-SPV method is that it allows customization of the interface. In the case of a Web site, all the search criteria can be grouped on the same page as the "Book tickets" and "Ask more information" tasks, and the user may choose the appearance of each search criteria (Fig. 26.20). Abstract tasks remain the same in all cases; what is changing is the interaction. Customization can also be applied to the device or the context of use. About the device, the organization of tasks is completely different for a Web site and a Wap phone (Fig. 26.21) because of the need to optimize the download time in this second device. As for the context of use, the system could detect if the room is noisy and adjust the volume of the vocal assistance, and in addition it could detect if the user is in a hurry but the user could also say "Faster" or "Hurry up!" (by the user's tone of voice), and respond by providing only the minimal information (e.g., the short version of the abstract).

26.5 CONCLUSION

This chapter has presented a task-based design methodology that uses the user's point of view (UPV) and the system's point of view (SPV) to specify abstract tasks that are independent of target platforms. With proper respect for the user's goal and ergonomics criteria, it allows the tasks to be adapted to different devices, taking into account underlying hardware and software constraints and the user's habits and the context of use.

REFERENCES

Artim, J., van Harmelen, M., Butler, K., Gulliksen, J., Henderson, A., Kovacevic, S., Lu, S., Overmyer, S., Reaux, R., Roberts, D., Tarby, J.-C., & Vander Linden, K. (1998). Incorporating Work, Process and Task Analysis Into Commercial and Industrial Object-Oriented Systems Development, CHI '98 Workshop, Los Angeles. Papers published in *SIGCHI Bulletin, 30*(4).

Barthet, M.-F. (1988). *Logiciels interactifs et ergonomie: Modèles et méthodes de conception.* Editions Dunod Informatique.

Bonifati, A., Ceri, S., Fraternali, P., & Maurino, A. (2000). Building multi-device, content-centric applications using WebML and the W3I3 tool suite. In *Conceptual modeling for e-business and the Lecture* Notes in Computer Science, Vol. 1921, p. 64. New York: Springer-Verlag.

Calvary, G., Coutaz, J., & Thévenin, D. (2001a). A unifying reference framework for the development of plastic user interfaces. In *Proceedings of Engineering Human Computer Interaction 2001*, Lectures Notes in Computer Science, Vol. 2256, p. 173. New York: Springer-Verlag.

Calvary, G., Coutaz, J., & Thévenin, D. (2001b). Supporting context changes for plastic user interfaces: A process and a mechanism. In A. Blandford, J. Vanderdonckt, & P. Gray (Eds.), *Proceedings of Interaction Homme Machine-Human Computer Interaction 2001* (pp. 349–363).

Constantine, L., & Lockwood, L. (1999). *Software for use: A practical guide to the models and methods of user-centered design.* Reading, MA: Addison-Wesley; New York: ACM Press.

Hackos, J., & Redish, J. (1998). *User and task analysis for interface design.* New York: Wiley.

Horrocks, I. (1999). *Constructing user interface with Statecharts.* Reading, MA: Addison-Wesley.

Jacobson, I., Christerson, W., Jonsson, P., & Övergaard, G. (1992). *Object-oriented software engineering: A use case driven approach.* Reading, MA: Addison-Wesley.

Paris, C., Tarby, J.-C., & Vander Linden, K. (2001). *A flexible environment for building task models.* In A. Blandford, J. Vanderdonckt, P. Gray (Eds.), *Proceedings of Interaction Homme Machine-Human Computer Interaction 2001*, BCS conference series, springer, New York.

Paternò, F., & Santoro, C. (2002). One model, many interfaces. In *Proceedings of 4th International Conference on Computer-Aided Design of User Interfaces 2002* (pp. 143–154). Kluwer Academic Publishers, Dordrecht.

Paternò, F., Mancini, C., & Meniconi, S. (1997). *ConcurTaskTrees: A diagrammatic notation for specifying task models.* In *Proceedings of Interact '97* (pp. 362–369). Chapman & Hall. 6th IFIP International Conference on Human Computer Interaction, Boston MA: Kluwer Academic Publisher.

Rosson, M. B., & Carroll, J. M. (2001). *Usability engineering: Scenario-based development of human-computer interaction.* San Francisco: Morgan Kaufmann.

Tam, R. C., Manlsby, D., & Puerta, A. R. (1998). *U-TEL: A tool for eliciting user task models from domain experts.* In *Proceedings of International Conference on Intelligent User Interfaces IUI '98* (pp. 77–80). New York: ACM press.

Tarby, J.-C. (1994). The automatic management of human-computer dialogue and contextual help. In *Proceedings of the 1994 East-West International Conference on Human-Computer Interaction (EWHCI '94).* Lecture Notes in Computer Science, Vol. 876, New York: Springer-Verlag.

Tarby, J.-C. (1997). *Merge tasks and objects: Dream or reality?* Paper presented at Object Oriented User Interfaces, CHI '97 Workshop, Atlanta, GA.

Tarby, J.-C. (1999). *When the task's world meets the object's world.* Paper presented at WISDOM '99, Workshop on User Interface Design and Object Models, ECOOP 99, Lisbon.

Tarby, J.-C., & Barthet, M.-F. (1996). *The* Diane+ method. In *Proceedings of the Second International Workshop on Computer-Aided Design of User Interfaces* (pp. 95–119). Namur, Belgium: Presses Universitaires de Namur.

Tarby, J.-C., & Barthet, M.-F. (1998). From tasks to objects: An example with Diane+. SIGGIH Bulletin, *30*(4), 33–36.

Tarby, J.-C., & Barthet, M.-F. (2001). Analyse et modélisation des tâches dans la conception des systèmes d'information: Mise en œuvre avec la méthode Diane+. In C. Kolski (Ed.), *Analyse et conception de l'IHM: interaction pour les systèmes d'information* (Vol. 1, pp. 117–166). Editions Hermes, Paris.

Traetteberg, H. (2002). Using user interface models in design. In *Proceedings of 6th International Conference on Computer–Aided Design of User Interfaces, 2002.* Kluwer Academic Publishers, Dordrecht.

van Harmelen, M., Artim, A., Butler, K., Henderson, A., Tarby, J.-C., Roberts, D., Rosson, M. B., & Wilson, S. (1997). *Object models in user interface design, A CHI '97 Workshop. Papers published in SIGCHI Bulletin, 29*(4).

Vanderdonckt, J., Derycke, A., & Tarby, J. C. (1998). Using data flow diagrams for supporting task models. In P. Markopoulos & P. Johnson (Eds.), *Supplementary Proceedings of 5th Eurographics Workshop on Design, Specification, Verification of Interactive Systems, (DSV-IS '98)* (pp. 1–16). Eurographics Association, Aire-la-ville, Switzerland.

27

Linking Task and Dialogue Modeling: Toward an Integrated Software Engineering Method

Chris Scogings
Massey University

Chris Phillips
Massey University

To date, software development methods and models have largely ignored the user interface. This is despite its increasing importance over the past decade as graphical user interfaces (GUIs) have become dominant, and as more attention has been devoted to usability aspects of interactive systems. This chapter initially reviews task and dialogue modeling, and examines task representation in UML. Lean Cuisine+, a notation incorporating both dialogue and task modelling capabilities, is then introduced. A method for the early stages of interactive systems design is proposed that integrates the two notations. The method is presented via a case study.

27.1 INTRODUCTION

Over the past 20 years a plethora of methods, models, and notations have been created to assist with software development, often supported by the use of computer-aided software engineering (CASE) tools (see chaps. 1 and 23). Some convergence is currently occurring through initiatives such as the Rational Unified Process (Jacobson, Booch, & Rumbaugh, 1999) and the Unified Modeling Language (UML; Rumbaugh, Jacobson, & Booch, 1999; chaps. 7, 11, 12, 19, 22, 23, and 24).

To date, software development methods and models have largely ignored the user interface (Kemp & Phillips, 1998; Kovacevic, 1998). The primary focus has been on support for functional decomposition of systems and on the mapping of the functionality to software architectures. This has been referred to as the design of the "internal" system, as opposed to the "external" or visible system (Collins, 1995). There is a need to look at the overall software engineering design process and at how user interface design can best be incorporated.

The design of the external system has assumed increasing importance over the past decade as graphical user interfaces (GUIs) have become dominant and as more attention has been devoted to usability aspects of interactive systems. It has been said that, to the user of an interactive application, "the interface *is* the system" (Collins, 1995, p. 49). User interface design requires good descriptive systems and the development of tools to support it (Guindon, 1990). Models and notations are required for describing user tasks and for describing the structure of the human-computer dialogue used to support these tasks.

Task models, which focus on task decomposition and/or task flow, form an important input to user interface design. Their primary purpose is to define the activities of the user in relation to the system as a means of uncovering functional requirements to be supported. The frequency of tasks and the objects manipulated by tasks can have a direct bearing on the structure and appearance of the user interface. The use of task models in user interface design is currently limited by a lack of tools to support their definition and a weak linkage to dialogue (*Tool Support,* 1999), notwithstanding the reviews in chapters 23 and 24 and the examples in part IV.

Dialogue models have been developed as a means of capturing information about the behavior of the user interface at an abstract level. They are important because they provide the interface designer with a means of describing the structure and behavior of the user interface at a level removed from the surface "look and feel." Dialogue models should provide for analysis of the dialogue in relation to the tasks that must be supported, especially their efficiency and completeness. However, a criticism of existing dialogue notations is that the linkage to tasks is often obscure (Brooks, 1991): They define the structure and behavior of the dialogue but do not always reveal the interconnections between dialogue components during the execution of higher level tasks. That is, they do not represent tasks within the *context* of the dialogue structure (the collection of dialogues and subdialogues that make up the interaction).

This chapter first reviews task and dialogue modeling and then examines task representation in UML. Next, it introduces Lean Cuisine+, a notation incorporating dialogue- and task-modeling capabilities. Finally, it proposes a method for the early stages of interactive systems design that integrates the two notations. The method is presented via a case study.

27.2 TASK MODELS

A variety of task description methods have been developed, and many of these involve graphical techniques. Some of them focus primarily on task decomposition and are top-down, such as Hierarchical Task Analysis (HTA; Shepherd, 1989; chap. 3). The outputs of HTA are a hierarchy of tasks and subtasks and plans describing in what sequence and under what conditions subtasks are performed. Other task models are concerned primarily with task flow and are bottom-up, such as use case models (Constantine & Lockwood, 1999). Many of them describe the activities of both the user and the system. Some, like GOMS (Card, Moran, & Newell, 1983; chap. 4), include goal-oriented cognitive aspects of the interaction. Graphical notations applied to task description include input-output diagrams, process charts, decision-action diagrams, Petri nets, and flow graphs (Kirwan & Ainsworth, 1992, provide a comprehensive review).

Use cases, which are abstract task scenarios (see chap. 24), were originally developed in connection with object-oriented software engineering (Jacobson, Christerson, Jonsson, & Övergaard, 1992) as a way of uncovering the functionality of a system. They model user actions and system responses and involve the manipulation of objects and the sequencing of interactions. A comprehensive review of the use of scenarios in design is provided in Filippidou (1998). Scenario-based development methods are still evolving, and the relationship between task scenarios and object-oriented software development has been explored in other research (Rosson & Carroll, 1995; chap. 5).

27.2.1 Task Modeling in UML

UML is a modeling language consisting of a collection of semiformal graphical notations, including use case diagrams, activity diagrams, class diagrams, interaction diagrams, and state diagrams. These notations can be used to support different development methodologies (Fowler &

Scott, 1997; Quatrani, 1998). Tasks in UML are described using use cases and scenarios (a scenario is defined as one thread or path through a use case). Associated with UML is the Rational Unified Process, which consists of a set of concepts and activities needed to transform requirements into a software system. It is not a detailed, step-by-step software design method but rather a framework that can be adapted to suit local conditions. A number of other UML-based methods have been developed (Eriksson & Penker, 1998; Fowler & Scott, 1997; Richter, 1999).

All the methods begin with the definition of use cases. Use cases capture the relationship between tasks and objects. Useful guidelines for establishing use cases are provided in Richter (1999). It has been pointed out that use cases can represent different levels of task detail and may subsume other use cases (Fowler & Scott, 1997). Determining the level of detail remains a challenge for today's system developers. Use cases can be represented at a system context level in use case diagrams, which show the boundaries of the system, the external actors, and any structural relationships existing between use cases.

The internal behavior of a use case can be described graphically in UML using either activity diagrams or interaction diagrams, depending on the perspective required and the complexity of the task. Activity diagrams can be used to show the flow of control through use cases, depicting an abstract workflow view of the activities of each use case. Interaction diagrams describe sequences of message exchanges between the objects involved in individual scenarios (threads through use cases). Several commentators (e.g., Jacobson et al., 1999; Richter, 1999) support the need for detailed textual use case descriptions, not least so that clients (end users) can fully understand and agree on the details. A structured format describing the primary flow of events, as well as details of any possible variations, is the norm (Quatrani, 1998). Textual use case descriptions are used in Step 1 of the method proposed in this chapter.

Various techniques exist for identifying object classes (Overmyer, 2000; Stevens & Pooley, 2000), including derivation from use cases. Static object relationships are captured in UML class diagrams, and the behavior of object instances can be described using UML state diagrams. Dialogue modeling is not directly supported in UML but might be attempted using state diagrams. More significantly, there is no direct means of relating tasks, captured as use cases, to dialogue structure.

27.3 DIALOGUE MODELS

At the early stage of user interface design, the designer requires models and notations that can assist with analysis of dialogue structure and are unconstrained by implementation issues. Dialogue models embody information about the interface in terms of objects, actions, states, and relationships, including both pre- and postconditions associated with the actions (chap. 19). Dialogue models can be applied both in the design of new interfaces and the analysis of existing interfaces, and they can be used in association with user interface prototyping.

A variety of dialogue models at various levels of abstraction have been employed in the early stages of interface design, and these have used both graphical and textual notations. They have included statecharts, Petri nets, user action notation (UAN; chaps. 22–24), event response systems (ERSs), communicating sequential processes (CSPs), and Lean Cuisine. Four graphical notations are briefly reviewed below (see Phillips, 1994, for a fuller treatment).

State transition diagrams (STDs) provide an easily displayed and manipulated network representation based on a small number of constructs. They predate direct manipulation interfaces and suffer from a number of shortcomings in this respect, notably that they are based on a "single event/single state" model, which can lead to an exponential growth in states and transitions in asynchronous dialogues and consequently to large and very complex diagrams. In particular, they do not handle unexpected user actions well. They are inherently sequential in nature and well suited to describing individual task sequences.

Statecharts (Harel, Pnueli, Schmidt, & Sherman, 1987) are a higraph-based extension of the STD formalism and are state based. Areas represent states, and labeled directed arcs represent transitions. Statecharts were developed for describing the behavior of complex reactive systems. The concepts that statecharts add to basic STDs—hierarchy, orthogonality, and broadcast events—provide expressive power and reduce the size of the resulting specification, although statecharts have been criticized as being too difficult for the designer. The need to show transitions explicitly remains and is compounded to an extent by the requirement for "null" states to be shown.

Petri nets (Peterson, 1981) are abstract virtual machines with a well-defined behavior, and they have been used to specify process synchrony during the design phase of time-critical applications. Like STDs, Petri nets define possible sequences of events in a modeled system. However, the need to show all transitions explicitly clutters and detracts from the diagrams. This is especially problematic where a number of transitions relate to a particular state.

Lean Cuisine (Apperley & Spence, 1989) is a graphical notation based on the use of tree diagrams to describe systems of menus. A menu is viewed as a set of selectable representations, called *menemes*, of actions, parameters, objects, states, and other attributes in which selections may be logically related or constrained. A meneme is defined as having just two possible states, "selected" and "not selected." Lean Cuisine offers a clear, concise, and compact graphical notation for describing the *structure* of menu-based dialogues, but it is not able to describe task sequences. Events are not directly represented but are implicitly associated with menemes.

A distinction can be drawn between STDs that are based on a single threaded model (single event/single state) and the other notations, all of which can describe concurrent states and therefore multithreaded dialogue. Shortcomings were uncovered in all these notations in relation to describing direct manipulation GUIs (Phillips, 1994), in that none could describe satisfactorily both asynchronous and sequential aspects of the dialogue. In particular, these notations had difficulties either capturing event sequences or relating event sequences to (higher level) tasks.

27.4 THE LEAN CUISINE+ DIALOGUE NOTATION

Lean Cuisine+ (Phillips, 1995), a semiformal graphical notation for describing the behavior of event-based direct manipulation GUIs, is an extension of Lean Cuisine. Lean Cuisine was only a dialogue-modeling notation. Lean Cuisine+ introduced the concept of task modeling within the context of the dialogue. It also added various minor improvements, such as using the fork to represent a homogenous group. Further incremental changes to Lean Cuisine+ have since been made. A current Lean Cuisine+ definition is provided in Scogings (2000) and is available at www.massey.ac.nz/~wwiims/rlims/page3.html.

The purpose of dialogue modeling is to analyze the dialogue in relation to the tasks that must be supported. It is not only a question of what a menu should look like but whether the menu should exist. Dialogue modeling has been going on for some time, using notations such as STDs, Petri nets, and so on. Traditional dialogue-modeling notations have two problems: (a) They do not show the tasks (e.g., an STD shows every state the system can possibly be in but no tasks), and (b) they do not connect to any programmer-approved method of constructing software. Lean Cuisine+ solves both of these problems because (a) it models both dialogue and tasks, and (b) it can be linked to use case and class construction in UML.

27.4.1 A Method for the Early Stages of GUI Design

This section uses a case study to explicate a method that supports the mapping of tasks to dialogue. This method, which builds on the strengths of UML and Lean Cuisine+, first appeared in Scogings and Phillips (2001). The case study concerns the development of part of an online

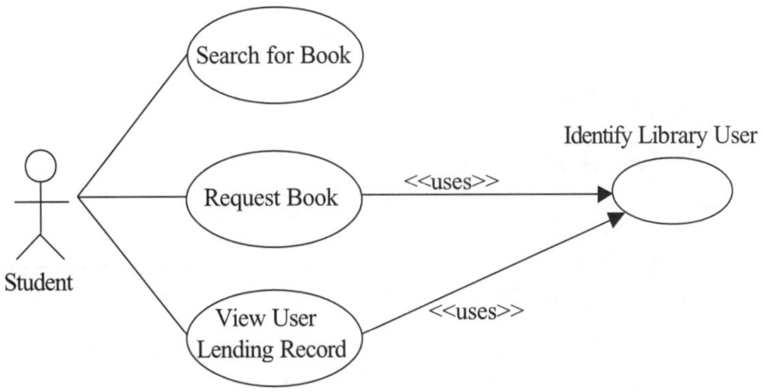

FIG. 27.1. UML use case diagram for the Library Catalogue.

library catalogue system. Library users can search for books, request that books be reserved, and display their lending record. The system requires user identification for the reservation of books or the display of a lending record.

This case study is a particularly good one to use to illustrate the combination of UML and interface design, as the use cases and class diagram were constructed as an earlier, separate project (Phillips, Kemp, & Kek, 2001). The case study thus confirms that Lean Cuisine+ dialogue can be extracted from independently produced UML use case specifications.

Four use cases are defined for the library catalogue system, and they are illustrated in a UML use case diagram (Fig. 27.1). The functionality of the *Identify Library User* use case is common to both *Request Book* and *View User Lending Record.* This is represented in the use case diagram by the UML <<uses>> relationship.

The proposed method is now presented as a series of steps illustrated by examples drawn from the library catalogue case study and is summarized at the end of this section.

27.4.1.1 Step 1: Identify Use Cases

The details of each use case are provided in the form of a textual description for the use case (Figs. 27.2 to 27.5). The style of textual description used here is from Phillips et al. (2001). Textual descriptions of use cases are recommended by the unified process and used by most UML-based methods.

27.4.1.2 Step 2: Construct a Class Diagram

Having established the use cases, the designers proceed to establish the classes for the main objects in the system. At this point, it is only necessary to establish classes for the *domain objects* (e.g., semester and course), as opposed to *system objects* (e.g., window and menu), which may be added later. These domain objects can be represented as a UML class diagram (Fig. 27.6).

Because the method is concerned only with the design of the user interface (and not the design of the underlying system), the entire class diagram is not required. In Fig. 27.6, the relationship between Library User and Book Copy is annotated to indicate that *each* library user is associated with zero or more (**0...** *) book copies. This means that each student can have several books (possibly none) checked out of the library. Likewise, each catalogue entry is associated with one or more (**1...** *) book copies to indicate that any book that appears once in the catalogue may have several physical copies on the shelves. However, relationships between classes are not usually required for the early stages of interface design. Class diagrams are developed incrementally as the system is designed (Paech, 1998). Note that any attributes and operations required for user selection *must* be included at this stage.

Use Case:	*Search for Book*
Actor:	Library User (Student)
Goal:	The library user can search for a book.
Preconditions:	None.
Main flow:	The library user activates the use case and then selects the search criterion, which can be one of title, author, ISBN/ISSN, call number, keywords, subject or date of publication. The system searches the catalogue using the search criterion and then displays a list of books which meet the search criterion. This list could be empty. The library user selects a book from the list. The system displays all the information about the selected book. The use case ends.
Exceptions:	None.
Postconditions:	The details of a particular book are visible.

FIG. 27.2. Textual description for the *Search for Book* use case.

Use Case:	*Request Book*
Actor:	Library User (Student)
Goal:	The library user can request a book.
Preconditions:	The details of a particular book are visible.
Main flow:	The library user selects a particular copy of the book. (**E1**) *Identify Library User* is activated. (**E2**) The system places the request in the queue for processing. The use case ends.
Exceptions:	**E1** If the selection is invalid, the system prompts the user to re-select. The use case continues. **E2** If *Identify Library User* fails, the use case ends.
Postconditions:	None.

FIG. 27.3. Textual description for the *Request Book* use case.

27.4.1.3 Step 3: Construct the Initial Lean Cuisine+ Diagram

The basic Lean Cuisine+ diagram consists of a collection of selectable menemes. These menemes have some correspondence with the domain object classes of the system under construction. Indeed, the major purpose of a user interface is to provide the user with access to objects in the domain of interest, and this is particularly so in the case of *direct manipulation* GUIs (Dix et al., 1993). Thus it is appropriate to base the initial Lean Cuisine+ menemes and

Use Case:	*View User Lending Record*
Actor:	Library User (Student)
Goal:	The library user can check details of checked-out books, requested books or unpaid library fines.
Preconditions:	None.
Main flow:	The library user activates the use case. *Identify Library User* is activated. (**E1**) The library user selects one or more of: **1.** Display checked-out books :- the system displays the list of books. **2.** Display requested books :- the system displays the list of books. **3.** Display fine information :- the system displays the fine information. The use case ends.
Exceptions:	**E1** If *Identify Library User* fails, the use case ends.
Postconditions:	None.

FIG. 27.4. Textual description for the *View User Lending Record* use case.

Use Case:	*Identify Library User*
Actor:	Library User (Student)
Goal:	The library user must be identified as a registered user of the system
Preconditions:	None
Main flow:	The library user enters identifying information. The system verifies that this information is correct. (**E1**) The use case ends and reports successful identification. (**E2**)
Exceptions:	**E1** If the information is invalid, the system may request that it be re-entered. The use case continues. **E2** If the information is persistently invalid or the user abandons the system, the use case ends and reports a failure.
Postconditions:	The identification check reports success or failure.

FIG. 27.5. Textual description for the *Identify Library User* use case.

their relationships on the domain object classes of the system. The basic Lean Cuisine+ tree structure is constructed as follows:

- A subdialogue header meneme corresponding to each domain class name is created. At this stage, these headers are created as *virtual menemes*—subgroup headers that are not available for selection in the final interface. A virtual meneme is indicated by placing the meneme name within braces $\{\ldots\}$.

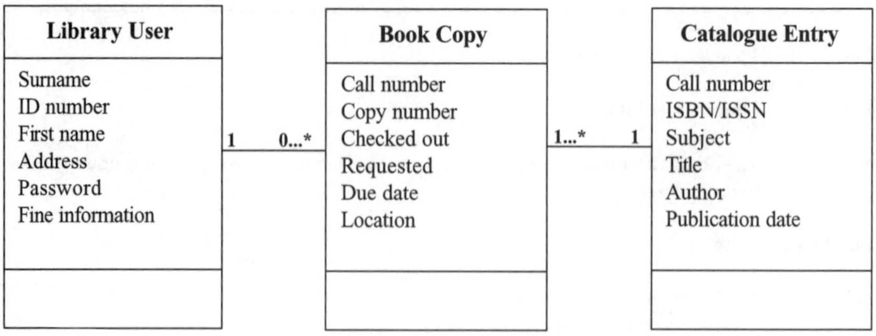

FIG. 27.6. The domain classes for the Library Catalogue (this is not a complete class diagram).

- A meneme for each *selectable* class attribute in the appropriate subdialogue is created. A selectable attribute is one that may be selected by a user. For example, library users in the case study are never selected by first name. Thus, no meneme is created to correspond to the first name attribute in the class diagram. The knowledge of which attributes can be selected is obtained through domain analysis. If there is uncertainty, an attribute should be included, as all unused menemes are removed later on.

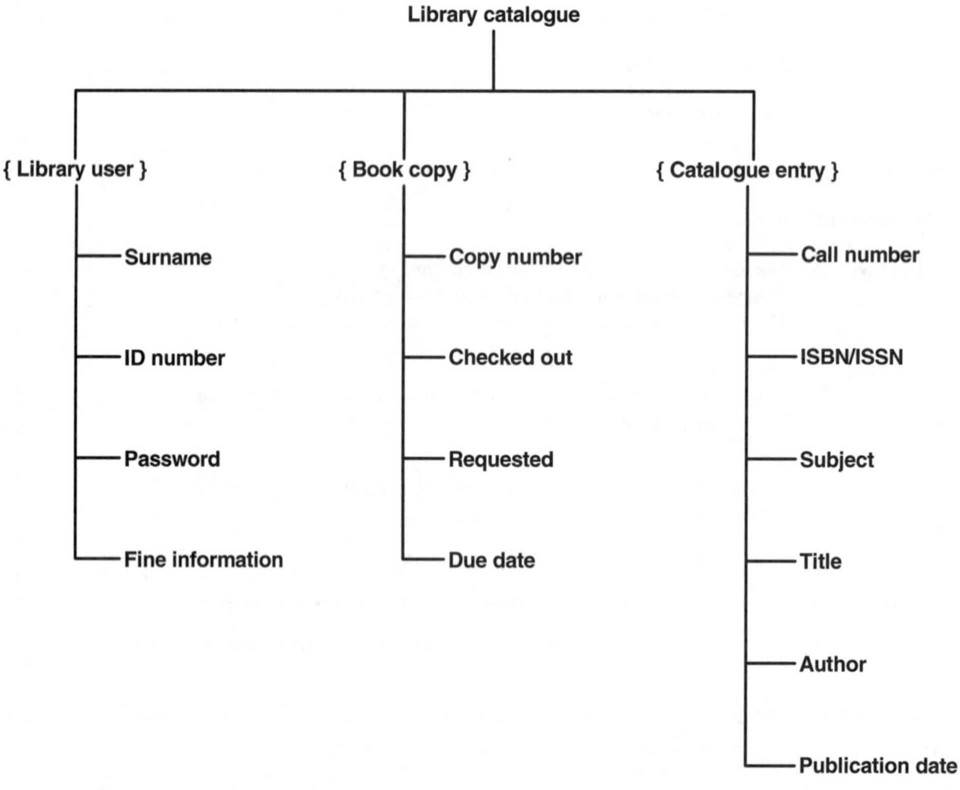

FIG. 27.7. The first draft of the basic Lean Cuisine+ diagram for the library catalogue.

Figure 27.7 shows the basic *dialogue tree* structure for the library case study. The meneme names are a subset of those appearing in the class diagram in Fig. 27.6. *Nonselectable* attributes have not been included as menemes. For example, books are never selected by location, so Location is omitted as a meneme in the {**Book copy**} subdialogue. It must be emphasized that this is the starting point for an iterative process and that the diagram is not intended to be complete at this point.

All known classes have been included in the basic tree for this example. However, if classes obviously have nothing to do with the user interface, they could be omitted from the basic tree. For example, a class may deal with low-level communication protocols. Again, if uncertainty exists, the class should be included, as all unused menemes are removed at a later stage.

The way menemes are arranged within a Lean Cuisine+ subdialogue indicates whether that group of menemes is mutually exclusive (1 from *n*) or mutually compatible (*m* from *n*). At this early stage of development of the diagram, it may be difficult to decide which type of structure is required for each subdialogue, and the initial structure may require revision at a later stage. Meneme groupings within subdialogues are assumed to be mutually exclusive until known to be otherwise.

27.4.1.4 Step 4: Construct Lean Cuisine+ Task Action Sequences

Lean Cuisine+ task action sequences, which are the means by which Lean Cuisine+ models tasks, are now overlayed on the initial base diagram. These task action sequences are directly derived from the use cases. A task action sequence may correspond to an entire use case or to a scenario (one path through a use case). Figure 27.8 shows the initial task action sequence for the *Search for Book* use case superimposed on the initial tree of Fig. 27.7.

Four *floating* menemes—**Initiate search**, **Display book list**, **Select book**, and **Display book details**—have been added to the diagram. During the construction of the task action sequence, it was discovered that these menemes were required but had not been provided.

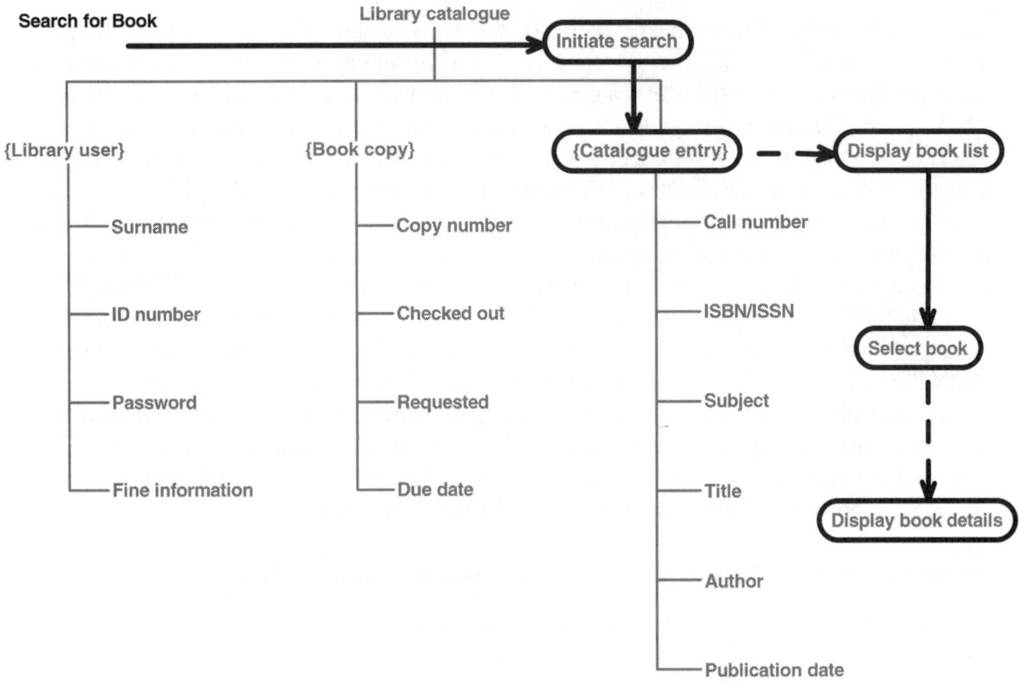

FIG. 27.8. The first draft of the task action sequence for *Search for Book*.

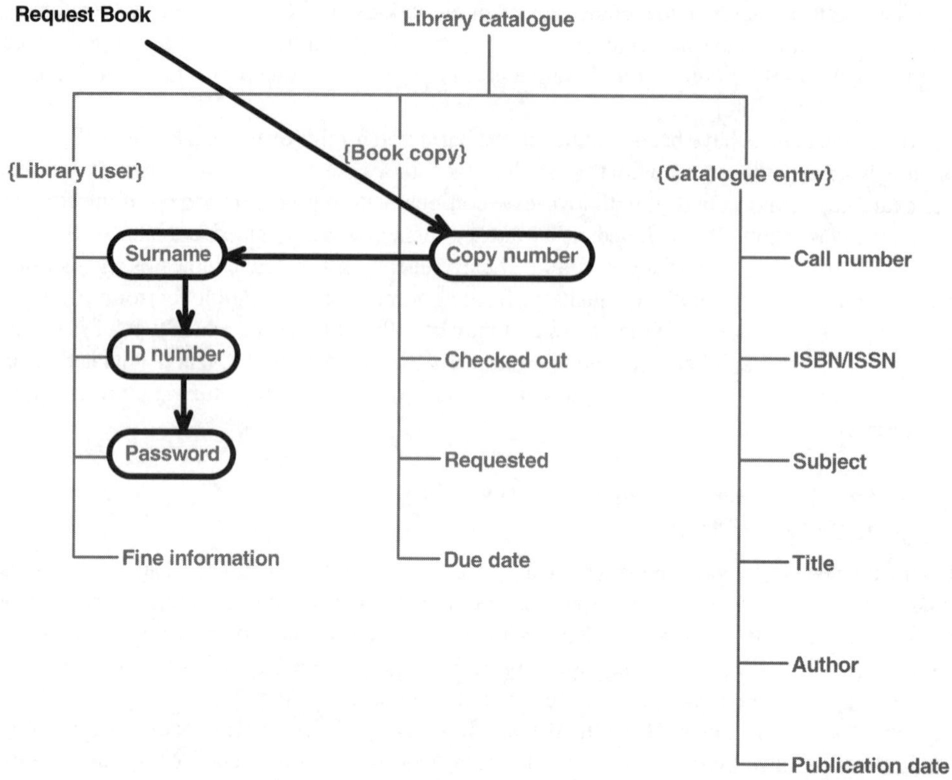

FIG. 27.9. The first draft of the task action sequence for *Request Book*.

They will be included in the next iteration of the tree structure. The dashed link to **Display book list** indicates that this is a system action. This is derived from domain knowledge; in this case, the system (not the user) causes the list of books to be displayed. It has also become obvious that {**Catalogue entry**} is not a good name for the subdialogue that allows the user to select the search criteria. Several names will be changed during Step 5 as part of the transition from class diagram to interface design. Virtual menemes cannot be selected, and the "selection" of the virtual meneme {**Catalogue entry**} indicates that any one of the menemes within that subdialogue may be selected at that point.

One or more task action sequences are constructed for each use case. Figures 27.9 and 27.10 illustrate the task action sequences constructed for the *Request Book* and *View User Lending Record* use cases respectively. Each task action sequence introduces new floating menemes that will need to be included in the next iteration of the Lean Cuisine+ dialogue tree.

A Lean Cuisine+ task sequence has not been constructed for the use case *Identify Library User* since this use case is used by *Request Book* and *View User Lending Record* (see Fig. 27.1) and has been included in the task sequences for those two use cases. It could easily be represented as a separate task sequence if that were deemed to be useful.

27.4.1.5 Step 5: Refine the Lean Cuisine+ Dialogue Tree

The initial task action sequences are now reviewed to identify

- menemes that are required but are not yet present in the diagram (called floating menemes)
- menemes that are present but are never used

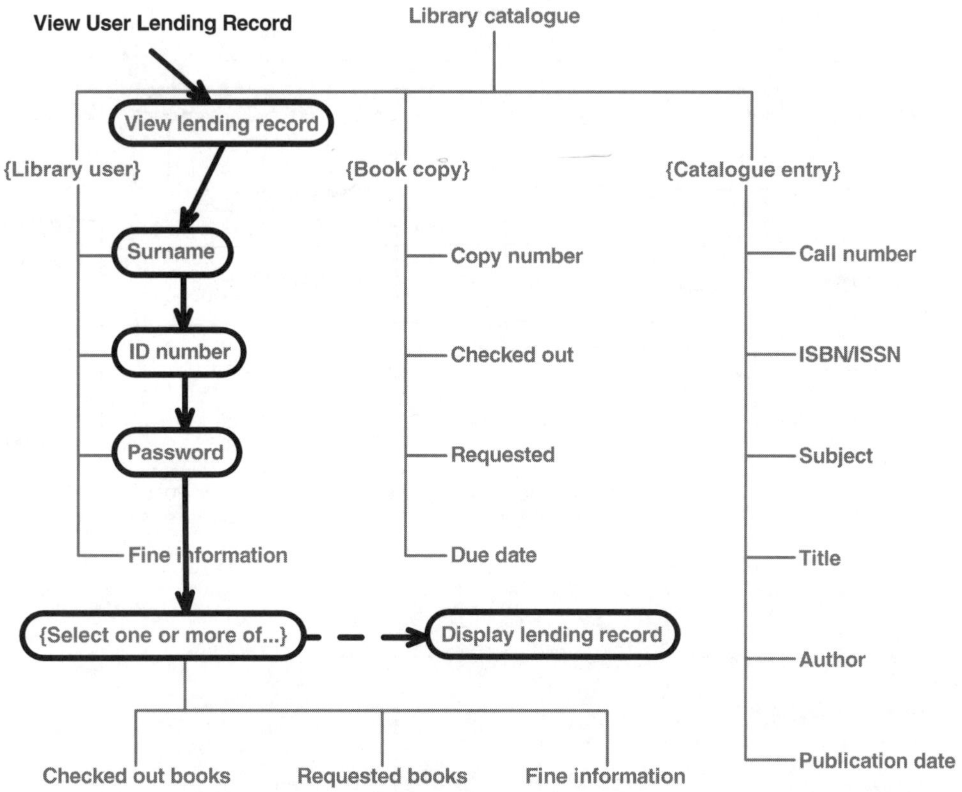

FIG. 27.10. The first draft of the task action sequence for *View User Lending Record*.

- changes to the tree structure itself – decisions are required on which meneme groupings should be mutually exclusive or mutually compatible, which menemes represent homogeneous groups and thus require *forks* in the diagram, which subdialogue headers should become virtual menemes, and so forth.

Significantly, this is the point where the design process changes from a *system-centered* approach to a *user-centred* approach. Thus menemes that were constructed from class attributes (internal to the system) may now change (in name) to reflect the user's view of the system.

Figure 27.11 illustrates refinements made to the Lean Cuisine+ tree structure for the library case study. This refined dialogue tree should be compared with the initial tree structure in Fig. 27.7.

Menemes that were not used by any task action sequence have been removed (e.g., **Due date**). The floating menemes from the various task sequences have all been added to the diagram (e.g., **Display lending record**). The names of subdialogues have been changed to provide a user's perspective (e.g. {**Catalogue entry**} has been changed to {**Search options**}), and some regrouping of menemes has taken place.

In Fig. 27.11, it is known that the library user can select one, two, or three menemes in the {**View options**} subdialogue so the menemes in this subdialogue are arranged to indicate a mutually compatible group (the menemes in the group are listed horizontally instead of vertically). The *fork* symbol has been added to represent homogeneous groups. For example, **Copy no.** is a mutually exclusive group, as several copies of a book may be listed but the library user can select only one copy, by number, when requesting it from the library.

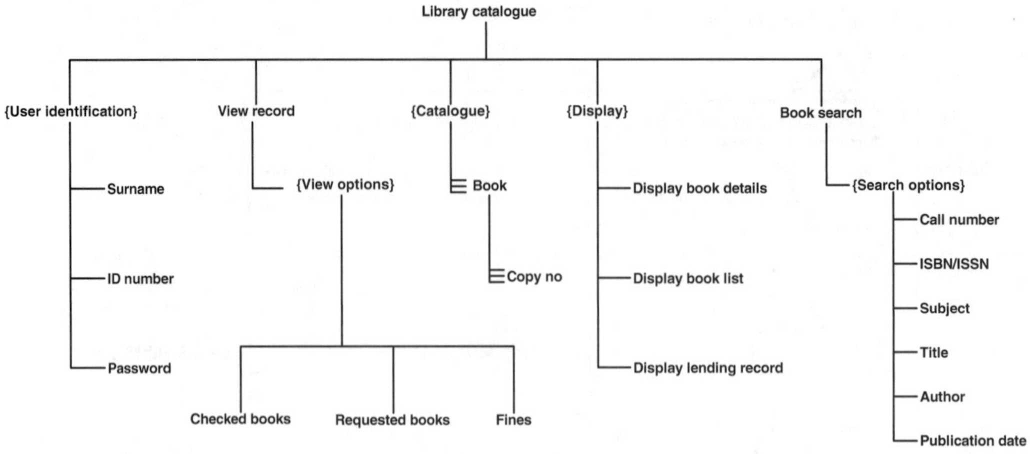

FIG. 27.11. The refined tree structure for the Library Catalogue.

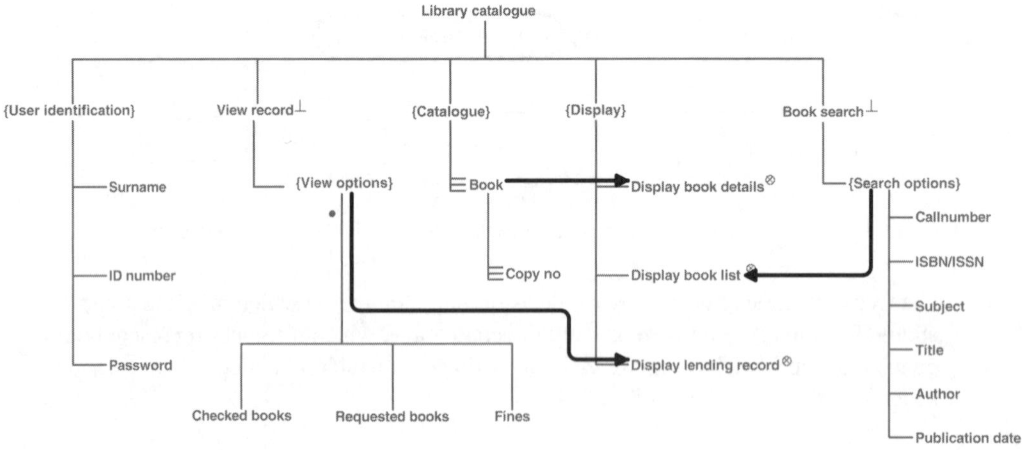

FIG. 27.12. Lean Cuisine+ diagram for the Library Catalogue showing selection triggers and meneme designators.

27.4.1.6 Step 6: Enrich the Lean Cuisine+ Diagram

Selection triggers are now added to the diagram. These correspond to the system actions in the task action sequences. For example, in Fig. 27.12, the trigger from **Book** to **Display book details** indicates that the selection of **Book** by the user will trigger the selection of **Display book details** by the system.

Meneme designators are also added to the diagram at this stage. Most menemes can be selected or deselected by the user, and the designators indicate the exceptions to this rule. A *passive* meneme (⊗) cannot be selected or deselected by the user but can only be selected by the system. A *monostable* meneme (−), such as {**View record**}, can be selected by the user but reverts to an unselected state on completion of the subdialogue that it represents. A dot next to a subdialogue (see {**View options**}) indicates an *unassigned default* choice that takes on the value of the last user selection from that group.

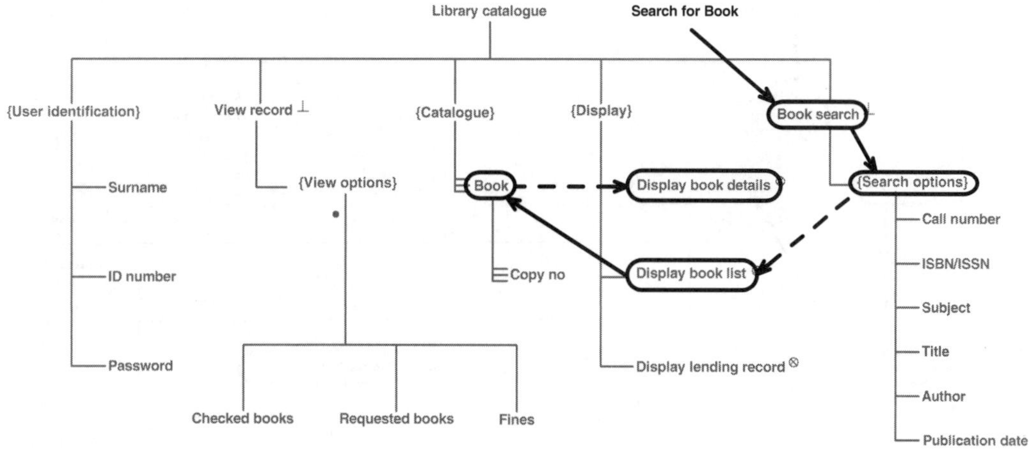

FIG. 27.13. The revised Lean Cuisine+ task action sequence for *Search for Book*.

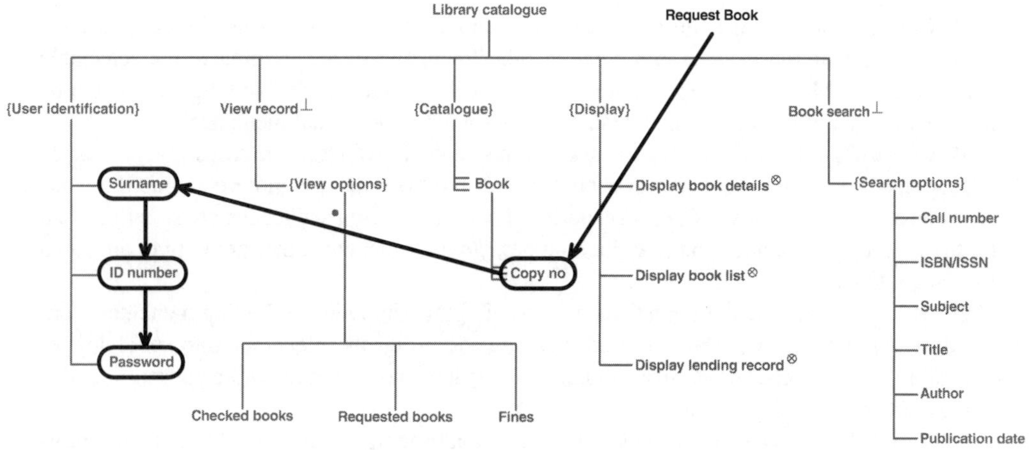

FIG. 27.14. The revised Lean Cuisine+ task action sequence for *Request Book*.

27.4.1.7 *Step 7: Remap the Task Action Sequences*

The Lean Cuisine+ task action sequences (first constructed in Step 4) are now remapped on to the refined dialogue tree. The revised task action sequence for *Search for Book* shown in Fig. 27.13 is derived from the initial task action sequence shown in Fig. 27.8. Likewise, the revised task action sequences in Figs. 27.14 and 27.15 are derived from the initial task action sequences in Figs. 27.9 and 27.10 respectively.

27.4.2 Summary of the Method

The method described above can be summarized as follows:

1. Define a set of detailed UML use case descriptions.
2. Identify the important domain classes and construct a UML class diagram, which does not have to be complete at this stage but must include any attributes that may be selected by a user.

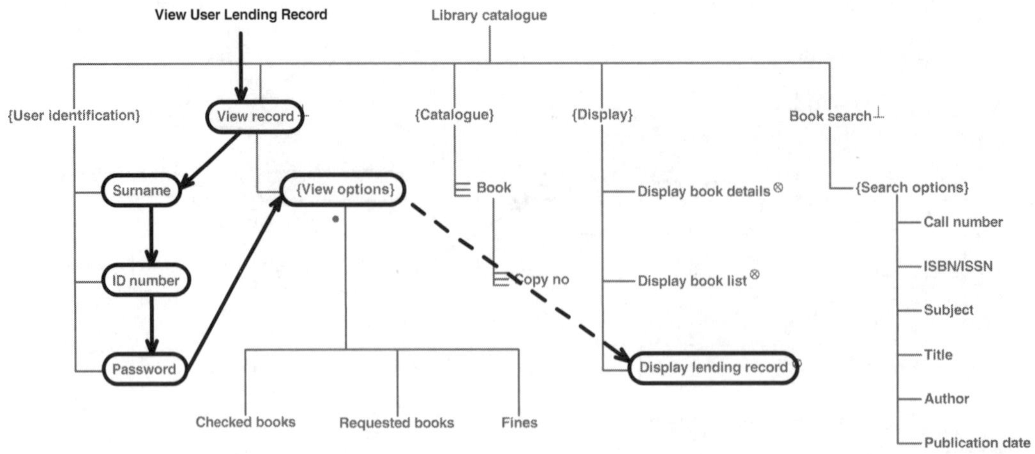

FIG. 27.15. The revised Lean Cuisine+ task action sequence for *View User Lending Record*.

3. Construct an initial Lean Cuisine+ dialogue tree structure by using the domain class names as subdialogue headers and creating other menemes corresponding to the selectable class attributes. Meneme groupings are assumed to be mutually exclusive unless known to be mutually compatible. Subdialogue headers are assumed to be virtual menemes.

4. Construct initial Lean Cuisine+ task action sequences from the use cases. There must be at least one task action sequence for each use case, but there may be more than one if the use case contains significantly different scenarios. These task action sequences are superimposed on the initial tree structure and may include new *floating* menemes that are not yet allocated to a subdialogue.

5. Review and refine the Lean Cuisine+ tree diagram. Include the floating menemes identified in step 4 and remove any unused menemes. Regroup the menemes into subdialogues according to functionality. Modify subdialogue headers and structure where appropriate. Rename menemes to reflect the user point of view.

6. Enrich the Lean Cuisine+ model. Add selection triggers to correspond to system actions in the task action sequences. Add meneme designators as required.

7. Remap the task action sequences on to the modified tree structure.

Steps 5, 6, and 7 may need to be repeated as part of an iterative design process. At each stage, the task action sequences need to be analyzed to ensure the following:

- All the specified use cases are covered in the model.
- No required menemes are missing.
- No menemes are superfluous.
- The flow of each task action sequence is logical and efficient.

It should be noted that OVID (Object View and Interaction Design) is another method for using existing UML notations to perform user interface design. The first two steps of OVID are exactly the same as those of the method described here: the identification of use cases and the construction of class diagrams. OVID then proceeds to describe tasks using UML sequence diagrams and statecharts (Roberts et al., 1997). It thus differs from the Lean Cuisine+ method in that the representation of tasks is divorced from the dialogue structure. It should be noted that the UML notations used by OVID were not specifically developed for use in user interface design.

27.5 CONCLUSION

This chapter has examined task modeling in UML and dialogue description in Lean Cuisine+ and described a method for the early stages of interactive systems design that incorporates both notations. This provides a means of representing tasks in the context of the structure of the user interface, that is, of explicitly showing the transformation of tasks into dialogue and hence the appropriateness of the dialogue structure to the tasks.

The method involves development of UML use cases and class diagrams, the definition of an initial Lean Cuisine+ dialogue tree, and the subsequent refinement and modification of this structure in order to better support task action sequences. This refinement process is repeated for each task action sequence, and any conflicts are resolved. The final dialogue structure is thus arrived at via an iterative process driven by an analysis of task action sequences. The refined Lean Cuisine+ model captures the behavior of the interaction at a level above the "look and feel" and establishes an overarching dialogue structure. The method needs to be further developed and tested. Work in progress includes

- an analysis of the level at which task action sequences are most useful. For example, should a task action sequence represent a use case or one scenario within a use case? Should a task action sequence include alternative paths or optional actions?
- further development of a software environment to support the method. A prototype Lean Cuisine+ support environment already exists, and it provides for the generation of dialogue trees and task action sequences. The hope is that it will be able to at least partially automate the production of initial Lean Cuisine+ diagrams and task action sequences from UML use cases and support the execution of the notation.
- further exploration of the use of Lean Cuisine+ as an analytical tool in order to check that all tasks uncovered at the requirements stage are supported by the system and that they can be carried out in an efficient manner (in regard to key strokes and selections). It may also be possible to partially automate this.

The next stage is to translate the Lean Cuisine+ model into a visible interface. This requires decisions to be made on the interaction style to be adopted, any metaphors to be employed, and ultimately the spatial grouping of objects.

REFERENCES

Apperley, M. D., & Spence, R. (1989). Lean Cuisine: A low-fat notation for menus. *Interacting with Computers, 1*(1), 43–68.

Brooks, R. (1991). Comparative task analysis. In J. M. Carrol (Ed.), *Designing interaction: Psychology at the human-computer interface* (pp. 50–59). Cambridge: Cambridge University Press.

Card, S. K., Moran, T. P., & Newell, A. (1983). *The psychology of human computer interaction*. Hillsdale, NJ: Lawrence Erlbaum Associates.

Collins, D. (1995). *Designing object-oriented user interfaces*. Redwood City, CA: Benjamin/Cummings.

Constantine, L. L., & Lockwood, L. A. D. (1999). *Software for use: A practical guide to models and methods of usage-centered design*. Reading, MA: Addison-Wesley.

Dix, A., Finlay, J., Abowd, G., & Beale, R. (1993). *Human-computer interaction*. Upper Saddle River, NJ: Prentice Hall.

Eriksson, H.-E., & Penker, M. (1998). *UML toolkit*. New York: Wiley.

Filippidou, D. (1998). Designing with scenarios: A critical review of current research and practice. *Requirements Engineering, 3*, 1–22.

Fowler, M., & Scott, K. (1997). *UML distilled: Applying the standard object modeling language*. Reading, MA: Addison-Wesley.

Guindon, R. (1990). Designing the design process: Exploiting opportunistic thoughts. *Human-Computer Interaction, 5*, 305–344.

Harel, D., Pnueli, A., Schmidt, J. P., & Sherman, R. (1987). On the formal semantics of Statecharts. In *Proceedings of the Second Symposium on Logic in Computing Science* (pp. 54–64). New York IEEE Press, Itaica.

Jacobson, I., Booch, G., & Rumbaugh, J. (1999). *The unified software development process*. Reading, MA: Addison-Wesley.

Jacobson, I., Christerson, M., Jonsson, P., & Övergaard, G. (1992). *Object-oriented software engineering*. Reading, MA: Addison-Wesley.

Kemp, E. A., & Phillips, C. H. E. (1998). Extending support for user interface design in object-oriented software engineering methods. In *Proceedings of 13th Conference on Human Computer Interaction, British Computer Society* (pp. 96–97). Sheffield UK: Springer-Verlag.

Kirwan, B., & Ainsworth, L. K. (1992). *A guide to task analysis*. London: Taylor & Francis.

Kovacevic, S. (1998). UML and user interface modeling. In J. Bezivin & P.-A. Muller (Eds.), *Lecture notes in computer science 1618* (pp. 253–266). Berlin: Springer-Verlag.

Overmyer, S. (2000, May 8). *Conceptual modeling through linguistic analysis using LIDA*. Seminar at Massey University, Auckland, New Zealand.

Paech, B. (1998). *On the role of activity diagrams in UML*. In J. Bezivin & P.-A. Muller (Eds.), *Lecture notes in computer science 1618* (pp. 267–277). Berlin: Springer-Verlag.

Peterson, J. L. (1981). *Petri net theory and the modeling of systems*. Englewood Cliffs, NJ: Prentice Hall.

Phillips, C. H. E. (1994). Review of graphical notations for specifying direct manipulation interfaces. *Interacting With Computers, 6*, 411–431.

Phillips, C. H. E. (1995). Lean Cuisine+: An executable graphical notation for describing direct manipulation interfaces. *Interacting With Computers, 7*, 49–71.

Phillips, C. H. E., Kemp, E. A., & Kek, S.-M. (2001). Extending UML use case modeling to support graphical user interface design. In *Proceedings of Australian Software Engineering Conference 2001* (pp. 48–57). Los Altimos, California: IEEE Press.

Quatrani, T. (1998). *Visual modeling with Rational Rose and UML*. Reading, MA: Addison-Wesley.

Richter, C. (1999). *Designing flexible object-oriented systems with UML*. Indianapolis, IN: Macmillan Technical Publishing.

Roberts, D., Berry, D., & Isensee, S. (1997). OVID: Object view and interaction design. *Proceedings of interact* IFIP International Conference (pp. 663–664). London, UK: Chapman & Hall.

Rosson, M. B., & Carroll, J. M. (1995). Integrating task and software development for object-oriented applications. In *Proceedings of CHI '95*. (pp. 377–384). New York: ACM Press.

Rumbaugh, J., Jacobson, I., & Booch, G. (1999). *The Unified Modeling Language reference manual*. Reading, MA: Addison-Wesley.

Scogings, C. J. (2000). The Lean Cuisine+ notation revised. *Research Letters in the Information Sciences* (Massey University), *1*, 17–23.

Scogings, C. J., & Phillips, C. H. E. (2001). Linking tasks, dialogue and GUI design: A method involving UML and Lean Cuisine+. *Interacting With Computers, 14*, 69–86.

Shepherd, A. (1989). Analysis and training in information technology tasks. In *Task Analysis for Human-Computer Interaction*. Diaper, D. (Ed). (pp. 15–55). Chiehester: Ellis Horwood.

Stevens, P. and Pooley, R. (2000). *Using UML: Software Engineering with Objects and Components* [Updated Edition]. Harlow, England: Addison-Wesley.

Tool support for task-based user interface design (Workshop 7). (1999). CHI '99. Pittsburgh, PA.

V

Today and Tomorrow

Cooperation or Conflict? was the subtitle of Easterbrook's (1993) book on computer-supported cooperative work (CSCW). Part V of this handbook represents a conflict resolution between its editors. Neville Stanton wanted a final chapter summarizing and integrating the handbook's contents; the other editor wasn't convinced that further summary was needed, nor that integration, at this stage in the history of task analysis, was feasible within the time frame of the handbook's production. Chapter 28 is their compromise solution—one that both editors are entirely happy with. It focuses primarily on the many, complicated psychological issues associated with task analysis.

Both editors firmly agree that there is a need to highlight the outstanding problems with task analysis and propose how these might be resolved in due course. Hence, the final chapter, chap. 30, which offers suggestions for future directions.

Sandwiched between chapters 28 and 30, chapter 29 presents a persuasive argument for the need to extend task analysis beyond the confines of traditional IT-supported work. The editors can only hope that the chapter's suggestions will be picked up by researchers in the task analysis community.

Chapter 28
The Psychology of Task Analysis Today
Neville A. Stanton

Task analysis without psychology is like peaches and cream without the peaches: thin stuff. Chapter 28 reviews the psychological basis of most of the handbook's chapters, along with related issues and problems. It presents an emergent analysis of four themes that arise across the chapters reviewed: (a) data collection, (b) psychological underpinnings, (c) task representation, and (d) the accessibility of task analysis. By providing an introduction to contemporary psychological issues in task analysis, the chapter should be particularly helpful to anyone new to the field.

Chapter 29
Experiences People Value: The New Frontier for Task Analysis
John Karat, Clare-Marie Karat, and John Vergo

Neville Stanton's first reaction upon reading the draft of chapter 29 was that it was "anti–task analysis," which reflects just how important is this chapter. The authors accuse task analysis of being too firmly rooted in the world of jobs and assert that satisfactory work involves more than the effective and efficient performance of tasks to achieve goals (chap. 1). People treat their leisure and social activities differently from their work ones, and they have different types of goals and very different requirements from those characteristic of the workplace, although they may spend years perfecting their skills at leisure activities. The chapter's authors are undoubtedly correct that there is a current problem with task analysis in these IT-supported non-job areas of people's lives, although the editors believe that the problems are not insurmountable but just haven't yet been adequately addressed by the task analysis community. It is certainly worth looking beyond task analysis, as good design needs to resolve issues other than usability and performance. The authors of this chapter recommend investigating methods normally associated with the arts (such as "the Crit," as used in the exhibition of works of art) as a way to take into account human desires, needs, and aspirations. This is relatively unfamiliar territory for HCI, but we should be open to new challenges and explore places we have not visited yet.

Chapter 30
Wishing on a sTAr: The Future of Task Analysis
Dan Diaper and Neville A. Stanton

"What a mess!" Although we are utterly confident that the long-term future of task analysis is assured, whatever task analysis might eventually be called, there is a great diversity of theories, concepts, and definitions and of task analysis methods appropriate for different purposes in myriad application domains and project environments. Chapter 30 argues that, across the whole field of task analysis, its practitioners need to come to agreement about basic concepts, terms, theories and so forth, but without compromising the range of application and the flexibility of the methods described in the chapters of this handbook. Development of a task analysis meta-method is suggested as one way forward. Radically, the chapter argues that the core concept of a goal should be abandoned by the task analysis community.

Chapter 30 then addresses the other, equally important, problem with much current task analysis, that of delivering task analyses in understandable, usable, effective, and efficient forms to those in the computing industries who are responsible for designing, producing, and maintaining computer systems.

REFERENCE

Easterbrook, S. (Ed.). (1993). *CSCW: Cooperation or conflict?* New York: Springer-Verlag.

28

The Psychology of Task Analysis Today

Neville A. Stanton
Brunel University

Throughout this handbook, the authors of chapters have either implicitly or explicitly made reference to psychological theory in development of their task models of human performance. The purpose of this chapter is to review the psychological themes that run through the handbook with the aim of identifying commonality and differences. The psychological theory is definitely the most complex aspect of the book, because the field of research is still developing and there are alternative perspectives. One of the main conflicts identified in the book is the tension between the academic need to present a high fidelity psychological model of human performance and the commercial need develop a just-good-enough model within time and resource constraints. The key to resolving this conflict comes from a hierarchical model of system performance, where the high-fidelity representations of the system may be subsumed under lower fidelity models.

28.1 CHALLENGES FOR TASK ANALYSIS

The purpose of this chapter is to review the state of the art of task analysis from a psychological perspective and based on the contributions to this handbook. As has already been pointed out in the preface, task analysis can mean different things to different people. This has led to a broad range of approaches to the analysis of tasks and the development of underpinning theories. By drawing the essential elements from this range, it is hoped that a coherent overview and guide to the main issues in task analysis will emerge. At the very least, this chapter should provide the reader with a comprehensive assessment of the main issues.

Simplistically, most task analysis involves identifying tasks, collecting task data, analyzing this data so that the tasks are understood, and then producing a documented representation of the analyzed tasks suitable for some engineering purpose. Tasks are analyzed according to some dogma that involves both a theoretical model and a method of analysis. Task analysis in the published literature normally focuses on the collection of data, the dogma, and the form of

TABLE 28.1
Comparison of HTA, GOMS, and ICS-CTA

	HTA	*GOMS*	*ICS-CTA*
Data collection example	Interviews, observation, manuals, equipment	Interviews, observation, manuals, equipment	Interviews, observation, manuals, equipment
Dogma	Hierarchy of subgoals based on control theory	Model of human processor for procedural tasks	Nine interacting cognitive subsystems
Documentation	Numerical hierarchical diagrams and tabular format with notes	Numerical hierarchy of text lists with approximate time data	Tables of cognitive activity within each subsystem

documentation, although the border between latter two facets may be blurred. Different task analytic techniques might share the same data collection methods but differ in their dogma and/or documentation. This is illustrated in Table 28.1, which compares hierarchical task analysis (HTA; chap. 3), Goals, Operations, Methods, and Selection rules (GOMS; chap. 4), and Interacting Cognitive Sub-systems for Cognitive Task Analysis (ICS-CTA; chap. 15).

As shown in the table, these three task analysis approaches share data collection methods, but differ in their dogma and style of documented task representation. Annett (chap. 3) argues that the process of analysis is the most important aspect because it provides the analyst with new insight into the nature of the tasks and the ways in which the goals can be achieved. Both Kieras (chap. 4) and Annett stress the importance of the judgment of the analyst when interpreting task data. This is a skill that is acquired through experience in applying the techniques, although the authors of the methods usually do pass on some helpful advice to readers. May and Barnard (chap. 15) offer a different kind of analysis that, on the face of it, appears to be more closely linked to psychological mechanisms. They identify the individual cognitive subsystems that are likely to be implicated in the performance of a task and build the analysis up from this standpoint. They also distinguish between novice and expert performers of a task. It is difficult to recommend any one approach over another, as all of the techniques are likely to provide useful information in the hands of an expert analyst. Rather it is recommended that the reader explore each of the approaches and select the one that best suits the purpose to which it is intended to be put.

Given broad definitions of human-computer interaction (HCI) and task analysis, task analysis is nigh universal in HCI because it is concerned with task performance—what people and other things do to achieve work (Annett & Stanton, 2000; Diaper, 2002a, 2002b; chap. 1). Task analysis is complicated because the many methods available vary in so many ways. This variety is well illustrated by the differences found in the taxonomies of task analysis in chapters 6 and 22–25.

Stanton and Annett (2000) noted that task analysis has come a long way since the days of Taylor (1911) and Gilbreth (1911), but the challenges we face are bigger than they could have imagined. There are several reasons for this. First, the emphasis on cognitive rather than physical, work. Second, the dominance of the information technologies. Third, the multi-agent systems perspective (with systems comprising people and machines). Finally, the realization of the importance of understanding the cultural and contextual ecologies when analyzing system performance. These challenges are addressed to different degrees by the contributors to this handbook. Again, as Stanton and Annett (2000) pointed out, the challenges are perhaps too great for any one method. In their review of contemporary approaches, they identified at least six different demands for task analysis: determining knowledge requirements, understanding how devices are used, defining training needs, predicting human error, describing multi-agent systems, and designing computer interfaces.

In reviewing the contributions in this handbook, the editors undertook a thematic analysis. Four main themes emerged:

1. collection of data about the world
2. theoretical underpinnings of task analysis
3. representation of task analysis
4. accessibility of task analysis.

Each of these will be discussed with reference to the relevant chapters.

Task analysis is the central method for the analysis and design of system performance. The data, methods, analysis, and documentation can feed through virtually every aspect of the design cycle. Everything from concept, demonstration, specification, system development, and operation and maintenance to decommissioning is, or should be, touched on in one way or another by task analysis techniques. Tasks analysis is not undertaken for its own sake; rather, tasks are analyzed for various purposes. These purposes become clear when talking about system design and development (e.g., how tasks are allocated between individuals and technology). This process of functional allocation (see also chaps. 6, 22, and 24) requires that tasks are first defined before they can be allocated. The same is true of all other aspects of system design, such as procedure design, training design, interface design, and so on. All of these aspects of system design are interrelated and all feed on task analysis.

28.2 COLLECTION OF DATA ABOUT THE WORLD

Many of the contributions to this book underline the importance of accurately representing work in the analysis. The accuracy of this representation is necessary for evaluating systems and for producing future generations of systems (chap. 1). The starting point for designing future systems is a description of a current or analogous system. It might be fair to suggest that inaccuracies in this description could compromise the development effort. The general rules regarding data collection are (a) not to rely on one source of information and (b) to beware of potentially misleading information. Measures for reducing the sources of error in data collection include clear specification of stakeholders in the tasks to be analyzed, use of a wide variety of sources of data, cross-checking of data and task logic, and presentation of analysis back to the stakeholders involved. Most of the chapters in part II point out that there are gaps between realistic and ideal data collection activities in commercial projects where resources are extremely limited. These gaps could compromise the accuracy of the analysis and have put pressure on the tasks analysis community to develop quick methods. The range of methods typically used include observation, interviewing, analysis of artifacts, usability metrics, analysis of the context of task performance, and reference to generic cognitive theory and models. Each of these is briefly discussed below.

28.2.1 Observation

Observing system performance seems at first glance to be the most obvious way of collecting data for task analysis. This assumes, of course, that there is an existing system, prototype, or simulation that can be observed. Essentially the method simply requires one to observe users performing tasks, although Diaper (1989) warns that such observation itself is not easy. There is a considerable complexity of potentially interacting and confounding variables. It is necessary to choose what to observe and record. Observing people affects what they do, the type and number of people and tasks observed may bias the results and make them unrepresentative, and the way in which the data are recorded could compromise the reliability and validity of the observations. Overcoming these potential problems requires careful planning, and preparation, including, ideally, a pilot observational study.

28.2.2 Interviewing

Interviewing people (e.g., chap. 16) may enable a task analyst to capture data on the unobservable decision-making processes, particularly if the interviewee is describing a recent event and how he or she dealt with it. However, it has been found that when people describe past events, they mistake what they were actually doing at the time. A solution to this is to get them to "think aloud" (concurrent verbal protocol) while performing the task. This is not without its problems, as people may actually change the way they do things to make them easier to describe, the psychological processes producing the protocols may interfere with the task, and people may not wish to disclose how they really do the task to an analyst (although chap. 1 notes that workers often relish the opportunity to talk about their job).

Greenberg (chap. 2) proffers the idea that the task analyst can discover what tasks people perform and develop task descriptions through observation and then validate the descriptions through interviews with the same people. Annett (chap. 3) argues that data collection should comprise observation and interviewing at the very least. Most importantly, Annett (chap. 3) and Kieras (chap. 4) both warn of the dangers of relying on verbal reports of task performance alone and both argue for the least intrusive method of observation that circumstances permit. Observation by the analysts could be supplemented by documentation (e.g., training and procedure materials and system manuals), historical records, focus groups, participant observation, and so on.

28.2.3 Analysis of Artifacts

The role of artifacts in task performance emphasizes the importance of a systems-based analysis. Artifacts may be as humble as a pen and Post-it note (e.g., chaps. 5 and xxx) all the way up to very realistic, complex environments such as in air traffic control (chaps. 13 and 25) or the bridge of a large ship (chap. 21). It is the interaction between people and artifacts that provides the basis for the task description. Spillers (chap. 14) points out that the artifacts serve a very important role in the sequencing of tasks. Artifacts can prompt the start of a task, determine the order in which tasks are carried out, and identify when the tasks have come to an end. The role of artifacts and the environment as a source of data are noted by both Spillers (chap. 14) and Dix et al. (chap. 19).

Artifacts serve as task cues, prompts, and placeholders. Harris (personal communication, 2002) reported that flight crews have been observed covering the cockpit yoke with an empty polystyrene coffee cup to remind them not to take off because some versions of flight management software require the pilot to enter the final take-off weight before the rest of the flight plan details can be entered. Unfortunately, these details are often not available until all passengers and cargo are loaded. Pilots have found a way out of this problem by entering a nominal figure (e.g., 999) for the final take-off weight and then either covering the yoke with their coffee cup (having first consumed the coffee) or clipping their tie to the yoke before going on with entering the rest of the flight plan data. Similarly, Spillers (chap. 14) reports that some recreational vehicle (RV) drivers attach a hairclip to the steering wheel to remind them that the television antenna is still erected and to prevent them from driving off before taking it down, so avoiding damage to it. This makes us wonder what happens if the pilot forgets to put a coffee cup over the control yoke on the flight deck.

Dix et al. (chap. 19) describe how air traffic controllers are required to mark their paper flight strips to hold their place in the task. For example, the sequence may be as follows: The new flight level and directional arrow is written on the strip when the pilot is informed; the old level is crossed out when the pilot confirms the new flight level; and when the pilot confirms that the new flight level has been reached, it is ticked on the flight strip. This approach emphasizes the importance of communication in the person-artifact interaction cycle. It is suggested that the communication can break down in three main ways: through gaps in the action-effect loop (Norman's [1986] gulf of evaluation), through gaps in the stimulus-response loop (Norman's gulf of execution), and through lack of an effective stimulus. Dix et al. (chap. 19) broaden the scope of triggers to include time, previous tasks, external events, and working memory as well as cues provided by artifacts.

28.2.4 Context of Task Performance

The context of system performance has been increasingly acknowledged as an important source of data. Task analysts are required to understand both the ways in which people adapt to their environment and the ways in which they adapt their environment to themselves for a comprehensive insight into the task. Baber and Stanton (chap. 18) present a model of the world that proposes that human activity is very fluid and people develop plans on the fly in response to the image presented by a system at any point in time. Each step in an interaction with a computer can potentially trigger revisions of the planned sequence of activities directed toward a particular goal. This view at its most extreme might suggest that planning really only exists for the next step and that there is only a very loose connection between goals, plans, and actions.

In fact, all of the task analysis methods address context of performance to a greater or lesser extent. It would be nonsense to analyze a task independent of the context in which it was to be performed. Annett (chap. 3) has made this point repeatedly over the past 30 years. Techniques such as trigger analysis (chap. 19) and cognitive archeology (chap. 14) emphasize the role of things in starting and stopping task sequences in particular situations. Scenario-based task analysis (chap. 5) explicitly creates a context for task exploration in the form of situationally specific scenarios. Windsor (chap. 13) indicates the variety of air traffic control tasks carried out by controllers as well as their frequency and the conditions under which they are performed. These are just a few of the examples that demonstrate that contextual analysis is central to task analysis. The reader is challenged to think of an occasion when context is not important to task analysis.

28.3 THE PSYCHOSOCIAL UNDERPINNINGS OF TASK ANALYSIS

Whatsoever one's brand of metaphysics, HCI and task analysis, as engineering disciplines, must assume the existence of an external world, even if, following the early British empiricists, it is conceived of as a universe made of stuff. The role of theory in task analysis, and in much else, is to select some aspects of all this stuff, which we can call data; to organize the data into information; to analyse the information so that it becomes knowledge; and to apply the knowledge, which, when fruitful, is considered wisdom. Beyond the crude empiricist view of a wholly physical world, psychological, sociological, and systemic views, among others, recognize that there are properties of the world that are as real as the tangible, physical things in that their recognition, description, understanding, and so forth, depend on a theoretical basis (see chap. 1). The types of theory associated with task analysis, however, show great diversity of content and detail and vary in their explicitness. Most people will probably accept that explicit theories are to be preferred because they allow public scrutiny and discourse and aid those attempting to learn about particular task analytic approaches. Nonetheless, many theories, such as the psychological ones, are still nascent and cannot be entirely explicit. They therefore require craft skill for their application in particular task analysis methods.

Task analysis theories generally concern how to articulate a model of the world (that is, how to divide it up, classify it, etc.). Although dealing with physical things, including human behavior, is still potentially controversial, the truly difficult and controversial theories focus on how to model the people who perform tasks. Part I reviews a wide range of theories of this kind. After introducing HTA as a benchmark psychological model for task analysis, it presents other models, with particular attention to their level of detail and probable validity. Part I ends with a consideration of models that take a wider view than that of individual human cognition, although it is important to realize that models that describe a single person's cognition can often also be applied to groups, as Annett (chap. 3) discusses with respect to HTA.

28.3.1 Hierarchical Task Analysis

Annett's presentation of hierarchical tasks analysis in chapter 3 shows that the methodology was derived from developments in control systems and feedback theory. Annett points out that the initial development effort was in response to the need for greater understanding of cognitive tasks. He also shows that HTA predicated the proliferation of cognitive tasks analysis methods (e.g., chaps. 15 and 16) by some 10 to 15 years. With more automation in industrial work practices, the nature of worker tasks was changing in the 1960s. Annett argues that, as these tasks involved significant cognitive components (e.g., monitoring, anticipating, predicting, and decision making), a method of analyzing and representing this form of work was required. Existing approaches tended to focus on observable aspects of performance, whereas HTA sought to represent system goals and plans.

In the late 1960s, this was a radical departure from contemporary approaches. The cognitive revolution had yet to spread to applied psychology and ergonomics, where the behaviouristic paradigm was still dominant, although it had been on the wane in experimental psychology throughout the 1960s. In applied psychology, it was still considered unscientific to infer cognitive processes, and the focus remained principally on observable behavior. HTA, however, offered a means of describing a system in terms of goals and subgoals and feedback loops in a nested hierarchy. Its enduring popularity can be put down to two key features. First, it is inherently flexible: The approach can be used to describe any system. Second, it can be used for many ends, from person specification, training requirement specification, error prediction, and team performance assessment and to system design. Despite the popularity and enduring use of HTA and the fact that the analysis is governed by only a few rules, it still needs craft skills to apply effectively. Analysts can be trained in the basic approach relatively quickly. Stanton and Young (1998) show that initial instruction in the method takes only a couple of hours. It is generally acknowledged, however, that sensitive use of the method requires months of practice under expert guidance, and it may take years to achieve mastery.

28.3.2 Interacting Cognitive Subsystems

At the other end of the spectrum of theoretical complexity is the theory that May and Barnard (chap. 15) present to explain how data might be processed within individuals. This approach forms the basis of a type of cognitive task analysis. The idea is that there is a close coupling between the underlying cognitive subsystems and their representation in a task analysis. By way of comparison, HTA makes no claims to represent cognitive systems; rather, it presents a functional analysis of goals, subgoals, and plans.

The theory proposed by May and Barnard postulates nine interacting cognitive subsystems: the acoustic subsystem (what we hear in the world), the visual subsystem (what we see in the world), the morphonolexical subsystem (what we hear in the head), the propositional subsystem (what we know of the world), the implicational subsystem (what we feel about the world), the object subsystem (what we see in the head), the articulatory subsystem (what we say to the world), the body state subsystem (how we sense the physical world), and the limb subsystem (how we move through the world). May and Barnard explain how activities may be processed by each of the subsystems. In one example, they show the subsystems and processes involved in the response of an operator to a visual alert, and in another example, they show the subsystems and processes involved in pointing to a diagram while speaking simultaneously. May and Barnard offer the Interacting Cognitive Subsystems approach as a way for analysts to understand covert mental operations and thus become better prepared to design cognitive artifacts.

28.3.3 Goals, Operators, Methods, and Selection Rules

Kieras (chap. 4) presents a review of the Operators, Methods and Selection Rules (GOMS) methodology in a human-computer interaction context. Kieras points out that GOMS models make up a family that includes a complex model that maps the critical path between cognitive processes using a schedule chart, a midlevel model that represents cognitive activities in a form rather like natural-language programming, and a simple model that represents units of keystroke-level actions required to perform a task. It is the midlevel model that Kieras focuses on in his chapter.

The GOMS model is based on the model of the human information processor originally specified by Card, Moran, and Newell (1983). Kieras shows how the GOMS approach may be used to structure and interpret task data to produce a "programming language of the mind." Despite the apparent structure in the approach, Kieras warns that there are many judgment calls and required simplifications and that some degree of expertise and care in needed for producing the most sensitive description of a task. One of the most compelling aspects of the GOMS analysis is that it can be undertaken before a system has been constructed. The design specification and the model human processor can be used to simulate system performance before any programming code has been written.

28.3.4 Human Ability Requirements

The Human Ability Requirements (HAR) approach of Cockayne and Darken (chapt. 20) represents an incomplete cognitive theory in that it treats behavior as independent atomic units, but the necessary theory to combine these atoms is not available. Cockayne and Darken recognize this but argue that identifying HARs is a necessary first step and that the approach has been found to be useful. Although differently specified from the units analyzed in GOMS (chap. 4), human ability requirements might be combined in a similar way, that is, by summation, as GOMS itself makes the theoretical assumption that its units of behavior and thought are independent. This assumption is known to be false in most cases, as Sternberg's (1969) reaction time additivity is rare. Sternberg shows that experimental manipulations that affect perceptual or response proccsses (e.g., stimulus degradation and response compatibility, respectively) can be added to predict performance, indicating that these mental processes are independent. Such additivity, however, does not occur with most experimental manipulations. On the other hand, GOMS has been used to predict performance accurately, which implies that a simplified cognitive theory of independent cognitive events is usually good enough for engineering purposes. Where GOMS fails to predict performance, more sophisticated interactive models of cognition, such as May and Barnard's Interacting Cognitive Subsystems (Chapter 15), are likely to be essential.

28.3.5 Norman's Cognitive Engineering Model

Norman (1986) produced a classic model of a cycle of human cognition in task performance. In this model, users have a goal, which they convert into a specific intention, which leads to the organization of behaviour, which is then executed. The results of the behavior are then perceived and compared with the intention, possibly leading to further behavior. In methods such as TOOD (chap. 25) and trigger analysis (chap. 19), Norman's model is reduced to a psychological model in which mental things are classified as primarily involving perception, cognition, or action (chap. 1's 3Ps, perceive, process, and perform). Although these methods involve a gross simplification and depend on the codicil "primarily involving," as in most cases all three of the 3Ps are involved, the argument is that this scheme provides a sufficient classification for engineers, who are not psychologists. Undoubtedly there will be times when this sort of approach is not sufficient and more sophisticated psychological models will be necessary. On the other hand, the scheme does provide placeholders in an engineering model that can potentially identify where a more sophisticated psychology is necessary. It should be noted, however, that where performance predictions fail, the solution does not necessarily involve changing human performance, say, by training, although it might. Often some other aspects of the system might need changing, with the changes influencing the human parts of the system.

28.3.6 Scenario-Based Design

Carroll (chap. 5; Carroll, 2000) eschews an explicit psychological model of how people read and understand his storylike scenarios. He seeks to use people's expert skills as readers and to exploit their creative interpretation of scenarios. In this approach, the psychological model of the people portrayed in scenarios is that of the scenario readers. From an engineering point of view, there is much merit in this approach, particular in the early stages of a project.

The approach, however, has raised considerable discussion in the journal *Interacting with Computers* (Carroll, 2000; Carey, 2002; Diaper, 2002a; Paternò, 2002), particularly over the use of scenario-based design throughout the software life cycle, although Benyon and Macaulay (2002) describe a method of using scenarios of different types for such purposes. In the published discussion, Diaper (2002b) proposed a scenario-based cognitive theory and made the claim that all human thought can be characterized as being episodic and that internal, psychological representations can be described as envisioned scenarios. Considerable further research would certainly be necessary before such a psychological model is sufficiently explicit that it can support a scenario-based design process.

Carroll is actually using the old human cognitive psychologists trick of using people as sophisticated information processors without understanding the cognitive mechanisms. Although this is worth recognizing, what he is doing is no different from many parts of methods in HCI and software engineering that exploit people's craft skills, and there is undoubtedly considerable mileage in the scenario-based design approach even without an explicit psychological model of cognition.

28.3.7 Cognitive Development Method

Wong's (chap. 16) approach to cognitive modeling is based on the assumption that people do have access to their own thoughts and that they can describe the basis of their decision making in situations they have experienced. This assumption is undoubtedly false. Indeed, a science of psychology would be unnecessary if it were true. Experimental psychology started in the 1880s in Liepzig and Cambridge, and the phenomena discovered since clearly indicate that there are properties of the human mind that cannot be inferred by introspection, which had been the method of the philosophy of mind for several thousand years in the West. Furthermore, memory is labile and can appear to be subject to systematic distortions, although it involves a constructive process dependent in part on the current environment of the person attempting to recall previously experienced events. The act of introspection is likely to change current cognition, as introspection itself requires cognitive processing resources (a Hiesenberg effect), and the requirement to represent mental states in words, pictures, or other forms will lead to mismatches between the mental states and how they are described.

Despite these problems, Wong's type of approach to psychology is still capable of producing results that have engineering utility. Perhaps essentially a "broad brush" view of behavior, it nonetheless has the advantage of basing analysis on complex, real events that have occurred, because it uses a popularist psychological perspective, communicating its results and conclusions is easy.

28.3.8 Task Analysis for Error Identification

Baber and Stanton (chap. 18) propose a dynamic and interactive theory of rewritable routines to support the fluidity of cognitive activities in task performance. Their ideas suggest that human interaction with artifacts is embedded in the environmental context and is a process of continual adaptation, revision, and refinement of next-step planning based on the current representation of the system image (Norman, 1986; see section 28.3.5) and the current goal the person is pursuing. They propose that even novice users of a device may behave as if they are experts because they are able to draw on a vast repertoire of previous experience. People's success or failure in using a device depends largely on the design of the system image cueing the appropriate routine. Mismatch between the routine afforded by the system image and that actually required to operate the device is highly likely to lead to errors of the kind that the TAFEI method seeks to identify.

28.3.9 Activity Theory

A different and much broader theoretical framework that covers both individual and collective activities is provided by activity theory (Turner & McEwan, chap. 21), which is based on a Russian approach to psychology first developed in the 1930s. Activity theory looks at the relationship between the individuals performing tasks, the tools they are using, and the purposes that they are pursuing. Consistent with a systems theoretic perspective, analysis can be performed at the level of a single individual or at the level of a group of individuals. It can also be performed at the level of an individual operation (e.g., a keystroke), an action (e.g., communicating a message), and a whole activity (e.g., teaching a set of skills). Of particular relevance to task analysis in human-computer interaction is the idea of contradictions. The contradictions of concern are apparent mismatches between the objects of interaction and/or the levels in the system. Turner and McEwan describe four potential types: primary contradictions (contradictions within a single node of activity), secondary contradictions (contradictions between nodes of activity), tertiary contradictions (contradictions between old and new ways of working), and quaternary contradictions (contradictions between concurrent or coexisting activities). Systematic evaluation of tasks using this framework of analysis should help reveal potential points of conflict and even identify where errors in the system are likely to occur. Removing contradictions from the system or at least reducing their number and impact is likely to have beneficial effect on performance.

28.4 REPRESENTATION OF TASK ANALYSIS

The task representations presented in this handbook can be broadly divided into five basic types: hierarchical lists (e.g., HTA and GOMS), narrative descriptions (e.g., the Crit and Cognitive Archaeology), flow diagrams (e.g., TAFEI and Trigger Analysis), hierarchical diagrams (e.g., HTA, DUTCH, and CCT), and tables (e.g., Task-Centered Walk-though, ICS-CTA, HTA, SGT, and TAFEI). Some methods have multiple representations, such as HTA, which employs hierarchical text lists, hierarchical diagrams, and tables. A summary of the types of representation used by task analysis methods is presented in Table 28.2.

Of the different forms of representation, tables are the most popular (with 11 examples), and the remaining types each occur in five to seven methods. Of the 22 task analysis methods listed in Table 28.2, 13 rely on primarily one form of representation, 6 have two forms of

TABLE 28.2
Task Analysis Methods and Representations

	Types of Representation				
Task Analysis Method	*List*	*Narrative*	*Flow Diagram*	*Hierarchy Diagram*	*Table*
Critical Decision Method (SA-ETA)	√	√			√
Cognitive Archaeology		√		√	
ConcurTaskTrees				√	
The Crit		√			
Diane+			√	√	
DUTCH			√		√
Goals, Operations, Methods, Selection Rules (GOMS)	√				
Hierarchical Task Analysis	√			√	√
Cognitive Task Analysis (ICS)	√				
Isolde					√
Just in e-Time			√		√
Lean Cuisine+				√	
Scenario-Based Requirements				√	
Sub-Goal Template					√
Task Analysis for Error Identification			√		√
TAEUS		√			
Task-Centered Walk-through		√			√
Task Layer Maps			√		
Trigger Analysis					√
TOOD			√	√	√
User Task Metrics					√
User Interaction Scripts	√				

representation, and 3 have three forms of representation. It should be noted that, in methods with multiple representations, one representation may be an embellishment of another, but in other cases different representations are used for different aspects of the analysis.

28.5 ACCESSIBILITY OF TASK ANALYSIS

Some researchers have recognized that considerable effort is needed to acquire sufficient skill and expertise to conduct task analysis in a competent manner. Indeed, they have gone further and developed ways in which conducting a task analysis might be made more accessible. Two main streams of thought reign. The first way to make task analysis more accessible is to simplify the process of tasks analysis. This approach is exemplified by the work of Greenberg (chap. 2), which presents a stripped-down version of task analysis that is quite easy to grasp. The second way to make task analysis more accessible is to provide tools that guide people through the task analysis process. This approach is exemplified by the work of Omerod and Shepherd (chap. 17), which presents the subgoal template method to support the novice analyst. A byproduct of making task analysis more accessible is that the methods may be acquired and applied more quickly. In many ways the issue of accessibility is tied up with the issues surrounding commercial pressure. The outputs of task analysis have to be understood not only by those undertaking the analysis but often by those who are basing a commercial decision on the results.

28.5.1 Scenario-Based Analysis

The introductory student exercise in chapter 2 uses a task simulation scenario, and Greenberg argues that this offers a cost-effective method for analyzing a system slated to be redesigned. The Task-Centered System Design approach he advocates involves imagining users of three different types and anticipating their responses at each task step in the interaction within a new system. In the introductory exercise, the three users are an elderly customer, a novice customer, and a trained assistant helping a customer. Each user is accompanied by a brief biography and a description of his or her motivation for using the system. Greenberg argues that, although the approach is likely to miss some tasks and some user groups, it will be quick to execute and relatively cost-effective. He notes that the approach could be supplemented by other methods to make it more comprehensive.

Greenberg presents a scenario-based, task-centered approach to system design and evaluation. He claims that the approach can be applied after as little as 4 hours of instruction and falls within the tradition of discount usability engineering. The method has four principle parts. In the first part, a task description is developed. In the second part, some tasks are selected for the analysis, and a specification of three diverse user types is developed. In the third part, the scenarios for analysis are developed. In the final part, the scenario is simulated step by step, and the likely user response to each system image for each task step is recorded. A tabular format is used to log the anticipated users' knowledge about and motivation for the interaction, potential disruptions to task performance, and potential errors by the users. Comments on the interaction and design remedies are also noted. Greenberg demonstrates that this analysis could lead to an interface redesign that has the potential to improve the interaction.

The analysis presented by Go and Carroll (chap. 5) uses the scenario as a mechanism for eliciting questions about and prompts for tasks. Go and Carroll argue that scenarios can be constructed to represent stories about past, present, and future use of technology. As they build up a scenario, they use a number of what-if questions. Go and Carroll show how a scenario-based task analysis may be proceduralized to make the process rather more structured. As with Greenberg (chap. 2), the focus is on the task the user is to perform.

Anticipating potential problems that the user might have with a technology by using structured scenarios offers a degree of ecological validity that may help persuade system designers to be more human-centered in their design processes. Scenario-based analyses may be supported by more formal, reductionist approaches. As Greenberg (Chapter 2) also points out, scenario-based analysis can be extremely cost-effective, as it does not necessarily involve user trials. This does not rule out task simulation with real end users though, when the design specification becomes clear. Indeed, Go and Carroll (chapter 5) caution that representative end users should be involved in focus groups as part of the process of developing realistic task scenarios.

A less formal approach to scenario-based analysis is provided by Spencer and Clarke (chap. 10). They argue that some task analysis techniques require considerable effort, but designers need task analyses promptly for decision making. By way of a solution, they have developed an approach called User-System Interaction Scripts, which, it is claimed, can be learnt in less than 30 minutes and requires little effort to apply. An interaction script is created by describing the system goal and listing all the task steps and the corresponding step responses. Spencer and Clarke cite two informal evaluation studies in which project managers spent between 6 and 30 minutes creating a script. The results suggest that feedback from the managers was favorable, but Spencer and Clarke note that they had some occasional difficulty in getting a level of detail appropriate for the projects.

28.5.2 Tools for Support

Omerod and Shepherd (chapter 17) present the subgoal template method as a way of making task analysis simple enough for the novice analyst. Despite the fact that HTA is one of the most widely used and prominent methods and has a pedigree of more than 30 years, it is far from the easiest to grasp. Omerod and Sherpherd include the long learning curve in their list of inhibitory factors that make people unwilling to adopt more formal approaches to task analysis. Their solution to this problem was to devise a classification scheme for tasks that would offer a parsimonious account of operations that people could perform. They also argue that the provision of templates for task analysis would remove many of the inhibitory factors and encourage people to adopt HTA as part of a human-centered design approach. For example, there are subgoal templates for actions (e.g., activations, adjustments, and deactivations), exchanges (e.g., entering and extracting data), navigation (e.g., locating, moving, and exploring information), and monitoring (e.g., detecting abnormalities, anticipating changes, and observing rates of change during state transitions). With each of the subgoal templates, the analyst may choose from a range of plans to construct the analysis. The four plan types available are fixed sequence, contingent sequence, parallel sequence, and free sequence. With these two relatively unsophisticated tools—templates and plans—the analyst may construct quite elaborate and complex task analysis. Omerod and colleagues have evaluated the effectiveness of the training the subgoal template method elsewhere. Suffice to say that the method is easier to learn than unconstrained HTA.

Van Dyk and Renaud (chap. 12) offer a usability metric approach intended to help system developers understand what usability is and how to quantify it. In their approach, the tasks are first decomposed into their subgoals, and the metrics are then used to examine the degree to which the user interface supports the subgoals. At the heart of the method are a series of questions that the analyst has to ask of the system while pursuing each subgoal. A score is assigned to each aspect of the system and weighted prior to the final evaluation. In this way, poorer aspects of the system may be identified with relatively little training in task analysis and usability assessment. Obviously, the resulting assessment will not be as rich in information as some of the more involved methods in this handbook, but it will help justify further development effort and might even encourage software developers to undertake more rigorous analyses.

Wong (chap. 16) reports on two tools designed to assist in the interpretation of data generated by the Critical Decision Method. The Critical Decision Method is an update and extension of the Critical Incident Technique (Flanagan, 1954). The Critical Decision Method structures the interview in an incident analysis by asking the interviewee to review critical decision points that occurred as the event unfolded. A series of questions are then presented to probe each decision, such as the following:

- What information cues were attended to?
- What were the situation assessments?
- What information was considered?
- What options were considered?
- What basis was used to select the final option?
- What goals were to be achieved?

Wong argues that this approach is extremely productive of data, but tools are needed to draw out themes and structure the data if useful insights are to be achieved. To this end, he offers a structured approach for organizing the data and an emergent themes approach for searching for common themes in the data. Wong proposes that first the emergent themes should be extracted from the critical decision data by using an indexing and classification procedure to identify ideas and relationships between concepts. When the emergent themes are exhausted on the first analysis, the structured approach may be used to look top-down at the data and explore each emergent theme in turn. This may lead to further classification and reanalysis of the data. The structured and emergent themes approaches could be used iteratively on the data until no further reclassification is required. Wong argues that they are complementary techniques and can help overcome some of the shortcomings of critical decision data, but they will not compensate for poor interviewing techniques.

28.5.3 Commercial Pressure

All of the chapters in part II, on industry perspectives, emphasize the importance of easy-to-conduct tasks analyses and transparent final documentation. These chapters cite problems such as the difficulty of conducting a task analysis, the costly expense in time needed to produce an adequate analysis, and the difficulty of reading the resultant analysis by anyone other than the analyst.

Arnowitz (chap. 11) addresses these problems by offering a new task analytic method, called *Task Layer Maps*. This method offers a way of parsimoniously describing the flow between task units after first identifying the logical sequence. The method requires that all of the tasks and their contingencies first be identified. This part of the method is common to all task analysis methods. In this particular method, the analyst is required to distinguish between tasks performed in parallel and tasks performed in series. This information is used to identify the critical path taken by tasks in the second part of the method. Only essential and contingent paths between tasks are left. Arnowitz calls this process *peeling*, as all nonessential links between tasks are removed. In the third and final part of the method, the processes of reviewing the tasks and removing any redundant dependencies are undertaken. As the links between tasks in a timeline are stripped to their minimum, a much clearer picture of the essential task features emerges. Arnowitz argues that this simplified analysis is much more accessible to all of the stakeholders involved in system design and that such an approach will improve the uptake of task analysis.

Paternò (chap. 24) develops a hierarchical description of the human-computer system that links both elements within the same description. He calls this method ConcurTaskTrees. As a systems-based approach, it makes perfect sense not to discriminate between the parts of the task that are performed by people and the parts that are performed by technology (chap. 1). Indeed, the design of the system could change any assignment given in the description. The approach seems to have a resemblance to HTA (chap. 3), although the actual notation differs and the plans that guide the sequence of activities are placed between each task. It also seems to bring the whole argument about the readability of task analyses full circle, as the ConcurTaskTrees diagrams are just as readable as HTA.

Windsor (chap. 23) argues that, in the design and development of Swanwick Air Traffic Control Centre, a detailed tasks analysis would have been prohibitively expensive. Instead, the designers opted for a coarse task list for the analysis and simulation of the critical tasks. In building up their picture of the air traffic control (ATC) tasks, they asked many questions of the current and envisaged system, including the following: What are the information exchanges? What are the temporal characteristics? What are the task triggers? How do tasks overlap? What happens if a task is interrupted?

Their analyses revealed four main types of tasks in ATC: scanning tasks (e.g., scanning monitors continuously to monitor aircraft profiles and identify potential conflicts as well as making decisions), executive tasks (e.g., recording and communicating decisions to pilots and ATC staff), back office tasks (e.g., rerouting a flight), and systems tasks (e.g., managing configuration of the system). Their analyses also revealed that these tasks varied in their duration, their frequency, and their importance.

Windsor states that the analyses led to the use of a user-centered approach in designing the new control center and that the analyses went through several iterations to refine the concepts for final specification. The process of refinement used scenario-based analysis techniques to consider potential problems and failures and the system's effect on the ATC workload (see also chap. 14). Two pertinent points to arise from the ATC study were that task analysis does not necessarily need to go down to the level of fine detail for every task (the level of detail will depend on the safety-critical nature of the activity) and that task analysis is best integrated into the design process rather than performed as a stand-alone activity.

Coronado and Casey (chap. 8) point out that, although task analysis might be one of the most effective methodologies in system design, commercial pressure to adhere to software release deadlines might not always accommodate it in the first generation. They state that experience shows that software development often precedes structured task analysis and formal design. The formal "textbook" approach to design is seen in subsequent generations of a product, according to them, but the luxury of using such an approach is often not afforded in the first generation, which might instead use less resource-intensive techniques such as brainstorming. Coronado and Casey agree that task analysis is extremely useful for constructing a picture of task flow. It also provides a common reference for a shared understanding of product and user requirements. The main lesson they have learned from commercial software projects is that, in task analysis, flexibility and speed are of the essence.

One possible stance toward the problem of conducting a task analysis is not to bother. Along this line of thinking, Karat, Karat, and Vergo (chap. 29) propose using critical peer review (cf. the Crit as used in the artistic and performance disciplines) rather than the more formal scientific evaluations normally associated with human-computer interaction. They argue that, for some aspects of design and evaluation in human-computer interaction, it is not possible to apply the traditional methods and metrics. In particular, they point out that traditional human-computer interaction techniques do not reveal the extent to which individuals identify with a device image, the degree to which they engage with a device, or the social value of a technology. They argue for an approach that taps into people's value and belief systems. This approach focuses attention on the macro, not the micro, issues of device use and the purposes to which devises are put. They identify three key issues: the quality of the content that a device gives the person access to, the degree to which the device supports social activities such as communication with fellow human beings, and the degree to which the device supports the mastery of a skill or the development of expertise. Karat, Karat, and Vergo argue that these metrics may provide more important qualitative insights into the usefulness of technology than some of the traditional ease-of-use metrics. Certainly the purpose to which a technology is put will be an important issue for system designers. Matching this to the values, ideology, beliefs, and goals of people is likely to lead to a greater acceptance, and use of the technology in the future.

28.6 TASK ANALYSIS: THE STATE OF THE ART

This review of the state of the art of task analysis reveals some lines of tension. The first line of tension involves the underlying psychological theory of a particular task analysis method. This has some obvious effects, as it ultimately determines the academic credibility of the method. Some theorists prefer to talk in terms of general psychological functions, whereas others prefer to talk in terms of specific psychological mechanisms. This divide is unlikely to be resolved in the short term, but the competition between approaches is likely to enhance our understanding.

Another line of tension concerns the degree of richness of the underlying psychological model. Some authors have opted for a very rich and detailed model of human action, whereas others favor a lite model. A pragmatic fitness-for-purpose approach would suggest that we adopt a model that delivers the level of detail required by the situation. We might view the models in terms of a hierarchy of description, with the more complex approaches subsumed under the more general ones (see chap. 30).

A third line of tension concerns the constraints that the analysis is performed under. The chapters written from an industry perspective identify the need to analyze complex tasks with very limited resources. This often means taking sophisticated techniques and applying them to difficult problems and arriving at a solution that nontechnical personnel can understand, all within a limited time frame. It is no wonder that some analysts have sought to simplify the process of analysis and make the output more accessible. The commercial pressure to develop cost-effective methods has to be aligned with the academic pressure to devise methods that are more theoretically sound. It is hoped that both parties can understand the perspective of the other. What all of the book's contributors seem to agree on is the importance of understanding tasks in the context in which they are performed. The final take-home message of this chapter is that context analysis is the key to successfully undertaking task analysis.

REFERENCES

Annett, J., & Stanton, N. A. (2000). *Task analysis*. London: Taylor & Francis.

Benyon, D., & Macaulay, C. (2002). Scenarios and the HCI-SE design problem. *Interacting with Computers, 14*, 397–405.

Card, S. K., Moran, T. P., & Newell, A. (1983). *The psychology of human-computer interaction.* Hillsdale, NJ: Lawrence Erlbaum Associates.

Carey, T. (2002). Commentary on "scenarios and task analysis" by Dan Diaper. *Interacting with Computer, 14*, 411–412.

Carroll, J. M. (2000). *Making Use: Scenario-Based Design of Human-Computer Interactions.* MIT Press.

Diaper, D. (1989) *Task analysis in human-computer interaction.* Chichester, England: Ellis Horwood.

Diaper D. (2002a). Scenarios and task analysis. *Interacting With Computers, 14*, 379–395.

Diaper, D. (2002b). *Task scenarios and thought. Interacting With Computers, 14*, 629–638.

Flanagan, J. C. (1954). The critical incident technique, *Psychological Bulletin, 51*, 327–358.

Gilbreth, F. B. (1911). *Motion study.* Princeton, NJ: Van Nostrand.

Norman, D. (1986). Cognitive engineering. In D. Norman & S. Draper (Eds.), *User-centered system design.* Hillsdale, NJ: Lawrence Erlbaum Associates.

Paternò, F. (2002). Commentary on 'Scenarios and task analysis' by Dan Diaper. *Interacting with Computers, 14*, 407–409.

Stanton, N. A., & Annett, J. (2000). Future directions for task analysis. In J. Annett & N. A. Stanton (Eds.), *Task analysis.* London: Taylor & Francis.

Stanton, N. A., & Young, M. S. (1998). Is utility in the mind of the beholder? A study of ergonomics methods. *Applied Ergonomics, 29*, 41–54.

Stanton, N. A., & Young, M. S. (1999). *A guide to methodology in ergonomics: Designing for human use.* London: Taylor & Francis.

Sternberg, S. (1969). The discovery of processing stages: extensions to Donder's method. In: W. G. Koster (Ed). *Attention and Performance II.* Elsevier: Amsterdam.

Taylor, F. W. (1911). *Principles of scientific management.* New York: Harper & Row.

29

Experiences People Value: The New Frontier of Task Analysis

John Karat
IBM TJ Watson Research Center

Clare-Marie Karat
IBM TJ Watson Research Center

John Vergo
IBM TJ Watson Research Center

A major challenge in human-computer interaction is making design decisions, which will have a positive impact for a user in a task context. The focus of almost all evaluation in human-computer interaction has been on how well someone can complete some specified task using the technology being evaluated. This chapter expands the scope of task analysis to include the design and evaluation of affective components of use where there is no specified user task. A framework for the design and evaluation of experiences that people value is presented, followed by a discussion of what and how to measure these user experiences. Research involving the design and evaluation of immersive user experiences in streaming multi-media stories on the web and research on social communication are employed to illustrate concepts and methods. Recommendations on how to approach the challenge of this new area of task analysis are provided to guide further work in this domain.

29.1 INTRODUCTION

Evaluation is a fundamental part of human-computer interaction (HCI) and interface design. Designers attempt to understand the needs of some audience and then formulate system designs to meet those needs. The focus of almost all evaluation in HCI has been on how well someone can complete some specified task using the technology being evaluated. We can measure time on task, observe error rates, and note task completion rates to provide information for determining whether the technology is satisfactory for its intended purpose—as long as there is a clear intended purpose. Measuring the effectiveness, efficiency, and satisfaction of users trying to carry out tasks in particular contexts is how we define measuring usability (ISO 9241). Although there is an affective component of this measure—user satisfaction—it is generally regarded as a less powerful measure; it is "more subjective" and for the productivity-oriented end just not as critical.

However, it is increasingly clear that the uses of computing technology are reaching far beyond office productivity tools. Various authors have pointed out this trend (e.g., Furnas, 2001; Grudin, 1990) and suggested that HCI needs to adapt techniques to encompass a larger scope of human activities and interests. This includes moving beyond the desktop to considering less productivity centered aspects of human life, such as social interactions and the concerns of society as a whole. As this trend continues, HCI will shift its focus from people interacting with computers and information in order to accomplish tasks and goals to people interacting with people through technology in order to communicate, engage, educate, and inform one another and thus reap family, community, and societal benefits of unknown magnitude. (Chapter 5, for example, describes an example of future possible family-wear).

Looking beyond the role of technology in the workplace to its potential role in making a better world takes some expansion of the focus of the field of HCI. We need to be able to ask, "Will people value this technology?" and not just "Will people find this technology useful?" Measuring such value goes beyond economic notions. It should consider all aspects of a system that a user might feel makes owning and using it important. Such benefits can be identified, measured, and given a role in HCI design (Landauer, 1995, Putnam, 2000; Resnick, 2001).

As we move forward, the specific goals a user has in mind when approaching technology will become more varied and less easily identified. Topics such as appearance and aesthetics will play an increasingly important role in HCI design (Karvonen, 2000; Laurel, 1993). HCI professionals will need to be able to respond with something other than a blank stare when asked to assist in the design of artifacts that provide emotionally satisfying user experiences. Style (e.g., form factor, color, and materials used) is becoming increasingly important in the design of standard personal computers. Devices for entertainment that "connect to the Net" are becoming widespread. Newer uses of technology call for a better understanding of how people will view the technology and for approaches that are not completely "efficiency of task completion" oriented. Questions that we should consider include these:

- How can we design a user experience that is engaging?
- When will people buy something because of its image?
- How can we measure the social value of technology?

Two interesting trends have created new opportunities for behaviorally oriented work. One is the movement of the technology out of the workplace and into the home and everyday lives of people. Although people certainly do spend some resources on labor-saving devices for the home, like dishwashers and such, few customers consider such economic issues when making purchases of most of the things they surround themselves with outside of the work environment. Most people do not really consider the effectiveness and efficiency of lighting fixtures, furniture, and home entertainment equipment when purchasing these. As for the second trend, there is ever more consideration being given to widening the role of technology systems in the workplace beyond that traditionally assigned to workplace tools. Can technology enhance collaboration? Can it change what it means to work together on common goals? Can technology improve employees' quality of life?

These trends bring a new challenge to HCI research and practice. It just isn't possible to always find a measurable objective task toward which to orient HCI engineering methods with an eye to increasing ease and efficiency. It will take more than just asking if someone is "satisfied" with an experience to understand fully why they might "value" it, for valuing something can involve a number of considerations, including ethical issues (Nardi & O'Day, 1999). What makes someone like playing a game? Why do people spend hours in chat rooms?

If HCI professionals are to consider answering these questions—and they certainly seem to full within the compass of HCI—HCI professionals must become better equipped to address the affective side of value. Although the field might still have a long way to go in making user-centered design common practice, the basic tools for task analysis and performance measurement are already well covered, as can be seen in many chapters in this handbook and in the general focus on usability. It is possible that many in the field feel that there are aspects of personal choice and value that cannot be easily subjected to the usability engineering approaches of transaction-based systems. The argument presented here is not that furthering the practice of use-based evaluation is the only means for advancing such systems, only that it could benefit their design and development through "normal science" (Kuhn, 1996). One alternative would be to declare such issues as outside the scope of the field. This is certainly an option for anyone who feels that the difficulty of designing usable, functional systems is sufficiently challenging. Although extending evaluation techniques to cover affective areas certainly offers significant challenges, the field of HCI does seem to be the right home for these new challenges, and many within it will be interested in trying to meet them.

29.2 EXPERIENCES PEOPLE VALUE

Usability engineering has two primary features: (a) the use of observations of users and use situations as a data-gathering activity to inform design and (b) the use of measurable behavioral objectives in design iteration. These features pinpoint major areas of concern in extending usability engineering to environments in which there is no specific task. Although HCI professionals might not think of activities such as creating entertainment "content" or aesthetic design as being influenced by usability engineering, these activities generally involve a concern for the ultimate consumer. Test screenings are carried out for plays, movies, and television shows, and "user observations" are collected to inform the final product. The presentation of a design concept for an artifact (ranging from a product to a building) is done to judge the adequacy of the current design and improve the final version. The main difference is not whether user testing takes place—though this term might not be used in other domains—but in how the data-gathering activities are set up and what kind of observations are sought. "Experience this artifact and tell me if you like the experience" replaces "Try to carry out this task using this system" as the main guide for what the session will involve. Clear usability objectives, which are essential for usability engineering, are rarely present in non-task-oriented artifacts or design. Goal attainment in a non-task-oriented situation is more indirect. People do not watch a program or play a game just to complete it. People pursue such activities because they are engaged or entertained.

This chapter addresses *experiences involving technology that people value*. In this context, "experiences" is intended to have a wider connotation than "uses." All situations in which *computing technology delivers content that people experience* offer the potential for evaluation to inform the design of the technology. So if someone asks whether this includes the design of a television, the answer is that it does as long as changing the technology can change the experience in some way. It should be clear that the experience is a function of content, technology, and context.

HCI professionals standardly talk about users of systems, user goals, context of use, and usability. For the most part, the goals associated with using systems intended to provide pleasurable experiences rather than solve problems are not explicitly held by the users of the technology. HCI professionals might say that people use a technology because they have the goal of reaching a pleasurable state, but this is awkward and has not proven useful as a guiding approach in design. Partly this is because of the difficulty in objectively defining the goal state, and without this there isn't much the field can say about the path to the goal. The real problem is that, for usability engineering to work, it needs to be guided by models of goals and how people work to achieve them. Rather than pretending that valuable user experiences are something they understand in the same way they understand explicit goal-directed behavior, HCI professionals would tend to leave user goals undefined. The science of enjoyment is not capable of defining a goal-directed approach. What is clear is that people can value experiences that involve technology beyond the contribution of those experiences to task-oriented activities. The focus here is on the value people place on the experience with a system. Although people might view "usability" as just a name for the measure of the quality of interaction with a system, it is difficult for people to break from the task-oriented focus unless a different terminology is employed. Therefore, different terms are used here. The act of experiencing content, whether it is looking at an image, reading some text, or watching a movie, is defined as "valuing" the content. Unless a particular mode is identified, "systems that people value" include systems valued for a wide range of possible reasons (sensory or cognitive).

29.3 SYSTEM CATEGORIES OF EXPERIENCES PEOPLE VALUE

We identify several general classes of systems for which different types of HCI activities might be useful in the design of systems that people value. The framework may grow as HCI work in the area proceeds. First, there are systems for which technology can deliver content that people find valuable. At the most general level, we can view a web browser as this sort of system. Although the Web can certainly be viewed as a system that helps people complete "tasks" such as obtaining driving directions or finding out what the weather will be like this week in Paris, it also can be viewed as providing content that people just enjoy browsing. For these types of systems, the two major features that one can explore with HCI methods are the simplicity of the interactions and the quality of the content. Simplicity of interaction is a fairly common theme in HCI and will not be covered in this chapter. Quality of content—from the perspective of providing valuable experiences for the reader or viewer—is a different issue. Quality of content is not generally explored within task-based HCI (outside of education applications) but is something that is easily accommodated within behavioral observation methods employed in HCI. Here it seems that the value comes primarily from *what* the system gives the user access to, and such systems will be called *content driven*.

Second, there are systems that support activities that people find valuable and will use without strong task-based requirements. Communication is an example of this type of activity, and e-mail and chat systems will be addressed. Certainly, HCI professionals can look at e-mail and chat systems from a task-oriented perspective. However, in doing so they might miss the point that these systems have a great deal of use outside of the task perspective. E-mail has enabled new forms of social communication and provided uses that cannot be accounted for in terms of goal achievement. Chat systems have become wildly popular, and although people are certainly finding business uses for them, their popularity arises from their support of social activity. A system that is developed to support social communication might be a lot more powerful than one designed to support a particular work communication task. Systems that can fulfill other basic human needs could be researched in similar ways. Here it seems that the value comes primarily from *WHOM* the system gives the user access to, and such systems will be called *communication driven*.

Third, there are systems that derive value from interaction and the results the interaction can produce. Games are the most common technology artifact, but the piano or another musical instrument may be a more suitable prototype. It is important to note here that the goal might not necessarily be ease of use: There is something in acquiring a skill that many people value. Mastery of something complex can be a goal people will work for, whether it is being able to execute a triple axle, play a Beethoven concerto, or blast video-game enemy to smithereens.

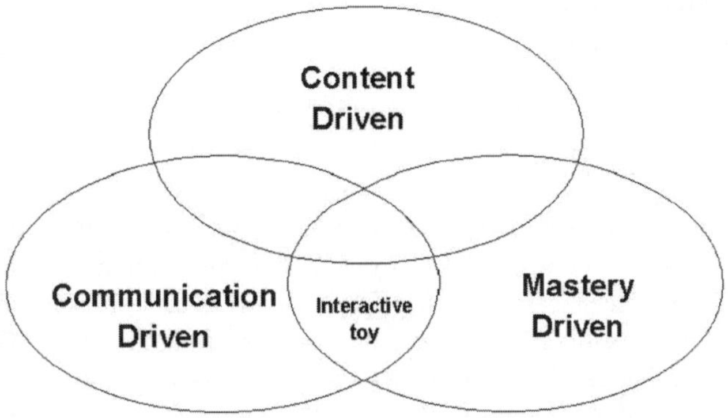

FIG. 29.1. The location of an interactive toy within the Content, Communication, and Mastery-Driven domains.

Again, although HCI techniques might not generally be employed in the design of such systems, it is easy to see how they might be. The value of the experience comes primarily from the *Interactive experience* the system gives the user access to, and such systems will be called *mastery driven*.

All three of these domains offer challenges appropriate to the skills of an HCI practitioner or researcher, whether the objective is (a) to understand why these systems are valued when their purpose is less obvious than our transaction-oriented approaches might assume or (b) to iteratively design and develop systems that "meet needs." It is not important that the classification presented here be complete or nonoverlapping (i.e., it may be heterarchical; see chap. 1). For example, interactive dolls or animals can be seen as artifacts that are valued as surrogate communicative playmates or as interactive toys (see Fig. 29.1). An interactive toy gives a child access to communication with a valued playmate (communication driven), and if the toy provides the child with more satisfying communication and interaction as the child develops better interaction skills, then the child is experiencing a mastery-driven toy.

Advice will not be presented in separate sections for each of the three categories because the HCI field has not yet sufficiently differentiated HCI methods that would call for applicability to one category (e.g., content-driven systems) over another (e.g., communication-driven systems). In all cases the advice is similar: Consider what the value might be, observe people in the appropriate class of interaction, and use the observations to inform future design decisions.

29.3.1 Content-Driven Systems

Content-driven systems are discussed using the entertainment domain as an example. (These systems are built for education and other purposes as well.) Content driven systems are those that are valued for the content that they can provide access to more than for their own features. The paradox of the VCR helps illustrate a point about the distinction between usability and value. People in the HCI field have joked for years about the difficulty of using VCRs. Still people buy them in large quantities. This is clearly not for the joy of playing with the buttons or trying to figure out how to use the features. People own VCRs to receive access to content that they enjoy. Although there are certainly educational videos that might provide task-oriented training, VCR purchases are probably driven far more by entertainment goals than by practical task-oriented goals.

There are three primary aspects that need to be considered: the quality of the content itself; how the content might be accessed; and the experience of browsing, viewing, or reading the content in a possibly new context. These are the primary aspects that need to be considered from a design and evaluation perspective in order to understand why people would value a content-driven system.

29.3.1.1 Content Quality

In the world of such new technologies as electronic books and multimedia computers, it is perfectly reasonable to ask what makes good content for different viewing technologies. There seem to be two basic questions. First, HCI practitioners can evaluate content, answering the question whether something is perceived as worthy of attention or not. Clearly there are tools utilized in HCI that can be useful for this purpose. These include field and laboratory observation methods and techniques such as questionnaires and focus groups. In selecting the method and what to look for, the first piece of advice is to consider how one wants to operationalize "quality" or "value." HCI practitioners can measure the time someone spends looking at something as a measure of interest or entertainment, they can count the number of laughs or of tears shed during a movie, and they can ask directly if an experience was enjoyable. Both contextual observation and questionnaire methods of data collection have value for exploring content quality. Generally contextual methods have a greater role in early evaluation, and survey methods in later stages.

Second, HCI practitioners can ask whether there is design advice based on HCI research that can be given to content designers. Here the focus is on practical theories related to human emotional states. Rather than detailing the research on this topic, the authors summarize by saying that this research is a long way from being in a form that would provide practical guidance for content designers. Laurel's (1993) work is one exception in this area.

Defining entertainment value is by no means an easy task. Although the term *entertainment* is used all the time, it is actually difficult to define (see Dyer, 1992, for a discussion of this issue). Entertainment has been described as "any activity without direct physical aim, anything that people attend to simply because it interests them" (Langer, 1954). For the purposes of the discussion here, we can say that people are entertained when they are undergoing an experience that interests them and gives them some amount of satisfaction.

A seemingly important distinction regarding entertainment is whether an entertainment-related artifact is interactive or primarily passive (see also chaps. 1, 18, and 26). The distinction is not pure: Content such as a DVD movie can certainly be interacted with as part of the experience, and games can be appealing because of the graphics. But it is helpful to make the distinction between "watching" appeal and "interactive" appeal in examining the role of HCI in entertainment applications.

There has been some effort to advance engagement (holding the attention of the viewer) as an important concept for the design of multimedia systems, particularly in education (e.g., Chapman, Selvarajah, & Webster, 1999; Jacques, Preece, & Carey, 1995; Malone & Lepper, 1987; Webster & Ho, 1997). Much of this work is aimed at tapping into the effort that people are willing to give to activities such as games. Certainly if HCI practitioners could make learning in school more enjoyable, they might find students learn better. These efforts have met with mixed success. For the most part, their goal is to provide more attractive interfaces for task-oriented systems. Although they might succeed partially they also might fail by mixing "business and pleasure." If there are task-related benefits that occur in entertaining activities, such as incidental learning that might take place during the viewing of a movie or play, the task behaviors tend to become the major focus. For the study and advancement of positive experiences with technology, the role of engagement in education must be a major focus.

When we talk about passive experience of content, such as in reading or viewing content, we mean that reading and viewing are passive in the sense that the reader or observer does not get up and direct the experience (that is another person or group is in charge of "telling the story"). We invite HCI professionals to consider this topic as open to contributions and to view content as something that is not outside the design of the total user experience.

29.3.1.2 Access to Content

At some level, it is easy to imagine that the role of HCI in content access is simply to make the task of getting the content as usable (meaning effective, efficient, and satisfactory) as possible. Accessing the content is the assumed goal, and making it easy is the usability engineer's task. Certainly there might be times when this is an appropriate approach to take. After all, we are primarily interested in the music, not the player, right? The answer is "not always." Certainly the designer doesn't want the access device to get in the way of user enjoyment, but sometimes it can be a part of it. When all mechanisms for access are essentially equally easy and functional—and even sometimes when they aren't—aesthetic attributes of the content player can become distinguishing features that people value. The appropriate metric for looking at access for nontask systems is whether the interaction seems engaging to the user rather than whether it is easy.

An example from the audio product world illustrates this point. Stereo equipment is not highly differentiated by ease of use. Although there is a fair amount of product differentiation in the labeling and placement of controls, the fact is that all systems have very similar controls for playing tapes, CD discs, and the like. People are also attracted to a particular device by other factors such as price, technical performance, and design. There are product lines, for example, that do differentiate themselves with respect to design in general and user experience in particular, such as the products of Bang and Olafeson. The point isn't that everyone demands this particular brand; price factors keep this from being the case at a minimum, and the design style might not be for everyone. The point is that this is a product line that offers a different user experience and that some people value enough to pay a substantial premium on the base value of the technology. Products of this type do not emerge from a focus on efficiency or traditional usability. They emerge from a culture aimed at producing things that people will value. Does Bang and Olafeson use user-centered design? Probably not by that name—actually it operates largely through the designs of a single individual. As an organization, it is aware of the value it is trying to create and takes steps to advance the products' value—steps that can be seen as "evaluation methods," including the holding sessions in which user reactions to design ideas are examined in much greater detail than provided by a check on a satisfaction scale.

There are lots of ways to structure such evaluations—in focus groups, in home settings, in product laboratories—but all require two leaps for usability engineers. First, the engineers must get over a fundamental mistrust of subjective data. Second, they must invest in more than a surface picture of satisfaction. It is difficult to iterate on a design to improve it if all the engineers have is a number that represents a user's satisfaction with a particular version of the design. What is needed is a "craft" view of the design process to supplement the engineering view (Wroblewski, 1991).

Aesthetics should not play the only role in designing systems for access to content, but affective components do have a role here beyond the normal usability perspective. Work is work, and play is a place where people are generally freer to make decisions based on feelings rather than efficiency. This doesn't mean that people will give priority to other design features, but it does mean that they are more likely to than in situations in which a specific goal is the reason for using a system.

29.3.1.3 Content in the Context of Experience

If everyone could see or hear all of the music, plays, sporting events, and cultural exhibits that they wanted in the comfort of their home, would anyone go to the movie, to a museum, to a stadium to experience these things? Is smelling the popcorn or hearing the crowd roar part of the experience that makes an event seem remarkable?

Any technology use takes place in some context. In the past, the most pervasive assumption of HCI practitioners was that the tools they designed were used by someone in a nice quiet office. As the HCI field moved from individual work tools to organizational work environments, the focus on computer-supported cooperative work (CSCW; e.g., chaps. 7 and 19) expanded its boundaries to include other people in a work environment as part of the context of use. Pervasive devices used in work and leisure situations have enlarged the context of use to the point where it encompasses everywhere. Consider the various contexts in which people like to spend their leisure time. In some cases, solitude is what people value and seek. For example, there are situations in which someone might view having other people around to be a nuisance—when reading a book or enjoying a painting in a museum. However, there are times when reading together with others or being part of a museum tour might be a preferred experience.

Although it might not be possible to determine a single best possible viewing context for any particular content, it is possible to consider whether or how a system might be used in solitary or social situations and to include evaluations of user experiences with the system in these different contexts (Karat, Pinhanez, Karat, Arora, & Vergo, 2001; Pinhanez et al., 2001). The Karat et al. (2001) study found that users enjoyed multimedia streaming Web tours (an example of a content-driven experience) both individually and in a small group settings where pairs of users were able to talk about the experience, share their perceptions and questions about it, and make joint decisions about what topics to view in more detail. Although sitting in front of a computer is generally a solitary activity, users found entertaining cultural Web tours to be an engaging experience to have both as individuals and in small groups.

The advice to observe current and proposed practice *in context* is probably as worthy as any. Although there is a good deal of emphasis on techniques such as heuristic and laboratory evaluation in usability engineering, these techniques are simply not adequate for discovering the value of social context, particularly for casual use systems (Putnam, 2000). Not everyone values social contact to the same extent, but in general people are social creatures who do value being around other people at times. In these evaluations, it is important to observe the interactions between the users and their environment and how the technology might enhance or diminish such interactions. Noting such interactions—even in a simple list of pluses and minuses—can help in gaining a perspective on how the technology will fit in a real context.

29.3.2 Communication-Driven Systems

What do you need a cell phone for, really? Are there tasks that we could have identified and used to predict that this technology would be as pervasive worldwide as it is becoming? What is the value of communication limited by small screen displays and numeric keypad message entry? What makes a game fun? Why are many successful games minor variations on a basic hunt-and-destroy theme? HCI practitioners need more evaluation-based information to be able to answer these questions. They need to consider the domain and the technology, make guesses about how to apply the technology, and then look to be informed by people attempting to use it. For designing new communication technologies that people would value, there is even less formal guidance than for designing content-driven systems. There are schools and design traditions for content areas such as film, theater, literature, and so on. Game design schools are beginning to emerge, but there are no chat room design schools. For these reasons, the advice offered here is more basic and subject to development by HCI practitioners based on context. If you think you have a technology that supports communication, ask questions to uncover why people might value it.

One way to view the goals of communication-driven systems is simply "to provide a way to communicate with other people." This is a goal that people seem quite willing to accept as a valuable part of their lives. It is perhaps useful to think of communication as a goal with a high value rather than as a task that needs to be accomplished efficiently. People do care about how easy it is to interact with a technology to communicate, but they are also willing to adapt to a technology if improved communications are possible (e.g., many are willing to learn complex schemes for entering text messages on cell phones).

Using the scheme developed above, what guidance can be offered by considering content quality, issues of access and interaction, and the context of experience? The example of "chat systems" is considered briefly below. These are systems, like AOL's Instant Messenger, that enable people to see when someone is currently connected to the Internet and to send the person messages directly. They also allow for communication between more than two parties at a time.

29.3.2.1 Content Quality

Very little has been done to improve the quality of communication in chat systems. Although functions such as spelling checking and grammar checking are available for such systems, there has not been a strong sense to date that they are requirements for a successful system. Just as people are tolerant of a wide range of flaws in normal spoken communication—which is often ungrammatical and populated with disfluencies—they seem to value the informality of imperfect exchanges in chat. The success of chat systems despite the scant attention paid to content quality does not mean that it is unnecessary (or possibly valuable) to consider content quality in communication systems. Instead, it might be that designers have not really thought much about what it would mean to improve the quality of a message used primarily for interpersonal communication.

Improving the quality of the content of a chat system might be taken as a challenge to improve the expressiveness of the messages that people can send. Viewed this way, it is possible to consider the emergence of "emoticons" (expressive faces created through combinations of various special characters on the keyboard) as an effort to improve the quality of text messages. Are other improvements possible? Certainly a text message lacks much of the expressive content present in face-to-face communication. Some of what is lacking people learn to compensate for when they learn written forms of language. For example, the forms of emphasis available to a speaker through voice changes can be conveyed through manipulations of font characteristics in even the most basic chat systems. Although we might try to fall back on considering "expressing a thought" as a task and then ask how well the technology enables "thought expression," we suspect that it will be more productive to simply try to include more expressiveness over time.

29.3.2.2 Access and Interaction

Behavioral science has long pointed out that humans are fundamentally social creatures. People value contact with others. If they are given access to a technology that enables them to have contact with others, they will use it. Beyond economic constraints, there does not seem to be any reason to expect that the more access technology provides, the more people will grow to value it. In fact, current technology provides an interesting illustration of how much people will forsake "easy-to-use" methods in order to learn new techniques for communicating. Learning to enter text messages on keyboards designed for entering numbers—the typical phone keypad—is something traditional task analysis would have a hard time justifying.

What guidance can we offer when considering access and interaction for communication systems? Although not denying that easier is better, we would point out that a technology that fulfills a need to communicate will be something that people will be willing to exert an effort to learn to use. Certainly phones that can recognize speech to determine who you are trying to reach and that correctly navigate through myriad possible work and home numbers to reach someone would make the task of accessing an individual easier. However, we need to keep in mind the relative value of ease of use compared with making something possible. Making something possible can provide a significant current value even if one can imagine better ways to use it in the future.

29.3.2.3 *Context of Experience*

Is communication only about one person talking to another individual? Although there is certainly a great deal in one-on-one conversation, there is a lot more to communication than this restricted version might offer. Again, thinking about the social nature of humans and the way in which they communicate face to face, it is easy to notice that the messages are generally delivered in a broader social context. Who is in the room can dramatically alter what is said and what it means. Similarly, awareness of who is in the audience for a chat can provide a great deal of useful information for the speaker. Are people listening attentively? Do they seem to wish that I would stop talking?

Some of these ideas have been explored within HCI. Erickson and Kellogg (2000) address trying to make the social context of a dialog involving more than two people more apparent. Through a variety of information visualization techniques, they have attempted to bring in information about the active participants in a discussion and have found this to be of significant value in facilitating remote discussions. These techniques can enhance a feeling of presence of individuals in a group—something not easily achieved on a standard conference telephone call.

HCI designers are encouraged to think of communication as more than just a task to be facilitated in a functional system. The functional view cannot easily account for the proliferation of communication devices (from cell phones to pagers to various e-mail and chat programs) or for the preferences that people have for different modalities of communication. The development of communication-driven technologies needs to consider value issues.

29.3.3 Mastery-Driven Systems

If the primary value a user achieves from a technology system is the experience of mastering a particular set of skills or competing with other people, the system is a mastery-driven system. The user actively interacts with the system and the user's skill level interacts with the content to determine the quality of the experience (e.g., playing a multimedia game). The user is attempting to master a skill, solve a problem, create a strategy, or compete with other people. The value to the user is derived from mastering some aspect of the game (in other words, meeting a challenge). As the user's skill level increases, his or her perception of the value of a particular game, the *content quality*, may change. Users may have mastery-driven experiences individually or in social groups.

Research on a class of popular experience-driven applications, games, suggests that the human-computer interaction can be divided into three areas (Clanton, 1998): the game interface (the interface at the perceptual and motor level), game mechanics (the "physics" of the game, such as flight simulation, motion, virtual environments, etc.), and game play (the goal of the game and the challenges presented to a user in the course of attempting to accomplish the goal). According to Clanton (1998), game play is the most important ingredient for a good game, and the keys to engaging a user and creating a highly valued experience are as follows:

- Set the goal of the game. It must be easily stated and understood for all classes of users (novice through expert, all targeted age ranges, etc.).
- Make sure the game is easily learnable.
- Allow users to progress at their own rate.
- Provide "pressure" (e.g., enact time limits for certain functions). Pressure can be fun.

- When a user is having trouble, provide hints, not answers. It should never be too easy to achieve the goal of the game.
- Make obstacles appear harder than they actually are in order to increase the sense of accomplishment on the part of the user.

There are some obvious parallels and differences between traditional systems that have been the focus of the HCI community ("tools") and such mastery-driven systems ("toys") (Malone, 1984). In both cases, the user is engaged in goal-directed behavior, but in the game world there is no "external," user-defined goal. The goal is provided by the system. Some games allow the user to define a goal, but the goal only exists in the context of the game. We still classify such a goal as "nonexternal," meaning the user is not attempting to achieve a goal that exists independent of the game.

Other strong parallels in the design heuristics between games and tools are that they should be easily learnable and should help users when they are having trouble. The heuristics take a divergent path owing to one simple but critical difference between tools and games: The goal of the tool designer is to make the system easy to master whereas the goal of the game designer is to make the game difficult to master (but not so difficult that the player gives up in despair!). When game designers confront the issue of helping a user, they construct the system to provide "hints, not answers," an idea that is antithetical to tool designers, whose object is to get the user past a problem easily and quickly and on to the next task.

29.3.3.1 Access and Interaction

Another popular method of creating the "right" level of difficulty for a range of users with different skill and experience levels is to provide different playing levels within a game. The analog in the tool world is novice- versus expert-level interfaces. Although the two concepts appear similar, the design goals could not be more different. The game designer creates "expert" levels of play to challenge the user—to make the game difficult enough to keep the user engaged. Game players actually strive to play at the most difficult levels of play! In contrast, a tool user who is attempting to achieve an external goal will typically use the simplest method available as long as it allows efficient completion of the goal.

The notion that "pressure" adds to the value of a game-playing experience is well understood and clearly marks a difference in the design of tools versus toys. Many types of games implement scenarios in which the user has a fixed period of time to accomplish a task. The consequences for failure can be quite severe (e.g., the user is forced to start the game over from scratch). Imagine if a word processor regularly challenged a user with an artificial task and threw away the user's work if the user failed to complete the task!

An element of luck makes many games intriguing but would likely serve to enrage the user of a tool who is trying to accomplish an external task. It can be argued that humor might have a positive impact on both toys and tools, but too much humor might detract from the ability of a tool user to complete tasks, resulting in an overall degradation of the user experience. Following are obvious questions: Is humor ever appropriate in tool design? If so, what is the right level of humor? What is the right type of humor? Does the humor level of type need to change based on the individual characteristics of the user? How do HCI professionals measure and test all of this?

29.3.3.2 Context of Experience

A peek into a video arcade will provide some interesting information for anyone interested in game design. Relative to the population as a whole, the population in the arcade is younger and composed of a higher proportion of males. Does this mean that games are only popular with (or possibly only interesting to) young males? It certainly suggests that games of a certain type—those in which someone is blasting something on a mission to find or save something—seem attractive to a particular audience. This is evidence that we have some knowledge of how to provide a valuable user experience for some segment of the population at least, but we should stop short of concluding that this is the only audience interested in game experiences. Brenda Laurel (1993), among others, has looked at gender and game preferences and has concluded that there are fundamental differences, possibly genetic, in the preferences exhibited by males and females.

Purple Moon interviewed 2000 girls and found the following play patterns (Gorriz & Medina, 2000):

- Girls prefer collaboration over competition.
- Girls are more comfortable with lack of closure and enjoy exploratory games more than boys.
- Girls prefer puzzle solving more than games that involve eye-hand coordination.
- Girls enjoy more complex social interaction.
- Girls have a greater preference for well-developed characters.

Girls prefer subtle forms of competition; boys prefer scoring and fighting. Girls prefer games with narratives and adventure games that include chat and collaboration. The content of a game must be gender appropriate. Barbie Fashion Designer, released in 1996, was an enormous success because it offered gender-specific content. It single-handedly proved the viability of the girl's game market, which has been steadily growing ever since.

The research shows that there are clear differences in how males and females value games. The results of the Purple Moon study reveal interesting gender-specific patterns that can guide game design, but they only scratch the surface. What are the causes of the differences? Exactly how different are males and females on these dimensions? What is the relative weight of the various factors and how do they interact with one another? Are girls inherently less "task oriented" than boys? The science of game design is in its infancy and provides a rich area for future HCI research.

The social context of mastery-driven systems can be broken into four categories: individual, one to one, one to many, and team. Individual-use mastery-driven systems are quite common and have been around for decades. The computer game Pong is an early example of such a system. More recently, rich simulation environments such as SimCity (www.simcity.com) challenge users to master very complex interactions. The capabilities of computer-playing chess systems are well known and extremely popular. The Internet has resulted in the creation of online communities that enable one-to-one, one-to-many, and team play systems. Many "classic" games such as chess, backgammon and cribbage are now being played in one-to-one social contexts (that is, two people playing against each other, with the game being mediated by the system). Other real-world games such as bridge, spades, and hearts are being played online as team games. The online multiplayer game world has grown exponentially in recent years, with people playing one-to-one and one-to-many games such as Starcraft (www.blizzard.com).

The social context of use has had an important impact on the design of mastery-driven systems. For example, most online games provide chat capabilities. Even without built-in chat, the pervasiveness of instant messaging allows for chat conferencing as well as private channels of communication (sometimes in parallel). As bandwidth increases, we can expect to see video- and audioconferencing features become ubiquitous. The result will be the emergence of many new HCI and social computing issues. Many of these, including issues involving privacy, engagement, entertainment, and the use and availability of social cues, represent an expansion of traditional task-based design and evaluation and will fall within the purview of the HCI community.

29.4 THE CHALLENGE: MEASUREMENT TO INFORM THE DESIGN OF VALUED EXPERIENCES

An assumption carried forward into the discussion that follows is that we will look for ways in which the design of systems intended to be valuable to people can be informed by data gathered from potential users of the technology. This is the central goal of user-centered design (Karat, 1997a), and its main concern is not whether to use potential users during the design process but how to use them and how to use measurement to inform design. One question to be addressed is what kind of guidance can be given to the developers and designers of the next generation of technology to help them deal with issues other than those related productivity. At issue is whether we can find an analogue for tasks in which there is no specific goal to be achieved and for which eliminating user effort might not be desirable.

It is important to distinguish between not having clear objectives and having no clear objectives possible. Not having clear objectives is often associated with design failure (we can't design successfully because we are not sure what we want to design). It suggests a lack of effort in trying to identify objectives. For systems that people value, it is not that we are unclear about the objective, it is that we are unsure how to go about reaching it. We can learn a lot by looking. A closer examination of people and what they value in life can lead to a better sense of how to address their needs. This might be thought of as a "task analysis will prevail" view: If we look closely enough at what people think of as contributing to their quality of life, we can figure out how to measure it, perhaps creatively, in ways that will facilitate informing design in much the same way as we track productivity in design now. There is a bit of experience and a bit of blind faith in this statement. In domains in which productivity is not the primary objective, basic user centered design processes have proven useful (Vogel, 1998). There have been failures and uncertainty about how to proceed in such areas. The advice of contextual design (Holtzblatt & Beyer, 1998) is even more important in these domains, and the contributions of standards, guidelines, and heuristics may be even more limited.

How can HCI professionals measure the value of a user experience? The problem is related to how practitioners measure any sort of behavioral variable. There are aspects of human behavior that lend themselves to measurement because they can be directly observed (e.g., task completion or the amount of time spent on a task). Affective measures, such as user satisfaction or the value of an experience, are not among these. Measurement relies on indirect measures obtained through techniques such as interviews or surveys or on psychometric measures that are either known to be or can be shown to be associated with the quality we seek to measure. Just because some reactions are not easily measured does not mean that they are not real or reliable. Emotional reactions in humans are very real and they can be very reliable. There are stimuli (e.g., a laughing or a crying baby) that will produce consistent responses in the population. Like in the case of more easily measured attributes of behavior, there are individual differences in reactions or evaluations of value. Again, this variability should not be confused with randomness or unpredictability.

Although some work has been done in measuring or monitoring arousal states through various physiological measures such as heart rate or skin resistance, these will not be discussed in detail here (see Vogel, 1998). The focus is on measures that in one form or another rely on self-reports from the holder of the experience, whether obtained through questionnaires, context observations, or interview sessions. The relative merits of the various techniques will not be discussed, as there is evidence that all can be useful for measuring the value of a user experience. It is suggested that practitioners select a particular tool based on the demands of the situation and, if possible, collect data using two different methods in order to achieve cross-validation. Observing a few people in real-life settings coupled with other methods of data collection might be the cost-effective compromise in many situations.

Another place to look for knowledge about evaluating user experiences is the entertainment industry. The question of what people enjoy is certainly key here, and this industry looks not only at the content (e.g., the movies) but also at the setting (e.g., the theaters) as components of user experiences. A common technique is to screen a movie before it is released. People are invited to see the movie and are asked to provide feedback (generally by filling out a questionnaire) following the viewing. Such methods can actually provide information about the viewer's experience, in contrast to TV monitoring systems, which simply record what people are watching and draw conclusions from the data about what they might have enjoyed.

In our research on creating engaging and entertaining experiences for users on the Web, we found that focusing on the amount of time people spend with a technology system that provides experiences that people value is an insightful way to gather design information. Also very useful in the design and evaluation of these types of systems is to observe the amount of time and number of interactions that the user has with a system as well as the pattern of these interactions (Karat et al., 2001; Vergo et al., 2001).

Other domains to consider include that of complex games (like chess) and that of musical instruments, for in both of these people spend years (even decades) mastering hard-to-learn devices that might deliver little functional value. Understanding the balance between ease of use and the value of possessing a skill poses an interesting challenge to HCI researchers.

As long as the technology we are considering has some form, from a desktop system to a computer embedded in jewelry (Mountford, 1994; Picard, 1997), another factor to consider is the value of the aesthetics of the form. Visual appearance is the most obvious example, but people have also considered other sensual dimensions such as sound, touch, and even smell. That form is important is not new—functionally oriented tools are designed with appearance as a part of their usability—even if it is mostly considered in the second-class satisfaction measurements. One important point, however, is that the aesthetic elements of a system (beyond those subject to industrial engineering guidelines or brand appearance constraints) are rarely addressed in the same design tradition as usability issues. One comparison that has been made between the HCI design tradition and the visual design tradition is in the area of accountability: How do we judge if it is good? William Gaver (2000) has referred to HCI design as driven by "empirical accountability"—evaluation by data of the fit of the system to the task. He goes on to describe visual design as driven by "peer accountability"—evaluation by reference to what other experts think of the design. In some way, of course, all product design is driven by marketplace accountability—what people will eventually purchase.

There are fields with rich design traditions that are far closer to craft than to engineering (e.g., see the discussion in Schon, 1983). One point to keep in mind in discussing different approaches is that design as craft and design as data-based usability engineering are not necessarily in opposition. It is completely possible that training in "sensitivity to the form" is a partner to training in "the ability to observe" (Schon, 1983). Whether design happens in the head of one creative individual or in a team, it represents a synthesis of data observed in the world. Whether the data were explicitly sought (as in UCD) or not (as perhaps in artistic design), the designer still uses observation to drive design. There is still a great deal to be learned about what makes for good design and good designers.

A final issue that merits serious consideration is the measurement of the impact of a technology on the overall quality of life. This is an issue often considered beyond the scope of a design project because there is always an economic incentive to make work tasks easier and faster. Approaching the development of discretionary technology with the assumption that "if people will buy it, it must be good" is simplistic. HCI practitioners should debate and consider seriously quality-of-life impacts related to new technology (Kubey & Csikszentmihalyi, 1990).

29.5 RECOMMENDED TOOLS FOR THE HCI EVALUATOR

A set of tools and techniques is recommended for the evaluation of user experiences with a system. These are drawn from the wide range of techniques described in this chapter. The goal is to give some guidance on what tools to think of drawing from in planning a project. As always in the HCI field, new twists or applications might call for creatively modifying the general tools to suit the particular situation. This is not weakness in the definition of the tool sets or methodologies but is a consequence of the kind of innovative environment HCI practitioners work in.

The basic areas for evaluation are content, access and interaction, and context of experience. The assumption is that the value of an experience is a function of these—not necessarily a linear function, but at least the value is likely to increase with increases in the values of the components. Therefore, all other things held equal, we can expect the value of an experience to be better if the content is improved, the interaction and access are made more engaging, or the context is made more appealing.

Table 29.1 presents an overview of the experience component (content, access and interaction, and context of experience), the primary evaluation domain target for each component, and the high-level evaluation technique appropriate to that target. For content evaluation, subjective measures like enjoyment and engagement are primary targets. Measures used for evaluating these components include objective measures (e.g., how long people will spend watching a video when they are free to do something else) and self-report measures from interviews or questionnaires. In general, there is a trade-off between questionnaire methods and interview methods. Questionnaires generally offering the possibility of greater response numbers (they can more easily be analyzed and administered outside of face-to-face situations), and interviews offer greater depth of information about the possible reasons for positive or negative evaluations. As with all of the guidelines here, it is not possible to recommend a single best approach for all situations. Using a mix of approaches in any project is generally considered "best practice," not just a way covering lack of specific guidance.

How might HCI professionals go about evaluating the aesthetic appeal of a system and its contribution to how a system is valued? The answer here is to give advice that is not very satisfactory to usability engineers or to HCI researchers but is very familiar to individuals who have been involved in many design projects. Although the efficiency aspects of usability are

TABLE 29.1
System Category by Value Aspect Matrix

	Content Driven	*Communication Driven*	*Mastery Driven*
Content quality	Seeking ways to evaluate context is of primary importance.	Content generally assumed to be developed by users, but technology may be able to impact quality of message.	Experience of value interacts with skill level.
Access and interaction	Ease of use a secondary issue.	Explore pervasive versus context-specific use.	Challenge level can be a major factor.
Context of experience	Explore individual and group use.	Availability a major issue. Explore synchronous and asynchronous use.	Explore individual and group use.

suitable for empirical evaluation, the aesthetic aspects are not part of our known science base. Two metrics of appeal are market success and a favorable rating in a large-sample poll, but neither are particularly useful for informing iterative design cycles (though market information can be useful in narrowing initial design ideas and directions). Commonly used in design fields like graphic design, architecture, or even music composition is expert or peer evaluation (Schon, 1983). Although perhaps "unscientific," extended critical reviews by people with experience in the particular domain (often termed "crits" for short) is the accountability metric most often employed in evaluating design.

How might this be done in a usability engineering environment? Experienced designers, even when they are not experienced in evaluating interactive systems, can provide input and guidance where no other source will be directly useful. Forming a design panel and reviewing designs periodically with them is a reasonable strategy for the development of systems with important user experience characteristics. If this can be done in an atmosphere of learning and cooperation, it is more likely to succeed. This is not a provable position: The reader cannot be provided with a study of *n* design projects in which advice offered by "experts" was found to be of significant value. Rather it is based on the experience of the authors and others in the design process community (Boyarski & Kellogg, 2000) who have attempted to understand different design traditions and their applicability to the design of computational devices. One conclusion, possibly controversial, is that aesthetic-based design has more to offer discretionary-use systems than task-oriented systems.

One final area to mention involves methods of evaluating the value of a system based on its possible contribution to the quality of life. This means both how a system might enhance the well-being of an individual (e.g., through stress reduction or making life for the family better; see chap. 5) and how it might benefit society as a whole (e.g., through facilitation of worldwide communication). Such factors are often viewed as intangible and are not subject to careful evaluation either because HCI practitioners do not know how to approach the evaluation or because they don't think they can measure the benefit economically. The advice here is simply to attempt to formulate best guesses. Because many people in HCI come from behavioral science backgrounds, they have little experience in making suggestions based on evidence known to be incomplete. In the engineering context of building systems, people make judgments based on incomplete knowledge all the time. There is no deceit in this as long as one is open about the limitations of the evaluations. Taken as a suggestion, the proposition that a system might have quality-of-life value can become a topic of discussion rather than something nobody dares talk about. As computing continues to make its way into all aspects of human life, it is becoming increasingly important to take up the challenge of determining how it impacts quality of life more seriously. The behaviorally oriented usability engineer is the best person to take up the challenge and advance evaluations in this area.

29.6 CONCLUSION

Designing systems to provide users with valuable experiences differs in a number of ways from designing systems that are usable for some task set. However, are the goals necessarily orthogonal? Is it a question of either/or—either usable or entertaining? As discussed above, usability does play an important role, at least in the access to content that people find valuable. Can the role go beyond that to include making tools more entertaining (making people's experiences with them more valued) or making games more productive? This is yet another design challenge to consider, one that may become more realistic once we get a better sense of what contributes to valuable experiences in nontask systems. There is great potential in, but not a quick path to, merging the two.

Perhaps it is asking too much to demand that the science of human-computer interaction reach out beyond work on functional tools to the design of systems that people value. The skills involved in the design of aesthetically pleasing or entertaining systems might best be left to people who focus on these areas. The only problem with this is that, as our technology becomes more pervasive, the distinction between tool and valued artifact becomes terribly blurred. This raises two related questions. First, does the design of technology artifacts need to become increasingly interdisciplinary (meaning that additional skills need to be brought to it)? Second, do the people responsible for the usability of a system need to broaden their perspectives to better accommodate human enjoyment? Until the HCI field develops a good sense of what it means to design valued experiences, these questions will be difficult to answer. For now, we simply encourage human-computer interaction designers to take them seriously.

REFERENCES

Boyarski, D., & Kellogg, W. (Eds.). (2000). *Designing interactive systems: Processes, practices, methods and techniques* (Proceedings of the DIS 2000 Conference). ACM Press: New York.

Chapman, P., Selvarajah, S., & Webster, J. (1999). Engagement in multimedia training systems. In *Proceedings of the 32nd Hawaii International Conference on System Sciences* (pp. xxx).

Clanton, C. (1998). An interpreted demonstration of computer game design. In *Proceedings of ACM CHI '98 Conference on Human Factors in Computing Systems* (pp. xxx). New York: ACM Press.

Dyer, R. (1992). *Only entertainment.* London: Routledge.

Erickson, T. D., & Kellogg, W. A. (2000). Social translucence: An approach to designing digital systems that support social processes. *ACM Transactions on Computer-Human Interaction,*

Furnas, G. W. (2001). Designing for the MoRAS. In J. M. Carroll (Ed.), *Human-computer interaction in the new millennium* (pp. xxx). Reading, MA: Addison-Wesley.

Gaver, W. (2000). Looking and leaping. In D. Boyarski & W. Kellogg (Eds.), *Designing Interactive Systems: Process, practices, methods and Techniques* (pp. xx). New York: ACM Press.

Gorriz, C., & Medina, C. (2000). Engaging girls with computers through software games. *Communications of the ACM, 43*(1),

Grudin, J. (1990). The computer reaches out: The historical continuity of interface design evolution and practice in user interface engineering. In *Proceedings of the ACM CHI '90 Conference on Human Factors in Computing Systems* (pp. 261–268). New York: ACM Press.

Holtzblatt, K., & Beyer, H. (1998). *Contextual design.* San Francisco: Morgan Kauffman.

Jacques, R., Preece, J., & Carey, T. (1995). Engagement as a design concept for multimedia. *Canadian Journal of Educational Communication, 24,* 49–59.

Karat, C.-M., Pinhanez, C., Karat, J., Arora, R., & Vergo, J. (2001). Less clicking, more watching: Results of the iterative design and evaluation of entertaining Web experiences. In M. Hiro se (Ed.), *Human-Computer Interaction: Interact '01* (pp. 455–463). Amsterdam: IOS Press.

Karat, J. (1997a). Evolving the scope of user-centered design. *Communications of the ACM, 40,* 33–38.

Karat, J. (1997b). Software valuation methodologies. In M. Helander (Ed.), *Handbook of human-computer interaction* (pp. 689–704). Amsterdam: North-Holland.

Karvonen, K. (2000). The beauty of simplicity. In *Conference on Universal Usability (CUU 2000)* (pp. 85–90). New York: ACM Press.

Kubey, R., & Csikszentmihalyi, M., (1990) *Television and the quality of life.* Hillsdale, NJ: Lawrence Erlbaum Associates.

Kuhn, T. S. (1996). *The structure of scientific revolutions.* Chicago: University of Chicago Press.

Landauer, T. K. (1995). *The trouble with computers: Usefulness, usability, and productivity.* Cambridge, MA: MIT Press.

Langer, S. K. (1954). *Feeling and form.* New York: Scribner's.

Laurel, B. (1993). *Computers as theatre.* Reading, MA: Addison-Wesley.

Malone, T. (1984). Heuristics for designing enjoyable user interfaces: Lessons from computer games. In J. C. Thomas, & M. L. Schneider (Eds.), *Human factors in computer systems* (pp. 1–12). Norwood, NJ: Ablex.

Malone, T., & Lepper, M. (1987). Making learning fun: A taxonomy of intrinsic motivations for learning. In R. E. Snow & M. J. Farr (Eds.), *Aptitude, learning and instruction: Cognitive and affective process analyses.* Hillsdale, NJ: Lawrence Erlbaum Associates.

Mountford, J. (1994). Constructing new interface frameworks. In B. Adelson, S. Dumais, & Olson, J. (Eds.), *Proceedings of ACM CHI '94 Conference on Human Factors in Computing Systems* (Vol. 2, pp. 239–240). New York: ACM Press.

Nardi, B., & O'Day, V. (1999). *Information ecologies: Using technology with heart*. Cambridge, MA: MIT Press.

Picard, R. (1997). *Affective computing*. Cambridge, MA: MIT Press.

Pinhanez, C., Karat, C.-M., Vergo, J., Karat, J., Arora, R., Riecken, D., & Cofino, T. (2001). Can Web entertainment be passive? In *Proceedings on Multimedia and the Web* (pp. xxx).

Putnam, R. (2000). *Bowling alone: America's declining social capital.* New York: Simon & Schuster.

Resnick, P. (2001). Beyond bowling together: Sociotechnical capital. In J. M. Carroll (Ed.), *Human-computer interaction in the new millennium* (pp. xx). Reading, MA: Addison-Wesley.

Schon, D. (1983). *The reflective practitioner*. New York: Basic Books.

Vergo, J., Karat, C., Karat, J., Pinhanez, C., Arora, R., Cofino, T., Riecken, D., & Podlaseck, M. (2001). Less clicking, more watching: Results from the user-centered design of a multi-institutional Web site for art and culture. In *Museums and the Web 2001: Selected papers from an international conference* (pp. 23–32).

Vogel, H. L. (1998). Entertainment industry economics (4th Ed.). Cambridge: Cambridge University Press.

Webster, J., & Ho, H. (1997). Audience engagement in multimedia presentations. *The DATA BASE for Advances in Information Systems, 28*(2), 63–77.

Wroblewski, D. (1991). Interface design as craft. In J. Karat (Ed.), *Taking software design seriously: Practical techniques for HCI design* (pp. xx). New York: Academic Press.

30

Wishing on a sTAr: The Future of Task Analysis

Dan Diaper
Bournemouth University

Neville A. Stanton
Brunel University

Today, task analysis is a mess. The chapter considers what might be the future of task analysis and how it might be improved. It is suggested that there is a need for greater agreement about theories, concept, methods, notations and nomenclature across the many different, current approaches to task analysis. Although some form of meta-method is proposed as a possible way forward, the desirable things that current task analysis methods provide collectively is a long and daunting requirements list. One area that we think requires immediate attention is more agreement on how task analysis should represent alternative ways of doing tasks, and parallel and collaborative ones as well. Radically, the chapter suggests that the whole concept of goals should be abandoned as unnecessary and thus one of the major sources of confusion about task analysis would be removed. The main end users of task analysis methods should be those who work in the commercial computing industries and their requirements, it is argued, have generally not been adequately considered by those who have been involved in the development of task analysis methods. In summary, as school teachers are wont to write on their pupils' reports "promising, but could do better."

30.1 INTRODUCTION

People will always be interested in task analysis, for task analysis is about the *performance* of work (chap. 1). Less certain is whether it will be called task analysis in the future (Diaper, 2002a), or even whether what we have today will survive in some recognizable form. What we have today is a mess.

Chapter 6 identifies as a current shortcoming of task analysis a "Lack of understanding of the basic contents of each task model, including the rationale behind the analysis method, the concepts, their relationships, their vocabularies, and the intellectual operations involved." In summary, nothing is well understood about task analysis, and one of the consequences, as chapter 6 points out, is that analysts, methods and computer-aided software engineering

(CASE) tools are isolated from each other intellectually and in practice. One remedy suggested by chap. 6 is to let task analysis methods vary in "focus," by which the authors mean the size of the world of the task model, which in some cases should encompass "parameters external to the task, yet relevant to the context of use."

Before one attempts to address current problems with task analysis, however, it is surely worth asking, is task analysis really a field, is it a general subject area in which some people can claim competence, or is it just a collection of separate HCI approaches that have been arbitrarily thrown together? One reply is that no one suggested it is not a field when this handbook was advertised on various human-computer interaction (HCI) and ergonomic e-mail distribution lists, nor has it ever been suggested to the editor of this handbook's predecessor (Diaper, 1989a). Furthermore, task analysis is not an uncommon topic in HCI conference calls for papers, and, for example, the journal *Ergonomics* recently had a special issue on task analysis (Annett & Stanton, 1998). Long (1986) argued that defining a discipline involves some degree of agreement among its practitioners as to common goals, problems, and approaches. There would seem to be a consensus in the HCI community about task analysis, but social agreement is not sufficient for science and engineering, which require some theoretical basis, supported by empirical evidence, to justify a taxonomic system that makes task analysis a distinct thing within HCI. The differences in the taxonomic systems presented in chapters 6, 22–25, and 28 demonstrate that, despite the general agreement that task analysis is a distinct field, there is plenty of disagreement about exactly what it is.

Even if one could identify a common theoretical basis underlying all the work that purports to be on task analysis, the task analysis community would still need to agree on it. A standardized vocabulary might aid in the achievement of such agreement, but it is the acceptance of the theory that is critical. Perhaps one reason for the failure of previous calls for HCI to agree on a vocabulary, such as by Thomas Green in 1991 (reported in Diaper & Addison, 1992a), is that agreement on the semantics for a revised vocabulary (the theories on which the terminology is based) is lacking. Stammers (1995) made a similar point in describing the lack of development in the field of task analysis. So far the various standards bodies do not seem to have succeeded in imposing a standardized vocabulary for HCI either, perhaps for the same reason. Furthermore, even if general agreement could be obtained about task analysis and the meanings of its technical terms, there are still several directions in which task analysis could progress, and these ultimately depend on the use made of task analysis in the commercial computing industries. Thus it is also necessary to consider the "delivery-to-industry" issue (e.g., Diaper, 2002b) when contemplating the future of task analysis (section 30.3).

30.1.1 Predicting Futures

Many people seem to believe in *the* future. The sorts of vastly complicated systems that are of interest in HCI work are almost certainly all chaotic (Lorenz, 1993; for a nonmathematical introduction to chaos theory, try Peterson, 1998, chap. 7) in that occasionally some small change in the current state will lead to large, subsequent state changes. Similarly, predicting *the* future of task analysis is equally impossible, but predicting *futures* is not. Diaper (2002c) proposed that people think by envisioning scenarios (chap. 5), whether they are primarily perceiving the current world, recalling some past state, or contemplating the future or some other fantasy. Using their Process Logic–based approach (Diaper & Kadoda, 1999), forward scenario simulation (FSS), a.k.a. the human psychological processes involved in predicting future states, starts with a description of the current state of the world (see chaps. 1 and 7) and tends to produce clusters of similar scenarios of future states.

Starting with the mess today, when virtually nothing is agreed on, and the practitioners and researchers, the methods and CASE tools, and the results from task analysis are isolated, we

TABLE 30.1
Future Scenario Clusters

		Task analysis methods integrated	
		No	Yes
Agreed theory and vocabulary for task analysis	No	① No change	④ No explicit task analysis
	Yes	② Separate methods but a common terminology	③ A new, inclusive task analysis

Note. Four clusters of simulated future scenarios for task analysis organized post hoc by whether an agreed theory, vocabulary, etc., for task analysis emerges and whether task analysis methods become more integrated in the future.

think that there are four clusters of possible futures for task analysis (Table 30.1). This table separates these using two concepts, induced after a FSS exercise, that might help categorise the possible futures for task analysis. The concepts are: (a) whether an agreed theory and vocabulary does become established and (b) whether greater integration among the various task analysis methods will occur.

Cell 1 in Table 30.1, represents the situation where nothing really changes. That is, a few new methods and tools will be developed, but each approach to task analysis will have its own clique of adherents with their own specialized theories, ways of working, vocabularies, and so on. Cell 4 represents the situation where there is no agreement about the basis of task analysis but somehow task analysis becomes more integrated. In this situation, task analysis would likely cease to stand as a distinct discipline and would never be chosen at the main topic of any book, journal issue, or conference session. Instead, it would probably be absorbed into HCI or some other overarching discipline yet to be developed. It can't actually disappear in practice, of course, as it is fundamentally about work performance. The future scenarios represented in cell 4 may actually be considered highly desirable ones for task analysis. If the term task analysis is lost because task analysis itself is entirely integrated with all HCI activities, isn't this an ideal outcome from the perspective of today's champions of task analysis?

Cells 2 and 3 represent possible future scenarios in which some measure of agreement about theories, vocabularies, and so on, is achieved. In a cell 2 scenario, there is agreement about general theory and vocabulary, but the task analysis methods remain separate and would probably be selected for different types of computing projects or in different circumstances. One plus would be that producing a classification of methods and tools would become easier (and undoubtedly a new classification would replace the various ones offered in this handbook, such as occur in chapters 6, 22–25, and 28). In addition, selecting an appropriate method, most importantly by industry practitioners, would become easier as well.

Cell 3 is the task analysis researcher's dream. Your new theory, methods, notation, terminology, and so on, are widely accepted, and all the current approaches to task analysis are entirely subsumed without loss of analytical power, scope of application, and so on, within the new task analysis, which is easy to understand and apply.

Missing from Table 30.1 is any indication of how much time the scenarios will need to be realized. This may not be a meaningful issue, as cell 1, for example, represents a continuation of the status quo. A great number of timelines for the four clusters of future scenarios are possible. There are 17 plausible ones. At one end of the range of possibilities, a scenario from each cluster could come about by itself without involving any of the others. This is obviously true for cell 1, but it is equally easy to envisage the loss of explicit task analysis (cell 4) and the development of common theories, and so on, (cells 2 and 3). At the other end, scenarios might follow each other in the same order as the cells: (1) Nothing changes for a while, then (2) analysts agree on a common theory, vocabulary, and so on. This leads to (3) a new, fully integrated approach to task analysis that replaces all previous approaches. The approach is so wildly successful that (4) task analysis is completely integrated into all HCI work. All the other plausible sequences also follow this numeric order, except that in some a cell 4 scenario precedes a cell 3 scenario, as would occur if task analysis as we know it today vanishes and then reemerges some time later. Sequences in which a cell 4 scenario precedes a cell 1 or 2 scenario are not plausible, as task analysis, once lost, would be unlikely to be resurrected and return to its current state, not even with a new terminology and theory (a cell 2 scenario).

The FSS approach using the Process Logic is multi-teleological (chap. 1) in that, rather than having a single future state as a goal, it advocates designing processes that will change the current state of the world so that some future scenarios that are judged, by an evaluation process, as more desirable than others have an increased chance of happening. Importantly, the future is not something that is fixed and that we are trying to predict but a theoretical set of clusters of scenarios that are more or less well specified and that, by being foreseen, allow our current actions to make some possible futures more likely to become, for an atomic instant, reality.

The "paperless office" is an example of a prediction that shows just how wrong predictions can be, although Sellen and Harper (2002) suggest that it was the mass media of the mid-1970s who, wanting a "big story" about the computer revolution, made this prediction popular, not the researchers of the time, who "might not have thought they were in the business of inventing the paperless office" (p. 5). Sellen and Harper describe a two hundred year technological history behind the idea of the paperless office and analyze the roles of paper in knowledge work, reading, and collaborative work. Although task analysis is not listed in the index of their book, Sellen and Harper's approach is task analytic, for they consider what people do with paper, how they use it, and so on. Their sophisticated analysis of the paperless office scenario well illustrates the problem of predicting one simple scenario that is judged desirable and then attempting, and failing, to push the necessary technological development to achieve it. The goal of a paperless office at least had the advantage of providing a specific future whose achievement could subsequently be measured, which Sellen and Harper (2002) do using a variety of statistics concerning paper production, sales, and use. Indeed, they show "that paper manufacturers can take heart; paper consumption will not wane any time soon" (p. 209).

The history of IT over the last few decades is littered with predicted futures, usually tied to some "latest bandwagon" technology such as expert systems, hypertext, groupware, neural nets, and so forth. Process Logic provides a radical alternative to trying to predict *the* future and then find some means to make such a future come about, because it concentrates on the processes by which the current world can be changed so as to make the current state, sometime later, better than it might otherwise have been predicted to be.

The purpose of this chapter is to encourage debate within the task analysis community. Its authors believe that there is something identifiable as task analysis; that it is worth preserving as something separate from other HCI approaches, at least in the near future; but that change in the area of task analysis is desirable.

30.2 WHAT IS TASK ANALYSIS?

It is perhaps interesting that only chapter 1 attempts an extensive definition of task analysis; most chapters offer only a sentence or two about the nature of the beast. Does this indicate that everyone agrees what task analysis is? Or is fashioning a comprehensive definition too well recognized as such a hard problem that it is best left alone, a "can of worms" not to be opened? If indeed the chapters in this handbook constitute a reasonably fair sample of current work in task analysis, the answer to the first question must be no. Although the abbreviated definitions of task analysis offered in the chapters do share a superficial resemblance, the basic things modeled in the various approaches to task analysis indicate that task analytic approaches fall into two major classes depending on their "focus" and on whether they adopt a broad or a narrow definition of HCI. Ideally, the narrow focused approaches (that is, those that are user interface design oriented) should be a subset of the more general ones, which should cope with user interface design problems as well as other types of problems by training, help systems, organizational and business-process reengineering, and so on. Hierarchical task analysis (HTA; chap. 3), for example, is a method that can be applied to a broad range of HCI problems, not just those that arise in user interface design.

30.2.1 Three Axioms for Task Analysis

An axiom is a theoretical assumption that cannot be questioned from within the closed epistemological space that a set of axioms creates. Such an epistemological space, however, can contain many alternative models of the world that share the identical set of axioms but differ in other ways. Furthermore, although justifications for selecting an axiom can be made, they can only be made logically from within a larger axiomatic space, which leads to recursive problems, making it common for axioms to be given and then accepted by the experts concerned as so self-evidently true that no further justification is necessary. On this basis, this chapter proposes the following three axioms and invites its readers to dare to disagree with them:

Axiom 1: The purpose of task analysis is to solve problems.
Axiom 2: All task analyses are concerned with performance issues.
Axiom 3: Keep it as simple as possible.

Note that we do not claim that this list is complete (that it is sufficient) but we merely challenge readers to suggest that these are not true and utterly necessary.

Axiom 1 has a basis in the history of HTA. Professor emeritus John Annett, who is one of HTA's inventors (e.g., Annett & Duncan, 1967), makes the case for this axiom again in chapter 3. Furthermore, on the assumption that HCI is an engineering discipline and not a science (Diaper, 1989b, 2002b; chap. 1), its purpose had better be to solve, or at least contribute to solving, problems, as it would not seem to be good for much else.

A strong case for Axiom 2 is presented in chapter 1. Only one chapter in this handbook, chapter 20, does not deal with some aspects of how tasks are performed, and Chapter 20 is a special case, for it discusses an input device so novel that even the human physical task elements are not understood. When the Chapter 20 authors' work develops to the point that the task elements, both physical and intangible ones (e.g., mental or software processes), can be sequenced, eventually in real time, then this work should be properly task analytic and concerned with performance rather than only with classification. Perhaps the essential point about performance problems, which may involve measuring failure, errors, task alternatives, or time, is that they only occur when the system being modeled is operating, that is, when events are occurring in time. In other words, performance problems always occur in a context in which some things are changing.

Axiom 2 does not preclude task analysis involving the simulation of performance, which could be of very low fidelity if a static, nonperformance model of the world is run in the analyst's mind (chap. 1). Many uses of scenarios (Carroll, 2000; Diaper 2002a; chap. 5), for example, concerning ideas about a future system, can only be low-fidelity simulations because the future system exists only in the written scenario and in the analysts' minds.

Axiom 3 encapsulates the numerous issues that need to be considered to get task analysis methods more widely used in the commercial computing industries. This is discussed further in section 30.3.

30.2.2 Reinventing Task Analysis

If we started with a *tabula rasa*, a blank slate, and had to invent task analysis for HCI, what would we produce? Obviously we would start with the three axioms (section 2.1) and an extremely broad view of HCI that would involve everything to do with people and computers (chap. 1), as a broad definition of HCI can't do any harm and should not prevent users of our new version of task analysis from working in specialized aspects of HCI, like user interface design.

The role of this chapter is not to solve the current problems with task analysis but to set the stage for the whole task analysis community to get their act together. In any case, when the authors attempted to reinvent task analysis themselves, they simply produced a version of Systemic Task Analysis (STA; chap. 1), which is hardly surprising, as STA is still under development by the first author.

We think that what is needed, ideally, is the development of a meta-task analysis theory and method under which all the existing task analysis methods and notations can be subsumed, as also proposed in chapter 28. What must be preserved is the vast range of applications and purposes, the flexibility of task analysis use, the ability to use many sources of data, and the ability to vary the scope and level of detail. There must be room, in other words, for extra-lite-weight methods, such as User-System Interaction Scripts (USIS; chap. 10), that can be used by the HCI naive as well as highly detailed, thorough methods, such as Task Object-Oriented Design (TOOD; chap. 25) or ConcurTaskTrees (CTT; chap. 24), that require considerable expertise and are supported by sophisticated CASE tools. It is also necessary to preserve the theoretical scope of task analysis, particularly with respect to its psychological models, so that both sophisticated cognitive models such as Interacting Cognitive Sub-systems (ICS; chap. 15) and very simple psychological models (chap. 28) can be accommodated. Finally, task analysis should better integrate with software engineering methods than hitherto, notwithstanding the successes reported in chapter 7 and in parts II and IV.

Our idea is that a meta-task analysis method would allow each current task analysis method to be broken down and the components of the method, its theoretical constructs, and so on, would be linked to the meta-task analysis method. Chapter 6 uses the common notation of entity-relationship-attribute (ERA) modeling to show how nine well-known task analysis methods differ. One option for the development of a meta-method might be to produce a single ERA model that covers these and other task analysis methods and then use a reductive method to simplify the model. The reductive method might be based on traditional relational database design, which is the most common use of relational models in software engineering, or it might use some other method, more akin, for example, to the task layer maps approach (chap. 11). One can see a promising start towards an ERA meta-model for task analysis in the nine diagrams in Chapter 6 dealing with "task" ("operation" in Diane+; see also chap. 26) in that each entity has a relationship to some form of task decomposition. A great more would need to be done if this approach were to succeed, and people may be more comfortable with a different style of meta-model.

If a meta-task analysis theory and method could be developed and become widely accepted, then an agreed terminology might be possible, but, as suggested in this chapter's introductory section, the agreement on a theoretical framework must precede the agreement on terminology if widespread acceptance is to be likely. It is for this reason that we abandoned any attempt to impose a standardized terminology on the chapters of this handbook.

Naturally, Axiom 3 must not be forgotten, and the meta-task analysis method itself needs to be designed so that it is reasonably easy to understand, and perhaps even use, by people who are not task analysis experts, whether members of computing project teams, users, clients, managers, or other involved people. One way to help ensure the meta-method meets such comprehensibility and usability criteria would be to design it to have levels of abstraction, perhaps arranged hierarchically, so that it could be decomposed to the required level of detail. There are, of course, other ways of keeping the meta-method as simple as is possible.

30.2.2.1 Representing Task Sequence and Time

One of the current tensions in task analysis, noted in chapter 1, arises from the issue of whether a task analysis should represent specific task instances or a single, generic, ideal task. The problem with the latter strategy is that either the task analysis represents a single way of doing a task, and can thus be at any level of abstraction, or it represents a task doable in a number of different ways but the analysis is at a relatively high level of abstraction and cannot be reduced to the lower levels needed for detailed software design work.

If a task analysis models specific task instances, as would occur in the evaluation of users' performance with a prototype, it is difficult to combine the different ways that individual users complete the tasks. The problem with an activity list (also sometimes called a task protocol or interaction script; chap. 1) representation of a task as a sequence of behaviors or goal states is that the users will produce activity lists that differ in their behaviors or goals and in the order of occurrence. Although Diaper (2002a) was mildly critical of Carroll's (2000) method of dealing with such alternatives, which involves just noting them at the bottom of an interaction script, at least Carroll recognizes the problem, unlike most of the chapters of this handbook that deal with the analysis of specific task instances.

The advantage of activity list–like representations is that task sequence or time is represented on one of the two dimensions on a page, generally downwards. One way to represent task alternatives is to have columns of activity lists with conditional branching across the columns. The problem with this approach is that, if sequence or time continues to be represented on a physical dimension, one can get ludicrous repetition in cases where subtasks can be done in any order or some things can be done at any time during the task.

For example, when setting up the printing of a document on the first author's PC, which for special printing jobs may involve half a dozen or so button clicks across several temporary windows, it does not matter when the printer is turned on, provided it is before the print execute command is issued. Waiting to turn the printer on until after the command, however, has serious consequences, as, owing to a bug, the only way to reenable the printer software is to reboot the PC, which takes about 5 minutes. The big problem is how to represent turning on the printer when it can be done by the author anywhere in the task. Furthermore, the subtasks using the printer drivers' temporary windows to select special options can be done in any order, with no difference in effect, before the main task of issuing the print command is completed by the user. Even here, there is not an agreed way of representing the alternative subtask sequences in an activity list format.

The other alternatives require abandoning using one of the physical dimensions in the model to represent time or sequence. Task models that use this technique will basically look like a flowchart irrespective of the actual notation used. Most obviously, subtasks are represented as entities on such a model, and these are linked via arrows and decision boxes so as to represent alternative traversals of the network so produced. A complaint against HTA's diagrammatic representation is that it mixes a spatial layout representing sequence with a second one involving plans that allows the sequence of a particular task to dance around the diagram sometimes. This may facilitate HTA's flexibility, but there must be a cost in using a mixed notation to represent one of the most fundamentally important things about tasks.

There is a place in computing projects for idealized task models, usually at some relatively high level of abstraction, that do not represent task sequence alternatives, as such models are reasonably device independent (Benyon, 1992a, 1992b; Benyon & Macaulay, 2002; Diaper 2002c; Diaper & Addison, 1992b) and can therefore support high-level design. Such models can also be used for evaluation by identifying where real task performance deviates from ideal performance, which is the approach of the task analysis for error identification (TAFEI) method (chap. 18). They fail, however, when tasks can be accomplished in a range of different ways that do not differ significantly when measured in terms of either performance effectiveness or efficiency. Air traffic control (ATC; see chap. 13) is a domain where tasks are brief, frequent, and, at a high level of abstraction, routine. At a lower level of analysis, however, they are often far from routine in that there are numerous exceptions, special circumstances, and so on, requiring the ATC officers to make up and continuously modify plans as they go along, which is in line with the psychology proposed in chapter 18. In such circumstances, there are many possible sequences of tasks and subtasks, and the existing idealized task models cannot cope with the complexity.

We do not think it is necessary for the task analysis community to wait to come to some more general agreement about task analysis, if it ever does, before it addresses the issue of representing alternative task sequences and also parallel, related and unrelated, and cooperative tasks. We think this issue is one major part of today's task analysis mess and thus a major source of confusion about task analysis. Related to this is the more detailed but still essential identification of pre- and post-task conditions and that these are not the same as triggers or interrupts (chaps. 14 and 19). One concern we do have, however, is the effects of any such changes on the currently available CASE tools, which may need significant reprogramming.

30.2.3 Goals and Functionality: Two Shibboleths

Why is the concept of a goal necessary in task analysis? And do we need the concept of computer system functionality? See chapter 1 for some discussion of both these questions. Without wishing to be radical just for the sake of it in a future looking chapter, we believe that task analysis would be much simpler and easier to understand if we abandoned both the goal and functionality concepts. The issue, of course, is whether we would lose anything in such conceptual abandonment.

30.2.3.1 Functionality

The case against functionality is relatively straightforward. It is a concept only applied to computer systems and not to the other main type of agents in HCI and task analysis, people (i.e., we do not usually discuss the functionality of a person, although we might discuss the functionality of an abstract role). The concept of functionality as what application software is capable of doing to transform its inputs into its outputs may well have been a reasonable one in computing's infancy, when programs were small and did very little, but today computer systems, particularly when networked and/or involving some artificial intelligence (AI), have outgrown the utility of the functionality concept. We believe that future computer systems will become further empowered as agents and that either we should happily apply the functionality concept to both human and nonhuman agents or drop it completely and use the same concepts we use for people when addressing the behavior of complex computer systems. Chapter 1 (see also chap. 26) illustrates this approach by allowing that goals, which throughout most of this handbook are characterized as possessed by people, can be possessed by other things, such as computer systems and organizations.

30.2.3.2 Goals

What are goals for in task analysis? Chapter 1 identifies two roles for goals in task analysis: to motivate behavior and to facilitate abstraction away from specific tasks and thus promote device independence. Concerning the first, chapter 1 also rails against the mono-teleology of nearly all task analysis methods—the idea that a specific human behavior is caused by a single goal rather than a rich, ever shifting melange of goals. Whereas the first role is generally restricted to goals possessed by people, the second role can be played by goals possessed by nonhuman things as well (chaps. 1 and 26).

Throughout the history of cognitive psychology, psychologists have frequently borrowed the basis of their models from other areas of science and engineering. Since the 1960s, most of the borrowing has been from computing (Diaper, 1989c), but an earlier loan, famously exemplified in Freudian psychoanalysis, was from hydraulics, which was one of the most advanced of the Victorian engineering technologies. The basic idea is that there is some sort of psychological energy that can flow, be blocked, diverted, and so forth. Goals as motivators of behavior would seem to be a part of this type of psychological hydraulics. Given that there is no empirical evidence of any physical substrate that could function in such a hydraulic fashion, perhaps we do not need the concept of goals as behaviour motivators. The alternative claim is that people and other complex things behave continuously, as even physically doing nothing is a behavior, and that there is no need to posit some special thing that causes, somehow energizes, behavior. This is essentially a chaos theoretic view, for it treats the mind as being no different, in essence, from complex systems that exhibit organized behaviour but do not have goals. An example would be a natural system such as the weather (unsurprisingly, Lorenz, the father of chaos theory (1993), is a meteorologist). On this view, people, including task analysts, merely use goals as a post hoc explanation of system performance. Furthermore, attention, both selective and divided (see chaps. 14 and 15), is an existing cognitive psychological mechanism that controls the allocation of the massive, but still ultimately limited, mental information-processing resources, and we think that attentional mechanisms can entirely replace the motivational role of goals without resorting to the physically dubious hydraulic metaphor.

It became clear in the 1960s from a wide range of empirical research in social psychology that explanations of people's behavior were most frequently based on a post hoc justification of their previous behavior. For example, Festinger's (1957) cognitive dissonance theory was tested by Brehm (1966; reviewed by many, including Reich & Adcock, 1976). In his study, subjects subsequently rated a household item that they accepted in payment for taking part in the ratings experiment more highly than an item they had rated earlier and then rejected as payment. The general explanation of such post-choice dissonance effects is post hoc justification (i.e., people like a thing more after they have selected it, because they have selected it and vice versa for rejected things). In agreement with the explicit psychological theory in chapter 18, we suspect that goals in task analysis are similarly post hoc, whether elicited from task performers or inferred in some other way by task analysts. That is, we believe that goals are not part of the psychology that causes behavior but are used to explain it. If we are correct, then this is another reason for abandoning the concept of a goal in task analysis. We can only confuse ourselves by explaining behavior by means of something that is not actually there in any such form during behavior. Again, we think that the concept of attention is prefable, for attention is a mechanism, and, whatever the form of the cognitive mechanism, the empirical evidence for it in human task performance is overwhelming.

Extending the concept of a goal to allow goals to be possessed by things other than people (chaps. 1 and 26) also will not save the concept. Although Annett, Cunningham, and Mathias-Jones (2000) suggested that a naval warship (viewed as a system and not a mere lump of steel) possesses high-level goals such as float, maneuver, and fight and then proposed that such a system's goals will remain even if the human goals involved are fluid, these goals are just as plausibly the post hoc rationalizations of the task analyst rather than real things that actually determine whether a warship continues to achieve, moment to moment, these claimed goals. It is important to note that this is not a general argument against all emergent systemic properties (Patching, 1990; chap. 1) that do not have a physical substrate to support them. For example, we are still happy for a system to be described as honest or not. The distinction being made is between things like honesty and goals. Honesty is a property of a system, however judged, whereas goals as human behavior motivators are here being used as a mechanism that cause rather than merely explain behavior.

The suggestion that the concept of a goal should be abandoned in task analysis is not new. Diaper and Addison (1991) supported Atkinson's (1988) philosophical behaviorism (chap. 1) and in their task analysis approach used the physical behaviors of users to stand for relevant but unknown psychological entities, including goals. Diaper and Addison claimed that nothing was lost in their task analyses by doing this, although Diaper (2001a) noted that there has been little uptake of their idea within the general task analysis community. Chapter 1 suggests that one cause of confusion, notably in HTA but not uncommon in other task analysis methods, is that the language used to describe goals is hardly if at all distinguishable from that used to describe physical behavior. It is suggested that what is needed are clearly separate terms for describing the physical and the psychological. On such grounds, Diaper (2002c) proposed that all thought can be considered as envisioned scenarios and can provide a psychologically based language separate from that used to describe behavior. In this psychological model, there is no need for the representation of goals; instead, the model posits attentional mechanisms that devote cognitive processing resources to bias a person's currently envisioned scenario toward perception, memory, or fantasy.

One concern regarding the language used to describe goals is that the vaunted abstraction obtained by specifying a goal rather than what to do to achieve it is often and perhaps always illusory and that task-specific, device-dependent, premature design commitments are unconsciously made by analysts (Benyon & Macaulay, 2002; Diaper, 2002a). What we think is common is that, although many people claim that they are dealing with goals in their analyses, really they are dealing with descriptions of behavior, as Diaper and Addison (1991) made explicit. We believe that it is entirely possible to have high-level abstractions of tasks that are not couched in terms of goals and that these are more honest as well as easier to deal with. For example, a high-level task description such as "create a new document" could be claimed to be a goal specification, but it could equally be a high-level description of behavior that can be decomposed into more detailed descriptions of behavior without becoming utterly device dependent. For example, "create a new document" might decompose into "choose type of document"; "trigger the creation of the new, empty document"; "name document"; "open document." Clearly, at this level of decomposition, the descriptions are not completely device dependent, as on most computer systems, for example, there are alternative ways of naming a new document.

More recently, Diaper and Kadoda (1999) argued for goal abandonment, along with targets, aims, intentions, and so on. As a general logic, the range of applications to which the Process

Logic can be applied is vast, if not total. It replaces the specification of some future state, a goal, with a model of the current world and a set of processes that operate on the world. In a process thinking approach, goals are effectively abandoned, although for those people who can't leave the entrenchment of goal-orientated, Western thought, then goals become extremely general, abstract things (e.g., "Make the world a better place"). One class of processes are those involved in FSS, and another class of processes evaluate the relative merits of the envisioned future scenarios. With this approach, planning to achieve some future goal state is replaced by planning how to change the current processes that operate on the world so that desirable simulated future states are more likely to come to pass than less preferred ones (section 1.1).

To exemplify the use of the Process Logic, first consider Norman' (1986, his fig. 3.4) example of a document, a letter, that currently does not "look right" to its author and so induces a Level 1 intention (a type of goal) in the author to make the letter "look better" and consequently a lower level intention to "block paragraph" to achieve the higher level goal. Norman's model then goes through some lower level mechanisms that become increasingly specific to task performance. Once the task is executed, the results are perceived by the person and then evaluated against the expectations, which were also generated by the intentions. Like Norman's model, the Process Logic starts with a model of the current world and evaluates it. Unlike Norman, the Process Logic, and using Diaper's (2002c) "all thought is just envisioned scenarios" hypothesis, evaluates the current state of the letter by FSS. That is, if the letter is sent in its current state, the recipient's reaction will probably be X, Y, or Z or a combination of these, and these consequences are not judged by the author as equally probable or desirable, individually or collectively. Where Norman and the Process Logic diverge is that the latter then attempts to establish a range of processes (behaviors, actions, mechanisms, or whatever one wishes to call them) that can change the current situation (e.g., alter the letter so that it is more likely to look better later to its recipient). Norman's model magically produces a possible solution, a single "better" future state, a goal, an intention to "block paragraph," which might be a process selected from a range of processes considered by the Process Logic, but it is not in any sense a goal within Process Logic. Instead, it is a process to be executed based on the prediction that it will improve the world more than some other process or combination of processes. The subtle (or blatant) difference between Process Logic and traditional Western approaches, including Norman's, is that the effort spent on identifying specific desirable future states, a.k.a. goals, is replaced by an effort on the processes to change the current state. Perhaps the ultimate raison d'etre of the Process Logic is its avoidance of the utopians' fallacy, of knowing what a better future world would be like but not having the means to achieve this idyll. The whole emphasis of the Process Logic is on means, applied to the current world, not ends. Admittedly the Process Logic is still being developed, but it does demonstrate that alternatives are possible that do not need the concept of a goal.

In summary, goals as motivators of behavior are an unnecessary device based on a discredited hydraulic theory of the mind used to fallaciously explain behavior that will occur in any case. Such goal orientated explanations are generally post hoc in that they are not how the mind works, but only a way of explaining it. Goals are often confused with descriptions of task behavior, and because at some point task analysis must be able to describe behavior, the issue is whether goals are a necessary abstraction mechanism. We say they are not and that other mechanisms are available to task analysis. Note that we are in no way advocating ignoring a massive and necessary contribution from psychology to task analysis, we are merely recommending abandoment of the concept of a goal, whether applied to people or to any other aspect of systems.

In conclusion, task analysis does not need the concept of a goal, and nothing will be lost if the concept is abandoned entirely. If we do abandon it, task analysis will be very much simpler and thus easier to understand and use, and the world will indeed become a better place, in the future, we think.

30.3 DELIVERY TO INDUSTRY

There must be a smidgen of truth in the view that most of the major task analysis methods have been developed by academics and are unsuited for the frenetic world of the commercial computing industries, which are still failing to cope with the human productivity problem, a.k.a. the software crisis (chap. 1), and whose projects are always constrained by resource costs and availability and a severe shortage of time. Chapters 8 and 9 provide graphic examples of such situations. The journal *Interacting with Computers*, since its inception in 1989, has advertised for papers about failures, and Hewett (1991) argued in an editorial for the importance of gaining an understanding of failures in engineering of all sorts, an understanding, he suggested, that is still absent from software engineering and HCI. Indeed, year after year, virtually no papers in this category are submitted to the journal.

Task analysis methods are a case in point. They are like old soldiers that "just fade away" rather then being reported on and evaluated, preferably in commercial applications, and perhaps sometimes rejected, if not severely modified, when they fail in whole or in part. The one exception is Task Analysis for Knowledge Descriptions (TAKD), which its academic coinventor and main champion since the beginning of the 1980s has recommended to be abandoned as a contending task analysis method (Diaper, 2001a). There was much that was less than ideal about TAKD, even though it was an extremely powerful and flexible method and better specified than most others (e.g., Diaper, 1989d). At the root of nearly all its problems was that it was just too complicated, even when supported by a CASE tool (LUTAKD). We doubt that if any of the other major task analysis methods, such as those discussed in the taxonomies of chapters 6, 22–25, and 28, would stand up to a similar level of expert, detailed, critical scrutiny. The problem, of course, is that researchers and their backers have invested great amounts of time, effort, money, and other resources in their particular approach to task analysis, and people's reputations and careers, as well as the work itself, are at stake if serious shortcomings and failures are admitted.

A suspicion that may also have its smidgen of truth is that methods in software engineering and HCI and in task analysis are not developed by expert method designers. Do such experts exist? We doubt it but feel that there is a need for the requisite expertise because methods are very complicated. Diaper and Kadoda (1999; see also Diaper 2001a) identified a major difficulty with describing methods: They involve describing processes, and most languages, along with virtually all of Western thought, are inherently declarative and unsuited for this purpose. That is, they are good at describing things but not how things and the relationships between things change over time. Another cause of method complexity is that the particular representations and notations used affect how people think about what is being represented. Diaper (2001b), for example, analyzed the differing possible effects on constructing and reasoning with systems models that were represented either using tree diagrams and tables or his Simplified Set Theory for Systems Modelling (SST4SM; Diaper, 2000; chap. 1).

We don't think people sit down and design a method from scratch. What usually happens is that some individual or a small team of people expert in some aspect of software engineering, HCI, task analysis, and so on, decide to codify how they achieved their previous successes. What they produce is a method that works for them, but what they find hard to see is what will work for others. Diaper (2001a) reported that even though much effort had gone into specifying the TAKD method, building a CASE tool to support it showed numerous gaps in the method—gaps that were bridged by the TAKD expert's craft skill. In many cases the analyst was unaware of these gaps. The point is that being an expert at doing something such as task analysis doesn't automatically qualify one for designing a method that others can use. Even Michael Schumaker, as an expert racing driver, doesn't design his F1 Ferrari; his input to design as an expert car user is undoubtedly invaluable, but the car is designed by car designing experts. What is really needed for task analysis, we believe, is some method designing experts, supported, of course, by task analysis experts.

It is an adage that if you want to understand something, then write it as a computer program, because the rigor required for programming forces a very careful analysis. One of the strong messages from Diaper's (2001a) analysis of TAKD's failure is that, when designing a method, one should design a CASE tool in parallel with the method's development. The advantages of parallel development are that inevitably the method will be much better specified and, even more importantly, those places in the method where analysts will have to employ their craft skills will be identified because they will not be programmable without resorting to AI. Parallel method and CASE tool development is not however, the current practice. Instead, the method is developed first; probably tested in a range of projects and different circumstances, with the steps of the method being done manually by the analyst; and only some time later may a CASE tool to support the analyst be constructed.

The concept of requirements capture, analysis, and specification is utterly common in software engineering. Although there usually are some computer-orientated requirements in a specification, most of the requirements fundamentally concern people, what they need, and their environment or context of use (e.g. Diaper, 2002b). In contrast to software development, method development, including the development of task analytic methods, does not seem to have been subjected to an equivalent requirements exercise. If method development is treated like software development, which is particularly plausible if a CASE tool is developed in parallel with the method, requirements can be divided into two major types, those that concern a method's users, their work, and their project environment, and so on, and those that concern the types of problems the method is supposed to be able to address. Our contention is that task analysis methods have been primarily developed on the basis of the second type of requirement and that the first has been largely ignored. We interpret the failure of most task analysis methods to be widely taken up by industry as being in large part due to the lack of balance between these two types of requirements. We also suspect that HTA's enduring popularity in industry is its superficial simplicity and genuine versatility, although, as noted in chapters 3 and 28, although the basic concepts might be taught in 20 minutes or so, it takes a long time to become truly expert at HTA. Ainsworth and Marshall (1998) reported that the quality of HTA analyses in industry was highly variable and suggested that the variability is caused by the differing skills of analysts.

We assume that the major end users of task analysis are people employed in the computing industries, not academics and other types of researchers. It seems obvious to us that there are fundamental requirements related to issues such as these: Who will be the method's users? Are they task analysis or HCI experts or are they unversed in HCI? Do the method's users basically work alone or are they part of a highly interactive team? Chapter 11, along with other chapters, makes the important point that such teams have members who need to understand a task analysis method's outputs and, better still, be able to contribute to the application of the method even though their expertise is not in task analysis or HCI. There are also project demands that need to be part of a method's requirements specification. All the chapters in part II suggest that there are usually time, money, and other resource constraints and that access to particular types of people is often limited (chaps. 2 and 9).

Axiom 3 (section 2.1) is "to keep it as simple as possible." The "it" primarily refers to task analysis methods intended for use by the commercial computing industries. TAKD was too complicated, and chapter 4 admits that there is a public perception that the GOMS method is complicated. The "as possible" is also important in Axiom 3. Task analysis is more complicated than methods that are not concerned with performance issues, and it is this complexity that needs addressing. There are some areas, such as with safety-critical systems, that require detailed, methodologically sophisticated and rigorous, perhaps even formally based, task analyses. In many cases, however, commercial computing projects must have a task analysis method that is lighter, sometimes very much lighter, than that provided by the more sophisticated and therefore more expensive task analysis methods. Our current recommendation, although we don't deny there are alternatives, is to develop a task analysis meta-method (section 2.2) that has levels of abstraction so that the lightweight and heavyweight methods could share the same high-level method and its concepts, terminology, and so on.

The development of any such task analysis meta-method will have to meet the same criteria of reliability and validity required by Stanton and Young (1999a,b) of particular task analysis methods. Reliability is a measure of stability, stability of the method over time and stability of the method across analysts. Ideally, it should be possible to demonstrate that the application of a task analysis method will result in the same results if it is used by different people or on different occasions by the same people (provided that the system being analysed hasn't changed). As for reliability, a method would be considered to have minimally acceptable reliability if the method's expert creator could achieve repeatable results on different occasions. At the other extreme would be a method that delivered the same results when used by anyone with even a little training. Between these extremes would be most of the methods used in the computing industries, and whether any one of these would be considered to have an acceptable degree of reliability would depend on the expertise of thosing using it; various project constraints such as time and resource availability; the type of project; and the problem for which the method was being used. Indeed, when a task analysis method is being used creatively (e.g., see chap. 6), then high reliability, either for an individual analyst or across different analysts, may be undesirable, as it could restrict the range of alternatives considered. In contrast, in large projects with a number of task analysts, a much higher degree of reliability is necessary, as the results achieved by the different analysts will no doubt need integrating at some stage during the project.

The proposed meta-method is not itself a task analysis method, so its reliability needs to be gauged using different criteria than those for specific task analysis methods. Critically, will the meta-method allow different people or the same person on different occasions to produce the same task analysis method for the same project situation? What will almost certainly happen in practice is that the perception of the meta-method's reliability will be largely based on the perceived reliability of specific methods related to it and applied on particular projects.

If method reliability is not a simple thing, then validation is even more complicated. Stanton and Young (1999a) suggested that there are four types of validity: construct, content, concurrent, and predictive (see Fig. 30.1). As with reliability, it will be the perceptions of those in the computing industries that will determine whether the proposed meta-method is considered adequately valid. Construct validity, for example, concerns the underlying theoretical basis of a method. If you are Phil Barnard, then the construct validity of Interacting Cognitive Subsystem is undoubtedly very high. It will be much lower for many nonpsychologists in the computing industries, and for some people, those who cannot understand its principles, its construct validity will be zero. Similarly, if the meta-method is to have construct validity, then not only must its theoretical underpinnings be sound, but they must also be understandable to a suitable range of people in its primary target audience in the computing industries.

Content validity, according to Stanton and Young, is concerned with the credibility that a method is likely to gain among its users. They suggest that, ideally, a method should use appropriate terminology and language and seem up to the job of task analysis if it is to be taken seriously. Obviously, such validity requires agreement among those in the computing industries. The content validity of the meta-method will to a great extent depend on its format, notation, and other presentational aspects. This is one area where our sought-after method design experts might have an extremely valuable contribution to make.

Finally, Stanton and Young's concurrent and predictive validity concerns the extent to which an analyzed task or task set is representative of the total set of tasks that might have

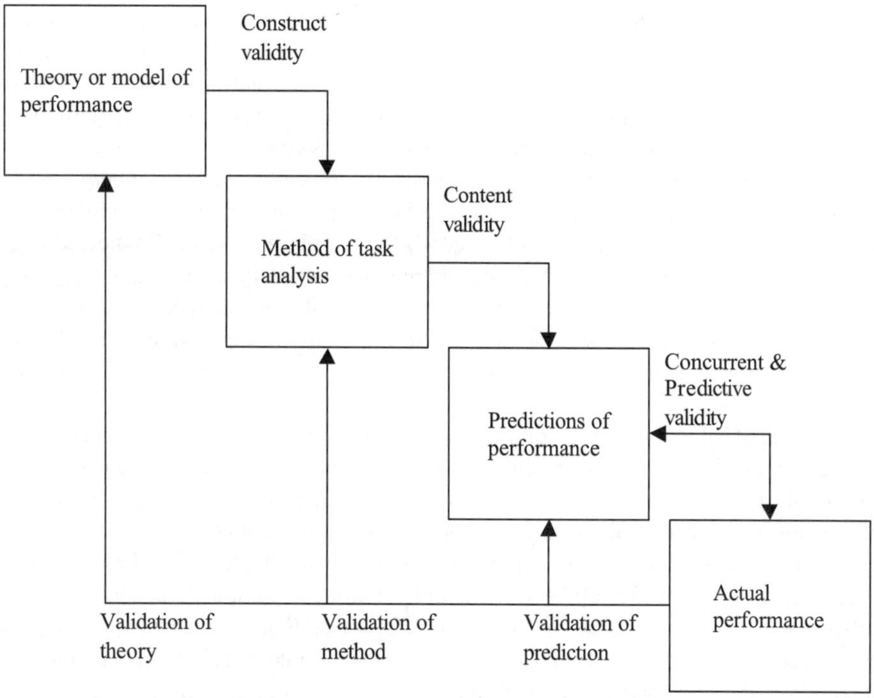

FIG. 30.1. Validation of task analysis methods.

been analyzed. The difference between concurrent and predictive validity is a matter of time: concurrent validity describes current systems sampled whereas predictive validity concerns the behavior of future, putative systems. Chapter 1 identifies two different types of "missed tasks", missed tasks within the range of a set of analyzed tasks and missed tasks outside the range. Although concurrent and predictive validity is absolutely vital for task analysis methods, the meta-method may not itself have this sort of validity. What is important instead is that the task analysis methods it produces do posses a level of concurrent or predictive validity suitable for their application.

There continues to be debate as to the role of validation in task analysis (Annett, 2002; Stanton, 2002; Stanton & Young, 1999b), and the issues are by no means resolved. Our proposed task analysis meta-method will need to meet both reliability and validity criteria. Although laboratory and other research work may be a desirable minimum, however, it is the perceptions of task analyses' ultimate users in the computing industries that will be most important.

30.4 CONCLUSION

HCI is always going to be concerned with the performance of work, and task analysis is that part of HCI devoted to the study of such performance. Indeed, the study of performance is what defines task analysis and differentiates it from other things in HCI and software engineering (chap. 1). As for the future of task analysis, we suggest that there are four general outcomes that may occur. Which outcome is realized will depend on whether eventually there is general agreement on a theory, method, and vocabulary for task analytic work and whether there is greater integration between the widely different current approaches to task analysis.

Our opinion is that the current state of task analysis is a mess and that the community and the theories, methods, results, and even vocabulary are fragmented. In this chapter, we proposed three core axioms for task analysis and then suggested that what is required is a meta task analysis theory and method that can subsume current task analysis approaches without losing the many and divers desiderata associated with existing approaches.

One issue in need of immediate attention is the difference between ideal task models and those that represent real task instances, as the latter need one or more techniques for the representation of equally good ways of successfully completing a task. Connected to this is how related and unrelated parallel tasks and shared tasks are represented. Many task analysis methods do have some ways of representing such things, but often the mechanisms are clumsy, and we are concerned that the representational forms used will not scale up to tasks that are highly parallel, involve complex cooperation between many agents, and can be done in many different ways.

The concept of a goal is the millstone around the neck of task analysis. We argued above that the concept should be abandoned forthwith, as goals are unnecessary as motivators of behavior and are really post hoc explanations of behavior. Our preference is to replace the concept of a goal, where necessary at all, with an attentional mechanism involved in the allocation of cognitive processing resources. We agree with statements in chapter 29 to the effect that task analysis will need to extend itself beyond the field of work as computers become increasingly prevalent in other areas of people's lives. Chapter 29 proposals are consistent with our proposal to abandon the concept of a goal, and we think such abandonment preferable to agonizing over what are a person's goals when playing a computer game, spending many hours in chat rooms, and so on. We are dismissive of the mono-teleology of most task analysis approaches (that is the association of a single goal with a particular behavior), but our preference is to abandon the whole concept of a goal rather than to tinker with a multi-goal approach.

Finally, we argued that the developers of the major task analysis methods have failed to adequately consider the needs of the ultimate users of task analysis methods, those who work in the commercial computing industries. We pointed out that there is a need for some people who are expert at method development to support task analysis experts in the development of future methods. We went so far as to suggest that the needs of commercial users should play as important a role in determining the nature of a task analysis method as the types of problems the method is designed to solve.

To close the circle and return to a point made in this handbook's preface, the field of task analysis for HCI is large but fragmented. Nonetheless, we would like to think that this final chapter will stimulate the task analysis cognoscenti so that we can open the next edition of this Handbook a few years hence with the claim that this time it really is the definitive work on the vitally important topic of task analysis for HCI.

REFERENCES

Ainsworth, L., & Marshall, E. (1998). Issues of quality and practicality in task analysis: Preliminary results from two surveys. *Ergonomics, 41*, 1607–1617. Reprinted in J. Annett & N. A. Stanton (Eds.). (2000). *Task analysis* (79–89). London: Taylor & Francis.

Annett, J. (2002). A note on the validity and reliability of ergonomics methods. *Theoretical Issues in Ergonomics Science, 3*, 228–232.

Annett, J., Cunningham, D, & Mathias-Jones, P. (2000). A method for measuring team skill. *Ergonomics, 43*, 1076–1094.

Annett, J., & Duncan, K. D. (1967). Task analysis and training design. *Occupational Psychology, 41*, 211–221.

Annett, J., & Stanton, N. A. (Eds.). (1998). Task analysis [Special issue]. *Ergonomics, 41*(11). Reprinted as J. Annett & N. A. Stanton (Eds.). (2000). *Task analysis.* London: Taylor & Francis.

Atkinson, M. (1988). Cognitive science and philosophy of mind. In M. McTear (Ed.), *Understanding cognitive science* (pp. 46–68). Chichester, England: Ellis Horwood.

Benyon, D. (1992a). The role of task analysis in systems design. *Interacting With Computers, 4*, 102–123.

Benyon, D. (1992b). Task analysis and system design: The discipline of data. *Interacting With Computers, 4*, 246–259.

Benyon, D., & Macaulay, C. (2002). Scenarios and the HCI-SE design problem. *Interacting With Computers, 14*, 397–405.

Brehm, W. J. (1966). *A theory of psychological reactance*. New York: Academic Press.

Carroll, J. M. (2000). *Making use: Scenario-based design for human-computer interactions.* Cambridge, MA: MIT Press.

Diaper, D. (Ed.). (1989a). *Task analysis for human-computer interaction.* Chichester, England: Ellis Horwood.

Diaper, D. (1989b). The discipline of human-computer interaction. *Interacting With Computers, 1*, 3–5.

Diaper, D. (1989c). Designing expert systems: From Dan to Beersheba. In D. Diaper (Ed.), *Knowledge elicitation: Principles, techniques and applications* (pp. 15–46). Chichester, England: Ellis Horwood.

Diaper, D. (1989d). Task Analysis for Knowledge Descriptions (TAKD): The method and an example. In D. Diaper (Ed.), *Task analysis for human-computer interaction* (pp. 108–159). Chichester, England: Ellis Horwood.

Diaper, D. (2000). Hardening soft systems methodology. In S. McDonald, Y. Waern, & G. Cockton (Eds.), *People and computers XIV* (pp. 183–204). New York: Springer.

Diaper, D. (2001a). Task Analysis for Knowledge Descriptions (TAKD): A requiem for a method. *Behaviour and Information Technology, 20*, 199–212.

Diaper, D. (2001b). The model matters: Constructing and reasoning with heterarchical structural models. In G. Kadoda (Ed.), *Proceedings of the Psychology of Programming Interest Group 13th Annual Workshop* (pp. 191–206). Bournemouth: Bournemouth University.

Diaper D. (2002a). Scenarios and task analysis. *Interacting With Computers, 14*, 379–395.

Diaper, D. (2002b). Human-computer interaction. In R. B. Meyers (Ed.), *The encyclopedia of physical science and technology* (3rd ed., Vol. 7, pp. 393–400). New York: Academic Press.

Diaper, D. (2002c). Task scenarios and thought. *Interacting With Computers, 14*, 629–638.

Diaper, D., & Addison, M. (1991). User modelling: The Task Oriented Modelling (TOM) approach to the designer's Model. In D. Diaper & N. Hammond (Eds.), *People and Computers VI* (pp. 387–402). Cambridge: Cambridge University Press.

Diaper, D., & Addison, M. (1992a). HCI: The search for solutions. In A. F. Monk, D. Diaper, & M. D. Harrison (Eds.), *People and computers VII* (pp. 493–495). Cambridge: Cambridge University Press.

Diaper, D., & Addison, M. (1992b). Task analysis and systems analysis for software engineering. *Interacting With Computers, 4*, 124–139.

Diaper, D., & Kadoda, G. (1999). The process perspective. In L. Brooks & C. Kimble (Eds.), *UK Academy for Information Systems 1999 Conference Proceedings* (pp. 31–40). New York: McGraw-Hill.

Festinger, L. (1957). *A theory of cognitive dissonance.* Stanford University Press.

Hewett, T. T. (1991). Importance of failure analysis for human-computer interface design. *Interacting With Computers, 3*, 3–8.

Long, J. (1986). People and computers: Designing for usability. In M. Harrison & A. Monk (Eds.), *People and computers: Designing for usability* (pp. 3–23). Cambridge: Cambridge University Press.

Lorenz, E. N. (1993). *The essence of chaos*. Seattle: University of Washington Press.

Norman, D. (1986). Cognitive engineering. In D. Norman & S. Draper (Eds.), *User-centered system design: New perspectives on human-computer interaction* (pp. 31–61). Hillsdale, NJ: Lawrence Erlbaum Associates.

Peterson, I. (1998). *The jungles of randomness: Mathematics at the edge of certainty.* New York: Penguin.

Reich, B., & Adcock, C. (1976). *Values, attitudes and behaviour change.* London: Methuen.

Sellen, A. J., & Harper, R. H. R. (2002). *The myth of the paperless office.* Cambridge, MA: MIT Press.

Stammers, R. B. (1995). Factors limiting the development of task analysis. *Ergonomics, 38*, 588–594.

Stanton, N. A. (2002). Developing and validating theory in ergonomics. *Theoretical Issues in Ergonomics Science, 3*, 111–114.

Stanton, N. A., & Young, M. S. (1999a). *A guide to methodology in ergonomics: Designing for human use.* London: Taylor & Francis.

Stanton, N. A., & Young, M. S. (1999b). What price ergonomics? *Nature, 399*, 197–198.

Author Index

M

N

O

Subject Index

D

E

F

U

V